Coastal and Estuarine Studies

Series Editors:
Malcolm J. Bowman   Christopher N.K. Mooers

Coastal and Estuarine Studies

54

Jörg Imberger (Ed.)

Physical Processes in Lakes and Oceans

American Geophysical Union

Washington, DC

Series Editors
Malcolm J. Bowman
Marine Sciences Research Center, State University of New York
Stony Brook, NY 11794, USA

Christopher N. K. Mooers
Division of Applied Marine Physics
RSMAS/University of Miami
4600 Rickenbacker Cswy.
Miami, FL 33149-1098, USA

Editor
Jörg Imberger
Department of Environmental Engineering
Centre for Water Research
University of Western Australia
Nedlands Western Australia 6907
AUSTRALIA

Library of Congress Cataloging-in-Publication Data
Physical Processes in lakes and oceans / Jörg Imberger, editor.
p. cm. — (Coastal and estuarine studies, ISSN 0733-9569 ; 54)
Includes bibliographical references.
ISBN 0-87590-268-5
1. Lakes. 2. Limnology. 3. Oceans. 4. Hydrodynamics.
I. Imberger, Jörg. II. Series
GB1603.2.P48 1998
551.48'2—dc21 98-41142
CIP

ISSN 0733-9569
ISBN 0-87590-268-5

Printed in the United States of America

CONTENTS

PREFACE
J. Imberger ix

PART I. INTRODUCTION
1. Flux Paths in a Stratified Lake: A Review
J. Imberger 1

PART II. SURFACE LAYER DYNAMICS
2. Air-Water Exchange Processes
M. A. Donelan 19

3. Turbulent Flux of Water Vapor in Relation to the Wave Field and Atmospheric Stratification
K. B. Katsaros 37

4. On the Structure of the Upper Oceanic Boundary Layer and the Impact of Surface Waves
A. Anis 47

5. Large Eddies in the Surface Mixed Layer and Their Effects on Mixing, Dispersion and Biological Cycling
M. Li and C. Garrett 61

6. Velocity, Temperature and Spatial Structure of Langmuir Circulation
D. Farmer, J. Gemmrich, and V. Polonichko 87

7. On Wavy Mean Flows, Langmuir Cells, Strain, and Turbulence
S. G. Monismith and J. J. M. Magnaudet 101

8. Modeling of Atmospheric Forced Mixing on the Shallow Shelf
I. D. Lozovatsky and A. S. Ksenofontov 111

9. Large Inflow-Driven Vortices in Lake Constance
E. Hollan 123

PART III. FORCED BASIN SCALE MOTIONS
10. Forced Motion Response in Enclosed Lakes
K. Hutter, G. Bauer, Y. Wang, and P. Güting 137

11. Excitation of Internal Seiches by Periodic Forcing
E. Bäuerle 167

12. **Thermohaline Transitions**
E. C. Carmack, K. Aagaard, J. H. Swift, R. G. Perkin, F. A. McLaughlin, R. W. Macdonald, and E. P. Jones 179

13. **Exchange Flows in Lakes**
P. F. Hamblin 187

14. **Gyres Measured by ADCP in Lake Biwa**
M. Kumagai, Y. Asada, and S. Nakano 199

15. **Circulation, Convection and Mixing in Rotating, Stratified Basins with Sloping Topography**
P. B. Rhines 209

PART IV. INTERNAL WAVE MOTIONS

16. **Internal Solitary Waves in Shallow Seas and Lakes**
R. Grimshaw 227

17. **Two Intersecting Internal Wave Rays: A Comparison Between Numerical and Laboratory Results**
A. Javam, S. G. Teoh, J. Imberger, and G. N. Ivey 241

18. **Breaking Internal Waves and Fronts in Rotating Fluids**
A. V. Fedorov and W. K. Melville 251

19. **A Laboratory Demonstration of a Mechanism for the Production of Secondary, Internal Gravity-Waves in a Stratified Fluid**
T. Maxworthy, J. Imberger, and A. Saggio 261

20. **Direct Numerical Simulation of Wave-Mean Flow and Wave-Wave Interactions: A Brief Perspective**
C-L. Lin, J. R. Koseff, J. H. Ferziger, and S. G. Monismith 271

21. **Momentum Exchange Due to Internal Waves and Wakes Generated by Flow Past Topography in the Atmosphere and Lakes**
P. G. Baines 285

22. **In Search of Holmboe's Instability**
G. A. Lawrence, S. P. Haigh, and Z. Zhu 295

PART V. TURBULENT MIXING

23. **Estimation and Geography of Diapycnal Mixing in the Stratified Ocean**
M. C. Gregg 305

24. **Special Closure for Stratified Turbulence**
J. Weinstock 339

25. **Turbulent Mixing in Stably Stratified Flows: Limitations and Adaptations of the Eddy Diffusivity Approach**
H. J. S. Fernando, J. C. R. Hunt, E. J. Strang, A. L. Berestov, and I. D. Lozovatsky 351

26. **Intermittency of Internal Wave Shear and Turbulence Dissipation**
C. H. Gibson 363

27. **Buoyancy Fluxes in a Stratified Fluid**
G. N. Ivey, J. Imberger, and J. R. Koseff 377

28. **Mixing Processes in a Highly Stratified River**
S. Yoshida, M. Ohtani, S. Nishida, and P. F. Linden 389

29. **Stratified Turbulence: Field, Laboratory and DNS Data**
H. Yamazaki and D. Ramsden 401

PART VI. INFLUENCE OF TOPOGRAPHY AND THE BENTHIC BOUNDARY LAYER

30. **Waves, Mixing, and Transports over Sloping Boundaries**
C. C. Eriksen 417

31. **Some Dynamical Effects of Internal Waves and the Sloping Sides of Lakes**
S. A. Thorpe 441

32. **Finescale Dynamics of Stratified Waters Near a Sloping Boundary of a Lake**
U. Lemmin, R. Jiang, and S. A. Thorpe 461

33. **Breaking of Super-Critically Incident Internal Waves at a Sloping Bed**
I. P. D. De Silva, J. Imberger, and G. N. Ivey 475

34. **Bottom Boundary Mixing: The Role of Near-Sediment Density Stratification**
A. Wüest and M. Gloor 485

35. **Turbulent Benthic Boundary Layer Mixing Events in Fresh Water Lakes**
C. Lemckert and J. Imberger 503

PART VII. IMPORTANCE OF TRANSPORT AND MIXING ON ECOLOGICAL PROCESSES

36. **Using Measurements of Variable Chlorophyll-*a* Fluorescence to Investigate the Influence of Water Movement on the Photochemistry of Phytoplankton**
R. L. Oliver and J. Whittington 517

37. **Plants in Motion: Physical - Biological Interaction in the Plankton**
C. S. Reynolds 535

38. **Turbulent Mixing and Resource Supply to Phytoplankton**
S. MacIntyre 561

39. **The Influence of Biogeochemical Processes on the Physics of Lakes**
D. M. Imboden 591

40. **Hydrodynamic vs. Non-Hydrodynamic Influences on Phosphorus Dynamics During Episodic Events**
Y. P. Sheng, X. Chen, and S. Schofield 613

41. **Coupling of Hydrobiology and Hydrodynamics: Lakes and Reservoirs**
M. Straskraba 623

42. **3D Modeling of Water Quality Transport Processes with Time and Space Varying Diffusivity Tensors**
C. Dejak, R. Pastres, I. Polenghi, C. Solidoro and G. Pecenik 645

List of Contributors 663

PREFACE

Physical limnology is the study of motion, transport and mixing in a lake and so forms an exact parallel to physical oceanography. However, in comparison with the field of physical oceanography, the hydrodynamics of lakes has received little attention; what is known about lakes is the result of the efforts of a few individuals who have relied heavily on the research results from the ocean!

This is a very peculiar situation if it is remembered that lakes are logistically much easier and cheaper to work in than the ocean. The answer must lie in the fear that since in a lake the boundaries are usually complicated and close to the field of observation, the motion is correspondingly complicated and thus difficult to unravel. The second peculiarity which should be pointed out is that lakes supply probably 80% of the world's drinking water, constitute major tourist attractions and support important fisheries, yet the funding of fundamental research into the hydrodynamics of lakes, both small and large, has always been and still remains pitifully small. At present no universities have a major program in physical limnology, few research institutes focus on physical limnology, there is no textbook on the subject, nor is there a journal focused on physical limnology. No wonder we are only beginning to understand the fundamentals of the hydrodynamics of lakes.

Lakes also form important laboratories in their own right, and as models of the ocean, for the study of how biological, chemical and sedimentological systems behave in response to water transport and mixing. This is what I am calling the flux path problem. Lake research offers immense opportunities for understanding such reactive transport systems. This in turn has importance not only for developing predictive water quality models for lakes, but also for developing tools which allow lakes to be used as an integrated part of our whole water maintenance system. Further, such research has immense importance for developing an understanding of the response of lakes and the ocean to such diverse pressures as global warming, depletion of selected species, and introduction of nutrients, synthetic organic chemicals and organisms detrimental to health.

A symposium on physical limnology was organised by the International Union of Theoretical and Applied Mechanics to focus attention on this unsatisfactory situation, to highlight the close links which exist between the hydrodynamics of lakes and the ocean, and to alert colleagues to the challenges of the flux path problem. This volume is an outgrowth of this symposium and comprises contributions from practically all presently practicing physical limnologists and from a group of selected physical oceanographers who kindly donated their time to cross fields. I hope the cross linking was rewarding to both sides. Near the end of the book is a selection of contributions, again from leaders in their field, which bring out some of the challenges of the flux path problem. The volume thus brings together contributions from leaders in physical limnology, physical oceanography and hydrobiology, and as such is, to my knowledge, the first book of its kind.

I would like to thank the authors for providing their manuscripts, the international reviewers for their constructive criticisms of the papers and Lorraine Dorn for the layout of the book and the considerable typing and editorial work she has undertaken. Vicki Sly organised the symposium, held in Broome, Western Australia, during September 1995, and I am sure I speak for all the participants in thanking her for co-ordinating a notable meeting in an unforgettable place.

Jörg Imberger
Editor

1

Flux Paths in a Stratified Lake: A Review

J. Imberger

Abstract

This review focuses on clarifying the energy transfer mechanisms operating in a stratified lake. The energy imparted by the wind to the lake leads to basin scale waves which may be identified as Kelvin or Poincaré waves or simple gravitational seiching; the vertical model structure of these waves appears to depend on the wind forcing and the basin slope, but most of the energy resides in the first and second modes. Once established these basin scale waves steepen due to non-linear processes and due to shoaling forming a spectrum of internal wave motions within the metalimnion with frequencies from about 10^{-6} Hz to 10^{-2} Hz depending on the size of the lake and the severity of the stratification. The waves at the high frequency end of the spectrum are often highly non-linear and are best described as solitary waves, bores and hydraulic jumps.

It is shown that waves in the frequency range from 5×10^{-5} to 10^{-2} Hz are close to critical with respect to typical bottom slopes. A review of internal wave reflection showed that for such incident waves most of the incident internal wave energy is dissipated at the boundary. This is also true for isolated solitary waves. There are thus two energy sources for the maintenance of the benthic boundary layer. First, at the level of the metalimnion incident internal wave modes break giving up their energy, and second, the velocity field associated with the first mode heaving leads to shear production at the lake bottom in the deeper parts of a lake basin.

Recent evidence also suggests that the composite internal wave field gives rise to turbulent patches in the metalimnion and to a lesser extent in the hypolimnion. The great majority of the turbulent events have very small displacement scales and are thus inefficient at transporting mass vertically, even though the mean dissipation of turbulent kinetic energy may be as high as 10^{-8} m^2 s^{-3}.

It appears that mass is thus transferred vertically in two ways in a stratified lake. First, through the benthic boundary layer, and second, to a much lesser extent directly by intermittent turbulent mixing.

Introduction

The study of hydrodynamics of lakes commenced with the pioneering work of Wedderburn (Wedderburn, 1912). This work showed, for the first time, that the thermal

Physical Processes in Lakes and Oceans
Coastal and Estuarine Studies Volume 54, Pages 1-18

structure in a lake seiches at well-defined frequencies and that if the wind is of sufficient severity the deeper water from the hypolimnion may upwell to the surface. Quite a few years later Mortimer (1974) successfully identified these long, basin scale waves as Kelvin, Poincaré and gravitational waves with well-defined vertical modal structures.

As water quality in lakes started to deteriorate in the late 1960's the emphasis shifted to the fluid mechanics of selective withdrawal (Imberger, 1980), inflow dynamics and differential heating and cooling as well as the study of mixing. For a review of these investigations the reader is referred to Fischer et al. (1979), Imberger and Patterson (1990), Imboden and Wüest (1994), and Imberger (1994).

These investigations were, more recently, augmented with direct measurements of the turbulence in the surface layer (Imberger, 1985), the benthic boundary layer (Wüest et al., 1994) and the main water column (Imberger and Ivey, 1991).

From a water quality point of view the main question is how do these motions influence such things as primary production (Reynolds, 1994), nutrient and heavy metal cycling (Stumm, 1985) and the behavior of particles (Buffle and van Leeuwen, 1992). The majority of active biological and chemical particles are smaller than 10 μm. Now the typical magnitude of the Kolmogorov length scale in a turbulent patch in the main water body is of the order 3×10^{-3}, so that the scale of the smallest eddies is about one or two centimeters (Capblancq and Catalan, 1994). Thus, there is a huge difference in size between the particles and the smallest motion; the motion will appear as a simple shear flow to all such particles. How then does the fluid motion influence the water quality in a lake?

There appear to be two clear links. First, the dissipation of turbulent kinetic energy ε determines the rate of strain $\gamma = (\varepsilon/\nu)^{1/2}$ for the particular motion, where ν is the kinematic viscosity of the fluid. As discussed in O'Melia and Tiller (1993) the rate of strain determines the rate of particle amalgamation and breakup for chemically active particles. For plankton and other biological particles there is recent evidence (Thomas and Gibson, 1990) that γ once again strongly influences the rate of growth; a large value for γ inhibits growth.

Second, the water transports such small particles as it moves throughout a lake. Some particles have gas vasicules (Reynolds, 1994) and can rise or fall relative to the water movement and others may have a specific gravity greater than one and slowly sink. However, the rise and fall velocities of such small particles, which have not formed colonies, are very small compared to even the slowest fluid motion and so particles are generally advected with the water. Such advection may, for instance, take particles through different light regimes as is the case for Langmuir circulation (Plueddemann et al., 1996), or to different depths with different chemical properties (Hamilton et al., 1995).

Thus, from a limnological point of view, two quantities are of paramount importance, the distribution of the rate of dissipation of turbulent kinetic energy and the path along which particles travel, called the flux path. The importance of these parameters is illustrated in the schematic shown in Figure 1. If the biogeochemical reactions which active particles undergo are not only a function of the state of the immediate neighborhood, but are also dependent on the internal state of the particle then the rate of reaction will depend on the history of the particles. This means that the Lagrangian path of particle becomes important requiring a Monte Carlo method (Beckenbach, 1956).

In the present review, as a general introduction to the rest of the book, we focus on clarifying what is known about the distribution of ε and the most likely mean flux path particles follow as they migrate within the body of the lake.

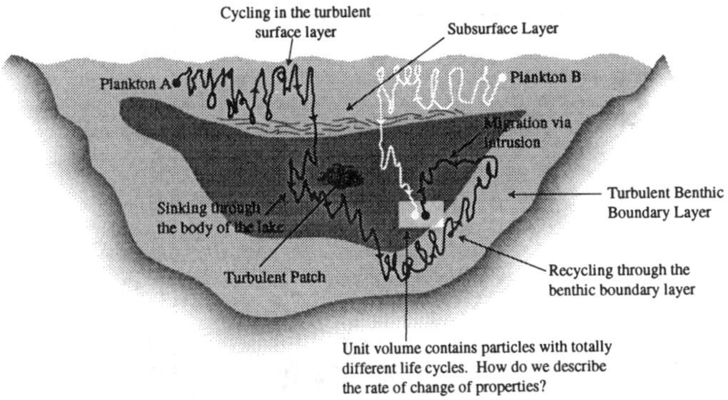

Figure 1. Illustration of the flux path problem.

Internal Wave Motions

By internal wave motions we understand motions of all scales and frequencies which cause isopycnals to oscillate with some well-defined frequency. Excluded from this are long term circulation patterns, intrusions and turbulent motions. Stratified lakes, by the very fact that they are stratified, rarely contain major circulation patterns; these are confined to large lakes where buoyancy plays a more minor role. The great majority of lakes fall into a category where the water motion is almost totally dominated by some type of internal adjustment (internal wave) since the forcing (wind stress) rarely is of a duration longer than a few periods of the longest internal wave periods.

As a concrete example, consider the spectrum of isotherm displacements in Lake Biwa reproduced in Figure 2 from Saggio and Imberger (1998). The spectrum is for the isotherm displacement during a period where the lake was strongly stratified, 28°C in the surface layer and 8°C at the bottom of the lake. The surface mixing layer had a mean depth of 15 m and the lake is about 80 km long, 30 km wide and 90 m deep at the deepest section. Over the time of the measurements (20 days), three major events occurred: the first produced winds to 10 ms^{-1}, the second was a relatively strong typhoon passing close to the lake leading to a peak wind speed of 20 ms^{-1}, and the last was due to a similar typhoon but one which passed some distance away so that the winds peaked at 15 ms^{-1}.

The spectrum shown in Figure 2 is typical for large lakes. As described by Saggio and Imberger (1998) there are four main features of the motion reflected in the spectrum. These are the discrete basin scale motions at low frequency, between zero and 3×10^{-5} Hz, a well defined peak around 3×10^{-5} Hz which is a mode directly forced by the wind, a range of motions from 3×10^{-5} Hz to 10^{-3} Hz where the spectrum decays as the square of the frequency, and two well-defined energy peaks between 10^{-3} Hz and 10^{-2} Hz.

The spectral properties outlined above have also been well documented in the oceanographic literature (Munk, 1981), and have been summarized in what is now known as the Garrett Munk Spectrum. The main features of the spectrum appear to be independent of the strength of the wind forcing, except that for increasing winds the peaks at the low frequencies become larger, more well-defined peaks form at the upper end of the basin scale range and encroach into the ω^{-2} range, the magnitude of the ω^{-2} range increases slightly and the two maxima immediately before the buoyancy frequency increase markedly (Saggio and Imberger, 1998).

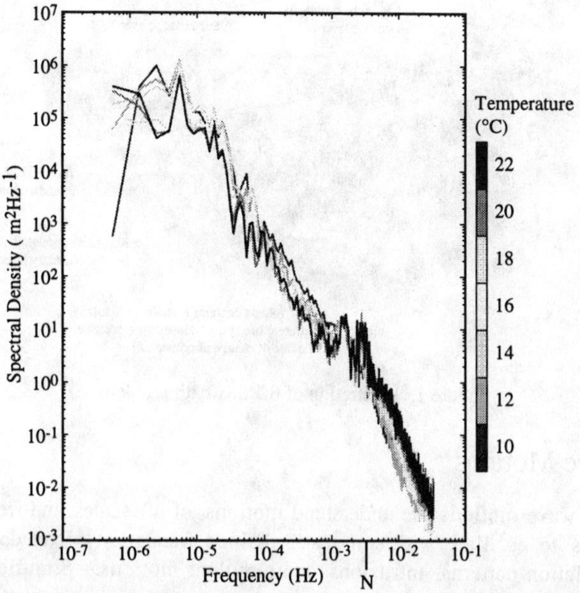

Figure 2. Spectrum of isotherm displacements from data collected in Lake Biwa.

In Table 1 the modes, corresponding to the isolated peaks in the low frequency band in Lake Biwa are summarized; each peak was identified by Saggio and Imberger (1998) as representing either a Kelvin or Poincaré wave with a definite vertical and horizontal modal structure. The direct wind forced wave occurs at a frequency of 3×10^{-5} and appears to be the response of water column to the wind forcing through the finite thickness surface layer; no evidence exists but it would appear likely that the frequency should depend on the thickness of the surface mixing layer. The horizontal wave length of the basin scale waves would naturally also be a strong function of the wind distribution over the lake but it is interesting that the rectangular basin model used by Saggio and Imberger (1998) was successful at reproducing all the observed frequencies of these basin scale waves. As numerical models improve and as detailed wind data becomes available it should be possible to successfully compute the whole range of basin scale waves excited by a particular wind history.

TABLE 1. Wave types present in Lake Biwa as computed from simulations in a rectangular basin.

Frequency Hz	Period hours	Vertical Mode	Horizontal Mode Width	Horizontal Mode Length	Type
2×10^{-6}	154	2	-	1	Kelvin
6×10^{-6}	46	1	-	1	Kelvin
1.2×10^{-5}	27	1	-	2	Kelvin
1.7×10^{-5}	16	1	1	3	Poincaré
2.3×10^{-5}	12	1	-	3	Kelvin
4.2×10^{-5}	6	1	-	-	Forced Wave
7 to 9×10^{-5}	3 to 4	higher	-	-	Various

The ω^{-2} subrange (Figure 2) deserves some further attention, since as discussed below, this frequency range (broadly from 5×10^{-5} to 10^{-3} Hz) matches the conditions for critical reflection from the sloping boundaries where the metalimnion borders the lake bottom. Saggio and Imberger (1998) band passed the isotherm displacement signals so that only energy in this band was retained. The resulting figure is reproduced in Figure 3 and it is seen that the waves in this frequency range are not continuous, but occur in groups.

Some of this groupiness is due to vertical sweeping of the higher frequency modes past the sensors by the vertical advection from the lower frequency waves, but even when this is removed, by referencing the signal to the long period isotherms, the groups remain. The overall conclusion reached by Saggio and Imberger (1998) was that these waves belonged to a complex field of multi-modal, strongly grouped, irregular waves being distorted by both higher and lower frequency waves.

The origin of these waves is still unknown, but most likely they are the result of the evolution of the various modes initiated by the wind forcing steepening due to shoaling and non linear processes. Ample laboratory evidence exists were mode one waves are generated by such processes (Chu and Chou, 1989) and recently Maxworthy et al. (1998) have shown how mode two waves can also be generated from mixing events triggered by basin scale activity.

The last band is the range 10^{-3} to 3×10^{-3} Hz where energy was observed to appear at the same time as the low period waves began to grow. These waves were identified as belonging to undular bores and solitary waves of large amplitude. The waves were predominantly of mode one and again occurred in groups, but in this frequency band the groups tended to be very isolated.

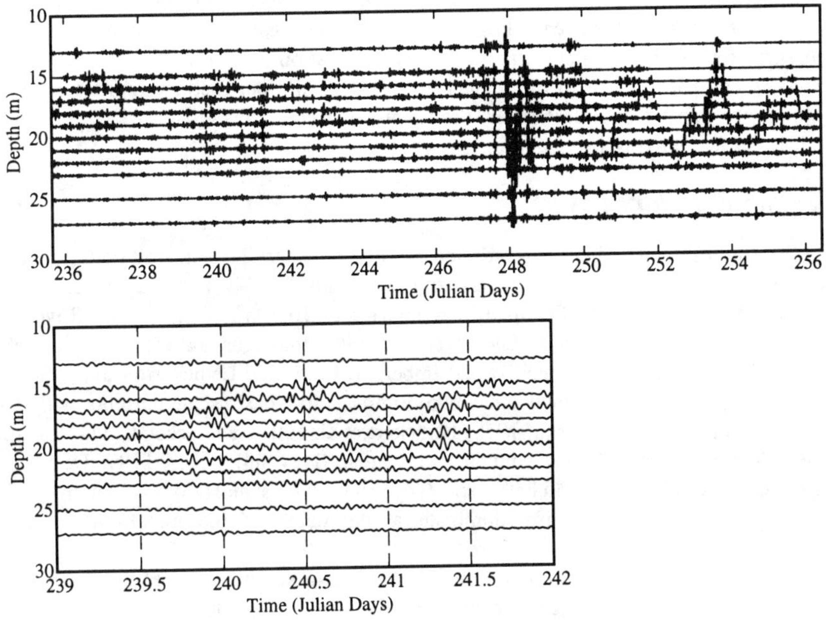

Figure 3. Thermistor signal from Lake Biwa: Band pass 1.5 to 4.0 10^{-4}.

The data from Lake Biwa thus showed that when a storm passes over a lake several classes of wave motion commence to gain energy. Time-frequency analysis was used by Saggio and Imberger (1998) to show that the time after the storm, for energy to appear was approximately equal for all frequencies recorded, from 5×10^{-7} Hz to 4×10^{-3}; no measurable time lag could be ascertained from the Lake Biwa data set. The wave activity, however, appeared at the measurement site nearly one day after the commencement of the strong winds associated with the passage of the typhoon which lasted about 6 hours. This would indicate that a storm generates large, long period internal waves as indicated by the experimental results of Stevens and Imberger (1996) and by numerical simulations (Saggio and Imberger, 1998). These waves intersect the bathymetry where they generate all the other frequencies and this mixture of waves then spreads into the interior, partly through its own propagation speed and partly by advection by the lower frequency, basin scale waves. However, Horn et al. (1998) have shown that non linear steepening is a more likely explanation for the appearance of the solitary waves and bores. The relative importance between the boundary interaction and the wave steepening as mechanisms for generation of high frequency waves should be revealed from the results of a recent experiment in Lake Kinneret where nine thermistor chains were placed around the lake. The important point to note is that the energy associated with internal wave modes in the metalimnion is likely to be dissipated on the first encounter with the boundary either by critical reflection (De Silva et al., 1998; Javam et al, 1998a) or by breaking of modal waves (Michallet and Ivey, 1998).

Vertical Turbulent Transport

Only a few measurements have been carried out of actual turbulent fluxes, either in the ocean or in a lake. However, a large number of investigations have concentrated on the measurement of the turbulent kinetic energy dissipation ε. In summary, these measurements reveal that in terms of turbulent activity there are four distinct regions in the water column, the benthic boundary layer, the main water column, the subsurface mixing layer, and the surface mixing layer.

Benthic Boundary Layer

In a stratified lake, there is a turbulent benthic boundary layer at the bottom, anywhere from a few centimeters to a few meters thick. In this benthic boundary layer dissipation levels vary, but generally fluctuate between 10^{-8} to 10^{-5} $m^2\ s^{-3}$ (Ivey and Boyce, 1982; Imberger and Patterson, 1990; Thorpe et al., 1990; Imberger and Ivey, 1993; Imberger, 1994; Wüest et al., 1994; Lemckert and Imberger, 1998; and Lemckert et al., 1998).

The source for this turbulence has two origins. First, basin scale currents will lead to a turbulent boundary layer (Armi, 1978; Lemckert et al, 1998), and second modes or internal waves impinging on a sloping boundary may break (Ivey and Nokes, 1989; De Silva et al. 1998; Javam et al, 1998a; Michallet and Ivey, 1998). It is most likely that internal wave shoaling and internal breaking dominate at the depth of the metalimnion and shear associated with basin scale waves predominate at depth.

At the depth of the metalimnion the evidence suggests that the greatest dissipation and largest flux Richardson numbers are present when the bottom slope is weakly supercritical (Eriksen, 1985; Thorpe et al., 1990; De Silva et al., 1998). Now the critical frequency is given by $\omega_c = N \sin \beta$ (Turner, 1973), where N is the buoyancy frequency and β is the bottom slope to the horizontal. Different lakes have very different slopes, but if we consider Lake Biwa and Lake Kinneret as representing two typical large natural basins then during the peak stratification, the maximum value of the maximum N equaled about 2×10^{2} Hz and the minimum value was about 8×10^{-3} and sin β ranges from 6.7×10^{-3} for the mildest slopes to about 4.0×10^{-2} for the steepest slopes. Thus the critical frequency ranges from about 6×10^{-5} Hz to 8×10^{-4}. Now this embodies almost the whole ω^{-2} range of internal wave groups as previously discussed. These wave groups thus serve a most important purpose, they appear to be the mechanism by which the wind energy is ultimately dissipated at the boundary of a lake.

As pointed out above, Ivey et al. (1998) have put forward a possible argument, based on laboratory experiments, that the thickness h_b of the benthic boundary layer is given by 0.15 λ, where λ is the wave length of the incident internal wave measured perpendicular to the slope. If it is assumed that all the energy of an incident internal wave is absorbed at the boundary when the slope is critical then Ivey et al. (1998) have shown that the dissipation in the benthic boundary layer should be given by:

$$\varepsilon = \frac{3\,\alpha^2 N}{4\pi} \sin 4\beta\,, \tag{1}$$

where α is the amplitude of the velocity of the incident internal wave field, N is the buoyancy frequency and β is the bottom slope. Again suppose we take values from Lake Biwa and Lake Kinneret then we find ε varies between $4.8 \times 10^{-3}\,\alpha^2$ and $3.2 \times 10^{-4}\,\alpha^2$. Now from direct velocity measurements in these two lakes with the portable flux profiler Etemad Shahidi and Imberger (1998) have shown that amplitude of the horizontal velocity is typically 5×10^{-2} ms^{-1} and the vertical wave length is about 10 m, leading to a general dissipation estimate ranging from 8×10^{-7} m^2 s^{-3} to 10^{-5} m^2 s^{-3} and a benthic boundary layer thickness of about 1.5 meters. These are values which are certainly typical of what is generally measured (Lemckert and Imberger, 1998). These comparisons thus suggest that the boundary layer thickness at the depth of the metalimnion is indeed determined by the characteristics of the incident wave field. Further, the assumption of almost complete dissipation of the internal waves as they approach a critical slope yields realistic estimates of the observed dissipation of turbulent kinetic energy. Last, the recent measurements of turbulent fluxes in the benthic boundary layer by Lemckert and Imberger (1998) show that the length scales of the turbulence is large enough, certainly for the larger Reynolds number events, so that the flux Richardson number $R_f = 0.25$ is a good approximation. The effective vertical turbulent eddy diffusion coefficient K_b (see Ivey et al., 1995) in the benthic boundary layer may thus be parameterized as:

$$K_b = 0.08\,\frac{\alpha^2}{N} \sin 4\beta\,, \tag{2}$$

and the thickness of the layer is given by,

$$h_b = \frac{0.15\,h_t}{n}\,, \tag{3}$$

where h_t is a measure of the thermocline thickness and n is the modal number of the most energetic wave group.

In the deepest part of the lake, at the bottom of the hypolimnion the water is only very weakly stratified and the benthic boundary layer is maintained by the friction velocity generated at the sediment-water interface by the motion induced by the mode one basin seiching (Lemckert et al., 1998). These authors showed that the benthic boundary layer followed the normal scaling laws of open channel flow where the turbulence is also energized by u_*^3.

Main Water Column

In the classical limnological literature (e.g. Wetzel, 1975) the main water column is separated into two regions, the metalimnion and the hypolimnion. Such authors also define the thermocline as the surface where N is a maximum; a surface embedded in the metalimnion. From a dynamical point of view such a precise separation is artificial because the gradual decay with depth of the buoyancy frequency means that the vertical extent of the involvement of the water column in any wave motion depends on the frequency of the motion or more generally the vertical modal structure of the waves excited by a particular storm. The water motion associated with mode one waves penetrates to the bottom, but the motion due to higher modes is confined predominantly to the wave guide formed by the metalimnion; the shear is thus higher in the thermocline but so is the stability so the local gradient Richardson number may have a similar variability throughout the water column.

The first direct flux measurements in the main water column of a lake recently became available (Etemad Shahidi and Imberger, 1998) from data collected in Lake Biwa and Lake Kinneret. These measurements have shown that the water column has an extensive number of turbulent patches with dissipation values log normally distributed and with a mean within patch value of about 10^{-8} m^2 s^{-3}, small displacement scales of around 1-5 cm and consistently negligible mass fluxes $\overline{w'\rho'}$ even though the turbulent Froude numbers $Fr_t \sim 1$, Reynolds numbers $Re_t \sim 10^3$ and strain ratios Fr_γ ranged from 10 to 10^2. Later experiments in Lake Kinneret, where all three velocity components were measured, showed that very few turbulent patches were associated with gradient Richardson numbers less than 0.25.

At first glance these results appear to contradict the generally accepted result for shear flow experiments (reviewed in Ivey and Imberger, 1991) that provided $Fr_\gamma > 15$, the fluxes should be non zero. These results were recently reinforced by direct numerical simulations (Ivey et al., 1998) lending confidence to this assertion for the case of shear, grid generated equilibrium turbulence.

However, as was shown by Imberger (1994), dimensional analysis implies for a simple shear flow in a fluid with a linear density gradient, the flow is stable to infinitesimal disturbances and finite disturbances grow without bounds, once introduced. This follows directly from the fact that such flows do not have any imposed length scales, other than $(\nu/N)^{1/2}$; the numerical results (see Ivey et al., 1998) confirm this conclusion. In the field, however, it is generally accepted that the length scale of the turbulence, as measured by some form of the displacement scale, is usually quite small, does not grow unboundedly, and is not always related to the Ozmidov scale $(\varepsilon/N^3)^{1/2}$. The reasons for this are still unknown, but there are three relatively obvious possibilities. First, the density (and velocity) profiles associated with the fine structure are seldom exactly linear and usually have variations which reflect a history of mixing events. They are marked by sharp transitions randomly distributed throughout the water column. Any fine scale straining is thus modified by these variations and their vertical scale will be imposed on the displacement scale. Second, in the metalimnion it is likely that mechanisms other than shear instabilities are responsible for the turbulence. Wave–wave interaction, as studied by McEwan (1973), Teoh et al. (1997), Javam et al. (1998b) and Lin et al. (1993) and is an example where the overturn length scale is fixed by the interaction mechanism imposed by the wave field. Other examples are critical layer absorption, reflection at a caustic and turbulence at a turning point (Javam et al., 1998b,c). In these examples, not only is the scale fixed by the instability mechanism, but also the patch size is fixed by the size of the interaction region. Third, even if both the density and velocity fields are relatively linear, they are rarely steady, so that the evolution of any turbulence generated will be modulated by time variation of the fine scale field.

Given these scenarios it is thus not difficult to imagine that any turbulence predominantly powered by the rate of strain of the fine scale internal wave field could be energetic, but small scaled. In such fields Fr_γ can be large; the rate of small scale strain is fast compared to the buoyancy frequency and the Ozmidov scale ℓ_0 is much larger than the Kolmogorov scale ℓ_k, yet the turbulence length scale does not grow with time as the scale is fixed by the instability mechanism. In such cases the transfer of mass and momentum will be small, even though the local dissipation can be moderately high.

Winters et al. (1995) have shown that it is important to differentiate between the reversible and irreversible contributions to the buoyancy flux $\overline{w'\rho'}$. They point out that the finer the grain of the turbulence the larger will be the rate of strain and thus diffusion is more important. Also the estimate $\overline{w'\rho'}$ is only an estimate of the irreversible buoyancy flux when the turbulence is in equilibrium with the background forcing; this is practically never the case in the metalimnion of a lake. These authors note that in order to estimate the irreversible flux we must either estimate the flux using the Osborn and Cox (1972) method based on the Cox number or a large number of estimates of $\overline{w'\rho'}$ must be averaged to extract the irreversible component. This theoretical result has recently been verified by Saggio and Imberger (1998) who showed that the turbulent diffusivity K_T is given by:

$$\frac{K_T}{\kappa_\rho} = 1 + 0.1\ Fr_\gamma{}^2, \tag{4}$$

where $Fr_\gamma = (\varepsilon/\nu N^2)^{1/2}$.

This indicates that estimates of the flux Richardson number based on measurements of $\overline{w'\rho'}$ may be distributed around zero with an equal number of positive and negative values (Etemad Shahidi and Imberger, 1998), but associated with these intermittent events is a local irreversible flux with a vertical diffusion coefficient ranging from 10^{-7} m^2 s^{-1} for the few negative events which occur in regions where the gradient Richardson number is maintained below 0.25.

A detailed study of the volume average value of ε does not appear to have been attempted, but given (see Imberger and Ivey, 1991; Etemad Shahidi and Imberger, 1998) that the occurrence of turbulent patches is very intermittent and the mean of the log normal patch dissipation is only 10^{-8} m^1 s^{-3}, the volume average of ε would be very small indeed; most likely much less than 10^{-9} m^2 s^{-3}.

Subsurface Mixing Layer

Recently it has been discovered (Imberger and Patterson, 1990 (see Figure 11); Wijesekjera and Dillon, 1991) that there exists a relatively thick layer immediately below the surface mixing layer (SML), called the subsurface mixing layer (SSML), where dissipation values are as high as in the SML and relatively constant throughout the water column with little intermittency, but where the stratification is very strong. This layer may be 2-5 m thick and be most active during periods of low wind immediately after strong thermal heating, before shear production at the base of the SML produces rapid deepening of the SML. In a Mediterranean climate this activity exists for three to four hours each day, before the strong afternoon wind causes the diurnal SML to deepen. Once again, the displacement scale in this region is usually a few centimeters, even though the dissipation itself ranges between 10^{-7} and 10^{-6} m^2 s^{-3}.

The source of energy for the internal waves which energize this turbulence is a mixture of direct wind forcing through the SML and energy leakage at the base of the surface layer (Imberger and Patterson, 1990; Saggio and Imberger, 1998).

Thus in the SSML plankton are exposed, for much of the daylight hours, to a high strain rate and small scale buffeting, but to only very small vertical excursions.

Surface Mixing Layer

The surface mixing layer is the region near the lake surface which is actively mixing and which is energized by the turbulence transferred at the air-water interface and the turbulence generated by the shear due to the momentum introduced by the wind stress.

The surface layer seems to be uncoupled from the main water column in terms of mass transport, although it is the layer through which momentum is transferred into the main water column.

The vertical transport in the SML has been amply documented and simple one-dimensional models based on a budget of turbulent kinetic energy (Spigel et al., 1986) appear to describe the diurnal and seasonal behavior of the surface layer rather well. However, recent direct measurements of the structure within the SML have revealed the importance of Langmuir's circulation (Plueddemann et al., 1996) during or after periods of relatively strong winds and some major discrepancies in the levels of dissipation of turbulent kinetic energy during periods of strong heating and weak winds. Both these features are most important for the modeling of plankton in the SML. Coherent motions are responsible for cycling plankton cells over the depth of the SML and this may influence the rate of production (Patterson, 1994) whereas larger dissipation levels, as indicated by the measurements, will produce larger strain rates and so inhibit primary production in the SML (Thomas and Gibson, 1990). This is an area requiring urgent attention.

Large scale spatial variability in the surface layer dynamics have repeatedly been documented and for a review the reader is referred to Imberger and Patterson (1990). However, until numerical models get to the point where they can be used to resolve this variability and field data is collected at the same high resolution, little can be done with this information.

Mean Vertical Transport

Many studies exist where the rate of change of the temperature or some tracer concentration at the bottom of a lake is equated to a mean vertical flux. Such studies do not reveal the path of the transport, but since a lake is a closed basin during periods of no in- or out-flow they do nevertheless yield concrete estimates of basin averages of the vertical heat and mass fluxes. Imboden and Emerson (1977), using radon and phosphorous, arrived at a net vertical diffusion coefficient of $0.5 \rightarrow 2.0 \times 10^{-5}$ m^2 s^{-1} in the Greifensee. Kullenberg et al. (1974) using the growth of dye clouds in Lake Ontario found a value ranging from 0.1 to 5×10^{-5} m^2 s^{-1}, whereas Robarts and Ward (1978) arrived at a value of the average diffusion coefficient of about 2×10^{-5} m^2 s^{-1} from temperature measurements in Lake McIlwaine. These values are all much larger than the values which follow from direct measurements in the turbulent patches that have an average value close to 10^{-7} m^2 s^{-1}.

Ward (1977) summarized much of the earlier work and suggested an empirical formula which implies that the vertical diffusion coefficient is proportional to the square root of the surface area of the lake and is inversely proportional to the buoyancy frequency. As reviewed by Imberger and Patterson (1990) the dependence on the buoyancy frequency varies greatly from one investigator to the next, with the dependence ranging from N^{-1} to N^{-2}.

None of these formulae, however, take into account the forcing acting on the lake or the lake geometry. As previously discussed, a lake quickly fills with internal wave energy as a wind event passes overhead and the energy contained in this distributed internal wave field is then dissipated through turbulence both in the interior and at the boundary.

Stevens and Imberger (1996) have recently shown that when a wind exerts a stress on the water surface, basin scale internal waves are generated in direct relationship to the magnitude of two non-dimensional numbers; the Wedderburn number W and the Lake number L_N (Imberger and Patterson, 1990).

If h_1 is the depth of the diurnal thermocline, g'_{12} is the reduced gravity across the diurnal thermocline, u_* is the water shear velocity induced by the wind and L is a representative length of the lake in the direction of the predominant winds, then

$$W = \frac{g'_{12} h_1^2}{u_*^2 L}, \tag{5}$$

and

$$L_N = \frac{S}{U_*^2 A_o^{3/2}} \left(\frac{1 - Z_T / H}{1 - Z_T / H} \right), \tag{6}$$

where the stability of the lake

$$S = \frac{g}{\rho_0} \int_0^H (Z_v - z) A(z) \phi(z) \, dz, \tag{7}$$

where A_o is the surface area of the lake, Z_T is the height of the seasonal thermocline, Z_V is the height to the center of volume and H is the total depth of the lake.

For the simple case of a rectangular basin with a two layer stratification

$$Z_v = H/2,$$

$$Z_T = (H - h_1),$$

$$A_o^{1/2} \sim L, \tag{8}$$

and

$$S = {}^1\!/_2\, g'_{12} L^2 (H - h_1) h_1,$$

so that

$$L_N = \frac{g'_{12}}{u_*^2} \frac{h_1^2}{L} \frac{H - h_1}{L},$$

$$\approx W, \tag{9}$$

for $h_1 << H$. Thus for the two layer case L_N and W have the same numerical value and the same physical interpretation. The stratification in most lakes, however, is better modeled by a three layer stratification with a surface layer of density ρ_1 and thickness h_1, an intermediate layer of density ρ_2 and thickness h_2 and a lower layer of density ρ_3 and thickness h_3. For such a stratification in a rectangular basin it is easy to show that

$$Z_V = H/2,$$

$$Z_T = h_3 + h_2/2$$

$$A_o^{1/2} \sim L \tag{10}$$

and

$$S = \frac{1}{2} L^2 \{g'_{32} h_3 (h_2 + H_1) + g'_{12} (h_2 + h_3) h_1\}, \tag{11}$$

so that

$$L_N = \frac{1}{u_*^2 LH} \left(h_1 + \frac{h_2}{2}\right) \{g'_{32} h_3 (H - H_3) + g'_{12} (h_2 + h_3) h_1\},$$

$$\approx \frac{1}{2} \frac{g'_{23} h_2^2 h_3}{u_*^2 L H}, \tag{12}$$

for the case where $h_1 << h_2, h_3$.

Stevens and Imberger (1996) showed, via a modal analysis of the response of the water column to an imposed wind stress, that the ratio of the mode 2 to mode 1 response is proportional to the ratio W/L_N. This implies that for a thin diurnal surface layer with a weak density jump at the base, the response will be predominantly mode 2, with the lower layer remaining essentially stationary (small W, large L_N). By contrast if the layer 2 is relatively thin and the density jumps g'_{12} and g'_{23} are roughly equal the response will be mode 1 (small L_N and W). This would imply that for the first case $W < 1$, $L_N > 1$ the turbulence in the lake remains confined to the surface layer and the water immediately beneath, whereas if $W < 1$, $L_N < 1$ then the lake as a whole is agitated.

Wüest (1987) was the first to try such a global parameterization of the turbulent properties of a lake. He used a simple two layer analogy (W parameterization) applied to Sempachersee to show that the vertical diffusion coefficient ranged from 10^{-6} m^2 s^{-1} for $1 < L_N < 200$ to 10^{-4} m^2 s^{-1} for $L_N < 0.5$. Imberger (1989) analyzed data from three different lakes and showed that the basin averaged vertical diffusion coefficient was a strong function of the Lake Number L_N (as computed by equation 6), and these results are shown in Figure 4. Again these results show that for $10^2 < L_N < 10^3$ the effective diffusion coefficient was about 10^{-6} m^2 s^{-1} and for $10^{-1} < L_N < 10$ this rose to 5×10^{-5} m^2 s^{-1}.

There is thus strong empirical evidence that the magnitude of the global parameter L_N determines not only the amplitude of the initial 1st mode response, but also seems to capture the basin averaged net vertical mass transport. Further, it follows directly from these observations that the net mass flux is not related to the stability, as embodied by the magnitude of N, but also by the net forcing at the surface of the lake; this is an obvious conclusion but one which has not, in the past, been highlighted.

By comparing the total energy in the internal wave field with the rate of dissipation in the water column, some headway can be made in answering the question of where, in the lake, the main transport takes place. If we take a very simple model of the energy per unit area contained in the first mode basin scale seiche to be equal to $1/4\ g'_{12}\ \rho_0\ a^2$, where it is assumed that the lake is modeled simply by a two layer system and the amplitude of the mode 1 wave is a. Again using the Lake Biwa data, with $g'_{12} = 3.2 \times 10^{-2}$, a = 5 m, and the average depth H = 40 m, the energy per unit mass = $6.5 \times 10^{-3}\ m^2\ s^{-2}$. Now, if it is assumed that the average dissipation in the water column is $10^{-9}\ m^2\ s^{-3}$ then the decay time for the waves would be about 75 days. This is a very large number, much larger than that which is observed. This can only lead to one conclusion and that is that the energy is predominantly dissipated at the boundaries near the depth of the metalimnion by internal wave modes shoaling and breaking and in the hypolimnion by bottom friction generated turbulence.

Once again consider the energy implications of this statement. The volume of Lake Biwa is about $2.7 \times 10^{10}\ m^3$ so that the total store of energy in the internal wave field, after a typhoon, would be about 1.8×10^{11} joules. Now suppose the thermocline has a thickness of 30 m, the perimeter at the thermocline depth is about 80×10^3 m and the slope is such that the wetted slope length subtended by the thermocline is 3 km. Then, if we assume a wave decay time of 10 days and a turbulent boundary layer thickness of 2 m the required dissipation in the benthic boundary layer would be about $4.2 \times 10^{-7}\ m^2\ s^{-3}$, a number representative of the measured values (as discussed above).

These arguments strongly suggest that the energy in the internal wave field is predominantly dissipated at the boundary.

Flux Path in a Stratified Lake

Given the above scenario and assuming that the groups of waves which bring the energy to the boundary are randomly distributed throughout the metalimnion then it is possible to apply the theory of Imberger and Ivey (1993) to estimate the volume flux induced within the benthic boundary layer.

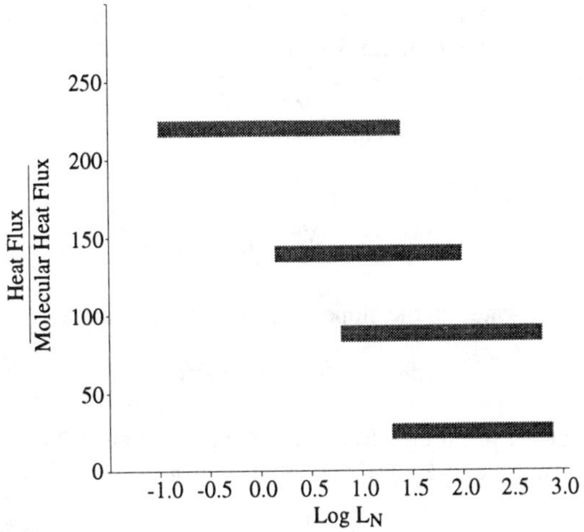

Figure 4. Basin average vertical flux as a function of Lake Number. Data from 3 lakes: Wellington Reservoir, Lake Rotonaigo, Canning Reservoir (after Imberger, 1989).

As an example let us take, once again, the data available from Lake Kinneret (Lemckert and Imberger, 1998). For this purpose assume the sin β is 10^{-2}, the thickness h of the benthic boundary layer is 2 m, the average buoyancy frequency $N = 5 \times 10^{-2}$ ms^{-1}, the internal wave velocity amplitude $\alpha = 0.05$ ms^{-1}, the slope length L is about 1000 m and the gravity jump across the whole thermocline $g' = 2.5 \times 10^{-2}$ ms^{-2}.

The vertical eddy diffusion coefficient within the benthic boundary is given by (2). Substituting the above values yields 1.6×10^{-4} m^2 s^{-1}. This may be compared to the estimate obtained from the direct measurements of the dissipation ε using (Ivey and Imberger, 1991):

$$K_\rho = \frac{R_f}{1 - R_f} \frac{\varepsilon}{N^2} \tag{13}$$

Using, as suggested by Lemckert and Imberger (1998), $R_f = 0.25$ and $\varepsilon = 10^{-6}$ m^2 s^{-3} yields a value for K_ρ of 1.4×10^{-4} m^2 s^{-1}, in good agreement with the above.

Now the volume flux induced by the gravitational circulation within the benthic boundary layer was shown by Imberger and Ivey (1993) to be given by:

$$q = 8.4 \times 10^{-6}\, Gr^2 A^3 K_\rho H^3 \left(\frac{1}{\Delta\rho}\frac{d\rho}{dz}\right)\left(\frac{1}{\Delta\rho}\frac{d^2\rho}{dz^2}\right), \tag{14}$$

$$Gr = \frac{g' h^3}{K_\rho^{\,2}} = 7.6 \times 10^6 \qquad A = \frac{h}{L} = 2 \times 10^{-3} \qquad K_\rho = 1.6 \times 10^{-4}\, m^2\, s^{-1}$$

$$H = 10\, m \qquad \frac{1}{\Delta\rho}\frac{d\rho}{dz} = 1.6 \times 10^{-1}\, m^{-1} \qquad \frac{1}{\Delta\rho}\frac{d^2\rho}{dz^2} = 4.9 \times 10^{-2}\, m^{-2}$$

Substituting these values into (14) leads to a value for

$$q = 4.9 \times 10^{-3}\, m^2\, s^{-1}. \tag{15}$$

Given that, at the depth of the metalimnion, the perimeter of the lake is about 35,000 m, the area of the lake is about 6×10^7 m^2 and the volume is about 6×10^8 m^3 we get a total discharge:

$$Q = 171.5\, m^3\, s^{-1}, \tag{16}$$

a cycle time T_c for the hypolimnetic water

$$T_c = \frac{V}{Q} = 40 \text{ days}, \tag{17}$$

and a travel time of the water up the slope in the benthic boundary layer of around

$$T_t = \frac{Lh}{q} = 4.7 \text{ days}. \tag{18}$$

If this flux is equated to an effective vertical lake averaged diffusion process then the effective diffusion coefficient K_ρ for the whole basin is given by (Imberger and Ivey, 1993)

$$K_\rho = \frac{HQ}{A} = 2.8 \times 10^{-5}\, m^2\, s^{-1}. \tag{19}$$

These various values all have magnitudes which are comparable with the known measurements during periods of low Lake Number. During calm day α and h would be less and also L is longer than 10^3 m for much of the slope.

This means the above numbers are an upper bound putting the overall estimates into the same range as the field measurements. The dependency of the energy transfer scenario on the Lake Number will require a series of detailed field experiments.

Conclusions

Examination of recent data from Lake Biwa and Lake Kinneret has allowed a scenario to be postulated for the way wind at the surface induces transport in a stratified lake. Initially the wind transfers momentum to the surface layer which then induces seiching of the thermal structure; the initial motion may be decomposed into various modes of seiching, Kelvin and Poincaré waves and forced gravity waves. The severity of the waves is inversely related to the magnitude of the Wedderburn and Lake Numbers. Much of this energy is transferred into higher frequency internal solitary waves, internally, by non-linear steepening and by shoaling at the boundary where the metalimnion intersects the lake boundary. These higher frequency waves propagate throughout the lake in groups and the spectrum of the isotherm displacement shows the well known ω^{-2} behavior. These waves are absorbed when they impinge on the boundary. The first mode basin scale seiches are thus damped by both boundary friction and non linear steepening. The wave breaking releases turbulent kinetic energy which drives a flux in the benthic boundary layer which in turn cycles the hypolimnetic water through the benthic boundary layer fast enough to fully explain the rate of change of temperature at the bottom of the hypolimnion of the lake.

Acknowledgment. This work is the result of field experiments funded by the Centre for Environmental Fluid Dynamics and forms contribution ED 1073 JI at the Centre for Water Research, The University of Western Australia.

References

Armi, L., Some evidence for boundary mixing in the deep ocean. *J. Geophys. Res., 83*, 1971-1977, 1978.

Beckenbach, E. F. (Ed.), *Modern Mathematics for the Engineer*. pp. 514, McGraw Hill, 1956.

Buffle, J. and H. P. van Leeuwen (Eds.), *Environmental Particles*. Vol 1 and 2. Lewis Publishers, Michigan, 1992.

Capblancq, J. and J. Catalan, in *Limnology Now, A Paradigm of Planetary Problems*, edited by R. Margalef, pp. 9-36, Elsevier Science BV., 1994.

Chu, C. K., and R. L. Chou, Solitons induced by boundary conditions, in *Advances in Applied Mechanics*, edited by J. W. Hutchinson and J. Y. Wu, *27*, 283-302, Academic Press, Cambridge, 1989.

De Silva, I. P. D., J. Imberger, and G. N. Ivey, Breaking of super-critically incident internal waves at a sloping bed. *Physical Processes in Lakes and Oceans*, edited by J. Imberger, pp. 475-484, AGU, this volume, 1998.

Eriksen, C. C. Implications of ocean bottom reflection for internal wave spectra and mixing. *J. Phys. Oceanogr. 15*, 1145-1156, 1985.

Etemad Shahidi, A. and J. Imberger, Anatomy of turbulence in thermally stratified lakes. Submitted to *Limnol. Oceanogr.*, 1998.

Fischer, H. B., E. J. List, R. C. Y. Koh, J. Imberger and N. H. Brooks (Eds), *Mixing in Inland and Coastal Waters*, pp. 483, Academic Press, New York, 1979.

Hamilton, D. P., S. G. Schladow, and I. H. Fisher, Controlling the indirect effects of flow diversions on water quality in an Australian Reservoir. *Environ. Intnl. 21*, 583-590, 1995.

Horn, D. A., J. Imberger and G. N. Ivey, The degeneration of mode one basin scale internal waves. To be submitted to *J. Fluid Mech.*, 1998.

Imberger, J., Selective withdrawal, A review. *Stratified Flows 2nd Int. IAHR Symposium*, Trondheim June 1980, 1, 381-400, 1980.

Imberger, J., The diurnal mixed layer. *Limnol. Oceanogr. 30*, 737-770, 1985.

Imberger, J. Vertical heat flux in the hypolimnion of a lake. *Proc. Tenth Australasian Fluid Mechanics Conf.*, Melbourne, Australia, December 1989. I, 2. 13-2. 16, 1989.

Imberger, J., Transport processes in lakes, A review, *in Limnology Now, a Paradigm of Planetary Problems*, edited by R. Margalef, pp. 99-194, Elsevier Science BV., 1994.

Imberger, J. and G. N. Ivey, On the nature of turbulence in a stratified fluid. Part 2, Application to Lakes. *J. Phys. Oceanogr. 21*, 659-680, 1991.

Imberger, J. and G. N. Ivey, Boundary mixing in stratified reservoirs. *J. Fluid Mech. 248*, 477-491, 1993.

Imberger, J. and J. C. Patterson, Physical Limnology, in *Advances in Applied Mechanics*, edited by T. Wu, Vol. 27, pp 303-475, Academic Press, Boston, 1990.

Imboden, D. M. and S. Emerson, Natural radon and phosphorus as limnologic tracers, Horizontal and vertical eddy diffusion in Greifensee. *Limnol. Oceanogr. 23*, 77-90, 1977.

Imboden, D. M. and A. Wüest, Mixing mechanics in lakes, in *Lakes*, edited by A. Lerman, Springer, New York, 1994.

Ivey, G. N. and F. M. Boyce, Entrainment by bottom currents in Lake Erie. *Limnol. Oceanogr. 27*, 1029-1038, 1982.

Ivey, G. I. and J. Imberger, On the nature of turbulence in a stratified fluid. Part 1, The energetics of mixing. *J. Phy. Oceanogr. 21*, 650-658, 1991.

Ivey, G. N. and R. I. Nokes, Vertical mixing due to the breaking of critical internal waves on sloping boundaries. *J. Fluid Mech., 204*, 479-500, 1989.

Ivey, G. N., P. De Silva and J. Imberger, Internal waves, bottom slopes and boundary mixing. Eighth *'Aha Huliko'a Proceedings: Flow-Topography Interactions*, January 1995, Hawaii, pp 199-206, 1995.

Ivey, G. N., J. Imberger and J. R. Koseff, Buoyancy fluxes in a stratified fluid. *Physical Processes in Lakes and Oceans*, edited by J. Imberger, pp. 377-388, AGU, this volume, 1998.

Javam, A., J. Imberger and S. Armfield, Numerical study of internal waves breaking on sloping boundaries. *J. Fluid Mech.*, (in press) 1998a.

Javam, A., J. Imberger and S. Armfield, Numerical study of internal wave-wave interactions. Submitted to *J. Fluid Mech*, 1998b.

Javam, A., J. Imberger and S. Armfield, Numerical study of internal wave-caustic and internal wave-shear interactions in a stratified fluid. Submitted to *J. Fluid Mech.*, 1998c.

Kullenberg, G., C. R. Murthy, and H. Westerberg, Vertical mixing characteristics in the thermocline and hypolimnion regions of lake Ontario. (IFYGL). *Proc. 17th Conf. Great lakes Res.*, Vol 1, pp 425-434. Intl. Assoc. Great Lake Res. Ann Arbor., 1974.

Lemckert, C. and J. Imberger, Turbulent benthic boundary layer mixing events in fresh water lakes. *Physical Processes in Lakes and Oceans*, edited by J. Imberger, this volume, pp. 481-494 1998.

Lemckert, C., A. Saggio and J. Imberger, Turbulent benthic boundary layers generated by shaoaling internal waves. *Limnol. Oceanogr.*, (in preparation) 1998.

Lin, C. L., J. H. Ferziger, J. R. Koseff, and S. G. Monismith, Simulation and stability of two dimensional internal gravity waves in a stratified shear flow. *Dyn. Atmos. Oceans*, 19, 325-366, 1993.

Maxworthy, T., J. Imberger and A. Saggio, A laboratory demonstration of a mechanism for the production of secondary, internal gravity-waves in a stratified fluid. *Physical Processes in Lakes and Oceans*, edited by J. Imberger, pp. 261-270, AGU, this volume, 1998.

McEwan, A. D., Interactions between internal gravity waves and their traumatic effect on a continuous stratification. *Boundary-Layer Met. 5*, 159-175, 1973.

Michallet, H. and G. N. Ivey, Experiments on mixing due to internal solitary waves breaking on uniform slopes. Submitted to *J. Geophys. Res.*, 1998.

Mortimer, C. H., Lake hydrodynamics. *Mitt. Int. Ver., Limnol., 20*, 124-197, 1974.

Munk, W., Internal waves and small scale processes, in *Evolution of Physical Oceanography*, edited by B. A. Wartren and C. Wunsch, pp 619, MIT Press, 1981.

O'Melia, C. R. and C. L. Tiller, Physico chemical aggregation and deposition in aquatic environments, in *Environmental Particles* Vol 2. edited by J. Buffle and H. P. Van Leeuwin, pp 353-386. Lewis Publishers, Boca Raton, 1993.

Osborn, T. R. and C. S. Cox, Oceanic fine structure. *Geophys. Fluid Dyn.* 3, 321-345, 1972.

Patterson, J. C., Modelling the effects of motion on primary production in the mixed layer of lakes. *Aquat. Sci. 53*, 218-238, 1991.

Plueddeman, A. J., J. A. Smith, D. M. Farmer, R. A. Weller, W. R. Crawford, R. Pinkel, and S. Vagle, Structure and variability of Langmuir circulation during the surface waves processes program. *J. Geophys. Res.* 101, 3525-3543, 1996.

Reynolds, C. S., The role of fluid motion in the dynamics of phytoplankton in lakes and rivers. *34th Symp. of the British Ecological Soc.*, edited by P. S. Giller, A. G. Hildrew and D. G. Raffaelli Aquatic Ecology, scales, pattern and process, pp 141-187, 1994.

Robarts, R. D. and P. R. B. Ward, Vertical diffusion and nutrients transport in a tropical lake (Lake McIlwaine, Rhodesia). *Hydrobiologia 59*, 213-221, 1978.

Saggio, A. and J. Imberger, Internal wave weather in a stratified lake. *Limnol. Oceanogr.* (in press) 1998.

Saggio, A. and J. Imberger, Mixing and buoyancy fluxes in the metalimnion of a stratified lake. To be submitted to *Limnol. Oceanogr.*, 1998.

Spigel, R. H., J. Imberger and K. N. Rayner, Modelling the diurnal mixed layer. *Limnol. Oceanogr. 31*, 533-556, 1986.

Stevens, C. and J. Imberger, The initial response of a stratified lake to a surface shear stress. *J. Fluid Mech., 312*, 39-66, 1996.

Stumm, W. (Ed.), *Chemical Processes in Lakes*. pp 435, John Wiley & Sons, New York, 1985.

Teoh, S. G., G. N. Ivey, and J. Imberger, Laboratory study of the interaction between two internal wave rays. *J. Fluid Mech.*, 336, 91-122, 1997.

Thomas, W. H. and C. H. Gibson, Quantified small scale turbulence inhibits a red tide Dinoflagellate Ganyaulox polyedra stein. *Deep Sea Res., 37*, 1583-1593, 1990.

Thorpe. S. A., D. Hall, and M. White, The variability of mixing at the continental slope. *Phil Trans. R. Soc. Lond. A331*, 183-194, 1990.

Turner, J. S., *Buoyancy Effects in Fluids*. Cambridge Press. pp 367., 1973.

Ward, P. R. B., Diffusion in lake hypolimnia. In *Proc. 17th Conv. Intl. Assoc. Hydraulic Research*, Baden Baden, IAHR. Vol 2, pp 103-110, 1977.

Wedderburn, E. M., Temperature observations in Loch Earn with a further contribution to the hydrodynamical theory of temperature seiches. *Trans. Roy. Soc. Edinburgh 48*, 629-695, 1912.

Wetzel, R. G., *Limnology*. W. B. Saunder Company, Philadelphia, 1975.

Wijesekjera, H. W. and T. M. Dillon, Internal waves and mixing in the upper equatorial Pacific ocean. *J. Geophys. Res. 96*, 7115-7125, 1991.

Winters, K. B., P. N. Lombard, J. J. Riley and E. A. D'Asaro, Available potential energy and mixing in density-stratified fluids. *J. Fluid Mech. 289*, 115-128, 1995.

Wüest, A., Ursprung und grosse von mischungsprozessen im hypolimnion naturlicher seen. *Rept. No. 8350*. Eidgenossischen Technischen Hoschschule, Zurich, 1987.

Wüest, A., D. C. Van Senden, J. Imberger, G. Piepke, and M. Gloor, Diapycnal diffusivity measured by microstructure and tracer techniques - a comparison. *Fourth International Stratified on Flows Symposium*, Grenoble, France, 3, B5 1-8, 1994.

2

Air-Water Exchange Processes

M.A. Donelan

Abstract

The significance of wind-generated waves in altering the coupling between air and water is discussed. Empirical growth formulae for wind waves and observed spectral forms are exploited to attempt to explain the wave age dependent wind stress, the enhanced kinetic energy dissipation in the wave zone and the strongly wind dependent mass transfer rates of slightly soluble gases.

Introduction

The exchange of momentum, energy and mass across the air-water interface to a large degree controls the weather, climate and the progress of life in the deep oceans. On a much smaller scale, physical limnology begins with the passage of energy across the boundaries of the water body, of which the air-water interface is paramount. The wind stress at the surface creates surface waves and drives the internal circulation of lakes. Radiative and turbulent heat exchange at the surface create and destroy the density stratification with seasonal regularity and less predictably on shorter time scales. In large shallow lakes, the wind-generated 'surface' waves, determine not only the roughness of the surface and the enhanced turbulent intensity resulting from their breaking, but also can reach down to interact directly with the bottom boundary layer and its complex community of benthic life, detritus and misplaced pollutants. We are concerned here with the turbulent exchange processes at the air-water interface, in particular the transfer of momentum, kinetic energy, heat and gases. The wind generated surface waves on scales of millimeters to tens of meters play important roles in all these processes, and this paper emphasizes these roles and draws attention to the significant gaps in our knowledge.

The appearance (or growth) of ripples on the surface follows an increase in the wind speed with an immediacy that leads to the identification of the one with the other. These first-formed short steep waves contribute to the aerodynamic roughness of the surface and, as the wind increases, they grow and break until, in moderate and strong winds (above 8 m s^{-1}), the wind stress appears to be supported principally by waves of various lengths, the viscous drag being relegated to an ever-diminishing role. The consequences of this are the following:

(i) The aerodynamic roughness of the surface depends on properties of the wave field;
(ii) At wind speeds above about 8 ms^{-1}, momentum is believed to be transferred between air

Physical Processes in Lakes and Oceans
Coastal and Estuarine Studies Volume 54, Pages 19-36

and water principally through waves and not by viscous drag on the surface. Thus, the kinetic energy is transferred at a rate commensurate with the phase velocities of the waves rather than the much smaller surface drift current. The resultant enhanced kinetic energy dissipation in the surface layer has important consequences for mixing and thermocline development, *inter alia.*

(iii) Gas transfer is restricted by thin diffusive layers near the surface. The breaking of waves of all scales acts to 'turn over' these thin layers and hence ventilate the boundary, and adds kinetic energy to the fluid, which erodes the thin boundary layer from below. The distribution of breaking with scale determines the efficiency of breaking-enhanced gas transfer.

In the following sections, these issues are dealt with in turn. Waves, their generation and dissipation, are critical in all cases.

The Aerodynamic Roughness of the Surface

Considerable experimental evidence has been accumulated on the effect of wave development on the roughness length, z_0, or drag coefficient, C_D - related through the logarithmic velocity profile characteristic of wall boundary layers:

$$U(z) = \frac{u_{*a}}{\kappa} \ln \frac{z}{z_0} \tag{1}$$

$$z_0 = z \exp\left(-\kappa / \sqrt{C_D}\right) \tag{2}$$

$$C_D = \frac{\underline{\tau}}{\rho_a U^2} = \left(\frac{u_{*a}}{U}\right)^2 \tag{3}$$

where U is the mean wind speed at height above the mean water level, $\kappa = 0.4$ is the von Kármán constant, $\underline{\tau}$ is the wind stress, ρ_a is the air density, and u_{*a} the friction velocity in the air.

For example, Donelan et al. (1993) relate the roughness length for fully rough flow ($u_{*a} z_0/\nu_a > 2.3$), normalized on rms wave height, σ, to the inverse wave age U_{10}/C_p:

$$\frac{z_0}{\sigma} = 6.7 \times 10^{-4} \left(U_{10} / C_p\right)^{2.6} \tag{4}$$

where U_{10} is the wind speed measured at 10 m height and C_p is the phase speed of the waves at the spectral peak, ν_a is the kinematic viscosity of the air. The lacustrine and oceanic data on which this is based are shown in Figure 1 (from Donelan et al., 1993). σ and C_p may be obtained through various wave growth formulae for fetch- or duration-limited conditions, e.g. Bishop and Donelan (1988), Donelan et al. (1992), and Hasselmann et al. (1973). As waves develop with time or fetch in a steady wind the frequency of the spectral peak (most energetic waves) decreases (see Figure 4). Correspondingly, the phase (propagation) speed of the peak, C_p, increases, until at "full development" it becomes about 20 % larger than the wind speed (Pierson and Moskowitz, 1964). Thus the ratio of phase speed at the spectral peak to wind speed, C_p/U_{10}, is a measure of the "age" of the wave field. The inverse of this "wave age" is a measure of the degree of wind forcing or transfer rate of momentum from the wind to the various wave components.

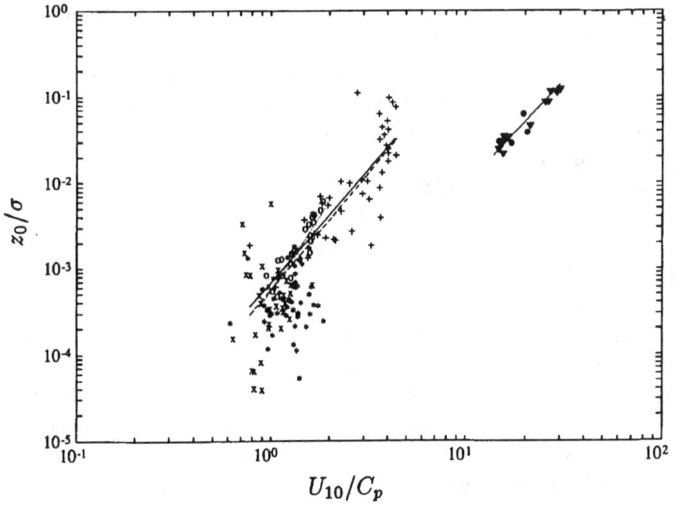

Figure 1. The ratio of measured roughness length z_0 to rms wave height σ versus inverse wave age U_{10}/C_p Roughness Reynolds number, $u_{*a}\, z_0/\nu_a > 2.3$, where ν_a is the kinematic viscosity of air. Symbols: Lake Ontario +; HEXOS (Smith et al. 1992) o; Atlantic Ocean, long fetch *; limited fetch ×; Donelan (1990) wave tank ∇; Keller et al.(1992) wave tank •; (From Donelan et al., 1993). Fetch is the over-water distance to the upwind shore.

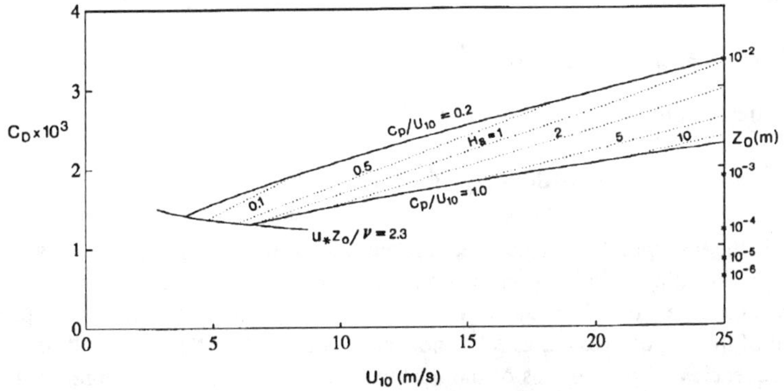

Figure 2. The relationship between drag coefficient and wind speed for various wave ages. The solid lines are for wave ages of 0.2 and 1.0 and the fully rough limit $u_{*a}\, z_0/\nu_a > 2.3$. The dotted lines are lines of constant significant height; the values indicated in meters. The corresponding roughness lengths are shown on the right-hand ordinate. (From Donelan et al., 1995).

Hsu (1994) has verified Eq. (4) using an independent data set. As U_{10}/C_p increases, corresponding to increasingly young waves, a larger fraction of the spectrum travels at slow speeds compared to the wind and therefore contributes to the roughness. The extent to which this affects the wind stress for a given wind is illustrated in Figure 2 (from Donelan et al., 1995). For a given wind speed, increasing U_{10}/C_p from 1 (almost fully developed) to 5 (very young waves) causes an increase in C_D (or $\underline{\tau}$) of $\approx$ 40 %. Large values of U_{10}/C_p correspond to short fetch or sudden increases in the wind, and thus significantly wave enhanced drag is a pervasive feature of limnology. This has been noticed since the early attempts at storm surge modelling on the Great Lakes (Platzman, 1963).

An observer travelling against a constant wind towards a lee shore – equivalent to moving upwards along a vertical line in Figure 2 – experiences increasing wind stress while encountering diminishing waves. The smaller waves at short fetch are rougher, apparently because they are steeper and white capping is more widespread. With increasing wind forcing (U_{10}/C_p) the steepness of waves near the peak increases uniformly. This is most clearly seen in the enhancement of the peak of the spectrum – first modelled by Hasselmann et al., 1973 – when spectra for different U_{10}/C_p are whitened ($\times\omega^4$) and plotted with their equilibrium ranges superimposed (Figure 3).

As waves develop with increasing fetch in an offshore wind, the spectral peak moves to lower frequencies and the frequency of the former peak moves to a position on the quasi-saturated tail ('equilibrium region') at a lower spectral energy level. Thus the energy at a particular frequency rises to a maximum with increasing fetch and falls again to a nearly constant level as illustrated in Figure 4. This 'overshoot' phenomenon becomes less pronounced as the waves move towards full development ($U_{10}/C_p \approx 1$, Figure 3). At the same time, however, the spectrum widens to include more longer waves (lower frequencies) while retaining the short waves in the equilibrium region. It is these short waves, whose phase speed $c < 5u_{*a} \approx 0.2$ U, that support most of the surface stress in aerodynamically rough flow (Phillips, 1977). Is the enhancement of the peak at higher U_{10}/C_p and the consequent increase in stress supported by the peak enough to offset the reduced stress to a narrower high frequency region? To answer this completely we need to know: (a) the rate at which momentum is transferred to various components of the spectrum, and (b) the full wavenumber spectrum.

The Rate of Amplification of Waves by Winds

The stress supported by all waves, $\underline{\tau}_i$, is given by:

$$\underline{\tau}_i = \int_0^\infty \tau_i(\omega)\,d\omega = \rho_w \; g\int_0^\infty \int_{-\pi}^{\pi} \frac{\beta}{\omega} F(\omega,\theta)\; k\cos\theta \; d\theta \; d\omega \tag{5}$$

where k is the wavenumber magnitude, ω is the radian frequency, ρ_w is the water density, $F(\omega,\theta)$ is the directional spectrum and β/ω is the non-dimensional exponential amplification rate. $\theta = 0$ is taken to be the wind direction and the mean propagation direction of the waves. Note that $\tau(\omega)$ and (later) $S(\omega)$ denote the contributions at various wave frequencies of gain or loss of momentum/energy, while $\underline{\tau}$ and $\underline{S}$ represent the integral over all frequencies, i.e. the total gain or loss.

β/ω is quadratic in the wind speed or friction velocity over a wide range of scales (e.g. Plant, 1982) and here we use the form given by Donelan and Pierson (1987) based on both field and laboratory experiments:

$$\frac{\beta}{\omega} = 0.194\frac{\rho_a}{\rho_w}\left(\frac{U_{(\pi/k)}\cos\theta}{C(k)} - 1\right)\cdot\left|\frac{U_{(\pi/k)}\cos\theta}{C(k)} - 1\right| \tag{6}$$

where the wind speed at one half wavelength (π/k is taken to be the relevant forcing parameter for a component of wavenumber, k. Note that β/ω (Eq. (6)) is deduced from observations of wind generated waves and therefore includes the enhancement of the growth rates due to flow separation over breaking waves (Banner and Melville, 1976).

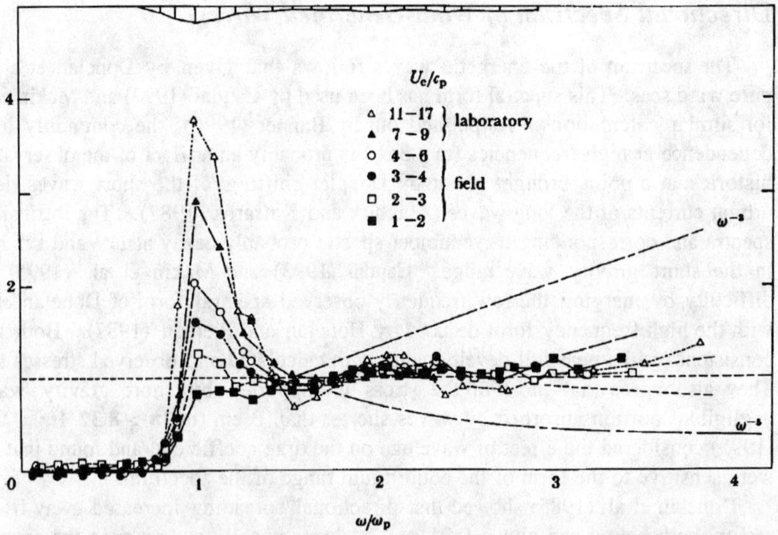

Figure 3. Frequency spectra times ω^4 normalized by the rear face $[\omega^4\ \phi(\omega)]_{r.f.}$ which is the average of $\omega^4\ \phi(\omega)$ in the region $1.5\ \omega_p < \omega < 3\ \omega_p$. The lines corresponding to ω^{-5} and ω^{-3} are also shown. The spectra are grouped in classes of U_c/C_p where U_c is the component of the 10 m wind in the propagation direction of the waves at the peak. (From Donelan et al., 1985). The ω^{-5} and ω^{-3} lines are drawn to illustrate the close adherence of the rear face to ω^{-4} (horizontal line). Relative to the normalized rear face, the "enhancement" of the peak with increasing forcing, U_c/C_p, is apparent.

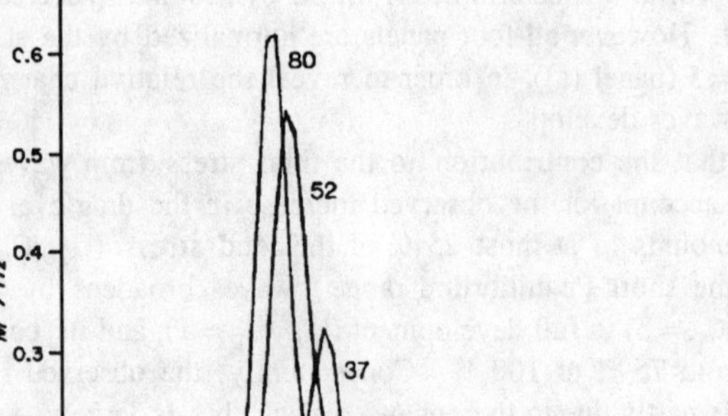

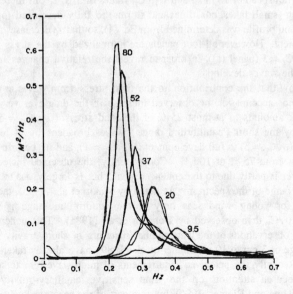

Figure 4. Evolution of wave spectra with fetch for offshore winds (11-12~h, Sept. 15, 1968). The spectra are labelled with the fetch in kilometers. (From Hasselmann et al., 1973).

Directional Spectrum of Wind-Generated Waves

The spectrum of the energetic waves follows that given by Donelan et al. (1985) for pure wind seas. This spectral form has been used by Caudal (1993) and Makin et al. (1995) for similar calculations. As pointed out by Banner (1990), the commonly observed ω^{-4} dependence at high frequencies ($\omega > 3\omega_p$) is probably an artifact of the observation of time histories at a point brought about by Doppler shifting of the short waves riding on the orbital currents of the long waves (Ataktürk and Katsaros, 1987). The intrinsic frequency spectra and corresponding wavenumber spectra probably decay at ω^{-5} and k^{-4} respectively in the short gravity wave range. Caudal (1993) and Makin et al. (1995) avoid this difficulty by merging the low frequency observed spectral form of Donelan et al. (1985) with the high frequency form deduced by Donelan and Pierson (1987). Both authors find consistent results near full development, with calculated and observed stresses about equal. They also agree that most of the stress is supported by short gravity waves with a negligible portion absorbed by waves shorter than 3 cm ($\omega/2\pi > 8.32$ Hz). Makin et al. (1995) considered the effect of wave age on the drag coefficient, and found that their results were sensitive to the form of the equilibrium range of the spectrum.

Donelan et al. (1985) showed that directional spreading increased away from the peak region, both below and above 0.95 ω_p, but were unable to determine the spreading much above the peak. Here we allow the spreading to be continuous and proportional to ω/ω_p in the range $0.95 < \omega/\omega_p < 3$. This is consistent with the observed ω^{-4} spectra in the energy containing region above the peak, Figure 3 (e.g. Donelan et al., 1985; Banner, 1990). From $3\underline{\omega}_p$ to high frequencies, we append an ω^{-5} tail (Terray et al., 1996) and further assume that the spreading in this quasi-saturated part of the spectrum is constant, having the value at $3\omega_p$. The calculated momentum distributed over wave frequency is indicated in Figure 5 for a wind speed of 10 m s^{-1} and U_{10}/C_p values of 5, 3, 2, 1 in increasing wave age from very young (small lakes, coastlines and storms) to fully developed (open ocean). In each case the wind profile was determined from Eq. (4), so that z_0 decreases as U_{10}/C_p tends to full development. However all four panels are normalized by the stress calculated from Eq. (4) for $U_{10}/C_p = 5$ (panel (a)), in order to reveal the relative changes in the calculated momentum as the waves develop.

It is clear now that the contribution to the total stress from waves at the peak of the spectrum does not account for the observed increase in the drag over very young waves. While this peak amounts to at most 25 % of the total stress ($U_{10}/C_p = 5$), the spectral range covered by the short ('equilibrium range') waves broadens by at least 50 % from young waves ($U_{10}/C_p = 5$) to full development ($U_{10}/C_p = 1$), and its contribution increases correspondingly from 75 % to 100 %. Consequently, the observed increase in the drag over young waves is partly due to the enhanced peak, but is largely due to the changes in the equilibrium range of the spectrum. As noted by Makin et al. (1995), their model yields increased drag for young wind seas only if the equilibrium range is more strongly dependent on U_{10}/C_p than observed by Donelan et al. (1985). This underscores the need for careful field observations of the wavenumber spectrum of short gravity waves – in the wavelength range of 3 cm to 10 m. Banner et al. (1989) have already taken some steps to fill this need. Recently the perceived requirements of microwave remote sensing and gas transfer have focused attention on the wind sensitive capillary-gravity range of the spectrum (e.g. Jähne and Riemer, 1990), but this part of the spectrum (wavelength < 3 cm) is almost irrelevant to the overall transfer of momentum and kinetic energy.

Various laboratory studies have demonstrated the pronounced effect of the artificially

introduced long waves on the spectral density of the short wind-generated waves (e.g. Mitsuyasu, 1966; Phillips and Banner, 1974; and Donelan, 1987), and recently Reid (1995) has shown similar effects from Southern Ocean data. Phillips and Banner (1974) suggest that the effect is due to the action of the wind drift and long wave orbital velocities in reducing the breaking amplitude of short waves. The amplitude modulation of the short waves by the oscillating current gradients of the long wave orbital velocities (Longuet-Higgins and Stewart, 1960; Longuet-Higgins, 1987) will produce breaking of the quasi-saturated short waves. Since the breaking process is generally much faster than the rate of regeneration by wind, the net effect is to reduce the amplitude of the short waves. This may very well be the principal process by which the surface becomes smoother as the waves develop – the long waves travel ever faster until they approach the speed of the wind and therefore do not themselves contribute to the drag; at the same time their orbital motions cause a reduction in the shorter waves that are the roughness elements.

How Much Momentum is Delivered to the Currents?

Of the total momentum delivered from air to water, $\underline{\tau} = \rho_a C_D U^2$, some remains in the wave field as the waves grow down fetch. By far the larger fraction is delivered immediately to the underlying water body either through direct viscous drag of one fluid on the other or through the intermediary of waves amplified and dissipated locally. This is generally believed to be greater than 90 % and increases to 100 % (by definition) as the wave field approaches full development. Hasselmann et al. (1973), based on the

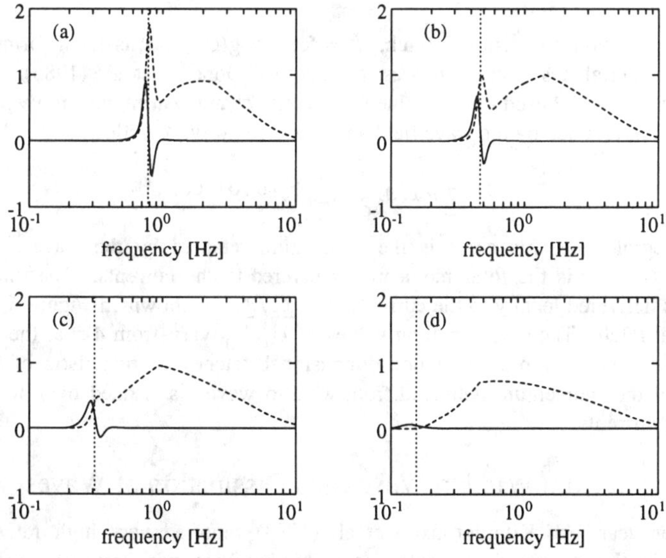

Figure 5. The spectral distribution of momentum input to waves from the wind $\tau_i(\omega)$ (- -), and of the momentum retained by the waves $\tau_r(\omega)$ (–). The four panels (a – d) are for values of the wind forcing, U_{10}/C_p of 5, 3, 2 & 1. The momenta have been normalized by the wind stress of panel (a) and multiplied by frequency (= $\omega/2\pi$), so that the curves are variance preserving. The vertical (dotted) line denotes the location of the frequency of the peak of the elevation (wave) spectrum.

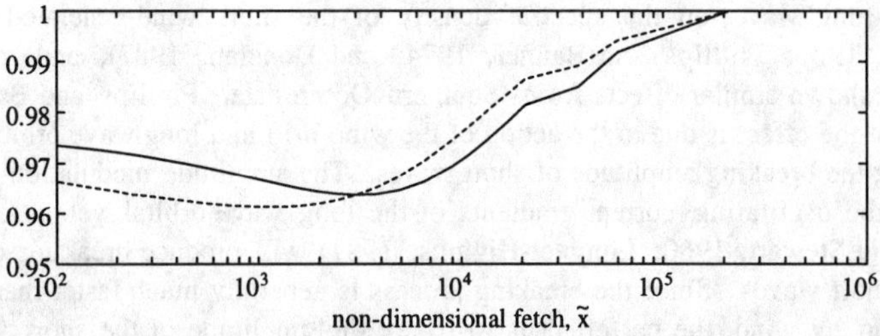

Figure 6. The fraction of momentum ($1 - \underline{\tau}_r / \underline{\tau}_i$), (- -); and of energy ($1 - \underline{S}_r/\underline{S}_i$), (–) _ from the wind that is delivered locally to the surface waters. The non-dimensional fetch, $\tilde{x}$ is related to the wind forcing parameter, U_{10}/C_p through Eq. (7).

JONSWAP empirical fetch formulae and an assumed drag coefficient of 0.001, found that 95 % of the momentum is delivered locally; while Donelan (1978), made direct measurements in fetch-limited conditions that were in keeping with the JONSWAP result.

The recent field study of Donelan et al. (1992) yielded an expression for the growth of waves from very short fetch to full development:

$$\tilde{x} = 4.0946\times10^4 \ln\left(\frac{U_{10}/C_p}{U_{10}/C_p - 0.8302}\right) - 3.3992\times10^4 \left(U_{10}/C_p + 0.4151\right)\cdot\left(U_{10}/C_p\right)^{-2} \tag{7}$$

where $\tilde{x}$ is the non-dimensional fetch, $\tilde{x}$ = fetch × g/U^2_{10}. This fetch formula, coupled with a full spectral description such as the modified Donelan et al. (1985) spectrum described above, may be used to determine the spectral changes in momentum with fetch; i.e. the momentum retained by the wave field and advected away, $\underline{\tau}_r$, illustrated in Figure 5:

$$\tau_r(\omega) = \frac{1}{2}\rho_w \; g\frac{\partial}{\partial x}\int_{-\pi}^{\pi} F\;(\omega,\theta)\;\cos^2\theta\;d\theta \tag{8}$$

The integral under this curve is the momentum retained by the waves, $\underline{\tau}_r$, and the difference $\underline{\tau}_i$ - $\underline{\tau}_r$) is the total momentum delivered to the currents. The fraction of the wind stress delivered locally to the currents ($1 - \underline{\tau}_r / \underline{\tau}_i$) is shown in Figure 6 versus non-dimensional fetch. The corresponding values of U_{10}/C_p vary from 4.5 at the left to 0.83 (full development). Even at short non-dimensional fetches characteristic of small lakes, virtually all the momentum delivered from wind to waves is handed over locally to the underlying currents.

Energy Input from Wind to Waves and Dissipation of Waves

Fourteen years ago Kitaigorodskii et al. (1983) reported very high rates of kinetic energy dissipation ε beneath strongly forced and breaking wind-generated waves in Lake Ontario. Observations of ε previous to that were in general accord with wall layer estimates; i.e. $\varepsilon = u^3_{*w}/\kappa z$ where u_{*w} is the friction velocity in the water. Since then enhanced (over wall layer) dissipation rates have been reported by several others; e.g. Gargett (1989), Agrawal et al. (1992), Anis and Moum (1992) and Terray et al. (1996). It seems likely that these very high dissipation rates near the surface are due principally to the release of wave energy in breaking, although no direct link between breaking and these field observations of enhanced dissipation has yet been established. Melville (1994) applied Phillips' (1985) estimates of dissipation in equilibrium spectra and his own estimates of breaking losses in coalescing laboratory groups (Rapp and Melville, 1990) to estimate the effect of breaking on the near surface layers. He concluded that enhanced volumetric dissipation rates should exist in the top layer to a depth of the order of the breaking wave height. Recognizing that nearly all the energy input to the wave field is delivered locally to the underlying water column, Terray et al. (1996) argued that the volumetric dissipation rate should be scaled on the overall energy input to the spectrum,

$$\underline{S}_i = \int_0^\infty S_i(\omega)\,d\omega = \rho_w\ g\int_0^\infty \int_{-\pi}^{\pi} \beta\ F(\omega,\theta)\ d\theta\ d\omega \tag{9}$$

Using the modified Donelan et al. (1985) spectrum described above and applying Eq. (6), they demonstrated that $\underline{S}_i$ may be 16 times the usual 'wall layer' estimate of $\underline{\tau}_i U_d$, where U_d is the surface drift velocity $\approx 0.5\ u_{*a}$ (Wu, 1975). (The amount of energy retained by the waves, $\underline{S}_r$,is a small fraction of $\underline{S}_i$ (see Figure 6), so that the rate of loss of wave energy to the water column $\approx \underline{S}_i$). Their observations indicate a decay rate of z^{-2} in an intermediate depth zone, which asymptotes to wall layer behaviour, z^{-1} at greater depths. Using the integral constraint (9), and ignoring the small amount of energy retained by the waves, they determine a depth z_b above which the dissipation rate must be relatively constant. The data yield $z_b = 0.6\ H_s$, where H_s is the significant wave height. This is in qualitative agreement with the laboratory results of Rapp and Melville (1990). This three-layer structure of the dissipation rate is illustrated in Figure 7 from Terray et al. (1996). An attempt to model the structure of energy dissipation in the upper layers, using an eddy viscosity model for the turbulence, has been proposed by Craig and Banner (1994).

It is of interest to examine the spectral distribution of dissipation in the wave field to see how this may produce the observed turbulent kinetic energy dissipation rates. We consider the energy balance equation (e.g. Komen et al. 1994) in terms of the frequency direction spectrum, $F(\omega,\phi)$:

$$\rho_w\ g\left\{\frac{\partial}{\partial t}+\mathbf{C}_g\cdot\frac{\partial}{\partial \mathbf{x}}\right\}\ F(\omega,\theta,\mathbf{x},t) = S_i + S_n + S_d \tag{10}$$

where $\mathbf{C}_g$ is the group velocity and the 'source functions' $S(F;\omega,\theta)$ are respectively: wind input, nonlinear wave-wave interactions and dissipation. The calculated spectral distribution of energy input to the wave field, $S_i(\omega)$ (Eqs. (6) and (9)) is plotted (dashed line) in Figure 8. Here, as for momentum, the fetch development formula (Eq. (7) permits the calculation of the energy retained by the wave field (solid line). The nonlinear wave-wave interaction term, $S_n(\omega)$ stabilizes the spectral shape and is believed to be largely responsible for the growth on the forward face (frequencies below the peak) of the spectrum. Exact calculations (Komen et al., 1994) indicate that $S_n(\omega)$ is largest near the spectral peak – having a positive lobe (spectral growth) at lower frequencies and a negative lobe at higher frequencies. Near full development the pattern shifts to higher frequencies so that the positive lobe coincides with the peak. The size of the energy retained by the waves, $S_r(\omega)$ on the forward face gives a measure of the maximum of $S_n(\omega)$ compared to $S_i(\omega)$. Clearly, over most of the spectrum $S_i(\omega)$ far exceeds $S_n(\omega)$ and the spectral balance must be essentially between wind input, $S_i(\omega)$ and dissipation, $S_d(\omega)$. The fraction of the energy given to the waves by the wind that is delivered locally to the upper mixed layer ($1 - \underline{S}_r/\underline{S}_i$) is indicated on Figure 6. As in the case of momentum, most of the energy passes through the wave field and leads to the enhanced dissipation levels observed near the surface.

Distribution of Dissipation Rate with Depth

The inferred balance between $S_i(\omega)$ and $S_d(\omega)$ over most of the spectrum may be used to determine the spectral distribution of $S_d(\omega)$. This is the rate of dissipation in the wave field per unit surface area. Assuming that the principal process of wave dissipation in gravity waves is through breaking, the depth of rapid mixing following the passage of a breaking event reveals the extent to which the flux of wave energy per unit surface area is distributed to volumetric dissipation of turbulent kinetic energy. The elegant laboratory experiments of Rapp and Melville (1990) indicate that breaking begins at slopes, ak of about 25 % and that the depth of initial mixing is about kz = 0.94 or z = 0.15 λ, where λ is the wavelength. We assume that in breaking, each wave deposits some energy and momentum mixed uniformly to a depth proportional to its length. This is consistent with mixing depths proportional to wave height, since breaking occurs in a relatively narrow range of slopes. These volumetric dissipation contributions from various wavelengths in the spectrum may be summed at each depth to yield the direct wave breaking-produced dissipation.

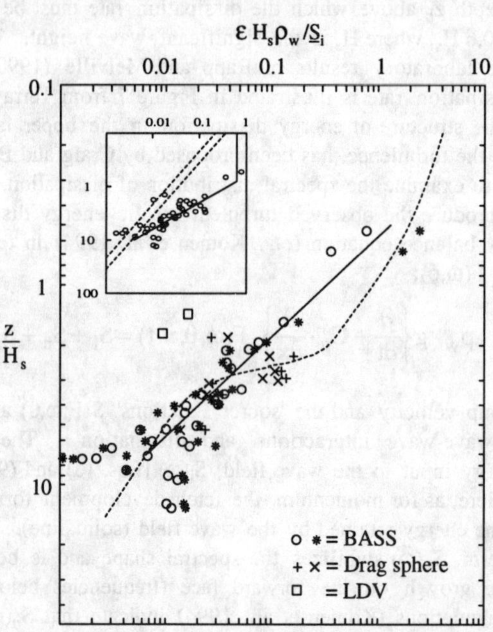

Figure 7. The normalized dissipation rate, $\varepsilon\ H_s\ \rho_w/\underline{S}_i$ versus dimensionless depth, z/H_s. The solid line is the regression line to the data, with a slope of z^{-2}; the dashed line is the calculated volumetric dissipation rate from the spectral distribution of wave dissipation per unit area. The inset shows the data denoted with o, the wall layer scaling for the data (dashed) and the Terray et al. (1996) proposed vertical structure: constant dissipation down to z_b, $(z/H_s)^{-2}$ scaling below that with a transition to wall layer scaling at greater depths. The two (dashed) wall layer lines show the possible range for the data in these coordinates. (From Terray et al., 1996).

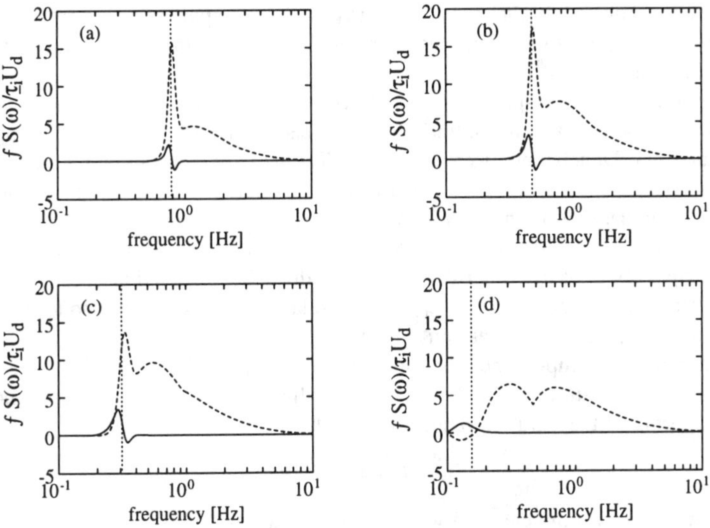

Figure 8. The spectral distribution of energy input to waves from the wind $S_i(\omega)$ (- -), and of the energy retained by the waves $S_r(\omega)$ (–). The four panels (a - d) are for values of the wind forcing, U_{10}/C_p of 5, 3, 2 & 1. The energy distributions have been normalized by $\underline{\tau}_i \times U_d$ where U_d, the surface drift velocity is set equal to $0.5u_{*a}$ The curves have been multiplied by frequency (= $\omega/2\pi$) so that they are variance preserving. The vertical (dotted) line denotes the location of the frequency of the peak of the elevation (wave) spectrum.

In addition, the release of momentum to the currents enhances the shear production of energy and further increases the total dissipation. This shear-produced dissipation is assumed to be $u^3_{*w}/\kappa z$, where the friction velocity in the water, u_{*w} increases with depth as the waves inject momentum into the upper layers. The calculated overall dissipation rate for the mean conditions of, and scaled according to, Terray et al. (1996) is shown (dashed) in Figure 7. Below 15 % of the wavelength of the longest waves contributing to the dissipation no kinetic energy is directly injected into the flow, and the calculated dissipation is assumed to follow the wall layer form, $u^3_{*w}/\kappa z$. Above this there is a layer of rapid increase of dissipation capped by a layer of enhanced dissipation having a weaker gradient. These calculations reflect only the direct effect of breaking in injecting kinetic energy to various depths. Subsequent diffusion of the injected energy will tend to produce a smoother depth dependence of the average dissipation. See Terray et al. (1996) for a detailed description of the observed kinetic energy distribution beneath breaking waves.

Field Observations of Wave Dissipation

Direct estimates of wave dissipation in the field are difficult to obtain and it is generally treated as a residual in attempts to establish the evolution of the wave field in response to the wind. Although there are some theoretical ideas on how dissipation should be modelled (see Komen et al., 1994), none has yet been put to the experimental test. As noted above, in strong forcing conditions the nonlinear transfer function $S_n(\omega)$ crosses zero at the spectral peak, the retained energy $S_r(\omega)$ is essentially zero at the peak and the calculated wind input, $S_i(\omega)$, is large. At steady state, therefore, in strong forcing conditions the balance at the peak is clearly between input and dissipation, $S_i(\omega)$ and $S_d(\omega)$. We can exploit this balance to calculate the form of the dissipation function near the spectral peak.

Observations of wave spectra in a shallow lake of about 20 km diameter (Quill Lake, Saskatchewan) were made in conditions of moderate and strong winds during June to October of 1993 (Donelan and Kenney, 1998). The measurements to be reported were made with a capacitance wave gauge on a tower in 1.2 m of water. The wind speed was measured by a cup anemometer at 3.9 m above mean water level. Wave dissipation due to interaction with the bottom is insignificant compared to that due to breaking when the wind is strong -- the principal effect of the water depth was to reduce the propagation speeds of the waves and thus keep the forcing rate (U_{10}/C_p) high at relatively long fetch. Bottom dissipation was calculated using the expression given in Komen et al. (1994). It is negligible compared to the wind input when the saturation spectrum (defined below) exceeds 0.02.

The wave dissipation is modelled, after Phillips (1985) and Donelan and Pierson (1987), in terms of the 'saturation spectrum', $B(\mathbf{k})$,

$$B(\mathbf{k}) = k^4 \Psi(\mathbf{k}), \tag{11}$$

where $\Psi(\mathbf{k})$ is the two-dimensional wavenumber spectrum:

$$S_d(\mathbf{k}) = -A_d \rho_w g \omega k^{-4} B^n(\mathbf{k}) \tag{12}$$

Si(wp) was calculated from (9) and plotted (Figure 9) against Sd(wp) (Eqs. (11) and (12)) at the spectral peak. There were 109 cases, each 20 minutes long. In the top panel $S_d(\omega_p)$ is calculated with n = 1, but as the wind input $S_i(\omega_p)$ increases, the dissipation rate is unable to keep up. Clearly a stronger dependence of $S_d(\omega_p)$ on wave steepness is called for. The value of n that yields the best correspondence between $S_i(\omega_p)$ and $S_d(\omega_p)$ is 5, somewhat stronger than Phillips' (1985) suggestion (n = 3), and in agreement with the choice of Donelan and Pierson (1987), which was selected to match radar reflectivity data in comparison with their spectral balance model. For n = 5, the constant A_d = 100.

Other models of the dissipation function, in which the nonlinearity appears as an integral property of the spectrum (e.g. Hasselmann, 1974), may fit this data also and a rigorous test of suitable dissipation functions will require a comparison above the peak as well. However, the Doppler shifting of point measurements due to the orbital velocities of longer waves renders them useless for calculating B(k) much above the peak. Here again the pressing need for wavenumber spectra over the entire gravity wave range is apparent.

The very strong dependence of $S_d(\omega_p)$ on degree of saturation underlines the nonlinearity of wave dissipation via breaking and its capacity for handling a wide range of wind forcing without much change in the spectral density – perhaps the principal reason for the existence of a quasi-saturated 'equilibrium range'.

Gas Transfer

The transfer of gases across the air-water interface is controlled by the resistance in the turbulent boundary layers and the molecular diffusive layers in direct contact with the interface. The turbulent diffusivities are far greater than the molecular diffusivities, and so the principal resistances are in the inner diffusive layers. Diffusion coefficients in air are about 10^4 times those in water so that gases of low solubility meet the largest resistance to transfer in the aqueous diffusive boundary layer.

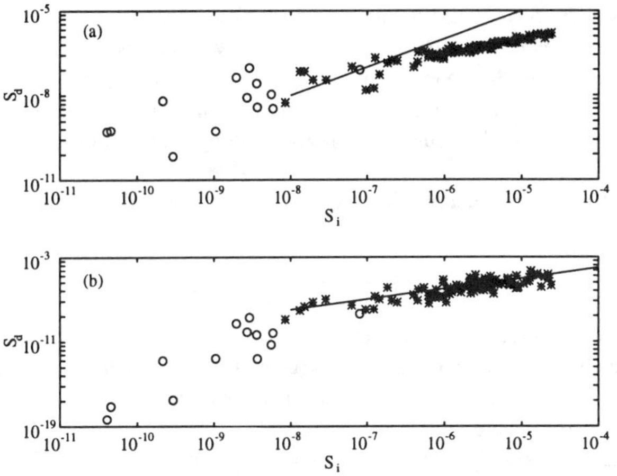

Figure 9. Comparison of calculated input $S_i(\omega_p)$ and dissipation $S_d(\omega_p)$ at the peak of the spectrum from wave records obtained in Quill Lake (Donelan and Kenney, 1998). Panel (a), $S_d(\omega_p)$ is assumed to be directly proportional to the spectral density; panel (b), $S_d(\omega_p)$ is proportional to the fifth power of the saturation spectrum, $B(\mathbf{k}) = k^4 \Psi(\mathbf{k})$. The straight lines represent a perfect balance between input and dissipation; (*) $B(\mathbf{k}) > 0.02$; (o) $B(\mathbf{k}) \leq 0.02$.

The transfer properties of gases that are controlled in the air boundary layer (e.g. H_2O. SO_2 and NH_3) have been successfully modelled using techniques borrowed from flow over solid walls: e.g. the surface renewal models of Brutsaert (1975) and Liu et al. (1979) and the mixing length model of Kitaigorodskii and Donelan (1984). The very large density difference between air and water allows one to treat the air flow in analogy to flow over rigid porous surfaces. On the other hand, the treatment of the diffusive properties of the aqueous boundary layer requires a different approach. Some of the reasons why the analogy with classical wall layers is inappropriate for the aqueous boundary layer are:

- horizontal velocities at the surface are essentially unconstrained;
- wave breaking produces intermittent injection of energy and momentum into the fluid;
- during the process of wave breaking the surface waters are renewed on a rapid time scale;
- intense wave breaking entrains air and bubbles may be carried to substantial depths.

In view of this one would expect the dissipation of surface waves to be of critical importance to the process of gas transfer of gases of low solubility; e.g. CO_2, O_2 and SF_6. In this section some expected effects of wave dissipation on gas transfer are discussed.

The mass transfer velocity or piston velocity, K is defined to be the ratio of gas flux to concentration difference between air and water. Brutsaert (1975) pointed out that the Schmidt number (Sc = the ratio of kinematic viscosity, ν to molecular diffusion coefficient, D) dependence of mass transfer should be proportional to $Sc^{-2/3}$ for smooth flow and $Sc^{-1/2}$ for rough flow, and this is in general agreement with wall layer experiments; e.g. Shaw and Hanratty (1977), Dawson and Trass (1972). Csanady (1990) showed that $Sc^{-1/2}$ is appropriate for a fluid boundary and the laboratory experiments of Jähne et al. (1987) are in general accord with this when the surface is disturbed by waves. They also showed significant enhancement of the mass transfer rates of water-phase limited compounds over solid wall results and suggested that the total mean square slope of the wave field was the relevant enhancement parameter. Comparison of mass transfer rates of water-phase limited compounds from various laboratory tanks (Ocampo-Torres et al., 1994) show systematic differences of a factor of two or so for winds above 6 m s^{-1}. The recent direct field measurements of the mass transfer rate of CO_2 by Donelan and Drennan (1995) indicate differences of 2 to 3 with the laboratory results of Ocampo-Torres et al. (1994), (Figure 10). This is possibly due to the higher energy dissipation in the field for the same wind speed.

Shear-free turbulence models (e.g. Lamont and Scott, 1970; Banerjee et al., 1968) yield:

$$K = a_1 \, Sc^{-1/2} \, (\varepsilon_0 \, \nu_w)^{1/4} \tag{13}$$

where ε_0 is the turbulent kinetic energy dissipation very near the surface, ν_w is the kinematic viscosity of the water, and a_1 is an empirical constant. Kitaigorodskii (1984) included the dissipation caused by wave breaking and showed that enhanced mass transfer rates could occur in these conditions.

The aqueous diffusive boundary layer is generally much less than 1 mm in thickness, so that an estimate of the dependence of ε_0 on the dissipation in the wave field per unit area may be obtained by distributing the wave dissipation over a depth commensurate with its wavelength ($z = 0.15\ \lambda$) as described above, and integrating over all wavelengths (frequencies) at depth ≈ 1 mm:

$$\varepsilon_0 = (0.15\ \rho_w)^{-1} \times \int_0^{\infty} (\lambda(\omega))^{-1} \int_{-\pi}^{\pi} S_i(\omega,\theta)\ d\theta\ d\omega \tag{14}$$

Figure 11 illustrates the spectral contributions to ε_0 from various wave ages and U_{10} = 10 m s^{-1}. This suggests that most of the enhancement of ε_0 comes from frequencies in the range of 1 Hz to 10 Hz or wavelengths 2 m to 2 cm. Increasing the wind speed to 20 m s^{-1} does not alter this range significantly.

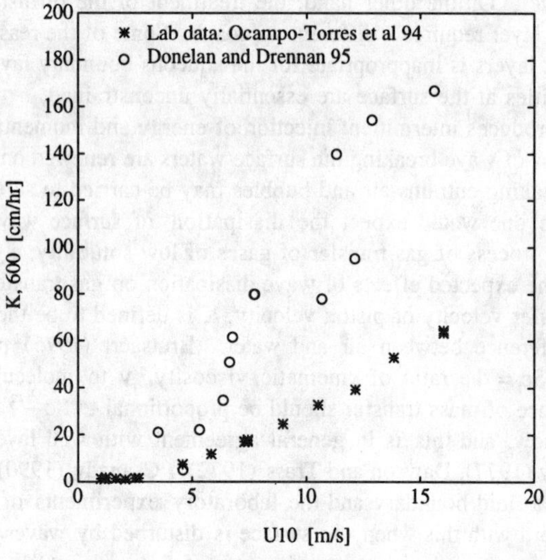

Figure 10. Mass transfer velocity ($K_{Sc=600}$) versus wind speed for the field measurements of Donelan and Drennan (1995) (o) and the laboratory experiments of Ocampo-Torres et al. (1994) (*). (From Donelan and Drennan, 1995).

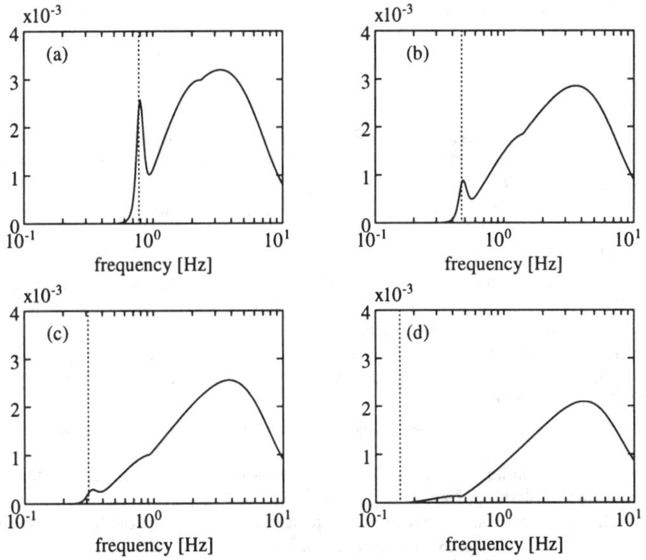

Figure 11. The spectral distribution of kinetic energy dissipation ($m^2 s^{-3}$) at the surface deduced from a balance between $S_i(\omega)$ and $S_d(\omega)$ in the spectrum of short gravity waves. The four panels (a – d) are for values of the wind forcing, U_{10}/C_p of 5, 3, 2 & 1. The curves have been multiplied by frequency (= $\omega/2\pi$) so that they are variance preserving. The vertical (dotted) line denotes the location of the frequency of the peak of the elevation (wave) spectrum.

Comprehensive reviews by Thorpe (1992), MacIntyre et al. (1995) and Melville (1996) have dealt with effects of enhancement of gas transfer due to chemical reactions and to bubble entrainment and super-saturation. Csanady (1990) has analysed the effect of wavelets in producing surface divergences that act to thin the diffusive boundary layer. In this paper the emphasis is on the direct role of all breaking waves in enhancing the turbulent kinetic energy dissipation levels beneath the surface.

Concluding Remarks

There is general agreement that wind-generated waves and their breaking alter the coupling between air and water, increase the mixing of near surface waters and enhance gas transfer. The details of the mechanisms involved remain to be clarified, but it is evident that further knowledge of wind-wave characteristics is a high priority. Among the important remaining questions for which there are yet no complete answers, must surely be included these:

- What is the complete wavenumber spectrum of wind-generated waves?
- What is the spectral (scale) distribution of wave dissipation?
- How much does wave breaking increase the aerodynamic roughness of the surface and the kinetic energy dissipation in the water?
- What is the balance of various breaking-related processes on the efficiency of gas transfer?

Acknowledgments. I thank W.M. Drennan for many helpful and enlightening discussions.

References

Agrawal, Y. C., E. A. Terray, M. A. Donelan, P. A Hwang, A. J. Williams III, W. M. Drennan, K. K. Kahma and S. A. Kitaigorodskii, Enhanced dissipation of kinetic energy beneath surface waves. *Nature, 359*, 219-220, 1992.

Anis, A. and J. N. Moum, The superadiabatic surface layer of the ocean during convection, *J. Phys. Oceanogr. 22*, 1221-1227, 1992.

Ataktürk, S. S. and K. B. Katsaros, Intrinsic frequency spectra of short gravity-capillary waves obtained from temporal measurements of wave height on a lake. *J. Geophys. Res., 92*, 5131-5141, 1987.

Banerjee, S., E. Rhodes and K. S. Scott, Mass transfer to falling wavy liquid films in turbulent flow. *Ind. Eng. & Chem. Fundamentals*, 7, 22-27, 1968.

Banner, M. L., Equilibrium spectra of wind waves. *J. Phys. Oceanogr. 20*, 966-984, 1990.

Banner, M. L., I. S. F. Jones and J. C. Trinder, Wavenumber spectra of short gravity waves. *J. Fluid Mech., 198*, 321-344, 1989.

Banner, M. L., and W. K. Melville, On the separation of air flow over water waves. *J. Fluid Mech., 77*, 825-842, 1976.

Bishop, C. T. and M. A. Donelan, Waves and wave forecasting. Chapter in Civil Engineering Practice, Vol. 3, Technomic Publishing, Lancaster, Basel, 653-695, 1988.

Brutsaert, W., A theory for local evaporation (or heat transfer) from rough and smooth surfaces at ground level. *Water Resource Res., 11*, 543-550, 1975.

Caudel, G., Self-consistency between wind stress, wave spectrum and wind induced wave growth for fully rough air-sea interface. *J. Geophys. Res. 98*, 22743-22752, 1993.

Craig, P. D. and M. L. Banner, Modeling wave-enhanced turbulence in the ocean surface layer. *J. Phys. Oceanogr. 24*, 2546-2559, 1994.

Csanady, G. T. The role of breaking wavelets in air-sea gas transfer. *J. Geophys. Res. 95*, 749-759, 1990.

Dawson, D. A., and O. Trass, Mass transfer at rough surfaces. *Int. J. Heat and Mass Transfer, 15*, 1317-1336, 1972.

Donelan, M. A., The effect of swell on the growth of wind waves. *Johns Hopkins APL Tech. Digest, 8*, 18-23, 1987.

Donelan, M. A., Air-sea interaction, in *The Sea*, edited by B. LeMéhauté, and D. Hanes, Volume 9A, pp. 239-292, Wiley,1990.

Donelan, M. A., F. W. Dobson, S. D. Smith and R. J. Anderson, On the dependence of sea surface roughness on wave development. *J Phys Oceanogr., 23*, 2143-2149, 1993.

Donelan, M. A., F. W. Dobson, S. D. Smith and R. J. Anderson, Reply to comments on *The dependence of sea surface roughness on wave development.*, *J Phys Oceanogr., 25*, 1908-1909, 1995.

Donelan, M. A., and W. M. Drennan, Direct field measurements of the flux of carbon dioxide, *in Air-Water Gas Transfer,* edited by B. Jähne and E. C. Monahan, pp. 677-683, AEON Verlag & Studio, 1995.

Donelan, M. A., J. Hamilton and W. H. Hui, Directional Spectra of wind generated waves. *Phil. Trans. R. Soc. London, A315*, 509-562, 1985.

Donelan, M. A., and B. C. Kenney, Observations of wave dissipation in wind-generated waves. (In preparation), 1998.

Donelan, M. A., M. S. Longuet-Higgins and J. S. Turner, Whitecaps. *Nature, 239*, 449-451, 1972.

Donelan, M. A. and W. J. Pierson, Jr., Radar scattering and equilibrium ranges in wind-generated waves with application to scatterometry. *J. Geophys. Res., 92, C5*, 4971-5029, 1987.

Donelan, M. A., M. Skafel, H. Graber, P. Liu, D. Schwab, and S. Venkatesh, On the growth rate of wind-generated waves. *Atmos. -Ocean, 30*, 457-478, 1992.

Gargett, A. E., Ocean turbulence. *Ann. Rev. Fluid Mech., 21*, 419-451, 1989.

Hasselmann, K., On the spectral dissipation of ocean waves due to whitecapping. *Bound. Layer Meteorol., 6*, 107-127, 1974.

Hasselmann, K., T. P. Barnett, E. Bouws, H. Carlson, D. E. Cartwright, K. Enke, J. A. Ewing, H. Gienapp, D. E. Hasselmann, P. Kruseman, A. Meerburg, P. Müller, D. J. Olbers, K. Richter, W. Sell and H. Walden, Measurements of wind-wave growth and swell decay during the Joint North Sea Wave Project (JONSWAP) . *Dt. Hydrogr. Z., A8 (Suppl.),* 12, 95 pp., 1973.

Hsu, S. A., A verification of two shear velocity equations for the wind-wave interaction in a lake environment. *Bound. Layer Meteorol., 71*, 205-209, 1994.
Jähne, B., K. O. Münnich, R. Bösinger, A. Dutzi, W. Herber and P. Libner, On the parameters influencing air-water gas exchange. *J. Geophys. Res., 92*, 1937-1949, 1987.
Jähne, B. and K. S. Riemer, Two dimensional wavenumber spectra of small-scale water surface waves. *J. Geophys. Res., 95, C7*, 11531-11546, 1990.
Keller, M. R., W. C. Keller, and W. J. Plant, A wave tank study of the determination of X-band cross sections on wind speed and water temperature. *J. Geophys. Res., 97, C4*, 5771-5792, 1992.
Kitaigorodskii, S. A., On the fluid dynamical theory of turbulent gas transfer across an air-sea interface in the presence of breaking wind waves. *J. Phys. Oceanogr., 14*, 960-972, 1984.
Kitaigorodskii, S. A. and M. A. Donelan, Wind-wave effects on gas transfer, in *Gas Transfer at the Water Surface,* edited by W. Brutsaert and G. H. Jirka, pp. 147-170, Reidel, 1984.
Kitaigorodskii, S. A., M. A. Donelan, J. L. Lumley, and E. A. Terray, Wave-turbulence interactions in the upper ocean. Part II: Statistical characteristics of wave and turbulent components of the random velocity field in the marine surface layer. *J. Phys. Oceanogr., 13*, 1988-1999, 1983.
Komen, G. J., L. Cavaleri, M. Donelan, K. Hasselmann, S. Hasselmann, and P. A. E. M. Janssen, Dynamics and modelling of ocean waves, Cambridge University Press, 560 pp., 1994.
Lamont, J. C., and D. S. Scott, An eddy cell model of mass transfer into the surface of a turbulent liquid. *J. Amer. Inst. Chem. Engrs., 16*, 513-519, 1970.
Liu, W. T., K. B. Katsaros, and J. A. Businger, Bulk parameterization of air-sea exchanges of heat and water vapor including the molecular constraints at the interface. *J. Atmos. Sci. 36*, 1722-1735, 1979.
Longuet-Higgins, M. S., The propagation of short surface waves on longer gravity waves. *J. Fluid Mech., 177*, 293-306, 1987.
Longuet-Higgins, M. S., and R. W. Stewart, Changes in the form of short gravity waves on long waves and tidal currents. *J. Fluid Mech., 8*, 565-583, 1960.
MacIntyre, S., R. Wanninkhof and J. P. Chanton, Trace gas exchange across the air-water interface in freshwater and coastal marine environments, In *Biogenic Trace Gases: Measuring Emissions From Soil And Water,* edited by. P. A. Matson and R. C. Harriss, pp. 52-97, Blackwell Science Ltd., 1995.
Makin, V. K., V. N. Kudryavtsev and C. Mastenbroek, Drag on the sea surface. *Boundary Layer Meteorol., 73*, 159-182, 1995.
Melville, W. K., Energy dissipation by breaking waves. *J. Phys. Oceanogr., 24*, 2041-2049, 1994.
Melville, W. K., The role of surface-wave breaking in air-sea interaction. *Ann. Rev. Fluid Mech., 28*, 279-321, 1996.
Mitsuyasu, H., Interactions between water waves and wind (1). Rep. Inst. Appl. Mech. Kyushu Univ., 14, 67-88, 1966.
Ocampo-Torres, F. J., M. A. Donelan, N. Merzi and F. Jia, Laboratory measurements of mass transfer of carbon dioxide and water vapour for smooth and rough flow conditions. *Tellus, 46B*, 16-32, 1994.
Phillips, O. M., The dynamics of the upper ocean, 2nd ed. Cambridge University Press, 336pp., 1977.
Phillips, O. M., Spectral and statistical properties of the equilibrium range in wind-generated gravity waves. *J. Fluid Mech. 156*, 505-531, 1985.
Phillips, O. M., and M. L. Banner, Wave breaking in the presence of wind drift and swell. *J. Fluid Mech., 66*, 625-640, 1974.
Pierson, W. J., and L. Moskowitz, A proposed spectral form of fully developed wind seas based on the similarity law of S. A. Kitaigorodskii. *J. Geophys. Res., 69*, 5181-5190, 1964.
Plant, W. J., A relationship between stress and wave slope. *J. Geophys. Res., 87, C3*, 1961-1967, 1982.
Platzman, G. W., The dynamical prediction of wind tides on Lake Erie. *Meteorol. Monographs, 4, 26*, 44pp., 1963.
Rapp, R. J. and W. K. Melville, Laboratory measurements of deep-water breaking waves. *Phil. Trans. R. Soc. London, A331*, 735-800, 1990.
Reid, J. S., Observational evidence of the interaction of ocean wind-sea with swell. *Aust. J. Mar. & Freshwater Res., 46*, 419-425, 1995.
Shaw, D. A., and T. J. Hanratty, Turbulent mass transfer rates to a wall for large Schmidt numbers. *AIChE J., 23*, 28-37, 1977.
Smith, S. D., R. J. Anderson, W. A. Oost, C. Kraan, N. Maat, J. deCosmo, K. B. Katsaros, K. L. Davidson, K. Bumke, L. Hasse and H. M. Chadwick, Sea surface wind stress and drag coefficients: the HEXOS results. *Bound. Layer Meteorol., 60*, 109-142, 1992.
Terray, E. A., M. A. Donelan, Y. C. Agrawal, W. M. Drennan, K. K. Kahma, A. J. Williams III, P. A Hwang and S. A. Kitaigorodskii, Estimates of kinetic energy dissipation under breaking waves. *J.*

Phys. Oceanogr., *26*, 792-807, 1996.

Thorpe, S. A., Bubble clouds and the dynamics of the upper ocean, *Q. J. Roy. Meteorol. Soc.*, *118*, 1-22, 1992.

Wu, J., Wind-induced drift currents. *J. Fluid Mech.*, *68*, 49-70, 1975.

3

Turbulent Flux of Water Vapor in Relation to the Wave Field and Atmospheric Stratification

K. B. Katsaros

Abstract

Field measurements of evaporation rate, momentum and heat flux together with mean meteorological quantities allow inferences to be drawn concerning the sheltering of air in the troughs of large waves and the possible differences between moisture and heat fluxes in very stable regimes over water. Interpretation of the measurements in terms of the processes near the surface are based on the Monin-Obukhov similarity theory. The results presented indicate that in the presence of water waves and under strong atmospheric stable stratification further detailed measurements are needed to fully explain the processes at work and that Monin-Obukhov theory may need modification.

Introduction

Turbulent fluxes of water vapor, heat and momentum measured directly by the eddy correlation or the inertial dissipation method in several field experiments, over a lake, in coastal waters and over the deep ocean are analyzed with the assumption of logarithmic profiles of humidity, temperature and wind to obtain the roughness lengths of these profiles, z_q, z_t and z_o respectively. The bulk aerodynamic method of estimating turbulent fluxes is also invoked to obtain the corresponding exchange coefficients for water vapor, heat and momentum: C_E, C_H and C_D (the drag coefficient). These exchange coefficients can be thought of as measures of the efficiency of the turbulence under particular conditions of mean meteorological observations at one height z, say 10 m. These efficiencies are strongly affected by the atmospheric stratification (parameterized as correction functions ψ_q, ψ_T and ψ_m to the logarithmic profiles of humidity, temperature and wind speed for neutral atmospheric stratification). We use Monin-Obukhov similarity theory as defined by Businger et al. (1971) and Dyer (1974). The exchange coefficients can also be expressed in terms of the roughness lengths. Using these parameters, we can explore the dependence of the flux of water vapor (or the efficiency of the turbulent process) on the mean wind speed (or the drag) on the sea state (roughness) and the role of negative buoyancy (stable stratification) in restricting evaporation from a water surface as compared to sensible heat transfer.

Physical Processes in Lakes and Oceans
Coastal and Estuarine Studies Volume 54, Pages 37-46

The issues raised in this presentation are derived from field studies reported on in detail in joint publications with several collaborators.

Observations from our field station on Lake Washington placed at the end of a 7 km over-water fetch along a rather narrow embayment are reported on by Ataktürk and Katsaros (1998). They show the effects of relatively low wave heights for a fixed wind stress (due to the narrowness of the water body, we suggest) on the vapor flux and provide cases of stable stratification. Measurements from the Humidity Exchange over the Sea (HEXOS, Katsaros et al., 1987) experiments obtained in 1986 at the Noordwijk tower, 10 km off the Dutch coast in 18 m of water, are reported on by DeCosmo et al. (1996). They show the possible effects of large and steep waves on the water vapor exchange, and compare well with the model developed by Liu et al. (1979), which suggests that sheltering in wave troughs is responsible for a reduction of z_e and z_t as waves grow. Similar results have also been deduced from observations in deep water with large fetch-unlimited waves in the Azores region during the SOFIA (Surface de l'Océan les Flux et leurs Interactions avec l'Atmosphère) and SEMAPHORE (Structure des Echanges Mer Atmosphère Propriétés des Hétérogénéties de l'Océan, leur Répartition) experiments (e.g. Dupuis et al., 1995).

Similarity Theory for the Bulk Exchange Coefficients of Momentum Heat and Water Vapor

Surface fluxes of momentum, τ, sensible heat, H, and latent heat, H_L, can be directly determined by the eddy correlation technique (Busch, 1973),

$$\begin{aligned} \tau &= -\rho\ \overline{u'w'} \\ H &= \rho\ C_p\ \overline{w'T'} \\ H_L &= \rho L_e\ \overline{w'q'} \end{aligned} \tag{1}$$

where u', w', T' and q' are the turbulent fluctuations of horizontal and vertical components of wind, air temperature and specific humidity, respectively, ρ is the air density, C_p is the specific heat of air at constant pressure and L_e is the latent heat of evaporation; the overbars denote a space average. One often applies the ergodicity principle which assumes that stationarity is achieved over a long enough averaging time such that a time average is equivalent to a space average.

Monin and Obukhov (1954) similarity theory predicts that the turbulence structure in the surface layer with approximately constant fluxes with height for horizontally homogeneous conditions and stationarity can be scaled with characteristic velocity, temperature and specific humidity scales (designated with * 's) that depend on the fluxes and with the Obukhov length scale L, as follows:

$$u_* = \left(\frac{\tau}{\rho}\right)^{1/2} \ ; \ t_* = -\frac{H}{\rho C_p} u_*^{-1} \ ; \ q_* = -\left(\frac{H_L}{\rho L_e}\right) u_*^{-1} \ ; \ L - \frac{T_v u_*^3}{\rho K \overline{w' T_v'}} \tag{2}$$

where T_v=T(1+0.61q) is the virtual temperature in °K and K = 0.40 is the von Kármán constant. L characterizes the height where the mechanical production and the buoyant energy production become equal. For a more complete development of the equations in this section, please see Ataktürk and Katsaros (1998). Taking into account the adjustments of the profiles of wind speed temperature and humidity due to stratification, z/L, determined by L above, we can write the profiles in their integrated form as:

$$U - U_s = \frac{u*}{K}\left[\ln\frac{z}{z_o} - \psi_u(z/L)\right],$$

$$T - T_s = \frac{t*}{K}\left[\ln\frac{z}{z_t} - \psi_t(z/L)\right], \qquad (3)$$

$$Q - Q_s = \frac{q*}{K}\left[\ln\frac{z}{z_q} - \psi_q(z/L)\right],$$

with the stratification corrections functions given by:

$$\psi_u = \psi_t = \psi_q = -7z/L; \qquad z/L > 0,$$

$$\psi_u = 2\ln\left[\frac{1+X}{2}\right] + \ln\left[\frac{1+X^2}{2}\right] - 2\tan^{-1}(X) + \frac{\pi}{2}; \qquad z/L < 0, \qquad (4)$$

$$\psi_t = \psi_q = 2\ln\left[\frac{1+X^2}{2}\right]; \qquad z/L < 0.$$

where $X = (1 - 16z/L)^{1/4}$.

These equations are not applicable very close to the surface where the turbulence is suppressed and the molecular transports through viscosity, conduction and diffusion become important.

Since the turbulent fluxes are often not measured directly, we resort to the so-called "bulk" estimation techniques employing measured mean meteorological values at one height and surface values rather than the turbulent fluctuations of equation 1.

For this purpose we must evaluate the bulk coefficients in the following equations empirically:

$$\frac{\tau}{\rho} = C_D\left[U_{(Z)} - U_s\right]^2 = u_*^2,$$

$$\frac{H}{\rho C_p} = C_H\left[U_{(Z)} - U_s\right]\left[T_s - T_{(Z)}\right] = -u_* t_*, \qquad (5)$$

$$\frac{H_L}{\rho L_e} = C_E\left[U_{(Z)} - U_s\right]\left[Q_s - Q_{(Z)}\right] = -u_* q_*.$$

where C_D is the drag coefficient, C_H is the Stanton number, and C_E is the Dalton number. Since the coefficients C_D, C_H and C_E are both height and stratification dependent, if we use measured values of U(z) T(z) and Q(z), we typically convert the measurements to equivalent values at 10 m height and neutral stratification using the flux profile relations iteratively obtaining expressions for the drag coefficient, the Stanton and Dalton numbers under neutral stratification and at 10 m height as follows:

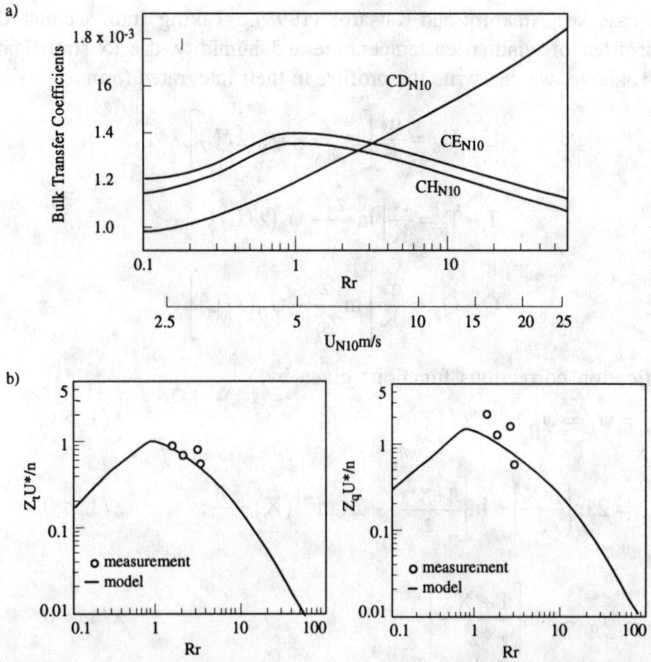

Figure 1. a) C_D, C_E and C_H as functions of Rr or U_{10}. and b) normalized z_T and z_q from the LKB model as functions of roughness Reynolds number, Rr, with a few field measurements from Lake Washington. (From Liu, Katsaros, Businger, 1979)

$$C_{DN10} = K^2\left[\ln\frac{10}{z_o}\right]^{-2},$$

$$C_{HN10} = K^2\left[\ln\frac{10}{z_o}\right]^{-1}\left[\ln\frac{10}{z_t}\right]^{-1} = KC_{DN10}^{1/2} / \ln\frac{10}{z_t}, \qquad (6)$$

$$C_{EN10} = K^2\left[\ln\frac{10}{z_o}\right]^{-1}\left[\ln\frac{10}{z_q}\right]^{-1} = KC_{DN10}^{1/2} / \ln\frac{10}{z_q}.$$

Liu, Katsaros and Businger (1979) incorporated effects of the molecular sublayers in air and water when solving these equations iteratively for the flux values, while starting with mean meteorological observables from the surface layer, and the "skin" rather than the "bulk" sea surface temperature. Classical empirical results obtained in the laboratory provide the foundation for this development. Figure 1 illustrates the results.

Experimental Results

Roughness Lengths: Lake Washington and HEXOS

The Liu et al. (1979) formulation gives values of C_E, C_H, z_q and z_t decreasing with increasing $\overline{U}$ beyond roughness Reynolds number equal 1. The measurements obtained at that time at a field station with 180° exposure to Lake Washington, where it is a wide body of water, were too few to fully validate or refute this model.

Later measurements obtained in the HEXOS program show a C_E value constant at $1.2 \pm 0.2 \times 10^{-3}$ over a wide range of wind speeds (6 - 18 m s^{-1}, DeCosmo et al., 1995), but because of the dependency of C_E on C_D – (6), this also implies a z_q decreasing with $\overline{U}$ where C_D has a linear increase with $\overline{U}$ (e.g. DeCosmo et al. 1996). The values of z_q obtained from measured $\overline{w'q'}$ and $\overline{q}$ at height z (Figure 2 after DeCosmo et al., 1996), assuming a logarithmic-linear profile, show an excellent fit with the Liu et al. (1979) formulation, albeit with large uncertainty limits on the measurements. It is only because we have an unusually large data set from one experiment that we can derive z_q at all. It is a

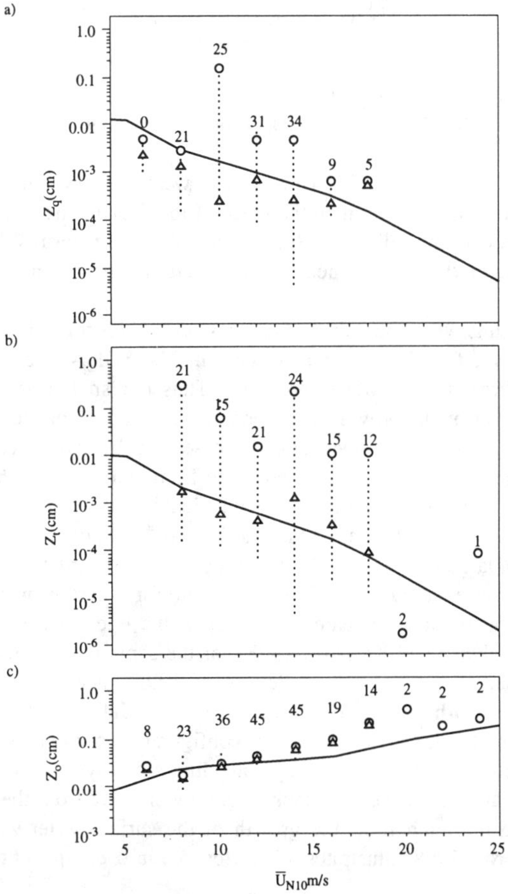

Figure 2. Log of (a) the water vapor roughness length z_q, (b) the temperature roughness length z_t, and (c) the roughness length z_o calculated from data obtained in the HEXOS experiment by the University of Washington team as a function of the mean wind speed. The solid lines are calculated with the surface layer model of Liu et al. (1979) as it is represented in the planetary boundary layer model of Brown and Liu (1982). The data have been combined in 2 m s^{-1} wind speed categories. Within each category, triangles represent the median value, dotted lines the interquartile range, and circles the mean; the numbers of points in each category is shown. (After the article by DeCosmo et al., 1996.).

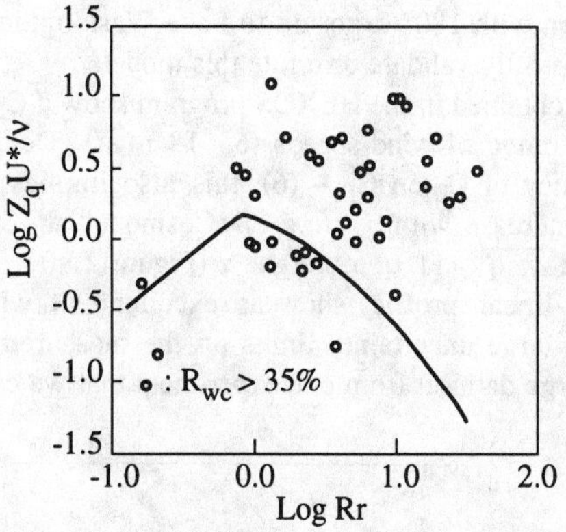

Figure 3. $z_q u_*/\nu$ against roughness Reynolds number, $Rr = z_o u_*/\nu$ where ν is the viscosity of air. The solid line is the prediction by Liu et al. (1979). (O) (After Ataktürk, 1991).

very difficult quantity to obtain. The HEXOS data were averaged into 2 m s^{-1} wind speed bins and we plot the median rather than the mean of the data in logarithmic form (to avoid the effects of a few extreme points). Similar corroborating results have been found in measurements of vapor fluxes (obtained with the inertial subrange method) (not shown) for the SOFIA experiment.

The HEXOS results were obtained on the Noordwijk platform in 18 m of water where the 300 m or longer waves feel the bottom and steepen. The SOFIA data were obtained on the deep ocean with large surface waves present. Thus the sheltering of air close to the interface in the wave troughs provide a reasonable explanation for the behavior of z_q; why the flux efficiency, C_E itself, does not decrease can be explained by the drag C_D simultaneously increasing as form drag is added to frictional drag (Eq. 6 for C_E.).

As a corollary, that does not prove the physical mechanism implied above, but is consistent with it, we present the result that no decrease in z_q or z_t with $\overline{U}$ (between 3 and 8 m s^{-1}) is seen in a large sample of Lake Washington data (Figure 3). This is reasonable since the waves measured at our Lake Washington site for a certain wind stress are smaller in amplitude by a factor of 2 compared to waves in the open ocean or on Lake Ontario (Ataktürk, 1991, Ataktürk and Katsaros, 1998) for the same mean stress.

For an even broader fetch Kahma and Petterson (1995) measured even larger waves (as represented by the growth factor, α). We believe with Kahma and Petterson (1995) that the differences in amplitude may be due to the configuration of the body of water. For Lake Washington absorption of the wave energy spreading sideways by the beaches parallel to the fetch direction, and the consequent lack of energy arriving from the directions of these shorelines all along the fetch limit the growth of the surface water waves. Figure 4 after Ataktürk and Katsaros (1998) illustrates this difference in α compared to data of Donelan et al. (see Donelan, 1990).

Effects of Stable Stratification on Water Vapor Flux Efficiency

The air flowing over lakes often have more extreme values of stratification than one finds over the sea due to the greater heating or cooling of the nearby land surfaces and due to generally lower wind speeds. C_{E10N} and C_{H10N} obtained at Lake Washington over a

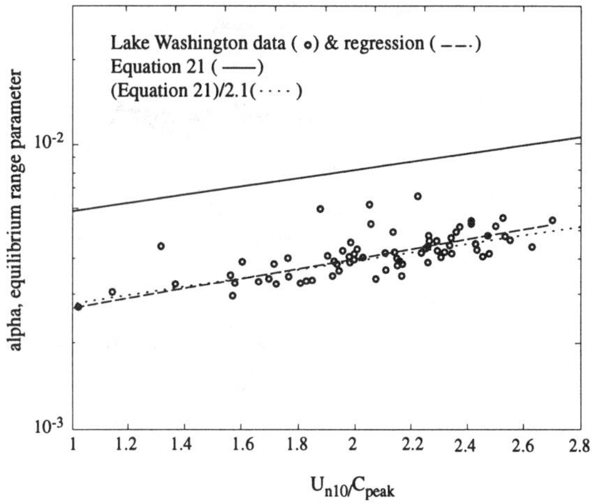

Figure 4. The equilibrien range parameter, α, versus wind forcing, showing that results from Lake Washington differ from Donelan et al., 1985 (solid line) in magnitude by a factor of 2.1 systematically (After Ataktürk and Katsaros, 1998).

period of years under very stable conditions (z/L as large as 3.0.) has been reported on by Ataktürk and Katsaros (1995). Figure 5 from that paper shows that the equivalent neutral Dalton number is lower than other values in the literature and is lower than the Stanton number for the same condition. However, such strongly stable conditions have not been measured previously. An attempt to explain why this difference between C_{H10N} and C_{E10N} occurs goes as follows: the buoyant plumes of water vapor enriched air rising from the water surface have little chance to reach all the way to 10 m height (or the level of mean humidity measurements, about 6 m on Lake Washington).

Downward sensible heat flux is produced under these conditions when intermittent breakdown of the stable density gradient due to temperature occurs. It is accompanied by downward momentum transfer. This process is likely to be more efficient than the vapor flux, since the momentum and heat transfer occur together. If some of the downward-directed bursts do not reach all the way down to the water surface, the low level moist plumes may never reach the flux measuring level (also about 10 m in height).

This argument implies, that our concepts of a constant flux layer and our eddy correlation technique for measuring turbulent fluxes are possibly not appropriate for very stable stratification; or, if our averages are really representative, that water vapor flux processes are less efficient than sensible heat flux processes in very stably stratified regimes. More detailed near surface measurements in the field are needed.

Summary and Discussions

The main conclusion is that sea state variability contributes significantly to the variations in the efficiency of the turbulence responsible for the vertical flux of water vapor from sea-to-air (for the same mean meteorological conditions). Large waves create sheltering in the troughs, where reduced turbulence (increased dependence on molecular transfer) results in net decrease of the flux—equivalent to a smaller roughness length (or a reduced mixing length as suggested by Donelan, 1990).

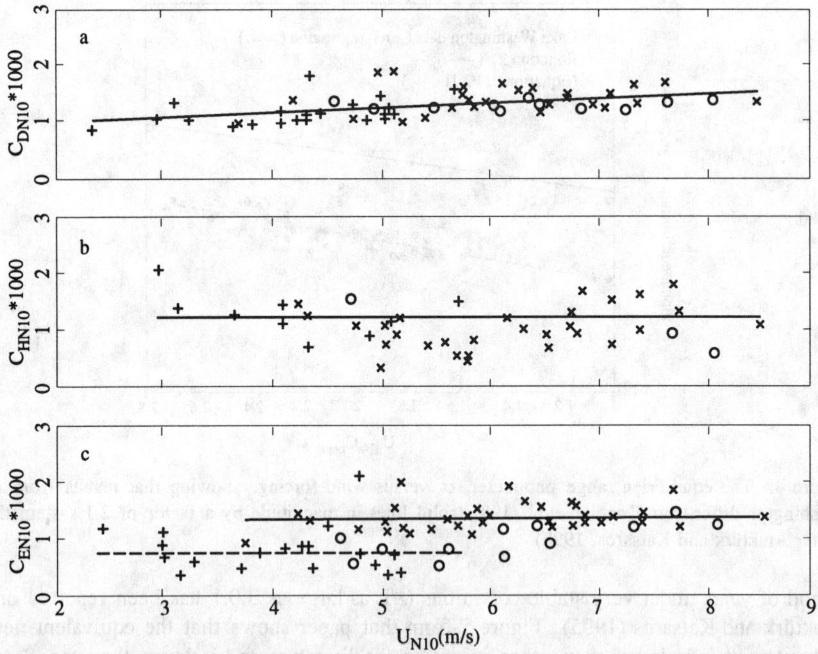

Figure 5. Bulk transfer coefficients obtained from atmospheric turbulence measurements at MSMAST on Lake Washington. With increasing wind speed, drag coefficient (a) becomes larger while Stanton (b) and Dalton (c) numbers remain unchanged. The data points with air-water temperature difference less than 1°C have been excluded from (b). Symbols indicate the atmospheric stratification associated with a data point, (x) unstable z/L<-0.1, (+) stable z/L>0.05, and (o) near neutral. Solid lines in top and middle plots indicate linear fit to the whole data set. Solid and dashed lines in the bottom plot are associated with data points from unstable and stable stratification conditions, respectively. Note that the neutral Dalton number in the latter case is reduced to half of its value for unstable stratification. Regression line and the root mean square error, erms, for each plot are:
(a) CdN10 = [0.08 UN10 + 0.84] x 10-3, erms = 0.35 x 10-3
(b) ChN10 = 1.24 x 10-3 erms = 0.69 x 10-3
(c) CeN10 = 1.39 x 10-3 for z/L < -0.1 (unstable), erms = 0.20 x 10-3
CeN10 = 0.67 x 10-3 for z/L > 0.05 (stable), erms = 0.24 x 10-3
(From Ataktürk and Katsaros, unpublished manuscript).

Our results on the Dalton number for z/L positive may indicate that flux-profile similarity relations cannot be extended to very stable conditions. z/L for many of these data points correspond to stabilities near the critical Richardson number. Other aspects of the flow, such as instabilities on the nocturnal jets and other upper level mesoscale or secondary flow effects may be responsible for variations in the flux (more so than in slightly stable or unstable regimes).

The issues raised here are only some of several indications that we need to look more carefully at the larger context and short term history of the wind and wave field when evaluating surface fluxes by direct measurements or parametrization schemes.

The interpretations of the measurements offered here have not been verified by near surface measurements, which are extremely difficult to make. They do not include the effects of evaporating sea spray (spume drops from the crests or bubble-mediated jet or film drops) on the net flux. For extremely high winds these effects may modify the profiles and, if these effects are important, the decrease in z_g with increasing $\overline{U}$ may simply reflect modification in the shape of the humidity profile due to evaporated spray (See for instance Andreas 1992, 1994, DeCosmo et al., 1996, Katsaros and DeCosmo 1990, 1993, Katsaros and de Leeuw 1994, Andreas et al., 1995, for discussion of this and related issues).

Acknowledgments. I am grateful to all my collaborators in the projects reported on here, in particular S.S Ataktürk, J. DeCosmo and H. Dupuis. This review has been supported by the National Science Foundation under Grant ATM-9024698 and by Institut Français de Recherche pour l'Exploitation de la Mer.

References

Andreas, E. L., Sea spray and the turbulent air-sea heat fluxes, *J. Geophys. Res.* 97, 11429-1141, 1992.

Andreas, E. L., Reply, *J. Geophys. Res.* 99, 14345-14350, 1994.

Andreas, E. L., J. B. Edson, E. C. Monahan, M. P. Rouault and S. D. Smith, The spray contribution to net evaporation from the sea, *Boundary-Layer Meteorol*, 72, 3-52, 1995.

Ataktürk, S. S., Characterization of roughness elements on a water surface. Ph. D. Dissertation, University of Washington, AK-40, Seattle, WA, 98195, 196 pp, 1991.

Ataktürk, S. S. and K. B. Katsaros, Wind stress and surface waves observed on Lake Washington. *J. Phys. Ocean* (in press) 1998.

Brown, R. A. and W. T. Liu, An operational large-scale marine planetary boundary layer model. *J. Appl. Met.,* 21, 261-269, 1982.

Bush, N. E., On the mechanics of atmospheric turbulence. *Workshop on Micrometeorology,* D. A. Haugen, ed., Am. Meteor. Soc., Boston, MA, 1-65, 1973.

Businger, J. A., J. C., Wyngaard, Y. K. Izumi, and E. F. Bradley, Flux-profile relations in the atmospherice surface layer. *J. Atmos. Sci., 28,* 181-189, 1971.

DeCosmo J., K. B. Katsaros, S. D. Smith, R. J. Anderson, W. A. Oost, K. Bumke and H. Chadwick, Air-sea exchange of water vapour and sensible heat: The Humidity Exchange over the Sea (HEXOS) results. J. Geophys. Res., 101, 12001-12016, 1996.

Donelan, M. A., Air-Sea Interaction. In *The Sea*: Ocean Engineering Science, edited by B. LeMehaute and D. Hanes, Vol. 9, pp 239-292, Wiley-Interscience, N. Y., 1990.

Dupuis, H., A. Weil, K. B. Katsaros and P. K. Taylor, Turbulent heat fluxes by profile and inertial dissipation methods: analysis of the atmospheric surface layer from shipboard measurements during the SOFIA/ASTEX and SEMAPHORE experiments. *Ann. Geophysicae, 13,* 1065-1074, 1995.

Dyer, A. J., A review of flux-profile relationships. *Bound. -Layer Met., 7,* 363-372, 1974.

Katsaros, K. B., S. D. Smith, and W. A. Oost, HEXOS - Humidity Exchange Over the Sea: A program for research on water vapor and droplet fluxes from sea to air at moderate and high wind speeds. *Bull. Amer. Meteor. Soc., 68,* 466-476, 1987.

Katsaros, K. B. and J. DeCosmo, Evaporation in high wind speeds, sea surface temperature in low wind speeds, examples of atmospheric regulation, in *Modelling the Fate and Influence of Marine Spray,* edited by P. G. Mestayer, E. C. Monahan and P. A. Beetham, pp 106-114, Whitecap Rep. No. 7, Marine Sciences Inst., U. of Connecticut, Groton, CT, 1990.

Katsaros, K. B. and J. DeCosmo, Water vapor flux from the sea at high wind speeds, in *Tropical Cyclone Disaster,* edited by J. Lighthill, Z. Zhemin, G. Holland and K. Emanuel, pp. 386-392, *Proc. ICSU/WMO Intl. Symp.*, Beijing, China, Oct. 12-16: 1992, Peking University Press, Beijing, 1993.

Katsaros, K. B. and G. de Leeuw, Comment on "Sea spray and the turbulent air-sea heat fluxes"; E. L. Andreas. *J. Geophys. Res., 97,* 14, 339-14, 343, 1994.

Kahma, K. K. and H. Pettersson, Wave growth in a narrow fetch geometry. *Global Atmos. and Ocean Sys.,* 2, 253-263, 1994.

Liu, W. T., K. B. Katsaros, and J. A. Businger, Bulk parameterization of air-sea exchanges of heat and water vapor, including the molecular constraints at the interface. *J. Atmos. Sci., 36,* 1722-1735, 1979.

Monin, A. S. and A. M. Obukhov, Basic laws of turbulent mixing in the ground layer of the atmosphere, *Tr. Aead. Nauk SSSR Geophys. Inst.* 151, 163-187, 1954.
Smith, S. D., R. J. Anderson, W. A. Oost, C. Kraan, N. Maat, J. DeCosmo, K. B. Katsaros, K. Bumke, L. Hasse, and H. M. Chadwick, Sea surface wind stress and drag coefficients: the HEXOS results. *Bound. -Layer Meteor.*, *60*, 10-142, 1992.

4

On the Structure of the Upper Oceanic Boundary Layer and the Impact of Surface Waves

A. Anis

Abstract

A detailed investigation of the upper oceanic boundary layer (OBL), under various atmospheric and sea conditions, was carried out. During convective conditions the vertical structure of the potential temperature, θ, and the turbulence kinetic energy (TKE) dissipation rate, ε, in the OBL mimic, in general, that in the convective atmospheric boundary layer (ABL). It is shown that the convective OBL exhibits a "slab" type behavior and from two independent estimates it is suggested that the heat flux profile is linear with depth.

In contrast to the apparent similarities between the OBL and ABL, there were instances when ε was greatly enhanced relative to wind stress and/or buoyancy production and exhibited an exponential depth decay. In these instances simple scaling laws predicted for turbulence near a solid surface severely underestimate turbulent mixing near the ocean's surface. Two different mechanisms, both invoking turbulence-wave interaction, are suggested to explain the behaviour of ε near the surface.

Introduction

The oceanic boundary layer (OBL), defined from above by the ocean's surface and by the top of the seasonal thermocline from below, is directly influenced by surface forcing such as heat flux, wind stress, and surface waves. Experiments in the convective OBL indicate that the vertical structures of the time-averaged potential temperature, θ, and TKE dissipation rate, ε, are similar to those in the convective atmospheric boundary layer (ABL) over land. Both systems include four distinct regimes (proceeding from the ocean's surface down and from the land's surface up):

(1) A "cool skin" (a few millimeters thick) near the surface of the ocean (e.g. Paulson and Simpson, 1981) which is analogous to the "hot skin" (or microlayer) near the land surface (Stull, 1988). Physical processes are governed by molecular diffusion

(2) A superadiabatic ocean surface layer (OSL) of O(10–20 m) in which $\partial\theta/\partial z \sim -2 \times 10^{-4}$ K m^{-1} (Anis and Moum, 1992) and a superadiabatic atmospheric surface layer (ASL) of O(10-100 m) in which $\partial\theta/\partial z \sim -1 \times (10^{-2} - 10^{-1})$ K m^{-1} (Driedonks and Tennekes, 1984; Stull, 1988).

(3) A well-mixed layer (ML) in which $\partial\theta/\partial z \sim 0$ and ε scales well on average with the

Physical Processes in Lakes and Oceans
Coastal and Estuarine Studies Volume 54, Pages 47-60

surface buoyancy flux J_b^0 (Shay and Gregg, 1986, for the OBL and Caughey and Palmer, 1979, for the ABL). Thicknesses of the ML are O(50–150 m) and O(1000–2000 m) for the OBL and the ABL, respectively.

(4) A stable thermocline in the OBL of O(100–200 m) and a stable inversion layer in the ABL of O(100–500 m) where $\partial\theta/\partial z > 0$. This regime is often referred to as the entrainment zone.

Other studies have shown that wind-driven near surface layers of oceans and lakes exhibit scaling laws consistent with constant stress layers over solid boundaries (e.g. Dillon et al., 1981; Soloviev et al., 1988). Lombardo and Gregg (1989) observed a range of conditions in a convective OBL. concluding that ε could be normalized very well by the sum of convective and SL similarity scalings.

In spite of the apparent similarities between the OBL and ABL there are many instances when the OBL exhibits large departures from the ABL scaling laws and parameterizations. In these cases parameterizations used for the OBL, most of which are based on results from the ABL and do not account for complications associated with the free surface of the ocean, may severely underestimate air-sea exchanges.

Evidence for enhanced turbulence and mixing in the upper part of the aquatic BL comes from a growing number of experimental field and laboratory studies. Field studies in the upper part of the OBL under different forcing conditions (Stewart and Grant, 1962; Shay and Gregg, 1984; Gregg, 1987; Gargett, 1989; Anis and Moum 1992) showed enhanced TKE dissipation rates much larger than predicted by SL and/or convective scalings. Gargett (1989) inferred a near-surface dissipation rate decaying as depth to the power −4, much faster than the −1 decay predicted by the law of the wall.

Kitaigorodskii et al. (1983) and Agrawal et al. (1992) reported enhanced dissipation rates beneath surface waves, observed during separate field experiments in Lake Ontario. Dissipation rates decayed with depth at a power between -3.0 and -4.6 (Drennan et al., 1992; Agrawal et al., 1992).

In a recent study, using data collected in Lake Ontario, Terray et al. (1996) propose a scaling for the rate of dissipation based on wind and wave parameters, and suggest a three stage depth dependence: very near the surface, within one significant wave height, dissipation is an order of magnitude greater than that predicted by the law of the wall and roughly constant; below this is an intermediate layer, the thickness of which is proportional to the energy flux from breaking normalized by ρu^3_*, in which dissipation decays with depth to the power of −2; at sufficient depth below this layer the dissipation rate asymptotes to values predicted by the law of the wall.

Thorpe (1984), using acoustic measurements of bubbles near the surface of the ocean, suggested the importance of turbulence generated by breaking waves. Both Kitaigorodskii et al. (1983) and Thorpe (1984) postulated a wave-affected layer of depth on the order of 10 times the wave amplitude.

In a laboratory experiment, Rapp and Melville (1990) showed that, as a result of wave breaking, mean surface currents in the range 0.02–0.03C (where C is the characteristic phase speed) were generated and took about 60 wave periods to decay to 0.005C. Turbulence rms velocities on the order of 0.02C were measured, decaying to 0.005C after more than 60 wave periods, and were still significant to a depth of k^{-1} (where k is the characteristic wavenumber). Melville's (1994) reanalysis of his earlier laboratory studies reveals that net dissipation levels in the SL are consistent with the measurements by Agrawal et al. (1992).

In another laboratory study, Cheung and Street (1988) carefully examined turbulence in the water at an air-water interface for different cases of surface gravity waves. They showed that although turbulence parameters followed wall layer scaling in the case of wind-generated waves, they behaved very differently in the case of wind-ruffled mechanically generated waves, for which 1) increased turbulence levels were observed, away from the surface to a depth of about 1/k (k is the wavenumber of the mechanically generated waves), and the depth decay of turbulence rms velocities followed the decay of wave rms velocities closely; 2) the phase difference between $\tilde{u}$ and $\tilde{w}$, the horizontal and vertical wave velocities, respectivley, was consistently less than 90°; 3) in both the wind-ruffled mechanically generated waves and the high wind speed wind wave experiments, the mean velocity profiles (as a function of depth) had slopes smaller than predicted for turbulent boundary layers near a solid surface, suggesting that the waves affect the mean flow. An ocean with swell and wind waves might closely resemble the laboratory wind-ruffled mechanically generated waves of Cheung and Street's experiment.

In a recent study Craig (1996) concentrated on the velocity profiles collected by Cheung and Street (1988). He used a one-dimensional, level 2.5 turbulence closure model (Craig and Banner, 1994), which showed that in the wave-enhanced layer the downward flux of TKE from the surface was balanced by its dissipation. The comparison between the model and the data indicate very good agreement, confirming that the model adequately describes the dynamics of the surface layer. The velocity profiles also allowed estimation of surface roughness length at five different wind speeds. A wall layer model, which ignores the flux of TKE, will, of course, not predict the enhanced dissipation near the surface.

Next we examine the details of the vertical structure of θ, ε, and shear of the horizontal current in the nighttime convective OBL and discuss the consequences. Following the discussion on the convective OBL we examine results from turbulence profiling measurements in the OBL, conducted under different meteorological and sea conditions. Possible effects of surface waves on TKE and consequently $\varepsilon(z)$, are discussed and scalings for $\varepsilon(z)$ are derived.

The OBL During Convection

Experimental Details and Background Conditions

A large data set was collected in the upper OBL of the Pacific Ocean between 13 and 20 March 1987 along 140° W from 17° N to 6° N. A total of 445 nighttime hydrographic and turbulence vertical profiles, collected with the Rapid-Sampling Vertical Profiler (RSVP; Caldwell et al., 1985), were analyzed. From the RSVP profiles θ and ε were calculated as functions of depth. Vertical profiles of horizontal velocity components were estimated from a shipboard acoustic Doppler current profiler (ADCP). Continuous shipboard measurements of meteorological data were taken throughout the experiment and are summarized in Table 1.

The Vertical Structure of Potential Temperature, TKE Dissipation and Velocity Shear

Our analysis was carried out for the period of each night during which J_b^0 varied little and the depth of the ML, D, traced the undulations in the depth of the seasonal pycnocline (about 11.5–13 hours). This part of the night, is referred to as the *quasisteady forcing*

TABLE I. Nighttime average values of meteorological parameters. The surface wind stress, τ_0; the difference between the air temperature and the sea surface temperature (SST), T_{air} – SST; the net surface heat flux, J_q^0; the net surface buoyancy flux, J_b^0 (positive for convective conditions); the Monin-Obukhov length scale, $L = -u^3_*/k\ J_b^0$ (where $u_* = \sqrt{\tau_0/\rho}$ is the ocean surface friction velocity and $L < 0$ during convection); the significant wave height, H_s for swell and wind-waves (from R/V *Wecoma*'s ship's officers' log).

	Night 1	Night 2	Night 3	Night 4	Night 5	Night 6	Mean
τ_0 (N m^{-2})	0.08	0.10	0.08	0.07	0.11	0.11	0.09
T_{air} - SST (K)	– 1.84	– 1.23	–1.53	– 0.88	– 0.73	– 0.86	–1.09
J^0_q (W m^{-2})	212	211	171	166	205	199	189
$10^7\ J_b^0$ (m^2 s^{-3})	1.61	1.66	1.37	1.31	1.54	1.61	1.48
L (m)	– 10.6	– 14.2	– 11.7	– 10.8	– 17.3	– 17.5	– 13.9
H_s swell (m)	2.4	2.7	2.0	1.8	2.2	2.1	2.1
H_s wind-waves (m)	0.8	0.9	0.7	0.6	0.9	0.9	0.8

phase. Individual profiles of $\theta(z)$ were referenced to the average value of θ in the ML (in the depth range $-D < z < 2L$) and then averaged over periods of 2 hours each during the quasi-steady convective forcing phase of each night Three distinct regimes are noticed from the profiles of θ (Figure 1): a superadiabatic OSL, where $\partial\theta/\partial z < 0$, in the upper 20–40% of the OBL (Anis and Moum, 1992); a relatively well-mixed layer beneath it in which $\partial\theta/\partial z \sim 0$ (the lower part of the ML often had a small positive temperature gradient similar in magnitude to that of the superadiabatic OSL: we attribute this to entrainment of cooler water below the ML base into the ML); a stable layer with a sharp positive gradient of θ capping the OBL from below and coinciding with the top of the seasonal pycnocline. The vertical structure of $\theta(z)$ was found to be constant within 95% confidence limits throughout the quasisteady convective forcing phase of each of the six nights of the experiment.

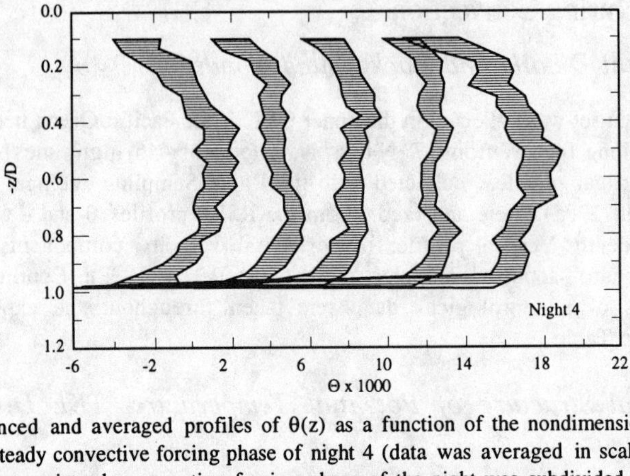

Figure 1. Referenced and averaged profiles of $\theta(z)$ as a function of the nondimensional depth, -z/D. during the quasisteady convective forcing phase of night 4 (data was averaged in scaled depth bins of -z/D = 0.04) The quasisteady convective forcing phase of the night was subdivided into 5 sequential periods of 2 hours. starting about 2 hours after J_b^0 changed sign from negative to positive and ending about 2 hours before J_b^0 changed sign again. Shading represents 95% bootstrap confidence intervals. The profiles are sequentially offset by 0.004 K.

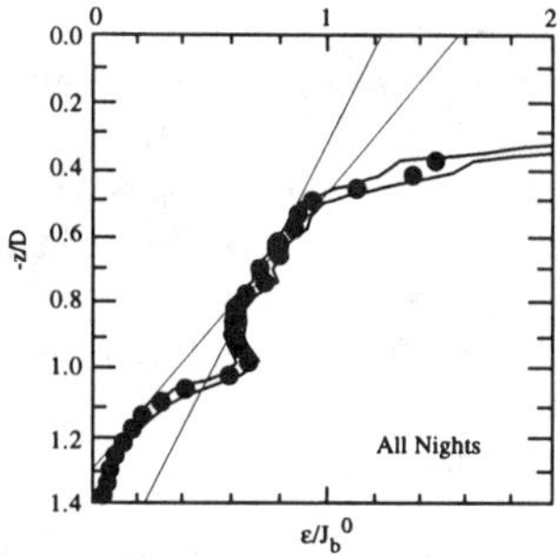

Figure 2. Scaled turbulence kinetic energy dissipation rate, ε/J_b^0, as a function of nondimensional depth, -z/D, for the average of all the nights of the expertment. The filled circles and solid lines represent the mean and the 95% bootstrap confidence envelopes, respectively. The broken line is the least-square fit of' the form $b_\varepsilon + a_\varepsilon z^*$ to the mean values (Eq. 10 and Table 2). Two straight solid lines were fitted by eye in the 95% confidence envelope such that their slopes represent the minimum and maximum values permissible by this envelope. This provides an uncertainty estimate on a_ε (see Table 2) which takes into account the 95% confidence intervals of the mean values of ε. Note also the increase in ε/J_b^0, near the base of the mixed layer (i.e. near -z/D ~ 1)

The vertical structure of ε was nondimensionalized as $\varepsilon^*(z^*) \equiv \varepsilon(z/D)/J_b^0$ for individual profiles, and ensemble averaged in depth bins of 0.04D for each night (as we did for the profiles of θ). A similar computation was made by ensemble averaging the individual profiles of all nights (Figure 2). A consistent feature of the averaged profiles was a rapid and roughly exponential decrease of values of $\varepsilon^*(z^*)$ in the upper 20–40% (~ 1 – 2L) of the convective OBL. In the lower half of the ML, roughly between $-z^* = 0.5$ and $-z^* = 0.9$, values of $\varepsilon^*(z^*)$ decreased linearly with z^*. Before dropping rapidly by several factors of ten, a slight increase in $\varepsilon(z^*)$ was observed near the base of the ML between $-z^* = 0.95$ and $-z^* = 1.05$. This is an indication of enhanced mixing in the entrainment zone, possibly a result of the enhanced shear levels observed near the base of the ML. Analysis of nighttime ADCP profiles showed that the estimates of $S^2 = (\partial U/\partial z)^2 + (\partial V/\partial z)^2$ in the bulk of the ML were equivalent, on average, to the detection limit of $O(1 - 2 \times 10^{-6}\ s^{-2})$ in estimating S^2. That is, the ML shear is not distinguishable from 0 s^{-1}. In the lower 20% of the ML a relatively sharp increase in shear was observed with a maximum of about $S^2 \sim 6 \times 10^{-5}\ s^{-2}$ near the base of the ML. The steadiness of the vertical structure of θ and the vanishing current shear in the bulk of the ML imply that a "slab"-type model might be an adequate representation of the convective OBL (see also Anis and Moum, 1995).

Estimates of the Vertical Heat Flux from θ Profiles and Heat Conservation

To isolate the effect of convection, we assume a purely l-D process so that the time rate of change of θ is solely due to the vertical turbulence heat flux, $\overline{w'\theta'}(z)$,

$$\frac{\partial\theta}{\partial t} = -\frac{\partial\left(\overline{w'\theta'}\right)}{\partial z} . \tag{1}$$

Integrating (1) vertically to the surface (z = 0),

$$\overline{w'\theta'}(z) = \overline{w'\theta'}(0) + \int_z^0 \frac{\partial\theta}{\partial t} dz' . \tag{2}$$

Equation (2), then, permits an estimate of the vertical heat flux at some depth z by vertically integrating the time rate of change of temperature between the surface and z and using the surface heat flux, $J_b^0 = \rho C_p \overline{w'\theta'}(0)$ (estimated from shipboard meteorological measurements; Table 1). In practice this is almost always impossible to do using oceanic data because of the relatively small temperature changes resulting from convection compared to those due to lateral temperature variability. (For example, a surface heat flux of ~ 200 W m^{-2} and a negligible heat flux at the base of the ML, a time period of Δt ~ 10 hours and a mean ML depth D = 65 m results in a cooling < 0.03 K.) A way to alleviate this problem is by taking the vertical derivative of (2), which results in

$$\frac{\partial}{\partial t}\left(\frac{\partial \theta}{\partial z}\right) = -\frac{\partial}{\partial z}\left(\frac{\partial \overline{w'\theta'}}{\partial z}\right) \tag{3}$$

From (3), the vertical structure of the turbulence heat flux can be found from the time rate of change of the temperature gradient, $\partial\theta/\partial z$. This can be accomplished more readily, since now we are not concerned with the *absolute* temperature differences between two profiles of $\theta(z)$, which are separated in both time and space, but only with the change in their *relative* vertical structure as a function of time. Because the latter does not change significantly throughout the quasisteady convective forcing phase of each of the six nights, we have $\partial(\partial\theta/\partial z)/\partial t \sim 0$.

By (3), this implies that the vertical heat flux due to turbulence in the convective OBL is linear in z, i.e.

$$\overline{w'\theta'}^*(z^*) = 1 + a_h z^*, \tag{4}$$

where $\overline{w'\theta'}^*(z^*) \equiv \overline{w'\theta'}(z)/\overline{w'\theta'}(0)$, and $a_h \equiv [1 - \overline{w'\theta'}^*(-1)]$. Field experiments in the convective ABL (e.g. Zhou et al., 1985; Young, 1988), numerical simulations (e.g. Deardorff, 1974 for a convective ABL; Andre and Lacarrere, 1985 for a convective OBL), laboratory experiments (e.g. Willis and Deardorff, 1974) and a field experiment in the convective aquatic boundary layer of a fresh water reservoir (Imberger, 1985) all indicate a linear dependence on z of the turbulence heat/buoyancy flux.

Estimates of a_h were made assuming a simplified form of the steady state TKE equation near the base of the ML in which the mechanical production of TKE, P, is balanced by TKE dissipation and by buoyancy destruction, J_b, i.e. $P = -J_b + \varepsilon$. The small, observed increase in ε at $-z/D \sim 1$ (Figure 2) and the increase in shear magnitude near the base of the ML seem to support such a balance during our experiment. Using the simplified TKE balance and a critical value for the flux Richardson number $R_f \equiv -J_b/P \sim 0.15$ (Osborn, 1980) results in an upper bound on the magnitude of the buoyancy flux given by

$$| J_b(-D) | \leq 0.2\varepsilon(-D). \tag{5}$$

Neglecting salinity effects, we have $J_b = g\alpha\overline{w'\theta'}$ (g is the gravitational acceleration and α is the thermal expansion coefficient of seawater) so that

$$| \overline{w'\theta'}^*(-1) | \leq 0.2\varepsilon^*(-1), \tag{6}$$

where $\varepsilon^*(-1) \equiv \varepsilon(-D)/J_b^0$. Estimates of the upper limit on the magnitude of $\overline{w'\theta'}^*(-1)$ ranged between 0.07 and 0.18 for the individual nights with an overall mean of 0.13 (Table 2). From (4) and the mean value of 0.13 we then have

$$\overline{w'\theta'}^*(z^*) \equiv 1 + 1.13z^*. \tag{7}$$

TABLE 2. Nighttime average values of mixed layer related quantities. D is the depth of the ML; $\varepsilon^*(-1) \equiv \varepsilon(-D)/J_b^0$ is the nondimensional TKE dissipation rate (calculated over the scaled depth range $0.98 < -z/D < 1.02$; values in parentheses are the 95% bootstrap confidence limits); $|\overline{w'\theta'}^*(-1)|$ is the upper bound on the nondimensional vertical heat flux near the base of the ML (calculated as $|\overline{w'\theta'}^*(-1)| \leq 0.2\,\varepsilon^*(-1)$; $a_h \equiv 1 - \overline{w'\theta'}^*(-1)$; a_ε and b_ε are the values resulting from a least square fit of the form $b_\varepsilon + a_\varepsilon z^*$ to $\varepsilon^*(z^*)$ (where $\varepsilon^*(z^*) \equiv \varepsilon(z)/J_b^0$ and $z^* \equiv z/D$) in the lower half of the ML (Figure 2; the ± values are the probable uncertainties in b_ε and a_ε resulting from the fit to individual nightly mean profiles and to the overall mean profile. For the latter, written in parentheses, are error estimates as determined from two graphically fitted straight lines in the 95% bootstrap confidence envelope (see Figure 2).

	Night 1	Night 2	Night 3	Night 4	Night 5	Night 6	Mean
D(m)	64.4	68.5	50.6	67.5	64.3	82.2	65.1
$\varepsilon^*(-1)$	0.36	0.65	0.61	0.47	0.90	0.67	0.63
	(.28, .44)	(.54, .77)	(.54, .69)	(.42, .53)	(.72, 1.12)	(.58, .76)	(.58, .68)
$\lvert\overline{w'\theta'}^*(-1)\rvert$	0.07	0.13	0.12	0.09	0.18	0.13	0.13
a_h	1.07	1.13	1.12	1.09	1.18	1.13	1.13
a_ε	.30 ± .10	1.26 ± 0.13	1.50 ± .14	1.25 ± .08	.32 ± .13	.57 ± .07	.93 ± 0.5
							(0.73, 1.23)
b_ε	.66 ± .07	1.64 ± .09	1.99 ± .10	1.49 ± .06	1.10 ± .10	1.15 ± .04	1.40 ± .04
							(1.25, 1.60)

Estimates of the Vertical Heat Flux from ε Profiles and TKE Balance

In the convective OBL, away from the ocean's surface and away from the ML base, TKE production is mainly by buoyancy forcing. For a steady-state, neglecting turbulence transport, and since the background shear is indistinguishable from zero, neglecting also mechanical production, we expect TKE dissipation to nearly balance buoyancy production such that

$$\varepsilon(z) \sim J_b(z). \tag{8}$$

Neglecting salinity effects and nondimensionalizing

$$\varepsilon^*(z^*) \sim \overline{w'\theta'}^*(z^*). \tag{9}$$

Equation (9) permits an independent estimate of the depth-dependent heat flux from the dissipation profiles. Within the ML we fitted a linear profile, by least squares, of the form

$$\varepsilon^*(z^*) = b_\varepsilon + a_\varepsilon z^*. \tag{10}$$

The coefficients a_ε and b_ε are equivalent via (9) to those in (4) and (7), i.e. if (10) is true, we expect $b_\varepsilon \sim 1$, $a_\varepsilon \sim a_h$. Results of least square fits to the observed $\varepsilon^*(z^*)$ profiles (e.g. Figure 2) show that a_ε is consistent with the average value a_h of 1.13 and b_ε is not significantly different from 1 (Table 2).

In spite of the simplifying assumptions we have made, the fact that the vertical structure of θ was constant throughout each night and that ε decreased linearly with depth in the lower half of the ML are solid and consistent results, both leading to the same conclusion of a linear heat flux profile. Moreover, the suggested slope of the heat flux profile (i.e. $a_h = 1.13$) is in general agreement with the range of 1.1–1.3 suggested by Stull (1976) based on a large number of published values. Simulations of the Wangara experiment by Deardorff (1974) resulted in $a_h = 1.13$ and a numerical study of a buoyancy driven OBL (André and Lacarrere, 1985) resulted in $a_h = 1.16$.

Surface Wave-Turbulence Interactions

In the next two sections we discuss the necessary theoretical background, followed by a description of experimental details and observational results. Two different mechanisms are proposed to explain the behavior of ε near the surface, leading to two new scaling schemes (see also Anis and Moum, 1995).

Turbulence in an Irrotational Wave Field

We assume an irrotational wave field, $\tilde{u}_i = \partial\Phi/\partial x_i$, and a rotational turbulence field $u'_i = \varepsilon_{ijk}\partial V_k/\partial x_j$, where Φ and V are, respectively, the scalar and vector velocity potentials (e.g. Kitaigorodskii et al., 1983). We consider, for simplicity, a statistically homogeneous flow in the horizontal plane with the x-axis aligned in the direction of the mean flow, so that $\overline{U}_i = [U(z,t),0,0], \overline{P} = P(z,t)$, and a wave field propagating in the x direction. The resulting kinetic energy equations for the mean, wave, and turbulence parts of the flow are, respectively,

$$\frac{\partial}{\partial t}\left(\frac{1}{2}\overline{U}\overline{U}\right) = + \nu\overline{U}\frac{\partial^2\overline{U}}{\partial z^2} - \frac{\partial}{\partial z}\overline{(w'u'U)} + \overline{w'u'}\frac{\partial\overline{U}}{\partial Z}, \tag{11}$$

$$\frac{\partial}{\partial t}\left(\frac{1}{2}\overline{\tilde{u}_i\tilde{u}_i}\right) = -\overline{\tilde{u}_i\tilde{u}_j\frac{\partial u'_i}{\partial x_j}} - \frac{\partial}{\partial z}\left(\overline{w'\frac{1}{2}\tilde{u}_i\tilde{u}_i}\right) - \frac{\partial}{\partial z}\overline{(w'u'_i\tilde{u}_i)} + \overline{u'_i u'_j\frac{\partial\tilde{u}_i}{\partial x_j}}, \tag{12}$$

$$\begin{aligned}\frac{\partial}{\partial t}\left(\frac{1}{2}\overline{u'_i u'_i}\right) = &- \frac{\partial}{\partial z}\left[\overline{w'\left(\frac{p'}{\rho} + \frac{1}{2}u'_i u'_i\right)}\right] - \frac{g}{\rho}\overline{w'\rho'} - \overline{w'u'}\frac{\partial\overline{U}}{\partial z} - \frac{\partial}{\partial z}\left(\overline{w'\frac{1}{2}\tilde{u}_i\tilde{u}_i}\right) \\ &- \frac{\partial}{\partial z}\left(\overline{\tilde{w}\frac{1}{2}u'_i u'_i}\right) - \overline{u'_i u'_j\frac{\partial\tilde{u}_i}{\partial x_j}} - \varepsilon .\end{aligned} \tag{13}$$

Turbulence in a Rotational Wave Field

Although in many cases the surface wave field is assumed to be irrotational, a number of field and laboratory studies have shown instances where the surface wave field departed from irrotationality (e.g. Shonting, 1964; Cavaleri and Zecchetto, 1985). These observations showed the vertical and horizontal wave orbital velocities to be out of quadrature leading to a downward momentum flux which decayed rapidly with depth. Laboratory studies by Cheung and Street (1988) found that in the case of wind-ruffled mechanically-generated waves the wave-induced stress was negative and larger in magnitude than the turbulence Reynolds stress, resulting in energy transfer from the wave field to the mean flow through the term $\overline{\tilde{w}\tilde{u}}\partial\overline{U}/\partial z$. Due to the apparent departure of the wave field from the classical notion of irrotationality, wave-turbulence separation can be achieved by the phase averaging (denoted here by $\langle\ \rangle$) technique (Hussein and Reynolds, 1970). This method is suitable for extracting the wave-induced motion when the wave field is characterized by a specific wavelength in the spectrum. An example is the laboratory case of wind-ruffled mechanically-generated waves (Cheung and Street, 1988), or the ocean in which the surface wave field is a combination of wind-waves and dominant swell.

Assuming, as before, statistical homogeneity in the horizontal plane, and taking the x axis in the direction of the mean current, results in the following equations for MKE,WKE, and TKE:

$$\frac{\partial}{\partial t}\left(\frac{1}{2}\overline{UU}\right) = +\nu\overline{U}\frac{\partial^2\overline{U}}{\partial z^2} - \frac{\partial}{\partial z}\left(\overline{w'u'U}\right) + \overline{w'u'}\frac{\partial \overline{U}}{\partial z} - \frac{\partial}{\partial z}\left(\overline{\tilde{w}\tilde{u}U}\right) + \overline{\tilde{w}\tilde{u}}\frac{\partial \overline{U}}{\partial z}, \tag{14}$$

$$\begin{aligned}\frac{\partial}{\partial t}\left(\frac{1}{2}\overline{\tilde{u}_i\,\tilde{u}_i}\right) = &- \frac{\partial}{\partial z}\left[\overline{\tilde{w}\left(\frac{\tilde{p}}{\rho} + \frac{1}{2}\tilde{u}_i\tilde{u}_i\right)}\right] + \overline{\nu\tilde{u}_i\frac{\partial^2\tilde{u}_i}{\partial x_j^2}} - \overline{\tilde{w}\tilde{u}}\frac{\partial \overline{U}}{\partial z} - \frac{\partial}{\partial z}\left(\overline{\langle w'u'_i\rangle\tilde{u}_i}\right)\\ &+ \overline{\langle u'_i\,u'_j\rangle\frac{\partial \tilde{u}_i}{\partial x_j}},\end{aligned} \tag{15}$$

$$\begin{aligned}\frac{\partial}{\partial t}\left(\frac{1}{2}\overline{u'_i\,u'_i}\right) = &- \frac{\partial}{\partial z}\left[\overline{w'\left(\frac{p'}{\rho} + \frac{1}{2}u'_i\,u'_i\right)}\right] - \frac{g}{\rho}\overline{w'\rho'} - \overline{w'u'}\frac{\partial \overline{U}}{\partial z} - \frac{\partial}{\partial z}\left(\overline{w'\frac{1}{2}u'_i\,u'_i}\right)\\ &- \overline{\langle u'_i\,u'_j\rangle\frac{\partial \tilde{u}_i}{\partial x_j}} - \varepsilon\,.\end{aligned} \tag{16}$$

Experimental Details and Observations

Vertical profiles were collected using the microstructure profiler, Chameleon (Moum et al., 1994), in water depths between 1000 and 2000 m off the Oregon coast during the summer of 1989. Chameleon, was lowered with the aid of an attached weight which upon release caused the instrument to freely rise, while taking microstructure measurements on its way to the ocean surface (in cases when the profiler was suspected to have surfaced anywhere near the wake, the profiles were not used in the analysis). Surface wind, heat, and buoyancy fluxes were calculated from continuous shipboard measurements of meteorological parameters. The significant wave height, Hs, and periods of swell and wind waves and their direction were estimated by the mate on watch every 15 min. Chameleon profiles for this experiment were collected during two nights. During the first night strong steady northerly winds (~ 13 m s^{-1}) produced an average surface stress of 0.25 N m^{-2} (Table 3). Sea state was dominated by two swells, one, heavy, from NW with a period of about 12 sec and 3.0 m significant wave height, and the other from NNW with a shorter period of 6 sec and significant height of 2.5 m. Intense breaking, mainly of wind-waves with significant height of about 1 m, took place throughout the night. Meteorological and sea state conditions, during the second night, were substantially different and relatively moderate compared to the rough conditions of the first night. Winds, decreasing from 10.0 to 8.0 m s^{-1}, produced an average surface stress of 0.11 N m^{-2}. A 2.0 m swell with a period of 6–8 sec and wind-waves of 0.6 m, breaking only sporadically, defined the sea state.

A comparison of $\varepsilon(z)$ to u^3_*/kz shows a clear distinction between the two nights: for night 2 $\varepsilon/(u^3_*/kz) \sim 1$ in the upper 10 m or so, as expected from SL scaling, while for night 1 ε was larger than expected from SL scaling alone. To quantify the excess in dissipation over u^3_*/kz, we calculated the ratio of the depth-integrated dissipation rate to the depth-integrated wind stress production in the depth interval -0.5 m > z > - D, $\int^{-0.5}_{-D} \rho\varepsilon\, dz \,/\, \int^{-0.5}_{-D} \rho(u^3_*/kz)dz$. The value of this ratio (Table 3) for night 1 is an order of magnitude larger than that of night 2, reflecting the significant difference between the two nights.

Scaling $\varepsilon(z)$ in a Wave Field

To determine if the observed $\varepsilon(z)$ profiles are consistent with energetics considerations estimates of the energy lost by breaking surface waves were made. The three independent estimates made show that the energy lost by breaking surface waves is much greater than

TABLE 3. Daytime and nighttime averaged values of the atmospheric and sea state conditions for the experiment (definitions of the meteorological parameters are given in the caption of Table 1). The significant wave height, H_s and period, T, are from R/V *Wecoma*'s ship's officers' log and are given for swell and wind waves. $\int \rho\varepsilon\, dz / \int (\rho u^3_* / kz) dz$ is the ratio of the depth-integrated dissipation rate to the depth integrated wind stress production and was estimated for the depth interval -0.5 m > z > -D.

	No. of profiles	τ Nm^{-2}	$10^7 J_b^0$ $m^2 s^{-3}$	L m	Swell H_s (m), T (s)	Wind-waves H_s (m), T(s)	$\frac{\int \rho\varepsilon dz}{\int(\rho u^3_* / kz)dz}$
				m			
night 1	29	0.25	1.4	– 67.7	3.0, 12	1.0, 4	13.0
					2.5, 6	–	
night 2	20	0.11	0.4	-63.0	2.0, 6 – 8	0. 6, 4 – 5	1.3

the turbulence energy generated by the wind stress as predicted by SL scaling. These estimates are, however, consistent with the observed vertically integrated turbulence dissipation rate, $\int \rho\varepsilon\, dz$ (Anis and Moum, 1995). Because high values of $\varepsilon(z)$ were observed deeper than e-folding scales of the wind-waves, a possible scenario is that breaking waves (mainly wind-waves) form a layer near the surface (on the order of the wind-wave height) of enhanced TKE. Turbulence is then transported downward by the swell, while decaying with time and diffusing spatially. This could be achieved via the term $-\partial(\overline{\tilde{w}\ 1/2\ u_i'u'_i})/\partial z$ (13) and (16) which describes the vertical transport of TKE, $1/2\overline{u_i'u_i'}$, by the wave velocity $\tilde{w}$. If a surface layer, with thickness of $O(k^{-1}$; k is the wavenumber), is governed by wave-turbulence interactions such that the main local balance is between $-\partial(\overline{\tilde{w}\ 1/2\ u_i'u_i'})/\partial z$ and the TKE dissipation rate, then

$$\varepsilon(z) \sim -\frac{\partial}{\partial z}\left(\overline{\tilde{w}\frac{1}{2}u_i'u_i'}\right) \sim -\frac{\partial}{\partial z}\left(\overline{\omega a\ \exp(kz)\ \sin(kx-\omega t)\frac{1}{2}u_i'u_i'}\right), \qquad (17)$$

where ω and a are the radian frequency and amplitude, respectively, and $\tilde{w} = \omega a \sin(kx - \omega t)\exp(kz)$. An upper bound on $\varepsilon(z)$ in the wave-dominated surface layer can now be found. Assuming, for simplicity of differentiation, a depth-independent (or in turn a depth-averaged) correlation coefficient, r, between $\tilde{w}$ and $u'_iu'_i$, and taking $1/2\ \overline{u_i'u_i'} \sim 3/2u^2_{rms} \sim 3/2(0.02C)^2$ (C is the phase speed of the waves) as an upper bound on the TKE (Rapp and Melville, 1990), we find from (17)

$$\varepsilon(z) \leq a\frac{1}{\pi}g^{1/2}k^{3/2}r\frac{3}{2}(0.02C)^2\exp(kz), \qquad (18)$$

where $\omega^2 = kg$, and averaging is performed over half the cycle of downward transport, i.e. when $\tilde{w} < 0$.

If on the other hand the wave field is not truly irrotational, the resulting wave stresses may indirectly enhance turbulence in the wave-influenced layer near the surface. For steady state, and neglecting molecular diffusion, the MKE equation (14) reduces to

$$-\overline{w'u'} - \overline{\tilde{w}\tilde{u}} = \text{constant} = \frac{\tau_0}{\rho}. \qquad (19)$$

Cheung and Street (1988) found that $\sim\overline{\tilde{w}\tilde{u}}$ and $-\overline{\tilde{w}\tilde{u}}\ \partial\overline{U}/\partial z$ were generally negative, decaying rapidly with depth. Thus, energy is transferred from the wave field to the mean field via the term $\overline{\tilde{w}\tilde{u}}\ \partial\overline{U}/\partial z$, which appears in (14) and (15) with opposite signs. If, in the wave-dominated layer near the surface, the wave stresses are much larger than the surface wind stress (e.g. Shonting, 1964; Cavaleri et al., 1978) such that $\tilde{w}\tilde{u} >> \tau_0/p$, then, to satisfy the balance in (19), one arrives at

$$-\overline{w'u'} \sim \overline{\tilde{w}\tilde{u}} >> \tau_0/p. \quad (20)$$

This balance does not violate the continuity of tangential stress across the air-water interface, but rather allows for much larger respective wave and turbulence stress terms than would be predicted from an estimate τ_0/ρ derived from surface winds.

If the wave stress is simply a result of $\tilde{u}$ and $\tilde{w}$ being out of quadrature then

$$\overline{\tilde{w}\tilde{u}} \approx \frac{1}{2}a^2 gk\exp(2kz)\sin\phi \quad (21)$$

where ϕ is the phase shift from quadrature. Assuming that the main balance in (16) is between the TKE dissipation rate, ε, and the shear production, $-\overline{\tilde{w}\tilde{u}}\ \partial\overline{U}/\partial z$ and using (20) and (21), results in

$$\varepsilon = \overline{-w'u'}\frac{\partial\overline{U}}{\partial z} \approx \frac{1}{2}a^2 gk\exp(2kz)\sin\phi\frac{\partial U}{\partial z}. \quad (22)$$

To compare the predicted scalings to our observations simple exponential forms, $\varepsilon(z) = \varepsilon_0 \exp(\alpha z)$, were fit to the observed dissipation rates in the upper part of the OBL (Figure 3). The observations reveal a change in slope of $\varepsilon(z)$ at a depth of ~ 6 m, with a more rapid depth decay in the upper 6 m. Within the context of our wave-based scalings, one interpretation is that the upper 6 m is dominated by wind waves while, below 6 m, the dominant effect is the longer wavelength swell. If dissipation locally balances the downward transport by swell of enhanced surface turbulence produced by wind-wave breaking, then vertically integrating $\varepsilon(z) = \varepsilon_0 \exp(\alpha z)$ to the surface should result in a value consistent with the energy lost by wave breaking. Vertical integration results in values of 0.45 and 0.11 Wm^{-2} for the fits in the depth intervals of 1.5–5.5 m and 5.5–14.5 m, respectively.

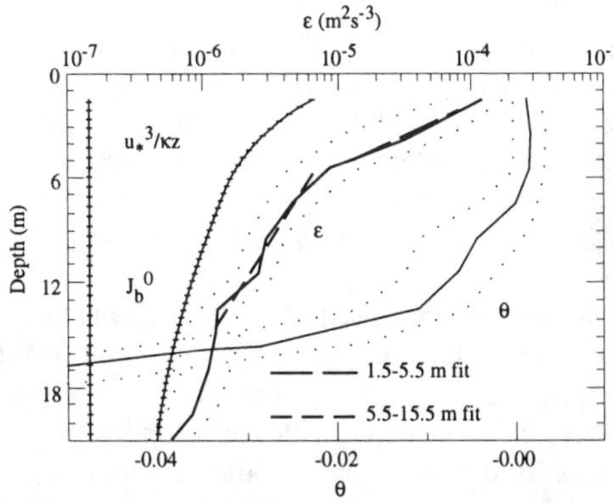

Figure 3. Referenced and averaged profiles of θ and ε for night 1. Also plotted are u^3_*/kz and the nighttime surface buoyancy flux, $J_b{}^0$. Dotted lines represent the 95% bootstrap confidence limits. Note that the dissipation rate in the neutral OBL is larger than the wind stress production, u^3_*/kz. Two fits, of the form $\varepsilon(z) = \varepsilon_0 \exp(\alpha z)$, to the estimated values of ε in the upper part of the OBL are shown. A fit in the depth range 1.5–5.5 m (long dashes) results in $\varepsilon_0 = 2.9 \times 10^{-4}\ m^2\ s^{-3}$ and $\alpha = 0.66\ m^{-1}$, and a fit in the depth range 5.5—14.5 m (short dashes) results in $\varepsilon_0 = 2.0 \times 10^{-5}\ m^2\ s^{-3}$ and $\alpha = 0.18\ m^{-1}$. Note that down to the base of the OBL (~ 13 m) $\varepsilon(z)$ is larger than predicted by constant stress layer scaling.

Both values are similar to that obtained by integrating the observed $\varepsilon(z)$ and are consistent with the estimates of the rate of energy lost by wind-wave breaking (Anis and Moum, 1995).

In the mechanism described by (17) and (18), by which TKE generated by breaking wind-waves is transported downwards by the swell, the depth decay of $\varepsilon(z)$ is dictated by that of the orbital velocity of the swell. However, if in this case downward transport of turbulence is dominated by wind-wave transport in the upper part and swell transport in the lower part, then $k_w = \alpha_1 = 0.66\ m^{-1}$ and $k_s = \alpha_2 = 0\ 18\ m^{-1}$, where k_w and k_s are the wind-wave and swell wavenumbers, respectively. These wavenumbers correspond, respectively, to $\lambda_w = 9.5$ m and $\lambda_s = 34.9$ m or to $T_w = 2.5$ sec and $T_s = 4.7$ sec, compared to the ship's officer's log estimates of $T_w = 4$ sec and $T_s = 6$ sec (for the shorter of the two swells observed). (We note that if the turbulence undergoes significant diffusion and dissipation over half a wave cycle, then the vertical decay rate will be greater than (18)). The correlation coefficient, r, need not be large for (18) to be an effective mechanism and estimates using our data resulted in $r_0 \sim 0.1$ Estimates of the velocity scale, u_t for turbulence resulting from wave breaking were made assuming an eddy length scale lt ~ H (H is the height of the breaking waves), and using the relation $\varepsilon = u^3_t/l_t$, with $\varepsilon = \varepsilon_0$, ($\varepsilon_0$ is the surface intercept value of the exponential fit). Expressed in terms of the phase velocity, C_w, of the wind-waves values of u_t were ~ $0.011C_w = 0.066\ m\ s^{-1}$. This is consistent with the range of $0.005–0.02C_w$ of rms turbulence velocities resulting from wave breaking observed by Rapp and Melville (1990). For comparison; if near the surface $u_t \sim u_*$ we find , $u_t \sim 0.016\ m\ s^{-1}$. The latter value of u_t is smaller by a factor of 4, compared with that estimated above, resulting in an underestimate of ε by almost two orders of magnitude.

Lets now consider the mechanism described by (22), by which wave-induced shear stresses in a rotational wave field interact with the mean flow and the turbulence. In this case the depth decay rate of $\varepsilon(z\ )$ is determined by the decay rate of the wave stress (21), such that $\varepsilon(z) \propto \exp(2kz)$ (or $\alpha = 2k$ for comparison to the data). Assuming that the change in slope of the exponential in Figure 3 is due to dominance of wind-waves near the surface and swell deeper, we find $k_w = \alpha_1/2 = 0.33\ m^{-1}$ and $k_s = \alpha_2/2 = 0.09\ m^{-1}$ corresponding to $\lambda_w = 19.0$ m ($T_w = 3.5$ sec) and $\lambda_s = 69.8$ m ($T_s = 6.7$ sec), respectively. A lower limit on $\sin\phi$, in the wave-stress dominated surface layer, results in $\sin\phi > 0.11$ ($\phi > 6°$), using wind-wave parameters, and $\sin\phi > 0.018$ ($\phi > 1°$), using swell parameters. In other words, to explain the observed TKE dissipation rates near the surface, only a small departure from quadrature (on the order of a few degrees) is needed. For comparison, phase shifts from quadrature as large as 20°—30° have been reported in the upper few meters of the OBL (Shorting, 1970; Cavaleri et al., 1978).

Although the two mechanisms proposed above to explain the observed high ε values in the near surface layer, namely the transport of TKE by swell and the indirect production via wave stresses, are physically different, both processes might affect turbulence and mixing in a surface layer of $O(k^{-1})$. Beneath this layer either SL or convective similarity scalings, or a combination of the two, may be more appropriate.

Acknowledgments. This work was carried out as part of my graduate studies at Oregon State University under the guidance of Jim Moum. Long and helpful discussions with Dave Hebert and Laurie Padman over the course of this work are gratefully acknowledged. I am grateful to the Captain and crew of the R/V *Wecoma* for their help during the experimental phase of this study. I would also like to thank the reviewers for their constructive and helpful comments.

References

Agrawal, YC., E. A. Terray, M. A. Donelan, P. A. Hwang, A. J. Williams III, W. M. Drennan, K. K. Kahma, and S. A. Kitaigorodskii, Enhanced dissipation of kinetic energy beneath surface waves. *Nature, 359,* 219-220, 1992.

André. J. C. and P. Lacarrere, Mean and turbulent structures of the oceanic surface layer as determined from one-dimenensional, third-order simulations. J. *Phys. Oceanogr. 15,* 121-132, 1985.

Anis, A., and J. N. Moum, The superadiabatic surface layer of the ocean during convection. *J. Phys. Oceanogr., 22,* 1221-1227, 1992.

Anis, A., and J. N. Moum, Prescriptions for heat flux and entrainment rates in the upper ocean during convection. *J. Phys. Oceanogr., 24,* 2142-2155, 1994.

Anis, A., and J. N. Moum, Surface wave-turbulence interactions: scaling $\varepsilon(z)$ near the surface of the ocean. *J. Phys. Oceanogr., 25,* 2025-2045, 1995.

Caldwell, D. R., T. M. Dillon and J. N. Moum, The rapid sampling vertical profiler: an evaluation. *J. Atmos. Ocean. Technol., 2,* 615-625, 1985.

Caughey, S. J., and S. G. Palmer, Some aspects of turbulence structure through the depth of the convective boundary layer. *Quart. J. Roy Meteor Soc. 105,* 8ll-827, 1979.

Cavaleri, L., J. W. Ewing. and N. D. Smith, Measurements of the pressure and velocity field below surface waves, In *Turbulent Fluxes through the Sea Surface, Wave Dynamics, and Prediction,* edited by A. Favre and K. Hasselmann, pp 257-272, Plenum, 1978.

Cavaleri, L. and S. Zecchetto, Reynold Stresses, in *'The Ocean Surface,* edited by Y. Toba and H. . Mitsuyasu, pp. 443 -448, Reidel, 1985.

Cheung, T. K., and R. L. Street, The turbulent layer in the water at an air-water interface, *J. Fluid Mech. 194,* 133-151, 1988.

Craig, P. D. Velocity profiles and surface roughness under breaking waves. *J. Geophys. Res., 101.* 1265-1277, 1996.

Craig, P. D., and M. L. Banner, modelling wave-enhanced turbulence in the ocean surface layer. *J. Phys. Oceanogr 24,* 2546-2559, 1994.

Lombardo, C. P., and M. C. Gregg, Similarity scaling of viscous and thermal dissipation in a convecting surface boundary layer. *J. Geophys. Res.,* 94. 6273-6284, 1989.

Deardorff, J. W. Three-dimensional numerical study of the height and mean structure of a heated planetary boundary layer. *Bound. -Layer Meteor., 7,* 81 - 106, 1974.

Dillon, T. M., J. G. Richman., C. G. Hansen, and M. D. Pearson,. Near surface turbulence measurements in a lake. *Nature 290,* 390- 392, 1981.

Drennan, M. M., K. K. Kahma, L. A. Terray, M. A. Donelan, and S. A. Kitaigorodskii. Observations of the enhancement of kinetic energy dissipation beneath wind waves. in *Breaking Waves,* edited by M. L. Banner and R. H. J. Grimshaw, pp 95- 10I, Springer-Verlag, New York, 1992.

Driedonks, A. G. M., and H. Tennekes, Entrainment effects in the well-mixed atmospheric boundary layer. *Bound. -Layer Meteor, 30,* 75-105, 1984.

Gargett, A. E., Ocean Turbulence. *Ann. Rev Fluid Mech., 21,* 419-451, 1989.

Gregg, M. C. Structures and fluxes in a deep convecting mixed layer. In *Dynamics of the Oceanic Mixed Layer,* edited by P. Müller and D. Henderson, pp. 1-23, Hawaii Institute of' Geophysics Special Publication, 1987.

Hussain, A. K. M. F. and W. C. Reynolds, The mechanics of an organized wave in turbulent shear flow. Part 2. Experimental results. *J. Fluid Mech., 41,* 241-25X, 1970.

Imberger, J., The diurnal mixed layer. *Limnol. Oceanogr.,* 30, 737-770, 1985.

Kitaigorodskii, S. A., M. A. Donelan, J. L. Lumley and E. A. Terray, Wave-turbulence interactions in the upper ocean. Part 2: Statistical characteristics of wave and turbulent components of the random velocity field in the marine surface layer. J. Phys. Oceanogr., 13, 1988-1999, 1983.

Melville, W. K., Energy dissipation by breaking waves. *J. Phys. Oceanogr, 24,* 2041-2049, 1994.

Moum, J. N., M. C. Gregg, R. C. Lien, and M. E. Car, Comparison of ϵ from two microstructure profilers. *J. Atmos. Ocean Tech., 12,* 346-366, 1994.

Osborn, T. R., Estimates of the local rate of vertical diffusion from dissipation measurements. *J. Phys. Oceanogr, 10,* 83-89, 1980.

Paulson, C. A., and J. J. Simpson, The temperature diffference across the cool skin of the ocean. *J. Geophys. Res. 86,* 11044- 11054, 1981.

Rapp, R. J. and W. K. Melville. laboratory measurements of deep-water breaking waves. *Phil. Trans R. Soc. Lond A331,* 735-800, 1990.

Shay, T. J., and M. C. Gregg, Turbulence in an oceanic convective layer. *Nature 310*, 282-285, 1984.
Shay, T. J., and M. C. Gregg, Convectively driven turbulent mixing in the upper ocean. *J. Phys. Oceanogr, 16*, 1777- 1798, 1986.
Shonting, D. H., A preliminary investigation of momentum flux in ocean waves. *Pure Appl. Geophys. 57*, 149-152, 1964.
Shonting, D. H., Observations of Reynolds stresses in wind waves. *Pure Appl. Geophys. 81*, 202-210, 1970.
Soloviev, A. V., N. V. Vershinsky, and V. A. Bezverchnii, Small scale turbulence measurements in the thin surface layer. *Deep-Sea Res. 35*, 1859-1874, 1988.
Stewart, R. W., and H. L. Grant, Determination of the rate of dissipation of turbulent energy near the sea surface in the presence of waves. *J. Geophys. Res.* 67, 3177-31 80, 1962.
Stull, R. B., The energetics of entrainment across a density interface *J. Atmos . Sci.* 33, 1260-1267, 1976.
Stull, R. B., *An Introduction to Boundary Layer Meteorology*. Kluwer Academic Publishers. Dordrecht, 666 pp. 1988.
Terray, E. A., M. A. Donelan, Y. C. Agrawal, W. M. Drennan, K. K. Kahma, A. J. Williams, R. A. Hwang, and S. A. Kitaigorodskii, Estimates of kinetic energy dissipation under breaking waves. *Nature,* 26, 792-807, 1996.
Thorpe, S. A., On the determination of K_V in the near surface ocean from acoustic measurements of bubbles. *J. Phys. Oceanogr,* 14, 855-863, 1984.
Willis, G. E. and J. W. Deardorff, A laboratory model of the unstable planetary boundary layer. *J. Atmos. Sci. 31*, 1297-1307, 1974.
Young, G. S, Turbulence structure of the convective boundary layer. Part 1: variability of normalized turbulence statistics. *J. Atmos Sci. 45*. 719-735, 1988.
Zhou, Ming Yu., D. H. Lenschow, B. B. Stankov, J. C. Kaimal and J. E. Gaynor, Wave and turbulence structure in a shallow baroclinic convective boundary layer and overlying inversion. *J. Atmos. Sci., 42*, 47-57, 1985.

5

Large Eddies in the Surface Mixed Layer and Their Effects on Mixing, Dispersion and Biological Cycling

M. Li and C. Garrett

Abstract

Large eddies, such as those associated with wind-driven Langmuir circulation, buoyancy-driven thermal convection and shear-driven Kelvin–Helmholtz billows, play an important role in mixing and deepening the surface layer of the ocean and lakes. This paper reviews recent progress towards understanding the physics of large eddies. Both Direct Numerical Simulation (DNS) and Large Eddy Simulation (LES) models have been developed. Together with new observations, they suggest that Langmuir circulation (1) dominates over thermal convection in mixing the surface water in windy conditions, (2) plays an important role in deepening the mixed layer and in mixing within it, and (3) is an effective mechanism for horizontal dispersion.

Introduction

The upper region of the ocean or lakes typically exhibits a surface mixed layer, or epilimnion, with a thickness of a few metres to several hundred metres, in which the density is rather uniform.

This mixed layer is a key component in studies of climate, biological productivity and marine pollution. It is the link between the atmosphere and deep ocean and directly affects the air–sea exchange of heat, momentum and gases. For example, the heat capacity of the top 2.5 m of the ocean equals that of the entire atmosphere so that, with turbulent mixing typically extending to a depth many times this, the ocean essentially acts as a "thermal flywheel" for the climate system, smoothing out both temporal (seasonal) temperature changes and, through advection, meridional gradients. Surface and mixed layer processes also play a vital role in the exchange, between the atmosphere and the ocean, of gases such as oxygen, nitrogen and carbon dioxide. The exchange occurs by diffusion across the sea surface and across the surface of bubbles created by breaking waves (e.g. Wanninkhof, 1992; Farmer et al., 1993).

Several small-scale processes, generated either by surface fluxes of momentum and buoyancy or by shear across the base of the layer, contribute to turbulent mixing in the mixed layer (Figure 1). Thermal convection can be generated when the ocean loses heat through longwave back radiation or evaporative cooling. The shear generated in wind-

Physical Processes in Lakes and Oceans
Coastal and Estuarine Studies Volume 54, Pages 61-86

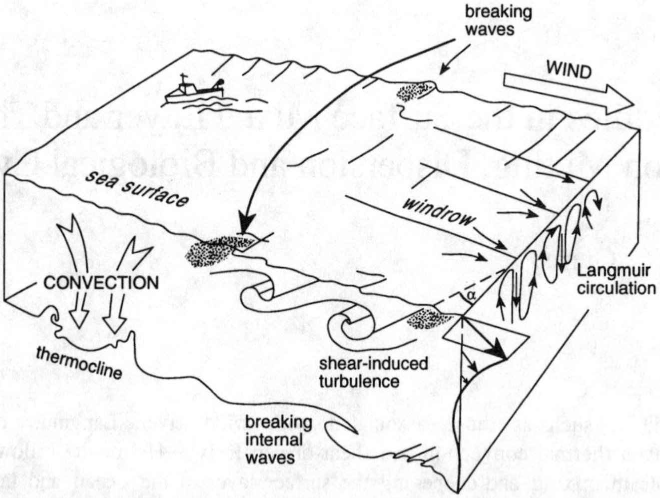

Figure 1. Schematic diagram showing small-scale processes taking place in the upper layer of the ocean. Redrawn from Thorpe (1985).

driven currents or in currents due to previous wind events can produce Kelvin–Helmholtz billows with their axes oriented roughly at right angles to the shear (Thorpe, 1985). These two turbulent processes bear a striking similarity to the buoyancy- and shear-driven turbulence found in the atmospheric boundary layer.

What distinguishes the ocean mixed layer, however, is the presence of surface waves. Wave breaking is a dominant source of turbulence near the surface and produces a greatly enhanced dissipation rate of kinetic energy in the top few metres (Agrawal et al., 1992). The interaction between the mean particle (Stokes) drift of surface waves and the wind-driven shear current also produces another coherent structure, known as Langmuir circulation (Langmuir, 1938), consisting of counter-rotating vortices with their axes roughly parallel to the wind direction.

Other processes, although not directly caused by local atmospheric forcing, can affect the properties of the mixed layer. Notable examples are internal waves, horizontal advection, and restratification driven by horizontal buoyancy gradients (Tandon and Garrett, 1995). These are major sources of uncertainty in the use of one-dimensional models to predict seasonal cycles of the mixed layer.

Turbulence in the ocean mixed layer, as in any boundary layer, may be dominated by organized coherent structures or so-called "large eddies". It is the theme of this paper to assess the role of large eddies in the deepening of the mixed layer, and in mixing and dispersion within the mixed layer.

Large eddies in the mixed layer can affect biological productivity by controlling both the supply of nutrients to the upper sunlit layer and the thickness of the mixed layer (Denman and Gargett, 1995). The eddies advect phytoplankton in an exponentially varying light field, with time scales comparable with those of light adaptation (Marra, 1978). Organic and inorganic matter is enriched in the surface films found on lakes and oceans so that when large eddies such as Langmuir circulation concentrate contaminants at convergence zones, harmful effects may be produced on organisms which feed in the surface microlayer (Sutcliffe et al., 1963; Barstow, 1983). Langmuir circulation has been frequently observed to distribute spilled oil into parallel bands (Thorpe, 1995a) and hence plays an important role in dispersing oil or similar pollutants.

Despite the apparent importance of large eddies, many existing mixed layer models have not explicitly taken their effects into consideration. In this review, we survey recent progress in understanding and modelling large eddies and discuss major open questions that warrant future investigations. The review will be biased towards our own work, and in many ways will only represent our personal perspectives. We apologize in advance for omitting many interesting papers. For more complete reviews, see Thorpe (1992, 1995b) and Large et al. (1994), though these have less emphasis on the phenomena we consider important.

One-Dimensional Mixed-Layer Models

Bulk Models

The simplest one-dimensional (1D) models are the so-called "bulk models" which capitalize on the small vertical density gradients found within the mixed layer. They parameterize changes in mixed-layer depth, and the averaged buoyancy and velocity of the mixed layer, in terms of the surface buoyancy flux and wind stress and/or the velocity and density difference between the layer and the water below it. These models are usually based on plausible physical arguments or analogy with laboratory entrainment experiments.

In bulk models, the mixed layer, with depth h, is assumed to have a sharp jump Δb in buoyancy $b = -g(\rho_w - \rho_0)/\rho_0$ at its base, where g is the gravitational constant, ρ_w is the water density and ρ_0 is a reference density. A key step is to prescribe the entrainment velocity $w_e = dh/dt$. One class of models (Phillips, 1977) gives

$$w_e = u_* F\left(\frac{h\Delta b}{u_*^2}\right) \tag{1}$$

where $Rb = h\Delta b/u_*^2$ is a bulk Richardson number based on the water friction velocity u_* and Δb, and is the only dimensionless parameter if u_* is the only relevant velocity scale. The function F is a decreasing function of Rb; if the rate of energy input to the mixed layer is assumed proportional to u_*^3, and a fixed fraction of this input is assumed to go into the increasing potential energy of the layer as dense fluid is entrained upwards, then $F \propto Rb^{-1}$ (Turner, 1973).

Another possible velocity scale is the magnitude $|\Delta \mathbf{u}|$ of the difference between the average velocity of the mixed layer and the velocity of the underlying fluid. Models for F(Rb), with $Rb = h\Delta b/(|\Delta U|)^2$, generally have a rapid decrease with increasing Rb (Phillips, 1977). A particularly simple form, suggested by Pollard et al. (1973), has $F = \infty$ for $Rb < 1$ and $F = 0$ for $Rb > 1$, so that the mixed layer depth adjusts to having Rb always equal to 1 during a deepening phase, though it is assumed not to "unmix" if $|\Delta \mathbf{u}|$ decreases. Price (1979) showed that an entrainment formula based on $|\Delta \mathbf{u}|$ was much more effective than one based on u_* in collapsing data from different laboratory studies. Thus the second type of bulk model, based on $|\Delta \mathbf{u}|$, has become more popular, particularly with the PWP model (Price, Weller and Pinkel, 1986) which stops the entrainments at $Rb = 0.65$ instead of 1 and also smooths the interface at the base of the layer to give a gradient Richardson number of 0.25.

In reality, however, the entrainment depends on the behaviour of turbulence in a stratified fluid and cannot be simply represented in terms of either u_* or $|\Delta\mathbf{u}|$, though possibly some simple parameterizations in terms of both velocity scales can be extracted from more thorough studies. In fact, Deardorff (1983) reviewed studies that cast doubt on the relevance of some laboratory studies because of the influence of side walls, and presented entrainment formulae involving bulk Richardson numbers based on u_*, $|\Delta\mathbf{u}|$, and a free convection velocity scale given by $w_* = (B_0 h)^{1/3}$ where B_0 is the surface buoyancy flux. Also, Davis et al. (1981) found that deepening formulae involving both u_* and $|\Delta\mathbf{u}|$, though in a rather *ad hoc* way, gave better results than formulae depending on u_* or $|\Delta\mathbf{u}|$ alone.

Turbulence Closure Models

Besides bulk models, turbulence closure models have been applied to the surface mixed layer of the ocean. The simplest of these assumes simple profiles of eddy viscosity and diffusivity, such as the nonlocal K-profile model recommended by Large et al. (1994). Many other mixed layer models are based on higher order closure schemes proposed by Mellor and Yamada (1982). However, as reviewed by Gaspar et al. (1990), the commonly used versions suffer from uncertainty in the prescription, or computation, of the "master length". Recently, Kantha and Clayson (1994) have extended the modified Mellor–Yamada second moment closure scheme of Galperin et al. (1988), but at the base of the mixed layer they use the same *ad hoc* Richardson number-dependent mixing formulae as employed by Large et al. (1994) in the transition zone.

A weakness in 1D models is that they do not explicitly incorporate key physical processes such as large eddies which are responsible for mixing the layer, and it is not at all clear that their parameterization implicitly models these processes correctly. For example, none of the existing 1D models have included Langmuir circulation, which is thought to be the most important large eddy process in the mixed layer (Thorpe, 1992). Even after careful calibration of empirical coefficients, the 1D models often over-predict the sea surface temperature in the summer and under-predict it in the fall owing to the difficulty in predicting the seasonal cycle of the mixed layer depth (Gaspar, 1988). When predicting mixing and diffusion within the mixed layer, turbulence closure models produce local down gradient diffusion while bulk models give implicit infinite diffusivity leading to all properties being instantaneously mixed throughout the layer. Neither of these predictions are reliable if there are nonlocal transports by large eddies.

Nevertheless, simple 1D mixed layer models are needed for a variety of applications since the processes involving small turbulent scales cannot be resolved except in very limited domains and for short time scales. The challenge to mixed layer modellers is to derive parameterizations of these processes for use in large-scale and long-term model integrations.

Two-Dimensional Modelling of Large Eddies

Forcing Functions

The surface of the ocean is forced by wind stress τ_w, with a corresponding friction velocity $u_* = (|\tau_w|/\rho_w)^{1/2}$ where ρ_w is the density of water, and by a surface buoyancy flux B_0 per unit area, given by

$$B_0 = -C_p^{-1} g\alpha\rho_w^{-1} Q + g\beta s\rho_w^{-1}(E-P). \tag{2}$$

Here, C_p is the specific heat of water, $\alpha = -\rho^{-1}(\partial/\rho/\partial T)_{p,s}$ is the coefficient of expansion of water at fixed pressure p and salinity s, Q is the net heat flux into the water, $\beta = \rho^{-1}(\partial\rho/\partial s)_{p,t}$, E is the evaporation rate and P is the rainfall rate (Gill, 1982). Positive B_0 corresponds to loss of buoyancy from the sea. The net heat flux into the sea is made up of insolation, which is distributed over an attenuation depth minus longwave back radiation and latent and sensible heat loss rates which act at the surface.

When the ocean loses heat, surface water becomes heavier and may sink. Buoyancy-driven thermal convection is a well known large eddy process and usually occurs at night in the ocean mixed layer. Convection may take the form of longitudinal rolls or various cellular patterns, although in a wind-driven shear current it organizes itself into rolls aligned in the wind direction (Kuo, 1963). Thermal convection in the atmospheric boundary layer has been well studied and most of the results can be directly carried over to the oceanic boundary layer (Thorpe, 1985).

The interaction between surface waves and the wind-driven shear current produces Langmuir circulation, however, which is unique to the ocean mixed layer. Due to the predominantly two-dimensional (2D) nature of these large eddies, and as a simplification, we have investigated a 2D Direct Numerical Simulation (DNS) model which has constant eddy viscosity and eddy diffusivity.

Dynamics of Langmuir Circulation

The prevailing theory for Langmuir circulation is the Craik–Leibovich model (Craik, 1977; Leibovich, 1977; hereafter the CL model) in which the Stokes drift of surface waves tilts the vertical vortex lines of a near-surface downwind jet to produce streamwise vorticity with surface convergence at the jet maximum. The jet is then reinforced by continued acceleration, by the wind stress, of the converging surface flow. The non-dimensional governing equations for Langmuir circulation in homogeneous water are given by Leibovich (1977) as

$$\frac{\partial u}{\partial t} + v\frac{\partial u}{\partial y} + w\frac{\partial u}{\partial z} = La\nabla^2 u, \tag{3}$$

$$\frac{\partial \Omega}{\partial t} + v\frac{\partial \Omega}{\partial y} + w\frac{\partial \Omega}{\partial z} = La\nabla^2 \Omega - \frac{du_s}{dz}\frac{\partial u}{\partial y}, \tag{4}$$

$$v = -\frac{\partial \psi}{\partial z}, \qquad w = \frac{\partial \psi}{\partial y}, \qquad \Omega = \nabla^2 \psi, \tag{5}$$

where u, v, w are the downwind, crosswind and vertical velocity components, respectively, Ω is the streamwise vorticity and u_s is the Stokes drift current. The dimensional Stokes drift is assumed to have an exponential profile, $\tilde{u}_s = 2S_0 e^{2\beta z}$ with the surface Stokes drift $2S_0 = 0.016U_w$ and the *e*-folding depth $1/(2\beta) = 0.12U_w^2/g$ estimated from the empirical Pierson and Moskowitz spectrum applicable only to fully developed seas. The controlling dimensionless number is the Langmuir number defined as

$$La = \left(\frac{\nu\beta}{u_*}\right)^{3/2}\left(\frac{S_0}{u_*}\right)^{-1/2} \tag{6}$$

where ν is the eddy viscosity.

Li and Garrett (1993, hereafter LG93) examined the 2D CL model and found that the predicted maximum downwelling velocity and downwind jet strength both depend on the magnitude of the Langmuir number. By adjusting the eddy viscosity within a plausible range and choosing a corresponding value of La ≈ 0.01, the predicted downwelling speed can be matched with measured values, but the jet seems to be weaker than observed. Using three-dimensional models, Skyllingstad and Denbo (1995) and Tandon and Leibovich (1995) have found that the pitch (downwind jet strength/downwelling velocity) is variable and closer to observations. This may be partly because their effective eddy mixing coefficients are rather large and give large La for which the 2D DNS model also predicts a larger pitch than at small La.

Two-dimensional simulations show that small counter-rotating vortices are generated near the surface and merge with each other to form bigger vortices. The maximum downwelling velocity w_{dn}, reached in these quasi-steady cells, increases as La decreases and is proportional to $La^{-1/3}$ at small La, i.e.

$$w_{dn} = 0.72u_*\left(S_0/u_*\right)^{1/3} La^{-1/3}, \tag{7}$$

as shown in Figure 2. The ratio of the surface Stokes drift to the water friction velocity is a function of sea state or wind fetch; for fully developed seas, $S_0/u_* = 4.6$ to 6.9 but is less in developing seas (see LG93).

By examining the flow fields in the quasi-steady cells, LG93 found that inner boundary layers developed within Langmuir cells. The flow field can be thus decomposed into a surface boundary layer, a narrow downwelling region, the corner where the two regions overlap, and an advective interior (Figure 3). By a scale analysis of the steady-state governing equations for the two sublayers, LG93 derived the $-1/3$ power law for w_{dn} and predicted that the thickness of the sublayers varies like $La^{1/2}$, both in agreement with the numerical results. The scale analysis also revealed that streamwise vorticity is generated in the narrow downwelling region. Thus it is expected that the 2D DNS results will depend on eddy viscosity unless, at very small Langmuir numbers (very high Reynolds numbers), unsteady small eddies are continuously generated in the sublayers so that the overall flow does not directly feel the effects of viscosity. Preliminary simulations at smaller La show that small cells can indeed be regenerated in the upwelling regions between the quasi-steady cells.

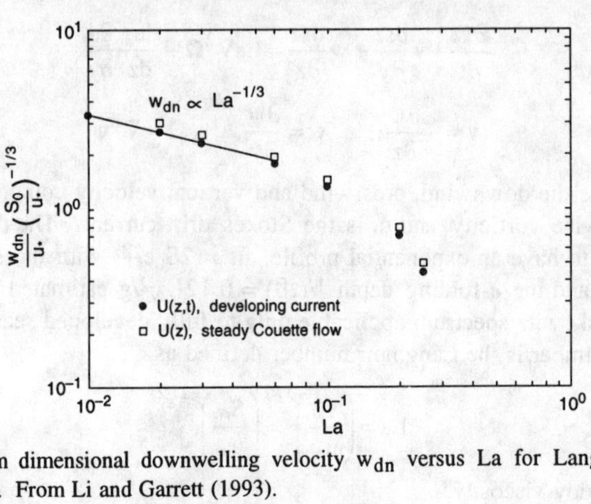

Figure 2. Maximum dimensional downwelling velocity w_{dn} versus La for Langmuir circulation in homogeneous water. From Li and Garrett (1993).

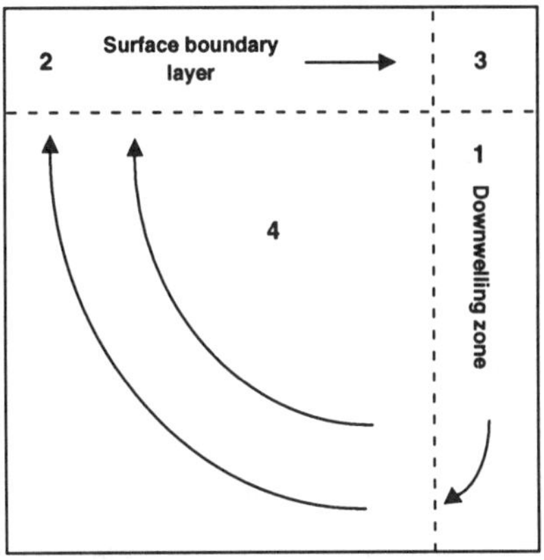

Figure 3. A schematic diagram of the four sub-regions within a cell. From Li and Garrett (1993).

Comparison between Langmuir Circulation and Thermal Convection

Wind-driven Langmuir circulation and buoyancy-driven thermal convection are two important large eddy processes found in the ocean mixed layer. To compare their roles in mixing the layer, Li and Garrett (1995, hereafter LG95) have used an extension of the CL model that includes the buoyancy force. Water is initially homogeneous but then is exposed to the wind and surface heat loss. LG95 have showed that this problem is characterized by three dimensionless numbers. The first is the Langmuir number La, the second is the eddy Prandtl number $Pr = \nu/\kappa$ and the third is the Hoenikker number

$$Ho = \frac{-\alpha g Q}{C_p \rho_w S_0 \beta u_*^2} \quad (8)$$

The latter dimensionless number can be expressed as the ratio of a length scale for the wave field to the Monin–Obukhov length, but it is more useful to recognize that HoPr represents the ratio, in the downwind vorticity equation, of convective forcing, by surface heat loss $-Q$, to wave forcing. The spectrum of the Stokes drift gradient $S_0\beta$ is flat so that Ho is fairly independent of fetch. From numerical solutions and scale analysis, LG95 have shown that, for Pr = 1, Ho must be as large as about 3 in order for convective forcing to be comparable to wave forcing at appropriately small values of La (Figure 4). The quantitative criterion used in this comparison is the maximum downwelling velocity w_{dn}, which is located between one-third and one-half of the cell penetration depth, although the Stokes drift gradient driving Langmuir circulation typically drops off more rapidly.

Realistic values for the surface heat flux, Stokes drift and wind stress give Ho significantly less than 0.1. It thus appears that the surface heat flux is insignificant compared with the vortex force, a conclusion that also holds for other values of Pr, for net heating Ho < 0 and for depth-distributed heating. The temperature then behaves as a passive scalar in these numerical experiments in which a heat flux is applied to previously homogeneous water.

For plausible values of the heat loss and other parameters, the predicted surface temperature difference $\Delta\theta$ from divergence to convergence is $0(10^{-2})$K, which is comparable to values reported by Thorpe and Hall (1982) and Weller and Price (1988).

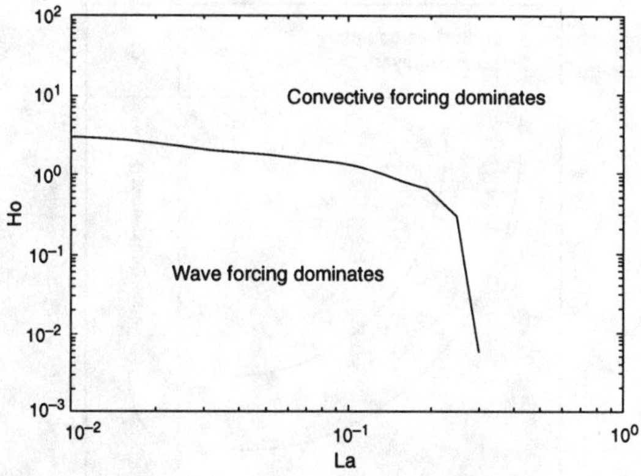

Figure 4. The separation of the wave-forcing-dominated regime from the convection-dominated regime in La, Ho space for Pr = 1. The solid line corresponds to W_{dn} being doubled from its value at Ho = 0. From Li and Garrett (1995).

Scale analysis, supported by numerical solutions, shows that

$$\Delta\theta = c\frac{S_0\beta u_*}{\alpha g}\left(\frac{S_0}{u_*}\right)^{-1/3} \mathrm{Ho}\mathrm{Pr}^{1/2}\mathrm{La}^{-1/6} \tag{9}$$

where c = 2 for cells of aspect ratio about 1. This implies weak dependence ($\nu/{-1/4}$) on the eddy viscosity ν, but more dependence on Pr.

Further questions remain regarding the validity of the 2D model and the appropriateness of using constant eddy viscosity and diffusivity. However, the constant $\mathrm{Ho}_c = 3$, for the transition from wave forcing to convective forcing, strongly suggests that the transition depends mainly on surface forcing unless the eddy Prandtl number is significantly different from unity, although convective plumes may produce a high turbulent eddy viscosity or a big La so that Langmuir circulation may be weakened or suppressed. We will return to this comparison when we discuss the LES model.

Role of Langmuir Circulation in the Deepening of the Mixed layer

Deepening due to Engulfment by Langmuir Circulation

To determine the role of wind-driven Langmuir circulation in deepening the mixed layer, Li and Garrett (1997, hereafter LG97) have investigated the interaction between Langmuir circulation and pre-existing stratification.

For initially linear stratification with buoyancy frequency N, the 2D DNS model shows that Langmuir cells erode the stratification in two stages. At first they engulf water and create a homogeneous surface layer. Later, they produce strong shear in a region beneath the downwelling jets, so that Kelvin–Helmholtz instability is facilitated. Figure 5 shows contours of stream function, vorticity, temperature and downwind current during the cell engulfment process for a given set of dimensionless parameters La, $R_{LN} = N^2\nu/(S_0\beta u_*^2)$ and Pr. One can see that the isotherms are raised at the upwelling sites as cold water is engulfed from below.

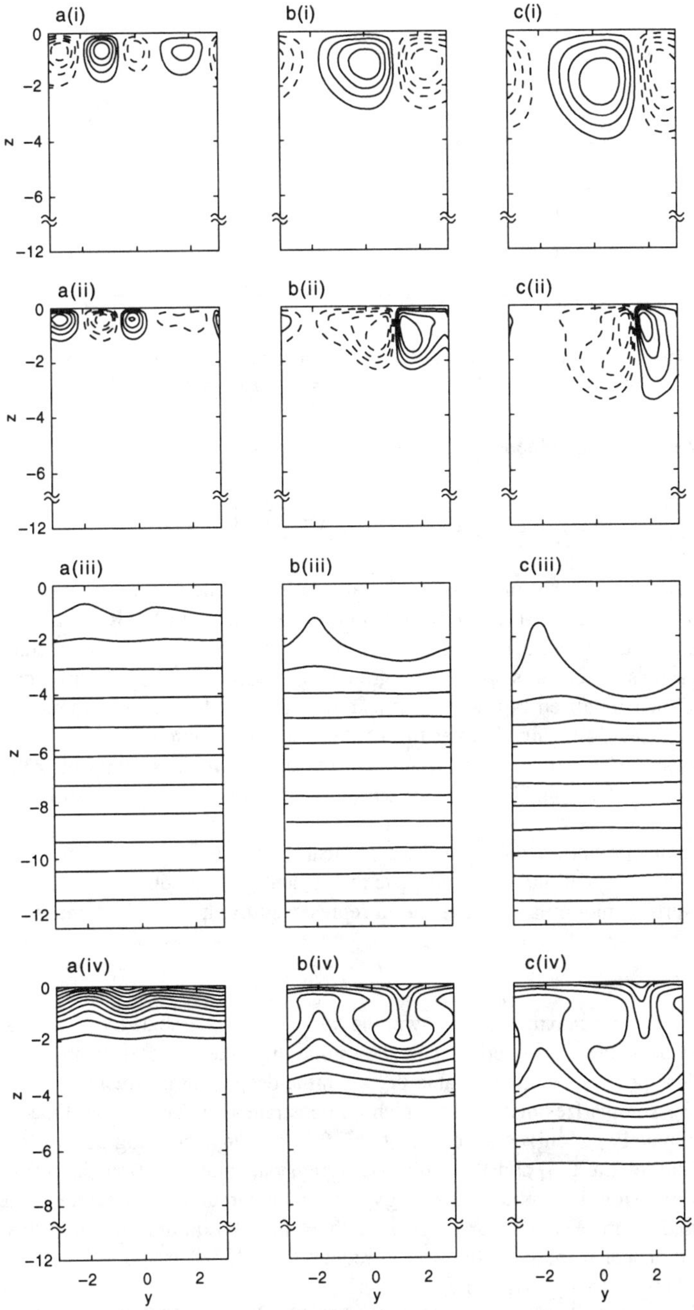

Figure 5. Contours of (i) stream function, (ii) vorticity, (iii) temperature and (iv) downwind current, for La = 0.03, R_{LN} = 0.05 and Pr = 1, at nondimensional times (a) t = 20, (b) t = 60, and (c) t = 160. The computational box has a nondimensional width L = 2π and a depth d = 4π.

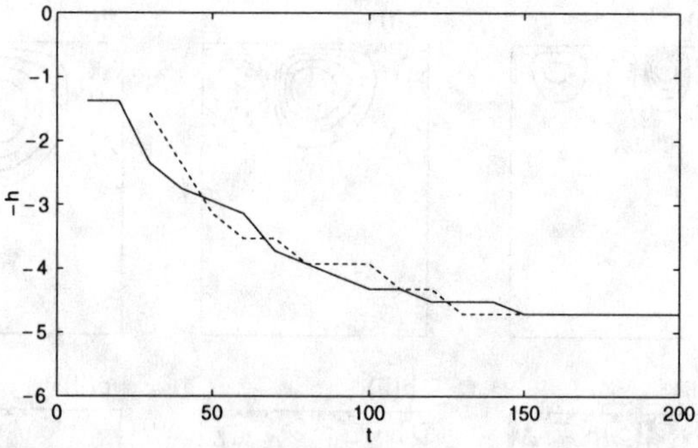

Figure 6. Nondimensional depth of the mixed layer generated by Langmuir cells in two computational boxes with $L = 2\pi$ (solid) and $L = 4\pi$ (dashed) for $La = 0.03$, $R_{LN} = 0.05$ and $Pr = 1$.

It is convenient to combine R_{LN} and La into a parameter

$$R_{LN}La^{-2/3} = \frac{N^2}{\beta^2 u_*^2}\left(\frac{S_0}{u_*}\right)^{-2/3} \tag{10}$$

which characterizes the pre-existing stratification without the involvement of eddy diffusivity. The model results can then be discussed in terms of La, $R_{LN}\,La^{-2/3}$ and Pr.

The mixed-layer depth h, defined to be at the maximum of the temperature gradient averaged across the cells, is shown in Figure 6 as a function of time for two numerical runs with nondimensionalised box width 2π and 4π. The depth increases rapidly as Langmuir cells grow in scale and engulf water from below, but then approaches an asymptotic limit. There appears to be virtually no difference between the two runs, suggesting that the mixed layer depth is determined by stratification rather than being constrained by lateral boundaries.

Langmuir circulation thus generates vertical velocities but with vertical penetration inhibited by stratification. The cell penetration depth h thus depends on the competition between vertical motion and stratification represented by the Froude number

$$Fr = \frac{w_{dn}}{Nh} \tag{11}$$

where w_{dn} is the maximum downwelling velocity generated by Langmuir cells in homogeneous water. Time series of Fr shows that the mixed layer deepens when Fr is high but that Fr approaches a critical value Fr_c when the deepening is arrested.

Figure 7 summarizes many runs which demonstrate that the critical Froude number Fr_c is approximately a universal constant of 0.6 for any oceanographically reasonable combination of La, R_{LN} and Pr. This has a physical interpretation in terms of kinetic energy conversion into potential energy; Langmuir circulation generates kinetic energy that is used to raise water particles from their initial equilibrium positions and stops penetrating deeper if the potential energy required ($\frac{1}{2} N^2h^2$) is more than a factor of 0(1) times the kinetic energy available ($\frac{1}{2} w^2$).

Using model results for w_{dn}/u_* for fully developed seas, LG97 obtained the mixed-layer depth produced by Langmuir circulation engulfment,

$$h = 10u_*/N. \tag{12}$$

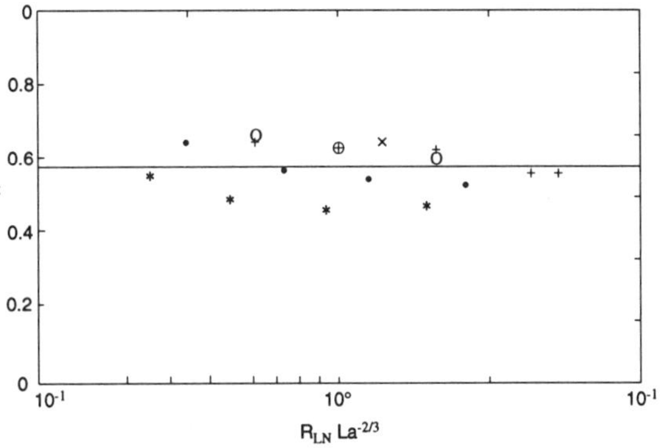

Figure 7. Summary of critical Froude number Fr_c in La and R_{LN} $La^{-2/3}$ parameter space. Symbol '*' corresponds to La = 0.1, '•' to La = 0.06, '+' to La = 0.03 and '×' to La = 0.02 at Pr = 1, while 'o' corresponds to La = 0.03 and Pr = 2. From Li and Garrett (1997).

Alternatively this means that Langmuir circulation will deepen the mixed layer until

$$\Delta b \geq c u_*^2 / h \tag{13}$$

where $c = 0.72\ S_0/(\nu\beta) \approx 50$ for fully developed seas, although it may be significantly smaller in developing seas.

In the bulk Richardson number criterion used in the PWP model (Price et al., 1986), mixed-layer deepening will not occur if

$$\Delta b \geq 0.65 |\Delta \mathbf{u}|^2 / h. \tag{14}$$

Comparison of (13) and (14) suggests that engulfment by Langmuir circulation dominates deepening if

$$|\Delta \mathbf{u}| \leq 9 u_* = 0.01 U_w, \tag{15}$$

i.e. if the velocity difference across the base of the mixed layer is no greater than 1% of the wind speed, though this fraction would be less in developing seas.

The buoyancy jump criterion (13) has been verified for a two-layer fluid (LG97) and found to be a robust formula suitable for use with any stratification profiles.

Enhanced Shear Instability Beneath Downwelling Jets

We have shown that Langmuir circulation rapidly produces a surface mixed layer through engulfment. Further entrainment due to shear instability can be possible, although the 2D model, which assumes no dependence in the downwind direction, cannot resolve this.

To detect the tendency for shear instability, we examine the distribution of the gradient Richardson number Ri_g. It exhibits a great variability in the crosswind direction, with variation by a factor of 3. The minimum value Ri_{gm} occurs beneath the surface convergence, where the downwelling jet carries down fast-moving fluid, making the shear much stronger. Figure 8 summarizes Ri_{gm} in parameter space. We observe that Ri_{gm} decreases with R_{LN} $La^{-2/3}$ defined in (10); when the water is less stratified, Ri_{gm} becomes smaller, falling below 0.25 for the realistic range of $0.1 < R_{LN}\ La^{-2/3} < 1$. Thus Langmuir

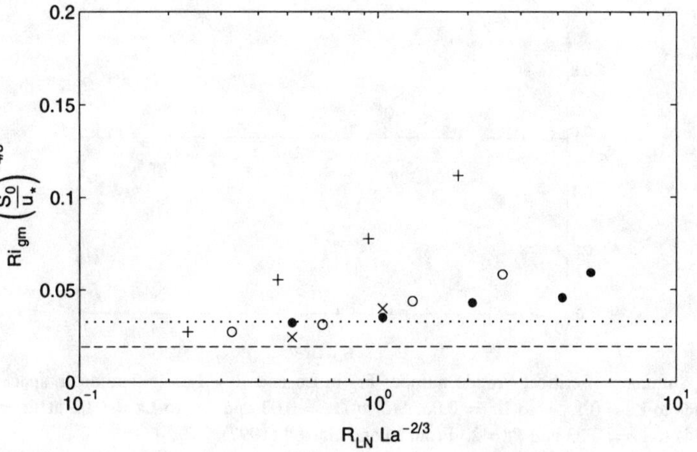

Figure 8. Summary of minimum gradient Richardson number $Ri_{gm}(S_0/u_*)^{-4/3}$ as a function of La, R_{LN} $La^{-2/3}$ and Pr. Symbol '+' corresponds to La = 0.1, 'o' to La = 0.06 and '•' to La = 0.03 at Pr = 1, while '×' corresponds to La = 0.03 and Pr = 2. The critical value Ri_g = 0.25 becomes a narrow band in $Ri_g(S_0/u_*)^{-4/3}$ because S_0/u_* = 4.6 to 6.9. From Li and Garrett (1997).

cells may facilitate shear instability in a horizontally confined region beneath downwelling jets, although allowance should be made for a different background shear due to previous wind events. Hence after the initial engulfment phase, shear and Langmuir circulation may act together to deepen the mixed layer with shear instability thickening the interface, the top of which can then be swept up by the cells, sharpening the interface again. Further work is required to learn how this second deepening phase should be parameterized and whether it can be described by a bulk Richardson number criterion, such as that used in the PWP model.

Test of Mixed-Layer Deepening Criteria Against Upper Ocean Observations

A commonly used test of mixed-layer models is to run the model over a seasonal cycle and then to check whether the predicted mixed-layer depth and sea surface temperature are in good agreement with observations. Disagreements are often ascribed to the errors of surface heat fluxes or the horizontal advection not accounted for in the 1D models. It is scientifically more appealing to directly evaluate the mixed-layer deepening criteria with upper ocean observations.

Li, Zahariev and Garrett (1995, hereafter LZG) have tested the bulk Richardson number and the new criterion representing Langmuir circulation against the upper ocean measurements obtained from the Long-Term Upper Ocean Study (LOTUS) experiment (Briscoe and Weller, 1984).

Using the estimated values of ΔT, $\Delta\mathbf{u}$ and h, LZG examined two normalized quantities:

$$Rb_n = 0.65\,|\Delta\mathbf{u}|^2/(h\Delta b), \tag{16}$$

$$LC_n = 50u_*^2/(h\Delta b). \tag{17}$$

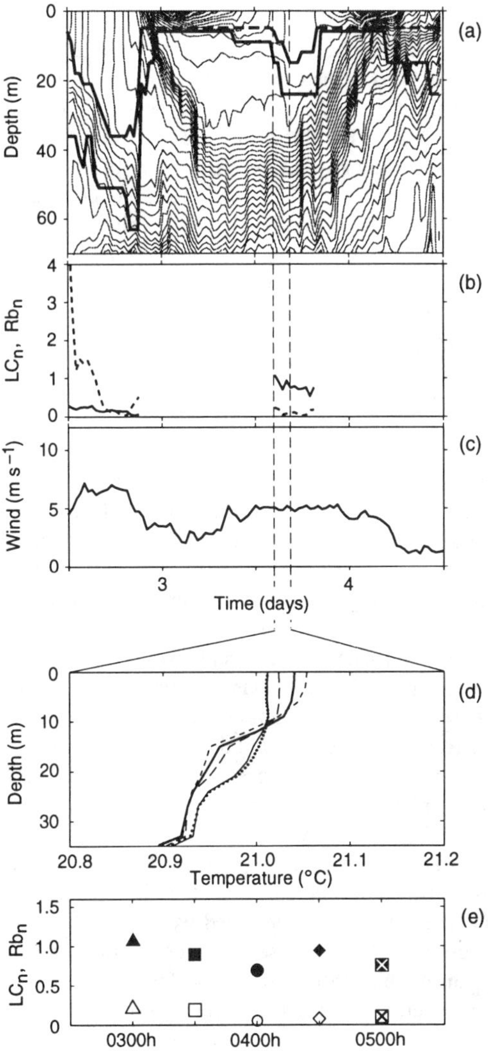

Figure 9. A mixed-layer deepening event dominated by Langmuir circulations. (a) Isotherms constructed from conductivity-temperature-depth (CTD) measurements are shown as thin contour lines. The thick upper line shows the depth of the surface mixed layer while the thick lower line shows the depth of the base of the transition layer. Mixed layer depths shallower than 5 m cannot be resolved by the data and are drawn as thick dashed lines at 5 m. They are excluded from the calculations of the flow indices. (b) Time series of the normalized stability indicators LC_n (solid line) and Rb_n (dashed line). $Rb_n > 1$ should cause mixed-layer deepening due to Langmuir circulation whereas $Rb_n > 1$ should cause deepening due to shear instability. (c) Times series of the wind speed. (d) Successive vertical profiles of temperature during a deepening event. (e) Corresponding indices LC_n (upper symbols) and Rb_n (lower symbols). The sequence used in the last two figures is: short-dashed line (triangles), thick solid line (squares), long-dashed line (circles), thin solid line (diamonds), and dotted line (boxed ×'s), representing profiles and indices, respectively, at half-hour intervals. From Li et al. (1995).

The PWP criterion and the proposed one representing Langmuir circulation suggest that deepening should occur if either Rb_n or LC_n exceeds 1. The possibility that deepening is dominated by convection is examined by checking changes in the heat content of the mixed layer. The event illustrated in Figure 9 is very suggestive of mixed-layer deepening caused by Langmuir circulation. For this event, the wind was approximately constant at 5 m s^{-1} and the mixed layer remained shallow prior to 0300h PST. Surface cooling then caused the mixed layer to deepen from 6 to 9 m between 0300 and 0400h. It also greatly reduced the buoyancy jump at the mixed-layer base. This precipitated a rapid penetrative deepening, though the heat content within the mixed layer was approximately conserved during this period. The Langmuir circulation index LC_n relaxed to about 1 as the mixed layer deepened. LZG also found mixed-layer deepening events dominated by shear instability and consistent with the PWP model.

Implications for Diurnal and Seasonal Cycles of the Mixed Layer

Although the proposed criterion for mixed-layer deepening by Langmuir circulation will only dominate occasionally, it may, nonetheless, be important to include in models of the mixed layer and will generally lead to a deeper layer than models without the criterion. In particular, running a model over a diurnal cycle shows how Langmuir circulation may delay daytime restratification of a mixed layer which is deep due to nighttime convection. Increasing insolation will heat the water near the surface, and can lead to a thin new surface mixed layer if only the shear criterion is used: if $Rb_n > 1$ (equivalent to $Rb < 0.65$), an increase in h conserves $h\Delta b$, but momentum conservation will decrease $|\Delta \mathbf{u}|$, decreasing Rb_n and allowing the new mixed layer to stabilize. However, increasing h while conserving $h\Delta b$ does not change the parameter LC_n, so that a shallower mixed layer cannot be formed unless the wind drops, and Langmuir circulation stops, for a time sufficient for insolation to provide a near-surface buoyancy content $h\Delta b$ of more than $50u_*^2$, where u_* is the friction velocity when the wind resumes. The time required for this is estimated to be about 15 hours for typical values of wind speed (water friction velocity $u_* = 0.01$ m s^{-1}) and surface buoyancy flux ($B_0 = 10^{-7}$ m^2 s^{-3}). Langmuir circulation may thus inhibit diurnal restratification that could occur in its absence, though the required time of calm could be much less if the later wind speed were less, the buoyancy flux greater and the sea state not fully developed. Over a seasonal cycle, the same argument suggests that in early spring Langmuir circulation may delay the formation of the seasonal thermocline in a deep winter mixed layer, unless the wind stress decreases significantly for a sufficient time.

Uncertainties in Using Bulk Models

Although the LOTUS experiment was a 2-year-long experiment (running from May 1982 to May 1984), we have not been able to find many well defined mixed-layer deepening events that can fit into the framework of bulk models. There are two major sources of uncertainty. One is horizontal advection of water with a different temperature. The other is internal waves which move isotherms up and down.

There are other difficulties in examining the buoyancy jump criteria used in bulk mixed-layer models. The estimates of mixed-layer depth, the thickness of the transition layer and the buoyancy jump across it are subjective and liable to errors, despite care being taken in examining individual profiles and in checking for consistency in time in the isotherms. Errors in these estimates will affect the evaluation of the buoyancy jump and velocity shear across the base of the mixed layer.

One may question the wisdom of treating the ocean surface layer as a bulk layer. An alternative way would be to parameterize turbulent stresses and fluxes, as in K-profile parameterizations (Troen and Mahrt, 1986; McPhee and Martinson, 1994; Large et al., 1994). In this way, intense mixing in the mixed layer can be represented by higher mixing coefficients, which depend on the properties of large eddies as well as local gradients, but this may not accurately represent the non-local mixing properties of large eddies.

Three-Dimensional Large Eddy Simulation (LES)

Subgrid-Scale Parameterization

The 2D DNS prediction for the strength of Langmuir circulation shows a sensitivity to eddy viscosity. To reduce this sensitivity, a three-dimensional (3D) LES model with a better parameterization of turbulence is required. LES models use a spatial filter to separate a turbulent field into large-scale and small-scale (subgrid) flows. The large-scale flows, which contain most of energy, are explicitly resolved, whereas small-scale flows are parameterized. If distance x_i is scaled with $1/\beta$, velocity u_i with u_*, temperature θ with the temperature jump ΔT across the mixed-layer base and pressure P with u_*^2, the nondimensionalized governing equations for the resolved-scale field are

$$\frac{\partial u_i}{\partial t}+(u_j+S_w u_{sj})\frac{\partial u_i}{\partial x_j}=\frac{\partial P}{\partial x_i}+S_w u_{sj}\frac{\partial u_j}{\partial x_i}+R\theta\delta_{ik}+\frac{\partial \tau_{ij}}{\partial x_j}, \tag{18}$$

$$\frac{\partial \theta}{\partial t}+(u_j+S_w u_{sj})\frac{\partial \theta}{\partial x_j}=\frac{\partial \tau_{\theta j}}{\partial x_j}, \tag{19}$$

$$\frac{\partial u_i}{\partial x_i}=0 \tag{20}$$

where $S_w = S_0/u_*$ and $R = \alpha g\Delta T/(u_*^2\beta)$. The spatial filtering operation used in deriving the resolved-scale equations is different from the usual temporal averaging used to derive the Reynolds stress (Leonard, 1974). This difference between the subgrid and Reynolds stresses is known as the "Leonard stress", though in practice it is swamped by numerical errors and thus often neglected (McComb, 1991). The subgrid stress and heat flux are usually specified by

$$\tau_{ij}=\nu\left(\frac{\partial u_i}{\partial x_j}+\frac{\partial u_j}{\partial x_i}\right), \tag{21}$$

$$\tau_{\theta j}=\kappa\frac{\partial \theta}{\partial x_j} \tag{22}$$

where the eddy viscosity ν and eddy diffusivity κ depend on the resolved flow.

The most widely used subgrid model is the Smagorinsky model (Mason and Brown, 1994) in which the nondimensionalized eddy viscosity and diffusivity are related to a mixing length ℓ and local strain rate S by

$$\nu = (\beta\ell)^2 S, \qquad \kappa = \nu/Pr, \tag{23}$$

$$S^2 = \frac{1}{4}\left(\frac{\partial u_i}{\partial x_j} + \frac{\partial u_j}{\partial x_i}\right)^2 . \tag{24}$$

Typically, ℓ is related to the grid size $\Delta = (\Delta x \Delta y \Delta z)^{1/3}$ through a constant $C_s = \ell/\Delta$. If the cutoff wavenumber in Fourier space lies in Kolmogorov's inertial subrange, one can adjust $C_s \approx (1/\pi)(3C_k/2)^{-3/4} = 0.18$ ($C_k = 1.4$ is the Kolmogorov constant) such that the ensemble-averaged subgrid kinetic energy dissipation is identical to the dissipation rate that matches the intensity of the turbulence (Lilly, 1967). Mason and Callen (1986) drew attention to the effect of C_s on numerical resolution and found that $C_s = \ell/\Delta = 0.2$ optimizes the tradeoff between achieving best numerical resolution and minimizing numerical errors.

This model is based on a 3D application of the mixing-length concept and assumes that the unresolved scales merely represent a one-way cascade to high wavenumber dissipation, as in a Kolmogorov spectrum, and do not have a backscatter effect on the resolved scale. One clear weakness of this model is that it only considers the ensemble-average stresses and ignores fluctuations of subgrid stresses which could influence the resolved-scale flows. It also produces too much dissipation near a solid boundary and cannot predict a logarithmic velocity profile. To overcome this problem, Mason and Thomson (1992) added a stochastic backscatter of energy to account for increased levels of subgrid motions. In thc so-called dynamic subfilter model, Germano et al. (1991) used the smallest resolved scales to explicitly calculate the mixing length and C_s in the Smagorinsky model, allowing for negative viscosities in regions that may show an inverse energy cascade.

A second model, proposed by Deardorff (1980), solves a prognostic equation for the subgrid-scale turbulent energy in which the closure assumptions are a down gradient diffusion assumption for the pressure term and the Kolmogorov hypothesis to evaluate the dissipation rate. The eddy viscosity is proportional to the product of a mixing length and a velocity scale determined by the local turbulent kinetic energy. Deardorff's model calculates a more complete turbulent energy budget, though it is computationally more expensive to implement.

The most interesting subgrid model is Kraichnan's spectral eddy viscosity model (Kraichnan, 1976). The flow is separated into low-wavenumber (resolved) and high-wavenumber (subgrid) components by a sharp spectral cutoff. In the equation for the kinetic energy spectrum for the resolved scale, the nonlinear energy transfer between the subgrid-scale and resolved-scale terms is parameterized by a spectral eddy viscosity. For 3D isotropic turbulence, this viscosity is constant for low wavenumbers except for showing cusp behaviour near the cutoff wavenumber. Fortunately, this singular behaviour is due to the sharp spectral cutoff and can be removed by using renormalization techniques (McComb, 1991). The spectral eddy viscosity model gives reasonable results even if the large-scale flows are neither isotropic nor homogeneous. Metais and Lesieur (1992) proposed a structure-function model that gives an equivalent spectral eddy viscosity in physical space. This model is computationally efficient and is as good as other subgrid models (Lesieur and Metais, 1996).

Intercomparison between different subgrid-scale models at fixed numerical resolutions suggests that the details of subgrid-scale parameterization do not matter (Nieuwstadt et al.,

1992). Subgrid terms are generally much smaller than the resolved-scale terms except near a boundary. Some high-resolution runs appear to have resolved part of the inertial subrange (Schmidt and Schumann, 1989). Nevertheless, convergence with increasing numerical resolution has not been convincingly demonstrated in LES simulations (Mason, 1994). This should be a key test of LES models, but it is hampered by restrictions of available computer resources. There is another resolution problem specifically related to the simulations of Langmuir circulation. The wave forcing drops off rapidly, with the e-folding depth of Stokes drift in the range of a few metres in fully developed seas. The mixing length (or the grid size) must be chosen to be much smaller than this depth in order to resolve the surface-concentrated Stokes drift forcing.

LES Simulations of Langmuir Circulation and Convection

Skyllingstad and Denbo (1995, hereafter SD) and McWilliams et al. (1997) have developed 3D LES models for the ocean surface layer. A snapshot of the Langmuir circulation flow field (Figure 10) obtained from SD's model shows vertical velocity distributions in a vertical cross-section perpendicular to the wind as well as on a horizontal plane. We observe that counter-rotating vortices are generated beneath the ocean surface and that at any instant there is co-existence of vortices of different sizes. Vortex amalgamation and regeneration occur simultaneously. On a horizontal plane, three bands of downwelling jets appear to be roughly aligned in the wind direction. Shorter downwelling bands are also seen embedded between the dominant bands. In some runs, Langmuir circulation convergence zones form downwind-directed Y-shaped junctions with remarkable similarity to patterns of bubble clouds observed in sidescan sonar images (Farmer and Li, 1995).

Convection cells can also align with the shear, but SD carried out a number of numerical experiments with different combinations of wind stress, wave forcing and convective forcing. They found that Langmuir circulation with surface heating produced vertical velocities, in the upper levels of the boundary layer, nearly as large as those produced with surface cooling. Thus both 2D DNS and 3D LES models agree that forcing associated with the Stokes drift dominates over thermal convection in driving large eddies. The LES simulations also showed that the addition of wave forcing to the wind-driven shear currents yields much stronger mixing than occurs with shear only, whether for surface cooling or surface heating. Deep in the mixed layer, different surface forcings give similar spatial distributions of vertical motions with scales proportional to the mixed-layer depth. These results are confirmed in a recent study by McWilliams et al. (1997), who show that the turbulent vertical fluxes of momentum and tracers are greatly enhanced by the presence of the Stokes drift.

Turbulent closure models such as the Mellor and Yamada (1982) model, or the Garwood (1977) bulk model, use estimates of the mixed-layer turbulent kinetic energy to determine vertical mixing and the growth of the mixed layer. SD showed that wave–current interaction may be an important source of turbulence which has not been included in vertical mixing parameterizations.

The fundamental difference between the DNS and LES models lies in the subgrid-scale parameterization. In DNS models the eddy viscosity and diffusivity are specified as constants. The eddy viscosity in the LES model depends on the local shear or turbulent kinetic energy and is normally a function of space and time. However, the eddy viscosity estimated from LES simulations is approximately a constant except close to the surface

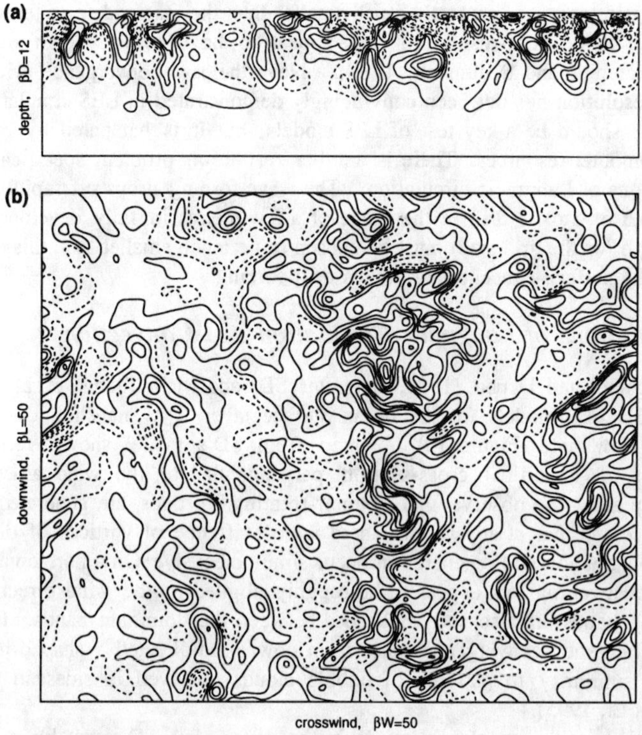

Figure 10. Vertical velocity distribution (upwelling indicated by dashed contour lines and downwelling by solid contour lines) obtained in the LES model. The computational domain has a nondimensional size of 50 (downwind length) by 50 (crosswind width) by 12 (depth) with the Stokes-drift e-folding depth $1/(2\beta) = 6.4$ m as the length scale. (a) In a crosswind section. (b) In the horizontal plane at a nondimensional depth of 4.

boundary, as found in the LES simulations of the atmospheric boundary layer (Mason and Brown, 1994). SD estimated the effective Langmuir number to be La = 0.1, an order of magnitude greater than the value 0.01 used in the 2D DNS model. The downwind-downward directed jets are wide and there appears to be less asymmetry between downwelling and upwelling flows (McWilliams et al., 1997). These features are broadly similar to those found in the 2D DNS model at a large value of La. It thus seems that small eddies resolved in the LES model produce high dissipation for the large eddies. This may represent the real physics of energy transfer between eddies of various sizes in turbulent flows. SD's eddy viscosity, however, was obtained from a formula involving the energy of all the motions other than the horizontally uniform mean shear flow. It thus involves the energy of the large eddies that one might wish to call Langmuir cells. A better comparison with the 2D models is hampered by the lack of a spectral gap between these and small-scale turbulence.

Nevertheless, it is interesting to note the similarity between the DNS and LES simulations in the interior regions. It does not seem unreasonable to view a LES simulation as a moderate Reynolds number DNS simulation of an interior flow patched to a high Reynolds number nonlinear boundary condition (Mason, 1994).

Outstanding Issues

Since the pioneering work of Deardorff (1972), LES has been used extensively in modelling the atmospheric boundary layer. A wealth of LES output has been documented, against which concepts and parameterizations can be tested. There is encouraging evidence that LES models of convection produce turbulence statistics in good agreement with atmospheric observations (Wyngaard, 1992). The application of LES to oceanography is just beginning and we are presently in the early stages of adapting LES techniques and of investigating uniquely oceanographic issues.

A challenging problem in the oceanic LES models is to represent near-surface turbulent processes such as breaking waves. Wave breaking, particularly plunging breakers, can introduce significant downward jets, with speeds locally comparable to or greater than those associated with Langmuir circulation. Breaking waves also produce high energy dissipation in the top few metres. Realistic LES simulations of the oceanic mixed layer may have to consider the non-uniform distribution of wind stress and the near-surface high dissipation. Craig and Banner (1994) showed that the enhanced energy dissipation by breaking waves can be accounted for in the Mellor–Yamada model by prescribing the turbulent kinetic energy flux to be proportional to u_*^3, although their results showed a sensitive dependence on the surface roughness length. It should also be borne in mind that the energy input by the wind appears to scale with the stress u_*^2 times the phase speed of rather short waves (Gemmrich et al., 1994). It may be possible, in the future, to prescribe this input via a subgrid model such as that of Deardorff.

It has not been established that the Craik–Leibovich equations can still be used in LES models. In the original derivation (Craik and Leibovich, 1976), a separation of time scales between turbulence, surface waves and currents was assumed. The vortex force was obtained after averaging over the time scale of many wave periods. In the LES model, the separation between Langmuir circulation currents and resolved turbulence does not exist. In a later paper Leibovich (1980) derived the CL equations by using the Generalized Lagrangian Mean (GLM) theory of Andrews and McIntyre (1978), though he made the crucial assumption that the eddy viscosity is constant. Additional terms can appear if eddy mixing coefficients vary in space and time; these presumably represent wave–turbulence interaction at small scales. Further work is required to transform the dissipative terms in the GLM formulation into their Eulerian counterparts. However, the subgrid-scale terms are usually small compared with the resolved-scale terms and it may not be necessary to seek more comprehensive representations. Nevertheless, it appears to be worthwhile to try to derive the LES governing equations that include the full effects of surface waves, and perhaps even investigate the effects of wave groupiness.

LES models have not yet been used in a systematic study of surface mixed-layer deepening by Langmuir circulation. We hope that it will prove possible to derive robust parameterizations of the initial engulfment, perhaps similar to the criterion of LG97, and also of the later deepening associated with shear instability. The need for such parameterizations is paramount due to the extreme and excessive computational demand of running a mixed-layer LES model for each grid square of a general circulation model of the ocean.

Horizontal Dispersion

We have seen some striking effects of large eddies in mixing and deepening the mixed layer. They also play a critical role in the horizontal dispersion of materials and pollutants at scales of O(10) m to O(1) km.

Surface windrows consisting of foam and flotsam are vivid demonstrations that Langmuir circulation can cause floating particles to congregate at convergence zones. It thus appears that Langmuir circulation is a mechanism for particle concentration. However, temporal evolution, including cell merging and disintegration, makes Langmuir circulation an effective mechanism for dispersion. Csanady (1974) was the first to recognize that a windrow with a life time T_c may split into many windrows spaced at ℓ_{cw}. A dimensional argument then gives the crosswind diffusivity as

$$K_y \propto \ell_{cw}^2 T_c^{-1}. \qquad (25)$$

By running an idealized model for a time-varying Langmuir circulation field, Faller and Auer (1988) found that windrow wandering and meandering is the main mechanism for dispersion. Their empirical fit into the model output showed that

$$K_y = 0.5\left(\frac{T_c v}{\ell_{cw}}\right)^{1/2} \ell_{cw}^2 T_c^{-1} \qquad (26)$$

where v is the crosswind velocity.

Recently, Thorpe and Curé (1994) and Thorpe et al. (1994) used sonar observations of convergence lines to infer the surface dispersion of particles. They found that K_y varies from 5×10^{-3} to 0.5 $m^2\ s^{-1}$ and depends on the Langmuir circulation life time T_c, the crosswind velocity v, and the excess downwind velocity Δu at the convergence lines and the mean wind drift $\bar{u}$. Their sonars scanned the water surface in fixed directions and hence could not directly measure the surface distribution of convergence zones.

Using rotating sidescan sonars, Farmer and Li (1995) obtained 2D images of the ocean surface at about 40-s intervals. These images were processed to produce patterns of convergence lines, which can be used to infer dispersion of particles floating at the ocean surface. Because there were no simultaneous measurements of velocity components, a horizontal flow field corresponding to the line pattern was prescribed. As a simplification, the particle is set to move crosswind at a constant speed v towards the closest convergence line if it lies between convergence lines. Because the instrument drifted with the prevailing current, it was assumed that no relative movement existed between the surface water and the instrument. Once a particle was trapped at a convergence line, it moved downwind at an excess speed Δu until it reached the end of that line, and then again moved to the nearest convergence at speed v. A background turbulent diffusion, as represented by a random walk

$$L = \sqrt{4K_t \Delta t} \qquad (27)$$

was added to the velocity field associated with Langmuir circulation.

Figure 11 shows the distributions of particles at the start of the dispersion experiment and at a half hour later. We see that the particles have moved apart from each other, indicating that Langmuir circulation is a dispersive mechanism. Figure 12 shows the time series of variances and eddy diffusivities in both the crosswind and the downwind directions for $v = \Delta u = 0.05$ m s^{-1} and Kt = 0.003 $m^2\ s^{-1}$. The absolute diffusivities were calculated using Taylor's approach. In this, assuming a homogeneous and stationary field, as when the wind is approximately steady, diffusivity can be calculated as the product of the variance of the downwind and lateral speeds and the Lagrangian integral time scales.

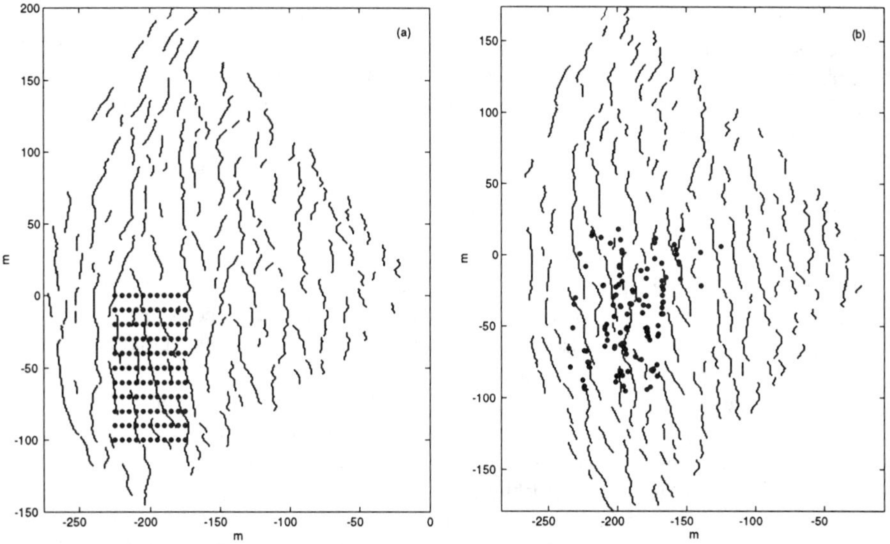

Figure 11. Horizontal particle dispersion by Langmuir circulation. A tally of 121 particles are seeded in (a). Their locations half hour later are shown in (b).

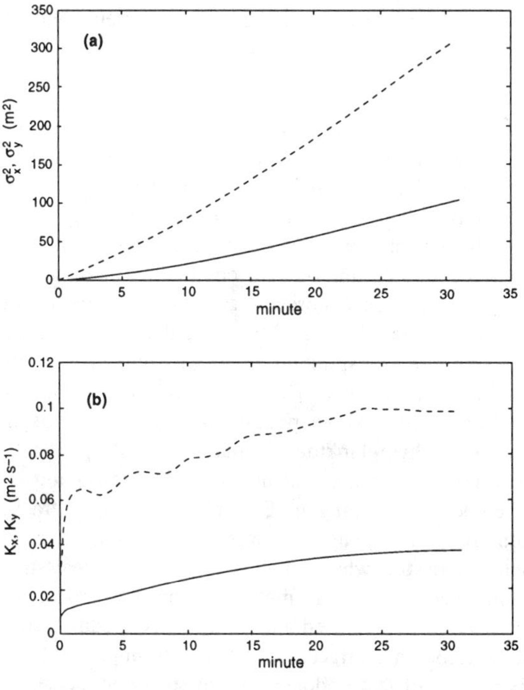

Figure 12. Time series of downwind (solid line) and crosswind (dashed line) variances (a) and diffusivities (b).

The horizontal diffusivities were found to be insensitive to the precise value of a small background turbulent diffusivity. Thus Langmuir circulation renders the incoherent turbulence irrelevant in dispersing particles. However, the diffusivities show a strong dependence on the crosswind velocity and downwind jet strength. This has yet to be understood and quantified.

Biological Cycles

Large eddies of the ocean mixed layer affect the marine biota community. A good example of such physical/biological interaction is the sinking of *Sargassum*, a pelagic plant found in substantial quantities in the Sargasso Sea, Caribbean Sea and Gulf of Mexico. Johnson and Richardson (1977) suggested that the cycling vertical motions in Langmuir circulation may be the cause of local elimination of *Sargassum* by collapsing their gas vesicles. Because of its proximity to the air–sea interface and the dominant nature of its role in carbon fixation and metabolism there, the *Sargassum* community may represent an important, though geographically limited, pathway for the incorporation of organic matter into the marine biosphere.

Successful survival of, and photosynthetic production by, phytoplankton require that they find and stay in regions with both sufficient light and nutrients (Denman and Marra, 1986). Thus large eddies in the mixed layer may affect biological productivity in two ways. First, large eddies advect phytoplankton in an exponentially varying light field, with a time scale comparable with that of light adaptation. Secondly, large overturns at the base of the mixed layer may control the supply of nutrients into the sunlit upper layer.

Phytoplankton Photosynthesis

The vertical cycling of phytoplankton by large eddies may enhance phytoplankton productivity by limiting photoinhibition that occurs if the phytoplankton stay near the surface and are exposed to intense light for too long. On the other hand, the cycling can be a disadvantage to phytoplankton if, in the lower depths of the mixed layer, the light intensity is too low for net photosynthesis to be achieved. As the depth of the mixed layer through which the phytoplankton are circulating increases, so the average light intensity to which the cells are exposed decreases, and consequently the total rate of photosynthesis by the whole phytoplankton population throughout the water column decreases. However, the rate of respiration of the phytoplankton population is relatively independent of the mixing depth. Thus there exists a critical depth below which respiratory carbon loss by the whole population exceeds photosynthetic carbon gain, resulting in a decrease in phytoplankton population (Sverdrup, 1953; Platt et al., 1991).

To quantify the effects of large eddies on phytoplankton photosynthesis, Woods and Onken (1982) developed a "Lagrangian Ensemble" model to track a large number of organisms individually or in groups. Large eddy effects were simulated by cycling individual organisms over the whole depth of the mixed layer in a given time, with additional small components of random motion. Other Lagrangian models, such as those of Kamykowski et al. (1994), have used a depth-dependent eddy diffusion coefficient.

Such treatment of turbulence effects can perhaps be improved by using more physically based DNS or LES models of large eddies. There seems to be no inherent difficulty in coupling these physics models with a phytoplankton photoresponse model such as that proposed by Denman and Marra (1986).

Supply of Nutrients

A more important interaction between large eddies and marine biota appears to be through the cycling of nutrients from deep water to the surface mixed layer. Based on results presented in sections 4.1 and 4.2, one might speculate that, coupled with the mixed-layer deepening due to enhanced shear instability beneath the downwelling jets, the engulfment of cold water at the upwelling sites by Langmuir cells may provide nutrients to the upper layer.

It has been established that nutrients such as nitrogen and phosphorus are essential elements required in phytoplankton photosythesis and can limit phytoplankton growth when depleted. Nutrients are usually abundant beneath the euphotic zone but can be exhausted in the shallow summer mixed layer due to heavy use by phytoplankton. This may cause a decrease in the phytoplankton population until the surface layer is replenished with nutrients in the fall when the mixed layer deepens again.

Many coupled mixed-layer and ecological models have been developed to investigate the physical/biological interaction in the mixed layer (see Denman and Gargett, 1995). Evans and Parslow (1985) were the first to show that the annual cycle of the mixed layer plays a critical role in the seasonal cycle of plankton populations. Fasham (1995) even suggested that the difference in mixed layer depth could be a cause of phytoplankton blooms occurring in the North Atlantic but not in the North Pacific. These coupled models employ either a bulk mixed-layer model or a turbulence closure model and frequently invoke rather arbitrary additional exchange coefficients to bolster nutrient supply into the mixed layer (Fasham et al., 1990). It is clear that further progress in ecological modelling will require physical models that incorporate the relevant mixing processes. In section 4.4, we alluded to the possible role of Langmuir circulation in disrupting diurnal and seasonal restratification. Thus mixed-layer models incorporating this effect may predict a deeper mixed layer and lead to more nutrient entrainment than those without it (with observations being the test of the validity of this). Other mixing processes, such as the breaking of internal waves, or upwelling caused by large-scale divergent flows, may also be important in nutrient supply. Nevertheless, it seems to be important to understand how turbulent mixed-layer processes interact with the mixing processes in the thermocline and how these processes jointly affect the entrainment of nutrients into the mixed layer.

Conclusion

In this paper we have presented our personal perspectives on mixed-layer physics. Our assessment is that energy-containing turbulent large eddies hold a central position in the surface mixed layer and should be a focus in future research. The 2D DNS models have helped advance our understanding of large eddy effects. Future research activities will concentrate on 3D LES models that can fully resolve three-dimensional turbulence, although concerns remain regarding the effective Reynolds numbers that can be achieved in LES models.

Acknowledgments. We thank Jörg Imberger for inviting us to attend the IUTAM symposium on physical limnology and two referees for thoughtful comments. We are grateful to Eric Skyllingstad for giving us his LES code. Rosalie Rutka provided editorial assistance. This work is supported by a Strategic Grant from Canada's Natural Sciences and Engineering Research Council.

References

Agrawal, Y. C., E. A. Terray, M. A. Donelan, P. A. Hwang, A. J. Williams III, W. M. Drennan, K. K. Kahma, and S. A. Kitaigorodskii, Enhanced dissipation of kinetic energy beneath surface waves. *Nature, 359*, 219–220, 1992.

Andrews, D. G., and M. E. McIntyre, An exact theory of nonlinear waves on a Lagrangian-mean flow. *J. Fluid Mech., 89*, 609–646, 1978.

Barstow, S. F., The ecology of Langmuir circulation: A review. *Mar. Envir. Res.*, 9, 211–236, 1983.

Briscoe, M. E., and R. A. Weller, Preliminary results from the long-term upper-ocean study (LOTUS). *Dyn. Atmos. Oceans, 8*, 243–265, 1984.

Craig, P. D., and M. L. Banner, Modeling wave-enhanced turbulence in the ocean surface layer. *J. Phys. Oceanogr., 24*, 2546–2559, 1994.

Craik, A. D. D., The generation of Langmuir circulations by an instability mechanism. *J. Fluid Mech., 81*, 209–223, 1977.

Craik, A. D. D., and S. Leibovich, A rational model for Langmuir circulations. *J. Fluid Mech., 73*, 401–426, 1976.

Csanady, G. T., Turbulent diffusion and beach deposition of floating pollutants. *Advances in Geophysics*, Vol. 18A, Academic Press, 371–381, 1974.

Davis, R. E., R. deSzoeke, D. Halpern, and P. Niiler, Variability in the upper ocean during MILE. Part I: The heat and momentum balances. *Deep-Sea Res., 28A*, 1247–1451, 1981.

Deardorff, J. W., Numerical investigations of neutral and unstable planetary boundary layers. *J. Atmos. Sci., 29*, 91–115, 1972.

Deardorff, J. W., Stratocumulus-capped mixed layers derived from a three-dimensional model. *Boundary-Layer Meteorol., 18*, 495–527, 1980.

Deardorff, J. W., A multi-limit mixed-layer entrainment formulation. *J. Phys. Oceanogr., 13*, 986–1002, 1983.

Denman, K. L., and A. E. Gargett, Biological–physical interactions in the upper ocean: The role of vertical and small scale transport processes. *Annu. Rev. Fluid Mech., 27*, 225–255, 1995.

Denman, K. L., and J. Marra, Modelling the time dependent photoadaptation of phytoplankton to fluctuating light, in *Marine Interfaces Ecohydrodynamics*, edited by J. C. J. Nihoul, pp. 341–349, Elsevier, 1986.

Evans, G. T., and J. S. Parslow, A model of annual plankton cycles. *Biol. Oceanogr., 3*, 327–347.

Faller, A. J., and S. J. Auer, The roles of Langmuir circulations in the dispersion of surface tracers. *J. Phys. Oceanogr., 18*, 1108–1123, 1988.

Farmer, D., and M. Li, Patterns of bubble clouds organized by Langmuir circulation. *J. Phys. Oceanogr., 25*, 1426–1440, 1995.

Farmer, D. M., C. L. McNeil, and B. D. Johnson, Evidence for the importance of bubbles in increasing air–sea gas flux. *Nature, 361*, 620–623, 1993.

Fasham, M. J. R., H. W. Ducklow, and S. M. McKelvie, A nitrogen-based model of plankton dynamics in the ocean mixed layer. *J. Mar. Res., 48*, 591–639, 1990.

Fasham, M. J. R., Variations in the seasonal cycle of biological production in subarctic oceans: A model senstivity analysis. *Deep-Sea Res., 42*, 1111–1149, 1995.

Galperin, B., L. H. Kantha, S. Hassid, and A. Rosati, A quasi-equilibrium turbulent energy model for geophysical flows. *J. Atmos. Sci., 45*, 55–62, 1988.

Garwood, R. W., An oceanic mixed layer model capable of simulating cyclic states. *J. Phys. Oceanogr., 7*, 455–471. 1977.

Gaspar, P., Modeling the seasonal cycle of the upper ocean. *J. Phys. Oceanogr., 18*, 161–180, 1988.

Gaspar, P., Y. Grégoris, and J. -M. Lefevre, A simple eddy kinetic energy model for simulations of the oceanic vertical mixing: Tests at station Papa and long-term upper ocean study site. *J. Geophys. Res., 95*, 16,179–16,193, 1990.

Gemmrich, J. R., T. D. Mudge, and V. D. Polonichko, On the energy input from wind to surface waves. *J. Phys. Oceanogr., 24*, 2413–2417, 1994.

Germano, M., U. Piomelli, P. Moin, and W. H. Cabot, A dynamic subgrid-scale eddy viscosity model. *Phys. Fluids A, 3(7),* 1760–1765, 1991.

Gill, A. E., Atmospheric–Ocean Dynamics. Academic Press, New York, 662 pp., 1982.

Johnson, D. L., and P. L. Richardson, On the wind-induced sinking of *Sargassum. J. Exp. Mar. Biol. Ecol., 28,* 255–267, 1977.

Kamykowski, D., H. Yamazaki, and G. S. Janowitz, A Lagrangian model of phytoplankton photosynthetic response in the upper mixed layer. *J. Plankton Res., 16,* 1059–1069, 1994.

Kantha, L. H., and C. A. Clayson, An improved mixed layer model for geophysical applications. *J. Geophys. Res., 99,* 25,235–25,266, 1994.

Kraichnan, R. H., Eddy viscosity in two and three dimensions. *J. Atmos. Sci., 33,* 1521–36, 1976.

Kuo, H., Perturbations of plane Couette flow in stratified fluid and origin of cloud streets. *Phys. Fluids, 6,* 195–211, 1963.

Langmuir, I., Surface motion of water induced by wind. *Science, 87,* 119–123, 1938.

Large, W. G., J. C. McWilliams, and S. Doney, Oceanic vertical mixing: A review and a model with a nonlocal boundary layer parameterization. *Rev. Geophys., 32,* 363–403, 1994.

Leibovich, S., Convective instability of stably stratified water in the ocean. *J. Fluid Mech., 82,* 561–585, 1977.

Leibovich, S., On wave–current interaction theories of Langmuir circulations. *J. Fluid Mech., 99,* 715–724, 1980.

Leonard, A., Energy cascade in large-eddy simulations of turbulent fluid flows. *Adv. Geophys., 18A,* 237–248, 1974.

Lesieur, M., and O. Metais, New trends in large-eddy simulations of turbulence. *Annu. Rev. Fluid Mech., 28,* 45–82, 1996.

Li, M., and C. Garrett, Cell merging and jet/downwelling ratio in Langmuir circulation. *J. Mar. Res., 51,* 737–769, 1993.

Li, M., and C. Garrett, Is Langmuir circulation driven by surface waves or surface cooling? *J. Phys. Oceanogr., 25,* 64–76, 1995.

Li, M., and C. Garrett, Mixed-layer deepening due to Langmuir circulation. *J. Phys. Oceanogr.,* 27, 121-132, 1997.

Li, M., K. Zahariev, and C. Garrett, The role of Langmuir circulation in the deepening of the ocean surface mixed layer. *Science, 270,* 1955–1957, 1995.

Lilly, D. K., The representation of small-scale turbulence in numerical simulation experiments. In *Proc. of the Tenth IBM Scientific Computing Symposium on Environmental Sciences,* pp. 195–210, 1967.

Marra, J., Phytoplankton photosynthetic response to vertical movement in a mixed layer. *Mar. Biol., 46,* 203–208, 1978.

Mason, P. J., Large-eddy simulation: A critical review of the technique. *Q. J. R. Meteorol. Soc., 120,* 1–26, 1994.

Mason, P. J., and A. R. Brown, The sensitivity of large-eddy simulations of turbulent shear flow to subgrid models. *Boundary-Layer Meteorol., 70,* 133–150, 1994.

Mason, P. J., and N. S. Callen, On the magnitude of the subgrid-scale eddy coefficient in large-eddy simulations of turbulent channel flow. *J. Fluid Mech., 162,* 439–462, 1986.

Mason, P. J., and D. J. Thomson, Stochastic backscatter in large-eddy simulations of boundary layers. *J. Fluid Mech., 242,* 51–78, 1992.

McComb, W. D., *The Physics of Fluid Turbulence,* 572 pp., Oxford University Press, 1991.

McPhee, M. G., and D. G. Martinson, Turbulent mixing under drifting pack ice in the Weddell Sea. *Science, 263,* 218–221, 1994.

McWilliams, J. C., P. P. Sullivan, and C. -H. Moeng, Langmuir turbulence in the ocean. *J. Fluid Mech., 334,* 1-30. 1997.

Mellor, G. L., and T. Yamada, Development of a turbulence closure model for geophysical fluid problems. *Rev. Geophys., 20,* 851–875, 1982.

Metais, O., and M. Lesieur, Spectral large-eddy simulations of isotropic and stably-stratified turbulence. *J. Fluid Mech., 239,* 157–194, 1992.

Nieuwstadt, F. T. M., P. J. Mason, C. -H. Moeng, and U. Schumann, Large-eddy simulation of the convective boundary layer: A comparison of four computer codes, in *Turbulent Shear Flows,* edited by F. Durst et al., Vol. 8, pp. 343–367, Springer-Verlag, 1992.

Phillips, O. M., *The Dynamics of the Upper Ocean,* 336 pp., Cambridge University Press, 1977.

Platt, T., D. F. Bird, and S. Sathyendranath, Critical depth and marine primary production. *Proc. R. Soc. Lond. B, 246,* 205–217, 1991.

Pollard R. T., P. B. Rhines, and R. O. R. Y. Thompson, The deepening of the wind-mixed layer. *Geophys. Fluid Dyn., 3,* 381–404, 1973.

Price, J. F., On the scaling of stress-driven entrainment experiments. *J. Fluid Mech.*, 90, 509–520, 1979.

Price, J. F., R. A. Weller, and R. P. Pinkel, Diurnal cycling: Observations and models of the upper ocean response to diurnal heating, cooling and wind mixing. *J. Geophys. Res.*, 91, 8411–8427, 1986.

Schmidt, H., and U. Schumann, Coherent structure of the convective boundary layer from large-eddy simulations. *J. Fluid Mech.*, 200, 511–562, 1989.

Sverdrup, H. U., On conditions for vernal blooming of phytoplankton. *J. Cons. Perm. Int. Explor. Mer, 18,* 287–295, 1953.

Skyllingstad, E. D., and D. W. Denbo, An ocean large-eddy simulation of Langmuir circulations and convection in the surface mixed layer. *J. Geophys. Res.*, 100, 8501–8522, 1995.

Sutcliffe, W. H., Jr., E. R. Baylor, and D. W. Menzel, Sea surface chemistry and Langmuir circulation. *Deep-Sea Res., 10,* 233–243, 1963.

Tandon, A., and C. Garrett, Geostrophic adjustment and restratification of a mixed layer with horizontal gradients above a stratified layer. *J. Phys. Oceanogr.*, 25, 2229–2241, 1995.

Tandon, A., and S. Leibovich, Simulations of three-dimensional Langmuir circulation in water of constant density. *J. Geophys. Res.*, 100, 22,613–22,623, 1995.

Thorpe, S. A., Small-scale processes in the upper ocean boundary layer. *Nature,* 318, 519–522, 1985.

Thorpe, S. A., Bubble clouds and the dynamics of the upper ocean. *Q. J. R. Meteorol. Soc.*, 118, 1–22, 1992.

Thorpe, S. A., On the meandering and dispersion of a plume of floating particles caused by Langmuir circulation and a mean current. *J. Phys. Oceanogr., 25,* 685–690, 1995a.

Thorpe, S. A., Dynamical processes of transfer at the sea surface. *Prog. Oceanogr., 35,* 315–351, 1995b.

Thorpe, S. A., and M. S. Curé, One-dimensional dispersion in a lake inferred from sonar observations, in *Mixing and Transport in the Environment,* edited by K. J. Bevan, P. C. Chatwin, and J. H. Millbank, pp. 17–28, Wiley, 1994.

Thorpe, S. A., M. S. Curé, and A. Graham, Sonar observations of Langmuir circulation and estimation of dispersion of floating particles. *J. Atmos. Ocean. Tech., 11,* 1273–1294, 1994.

Thorpe, S. A., and A. J. Hall, Observations of the thermal structure of Langmuir circulation. *J. Fluid Mech., 114,* 237–250, 1982.

Troen, I. B., and L. Mahrt, A simple model of the atmospheric boundary layer: Sensitivity to surface evaporation. *Boundary-Layer Meteorol., 37,* 129–148, 1986.

Turner, J. S., *Buoyancy Effects in Fluids,* 367 pp., Cambridge University Press, 1973.

Wanninkhof, R., Relationship between wind speed and gas exchange over the ocean. *J. Geophys. Res., 97,* 7373–7382, 1992.

Weller, R. A., and J. F. Price, Langmuir circulation within the oceanic mixed layer. *Deep-Sea Res., 35,* 711–747, 1988.

Woods, J. D., and R. Onken, Diurnal variation and primary production in the ocean – preliminary results of a Lagrangian ensemble model. *J. Plankton Res., 4,* 735–756, 1982.

Wyngaard, J. C., Atmospheric turbulence. *Annu. Rev. Fluid Mech., 24,* 205–233, 1992.

6

Velocity, Temperature and Spatial Structure of Langmuir Circulation

D. Farmer, J. Gemmrich and V. Polonichko

Abstract

Although Langmuir circulation is thought to play a significant role in vertical mixing there have been few detailed measurements of the resulting thermal and velocity flow field. Recent observations provide an opportunity for comparing measurements of the vertical and horizontal velocity field and the fine-scale temperature structure with various model predictions. Compared with results of 2-dimensional modeling, the observations show a broader downwind jet in the convergence zone and somewhat larger values of the pitch (the ratio of maximum downwelling to horizontal jet speed). While the 3-dimensionality in Langmuir circulation structure will undoubtedly increase the variance of these variables, we suggest that the observed discrepancy between the model predictions and measurements is related to the use of a constant eddy viscosity in the model. In order to determine overall turbulent diffusivity we use fine scale temperature measurements. The temperature structure at 6.5 m depth is consistent with numerical predictions, but close to the surface, where turbulent field is expected to be influenced by the surface wave effects, our observations show much greater variability than predicted by a constant eddy diffusivity model.

Introduction

Langmuir circulation consists of a near surface circulation pattern, which in its simplest 2-dimensional form is a succession of parallel counter-rotating vortices oriented in the wind direction (Langmuir, 1938). Interest in this topic has been stimulated by recognition of its role in the vertical flux of gas, heat and momentum through the surface layer (Thorpe, 1985). Recent analysis (Li et al., 1996) has further demonstrated the importance of Langmuir circulation in ocean surface mixed layer deepening.

While the mechanism for generating Langmuir circulation has been the subject of extensive theoretical analysis (c.f. Craik, 1977; Leibovich 1977, 1983), field observations, especially of the crucial dynamic variables associated with the surface convergence and downwelling, remain sparse. Sonar backscatter signals have been used to delineate the distribution of bubble clouds organized along convergence zones (Thorpe, 1986; Zedel and Farmer, 1991; Thorpe et al., 1994; Farmer and Li, 1995) and Doppler sonar has revealed the near surface velocity field (Smith et al., 1987; Smith, 1992). There have, however,

Physical Processes in Lakes and Oceans
Coastal and Estuarine Studies Volume 54, Pages 87-100

been almost no combined measurements of both vertical and horizontal velocities which would allow direct comparison with theoretical models except for a singular example provided by Weller and Price (1988).

Similarly there are few detailed measurements of the fine-scale temperature structure within the wind driven mixing layer. Thorpe and Hall (1982) have described measurements of temperature variability acquired with a towed mast fitted with sonar and thermistors. They found fluctuations coincident with bubble clouds ascribed either to Langmuir circulation or breaking waves. However, the link between variability in the temperature field due to Langmuir circulation and the air-sea heat flux, though modeled by Li and Garrett (1995), hereafter LG95, has yet to be unambiguously confirmed by field measurements. In the present report we describe recent measurements of the near surface velocity field, temperature and bubble cloud distribution and compare them with model results.

Recent 2-dimensional modeling has resulted in a number of predictions. These include the strength and structure of downwelling and downwind currents, the jet width/cell width ratio, the temperature anomaly caused by the circulation and corresponding dependency on parameters determining the circulation strength. Li and Garrett (1993), hereafter LG93, numerically integrated the Craik-Leibovich equation so as to examine both 2-dimensional vortex merging and the strength and structure of the cells. Farmer and Li (1995) discuss observations of cell merging which are far from 2-dimensional, and recent large eddy simulations (Skyllingstad and Denbo, 1995; Li, 1995) reveal some of the complexity to be found in 3-dimensional model results. However, at present 3-d models have not provided general predictions of the circulation parameters and their dependence on dimensionless parameters. Despite its limitations, the 2-dimensional Craik-Leibovich (CL2) model provides a useful reference point for comparison with observations and allows us to test the validity of this description.

Dimensional analysis of the averaged governing equations (c.f. Leibovich, 1983) leads to no less than 7 dimensionless variables (LG93). The most important in many situations is the Langmuir number,

$$La = \left(\frac{\nu_T \beta}{u_*}\right)^{3/2} \left(\frac{S_0}{u_*}\right)^{-1/2}, \quad (1)$$

representing the ratio of viscous to inertial forces. In (1) ν_T is an eddy viscosity, normally taken as a constant, $2S_0$ is the surface magnitude and $[1/(2\beta)]$ the e-folding depth of the Stokes drift, and u_* is the water friction velocity.

From dimensional reasoning the downwelling velocity is expected to be proportional to $La^{-1/3}$. The LG93 numerical results give :

$$w_{max} = 0.72\, u_* \left(\frac{S_0}{u_*\, La}\right)^{1/3}, \quad (2)$$

where w_{max} is the maximum downwelling velocity. They further define the 'pitch' of the circulation:

$$Pt = \frac{u_{con} - u_{div}}{w_{max}}, \quad (3)$$

where u_{con} is the surface down-wind velocity component in the convergent zone, and u_{div} the corresponding value in the divergent zone. For comparison with observations, the pitch is normalized as

$$PT = Pt\,(S_0/u_*)^{3/2}, \tag{4}$$

which LG93 found to be proportional to $La^{-1/3}$. LG95 added a buoyancy forcing term and the heat equation to the two-dimensional model so as to derive the temperature field of Langmuir circulation. They also estimated the role of buoyancy forcing in Langmuir circulation. They expressed the ratio between thermal and wave forcing through Stokes drift in terms of the Hoenikker number Ho as:

$$Ho = -\frac{\alpha g\, Q/(c_p \rho_w)}{S_0\, \beta\, u_*^2}, \tag{5}$$

where α is the coefficient of thermal expansion, g the gravitational constant, Q the surface heat flux, c_p the heat capacity and ρ_w density of sea water.

For small La thermal forcing balances wave forcing when the Noenikker number HO equals 3 and for HO << 3 the temperature field is dynamically inactive so that temperature can be regarded as a passive tracer. LG95 obtained the following expression for the temperature difference of the surface boundary layer between convergent and divergent zones.

$$\delta\tilde{\theta} = C\frac{S_0\, \beta\, u_*}{\alpha g}\left(\frac{S_0}{u_*}\right)^{-1/3} Ho\, Pr^{1/2}\, La^{-1/6}. \tag{6}$$

Here C is an empirical constant and Pr is the turbulent Prandtl number.

Our observations allow us to evaluate normalized pitch PT and temperature deviation $\delta\tilde{\theta}$ and thus make direct comparisons with the predictions.

Observational Approach

Our measurement approach has primarily depended on the use of self-contained instruments that drift freely with the surface layer. This choice is dictated by the difficulty of acquiring data from a ship in heavy seas, and by the desirability of minimizing advective effects which are particularly important in vertical flux studies.

The present report is concerned with measurements acquired from two instruments: a set of scanning and fixed orientation sonars mounted on a platform suspended at a depth of approximately 30 m by a rubber cord from a surface buoy and a surface tracking float supporting thermistors and conductivity cells.

The surface tracking float measures temperature at four fixed depths and also has an electrically driven vertically profiling thermistor which cycles through the upper 2m of the water every minute. A capacitance wire sensor is used to determine small departures of the sensors from their nominal depths. The acoustic sensing platform has both vertically and horizontally oriented sonars and represents a refinement of earlier versions of the same instrument (Farmer et al., 1990). The 4 horizontally directed sonars are of 'fan-beam' type (directivity is 1.5° in the azimuthal and 45° in the vertical plane and can be independently oriented by stepping motors. Two modes have been used: the sonars can sweep in azimuth, producing successive radar-like images of the surface backscatter over 360° every 30 s (Figure 1), or they can be programmed to point in fixed predetermined azimuthal directions, compensating in real time for small rotational motions of the platform thus providing measurements along four fixed paths (Figure 2). Both vertical and horizontal sonars transmit a broad band coded pulse (Trevorrow and Farmer, 1992) from which backscatte intensity and Doppler velocity are derived as a function of range. The intensity field reveals the bubble distribution, both vertically as the sonar passes beneath bubble clouds, and horizontally along the scanning path just beneath the sea surface Figure 3. A recording temperature sensor is attached to the rubber cord at a depth of approximately 6.5m. Except when masked by deeply penetrating bubble clouds the temperature sensor appears in the sonar image (bottom plate in Figure 3).

Results

Here we mainly focus on data acquired with the surface following float and subsurface sonar platform during two cruises aboard RV WECOMA in January 1995 off the coast of Oregon and April 1995 during the Marine Boundary Layer Experiment, off Monterey, California. The surface following float was only deployed on the second cruise. The sweeping side-scanning sonars image the 2-dimensional near-surface bubble cloud distribution out to a range of approximately 280 m (Figure 1). The clouds appear black in the image and are narrow, elongated structures approximately aligned with the wind. Although the bubble clouds can be coherent over at least some hundreds of metres, they also display considerable irregularity. When the sonar points upwind or downwind, the elongated structure is not observed due to irregularity in bubble penetration depth. Particularly deep and dense bubble clouds can also cast shadows, which will appear as swaths of light gray or white in the image.

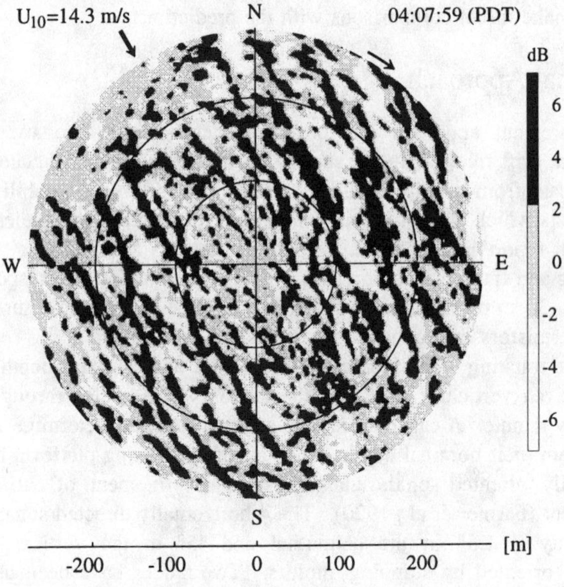

Figure 1. Scanning sonar image showing bubble cloud distributions (18 April 1995). Range rings are at 100 m intervals, U_{10} = 14.3 m s^{-1} at 330° N. Dark bands are bubble clouds organized by Langmuir circulation. The curved arrow shows the instrument rotation during the scan.

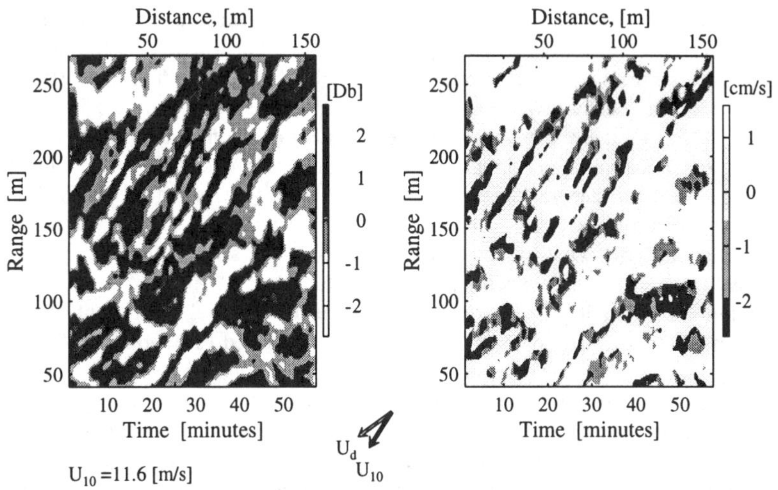

Figure 2. Filtered backscatter intensity (left) as a function of range and time when the sonar is pointing in a fixed orientation as the bubble clouds drift by. The arrows (lower center) marked U_{10}, U_d identify the wind direction and the direction of surface drift relative to the instrument respectively. The sonar heading is upward. Filtered Doppler velocity measurements (right) for the same period, corresponding to the areas of strong bubble scatter. Positive speed is away from the sonar. The downward flow within the bubble clouds is consistent with the concept of downwind jets within the convergence zones. (January 17, 1995)

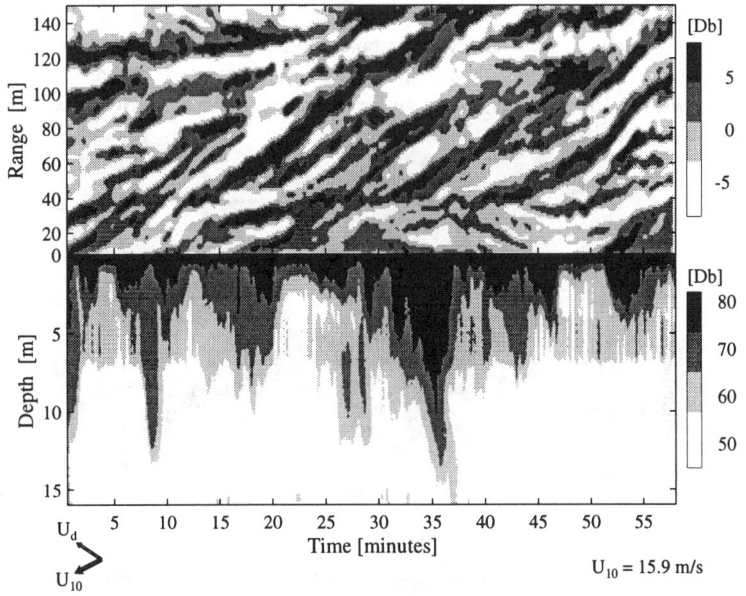

Figure 3. Simultaneous horizontal (top plate) and vertical (bottom plate) sonar backscatter intensity. Arrows marked U_{10}, U_d depict the wind and the surface drift direction respectively. (April 18, 1995).

Figure 2a, b shows time series of fixed orientation sidescans measuring the near-surface backscatter intensity and Doppler velocity. Successive images of this kind illustrate the complex temporal evolution of the bubble distribution and velocity field along four sidescan beams. While the overall pattern drifts away from the instrument at a speed of 0.028 m s^{-1} there also appears to be a smaller component moving in the opposite direction. Only the velocity signal present in the areas occupied by bubble clouds is shown Figure 2b. In general the velocity signal reveals an enhanced downwind surface flow coincident with the bubble clouds. However, occasional patches of more slowly moving fluid (the flow actually seems to be moving more slowly than the drifting buoy) are thought to be due to more deeply penetrating clouds.

The dominant velocity signal arises from the resolved orbital component of the surface waves from which the directional wave spectrum is recovered (Trevorrow, 1995). A more challenging task is to separate the weak and more slowly changing vertical and horizontal velocity field from the wave motion. We correct for small instrument motions and make use of the measured surface elevation and linear dispersion relations to remove the orbital component from our vertical current measurements. The time series of the horizontal Doppler velocity as a function of range are modified at close ranges by the slant angle effect, which we correct for.

The 2 dimensional Fourier transform is applied to this corrected time-range velocity field and the resulting field is expressed in terms of wave number (from the range dependence) and frequency (from the time dependence). Retaining only the portion near the origin and inverting the transform then serves as a convenient 2-d spectral filter which isolates the residual low frequency motions regardless of the mutual orientation of the waves and the sonar beam. An example of the raw 2-d Doppler velocity spectrum and the 3-dB filter contour are shown in Figure 4. The continuous curve in Figure 4 depicts the best fit of the resolved linear dispersion relation for the deep water surface gravity waves.

The vertical sonar shows numerous bubble plumes (Figure 5), some penetrating to about 15 m depth. The vertical velocity profile, which can only be derived from the higher backscatter within bubble clouds, is shown for one particular bubble cloud identified in the figure with a vertical dashed line. In this case the vertical velocity is close to zero within

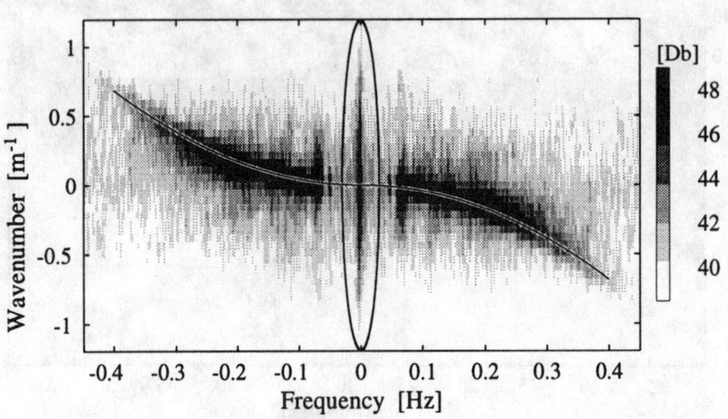

Figure 4. Two dimensional raw velocity spectrum shows both quasi-steady velocities (within ellipse) and a wave group. The wave energy has been fitted with a directionally resolved dispersion relationship (continuous curve). The ellipse shows the 3-dB contour of the low pass Fourier domain filter which is applied to the raw velocity spectrum so as to remove the wave signal.

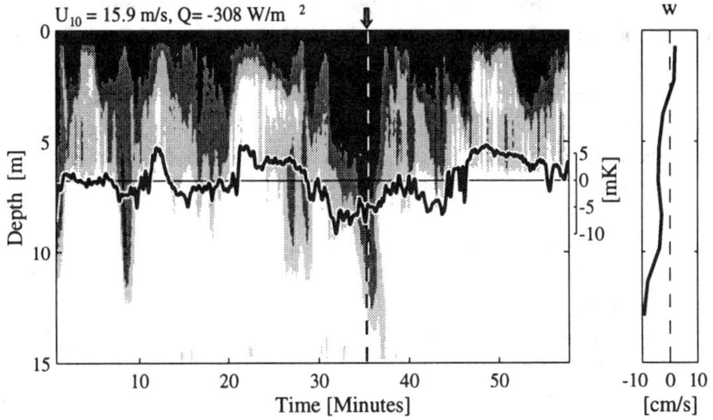

Figure 5. Backscatter intensity detected with an upward looking sonar (400 kHz), together with a single profile of the vertical velocity component evaluated from the Doppler measurement. The velocity profile corresponds to the time identified with a vertical dashed line. Also shown is a time series of the temperature deviation at 6.5 m depth. The vertical dashed line and arrow mark the time for which the velocity is drawn. (April 18, 1995). Average wind speed is 15.9 m s^{-1}, and the average heat flux shows cooling at a rate of 308 W m^2.

the upper 2–3 m, changing to -10 cm s^{-1} at 15 m. This is an unusually high vertical velocity; more typical values are in the range of 4–8 cm s^{-1}. Figure 5 also shows the corresponding temperature record, expressed here as departure from the mean value, measured by the temperature sensor attached to the cord above the sonar at a depth of 6.5 m. In general the temperature is below the mean value within the clouds and above the mean value between them, when there is strong surface cooling. This result is consistent with Thorpe and Hall (1982) and with the interpretation of cooler water being drawn down in Langmuir convergence zones when there is an upward heat flux, to which we return subsequently.

There is some ambiguity between the bubble clouds introduced by wave-breaking, and those associated with Langmuir convergence zones. Li and Farmer (1995) found a large number of short lived and small sized bubble clouds which they attributed to wave breaking, but identified the majority of those contributing to the fractional area coverage as being associated with convergence zones. An indication of the relative significance of wave breaking induced bubble clouds, as opposed to those due to Langmuir convergence effects, is provided by comparison of the simultaneous vertical and horizontal backscatter images (see Figure 3). The majority of the clouds penetrating deeper than 3.5 m can be traced to the corresponding horizontal bubble streaks. Analysis of images of this type suggest that more than 80% of bubble clouds detected on the vertical sonar coincide with persistent structures observed by the side looking sonar and are attributed to Langmuir circulation. In this report we limit discussion to these convergence zone features while recognizing that there are smaller scale features lying between them.

Discussion

Although our bubble cloud measurements (Figure 1) show the elongated structures associated with Langmuir convergence zones, the circulation pattern includes several 3-

dimensional features (c.f. Farmer and Li, 1995). Nevertheless we might expect that the 2-dimensional models would provide a suitable scaling for analysis of our data. Here we make a preliminary comparison with the results of LG93 numerical study of the Craik-Leibovich model. The comparisons most accessible from our data relate to the horizontal spatial structure and velocity and temperature fields, scaled by the Langmuir number which, as shown by the model analysis, is deduced from the measured maximum vertical velocity.

Figure 6 shows the distribution of convergence zone spacing or cell spacing, and the width of the downwind jet. The latter was determined in two ways, both of which gave similar results. First, zero crossings of the horizontal velocity gradient measured in cross-wind direction (Figure 2) were used to bound the convergence zones; second, the width of the downwelling plume (Figure 5) and the speed at which the instrument was observed to drift perpendicular to the Langmuir cell orientation was used to derive a plume width. As shown in Figure 7 the typical ratio of jet width to cell spacing is 0.1 corresponding to a significantly greater relative jet width than found by Li and Garrett in their numerical calculations (LG93). As discussed below, a limitation of the 2-d model is the use of a constant eddy viscosity and this could account for the discrepancy. This point is taken up later.

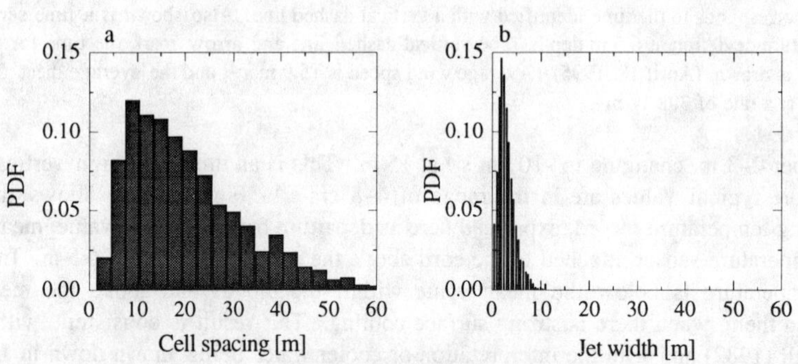

Figure 6. (a) Distribution of Langmuir cell spacing, i.e. spacing between successive convergence zones, and (b) downwind jet widths, evaluated from a sonar oriented at 92° N with $U_{10} = 11.7$ m s^{-1} from 167° N. (January 17, 1995)

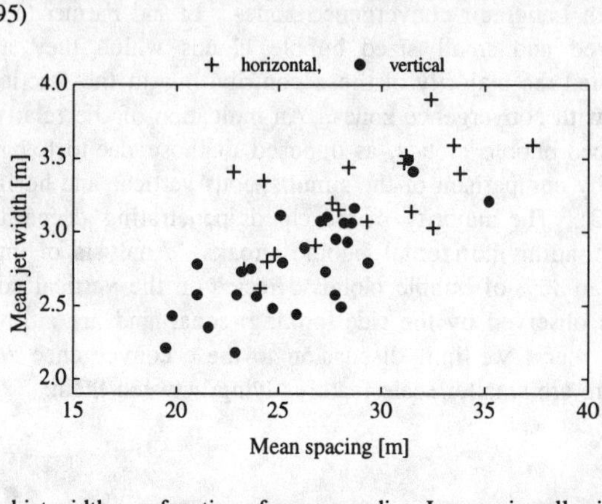

Figure 7. Downwind jet width as a function of corresponding Langmuir cell width. Crosses correspond to jet widths evaluated from zero-crossings in the cross-wind surface velocity gradient field, circles to jet widths evaluated from width of vertical velocity zones and measured cross-wind drift (Jan. 17, 1995).

The distribution of the residual vertical bubble velocities is shown in Figure 8 (vertical velocities are only measured where bubbles are present). Bubbles move downwards at higher speeds than upwards, consistent with an asymmetrical picture of the circulation. With measurements of both horizontal jet speed and vertical velocity we may calculate the normalized pitch as defined in (3)–(4) and express the result as a function of Langmuir number.

Comparison with LG93 model predictions (Figure 9) shows that most of our results lie significantly above the theoretical calculation but are also closer than the values inferred from Weller and Price (1988) data, which range from 5 to 8. As for the discrepancy in the jet width we again draw attention to the use of a constant eddy viscosity. If the model uses too low a viscosity the vertical velocity is overpredicted leading to an underestimated pitch.

Uncertainty in the role of turbulence motivates an analysis of the near surface temperature field. We have measurements both at 6.5 m from a recording thermistor placed above the imaging sonar,and from the thermistors mounted on the surface following buoy.

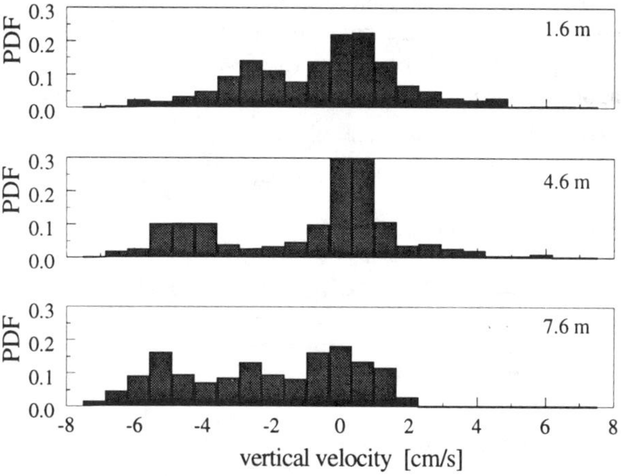

Figure 8. Distribution of measured residual vertical bubble velocities at different depth. (Jan 17, 1995).

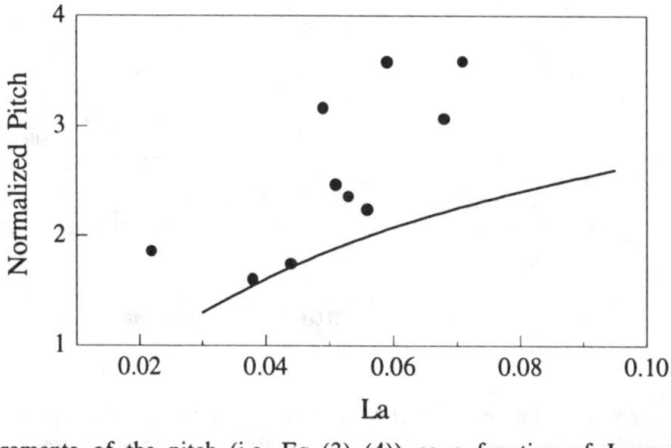

Figure 9. Measurements of the pitch (i.e. Eq (3)–(4)) as a function of Langmuir number. Line represents model predictions of LG93.

The heat flux during this night time observation ranged from -150 Wm^{-2} to -350 W m^{-2} leading to a Hoenikker number of 2×10^{-3} to 8×10^{-3} which is well below the critical value. We therefore interpret the temperature drops observed with our surface following float as well as the measurements at greater depth (Figure 5) as being associated with convergence zones and the magnitude of the fluctuation as the horizontal temperature difference dT across a Langmuir cell. For model comparisons the results are normalized as

$$\delta T = \frac{\delta \tilde{T} \rho c_p \kappa_T \beta}{Q}, \tag{6}$$

where κ_T is the eddy diffusivity estimated from La with the assumption of a turbulent Prandtl number Prt = 1. The normalized data are compared in Figure 10 with LG95 normalized CL2 model calculations. At 6.5m (solid circles), corresponding to the measurements illustrated in Figure 5, the data agree well with LG95. At shallower depths, the temperature differences across the cells exceed the prediction, a result attributed to suppression by the boundary of turbulent diffusivity below the constant value assumed in the model. This observed discrepancy motivates a closer analysis of the data.

The LG95 model calculations are based on a constant diffusivity assumption. To explore the depth dependence of the diffusivity we model the temperature field in a Langmuir circulation as a steady state 2-dimensional advective diffusion process with heat as a passive tracer:

$$\nabla \cdot (\kappa_T \nabla \theta) - \mathbf{U} \cdot \nabla \theta = 0, \quad \mathbf{U} = (u, w) \tag{7}$$

with following boundary conditions:

$$\begin{aligned} \frac{\partial \theta}{\partial z} &= \frac{1}{\kappa_T}, \quad z = 0; \quad \theta = 0, \quad z = \gamma, \\ \frac{\partial \theta}{\partial x} &= 0, \quad x = \pm 1. \end{aligned} \tag{8}$$

Nondimensionalization is achieved by

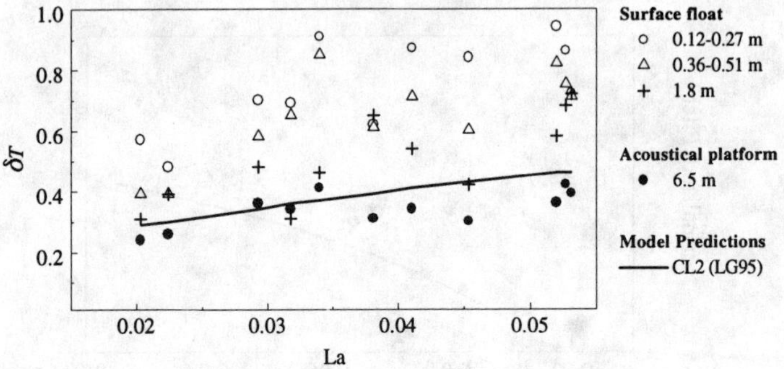

Figure 10. Normalized temperature difference between convergent and divergent regions of the Langmuir cells, for different depths, expressed as a function of Langmuir number. LG95 CL2 model predictions shown by a line. (April 18, 1995).

$$x = \frac{\tilde{x}}{L}, \quad \theta = \frac{\tilde{\theta}\rho c_p \tilde{\kappa}_T(\gamma\ L)}{QL}, \qquad (9)$$
$$U = \frac{\tilde{U}L}{\tilde{\kappa}_T(\gamma\ L)}, \quad \kappa_T(z) = \frac{\tilde{\kappa}_T(z)}{\tilde{\kappa}_T(\gamma\ L)}$$

where L, γ are cell width and cell aspect ratio, respectively. The diffusivity profile $\tilde{\kappa}_T(z)$ and velocity field $\tilde{U} = (\tilde{u}, \tilde{w})$ have to be prescribed.

We choose a circular flow pattern which allows for asymmetry in strength of upwelling and downwelling:

$$u = -\frac{\partial \psi}{\partial z}; \quad w = \frac{\partial \psi}{\partial x};$$
$$\psi = u_0 \sin(\pi x')\sin(\pi z/\gamma)$$
$$x' = \frac{1 - |x|}{2 - 2\alpha} \text{ for } |x| \geq \alpha; \qquad (10)$$
$$x' = \frac{|x|}{2\alpha} \text{ for } |x| < \alpha;$$

The asymmetry factor α, $0 < \alpha < 1$ determines the center of the cell and u_0 defines the maximum downwelling velocity. Hence, the flow pattern is set by four parameters, two to define the geometry of the cell and two for the strength of the circulation. These model parameters can be evaluated utilizing the acoustical observations and related data.

The temperature field within a Langmuir cell was calculated with this advective diffusion model (7) for the observed range of flow parameters and cell geometry. An example of the modeled temperature field for a deep cell ($\gamma = 1.4$) with strong downwelling ($\alpha = 0.4$) is given in Figure 11. Strong temperature anomalies, present in the near surface layer, are drawn down in the convergence zone, whereas the bulk of the upwelling region is nearly isothermal. In this calculation we arbitrarily choose a law of the wall diffusivity profile for $u_* = 0.02$ ms^{-1}, so as to illustrate the general behavior of the temperature field, although as discussed below our observations reveal substantially greater diffusivity near the surface.

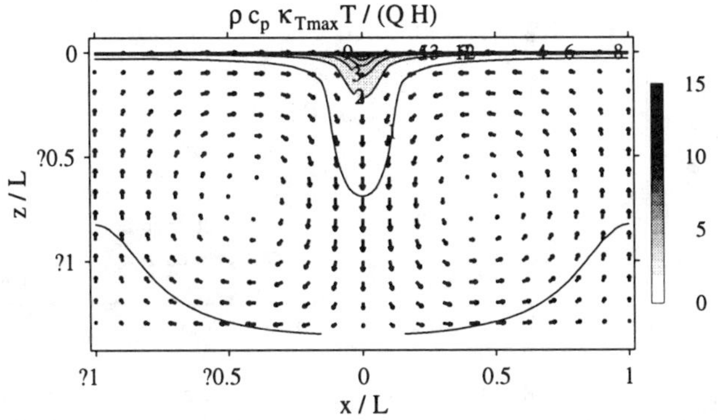

Figure 11. Non-dimensional temperature (grey scale) and velocity field (arrows) of Langmuir circulation. Cell asymmetry $\alpha = 0.4$, cell aspect ration $\gamma = 1.4$ sand maximum horizontal velocity $u_0 = 7.4$.

Horizontal temperature cross sections show pronounced fluctuations within the downwelling region, decreasing with depth, very similar to the fluctuations observed in the temperature time series (Figure 10). The modeled temperature difference

$$\Delta T(z) = T_{div}(z) - T_{con}(z) \tag{11}$$

between the line of maximum upwelling and maximum downwelling is compared to the magnitude of the observed temperature fluctuations at a given depth.

This is carried out for a range of model calculations utilizing the distribution of observed flow parameters and cell geometry and two different diffusivity profiles. The comparison is given in Figure 12. A wall layer type diffusivity results in unrealistically high horizontal temperature gradients close to the surface, whereas a constant diffusivity does not account for the decreasing gradient with increasing depth. The profile of the *apparent* diffusivity in the upper 2 m can be inferred from each measured temperature profile as

$$\kappa_T = \frac{Q}{\rho c_p}\left(\frac{dT}{dz}\right)^{-1}, \tag{12}$$

which includes effects due to the modification of the temperature gradient by advection within the cell. The mean profile obtained from an eight hour nighttime observation period is given in Figure 13 in wall layer coordinates. Close to the surface the diffusivity is significantly greater than the diffusivity in a wall bounded layer $\kappa_{WL} = \kappa u_* z$, where $\kappa = 0.4$ is the von Karman constant.

Recent measurements of energy dissipation in the surface layer (Drennan et al., 1992; Agrawal et al., 1992; Anis and Moum, 1992; Osborn et al., 1992) indicate enhanced turbulence levels close to the surface, attributed to wave breaking (Drennan et al., 1992), which implies a larger near surface diffusivity. This is also predicted in an attempt to model the wave stirred surface layer by Craig and Banner (1994), who modified a conventional 1-dimensional turbulence closure model to incorporate a boundary layer of wave-breaking enhanced turbulence matched to a law of the wall layer beneath. While detailed model

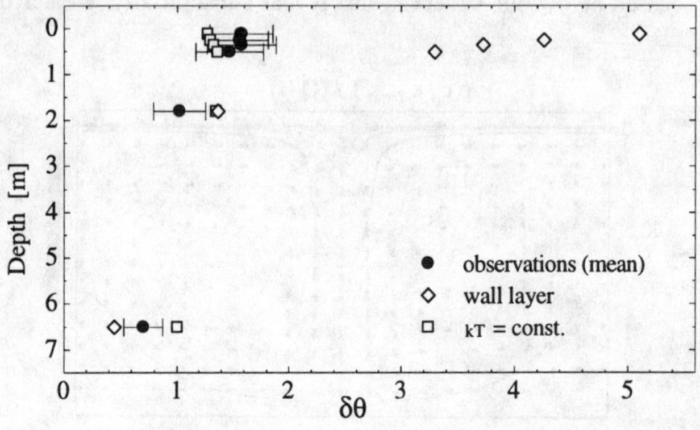

Figure 12. Comparison of observed (solid symbol) and modeled (open symbols) horizontal temperature differences associated with Langmuir convergences for different diffusivity profiles (diamonds for wall layer scaling and squares for constant diffusivity). The advective diffusion model was driven by observed flow parameters and cell geometry.

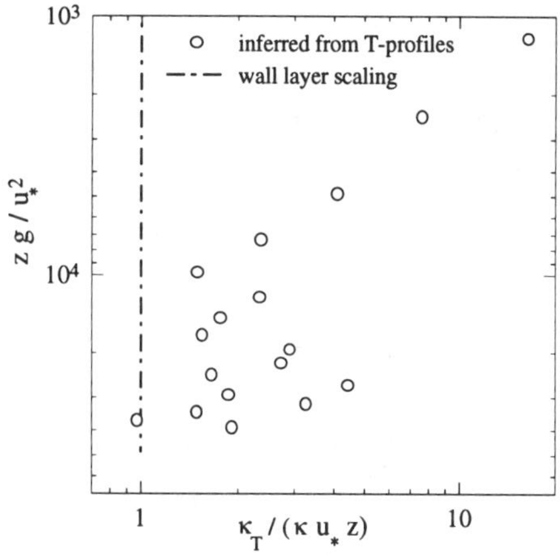

Figure 13. Diffusivity profile in the upper 2 m presented in wall layer coordinates. Circles are apparent diffusivities inferred from surface heat flux and vertical temperature gradients. The broken line represents wall layer scaling.

comparison remain a topic of further research, we note here that our observations are qualitatively consistent with a wave enhanced turbulent layer near the surface.

A further complication not included in existing model arises from the unsteady nature of the kinetic energy generation mechanism. If breaking waves are a primary source, this is inherently intermittent. Preliminary model results of the temperature field beneath breaking waves, utilizing the observed breaking frequency provide a further signal for investigation of near surface turbulence.

To summarize, the presented results show some consistency and some discrepancies with the CL2 model. The pattern of bubble clouds is generally aligned with the wind (Figure 1) but is rarely linear especially at higher wind speed, and shows evidence of 3-dimensional cell merging and other instabilities (Farmer and Li, 1995). We expect such 3-dimensionality to add greatly to the signal variance, although the mean effects are yet unclear. The ratio of jet width to cell spacing of 0.1 is significantly greater than the 2-d model predictions. Calculations of pitch are somewhat higher than LG93 predict. Finally, the measured temperature anomalies show remarkably close agreement at 6.5m but are generally greater than predicted nearer the surface. These discrepancies are almost certainly associated in part with the limitation of a constant diffusivity model. Analysis of the temperature observations supports the hypothesis of a boundary layer of wave breaking enhanced turbulence near surface, which is likely to have a significant influence on the Langmuir circulation structure. Enhanced turbulence near the surface will also add to the modification of the eddy viscosity. In any event it is clear that the model does not accurately incorporate the real turbulence field and we expect this simplification may in part account for the observed discrepancies.

Acknowledgments. This work was made possible with the help of technical staff at the Institute of Ocean Sciences. We are indebted to the crew of the RV WECOMA for assistance in deployment and recovery of the instrumentation and to J. Edson (Woods Hole Oceanographic Institution) for measurements of the surface heat flux. The program received financial support from the US Office of Naval Research through the Marine Boundary Layer program.

References

Agrawal. Y. C., E. A. Terray, M. A. Donelan, P. A. Hwang, A. J. Williams III, W. M. Drennan, K. K. Kahma and S. A. Kitaigorodskii, Enhanced dissipation of kinetic energy beneath surface waves, *Nature*, 359, 219-220, 1992

Anis, A. and J. N. Moum, The superadiabatic surface layer of the ocean during convection *J. Phys. Oceanogr., 22*, 1221-1227, 1992.

Craig, P. D. and M. L. Banner, Modeling wave-enhanced turbulence in the ocean surface layer. *J. Phys. Oceanogr., 24*, 2546-2559, 1994.

Craik, A. D. D., 1977, The generation of Langmuir circulations by an instability mechanism. *J. Fluid Mech., 81*, 209-223

Drennan, W. M., K. K. Kahma, E. A. Terray, M. A. Donelan and S. A. Kitaigorodskii, Observations of the enhanced of the kinetic energy dissipation beneath breaking wind waves. in *Breaking waves*, Banner, M. L. and R. H. J. Grimshaw, Eds., Springer-Verlag, 95-102, 1992.

Farmer, D. M. and J. R. Gemmrich, Measurements of temperature fluctuations in breaking surface waves. *J. Phys. Ocean., 5*, 816-825, 1996.

Farmer, D. M, and M. Li., Patterns of bubble clouds organized by Langmuir Circulation, *J. Phys. Oceanogr.*, 25, 1426-1440, 1995.

Farmer, D. M., R. C. Teichrob, C. J. Elder and D. G. Sieberg, Novel acoustical instrumentation for the study of ocean surface processes. *J. Atmos. Oceanic Technol., 12*, 603-616, 1990.

Langmuir, I., Surface motion of water induced by wind. *Science, 97*, 119-123, 1938.

Leibovich, S., On the evolution of the system of wind drift currents and Langmuir circulations in the ocean. Part 1. Theory and averaged current. *J. Fluid Mech., 79*, 715-743, 1977.

Leibovich S., The form and dynamics of Langmuir circulations. *Ann. Rev. Fluid Mech., 15*, 391-427, 1983

Li, M., Comparison between DNS and LES simulations of coherent flow structures in the ocean surface boundary layer. in *Computational Fluid Mechanics '95*, edited by P. A. Thibault and D. M. Bergeron, 433-440, 1995

Li, M. and C. Garrett, Cell merging and jet/downwelling ratio in Langmuir circulation. *J. Marine Res., 51*, 737-769, 1993.

Li, M. and C. Garrett, Is Langmuir circulation driven by surface waves or surface cooling? *J. Phys. Oceanogr., 25*, 64-76, 1995.

Li, M., K. Zahariev and C. Garrett, The role of Langmuir circulation in the deepening of the ocean surface mixed layer, *Science, 270*, 1955-1957, 1996.

Osborn, T., D. M. Farmer, S. Vagle, S. A. Thorpe and M. Cure, Measurements of bubble plumes and turbulence from a submarine. *J. Atmos. and Ocean., 30*, 419-440, 1992.

Skyllingstad, E. D. and D. W. Denbo, An ocean large eddy simulation of Langmuir circulations and convection in the surface mixed layer. *J. Geophys. Res., 101*, 1095-1110, 1996.

Smith, J. A., Observed growth of Langmuir Circulation, *J. Geophys. Res., 97*, 5651-5664, 1992

Smith, J. A., R. Pinkel and R. A. Weller, Velocity structure in the mixed layer during MILDEX, *J. Phys. Oceanogr., 17*, 425-439, 1987

Thorpe, S. A., Small-scale processes in the upper ocean boundary layer, *Nature, 318*, 519-522, 1985.

Thorpe, S. A., Measurements with an automatically recording inverted echo sounder: ARIES and the bubble clouds. *J. Phys. Oceanogr., 16*, 1462-1478, 1986

Thorpe, S. A. and A. J. Hall, Observations of the thermal structure of Langmuir circulation, *J. Fluid Mech., 114*, 237-250, 1982

Thorpe, S. A., M. S. Cure, A. Graham and A. J. Hall, Sonar observations of Langmuir circulation and estimation of dispersion of floating particles. *J. Atmos. Ocean. Technol., 11*, 1273-1294, 1994.

Trevorrow, M., Measurement of ocean wave directional spectra using Doppler side-scan sonar arrays, *J. Atmos. Oceanic Technol., 12*, 603-616, 1995

Trevorrow, M. V. and D. M. Farmer, The use of barker codes in Doppler sonar measurements. *J. Atmos. Oceanic Technol., 9*, 699-704, 1992

Weller, R. A. and J. F. Price, Langmuir circulation within the oceanic mixed layer, *Deep-Sea Res., 35*, 711-747, 1988

Zedel, L., and D. M. Farmer, Organized structures in subsurface bubble clouds: Langmuir circulation in the open ocean, *J. Geophys. Res., 96*, 8889-8900, 1991.

7

On Wavy Mean Flows, Langmuir Cells, Strain, and Turbulence

S. G. Monismith and J. J. M. Magnaudet

Abstract

In this paper we discuss several laboratory experiments examining interactions between a turbulent shear flow and surface gravity waves. In general, wavy flows differ markedly from similar wall flows: mean flow profiles are different and depend on wave behavior and turbulence structure and spectra can both be different. We attempt to fit these observations into a single framework by considering how wave strains can affect turbulence and produce Langmuir circulations. Wavebreaking notwithstanding, we argue that irrotational wave strain is the prime means by which waves, mean flows and turbulence interact. Using this perspective we derive several scales which determine the type of wave interaction possible: quasi-steady flows in equilibrium with the wave strain, non equilibrium unsteady flows forced by the wave strain, and large-scale motions, Langmuir cells, that represent the rectified effects of fast, elastic straining.

1. Introduction

There has been a good deal of work, both in the field and in the laboratory, on wind-wave turbulence. Recent reviews of this literature (emphasizing different aspects of the field) can be found in Melville (1994), Magnaudet and Thais (1995), and in Craig and Banner (1994). Early work concentrated on determining mean velocity profiles (Wu, 1973) as a function of wind stress (speed), while later work has explored turbulence properties like dissipation (Agrawal et al., 1992), Reynolds stress (Cheung, 1985; Jiang et al., 1990), wave-turbulence interaction (Jiang et al., 1990 and Magnaudet and Thais, 1994), and wave effects on turbulence statistics and spectra (Thais and Magnaudet, 1996).

In contrast to the body of experimental work that exists on wind-wave turbulence there has been comparatively little theoretical work. For example, Magnaudet and Masbernat (1990) analyzed purely periodic waves using an eddy viscosity type relation to derive a form for the wave stress induced by wave turbulence interactions, Their result, although in agreement with Cheung and Street's (1988a) data, is suspect because the turbulence distortion by waves is too rapid for an eddy viscosity-type closure to be applicable (Belcher et al., 1994). Indeed, the experimental results reported in Thais and Magnaudet

Physical Processes in Lakes and Oceans
Coastal and Estuarine Studies Volume 54, Pages 101-110

(1996) do not support the model of Magnaudet and Masbernat. Kitaigorodskii and Lumley (1983) assumed that the waves were irrotational and argued that the divergence of the turbulent transport of the wave kinetic energy could represent a significant source of turbulent kinetic energy.

A clearer connection has been made between waves and turbulence through wavebreaking: Elevated near surface dissipation rates observed by Agrawal et al. (1992) have been shown by Melville (1994) to be consistent with a best estimate of the rate of input of energy by the wind and that the dissipation was concentrated in a region of the order of the wave height, a. As modeled by Craig and Banner (1994), this results in a turbulent layer that has an extremely energetic near-surface layer riding on a turbulent wall layer having more or less the same degree or structure and order as seen in a typical wall flow. However, as shown by Craik and Leibovich (1976), it appears that wind-driven waves can also interact with wind-shear forced mean flows to create large-scale streamwise vortices known as Langmuir cells (Langmuir, 1938). After averaging the effects of the waves on the slowly-evolving non-wave flow a new forcing term appears in the momentum equation that takes the form:

$$\vec{F}_{CL} = \vec{u}_s \times \vec{\omega} \tag{1}$$

where F_{CL} is the extra force acting on fluid particles, $\vec{\omega}$ is the vorticity, and $\vec{u}_s$ is the Stokes drift defined in terms of a wave velocity field and the phase (or time) average $< >$ as

$$\vec{u}_s = \left\langle \left(\int \vec{u}_w \; dt \right) \cdot \nabla \vec{u}_w \right\rangle . \tag{2}$$

As described by Leibovich and co-workers, this term can be de-stabilizing (producing longitudinal vortices via the so called CL2 instability) when for uni-directional waves producing a Stokes drift current U_s the mean shear and the shear in the Stokes drift have the same sign (see e.g.. Leibovich et al., 1989).

Calculations of velocity fields using the so-called CL model generally produce reasonably smooth fields, although some increase in scale of circulations is seen (Li and Garrett, 1993). More recently, Skyllingstad and Denbo (1995) carried out Large Eddy Simulations (LES) of the upper ocean applying the forcing term directly to the resolved turbulence field. They found that inclusion of the CL forcing term dramatically modifies the dynamics of a modeled ocean mixed layer, greatly increasing vertical heat fluxes. In effect, as argued by Gargett (1989), it seems conceivable that one might view these structures as large-scale turbulent structures, and so their dynamic might be considered a part of wave-turbulence interaction. Thus, while theory for wave-turbulence interactions has been difficult to produce, the CL theory for Langmuir cells appears to give an explicit result for one form of interaction.

In the sections below we consider how these various mechanisms for turbulence production discussed above might be fit together. The picture that emerges is one that indicates mixing on the large scales by Langmuir-cell like structures driven in the fashion given by CL theory, with more complex behavior on smaller scales caused by the unsteady response of the turbulence to wave strains. In §2 we discuss how mean flows can be altered by waves, while in §3 we consider turbulence interaction in the light of rapid distortion theory (RDT); finally, we summarize our results in §4.

2. Mean Flows and Langmuir Cells

2.1 Wind Driven Shearflows

Generally, it has been claimed that mean flows, $\overline{U}(z)$, under wind waves satisfy a relation of the form

$$U^+ = \frac{\overline{U}_s - \overline{U}(z)}{u_*} = \frac{1}{\kappa} \ln\left(\frac{z}{z_0}\right) \quad (3)$$

where u_* is the shear velocity, κ is von Karman's constant $= 0.41$, z is the distance from the surface. U_s is the mean Eulerian surface velocity and z_0 is roughness length. Failing direct measurements of the shear stress, u_* and z_0 are chosen so that the observed data satisfy the wall relation, (3). In contrast, the detailed LDA measurements taken in the large windwave facility at Stanford and reported in Cheung and Street (1988a, 1988b) for wind waves and for wind-ruffled mechanical waves included direct measurement of u_*. Plotted in wall coordinates their measurements for windwaves appear to fail to satisfy (3) in that the slope of U^+ vs ln (z^+) is not $\kappa^{-1} = 2.5$. Here, the wall coordinate z^+ is defined by the relation

$$z^+ = z\, u_*/\nu \,. \quad (4)$$

In this case z_0 can only serve to shift the curves vertically in the U^+– ln (z^+) plane. However, examination of their data for large values of z^+ the slope asymptotes to the wall value. Taking this observation, we fit these measurements by eye to (3) using the points at the largest values of z^+ to calculate z_0 . The results are shown in Figure 1, where the collapse of the data in this fashion is seen to be quite good. More importantly, z_0 so determined is not arbitrary; instead, we find that

$$z_0 = (0.0042 \pm 0.001) y_\eta \quad (5)$$

where y_η is the inverse wavenumber (k^{-1}) determined by Cheung and Street from the entire wave spectrum. Comparing their model to the data presented by Agrawal et al. (1992), Craig and Banner (1994) argue for this form of roughness. Careful modeling by Craig (1996) seems to confirm this scaling, although to match data and calculations Craig needed to subtract an offset from the measured velocity, and, more importantly, the resulting roughness scale is 100 times larger.

Figure 1 indicates a velocity profile with a zone of rapid mixing of momentum which is 50 to 60 z_0 thick below which the law of the wall holds, albeit with a wavelength (not wave amplitude) -determined roughness. This thickness corresponds to a region 0.2 k^{-1} thick. There is also evidence of a highly sheared near-surface layer above, although the collapse of the data is less good here, possibly including effects of averaging velocities from different distances under the free surface as the waves pass by the LDA (Cheung and Street, 1988b).

One could argue that the well-mixed region for $10 < z/z_0 < 40$ is the result of mixing associated with wavebreaking (suggested to us by an anonymous referee). There are two factors arguing against this interpretation. Firstly, the data shown in Figure 1 span a wide range of breaking conditions. Secondly, in the present case dissipation rates inferred by Jiang (1990) using inertial subrange fitting seem to match wall values reasonably well and do not exhibit the enhancement that has been observed in the wave breaking region near the surface. These dissipation estimates are shown in Figure 2, where we use the depth scaling suggested by Terray et al. (1996), i.e., z/H_s, where H_s is the significant wave height. Terray et al. show that there exists a transition depth where dissipation rates predicted from the law of the wall and from decay of breaking generated turbulence match. We calculated this depth using their equation (11) with an estimate of the "effective phase speed" of waves responsible (in an integral sense) for the flux of energy from the wind to the water based on the measured dominant wave. Given that the measurements appear to be taken below the transition depth, the wall values of dissipation seen by Cheung and Street are consistent with other measurements taken below the high dissipation surface layer. We may conclude that the well mixed region seen in Figure 1 was not due to the direct effects of mixing induced by breaking.

It is natural to ask the question as to whether or not it is possible that Langmuir cells might be responsible for the velocity structure seen in Figure 1. Faller has carried out a variety of experiments indicating that Langmuir cells do form in laboratory flows (Faller and Cartwright, 1983). For one of the cases treated by Cheung and Street (U= 6.7 m s^{-1}),

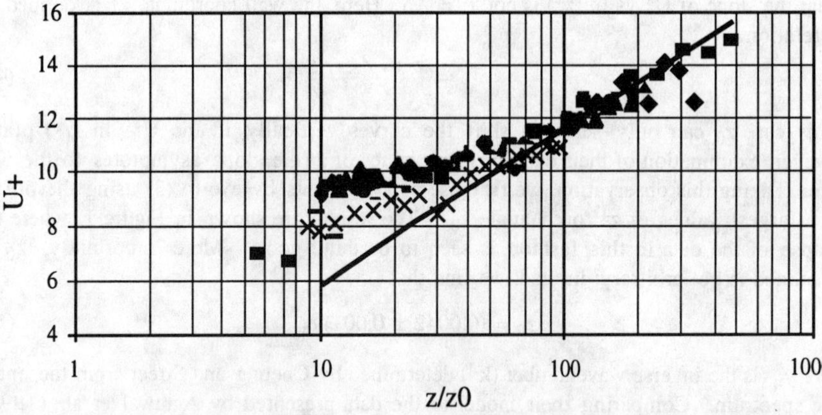

Figure 1. Cheung and Street (1988a) data replotted with z scaled by z_0 for (■) U = 3/2 m s^{-1}, (♦) U = 4.7 m s^{-1}, (▲) U = 6.7 m s^{-1}, (●) U = 9.9 m s^{-1}, (×) U = 13.2 m s^{-1} and (–) eq (3).

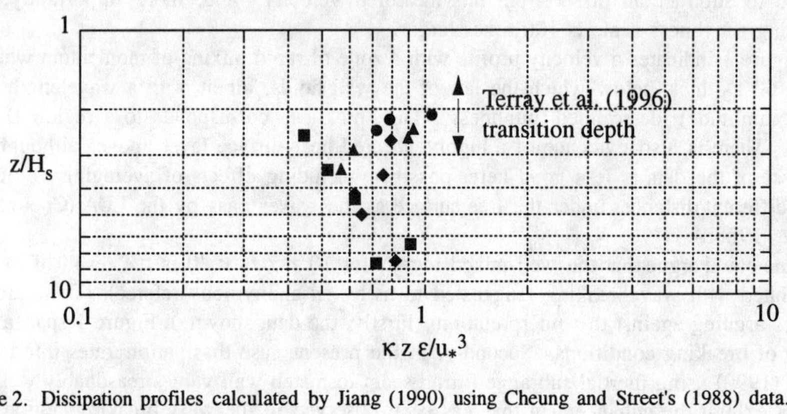

Figure 2. Dissipation profiles calculated by Jiang (1990) using Cheung and Street's (1988) data. The symbols are the same as in Figure 1.

Monismith (1990) carried out further experiments in the Stanford channel using a 3 component LDA and flow visualization. Although his results were somewhat limited, both methods showed the presence of relatively strong secondary flows that he interpreted as evidence of Langmuir cell formation. Indeed according to the CL theory, Langmuir cells would be expected for this wind speed.

The conclusion that Langmuir cells are important to turbulent mean flows is substantiated by calculation as well. Detailed 2D calculations of turbulent wind driven flows including the CL force and using a k-ε closure to model small scale turbulence are currently being made by M. Araujo, P. Maurel, D. Dartus, and L. Masbernat at the IMFT. They find that they can match with reasonable success the mean velocity profiles measured by Magnaudet and Thais (1995). Their prediction of the mean flow is much better with the CL force (and implicitly Langmuir cells) included than without. Importantly, they find that the shear stress vanishes a relatively short distance below the free surface rather than penetrating to the bottom of the channel as should occur for Poiseuille-Couette flow; it is something that is seen in the essentially identical experiments done at Stanford (Cheung, 1985). Thus, the CL case is clearly in better agreement with the data than is the non-CL case.

Since the Langmuir cell forcing is confined to a distance k^{-1} below the water surface, it would appear that we may conclude that the roughness effect is associated with Langmuir cell dynamics, and not with turbulence dynamics in the traditional sense of roughness effects at a rough wall (e.g. Townsend, 1976). However, the picture may be more complicated than indicated by the 2D model: Unlike 2D cells calculated using the CL model, the structures observed by Monismith (1990) were transitory and intermittent. The timescale of these fluctuations were generally quite long, giving spectra that were roughly twice as energetic at low frequencies as would be expected for flat plate boundary layers. Besides making characterization of secondary flow structures difficult, this unsteadiness strengthens the argument that Langmuir cells are turbulence. Firstly, if we imagine the Langmuir cell field to be an array of streamwise vortices, it seems likely that they will behave like all other similar flows (e.g. Dean vortices - Ligrani and Niver, 1988) and lose their simple periodic form through secondary instability to form more complicated structures. Indeed Li and Farmer (1995) present evidence that this is the case. Secondly, if we look at the vorticity equation including CL forcing for the case of a lD Stokes drift, the only stretching term that appears is the equation for the streamwise vorticity, i.e. $\omega_s\, (\partial U_s/\partial z)$, where ω_s is the streamwise vorticity. Thus if there is little spanwise variation in U, and little streamwise variation in V (the mean transverse velocity), the effects of the CL vortex force can only be transmitted to the flow through the turbulence itself, hence producing unsteady cells.

2.2 *Turbulent Channel Flows*

In an effort to isolate the mechanism for Langmuir cell formation and to avoid difficulties in working with the wind-driven flows, Nepf et al. (1995) carried out a series of experiments with waves propagating on turbulent channel flows. In order to produce conditions leading to Langmuir cell formation they employed curved screens upstream of the wavemaker (a surface penetrating cylinder) to impose extra positive (de-stabilizing) and negative (stabilizing) shears. By varying wave conditions (amplitude and frequency) they were able to produce structures that produced Langmuir cells, i.e., streamwise surface convergence zones and rapid mixing by vigorous vertical motion, as well as conditions for which cells did not form. Contrary to CL theory, the sign of the imposed shear had no effect on the formation of cells. However, wave steepness did appear to be important in that when the waves broke near the wavemaker, cells would form, if not, no cells would appear. Thus, in these experiments wave breaking was the causative agent for cell formation.

A reason for the possible divergence of the Nepf et al. (1995) data from CL theory emerges when mean velocity profiles are considered. For breaking waves, the velocity profile was typically uniform above the bottom boundary layer, indicative of the rapid vertical mixing of the cells. Unlike the wind stress case, in this case, there is no source for shear near the water surface, and hence the shear layer seen for $kz < 0.4$ in Figure 1 was absent. However, the velocity profile for non-breaking waves also has a remarkable property: a negative Eulerian shear develops such that the Lagrangian velocity is unchanged (Nepf, 1992). This results from the definition that the mean Lagrangian velocity is equal to the sum of the mean Eulerian velocity and the Stokes Drift (Andrews and McIntyre, 1979). Thus constancy of the mean Lagrangian velocity implies that when waves are added to the flow a change in the Eulerian flow equal and opposite to the local Stokes drift must appear. Figure 3 below summarizes a variety of experiments reported in Nepf's thesis in which we compare the change in Eulerian mean velocity to that which would be expected if the Lagrangian mean were to remain constant.

This result appears to hold not only for the configuration of Nepf et al., but for a variety of other mechanical wavemakers, including those used by Cheung and Street (1988b) (see Jiang and Street, 1991) and by Swan (1990). Applying the general theory developed in Andrews and McIntyre (1979), this negative Stokes drift effect appears to be a necessary result of the mean conditions that must prevail at the wavemaker itself. Based on further experiments, including some made with unsteady packets, Cowen (1996) argues that the waves generated mechanically appear to be Gerstner waves (Kinsman, 1984), vortical waves with closed orbits, and hence no Stokes drift. If indeed, Nepf et al.'s non-breaking waves were Gerstner waves, then the lack of cell formation is a necessary consequence of the lack of a Stokes drift. To explain the formation of cells within the context of CL theory, Nepf et al. argued as follows: When the waves break it is found that the Lagrangian mean velocity is not equal to the original Eulerian mean, suggesting that breaking has allowed a Stokes drift to form. Thus in addition to the stress imposed on the flow by breaking waves, the CL2 instability mechanism can operate, producing cells.

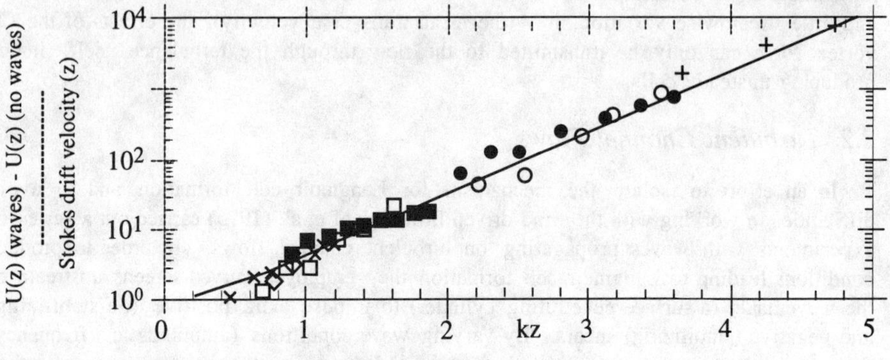

Figure 3. Velocity deficit data from Nepf (1992) (various symbols) compared to the value expected (solid line) if the Lagrangian velocity had remained constant. (taken from Nepf, 1992).

Finally, like the windwave measurements of Monismith (1990), time series measurements and observation of surface convergence behavior indicate that the cells, once formed are unsteady (Nepf et al., 1995), again suggesting the usefulness of considering Langmuir cells as a form of turbulence, albeit one with a rationally formed closure (S. Leibovich - presentation at 1994 ONR MBL workshop, La Jolla July 1994). The basis for this kind of closure is discussed in the next section.

3. The Importance of Strain: RDT and the Craik-Leibovich Force

It is well known that irrotational strains, like those provided by surface waves, can strongly influence turbulence. A productive framework for examining strain effects is provided by rapid distortion theory (RDT - see Hunt, 1979; Townsend, 1976; Reynolds, 1984), a linearization of the equations governing the perturbation fields in a turbulent flow. For example, excluding viscous effects, the RDT balance for the fluctuating vorticity is

$$\partial_t \omega_i + U_j \partial_j = \omega_j S_{ij} + \Omega_j S_{ij} \tag{6}$$

here we use lower case variables to mean fluctuations and upper case variables to refer to mean quantities. Ω_j is the jth component of the mean vorticity and S_{ij} is the ij^{th} component of the mean rate of strain tensor. The principal linearization involved in producing (6) is to neglect terms like $\omega_j \, s_{ij}$ which involve self-interactions of the turbulence while retaining terms which represent the effects of the mean strain field on the turbulence itself. In general, full RDT solutions are difficult to obtain (Hunt, 1979). Thus, RDT's utility lies in examining particular aspects of behavior, i.e. of single Fourier components in homogeneous turbulence.

In the present case we can take the mean flow to be the waves and the perturbation field to be the turbulence. Within this construct, $\Omega_j = 0$ and we must only consider the effects of strain. The fluctuating vorticity then satisfies the relation (see e.g. Reynolds, 1980)

$$\frac{D\omega_i'}{Dt} = \omega_j' \, S_{ij} \tag{7}$$

where we have used the Einstein summation convention, D/Dt is the material derivative, and S_{ij} is the local rate of strain tensor which we attribute to the waves alone (Leibovich, 1977). While numerical solution to this set of equations subject to appropriate initial conditions could be obtained, the important point to be made is that (7) is the fast (wave timescale) vorticity equation used to derive the CL equations. Thus, as originally derived by Leibovich, the fast time behavior of the vorticity field is nothing more than the rapid distortion of the slowly varying vorticity field by the wave strain. On the wave timescale turbulent eddies are distorted by the wave strain creating a periodic perturbation vorticity field; this field then interacts on the longer timescale with the periodic wave strain to create a rectified effect which is the CL force. Thus we can conclude that the wave turbulence interactions will be described by the CL forcing when the hypotheses of RDT theory (Hunt, 1979) are satisfied, namely: (1) Distortion timescale << turbulence timescale; and (2) distortion strain rate >> eddy strain rate.

These conditions can be easily (but not precisely!) examined via scaling arguments. The characteristic timescale of distortion is σ^{-1}, where σ is the wave frequency, while the rate of strain is $\alpha k\ \sigma^{-1}$. The turbulence timescale is (l/q) where l is the lengthscale of interest. If we assume inertial subrange dynamics (a necessary evil) $q \sim (\varepsilon l)^{1/3}$, where q is the turbulence velocity scale, we can identify three types of behavior depending on the length scales

$$L_N = \frac{\varepsilon^{1/2}}{\sigma^{3/2}}$$

$$L_L = (\alpha k)^{-3/2} \frac{\varepsilon^{1/2}}{\sigma^{3/2}} \tag{8}$$

where we have used the subscripts N to denote "nothing" and L to denote Leibovich. Either one of these scales can be taken to be analogous to the Ozmidov scale in stratified turbulence. Three ranges of turbulence can be identified (1) $1 < L_N$: Small-scale turbulence that is "fast" and has strains that are much bigger than those of the wave; m this case the waves do nothing. (2) $L_N < 1 < L_L$: Medium scale turbulence which is slow compared to the waves, but whose strains are larger than those of the waves. Turbulence with $1 \sim L_L$ would appear to be the most difficult to model in that wave strains are of equal importance to self strains, so one has neither the CL force, nor the usual cascade. (3) $1 > L_L$: Large scale turbulence that is not only slow relative to the waves, but is also weak compared to the waves. This turbulence is distorted elastically on wave timescales and evolves in response to the CL force it experiences on long time scales $t \sim (\alpha k)^2\ \sigma^{-1}$. These might be taken to be the Langmuir cell turbulence calculated in the LES model of Skyllingstad and Denbo (1995).

This approach can also be used to identify different regions below the water column where one might find different types of turbulence. If we let the integral scale of the turbulence $l_t \sim z$, and assume the wall scaling for ε, we can find the same three behaviors reflected in a 3 layer structure of an inner (Stokes) layer of $O(\sigma^{-1}\ u_*)$ in which an equilibrium approach is possible (Belcher et al., 1994), an intermediate layer of $O((\alpha k)^{-1}\ \sigma^{-1}\ u_*)$ which is not in equilibrium and the wave strains which might be important, and an outer layer for $z >> (\alpha k)^{-1}\ \sigma^{-1}\ u_*$ in which the turbulence field might take the form of a quasi-random set of Langmuir cells. This scaling could be modified to include the near-surface dissipation layer scaling, possibly giving a different thickness for the innermost layer. The outer layer location is not likely to be influenced since it should scale in thickness with k^{-1} (where the wave strain is), and is thus $(\alpha k)^{-1}$ thicker than the inner layer.

For the sake of definiteness, we can assume that order one constants apply to these scaling relations. We can take typical values of $\varepsilon = 10^{-6}\ m^2/s^3$, $\sigma = 0.6\ s^{-1}$, and $\alpha k = 0.2$ to find that $L_N \sim 0.002$m, while $L_L \sim 0.2$ m. Thus, for an O(100m) deep ocean mixed layer, or even an O(10m) deep lacustrine mixed layer, one would expect that Langmuir cell turbulence associated with the turbulence field produced in the presence of the CL force should be a decent model of wave-turbulence interaction outside of the near-surface region in which wave breaking is important. Given that even relatively small turbulence scales are described by this inviscid interaction, it seems entirely appropriate to use the CL force to model wave-turbulence interactions in an LES model like that of Skillingstad and Denbo (1995). However, it would also seem appropriate to separately model the effects of wave breaking, perhaps through a semi-random energy and vorticity flux through the top surface of the computational domain.

4. Summary

In summary, the data we discuss show that waves significantly modify the shear flow beneath. Observations and calculations made including the CL force derived by Craik and Leibovich (1976) show that the formation of Langmuir cells is likely and significantly modifies mean and turbulence flow structure. We argue that the means by which waves interact with turbulence is through the effect of irrotational wave strain on turbulence. To make this point concrete, we use RDT to argue that the CL force that drives Langmuir cell formation is the rectified effect of the rapid distortion of turbulence (RDT) by the wave strain. Accordingly, one can think of Langmuir cells as the result of an interaction of the large turbulence scales with the waves. By considering the type of interaction in terms of RDT we define three regimes of wave-turbulence interaction: (1) Small-scale turbulence that responds quasi-steadily to the wave strain; (2) Intermediate scale turbulence that interact nonlinearly with the wave strain; (3) Largescale turbulence that is distorted elastically but with a rectified effect described by the CL body farce.

Given the imprecise nature of the scaling arguments, it would be useful to carry out a direct numerical simulation of wave strain effects on turbulence to confirm the applicability of the CL force to model large-scale wave-turbulence interactions. This could be done by first considering initially homogeneous turbulence subjected to simple spatially and temporally varying irrotational strain fields. As long as these fields produce a "Stokes drift", the results of such a calculation should be applicable to wind-wave turbulence modeling. Unfortunately, it does not appear possible to test the scaling in laboratory flumes, at least with mechanically generated waves, because these appear not to produce a Stokes drift.

Acknowledgments. SGM is grateful for support from ONR (N00014-93-1-0377 monitored by S. Sandgathe and N00014-940190 monitored by E. Rood), for the many enjoyable discussions in which he has been engaged by his colleagues Bob Street, Jeff Koseff, Joel Ferziger, Todd Cowen, and Heidi Nepf, to JJMM, for the fine time he had visiting the IMFT, where this paper was written, and to Jörg Imberger for inviting him to participate in this symposium.

References

Agrawal, Y. C., E. A. Terray, M. A. Donelan, P. A. Hwang, A. J. Williams III, W. M. Drennan, K. H. Kahma, and S. A. Kitaigorodski, Enhanced dissipation of kinetic energy beneath surface waves. *Nature 359*, 219-220, 1992.

Andrews D. G. and M. E. McIntyre, An exact theory of waves on a Lagrangian mean flow. *J. Fluid Mech 89*, 609-646, 1978.

Belcher, S. E., J. A. Harris, and R. L. Street, Linear dynamics of wind waves in coupled turbulent air-water flow. Part 1. Theory. *J. Fluid Mech. 271*, 119-151, 1994.

Cheung, T., A Study of the Turbulent Layer in the Water at an Air-Water Interface. PhD Thesis, Dept. of Civil Eng., Stanford University, 1985.

Cheung, T. and R. L. Street, The turbulent layer in the water at an air-water interface. *J. Fluid Mech, 194*, 133-154, 1988a.

Cheung, T. and R. L. Street, Wave following measurements in the water beneath an air-water interface. *J. Geophys. Res. (Oceans) 93*, 14089-14907, 1988b.

Cowen, E. A., An experimental investigation of the near-surface effects of waves traveling on a turbulent current Ph. D. Thesis, Dept. of Civil Eng., Stanford University, 1996.

Craig, P. D. and M. L. Banner, Modeling wave-enhanced turbulence in the ocean surface layer. *J. Phys. Ocean. 24*, 2546-2559, 1994.

Craig, P. D., Velocity profiles and surface roughness under breaking waves. *J. Geophys. Res. (Oceans) 101*, 1265-1278, 1996.

Craik, A. and S. Leibovich, A rational model for Langmuir circulations. *J. Fluid Mech. 73*, 401-426, 1976.

Csanady, G. T., Vortex pair model of T Langmuir circulation. *J. Mar. Res., 52*, 559-581, 1994.

Faller, A. and R. Cartwright, Laboratory studies of Langmuir Circulations. *J. Phys. Oceanogr. 13*, 329-340, 1983.

Gargett, A. E., Ocean turbulence. *Ann. Rev Fluid Mech., 21*, 419-452, 1989.

Hunt, J. C. R., A review of the theory of rapidly distorted turbulent flows and its application. *Fluid Dyn., Trans. 9*, 121-152, 1979.

Jiang, J. Y., Wave-turbulence interaction and modulated flows beneath wind waves PhD Thesis, Dept. of Civil Eng., Stanford University, 1990.

Jiang, J. Y. R. L. Street, and S. P. Klotz, A study of wave-turbulence interaction by use of a nonlinear water-wave decomposition technique. *J. Geophys. Res. (Oceans) 95*, 16037-16054, 1990.

Jiang, J. Y. and R. L. Street, Modulated flows beneath wind-ruffled, mechanically generated waves. *J. Geophys. Res. (Oceans) 96*, 2711-2721, 1991.

Kinsman, B., Wind Waves: Their Generation and Propagation on the Ocean Surface. Dover, 1984.

Kitaigorodskii, S. A., and J. L. Lumley, Wave-turbulence interactions in the upper ocean. Part 1: The energy balance of interacting fields of surface wind waves and wind-induced three-dimensional turbulence. *J. Phys. Ocean., 13*, 1977-1987, 1983.

Kitaigorodskii, S. A., and others Wave-turbulence interactions in the upper ocean. Part II: Statistical characteristics of wave and turbulent components of the random field in the marine surface layer. *J. Phys. Ocean. 13*, 1988-1999, 1983.

Langmuir, I., Surface motion of water induced by wind. *Science, 87*, 119-123, 1938.

Leibovich, S. On the evolution of the system of wind drift currents and Langmuir circulations in the ocean. Part 1. Theory and averaged current. *J. Fluid Mech., 79*, 715-743, 1977.

Leibovich, S., Lele, S. A. K., and Moroz, I. M. Nonlinear dynamics m T Langmuir circulations and in thermosolutal convection. *J. Fluid Mech. 198*, 471-511, 1989.

Li, M. and D. Farmer, Patterns of bubble clouds organized by Langmuir circulation. *J. Phys. Oceanogr., 25*, 426-1440, 1995.

Li, M. and C. Garrett, Cell merging and the jet/downwelling ratio in Langmuir circulation. *J. Mar. Res. 51*, 737-769, 1993.

Ligrani, P. M. and R. D. Niver, Flow visualization of Dean vortices in a curved channel with a 40 to 1 aspect ratio. *Phys. Fluids 31*, 3605-3617, 1988.

Magnaudet, J. and L. Masbernat, Interaction des vagues de vent avec le courant moyen et la turbulence. *C. R. Acad. Sci Paris Serie 11 311*, 1461-1466, 1990.

Magnaudet, J. and L. Thais, Orbital rotational motion and turbulence beneath laboratory wind waves. *J. Geophys. Res. (Oceans) 100*, 757-771, 1995.

Melville, W. K., Energy dissipation by breaking waves. *J. Phys. Ocean. 24*, 2041-2049, 1994.

Monismith, S. G., A study of the influence of Langmuir cells on turbulence and secondary flows under wind-waves. *Proc. Intnl. Conf. on Physical Modelling of Transport and Dispersion*, Cambridge, UA, pp 3. 9-3. 14, 1990.

Nepf, H. M., The production and mixing effects of Langmuir circulations PhD Thesis, Dept. of Civil Eng., Stanford University, 1992.

Nepf, H. M, E. A. Cowan, S. J. Kimmel and S. G. Monismith, Longitudinal vortices under breaking waves. *J. Geophys. Res. (Oceans) 100*, 16211-16221, 1995.

Reynolds, W. C., Physical and analytical foundations, concepts, and new directions in turbulence modelling and simulation, in *Turbulence Models and Their Applications*, vol 2 pp. 148-294., Editions Eyrolles, 1980.

Skyllingstad, E. D. and D. W. Denbo, An ocean large-eddy simulation of Langmuir circulations and convection in the surface mixed layer. *J. Geophys. Res (Oceans). 100*, 8501-8522, 1995.

Swan, C., Convection within an experimental wave flume, *J. Hyd. Res. 28*, 273-282, 1990.

Thais, L. and J. Magnaudet, Turbulent structure beneath surface gravity waves sheared by the wind. *J Fluid Mech. 328*, 313-344, 1996.

Terray, E. A., M. A. Donelan, Y. C. Agrawal, W. M. Drennan, K. H. Kahma, A. J. Williams III, P. A. Hwang, and S. A. Kitaigorodski, Estimates of kinetic energy dissipation under breaking waves. *J. Phys. Ocean 26*, 792-807, 1996.

Townsend, A. A., *The Structure of Turbulent Shear Flows*, Cambridge University Press, 1976.

8

Modeling of Atmospheric Forced Mixing on the Shallow Shelf

I. D. Lozovatsky and A. S. Ksenofontov

Abstract

A one-dimensional non-stationary differential model of turbulent boundary layer is developed to analyze time-depth variations of the specific potential density σ_θ, the turbulent energy e_t, viscous dissipation rate ε, and turbulent buoyancy flux J_b on the shallow shelf in the north-western part of Black Sea. The boundary conditions were represented by the surface fluxes of momentum, buoyancy and turbulent energy, which were calculated using meteorological data obtained during seven days of continuous observations in late September and in the beginning of October 1989. Three mild storms accompanied by strong night-time cooling of the sea surface allowed to study turbulence variations in the upper layer produced by vertical shear and thermal convection. It was found that turbulent entrainment at the base of the mixed layer leads to sharp decreasing of the viscous dissipation rate in the transition zone between the quasi homogeneous layer and pycnocline. A local maximum of ε is formed in this transition zone, when the deepening of the thermocline completes and entrainment ceases. The results of modeling are compared with the hydrological measurements on the shelf.

1. Introduction

Mixing on a shallow shelf is mainly governed by convective and shear-induced turbulence in the near surface and near bottom boundary layers, which may occupy a large part of the water column. Deep sea turbulence far from the upper quasi homogeneous layer and from the bottom topography is commonly generated by K-H shear instability and random internal wave breaking, whereas atmospheric forcing is the major process providing vertical mixing from the sea surface down to the bottom on a shallow shelf. The turbulent energy and the kinetic energy dissipation rate, induced by wind stress and convection, determine the value of the vertical turbulent diffusivity and therefore responsible to the vertical transport of heat, oxygen and pollutant in the shallow waters.

An important aim of the present study is estimation of time-depth variations of the main turbulence parameters in the upper layer. To achieve this goal, we address to numerical modeling of time evolution of the vertical turbulent structure on the shallow Black Sea shelf at the beginning of autumn seasoning cooling. The main interest will be paid to the

Physical Processes in Lakes and Oceans
Coastal and Estuarine Studies Volume 54, Pages 111-122

mixing, induced by short period storm events. With regards to this problem, it is quite important to introduce an appropriate parameterization of intermittent mixing, associated with coupled influence of direct wind forcing and night convection at the sea surface. Therefore we will based on a one-dimensional model of Lozovatsky et al. (1993), which has shown a reasonable agreement between calculations and direct turbulent measurements in the oceanic upper boundary layer. The closure scheme was modified for this particular case.

The calculations were grounded on the meteorological and hydrological measurements, which were carried out in September-October of 1989 in the western part of the Black Sea shelf southern of the Bulgarian port Varna. This field experiment was performed in the beginning of autumn water cooling during the period of active air-sea interaction (Lozovatsky and Ozmidov, Ed., 1992).

2. Model

2.1. The Basic Equations and Closure Scheme

The complete system of equations consists of the equations of one-dimensional momentum balance (1), (2), the equation of buoyancy transfer (3), and the equation of turbulent energy balance (4):

$$\frac{\partial u}{\partial t} - fv = \frac{\partial}{\partial z} K \frac{\partial u}{\partial z}, \qquad \frac{\partial u}{\partial t} + fu = \frac{\partial}{\partial z} K \frac{\partial v}{\partial z} \qquad (1), (2)$$

$$\frac{\partial b}{\partial t} = \frac{\partial}{\partial z} K_b \frac{\partial b}{\partial z} \qquad (3)$$

$$\frac{\partial e_t}{\partial t} = KSh^2 + K_b \frac{\partial b}{\partial z} + \frac{\partial}{\partial z} K_e \frac{\partial e_t}{\partial z} - \varepsilon \qquad (4)$$

where u and v are the eastern and northern components of the mean velocity vector; $e_t = (3/2)\overline{w'^2}$ is the turbulent kinetic energy per unit mass; $\overline{w'^2}$- the variance of vertical turbulent velocity component; t - time, z - the vertical coordinate directed downward; $b = g(\rho - \rho_0)/\rho_0$ is buoyancy, ρ and ρ_0 are density and its reference value; g is gravity, f is the Coriolis parameter; K, K_b, and K_e, are the eddy diffusivities of momentum, buoyancy and turbulent energy; ε is the viscous dissipation rate, Sh^2 is the squared vertical shear. The turbulent dissipation rate and the turbulent eddy diffusivy are parameterized by well-known Kolmogorov's hypotheses (Kolmogorov, 1942; Rotta, 1951):

$$\varepsilon = \alpha_\varepsilon \frac{e_t^{3/2}}{l_t}\left(1 + \alpha_\varepsilon^0 \frac{\nu}{l_t\sqrt{e_t}}\right), \qquad K = l_t\sqrt{e_t} \qquad (5), (6)$$

where l_t is the turbulent scale to be introduced, ν - the molecular viscosity, α_ε and $\alpha_\varepsilon{}^0$ are the universal constants.

For the reasons of measurements lack we assume, following to Laikhtman et al. (1976), that the ratio K_e/K is equal to 0.73 and the turbulent Prandtl number $Pr_t \equiv K/K_b = 1$. In background of stable buoyancy stratification the turbulence length scale l_t is a function of e_t, db/dz, and Sh^2. Therefore it can be written in the following general form (Ksenofontov et al., 1990):

$$l_t = \phi(Ri)\sqrt{\frac{e_t}{Sh^2}}, \qquad \phi(Ri) = \frac{c_l}{1+\sqrt{Ri/Ri_{cr}}}, \qquad Ri \geq 0, \tag{7}$$

where $Ri = N^2/Sh^2$ is the gradient Richardson number and Ri_{cr} its critical value, which was taken as 0.25, $N^2 = db/dz$ is the squared buoyancy frequency. The function $\phi(Ri)$ was defined in Lozovatsky and Ozmidov (1979), Lozovatsky and Korchashkin (1984) and generalized in Ksenofontov et al. (1990) with the constant $c_l = 1.2$. Two asymptotes result from this formula. For strong density stratification ($Ri >> Ri_{cr}$), $l_t = 0.6(e_t/N^2)^{1/2}$, whereas for quasi homogeneous fluid ($Ri << Ri_{cr}$), $l_t = 1.2 \cdot (e_t/Sh^2)^{1/2}$.

The Eqs.(7) can not be applied for unstable stratification ($Ri < 0$), so we have to introduced a special closure. The negative vertical density gradients are used in the equation of buoyancy transfer along with the values of turbulence scale $l \equiv l_{inv}$ that are calculated by formula (Blakadar, 1962; Mellor and Yamada, 1974):

$$l_{inv} = c_l \int_0^{h(t)} \sqrt{e_t}\, z dz \Big/ \int_0^{h(t)} \sqrt{e_t}\, dz, \tag{8}$$

where $h(t)$ - is the thickness of the upper turbulent layer, $c_l = 0.4$ - an empirical constant. Such artificial procedure replaces convective flux by fictitious diffusive flux with the magnitude proportional to the density gradient, taken with the opposite sign. In spite of insufficient grounded of the convective parameterization, it has led to reasonable results in our previous calculations (Ksenofontov et al., 1990; Lozovatsky et al., 1993).

2.2. *The Initial and Boundary Conditions*

The following initial and boundary conditions were used to solved equations (1) - (8):

$$t = 0: \quad u = u^o(z), \quad v = v^o(z), \quad \rho = \rho^0(z), \quad e_t = e_t^o(z) \tag{9}$$

$$z = 0: \quad K\frac{\partial u}{\partial z} = -\frac{\tau_x}{\rho_0}, \quad K\frac{\partial v}{\partial z} = -\frac{\tau_y}{\rho_0}, \quad K_b\frac{\partial b}{\partial z} = -J_B, \quad K_e\frac{\partial e_t}{\partial z} = -P_e, \tag{10}$$

$$z = H: \quad u = v = 0, \quad \rho = \rho_H, \quad e_t = e_{tf} \tag{11}$$

To analyze time variations of the turbulence characteristics in the upper layer caused by atmospheric forcing, we set up the lower boundary of the calculation area $z = H$ over the bottom friction layer. We also assumed that there are no motion (11) at $z = H$. According to these restrictions, the calculations can not be applied to the near bottom boundary layer. The background turbulence energy, e_{tf}, was taken to be equal too small as 10^{-4} cm^2/s^2. If the turbulent eddy diffusivity is limited by the value of the molecular viscosity $K \geq K_f \equiv \nu$, the background kinetic energy dissipation rate $\varepsilon_f = 0.1e^2_{tf}/K_f$ should be equal to 10^{-7} $cm^2\, s^{-3}$ and a small background shear $Sh_f < 10^{-3}$ s^{-1} exists in the model.

The initial profiles u^o, v^o, ρ^0, e_t^o are usually taken from the measurements or from prognostic calculations as the stationary solution for a given density profile $\rho^0(z)$ with the constant wind stress and turbulent energy flux at the sea surface. To obtain the non-stationary solution, we must have the time series of the surface buoyancy flux, $J_B(t)$, the wind stress components $\tau_x(t)$, $\tau_y(t)$ and the turbulent energy flux, $P_e(t)$. The surface fluxes were calculated from meteorological data using standard aerodynamics bulk-formulae method following to Bunker (1979), Large and Pond (1981), Carroll and Noble (1982), Blanc (1985), and Ksenofontov (1992).

2.3. *The Grid System and Numerical Scheme*

A finite-difference scheme was developed to reproduce the main features of the basic system of the differential equations, satisfying to the conservation laws. We used absolutely stable conservative numerical schemes (Arakawa, 1966) with high order approximation for spatial and temporal derivatives. The original equations were reduced to a three point grid scheme by the balance method and computed by well-known iteration run method (Marchuk, 1977). Cranckle-Nickolson scheme of the second order of accuracy was used for integrating of the equations over time. We pitched upon the vertical step of 20 cm and time step of 2 minutes to achieve an appropriate resolution for small scale hydrophysical structures.

2.4. *The Limitations of the Model*

We would like to point out that present study is limited by modeling the evolution of mixing activity in the upper layer over the shelf at short time periods. A one dimensional model fails to simulate the upper layer processes on the Black Sea shelf during quite long time (a week or more). So we will consider only short separated periods of intensive air-sea interactions, such as storm mixing and night convection, providing vertical mixing as a local process. We do pay the main attention to formation and evolution of the vertical structure of the turbulent energy, the kinetic energy dissipation rate and turbulent buoyancy flux in background of local stratification.

It is evident that many processes are usually responsible for thermohaline structure formation on the shelf. For example, the present model works from the sea surface down to the lower boundary of the pycnocline only and does not resolve the near bottom mixed layer. Upwelling circulation, mesoscale eddies, trapped Kelvin waves, river inflow and other processes might affect mixing on a shelf, but they are not described by the one-dimensional model. The effects of the named processes should be significant when the simulation period is quite long compare with characteristic time of direct wind-induced mixing.

3. Boundary Fluxes and Initial Hydrology

3.1. *Observations*

The boundary fluxes were calculated from the meteorological data, obtained on September 28 - October 6, 1989 in the north-western part of the Black Sea shelf ($\varphi = 42^\circ$ 56' N, $\lambda = 28^\circ$ 01' E). Meteorological observations were carried out every 3 hours. Air temperature and humidity were measured by ordinary mercury thermometer and psychrometer from the pier end 15 - 20 meters off the coast line, 3 meters over the sea surface. The sea depth at that place was about 2 m. Sea surface temperature as a meteorological parameter was measured in the same place. Atmospheric pressure was recorded permanently by a barograph. An anemometer sensor for wind measurements was mounted at 10 meters beam located at the beach 50 meters from the coast line.

BAKLAN profiler was used for the hydrophysical measurements. The bottom depths at the main (almost flat) part of the shelf varied from 25 to 30 meters. Because in the present paper these data will be used only for a qualitative comparison with the results of numerical calculations, we refer to instrumentation details to Nabatov and Paka (1990), and Gibson et al. (1993). Preliminary analysis of these turbulent measurements were presented in

Lozovatsky and Erofeev (1994).

The hydrological measurements showed very small horizontal density gradients in the upper active layer at the main part of the testing area about 1 mile off the coast line. Therefore the evolution of the vertical structure and the thickness of the quasi homogeneous mixed layer was mainly caused by intensive convective and wind induced turbulence and only slightly affected by advection. Based on these reasons we believe that the presented one-dimensional turbulent model is a relevant tool to study turbulent processes in the upper layer over the shelf in this particular case.

3.2. Meteorological Background

Eastward and northward winds should induce off-shore surface current on this shelf, whereas southward and westward winds should produce on-shore current in the upper layer. From September 28 till October 7, wind had not southern component. Thus, during the first week of October 1989 winds were not favorable for the coastal upwelling circulation.

The transport of cold air masses to the Black Sea coast on October 1 and October 2 was responsible for rapid decreasing of the sea surface temperature and as a result to the origin of convective mixing near the coast. From October 1 till October 7 the cold air mass with minimal temperatures in early morning of 5.2 °C – 5.4 °C dominated over Bulgarian coast. The total heat balance for this period was negative and equal to - 14.9 MJ/m^2. A joint effect of low air temperature and the storm events on September 29, October 1 and 2 led to almost complete vertical homogeneity of the thermohaline structure in the upper 15 - 20 meters of the water column. On October 1 the heat losses from the sea to the atmosphere exceeded the total heat income during the next days. The diurnal heat balance remained negative due to high losses resulting from latent heat of vaporization, which was the main cause of water temperature decreasing during the first October decade.

3.3. Momentum and Buoyancy Fluxes

The time series of the boundary fluxes are shown in Figure 1. Wind stress variability shows three prominent storm events during the observational period. These storms were characterized by the peak wind stress $|\tau^{max}|$ in the range of 0.23 - 0.29 N/m^2, that was about an order higher than the background relatively low wind stress with the typical values of 0.02 - 0.04 N/m^2 during the calm periods. Three mild storms led to significant changes in vertical and horizontal thermohaline structure at the shelf. During the periods of stormy weather the wind stress increases sharply and produces intensive mixing in the shelf waters. Between the storms, when the momentum transfer was not so intensive, the inertial currents with the period about 17 hours dominated.

The surface buoyancy flux $J_B(t)$ had asymmetric diurnal variability (Figure 1b). Three times per day (in 9:00, 12:00, and 15:00) it was negative, while the rest five times of the meteorological observations (from 18:00 till 6:00) it was positive. Only once, in 9:00 on October 1, J_B had a small positive value owing to storm induced intensive vaporization. The peak values of day-time buoyancy flux were in the range of $J_B = -(1.6 - 2.9)\cdot 10^{-7}$ W/kg. The positive fluxes were more even in time $J_B = (0.5 - 1.0)\cdot 10^{-7}$ W/kg, except of storm periods on October 1 and 2, when J_B achieved $(2.1 - 2,3)\cdot 10^{-7}$ W/kg. The night-time positive buoyancy fluxes provided intensive turbulent convection on the shelf. This short analysis of thermodynamic air-sea interaction allows to assume that the one-dimensional turbulent model can be used to simulate time variations of thermohaline

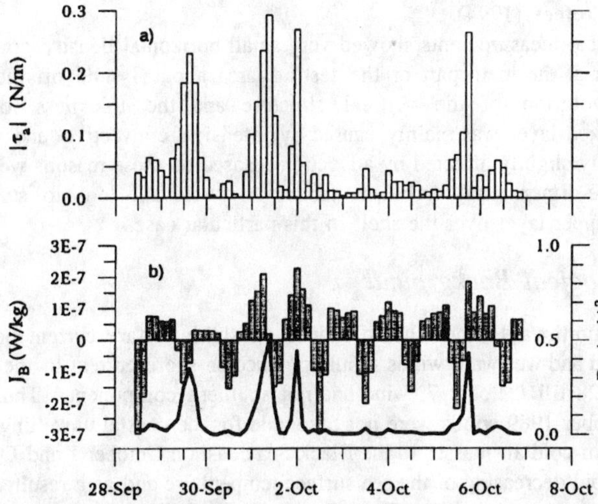

Figure 1. Fluxes at the sea surface: the module of wind stress $|\tau_\alpha|$ - (a), the buoyancy flux J_B, and the turbulent energy flux P_e.- (b).

profiles and turbulence parameters in the upper layer caused by coupled convective-wind-induced mixing.

3.4. *Initial Hydrology*

The vertical profile of specific potential density measured at 14:00 local time on September 28 is presented in Figure 2 (curve 1). The upper mixed 6 m layer was separated from the lower pycnocline by a thin but sharp density interface ($\Delta\sigma_\theta = 0.2$) caused mainly by salinity jump due to fresh water inflow from Kamchia River. The upper pycnocline was destroyed every night by convective mixing, therefore temperature profile was quasi homogeneous down to depth of 12 m. This structure was in a quasi stationary state resulting from permanent fine anticyclonic weather conditions during previous ten days. It allows to begin our prognostic numerical calculations without initial velocity and turbulence energy profiles, obtaining them from a diagnostic model.

4. Results

4.1. *Convective and Wind Induced Entrainment*

Moderate permanent wind ($V_a \approx 6$ m s^{-1}) and positive (downward) buoyancy flux with ($J_B \approx 7.5 \cdot 10^{-8}$ W/kg) led to formation of the mixed layer in the regime of penetrating convection, which started on September 28 once the J_B sign changed at the sea surface. The near surface turbulent layer was deepening to 10 m during next three hours (Figure 3b,c). The enhancement of negative (upward) buoyancy flux in the lower part of the upper layer was due to water entrainment from the pycnocline into the mixed layer. The negative buoyancy flux doubled during two hours after the sunset and reached $2.2 \cdot 10^{-7}$ W/kg, while

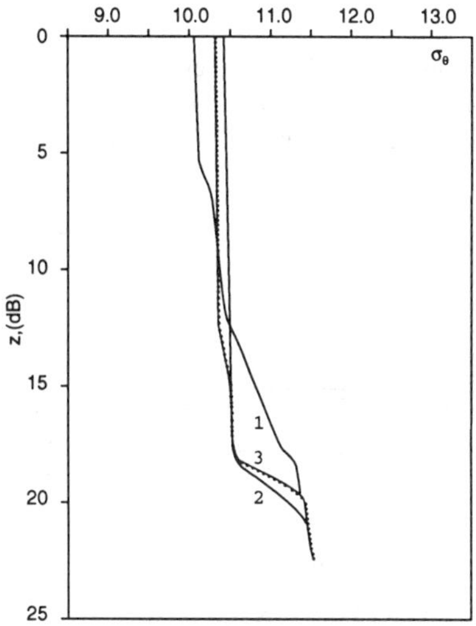

Figure 2. The measured specific density profiles. on September 28 - (1) and on October 5 - (3). The calculated profile (2) for October 5. The model density in the mixed layer is 0.15 units higher than the measured one resulting from a minor influence of fresh water advection.

it was $0.8 \cdot 10^{-7}$ W/kg at the surface. Time variations of the buoyancy flux in the mixed layer are shown in Figure 3d. A sharp vertical gradient of J_b at the bottom of the quasi homogeneous layer was formed at the first hours of the night cooling, associated with increasing of wind stress. The depth of the mixed layer boundary stabilized upon completion of the active entrainment in 21:00. The magnitude of buoyancy flux at the lower boundary became half as much as that at the surface. During the rest of the night the entrained waters were mixing through the whole turbulent layer; convection was continuing but in the non-penetrating regime that probably associated with decreasing of the wind stress. As the result, the isopycnal surface $\sigma_\theta = 10.2$ (Figure 3a) lifted up to the sea surface in 6:00 on September 29. (The following storm-induced variations, which are shown in the central part of Figure 3, will be discussed in sub-section 4.2). The mixing intensity during the period of penetrative convection led to increasing of the turbulence energy that is shown in the left of Figure 3b as the downward concave of $e_t(z, t)$ lines . In the late evening the turbulence energy at the sea surface reached 1.5 cm^2/s^2 that was correlated with the maximum values of the buoyancy flux. Later the wind stress and buoyancy flux decreased (Figure 1) and the isoline $e_t(z, t) = 1$ cm^2/s^2 lifted up to the sea surface in the early morning on September 29. The following storms resulted in enhancement of turbulent mixing and deepening of the pycnocline.

4.2. Storm-Induced Mixing

Three prominent storms occurred from September 28 till October 6. The first one began about 9:00 in the morning on September 29 and was continuing over 12 hours. On October 5 the third storm had approximately the same duration, whereas the second storm

(September 30 - October 2) was much longer. It consisted of two wind enhancements (up to 13 m/s) separated by 6 hours of calm weather. The storm actions led to irreversible deformations of thermohaline structure at the shallow shelf in the beginning of the autumn cooling season (Figure 2, curve 3). Our model allows to consider details of this process.

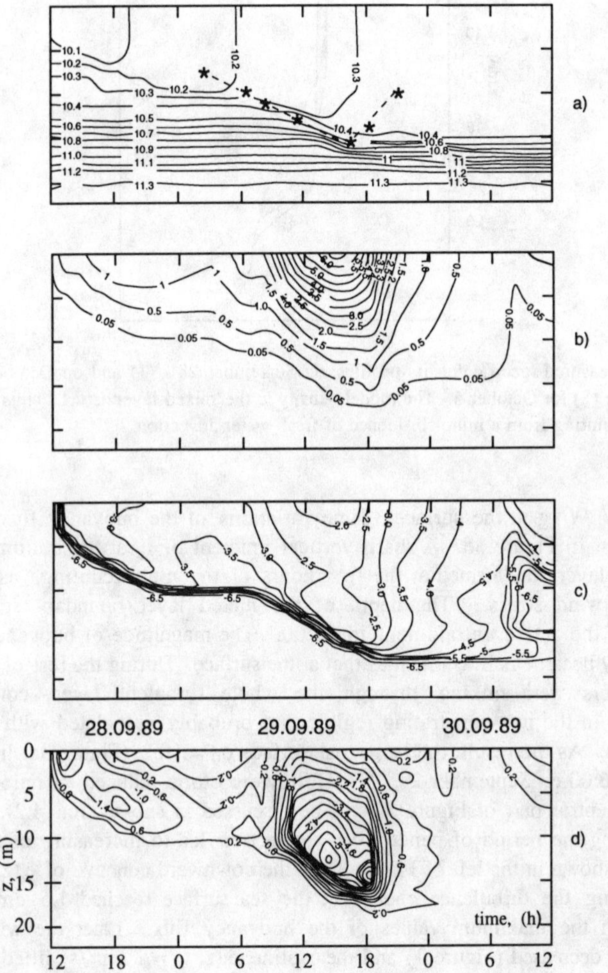

Figure 3. The model calculations of specific density - (a), the turbulence energy e_t (cm^2/s^2) - (b), the logarithm of the kinetic energy dissipation rate log ε (cm^2/s^3) - (c), and the buoyancy flux $J_b \cdot 10^{-7}$ (W/kg) - (d). During the night-time on September 28 - 29 the deepening of the pycnocline was caused by convective and wind induced entrainment. Turbulence intensification in day-time on September 29 resulted from the first storm action. The deepening of the mixed layer is marked by stars-dashed line at the upper panel in accordance with a relation $h_* = 0.1 \cdot u_*/f$.

During the first storm, the thickness of the upper mixed layer increased from 10 to 16 m and turbulence entrained down to the pycnocline to 18 m depth (see the central part of Figure 3). The density in the mixed layer increased on 0.15 units of σ_θ. The vertical density gradient in the underlying pycnocline became twice stronger, $\Delta\sigma_\theta/\Delta z = 0.25$ kg/m^4, compare with the beginning of the calculations. The rate of turbulent layer deepening varied as the wind increased. The deepening of the mixed layer continued until its thickness h became equal to $h_* = 0.1 \cdot u_*/f$, where $u_* = (|\tau|/\rho)^{1/2}$ is the friction velocity. The values of h_* are plotted by stars in Figure 3a. The mixed layer deepening ceases if the friction velocity decreases lower than 10hf. Owing to the active entrainment in the lower part of the homogeneous layer, the kinetic energy dissipation rate sharply decreased near the upper boundary of the pycnocline (Figure 3c). A linear dependence between e_{to} and u^2_* near the sea surface was found. The values of constant $c_* = e_{to}/u^2_*$ for three storms under consideration were equal to 2.7, 2.3, and 2.5 correspondingly, so the mean value of $<c_*> = 2.5$. The turbulent energy, generated by the first, and to a smaller extent, by the second storm, resulted mainly in entrainment and mixing of the cold water from the pycnocline so the isolines of $e_t(z, t)$ are slightly rarefied near the bottom of the upper layer. The storms produced active turbulence in the upper layer, so e_t was as high as 6 - 7 cm^2/s^2 near the sea surface and approximately linearly decreased with depth (Figures 3, 4). The second and the third storms generated turbulence in the already mixed quasi homogeneous layer. In theses particular cases the turbulent parameters became very sensitive to any atmospheric variations. That is why the turbulence energy e_t rapidly decayed within the mixed layer, when the wind became lighter, between two main enhancements of the wind stress during the second storm (see Figure 4b). The kinetic energy dissipation rate in the mixed layer, ε_{ml}, achieved 10^{-3} - 10^{-2} cm^2/s^3 for the storm periods and weakly depended on depth. In a quasi stationary state, the turbulent dissipation had a local maximum at the base of the mixed layer, which did not appear in the phase of entrainment. Storms induced buoyancy flux J_b was negative (downwards from the sea surface) in the upper part of the mixed layer and positive (upwards from the pycnocline) in the rest part of the water column. The upward flux J_b was several times larger during the wind entrainment than the flux caused by ordinary penetrating night-time convection.

Thus, three-layered density structure was formed after the last storm in the evening on October 5. It consisted of a 18 m quasi homogeneous layer, a 3.5 m pycnocline ($\Delta\sigma_\theta = 0.9$), and a near bottom weak stratified layer (Figure 1, curves 2 and 3). The measured density of the mixed layer was 0.15 units of σ_θ lower than the density resulted from the model calculations. It reflects an influence of along - shore fresh waters advection from Kamchia River. In spite of that discrepancy and taking into account complex coastal dynamics that is outside of the present model, a quite reasonable agreement between the calculated and measured profiles can be pointed out.

5. Summary

The comparison of numerical calculations with the field measurements shows that the proposed model can be used to predict short period evolution of the vertical thermohaline and turbulent structures caused by storm-induced mixing on the shallow shelf.

The thickness of the upper mixed layer with active turbulence h_* produced by wind stress action at the sea surface, is subjected to the condition $h_* = u_*/10f$. A simple approximate formula $e_{to} \approx 2.5 \cdot u^2_*$ can be introduced for a dependence between the turbulent energy at the sea surface e_{to} and wind-induced friction velocity u_*.

Turbulent entrainment from the pycnocline to the mixed layer accompanies by sharp decreasing of the kinetic energy dissipation rate near the lower boundary of the quasi homogeneous layer as for wind-induced as for convective mixing. The vertical profiles of ε(z) are characterized by a local maximum at the base of upper mixed layer when entrainment ceases.

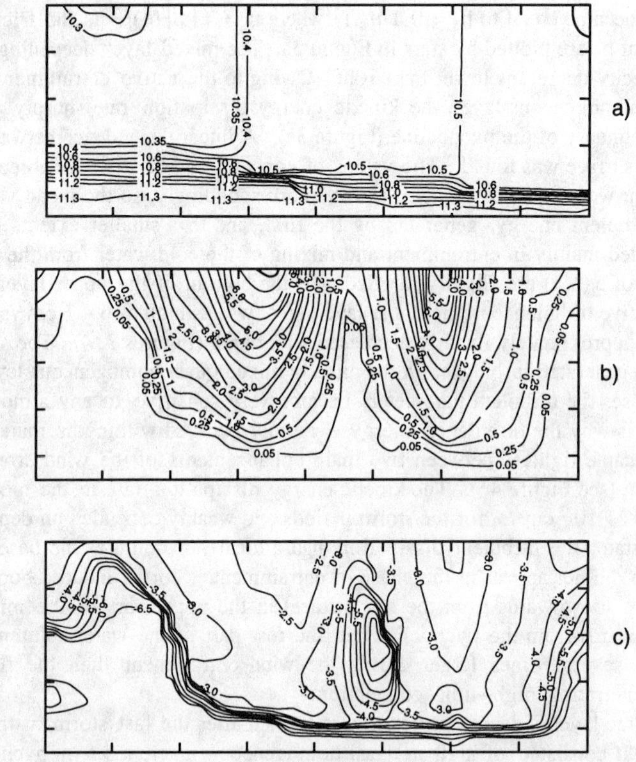

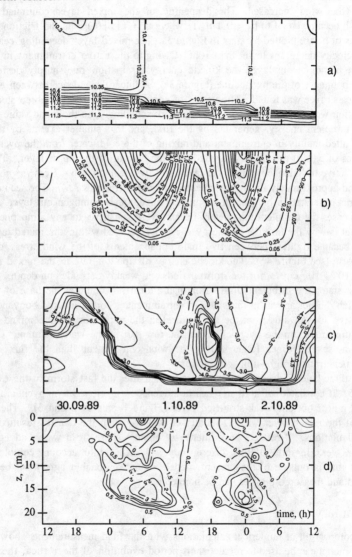

Figure 4. Storm induced variations of the mixed layer density and the turbulence parameters for two periods of moderate storm winds ($|V_a| = 13$ m/s) separated by 6 hours of wind ceasing ($|V_a| < 3.5$ m/s). The turbulence was generating in already quasi homogeneous layer. The turbulent energy and buoyancy flux in the upper layer are about five times larger compared with night convective mixing (Figure 3). The active turbulence is decaying rapidly between the periods of wind enhancements.

Acknowledgments. This work was partially funded by the Office of Naval Research, grant N00014-95-1-0786 and Russian Foundation for Fundamental Investigations, grant No. 94-05-16371. We are also indebted to the Bulgarian Academy of Sciences for support with the field measurements.

References

Arakawa, A., Computational design for long term numerical integration of the equations of fluid motion. *J. Compute. Phys. 1*, 119-143, 1966.

Blackadar, A. K., The vertical distribution of wind and turbulent exchange in neutral atmosphere, *J. Geophys. Res., 67*, 3095-3102, 1962.

Blanc, T. V., Variation of bulk derived surface flux, stability and roughness results die to the use of different transfer coefficient schemes. *J. Phys. Oceanogr., 15*, 650 -669, 1985.

Bunker, A. F., Computation of surface energy flux and annual air-sea cycle of the North Atlantic Ocean. *Mon. Weather Rev., 104*, 1122-1140, 1979.

Carrol, T. A., and R. D. Noble, Daily absorbed solar radiation at air-sea interface. *J. Environmental Systems, 11*, 289-294, 1981-1982.

Gibson, C. H., V. N. Nabatov, and R. V. Ozmidov, Measurements of turbulence and fossil turbulence near Ampere Seamount. *Dyn. Atmos. and Oceans, 19*, 175-204, 1993.

Kolmogorov, A. N., Equations of turbulent motion in non-compressible fluid, *Izvestiya of the Academy of Sciences of the USSR, ser. Physics, 6*, 56-58, 1942, (in Russian).

Ksenofontov, A. S., I. D. Lozovatsky, and R. V. Ozmidov, Diurnal variability of characteristics of the ocean upper turbulent layer, *Oceanological Res., 42*, 92-97, 1990, (in Russian).

Laikhtman, D. L., V. Ya. Rivkind, and T. A. Savelyeva, Upper ocean layer turbulence regime study by numerical methods. *Oceanology, 11*, 25-31, 1976, (in Russian).

Large, W. G., and S. Pond, Open ocean momentum flux measurements in moderate to strong winds. *J. Phys. Oceanogr., 11*, 324 -336, 1981.

Lozovatsky, I. D., and R. V. Ozmidov, Relation between small-scale turbulence characteristics and water stratification characteristics in the ocean, *Engl. Edition of Oceanology, 19*, 649-655, 1979.

Lozovatsky, I. D., and N. N. Korchashkin, On the structure of small scale turbulence in the equatorial zone of the Indian Ocean. *Oceanology, 24*, 34-41, 1984, (in Russian)

Lozovatsky, I. D., and R. V. Ozmidov, (Eds), *The variability of the hydrophysical fields in the coastal zone during autumn cooling. Data of Oceanological Study, 5*, Moscow, Nat. Geophys. Com. Publ., 212pp. (1992), (in Russian)

Lozovatsky, I. D., A. S. Ksenofontov, A. Yu. Erofeev, and C. H. Gibson, Modeling of evolution of vertical structure in the upper ocean by atmospheric forcing and intermittent turbulence in the pycnocline". *J. Mar. Systems, 4*, 263-273, 1993.

Lozovatsky, I. D., and A. Yu. Erofeev, Marine turbulence in the coastal zone. *OCEANS 94 OSATES Proceedings of Oceans engineering for today's technology and tomorrow's preservation conference*, Brest, France, *1*, 498-501, 1994.

Marchuk, G. I., *The methods of numerical mathematics*. Moscow, Nauka, 454 p., 1977, (in Russian).

Mellor, G. L., and T. Yamada, A hierarchy of turbulence closure models for planetary boundary layers. *J. Atmosph. Sci., 31*, 1791-1806, 1974.

Nabatov, V. N., and V. T. Paka, Metrology of the velocity shear pulsation measurements in the ocean, *Oceanological Res., 42*, 94-98, 1990, (in Russian).

Rotta, T.S., Statistische theorie nichthomogener turbulenze, *Z Phys., 129*, 547-572, 1951.

9

Large Inflow-Driven Vortices in Lake Constance

E. Hollan

Abstract

The formation of inflow-induced synoptic eddies in lakes is reviewed according to related general hydrodynamical treatments and compared on this basis with the conditions in Lake Constance. The possibility of generation is then estimated in terms of an idealized vortex geometry and simplified dynamics of the governing energy flux budget. In comparison with the stronger excitation by the vorticity of southeastern catabatic and northerly wind fields over the lake, the existing observations on lake-wide cyclonic motions (Figures 1 and 2) leave open, whether the main tributary, the Alpine Rhine (Figure 3), is strong enough to drive such large vortices as the maximum width of 12 km allows for. The assessment shows for a moderate discharge of 500 $m^3 s^{-1}$ that large eddies can be kept alive by the main inflow, in particular, when they become caught in a partially land-locked position. However, in the largest possible case of about 12 km diameter the response times increase so much and the corresponding maximum circular motions reduce to the low magnitude of 1 cm s^{-1} that the generation seems only realistic for events of extreme discharges of about 1,500 $m^3 s^{-1}$ and more. Thus persisting influence of outer circulations associated with the river jet is to be expected at most within a region of a few km adjacent to the inflow site.

Preliminary Remarks

The problem of synoptic eddies of lake-wide scale has been investigated in Lake Constance by observations since the 1920´s (Wasmund, 1927, 1928; Auerbach 1939). In particular, the generation of a large cyclonic gyre of maximum diameter of 12 km in the central part of Upper Lake Constance had been traced back by Elster (1938) to catabatic wind events with strong vorticity over the eastern part (Figure 1). He gave a diagnostic calculation in terms of the baroclinic geostrophical balance and explained thus the doming effect on the stratification measured just after such strong storms. For the first time Serruya et al. (1984) included the mechanism of horizontal shear in the forcing wind field as to Lake Constance and calculated also the leakage currents due to the main tributary, the Alpine Rhine, which enters the lake at its eastern end. However, this approach reflected mainly the joint effect of the earth´s rotation, bottom friction and variable topography, since the calculations were confined to vertically integrated steady two-dimensional transports in the homogeneous lake. Though conservation of potential vorticity seems to

Physical Processes in Lakes and Oceans
Coastal and Estuarine Studies Volume 54, Pages 123-136

govern certain structures of the observed leakage currents, the lack of lateral counter currents compared with the measurements from former times hints at a dynamically more complicated influence of stratification and spatially variable friction. These effects have been visualized to a certain extent by experiments with a rotating physical model of the lake (Stewart and Hollan, 1984). In particular, the river-induced onset of the large-scale cyclonic motion is qualitatively indicated in a few experiments on this subject which are documented by photographs in Stewart's (1988) final report.

The former investigators related the large rotatory motion also to the inflow of the main tributary, because its direction tangentially points toward the northern shore. Riverine water compounds they measured by water samples served as tracers of such motions and indicated the tendency of secondary lateral counter currents at the opposite shore, also when wind influence was vanishing. This interpretation is included in the map of the surface layer currents in Figure 2 drawn by Wasmund (1928) on the basis of numerous observations on the paths of drift nets.

The residence time of river water in Upper Lake Constance amounts to 4.5 years and implies at most a generally low river-induced secondary circulation. As outlined above, a satisfactory explanation is still pending and deserves to be carried on, since there is considerable significance of such persisting slow motions with respect to water management problems of the lake. However, with regard to the mean annual weather conditions over the lake, calm or weak wind periods prevail over those of pronounced wind forcing, thus rendering this question an important one. Moreover, in view of the permanent construction of the mouth of the Alpine Rhine since the early 1970's, there is a strong interest also in the remote effect of the man-made change of the site and direction of this inflow. In the following analysis a rough estimation is carried out in terms of different inherent time and spatial scales, whether the dynamical conditions exist that the main inflow can turn into a large vortex motion which occupies the greatest width of the lake.

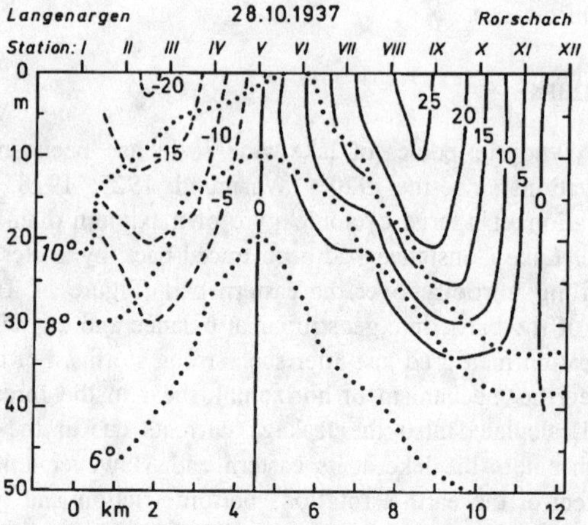

Figure 1. Transect of the temperature and current distribution in Lake Constance from Langenargen to Rorschach on 28 October 1937 just after strong Föhn-storms over the adjacent eastern region (after data from Elster, 1938). The isotherms are drawn by dotted lines in steps of 2°C, while the currents normal to the transect are represented by isotachs in steps of 5 cm s^{-1}. Solid lines mean southeastward, broken lines mean northwestward direction, resp. The currents of the cyclonic gyre have been diagnostically calculated by means of the dynamical method of physical oceanography (Defant, 1929).

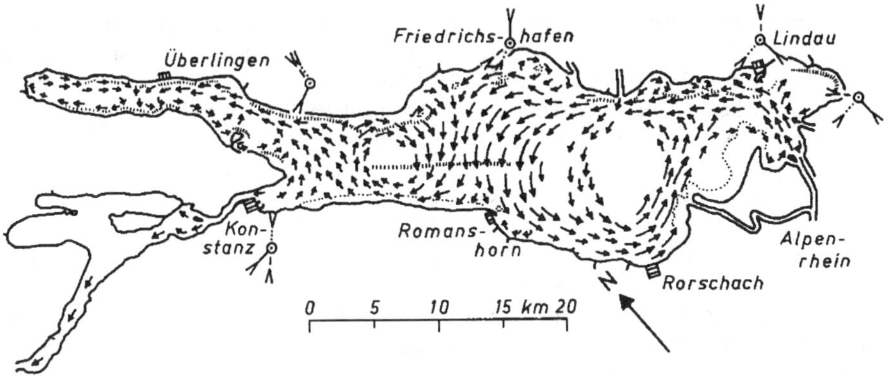

Figure 2. Map of typical surface currents in Upper Lake Constance after Wasmund (1928) according to numerous observations of paths of drift nets. The strength of the currents is given qualitatively. The annual wind frequency at five observational sites on the shore is shown as shafts by full, dotted and broken lines corresponding to most, second-most and third-most directions, resp.

Previous Results by Related General Calculations

A few theoretical treatments which include the effect of the earth's rotation shed light on the outer currents associated with inflow jets in an incompressible unstratified fluid. Despite the geometrical idealizations of these problems such as infinite or semi-infinite extent of the fluid, the results imply essential insight into the peculiar formation in the case of Lake Constance.

Savage and Sobey (1975) investigated the deflection of the path of a round jet and final development into a spiral in a rotating unstratified deep water reservoir with same density. A characteristic length scale, L, defined by $L^2 = \sqrt{\pi k J / 2\rho} / \Omega \alpha$ with J the momentum flux at the orifice, ρ the density of the fluid, $k = \ln 2$, Ω the angular frequency of the basin rotation and $\alpha = 0.095$ the width scale factor of the jet specified for 50% of the maximum centre-line velocity. Deep and shallow reservoirs are discerned by the ratio of $h/L \geq 0.21$ and ≤ 0.024, resp., where h is the mean depth of the reservoir. The evaluation of these relations for moderate high and average discharge of 500 and 230 $m^3\ s^{-1}$, resp., of the Alpine Rhine modified as a round jet into Upper Lake Constance yields L = 2,174 m and 1,680 m, resp. For h = 100 m and the orifice diameter of 24.7 and 19.0 m, resp., the classifying condition h/L results into 0.046 and 0.060, resp. If the neutral inflow is considered to be confined to the surface layer of h = 30 m depth, the ratio decreases to 0.014 and 0.018 resp. Thus Lake Constance has the character of a shallow lake as to this strong inflow and the adjustment to the Coriolis force will mainly take place by pressure redistribution and will bring about merely a slight deflection of the path, as is consistent with the observations.

The deep water solution displays certain interesting properties of the outer lateral flow directed towards the jet due to entrainment, as far as the jet has not been wrapped up by the formation of the spiral. However, the horizontal structure of the outer motion is better delineated in the more appropriate solution by Gadgil (1971) given in terms of a quasi-geostrophical two-dimensional jet entering from a vertical slit in a side-wall into the horizontally semi-infinite domain of the rotating water body bounded by rigid planes both at the bottom, z = h , and at the surface, z = 0. Due to the friction at these boundaries and at the sides of the jet, the inflow cannot penetrate beyond a certain point, X_E , on the centre-line. On the initial path the jet entrains water from the adjacent region and widens correspondingly (Figure 1 in Gadgil, 1971). This section is a kind of Schlichting jet according to Schlichting´s (1968) theory on the side-frictional jet without rotation. At $X_S < X_E$ there is a smooth transition into the second part of the jet which is dominated by bottom friction. In this regime water is laterally ejected from the jet and forms a branch of circulation away from the jet on both sides with increasing orientation normal to it as one moves away from X_S towards X_E. The bottom friction effect augments strongly by this horizontal structure of the current field and thus brings about the finite length X_E of the jet (Figure 4 in Gadgil, 1971). Considering the side circulation branches one would expect, heuristically, that in case of a finite lateral extent of the water mass, i.e. additional vertical walls in greater distance parallel to the x-direction, slow counter currents close to them would accomplish the outer circulation.

For a rough information, the inflow is assumed to occur in the surface layer with same density and same thickness, h. The lower boundary is represented by the steplike upper part of the pycnocline exerting interfacial instead of bottom friction, while the upper boundary is the free, frictionless surface. The evaluation of X_S and X_E for three different high water discharges, Q, the initial width of 200 m of the Rhine inflow and two thicknesses, h = 10 m and 20 m, yields the distances listed in Table 1. For comparison, the longitudinal extent, Z, of a planar jet without rotation and no outer circulation is added in the last two columns according to Fischer et al. (1979). This quantity is defined for a centre-line velocity of 2.5 cm s^{-1}, since the length of the jet reaches to infinity in this approach.

The X_E-values amount from 7.3 through 25.2 km and reduce merely secondarily, when the surface layer depth is doubled from 10 to 20 m. Compared to the flat initial structure of 200 m width, there is only a moderate effect of the increasing contribution by side friction. The range of the X_E-values for interfacial friction fits into the basin configuration and former observations as to the longitudinal distance between the river mouth and the region of Langenargen and Friedrichshafen, where the river plume is known to generally turn cross-lake towards the Swiss shore.

In the context of the existence of a large-scale outer circulation, a final reference to Sozou (1979) is worthwhile to be mentioned. This investigation concerns the laminar analogue of a closed lateral circulation induced by a round jet in three-dimensional infinite

TABLE 1. Longitudinal extent, X_E, of the Rhine jet in the surface layer of Upper Lake Constance due to interface and side friction (ν_v = 8 cm^2 s^{-1} and ν_h = 10^4 cm^2 s^{-1}, resp.) with rotation (after Gadgil, 1971) and for a planar jet without rotation, Z, after Fischer et al. (1979). X_S, length of the first section with horizontal entrainment according to Gadgil (1971).

Q [m^3 s^{-1}]	X_S [km]		X_E [km]		Z[v_{min}=2.5cms^{-1}] [km]	
	h=10m	h=20m	h=10m	h=20m	h=10m	h=20m
500	2.75	2.75	8.40	7.34	5.81	2.90
1000	5.50	5.51	16.80	14.68	23.23	11.62
1500	8.25	8.26	25.20	22.02	52.27	26.14

domain of an incompressible fluid without rotation, when a constant force is applied in the centre in terms of a jump function with time. Though the finite geometrical configurations, the turbulent viscosity as well as the rotation of the lake do not allow for a close comparison, the results give a hint to the possible existence and growing process of large-scale outer circulations also in respect of the two-dimensional case. The initial flow in the three-dimensional domain has the feature of a dipol with the centre at the impressed force and pointing into its direction. The closed loop is situated around the centre like a ring and migrates laterally to infinity in the linear solution establishing thereby the steady field of motion oriented on the whole in the same direction as the generating force. In earlier calculations Sozou and Pickering (1977) accounted for nonlinear terms in the governing equations and showed that the loop structure moves away laterally with a forward component as the orientation of the forcing.

If horizontal eddy viscosity of $\nu \sim 1\ m^2\ s^{-1}$ and the force corresponding to the discharge of the Alpine Rhine of 230 and 500 $m^3\ s^{-1}$ are tentatively inserted into the linear solution, similar relative conditions result as in the computed laminar cases. Desisting from any closer quantitative consideration because of the lower spatial dimension and rigid boundaries necessary for an approach to the lake, this feature of transient loop formation around the centre of generation may give an idea on corresponding structures just in question of Lake Constance.

Energetical Assessment of a River-Induced Large Vortex

Besides the effect of the pressure field which is associated with the inflow process and has been accounted for in the general solutions discussed above, there is an essential question, whether the mechanism of frictional generation by the inflow and dissipation allow for the response of such a large rotatory motion in the lake. For the purpose of an estimation, a rough approach of the complicated process is assumed as follows. The kinetic energy of the steady large vortex is considered to be balanced between the energy flux across its under water boundaries by shear of the rotational motion against the tangentially driving jet on one hand and the dissipation by the surrounding bottom and waters at rest on the other hand.

For the tractability the vortex is idealized by circular symmetrical shape and rotation like a rigid body. This latter simplification of cyclostrophic motions represents a rather rough, but physically reasonable approach, since the excitation of a vortex from its circumference and the following inward exchange of angular momentum occurs in relatively short time and is predominantly a two-dimensional process. According to Greenspan (1968) the characteristic time for this spin-up is $r/\sqrt{\Omega\nu}$, where r means the radius of the small cylindrical vessel in the laboratory experiment, Ω its forcing angular frequency and ν the kinematic viscosity of water. For large dimension of the process in nature a typical eddy viscosity coefficient has to be assumed instead of ν. Thus the vortex development and rotation variation is considered in terms of longer time-scales incorporating the spin-up process as a smaller period. Moreover, the influence of the earth's rotation is dropped in this context, since the process of excitation and dissipation by shear can be considered as dominant in terms of the search for the physical reality. A justification of this neglect is partly evident from the previous discussion and will be substantiated later by the scales of motion to be expected in Lake Constance.

The energy source of the main tributary is defined as kinetic energy flux by the pressure p that the discharge of the river exerts on the lake:

$$\frac{\Delta E}{\Delta t} = puF \tag{1}$$

where u is the mean inflow velocity of the river averaged over the area F of the cross-section at the river mouth. The pressure p is approximately related with u by the energetic equivalence $p = g\,\rho\,\Delta W = 0.5\,\rho\,u^2$, wherein g is the earth's gravity, ρ the density of water and ΔW the water level difference in the last part of the river before its mouth. In this relation the contribution of the turbulent energy flux into the lake at the river mouth has been omitted as minor contribution of roughly one magnitude smaller than that of the mean flow.

The dissipation of kinetic energy on the vortex boundary per unit time and surface amounts to

$$\frac{\Delta E}{\Delta t} = \pm \tau_b u_b \tag{2}$$

with τ_b the stress and u_b a typical tangential speed. While the plus sign designates energy flux from outside, the minus sign indicates this from inside the vortex. Introducing the marginal azimuthal speed $r \cdot \omega$ of a circular symmetric vortex like a rigid body with ω the angular frequency and r the radius to a point on its boundary, the energy flux balance of a large flat vortex, say in the surface layer of the lake with the upper side coinciding with the lake's surface, reads:

$$\frac{dE_{kin}}{dt} = \frac{1}{2}\frac{d\left(\theta\omega^2\right)}{dt} = \iint_{G_A} \tau_a (v - r\omega) d\sigma - \iint_{G_B} \tau_b r\omega d\sigma \tag{3}$$

The quantity θ represents the angular moment of the vortex and the surface integration is splitted into the range G_B of frictional loss of kinetic energy from the vortex and the range G_A of kinetic energy supply into it by the shearing contact with the inflow speed v at this site. The sum $G_A + G_B$ represents the whole underwater surface of the vortex. The left side of (3) consists of two terms, $(\omega^2/2)d\theta/dt + \omega\theta d\omega/dt$. The first one represents the effect of variable angular momentum and occurs, when the size of the vortex varies. The second one concerns the change of energy due to transient angular frequency. From the discussion in the second section it may be conceived that two such vortices will develop each on one side at the mouth of the inflow when this is starting from rest. This outer circulation of the jet will grow and be shifted to another site, prescribed by the configuration of the basin and the strength of the jet. While such a migration would explain the development of a lake-wide gyre, the local generation at the same site and for a constant size will be investigated as an alternative in the following, because it is sufficient for a first approach to estimate the conditions of the final spatial state. A proper treatment of the spatially transient case requires considerably more theoretical effort and is beyond the present scope. Thus on the left side of (3) the term containing dθ/dt is dropped and r is no longer a function of time. Introducing the quadratic law of the stress, $\tau_a = \rho\, k\, (v\text{-}r\omega)^2$ and, correspondingly, $\tau_b = \rho\, k\, r^2\, \omega^2$, into equation (3), the balance of the kinetic energy flux reads for constant size of the vortex:

$$\theta\omega\frac{d\omega}{dt} = \iint_{G_A} \rho\, k(v - r\omega)^3 d\sigma - \iint_{G_B} \rho\, k(r\,\omega)^3 d\sigma \tag{4}$$

Evaluating equation (4) for a disc-like cylindrical shape with $\theta_d = 0.5\,\rho\,\pi\,h\,r_m^4$ and distinguishing between interfacial friction at the bottom of the disc and usual bottom friction on its circumference integration on the right side yields the differential equation for the time-dependent angular frequency $\omega(t)$:

$$\frac{d\omega}{dt} = 2k_H\left[\frac{\varphi(v - r_m\omega)^3}{\pi\, r_m^3\,\omega} - \left(2 - \frac{\varphi}{\pi} + \frac{2 r_m k_i}{5hk_H}\right)\omega^2\right] \tag{5}$$

As new parameter appears the azimuthal angle φ which defines the length of the contact zone with the inflow jet on the circumference of the vortex with the radius, r_m. k_H and k_i represent the boundary friction coefficients in this zone and at the lower interface boundary, resp., due to the quadratic dependency of the stress on the velocity as applied in (4). The meaning of the terms on the right-hand side of (5) is easily discernible. While the first term in the main bracket yields an increase of momentum so long as the inflow speed v is larger than the rotation speed at the vortex rim, it may also result in a decrease, when this is vice versa. The second term represents the decrease by dissipation on the remaining boundary of the vortex. Integration of (5) yields the following solution, explicitly for t,

$$\omega_F \cdot t = \frac{1}{E}\left\{\ln\left[\frac{\sqrt{\tilde{\omega}^2 + 2q\tilde{\omega} + p}}{\sqrt{p}(1-\tilde{\omega})}\right] - \frac{q+p}{\sqrt{p-q^2}}\left(\operatorname{arctg}\left[\frac{\tilde{\omega}+p}{\sqrt{p-q^2}}\right] - \operatorname{arctg}\left[\frac{q}{\sqrt{p-q^2}}\right]\right)\right\} \tag{6}$$

where $\tilde{\omega} = \omega/\omega_0$ with $\omega_0 = (\tilde{y}_1 + c)\,\omega_F$. Herein ω_0 means the frequency of the final steady state of rotation and $\omega_F = v/r_m = \sqrt[3]{\eta}\cdot u / r_m$ is the forcing frequency established by the driving adjacent segment of the jet with its average velocity $u = Q/F$ given by the corresponding discharge Q and the cross-section area F of the jet. η means an efficiency coefficient. The remaining quantities in (6) are defined as follows:

$$\tilde{y}_1 = \sqrt[3]{c(1-c)^2} - \sqrt[3]{c^2(1-c)},\quad c = \varphi/(2\pi\, d),\ d = 1 + k_i\, r_m/(5k_H\, h),$$

$$q = (\tilde{y}_1 - 2c)/(2(\tilde{y}_1 + c)),\ p = c/(\tilde{y}_1 + c)^3 \text{ and } E = 4k_H d\left[2\tilde{y}_1 - c + c/(\tilde{y}_1 + c)^2\right].$$

A discussion of the solution is postponed to the following section, where two examples are evaluated for Upper Lake Constance. The main pattern of the time-dependency of $\tilde{\omega}(t)$ is the same, as compared with the solution derived below based on a more simplified process of energy uptake in the vortex.

Instead of the shear mechanism it is suitable to leave the supply process unspecified and to assume the energy flux from the jet into the fixed large-scale vortex as a rate in terms of a fraction η of the energy flux at the river mouth. The driving first term on the right side of (4) is free of ω in this case. The generation process will still be well approximated at low angular frequencies and will be physically reasonable as long as $r\cdot\omega$ will not exceed v. Hence, the explicit equation for $d\omega/dt$ assumes the form:

$$\frac{d\omega}{dt} = \frac{\rho\eta u^3 F}{2\theta\omega} - \frac{B}{\theta}\omega^2 \tag{7}$$

wherein B represents a geometrical factor depending on the friction coefficient k and the underwater boundary area of the vortex excluding the area of contact with the jet. The relation of k in B is linear and contains terms of k_i / k_H, if there are parts of the surface, where reduced interfacial friction occurs. For comparison with the solution (6) for a disc-like vortex in the surface layer the corresponding solution of (7) has been also calculated. With the same geometrical and physical quantities, as defined there, it reads:

$$\omega_F \cdot t = \frac{D}{12k_H\left(d-\frac{\varphi}{2\pi}\right)}\left\{ \ln\left[\frac{\sqrt{\tilde{\omega}^2+\tilde{\omega}+1}}{1-\tilde{\omega}}\right] - \sqrt{3}\left(\operatorname{arctg}\left[\frac{2\tilde{\omega}+1}{\sqrt{3}}\right]-\frac{\pi}{6}\right)\right\} \tag{8}$$

where $\tilde{\omega} = \omega/\omega_0$ with $\omega_0 = \omega_F / D$ the frequency of the final steady rotation and

$$D = \sqrt[3]{4\pi \ h \, r_m \, k_H \, (d-\varphi/(2\pi))/F} \ .$$

It has to be pointed out that the solution (8) is valid for the flux of all available energy through the size of the circumference segment defined by φ. If only a fraction of the total energy flux is provided there, because the complementary part is used up, for instance by synoptic eddies driven on the other side of the jet, then a corresponding reduction of ω_0 has to be introduced into (8). In the case of half the energy available for the large vortex, ω_0 has to be divided by $\sqrt{2}$. This modification has not to be accounted for in the solution (6), since the energy transfer is defined there as local process per unit area of the contact zone.

Application to Upper Lake Constance

At the longitude of the greatest width of about 12 km in Upper Lake Constance the maximum size of the vortex may be scaled by the radius of 6 km at the surface. The depth of the basin amounts to about 200 m in this region. The site of the vortex is indicated in Figure 3. It is still located in the eastern half of the elongated, about 60 km long basin. The distance to be passed by the inflow between the river mouth and the rim of the vortex amounts to approximately 8 km. The limit by the lake width implies that θ remains constant throughout the generation, if no less wide initial vortex geometry is admitted.

As first consideration, the magnitudes of spin-up times during the vortex formation process are of interest. From Greenspan's (1968) formula quoted in the preceding section the spin-up time T_f with respect to the forcing frequency $\omega_F = \sqrt[3]{\eta}\, u/r_m$ is given by $r_m / \sqrt{\omega_F \cdot \nu_H}$ for generation by side shear. Two examples of flat vortices situated in the surface layer are evaluated below to give an idea of the time scales. The radii assumed amount to r_m = 5,700 m and 400 m, while the thickness is the same, h = 20 m. For a discharge of the Alpine Rhine of 500 $m^3\ s^{-1}$, which represents a very moderate high water, with F = 500 m^2 and u = 1 ms^{-1}, the case of side shear may be characterized by a horizontal eddy viscocity of 1 and 0.1 $m^2\ s^{-1}$ and an efficiency coefficient η of 1 or 0.1. The corresponding spin-up times for the large gyre range from roughly 4 d in case of the stronger forcing and horizontal eddy viscosity through 23 d for the weaker conditions. Though the length parameter is certainly an overestimation, this kind of formation process proves to be a rather longterm one for this maximally possible dimension. The corresponding spin-up times of the smaller eddy range between 2.2 h and 10.3 h. Hence, medium scaled vortices seem to be favoured by river-induced generation.

In order to get a first idea on the variation the angular frequency $\omega(t)$ of a single large forced vortex displays in Lake Constance after spin-up, two examples of the solutions (6) and (8) have been evaluated for Q = 500 $m^3\ s^{-1}$. For orientation, the average and maximum discharge amount to 230 and, roughly, 2,000 $m^3 s^{-1}$, resp. As quoted above, the two disc-like vortices have been assumed in the surface layer of same thickness, h = 20 m. The large one represents the greatest possible dimension, while the small one is situated at the side of the jet near the river mouth (see Figure 3). The friction coefficients are chosen for internal friction by $k_i = 10^{-3}$ at the lower surface on the top of the thermocline and for bottom friction by $k_H = 2 \cdot 10^{-3}$ at the convex boundary. The azimuthal ranges, where the energy flux occurs at the convex boundary, have been chosen less wide for the small gyre because of the stronger curvature. In particular, the ranges amount to $\varphi = \pi/6$ and $\pi/2$. For the large gyre this parameter has been assumed greater, i.e. $\varphi = \pi/4$ and π.

Before the results are discussed in detail, a general remark on the solutions is in order concerning the parameters of the forcing in relation to the inflow, u = Q/F and the efficiency, η. They constitute the forcing frequency, $\omega_F = \sqrt[3]{\eta}\, u / r_m$, which appears as inherent quantity in solution (6) simply as factor with the time, t, and reciprocally with the normalized frequency, $\tilde{\omega}$. This is different in solution (8), though the time dependency has been expressed in the same way for the purpose of comparison. Originally, the steady state frequency, ω_0, is the characteristic quantity which includes the inflow parameters in this approach, as the transfer mechanism of energy has not been specified and ω_F is physically irrelevant in this context. However, the solution (8) has to be represented in the same coordinates as in (6) in order to show the differences. As consequence, the area of the cross-section, F, which is depending on Q, also separately appears in the quantity, D, which is a factor in the argument, $\tilde{\omega}$, and in the first term on the right side of (8). Thus the value of F = 477.3 m^2 corresponding to the discharge of 500 $m^3 s^{-1}$ has to be introduced, whereas such a specification is not necessary in the solution for the shear mechanism.

There are remarkable differences resulting from the two approaches of the forcing and the explanation yields elucidating insight. The four solutions computed for the small gyre are represented in Figure 4a. On first glance, the solutions of bulk forcing shown by solid

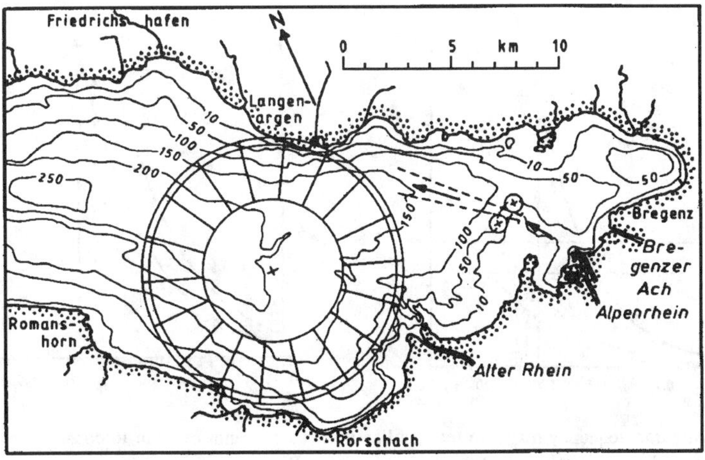

Figure 3. Map of the eastern part of Upper Lake Constance from the 1960's showing idealized circular vortices of conical shape in the central region and of disc form near the future site of the mouth of the Alpine Rhine. The flat cylindrical vortex of radius of 5,700 m considered in the text is indicated by the second largest circle in the central part. The jet of the Alpine Rhine is illustrated by hatched lines and arrows. Depths in m.

line are nearly of same variation. The maximum frequency is roughly three times greater than for the solutions resulting from shear, which are displayed by broken lines. Obviously, the energy flux prescribed by the quadratic stress law and the segment area given by φ on the convex surface of the gyre is considerably lower than in the bulk case, which is, by contrast, characterized by a constant rate of energy transmission. The difference is evident also in the shear solutions themselves. For three times greater azimuthal range, $\varphi = \pi/2$, there is finally a 25 % higher frequency than for $\varphi = \pi/6$. The response times T_h for the uptake of half the final rotational energy are additionally displayed in the diagram by vertical broken-dotted lines. In the case of generation by shear, the values grow proportional to the reduction of the final rotation rate, i.e. the final energy content of the gyre. This is related to the reduced total energy flux at the same time t for the solution with the smaller feeding window. In the approach by bulk forcing this difference is marginal and opposite, which is explained below in the context of Table 2. The corresponding results of the large vortex are represented in Figure 4b and display clearly the influence of the enlarged window size for feeding with kinetic energy. While the curve of the shear case for the smaller window, $\varphi = \pi / 2$, is always below the mostly coinciding pair of the bulk solutions, that one of the shear solution for large window, $\varphi = \pi$, is well above them. For this large size of the gyre, obviously the different formulations of the energy transfer mechanism do not result in such principal differences, as detected in the small gyre. The response times T_h vary similar to the final rate of rotation as in the preceding examples.

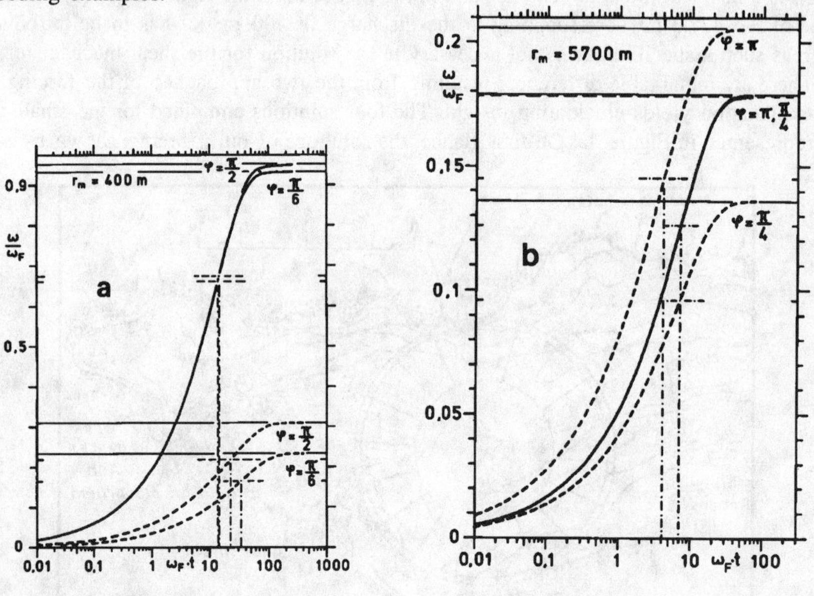

Figure 4. Angular frequency response for disc-like vortices of height of 20 m to constant forcing in the surface layer of Lake Constance by the jet of the Alpine Rhine. Solid lines: for horizontal bulk energy flux through a segment on the convex boundary defined by the corresponding azimuthal angle, φ. Broken lines: corresponding curves for energy flux by side shear on a segment of different size, φ . Vertical lines indicate half-energy response times. ω_F, forcing frequency (further explanations in the text); a) small gyre of radius of 400 m near the river mouth; b) large vortex of radius of 5,700 m (s. Figure 3).

Summarizing, the behaviour of the response due to shear means that the balance by dissipation occurs for higher rate of rotation, as the energy flux increases proportionally to the feeding window and the associated decrease of the area of dissipation. The latter influence is secondary, as the solutions for constant bulk energy flux show, where the same reduction of the area of dissipation brings about minimal increase of rotation.

As completion, the absolute values of the response times have been computed. For this reason, ω_F has to be specified by the forcing parameters, as quoted initially. The response times for the shear-induced and the bulk flux solutions are designated by $T_{h,s}$ and $T_{h,0}$, resp., and listed in Table 2 for $\eta = 1$. Values for lower efficiency are easily obtained by dividing by $\sqrt[3]{\eta}$, since the corresponding forcing frequencies ω_F have to be multiplied by this number. The response times $T_{h,0}$ listed in the last column of Table 2 have been determined unlike to the full energy transfer into a single vortex, instead, for half of this flux in order to account for the loss by a second vortex on the other side of the jet. Furthermore, the typical inflow velocity of u = 1.05 m s^{-1} resulting from the given values of Q and F has been reduced to half in a second evaluation for the response times of the shear cases thus accounting for a more realistic scale along the length of the feeding window in the vortex. These values of $T_{h,s}$ are listed in the third column for the parameter u = 0.52 m s^{-1}.

As expected, the shortest response time of ~1.8 h occurs for the bulk flux solution of the small gyre. The remaining values illustrate the physical interpretation given above in general terms. The response times of the small gyre after spin-up last up to 7 h for the shear case, while the corresponding maximum value for the large vortex is 21 h. The response for bulk energy flux amounts to 13.5 h in the latter case.

A detail of the T_h-values in Table 2 is interesting with respect to the different forcing mechanisms. The response times vary in opposite sense, when compared pairwise for increasing feed window in each vortex. Strong reduction in the shear cases is contrasted by slight increase in the bulk case. This behaviour is related to the different time history of the forcing energy flux. Due to the shear law this energy flux permanently reduces with increasing rate of rotation. Thus the energy uptake is much stronger in the first phase of reaction yielding an higher stage of kinetic energy earlier than in the other approach. By contrast, the latter process requires a little longer time during its constant rate of energy supply to arrive at the higher level, even if it is only a bit on the course to the slightly increased steady rotation in the end.

The calculated T_h-values become very significant, when they are converted to lower forcing velocities, v. Since ω_0 is proportional to ω_F, the response time depends

TABLE 2. Response times for half the maximum energy take up by the disc-like vortices as indicated in Figure 4 by vertical lines. $T_{h,s}$ for generation by shear and two different forcing velocities, u , by the inflow. $T_{h,0}$ for generation by half the maximum bulk energy flux on account of a second vortex driven on the other side of the jet. Efficiency in all cases $\eta = 1$.

r_m [m]	φ	$T_{h,s}$ [h] u = 0.52 [m s^{-1}]	$T_{h,s}$ [h] u = 1.05 [m s^{-1}]	$T_{h,0}$ [h] u = 1.05 [m s^{-1}]
400	$\pi/6$	7.09	3.55	1.82
	$\pi/2$	4.67	2.34	1.89
5700	$\pi/4$	20.87	10.43	13.45
	π	12.41	6.21	13.56

reciprocally on $v = \sqrt[3]{\eta} \cdot u$ and increases proportional to r_m. For instance, if v = 0.1 u with u = 1.05 m s^{-1} from Table 2, a more appropriate situation at the horizontal boundary of the large vortex would be given. The response times increase then by a factor of 10 and yield values of about 2.5 and more than 4 days for the excitation by shear, whereas the corresponding reaction by bulk energy flux needs about 5.5 days. These numbers make evident that the response is very gradual and may start only for more pronounced discharges or eddies of smaller size. This is elucidated by the example of the smaller gyre, when v = 0.1 u. The response time in the shear case amounts to roughly 1 and 1.5 d, resp., and rises to about 0.75d when the generation is by bulk flux. However, the long response times by the weak forcing considered refer to a reduction of the energy flux by $\eta = 1^0/_{00}$, which represents conditions unlikely for this kind of generation, at least when the large vortex and the moderate inflow are concerned.

Finally, the scales of the azimuthal velocities will be examined. For this purpose, the velocities, v_m, at the rim of each vortex have been evaluated for response times and listed in Table 3 accounting for different efficiencies. Very slow motions of v ~ 1 cm s^{-1} result for $\eta = 1^0/_{00}$ and show that this forcing by moderate discharge is extraordinarily weak and may hardly give rise to the largest outer circulation of the Alpine Rhine jet. However, it has to be mentioned that any large-scale motion which is induced by the pressure field associated with the inflow has been neglected. On the other hand, the possible generation of a second large-scale vortex below in the deep hypolimnion by virtue of the vorticity applied to the interface would delay the development of the directly forced vortex. Thus, the whole system requires stronger forcing to arrive at the same stage of rotation in the surface and is more probably to be expected on smaller horizontal scales in Lake Constance. As a further consequence from the low cyclonic azimuthal velocities, the doming due to the geostrophic balance is small compared to the assumed depth of the surface layer. Therefore it is subsequently justified to disregard the Coriolis effect.

The situation is different, when wind forcing with strong horizontal vorticity occurs over the lake, as observed by Elster (1938). Further interesting calculations, which concern the combined response of a surface and lower layer vortex, have been carried out by similar method and will be left to this latter subject of investigation for the sake of conciseness herein. It is worth mentioning that hydraulic experiments may serve a good deal to prove the theoretical estimation given above, but were beyond the scope of this work. Recent synoptic CTD-measurements from Lake Constance are presently evaluated in this context and will improve the knowledge on the structure and origin of lake-wide or partially land-locked eddies in this lake.

TABLE 3. Azimuthal velocity v_s and v_0 at the vortex rim for the response times $T_{h,s}$ and $T_{h,0}$ given in Table 2 for the examples of shear and bulk energy flux, resp. The mean inflow speed is u = 1.05 m s^{-1}; the low efficiency $\eta = 0.001$ corresponds to the long response times discussed in the text.

r_m	φ	$v_s(T_{h,s})$ [cm s^{-1}]			$v_0(T_{h,0})$ [cm s^{-1}]		
[m]		$\eta = 0.001$	$\eta = 0.1$	$\eta = 1$	$\eta = 0.001$	$\eta = 0.1$	$\eta = 1$
400	$\pi/6$	1.7	8.0	17.3	5.5	25.5	54.9
	$\pi/2$	2.3	10.7	23.0	5.6	26.0	56.0
5700	$\pi/4$	1.0	4.7	10.0	1.05	4.87	10.48
	π	1.5	7.1	15.2	1.05	4.89	10.53

Acknowledgements. The assistance with some evaluations by J. Kettner and P. Kirner as well as the final typing by K. Weih is gratefully acknowledged.

References

Auerbach, M., Die Oberflächen- und Tiefenströme im Bodensee, *Dt. Wasserwirtschaft*, 34, 193-202, and 358-366, 1939.

Defant, A., *Dynamische Ozeanographie*.- Naturwiss. Monographien u. Lehrbücher, Vol. 9, Einführg. in d. Geophysik III, 222 pp., Springer, Berlin, 1929.

Elster, H.-J., Einige Beobachtungen über das Verhalten der oberen Wasserschichten des Bodensees (Obersee), Schrr. VG Bodensee, No. 65, 1-34, 1938.

Fischer, H.B., E.J. List, R.C.Y. Koh, J. Imberger and N.H. Brooks, *Mixing in Inland and Coastal Waters*, 483pp., Academic Press, New York, 1979.

Gadgil, S., Structure of jets in rotating systems, *J. Fluid Mech.*, *47*, 417-436, 1971.

Greenspan, H.P., *The Theory of Rotating Fluids*., 327 pp., University Press, Cambridge, 1968.

Savage, B.S. and R.J. Sobey, Horizontal momentum jets in rotating basins, *J.Fluid Mech.*, *71*, 755-768, 1975.

Schlichting, H., *Boundary Layer Theory*, McGraw Hill, 1968.

Serruya, S., E. Hollan and B. Bitsch, Steady winter circulations in Lakes Constance and Kinneret driven by wind and main tributaries, *Arch. Hydrobiol., Suppl. 70*, 33-110, 1984.

Sozou, C., Development of the flow field of a point force in an infinite fluid, *J. Fluid Mech.*, *91*, 541-546, 1979.

Sozou, C. and W.M. Pickering, The round laminar jet: the development of the flow field, *J. Fluid Mech.*, *80*, 673-683, 1977.

Stewart, K.M., Tracing Inflows in a Physical Model of Lake Constance, *J. Great Lakes Res.*, *14*, 466-478, 1988.

Stewart, K.M. and E. Hollan, Physical model study of Lake Constance, *Schweiz. Z. Hydrol.*, *46*, 5-40, 1984.

Wasmund, E., Die Strömungen im Bodensee, Int. Rev. d. gesamten *Hydrobiol. u. Hydrographie*, *18*, 84-114 and 231-260, 1927.

Wasmund, E., Die Strömungen im Bodensee, Int. Rev. d. gesamten *Hydrobiol. u. Hydrographie* *19*, 21-155, 1928.

10

Forced Motion Response in Enclosed Lakes

K. Hutter, G. Bauer, Y. Wang and P. Güting

Abstract

We report experiences gained with two different models that allow computation of the wind induced barotropic and baroclinic motion in lakes. These are: Nonlinear dispersive two-layered wave model for finite depth fluids on the rotating Earth; Semi-spectral primitive equation model (SPEM) with implicit integration in time. The first is used to demonstrate that nonlinearities affect the fine structure of the thermocline response when the wind input is strong. This is shown by means of computations with a rectangular basin of constant depth and with Lake Constance. The implicit version of SPEM is used to study the Ekman problem and the baroclinic response in a rectangle and in Lake Constance. Both models concentrate upon the role played by diffusion in predicting the current and temperature distribution that is established in lakes under direct wind forcing.

Introduction

The most popular and probably also most successful tool for the interpretation of observations collected with moored instruments in enclosed lakes has so far been the theoretical-numerical analysis of barotropic and baroclinic linear waves by use of the one- and two-layer free shallow water equations; this has led to the identification (i) of the periods of Kelvin and Poincaré type eigenmodes of the gravity waves and (ii) very nearly also of topographic Rossby waves or higher order baroclinic gravity modes, which require a three- or many-layer approximation of the stratification; for a review, see e.g. Hutter (1993). Comparisons of measured time series of velocities and/or isotherm depths at selected positions in a (stratified) lake with corresponding time series deduced from the linear shallow water equations subject to direct wind forcing are much less frequent despite the existence of a formal analytical theory of forced oceanic waves (Fennel and Lass, 1989) and wind induced currents (Heaps, 1984; Kielmann and Simons, 1984; Simons, 1980; Csanady, 1984). The reasons are that (i) thermocline excursions are often large, i.e., a considerable fraction of the epilimnion depth and (ii) vertical diffusion of momentum (and energy) is not negligible when the current distribution is to be predicted. Thus, the linear formulations offer no advantage and are insufficient.

Nonlinear models have been developed on two levels: 1) two-layered large amplitude wave models for finite depth fluids and 2) full scale numerical implementations of the

Physical Processes in Lakes and Oceans
Coastal and Estuarine Studies Volume 54, Pages 137-166

Boussinesq-approximated shallow water equations. The former may give an adequate description only of the motion of the thermocline. Their advantage is that they have the least built-in dissipation, so persisting motions can be reliably followed for a reasonably long time; their simultaneous disadvantage is that they are prone to numerical instabilities. However, we shall demonstrate how much richer the response is to direct wind forcing of such a nonlinear model in comparison to its linear counterpart.

The numerical adaptation of the full scale shallow water equations, as popularized by oceanographers, is almost exclusively based on multi-layered time-explicit finite-difference (FD) approximations. In two detailed comparisons of the measured and computed storm-induced baroclinic motion in Lake Zurich (Hutter, 1983) and Lake Lugano (Hutter, 1991), in which isotherm-depth-time series were constructed, the following weaknesses became apparent: Numerical stability required excessively large Austausch coefficients, far bigger than physically suggested. Only four to five days after a heavy storm could reliably be reproduced by the model. Dissipation was too large and computed amplitude decay of the isotherm-depth-time series in the metalimnion too rapid. It was, and probably still is, unthinkable to reproduce a one-month episode, say, of a typical meteorologically forced summer situation. To this end the total diffusion of the numerical code must be substantially reduced and allowable time steps must be increased to make long time computations (over one to several months) economically feasible.

To bring down the required numerical diffusion to physically reasonable values time-implicit integrations are required. With such a model one may then show that the current distribution within the lake strongly depends on the numerical values and vertical distribution of the vertical turbulent diffusion coefficients. This becomes particularly manifest in the Ekman problem of a homogeneous lake. However, there is very little known about the appropriate functional selection of the diffusivities of momentum and energy, except perhaps that a Richardson-number dependence is suggested. One is necessarily led to introduce a higher order closure procedure in which the evolution of the turbulent Reynolds stresses is sought along with the other field variables of the problem. To our knowledge, this is still an open problem left largely untouched in the limnological literature.

The stability performance of a numerical code solving the shallow water equations depends also on the choice of continuity by which the lake shore is approximated. Equally sensitive are such codes to the form of the discretization of the field variables in the vertical direction. These points are quite obvious: A step-like approximation of the shore line introduces a multitude of edge boundary points, each a singular source from which information propagates inward prone to instabilities. On the other hand, if the choice of vertical diffusivities is so crucial, its contribution from discretizing the field equations should be kept minimal, which makes mandatory the vertical discretization of the field variables in as smooth a fashion as possible. This suggests to use semi-spectral models, i.e., function expansions of the field variables in the vertical and curvilinear coordinates in the horizontal directions such that the lake shore can be approximated by piecewise C1-continuous functions. Finite element representation would be an equivalent choice.

In this paper we shall report experiences gained with the following models:

- Nonlinear two-layered wave model for finite depth fluids that incorporates nonlinear advection, amplitude dispersion, gyroscopic effects due to the rotation of the Earth and variable depth (Diebels, Schuster Hutter, 1994; Bauer, Diebels Hutter, 1994; Lin, 1995).
- A semi-spectral primitive equation model, using a Chebyshev-Polynomial expansion in the vertical and orthogonal curvilinear coordinates in the horizontal directions. This

model, called SPEM was developed by Haidvogel, Wilkins and Young (1991) and has been extended/altered by us as follows: We use implicit temporal integration in the vertical and prescribe explicitly the vertical functional dependence of the vertical turbulent Austausch coefficient. This is a time-implicit zeroth order turbulent closure version of SPEM (Wang, 1995).

We shall demonstrate that the nonlinear two-layered wave model exhibits very small numerical diffusion so that initiated oscillations persist for very long times, making the model a likely candidate for computations of the meteorologically induced motion over several weeks (and perhaps months). Computations performed with the semi-implicit SPEM model, on the other hand, show that full scale three dimensional numerical models can operate only with relatively large numerical diffusion. To bring this diffusion down to a level that inertial, Kelvin and Poincaré waves (which are physically observable) are not damped out before they have practically developed, an excessively high numerical resolution is required. This makes realistic computations of wind induced current and temperature distributions on workstations virtually impossible and thus requires main frame (parallel) processing. This is economically expensive. In view of the significance of such computations for water quality measures this is unfortunate.

Nonlinear Finite Depth Two-layer Model with Amplitude Dispersion on the f-Plane

Preliminary Remarks

Models of large amplitude external and internal waves in fluid systems with a free surface have been available for many years. In the nondimensional finite-depth water-wave equations two non-dimensional parameters appear

$$\alpha_0 = A_0 / H, \qquad \varepsilon = H / \lambda, \tag{1}$$

where A_0 and λ are respectively the wave amplitude and horizontal length scales, and H is the constant quiescent water depth. Nonlinearity (wave steepening) and linear phase dispersion are measured by α_0 and ε respectively, and the Ursell number

$$U = \frac{3\alpha_0}{\varepsilon^2} = \frac{3A_0\lambda^2}{H^3} \tag{2}$$

gives the relative significance of these two effects. The shallow water regime applies to $\varepsilon << 1$; when the amplitudes are also small ($\alpha_0 << 1$), the governing equations can be solved via a perturbation expansion in α_0 and ε. To lowest order and in only one spatial dimension one obtains the familiar linear, nondispersive long wave equation, $\zeta_{tt} - c_0{}^2\zeta_{xx} = 0$, $\sqrt{gH}$ where c_0 = is the shallow water speed and ζ the surface displacement. To the next order, one finds, for waves travelling in the positive x-direction, the celebrated Korteweg-de Vries (KdV-) equation (in dimensional form)

$$\zeta_t + c_0\zeta_x + c_1\zeta\zeta_x + c_2\zeta_{xxx} = 0, \quad c_0 = \sqrt{gH}, \quad c_1 = 3c_0/2H, \quad c_2 = c_0H^2/6. \tag{3}$$

It possesses knoidal and solitary wave solutions; solitons are for and amplitude persistent and essentially preserved when two solitary waves interact. [This equation goes back to the doctoral thesis of de Vries (1894), see also Korteweg and de Vries (1895). For an historical review see Miles (1981) and for its mathematical properties Miura (1976). For an historical account of the development of the nonlinear equation see also Sander and Hutter (1991)]. The solitary wave solution of (3) was observed over a century ago by Russell (1844). We further mention that the dispersion term stems from the non-hydrostatic terms in the vertical component of the momentum balance and that ignoring it ($c_2 = 0$) will necessarily lead to wave steepening and eventually wave breaking, i.e instability.

For $c_2 = 0$ the general solution of (3) is

$$\zeta = F(X - t), \qquad X = \int_{x_0}^{x} (c_0 + c_1\zeta(\sigma))^{-1} d\sigma , \tag{4}$$

where $F(\cdot)$ is a differentiable function that can be obtained from the initial data $\zeta_0(x)$ at $t = 0$

$$\zeta_0(x) = F\left(\int_{x_0}^{x} (c_0 + c_1\zeta_0(\sigma))^{-1} d\sigma\right) . \tag{5}$$

Its speeds are $c_0 + c_1(\sigma)$, so points of F with large $c_1\zeta$ travel faster than points with small $c_1\zeta$, necessarily yielding wave steepening of any initial profile, for which $c_1\zeta > 0$.

Mysak (1984) reviewed and classified the unidirectional nonlinear internal wave theories in a finite constant depth fluid. To state the results (Benjamin, 1966), let λ, H, h_1 and h_2 be a horizontal length scale (wave length), total fluid depth and the epilimnion and hypolimnion depths respectively, (Figure 1). The shallow water theory is then classified as the limit $\lambda/H >> 1$, h_1/H, $h_2/H \leq O(1)$. A two-layer fluid with a rigid lid assumption incorporated again yields the KdV-equation as governing equation for disturbances of the interfacial surface ζ travelling in the positive x-direction, but now the coefficients c_0, c_1 and c_2 take the forms

$$c_o = \sqrt{g\frac{\Delta\rho}{\rho_2}\frac{h_1 h_2}{H}} , \quad c_1 = -3c_0\frac{h_2 - h_1}{h_1 h_2} , \quad c_2 = \frac{1}{6}\frac{c_0}{h_1 h_2} . \tag{6}$$

Note that for $h_1 = h_2$, $c_1 = 0$, the linear dispersive wave equation emerges that does not possess solitary wave solutions. Note, moreover, that $h_2 > [<]\ h_1$ implies $c_1 < [>]\ 0$ and invariance requirements of (3) dictate solitary wave solutions which are a depression [and a hump, respectively].

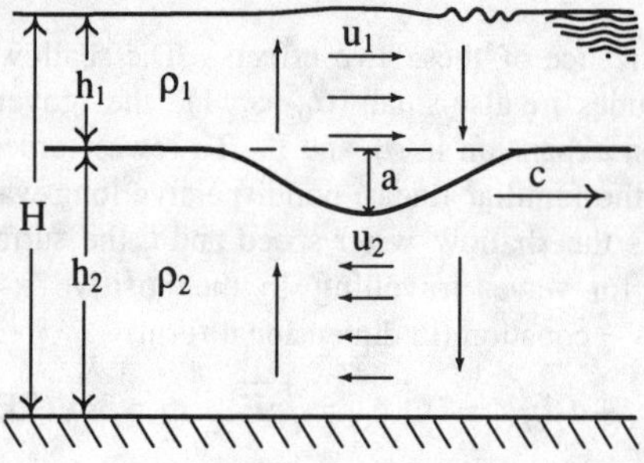

Figure 1. Internal solitary wave in a two-layer fluid with $h_2 > h_1$. Arrows indicate current pattern within the internal wave, resulting in the surface rip which leads the wave. When $h_2 < h_1$, the solitary wave is a wave of elevation, rather than a depression wave.

Model Equations and their Discretization

The above model has been generalized to also incorporate the second space direction, the rotation of the Earth and variable depth (Bauer, Diebels and Hutter, 1994). For completeness we should also mention that Beckers (1989) in a dissertation at the Federal Institute of Technology in Lausanne (EPFL) derived a two-layer model from the shallow water approximation of the Euler equations. In view of our previous remarks, that model must necessarily yield wave steepening and thus eventually become unstable. This has been corroborated by our own computations of wind induced waves in Lake Constance for which the nonlinear computer program of Beckers always failed but ran perfectly in its linear version. Nonlinear wave models for external and internal waves due to wind forcing were also derived from the full Euler equations on the rotating globe by Melville et al, (1984), Ostrovsky and Stepanyants (1990). In the equations of these authors the second Coriolis parameter $f = 2\Omega\cos\sigma$ (Ω = angular velocity of the Earth, σ latitude angle) does not arise. A priori, this is not justified. The equations in dimensionless form, read

$$\frac{\partial}{\partial x}\left(\left(\alpha_0\zeta_0 + h_1 - \alpha_1\zeta_1\right)\bar{u}_1\right) + \frac{\partial}{\partial y}\left(\left(\alpha_0\zeta_0 + h_1 - \alpha_1\zeta_1\right)\bar{v}_1\right) + \frac{\partial}{\partial t}\left(\frac{\alpha_0}{\alpha_1}\zeta_0 - \zeta_1\right) = 0, \quad (7)$$

$$\begin{aligned} &\frac{\partial \bar{u}_1}{\partial t} + \alpha_1\left(\frac{\partial \bar{u}_1}{\partial x}\bar{u}_1 + \frac{\partial \bar{u}_1}{\partial x}\bar{v}_1\right) - \frac{f}{R}\bar{v}_1 + \frac{\alpha_0}{\alpha_1}\frac{\partial \zeta_0}{\partial_x} \\ &+\varepsilon^2\left\{\frac{h_1}{2}\frac{\partial^3\zeta_1}{\partial x\partial t^2} - \frac{h_1{}^2}{3}\left(\frac{\partial^3\bar{u}_1}{\partial x^2\partial t} + \frac{\partial^3\bar{v}_1}{\partial x\partial y\partial t}\right)\right\} \\ &-\frac{\varepsilon\tilde{f}}{R}\left\{\frac{h_1}{2}\left(\frac{\partial^3\bar{u}_1}{\partial x} + \frac{\partial^3\bar{v}_1}{\partial y}\right) - \frac{\partial\zeta_1}{\partial t} + \frac{h_1}{2}\frac{\partial \bar{u}_1}{\partial x}\right\} - \tilde{T}_z = 0 \end{aligned} \quad (8)$$

$$\begin{aligned} &\frac{\partial \bar{v}_1}{\partial t} + \alpha_1\left(\frac{\partial \bar{v}_1}{\partial x}\bar{u}_1 + \frac{\partial \bar{v}_1}{\partial y}\bar{v}_1\right) + \frac{f}{R}\bar{u}_1 + \frac{\alpha_0}{\alpha_1}\frac{\partial \zeta_0}{\partial_\lambda} \\ &+\varepsilon^2\left\{\frac{h_1}{2}\frac{\partial^3\zeta_1}{\partial y\partial t^2} - \frac{h_1{}^2}{3}\left(\frac{\partial^3\bar{u}_1}{\partial x\partial y\partial t} + \frac{\partial^3\bar{v}_1}{\partial y^2\partial t}\right)\right\} \\ &-\frac{\varepsilon\tilde{f}}{R}\left\{\frac{h_1}{2}\frac{\partial \bar{u}_1}{\partial y}\right\} - \tilde{T}_y = 0 \end{aligned} \quad (9)$$

$$\frac{\partial}{\partial x}\left(\left(h_2 + \alpha_1\zeta_1\right)\bar{u}_2\right) + \frac{\partial}{\partial y}\left(\left(h_2 + \alpha_1\zeta_1\right)\bar{v}_2\right) + \frac{\partial}{\partial t}\zeta_1 = 0 \quad (10)$$

$$\begin{aligned} &\frac{\partial \bar{u}_2}{\partial t} + \alpha_1\left(\frac{\partial \bar{u}_2}{\partial x}\bar{u}_2 + \frac{\partial \bar{u}_2}{\partial y}\bar{v}_2\right) + (1-\delta)\frac{\partial\zeta_1}{\partial x} + \delta\frac{\alpha_0}{\alpha_1}\frac{\partial\zeta_0}{\partial x} - \frac{f}{R}\bar{v}_2 \\ &+\varepsilon^2\left\{\delta h_1\left(\frac{\partial^3\zeta_1}{\partial x\partial t^2} - \frac{h_1{}^2}{2}\left(\frac{\partial^3\bar{u}_1}{\partial x^2\partial t} + \frac{\partial^3\bar{v}_1}{\partial x\partial y\partial t}\right)\right)\right. \\ &\left.-\frac{h_2}{2}\left(\frac{\partial^3}{\partial x^2\partial t}(h_2\bar{u}_2) + \frac{\partial^3}{\partial x\partial y\partial t}(h_2\bar{v}_2)\right) + \frac{h^2{}_2}{6}\left(\frac{\partial^3\bar{u}_2}{\partial x^2\partial t} + \frac{\partial^3\bar{v}_2}{\partial x\partial y\partial t}\right)\right\} \end{aligned}$$

$$+\frac{\varepsilon\tilde{f}}{R}\left\{-\delta h_1\frac{\partial \bar{u}_1}{\partial x}+\frac{h_2}{2}\frac{\partial \bar{v}_2}{\partial y}-\left(\frac{\partial h_2\bar{u}_2}{\partial x}+\frac{\partial h_2\bar{v}_2}{\partial y}\right)\right\}=0, \tag{11}$$

$$\frac{\partial \bar{v}_2}{\partial t}+\alpha_1\left(\frac{\partial \bar{v}_2}{\partial x}\bar{u}_2+\frac{\partial \bar{v}_2}{\partial y}\bar{v}_2\right)+(1-\delta)\frac{\partial \zeta_1}{\partial y}+\delta\frac{\alpha_0}{\alpha_1}\frac{\partial \zeta_0}{\partial y}+\frac{f}{R}\bar{u}_2$$

$$\varepsilon^2\left\{\delta h_1\left(\frac{\partial^3\zeta_1}{\partial y\partial t^2}-\frac{h_1}{2}\left(\frac{\partial^3\bar{u}_1}{\partial x\partial y\partial t}+\frac{\partial^3\bar{v}_1}{\partial y^2\partial t}\right)\right)\right.$$

$$\left.-\frac{h_2}{2}\left(\frac{\partial^3}{\partial x\partial y\partial t}(h_2\bar{u}_2)+\frac{\partial^3}{\partial y^2\partial t}(h_2\bar{v}_2)\right)+\frac{h_2{}^2}{6}\left(\frac{\partial^3\bar{u}_2}{\partial x\partial y\partial t}+\frac{\partial^3\bar{v}_2}{\partial y^2\partial t}\right)\right\}$$

$$-\frac{\varepsilon\tilde{f}}{R}\left\{\delta h_1\frac{\partial \bar{u}_1}{\partial y}+\frac{h_2}{2}\frac{\partial \bar{u}_2}{\partial y}\right\}=0. \tag{12}$$

Here, α_0 and ε are defined in (1) and

$$\alpha_1 = A_1/H, \qquad \text{with } \alpha_0 < A_1 \ll 1 \tag{13}$$

is the dimensionless amplitude scale of the interface, and $\delta = \rho_1/\rho_2$ is the density ratio of the two layers. R is the *Rossby number*, a dimensionless form of the *Rossby radius* and given by $R = \sqrt{gH}/\Omega\lambda$. Moreover, ζ_0, ζ_1, $\bar{u}_1$, $\bar{v}_1$, $\bar{u}_2$, $\bar{v}_2$ are defined in Table 1. The equations are correct to $O(\varepsilon\alpha_1)$, not $O(\varepsilon^2\alpha_1)$ because of the occurrence of $\bar{f}$. No rigid lid assumption has been invoked.

TABLE 1. Dimensionless variables arising in equations (7)–(12), their names and the scales by which they have been nondimensionalized.

Dimensionless variables	Name	Scale
ζ_0	surface elevation	A_0, typical surface amplitude
ζ_1	interface displacement	A_1, typical amplitude of interface
h_1, h_2	upper and low layer depths	H, vertical characteristic length
$\bar{u}_1$, $\bar{v}_1$	mean horizontal velocity components in top layer	$\alpha_1\sqrt{gH}$, typical barotropic velocity with amplitude dependence
$\bar{u}_2$, $\bar{v}_2$	mean horizontal velocity components in bottom layer	$\alpha_1\sqrt{gH}$, typical barotropic velocity with amplitude dependence
$\tilde{\tau}_x$, $\tilde{\tau}_y$	components of wind-stress resp. layer acceleration	$(\alpha_1 gH/\lambda)$
t	time	$\lambda/\sqrt{gH}$, typical time for barotropic transport
$f = 2\Omega\sin\phi$, $\tilde{f} = 2\Omega\cos\phi$	Coriolis parameters	Ω, angular velocity of the earth

These equations were discretized by FD with a spatial discretization on a staggered Arakawa-grid (see Bauer et al. (1994) or Arakawa et al. (1977)). All linear terms were discretized by implicit time differences while the nonlinear terms are taken as explicit. The expression $\partial/\partial x((\alpha_0\zeta_0 + h_1)\, \bar{u}_1)$, for instance, is discretized as

$$\left\{\frac{h_1}{dx}\left((\bar{u}_1)_{i,j} - (\bar{u}_1)_{i-1,j}\right)\right\}^{n+1} + \frac{\alpha_0}{2dx}\left\{\left(\zeta_{0_{i,j}} + \zeta_{0_{i,j}}\right)(\bar{u}_1)_{i,j} - \left(\zeta_{0-1,j} + \zeta_{0_{i,j}}\right)(\bar{u}_1)_{i-1,j}\right\}^{n},$$

where n indicates time and i, j space. This stabilized the computations to the extent that in ideal basins with rectangular geometry no numerical diffusion is needed even when barotropic and baroclinic modes interact. However, for real lakes additional numerical diffusion was required. This diffusion was implemented by performing a simple smoothing if necessary after every hundred time steps (for details see Bauer (1993)).

A Rectangular Basin with Constant Depth Subject to Impulsive Wind

A rectangular basin of 17.0 km length and 7.0 km[1] width, 15 m and 85 m epilimnion- and hypolimnion depths and $\Delta\rho/\rho = 1 \times 10^{-3}$ density difference was subjected to a spatially uniform Heaviside-type wind-stress in time,

$$\tau(x, y, t) = \sqrt{\tau_x^{\,2}(x, y, t) + \tau_y^{\,2}(x, y, t)} = \tau^0 H(t) \tag{14}$$

with $\tau^0\, \rho_1 h_1 = \tau_x\, \rho_1 h_1 = 0.014$ Nm^{-2} for weak and $= 0.059$ Nm^{-2} for strong wind. The discretization was $\Delta x = \Delta y = 500$ m, $\Delta t = 60$ s, but computations were also performed in isolated cases with a grid of $\Delta x = \Delta y = 250$ m sidelength without visible changes in the results. A value of 47 degrees was used for the latitude ϕ. The internal shallow water wave speed in this case is $c_0 = 0.35$ ms^{-1} amounting to the fundamental seiche period of 26.7 h.

Excerpts of results covering a period of 30 hours from a computer animation are shown in Figure 2, but the steady state setup is only reached after approximately 4000 hours (!) as can be inferred from Figure 4 showing the interface-elevation-time series at the location shown in the inset. The shape of the linear response is insensitive to the wind amplitude. In Figure 2 (panels a) it is – besides the global factor 4 – nearly the same as that of the nonlinear response to weak winds (panels b), which is also 4 times weaker than the amplitudes for the strong wind and the nonlinear computation (panels c). We would like to emphasize the close similarity (of the signals shape not of its size) of the graphs in panels a) and b) as opposed to those of panels a) and c), which indicates a far richer small scale response in the case of strong forcing. These small scale features are likely not due to numerical instabilities because they are equally reproduced by the finer grid computations.

In fact it is those small scale features that bring out the nonlinearities often observed in measured isotherm-depth-time series. Figure 3 shows the 8°C-, 10°C-, 12°C- and 14°C-isotherm-depth-time series in the stratified lake Zürich obtained from measurements (solid curves) and calculations (using an older 3D finite difference model – not the SPEM described later in this paper – of the nonlinear shallow water equations, dotted) at the indicated stations 4, 9 and 10. Apart from demonstrating the relatively poor coincidence between the measured and computed results, we wish to point out the many short-periodic oscillations visible in the measured time-series but not discernible in the computed counterparts due to the high numerical diffusion, which had to be added for the purpose of stabilization. These small scale features are typical of nonlinear steepening effects due to advection and grow in importance compared to the dispersive terms with growing wind forcing. Such an interpretation of the results displayed in Figure 2 became particularly convincing in a video-film displaying the interface motion for the wind forcing of Figure 2 and has also been shown by Bauer et al. (1994).

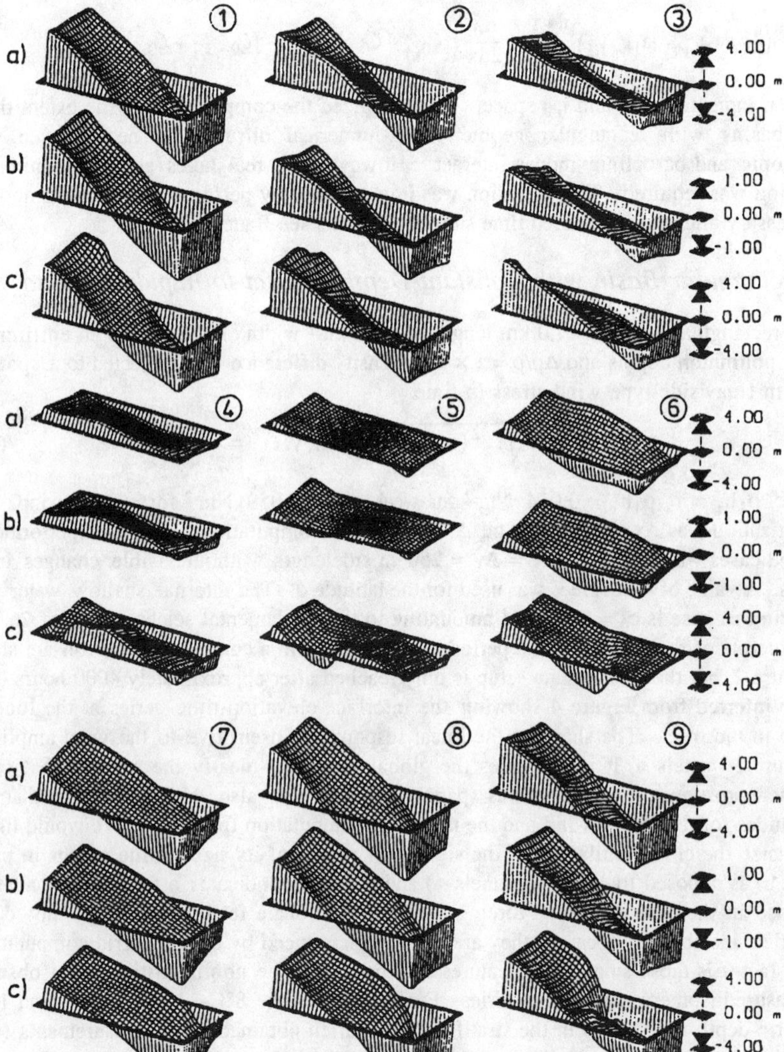

Figure 2. Temporal evolution of the interface motion in a two layer constant depth rectangular basin [17km × 7km, 15m/85 m layer depths] subject to constant uniform wind in the long direction The three panel sets show snapshots at the indicated time of the results of the linear theory (a), the nonlinear theory at weak (b) and strong (c) wind as indicated in the main text. Times of the snapshots are

Label	1	2	3	4	5	6	7	8	9
time (h)	14.75	18.5	22.25	26.0	29.75	33.5	37.25	41.0	44.75

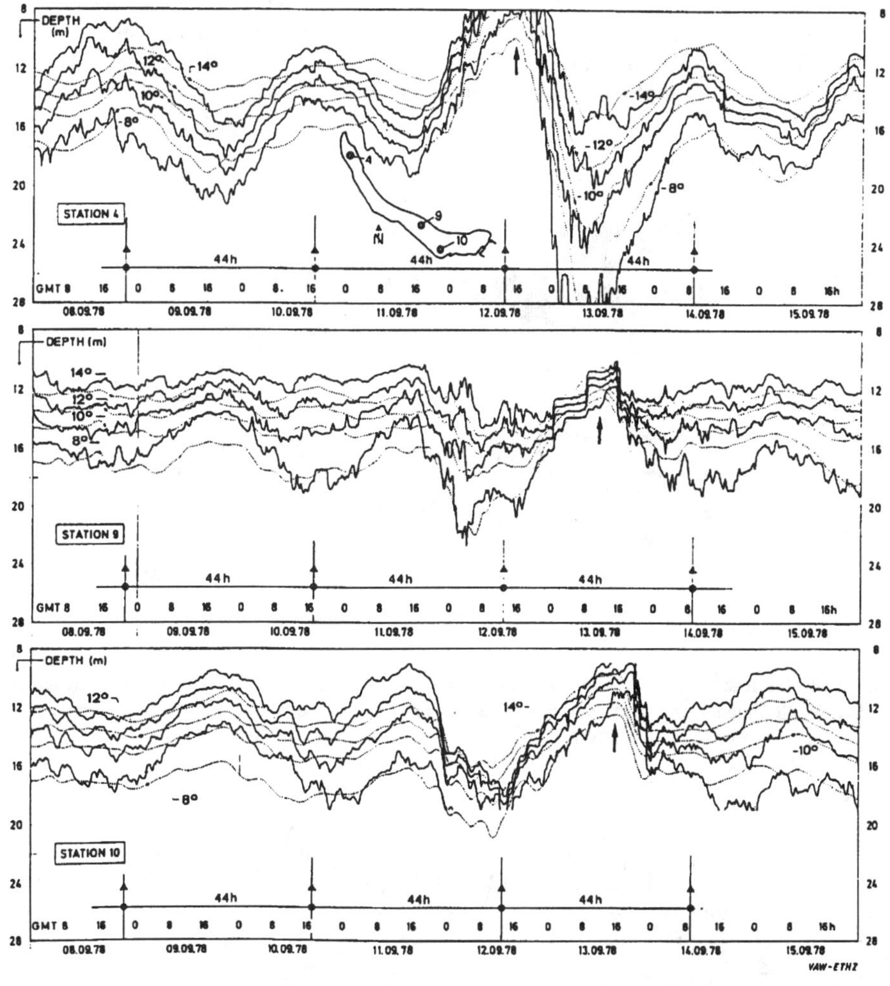

Figure 3. Lake Zürich comparison of time series of the 8°C-, 10°C-, 12°C- and 14°C- isotherm depths obtained from measurements (solid curves and model calculations (dotted) at the stations 4, 9 and 10 from 9 to 14 September 1978 (from Hutter, 1983).

Of particular interest is the performance with regard to dissipation. Figure 4 shows, how the system moves into steady state. Physically and analytically no dissipation is involved; so the long-time behavior ought to be a steady set-up plus a number of eigenmodes (dispersion should eliminate the smaller scale response). Computationally, implicit numerical diffusion - inescapable for the used finite-difference-scheme - dissipates all oscillations away in a time of approximately 4000 h. This is far better than what we have seen in the past with other models. The computational model possesses the potential of reproducing episodes in two-layer basins of several weeks duration. In Figure 5 the first 10 day period (240 h) of the elevation time series is displayed with stretched time axis for the linear (panel a) and the non-linear equations using weak (panel b) and strong (panel c) wind forcing. It is evident that with growing influence of the nonlinearities the signal shape changes significantly and discloses better resolution of the fine structure. We stress again these effects are physical and not numerical artefacts.

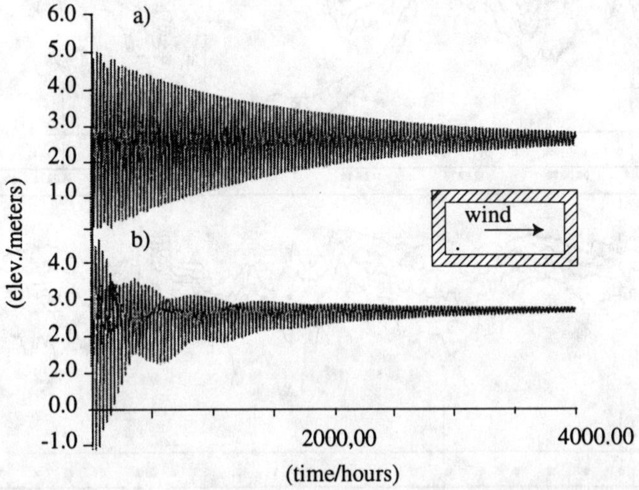

Figure 4. Interface elevation time series plotted against time for 4000h after the onset of an uniform wind. Wind direction and location of time series as indicated in the inset. (a) results as obtained with linear shallow water equations, (b) for the nonlinear equations and a strong wind.

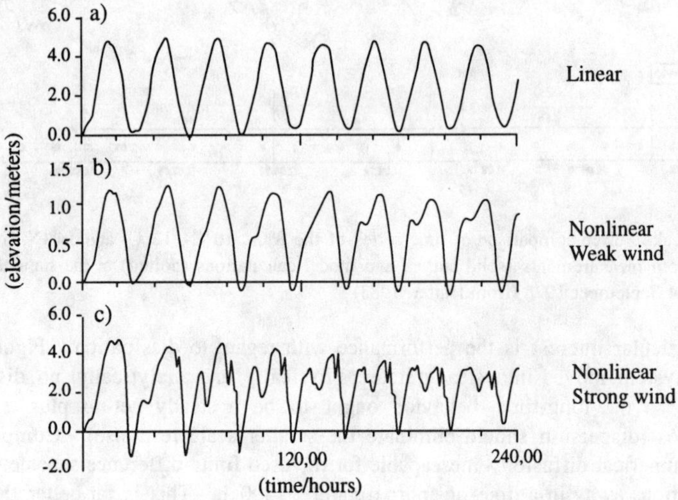

Figure 5. Interface elevation time series plotted against time for 240h after the onset of an uniform wind as shown in the inset to Figure 4, as obtained with the linear shallow water equations (a), with the nonlinear equations for weak (b) and strong (c) wind as described in the main text.

We also stress that equations (7) – (12) have no rigid lid assumption incorporated and so embrace both barotropic and baroclinic wave motion. More importantly, along with the baroclinic interface wave signal they also predict the associated nonlinear surface displacement. In view of the fact that internal waves in the ocean are now often observed via satellite imagery and thus trace the surface signal this is important (e.g Ademan Sea, Strait of Gibraltar, etc). More on this, see Diebels et al. (1994). Furthermore, the theory has shown good agreement with laboratory experiments (Maurer et al. 1996).

We have demonstrated many more features that can be described by this nonlinear two-layer wave model. Among these are (i) the back-curving of Kelvin-type wave crests due to the (nonlinear) amplitude dependence of the phase speed, (ii) the necessary nonlinear coupling of Kelvin-type waves with Poincaré-type waves, and (iii) the explicit demonstration that effects due to the second Coriolis parameter are generally overshadowed by the nonlinearities, see Bauer, Diebels and Hutter (1994).

Application to Lake Constance

Lake Constance, an Alpine lake bordering Germany, Austria and Switzerland consists of three basins: Obersee, Überlinger See and Untersee, but the Untersee is separated from the other two basins by the 5 km long "Seerhein"; we shall here only be concerned with the ensemble Obersee+Überlinger See, for brevity now simply referred to as Lake Constance. It is approximately 64 km long and 16 km wide, has a maximum depth of 250 m and an approximate mean depth of 100 m (Figure 6). The Überlinger See is a relatively deep and narrow arm, "separated" from the main upper lake by a sill just west of the island Mainau. The depth discretization we use in this paper is based on the bathymetric map of the 1990 survey and has in the following computations a square grid of 600 m side length corresponding to 109 × 34 nodal points.

For comparison, we performed similar computations as those for the rectangular basin with constant depth: Heaviside-type wind-stress (but weaker, now with $\tau_0 \rho_1 h_1 = 0.015$ Nm^{-2} and = 0.0037 Nm^{-2} respectively) 15 m upper layer depth and a density ratio of $\Delta\rho/\rho = 1 \times 10^{-3}$. The wind-stress was reduced as geometrical effects are responsible for larger amplitudes in some parts of the lake (especially in the Überlinger See). The induced reaction of the lake is shown in Figure 7, a snapshot taken after 24 hours. The upper layer is following the wind direction, the lower layer is moving in the opposite direction due to the pressure gradient built by the surface elevation. A series of snapshots would disclose a Kelvin-type motion counterclockwise around the basin. The effects of the nonlinearities

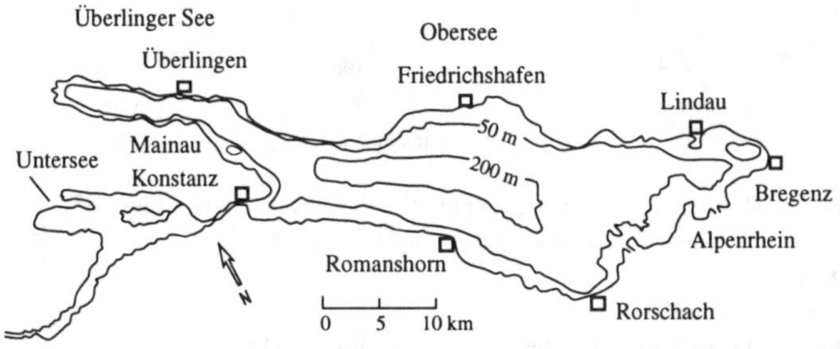

Figure 6. Map of Lake Constance: Untersee, Überlinger See and Obersee. Also shown are the 50 m and 200 m isobaths and a few main towns along the shore

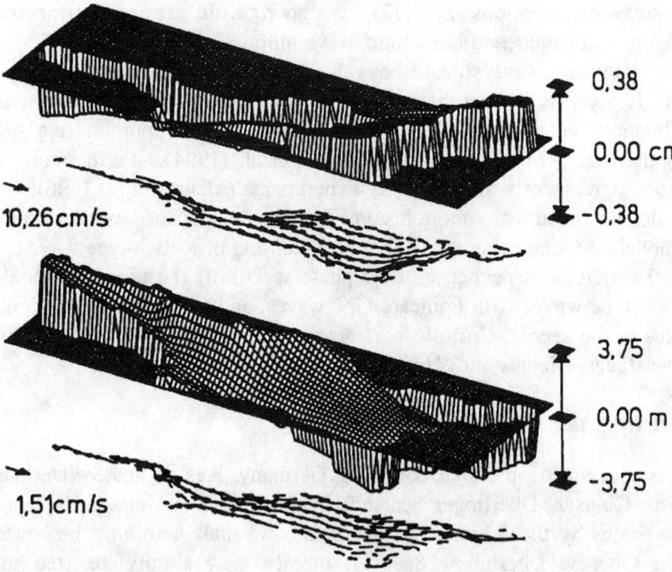

Figure 7. Response of lake Constance to a Heaviside wind stress. Free surface elevation, upper layer velocities, internal interface elevation and lower layer velocities after 24 hours.

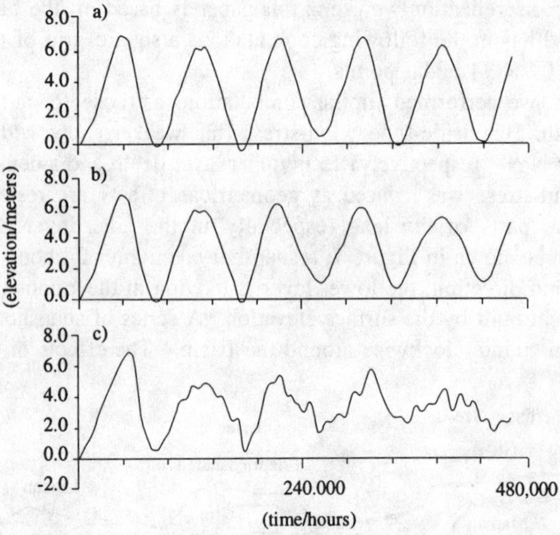

Figure 8. Interface elevation at the western end of "Überlinger See". Time series of the first 480 hours, computed without any smoothing, for the linear equations (a) and the weak (b) and strong (c) wind response in the nonlinear case.

can be made visible by looking at time series of the interface elevation at selected points as obtained with the linear and nonlinear theory, respectively. Figure 8 shows those at the western end of the "Überlinger See".

As in Figure 5 growing nonlinearities are responsible for a changing of the small scale response. Obviously, higher frequencies are determining the signals shape. Although the calculation up to this time is performed without any smoothing filter, energy dissipation is larger than in the linear case; the first large wave hump is much faster eroded than in the linear case. This is so, because with a Cartesian grid, the lake shore is approximated by a function exhibiting a multitude of edges and corners, each of which is the source of an ingoing wave prone to instabilities.

Semi-spectral Primitive Equation Model with Curvilinear Orthogonal Coordinates (SPEM)

We briefly describe here the numerical model as constructed by Haidvogel, Wilkin and Young (1990) and changed by us to allow large-time-step-temporal integration.

Original version of SPEM

Haidvogel et al. (1990) start from the hydrodynamic equations of mass, tracer mass, momentum and energy with the following simplifying assumptions invoked:

- Boussinesq assumption, i.e., density variations are only accounted for in the buoyancy term and balance of mass reduces to the continuity equation ($\text{div}\mathbf{v} = 0$).
- Shallow water approximation, i.e., the vertical momentum balance reduces to the hydrostatic equation balancing the vertical pressure gradient with the buoyancy force, and the effects of the eccentric rotation (second Coriolis parameter) are ignored.
- Rigid lid assumption, i.e., in the vertically integrated mass balance the influence of the motion of the free surface is ignored. This amounts to assuming the volume transport to be solenoidal and derivable from a streamfunction that can be determined independently of the baroclinic response.

A derivation of these equations from a proper scaling analysis is given by, e.g., Hutter (1993). We have argued above that inclusion of dispersion in integrated layer models is due to the inclusion of the acceleration terms in the vertical momentum balance and necessarily stabilizes those models both physically and computationally. This is not so for hydrostatic models and is one reason why hydrostatic models generally require a large amount of numerical diffusion.

In SPEM, smoothness in the representation of the discretization is the principle to minimize the necessary numerical diffusion; hydrostaticity is maintained. To this end in a *first step* the σ-transformation

$$x = \hat{x}, \quad y = \hat{y}, \quad \sigma = 1 + 2(z/H), t = \hat{t} \tag{15}$$

is implemented. Here (x,y,z,t) are Cartesian coordinates in physical space and (x,y,σ,t) are the σ-transformed coordinates, mapping $0 \le z \le H$ into $1 \ge \sigma \ge -1$, i.e., constant depth. In a *second step*, this is achieved by approximating the lake shore polygonally by a sequence of straight segments, and mapping the emerging irregular closed polygon with the aid of a sequence of Schwarz-Chrystoffel transformations onto a rectangle,

$$\hat{x} = \hat{x}\ (\xi,\eta), \hat{y} = \hat{y}\ (\xi,\eta) \Leftrightarrow \xi = \xi\,(\hat{x}, \hat{y}), \eta = \eta\,(\hat{x}, \hat{y}), \tag{16}$$

so that the computational domain is now a cube in (ξ, η, σ)-space. This step avoids discontinuous approximations of the lake boundary which could also be achieved by using

FE-approximations. In a *third step*, the vertical σ-dependence of the model variables is represented as an expansion in a finite polynomial basis set $\{T_k(\sigma)\}$; that is

$$\phi(\xi,\eta,\sigma,t) = \sum_{k=0}^{N} T_k(\sigma)\hat{\phi}_k(\xi,\eta,t), \tag{17}$$

where ϕ is any unknown field variable, $\{T_k(\sigma)\}$ are polynomials and $\{\phi_k\}$ are unknown coefficients; the only restriction placed on the form of the Tk is that $\int_{-1}^{1} T_k(\sigma)d\sigma = 2\delta_{k0}$, where δ_{k0} is the Kronecker delta - i.e., only the lowest order polynomial (k = 0) has a nonzero vertical integral. This isolates the depth average barotropic component of the field. $\{T_k(\sigma)\}$ are chosen to be Chebyshev polynomials, and the representation guarantees a very smooth representation of the variables in the -direction if only N is large enough. In the horizontal, i.e., the ξ, η-directions, finite difference representations with an Arakawa C-grid are used.

SPEM in its original version employs explicit FD-schemes for temporal integration and thus is numerically only conditionally stable, i.e., the allowable time step is restricted by the number N of polynomials taken into account in (17). This conditional stability also dictates which numerical values the horizontal and vertical diffusivities of momentum and energy can take. Because of the smoothness precautions taken in steps 1 to 3 these values are of the order of what would be physically suggested but the temporal steps of integration are still prohibitively small for many problems.

Implementation of Semi-implicit Integration in Time

A fully implicit integration in time is equally prohibitive because of the large nonlinear systems of equations that emerge. This fact suggested to us that we use ADI-procedures (Alternate Direction Implicit). Accordingly, the integration step from time level n to n+1 is subdivided as follows:

(i) $t^n \rightarrow t^{n+1/3}$, implicit in ξ-direction, explicit in η- and σ-directions,
(ii) $t^{n+1/3} \rightarrow t^{n+2/3}$, implicit in η-direction, explicit in ξ- and σ-directions,
(iii) $t^{n+2/3} \rightarrow t^{n+1}$, implicit in σ-direction, explicit in ξ- and η-directions.

It turned out that tripling the time step $\Delta\tau$ according to this procedure is less effective than employing a semi-implicit integration in the σ-direction alone with the full time step Δt. Furthermore, Lake-Constance computations with barotropic wind induced processes showed that time steps could be 200 - 400 times larger than with the explicit integration of the original SPEM. For baroclinic processes the gain is much less but still a factor of approximately 20, or more.

A detailed analysis of this, including a study of selection of the optimal horizontal and vertical diffusivities is given in a pending dissertation by Wang (1995).

Ekman Problem in Closed Basins

Rectangular Basin with Constant Depth

Consider a rectangular basin of 65 km × 17 km extent and 100 m depth; assume homogeneous water, initially at rest, and subject to external wind forcing. Let this wind blow in the long direction of the rectangle, uniformly in space, Heaviside in time and with strength 0.05 Nm^{-2} ($\simeq$4.7 ms^{-1} windspeed). This is the windforcing we apply throughout the paper. Integration starts at rest until steady state is reached. Let the discretization be implemented with 12 Chebyshev polynomials and $\Delta x = \Delta y = 1$ km, and let the numerical values of the horizontal and vertical diffusivities of momentum, ν_H and ν_v, be prescribed as follows: $\nu_H = 1\ m^2\,s^{-1}$ and

- Case (i):

$$\nu_V \ = \ \in [0.005, 1]\,m^2\,s^{-1}, \tag{18}$$

- Case (ii):

$$\nu_V \ = \begin{Bmatrix} A\sin\left(\dfrac{\pi z}{60}\right), & z > -45m \\ 0.01\ , & z \le -45m \end{Bmatrix} m^2\,s^{-1},$$

$$\text{where } A \in [0.01, 1]\,m^2\,s^{-1}\ , \tag{19}$$

- Case (iii):

$$\nu_V \ = \begin{Bmatrix} A_0 + (A_0 - 0.02)\dfrac{z}{10}, & z > -10m \\ 0.03\ , & z \le -10m \end{Bmatrix} m^2\,s^{-1},$$

$$\text{where } A_0 \in [0, 1]\,m^2\,s^{-1}\ , \tag{20}$$

In all three cases the horizontal diffusivity ν_H is kept constant, at the same value because it turned out that the numerical values of ν_H are not very crucial (Wang, 1995). In case (i) the vertical diffusivities are also spatially constant, but their values are varied over $2^1/_2$ log cycles. Case (ii) corresponds to a z-dependence of ν_V, where ν_V is large in the upper layer, grows first with depth, reaches a maximum in 30 m and then rapidly decreases to the value 0.01 $m^2\,s^{-1}$ at 45 m depth and below. Despite the analyticity this profile mimics realistic situations. Case (iii) assumes a linear variation of ν_V from the free surface to 10 m depth below which ν_V is kept constant at the value $\nu_V = 0.03\ m^2\,s^{-1}$. Choices like these are considered realistic and have to some extent been analysed with a linear theory in finite depth oceans, see Heaps (1984).

One typical result of the Ekman problem is the angle Φ between the direction of the wind and the surface current in steady state. In the original problem $\mathbf{v}_{surf}$ is 45° to the right of the wind. Results of our nonlinear calculations are summarized in Figure 9. In case (i) the maximum value of Φ is 42.5°, and Φ decreases rapidly with increasing vertical diffusivity to a value as low as 4° for $\nu_V = 1$. It can never reach 45° as it does (independent of the value of ν_V) in the infinite ocean, because of the finiteness of the rectangular basin, that induces a geostrophic flow which has the tendency to reduce Φ. This simply indicates how significantly the circulating motion in a finite basin can depend on the absolute values of the viscosities. Even more surprising are the results of case (ii). Figure 9 also displays the angle Φ between wind directions and $\mathbf{v}_{surf}$, now plotted against the amplitude A of (19). Here Φ varies nonmonotonously with A: For $A \le 0.12$ it decreases with growing A, while it grows with A when $A \ge 0.12$. Case (iii) presumes for most values of A_0 a very active turbulent near-surface layer and illustrates that for such a stiff situation the typical dependence of Φ can even be reversed. Not shown by a separate graph are the moduli of $\mathbf{v}_{surf}$ as functions of the diffusivities. As would be expected, they rapidly decrease with increasing diffusivities.

That the values of vertical diffusivities are significant can also be seen from Figure 10, which shows the Ekman spirals at the lake midpoint and a near shore midpoint as indicated in the inset and as computed with $\nu_V = 0.02$ m^2 s^{-1} (physically a large value, left panels) and $\nu_V = 0.005$ m^2 s^{-1} (physically a 4 × smaller, but still not realistic value, right panels). These Ekman spirals are considerably affected by the ν_V values. For the larger diffusivities (left panels) the boundary layers are "fatter", and the spirals are incomplete insofar as relatively strong transverse currents arise at the bottom. These currents are due to the finiteness of the basin; they essentially evolve from a geostrophic balance, a contribution that is independent of depth. For the smaller viscosities (right panels) the Ekman spirals are less fat and the transverse current at the bottom is less pronounced. The turning of the arrows making up the spirals also indicates that the surface Ekman boundary layer (if there is any) is thinner for the smaller values of the diffusivities than for the larger ones. All this is qualitatively as one would physically expect.

Even more striking are the differences obtained with $\nu_V = 0.02$ m^2 s^{-1} and $\nu_V = 0.005$ m^2 s^{-1}, when the transient response from initiation of the motion to steady state is analysed. We focus here attention exclusively to the time series of the horizontal velocities, u, v, at 10 m depth intervals in the midlake position. Figure 11 displays these for $\nu_V = 0.02$ m^2 s^{-1} (the larger value). Steady state conditions are approximately reached after 40 to 50 hours, but initially an oscillating motion is superposed on the monotonic trend into steady state, which is easily identified with the inertial wave of 16.3 hour period. This oscillating signal is damped away after less than two periods. By contrast in Figure 12 which shows the corresponding results as obtained with the small (physically more realistic) value $\nu_V = 0.005$ m^2 s^{-1} the inertial motions persist much longer. Of course, the above value for $\nu_V = 0.005$ m^2 s^{-1} is still somewhat large to realistically model vertical momentum diffusion. To reduce it further requires an increase of the number of Chebyshev polynomials to be accounted for. This, of course enlarges the necessary computer time. We shall demonstrate this below for the baroclinic case of wind-induced motion.

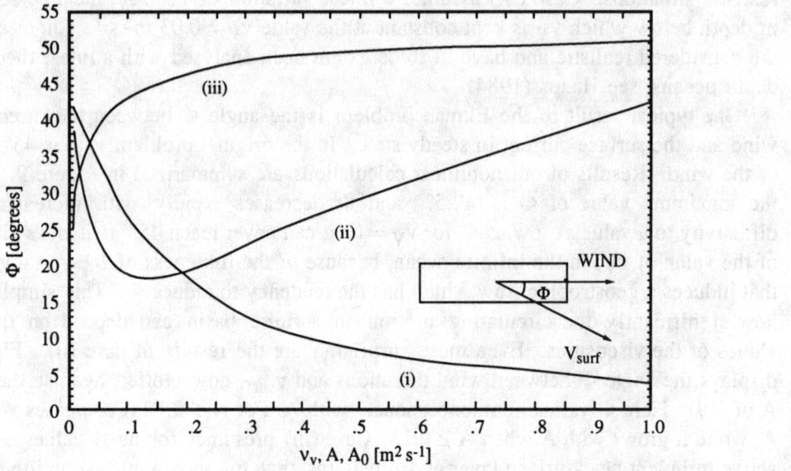

Figure 9. Rotation angle Φ between wind and surface current. In the infinite ocean problem $\Phi = 45°$. Here Φ is plotted against ν_v (i), A (ii) and A_0 (iii), respectively, corresponding to the cases (i), (ii) and (iii) in formulas (18), (19), and (20).

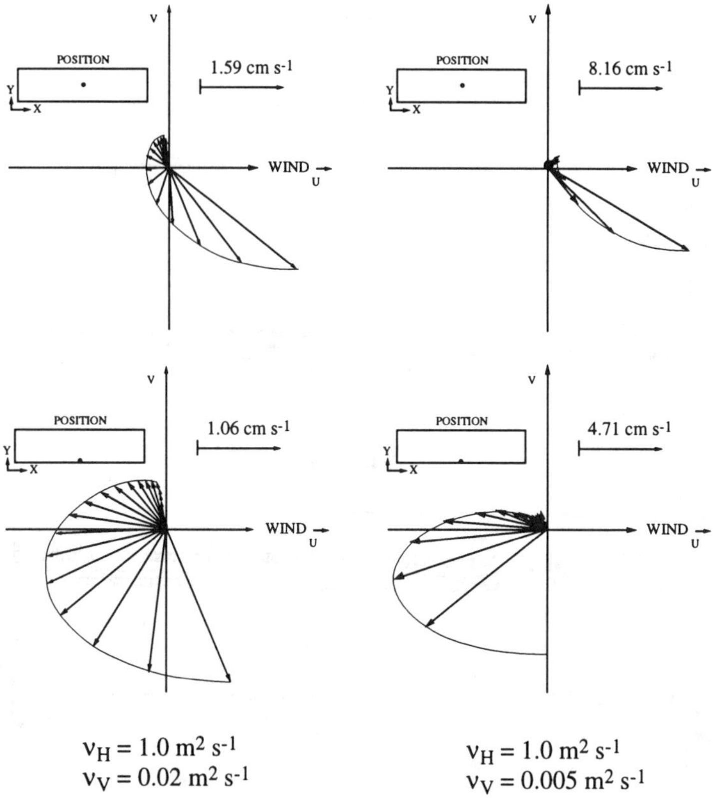

Figure 10. Ekman spirals in a rectangular basin with 65 km × 17 km side lengths and 100 m depth. The graphs show the horizontal velocities in a vertical profile at the position in the inset. The drawn arrows are 5 m apart starting from the free surface to the bottom. The graphs on the left are produced with $\nu_v = 0.02\ m^2\ s^{-1}$, those on the right for $\nu_v = 0.005\ m^2\ s^{-1}$.

It transpires that the distribution of the vertical diffusivities strongly affects the wind induced barotropic currents. The response can be so much counterintuitive that it appears almost impossible to prescribe the dependencies of the diffusivities such that Ekman-type currents (spinup and steady state) in a finite basin can reliably be predicted. We conclude from this: *The determination of the turbulent fluxes along with the solution of the governing equations becomes a mandatory prerequisite of future forced-motion software.*

Application to Lake Constance

An analogous study has been performed for Lake Constance (i.a. the Upper Lake) with 12 and 18 Chebyshev polynomials, respectively and a discretization as shown in Figure 13, with the same number of grid points (65 × 17) as for the rectangle. Computations were done for $\nu_H = 1.0\ m^2\ s^{-1}$ and $\nu_V = 0.02\ m^2\ s^{-1}$ and $\nu_V = 0.005\ m^2\ s^{-1}$, respectively. The larger number of polynomials was needed in the second case to achieve stable numerical integration.

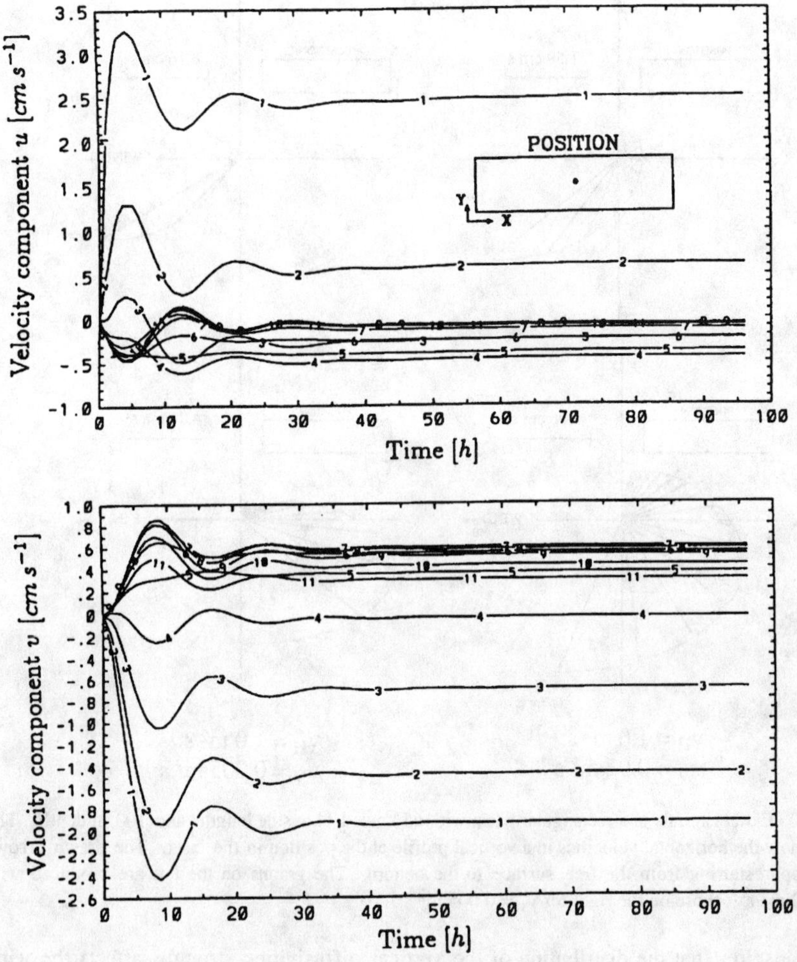

Figure 11. Temporal evolution of the horizontal velocity components u (top) and v (bottom) for a midlake position in the homogeneous rectangular lake of constant depth subject to a heaviside wind set up in the long direction. The symbols (1, 2, 3, ..., 11) refer to the depths (0, 10, 20, ..., 100) m ν_v = 0.02 $m^2\ s^{-1}$.

Of the very detailed computations by Wang (1995) we display in Figure 14 two steady Ekman spirals as they form for an impulsively applied spatially and temporally uniform wind from 35° NW (approximately in the long direction) and as obtained with the two different selections of the diffusivities. Qualitatively, the results are essentially the same as those obtained for rectangular basins. As an additional feature we see here in all four spirals the near-bottom backward turning Ekman spiral.

We could also show results as those displayed in Figures 11 and 12 with no new surprising features. More interesting is the comparison of the time series of the horizontal velocity components and for various depths in the midlake positions of Lake Überlingen and the Upper Lake as displayed in Figure 15. At both positions transient oscillations can

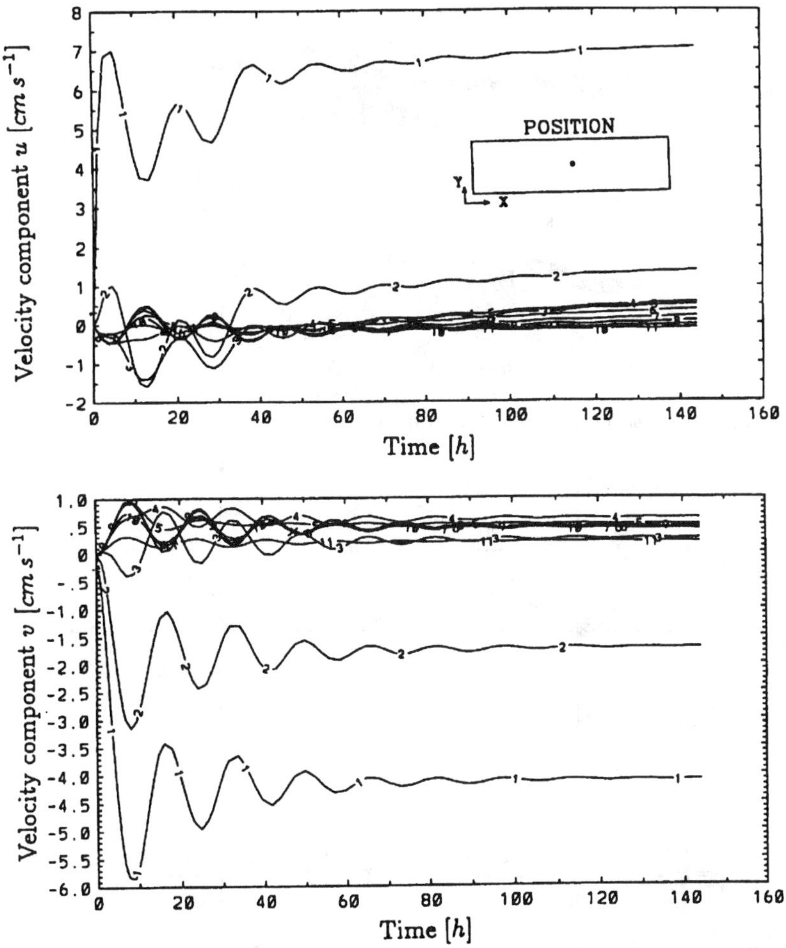

Figure 12. Temporal evolution of the horizontal velocity components u (top) and υ (bottom) for a midlake position in the homogeneous rectangular lake of constant depth subject to a heaviside wind set up in the long direction. The symbols (1, 2, 3, ..., 11) refer to the depths (0, 10, 20, ..., 100 m) $\nu_V = 0.005$ $m^2 s^{-1}$.

be discerned with a period of approximately 16 h, corresponding to the inertial period; steady state is reached after less than 4 days. The oscillations can be seen at all water depths, however with decreasing amplitude as the depth increases. Furthermore, they have died out after 2 days in Lake Überlingen, but only after 4 days in the Upper Lake. A likely reason is the smaller size of the basin for Lake Überlingen and therefore the enhanced frictional resistance at the lake bottom and the side shores. Figure 15 also shows that the boundary layer close to the free surface is thick in the transverse velocity component and thin in the along shore velocity component. For wind forcing in the transverse direction this is opposite, as expected.

We have performed many more computations but do not have the space to present any of these (see Wang, 1995).

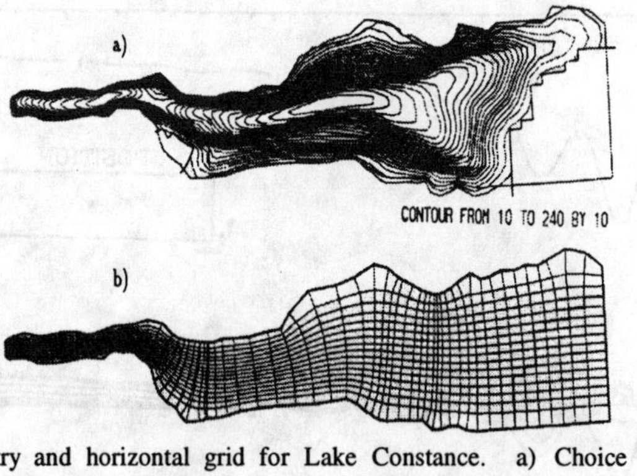

Figure 13. Bathymetry and horizontal grid for Lake Constance. a) Choice of polygon for the construction of the curvilinear coordinate net, choice of the shore line approximation and water depth in 10 m intervals, b) distribution of horizontal net with computational shoreline.

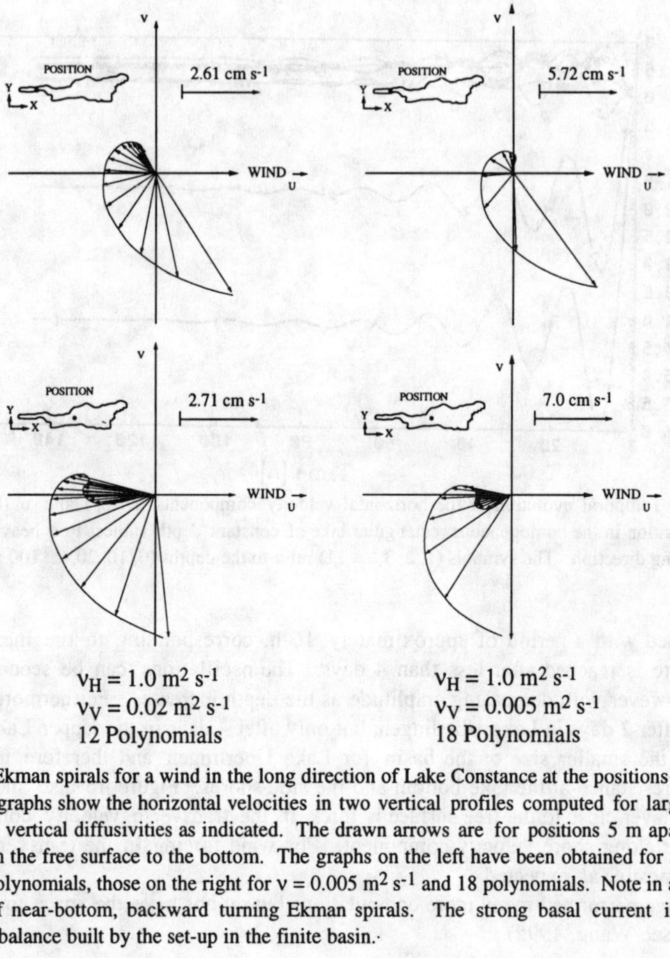

Figure 14. Ekman spirals for a wind in the long direction of Lake Constance at the positions shown in the insets. The graphs show the horizontal velocities in two vertical profiles computed for large (left) and small (right) vertical diffusivities as indicated. The drawn arrows are for positions 5 m apart from one another from the free surface to the bottom. The graphs on the left have been obtained for $\nu = 0.02\ m^2\ s^{-1}$ and 12 polynomials, those on the right for $\nu = 0.005\ m^2\ s^{-1}$ and 18 polynomials. Note in all graphs the formation of near-bottom, backward turning Ekman spirals. The strong basal current is due to the geostrophic balance built by the set-up in the finite basin.

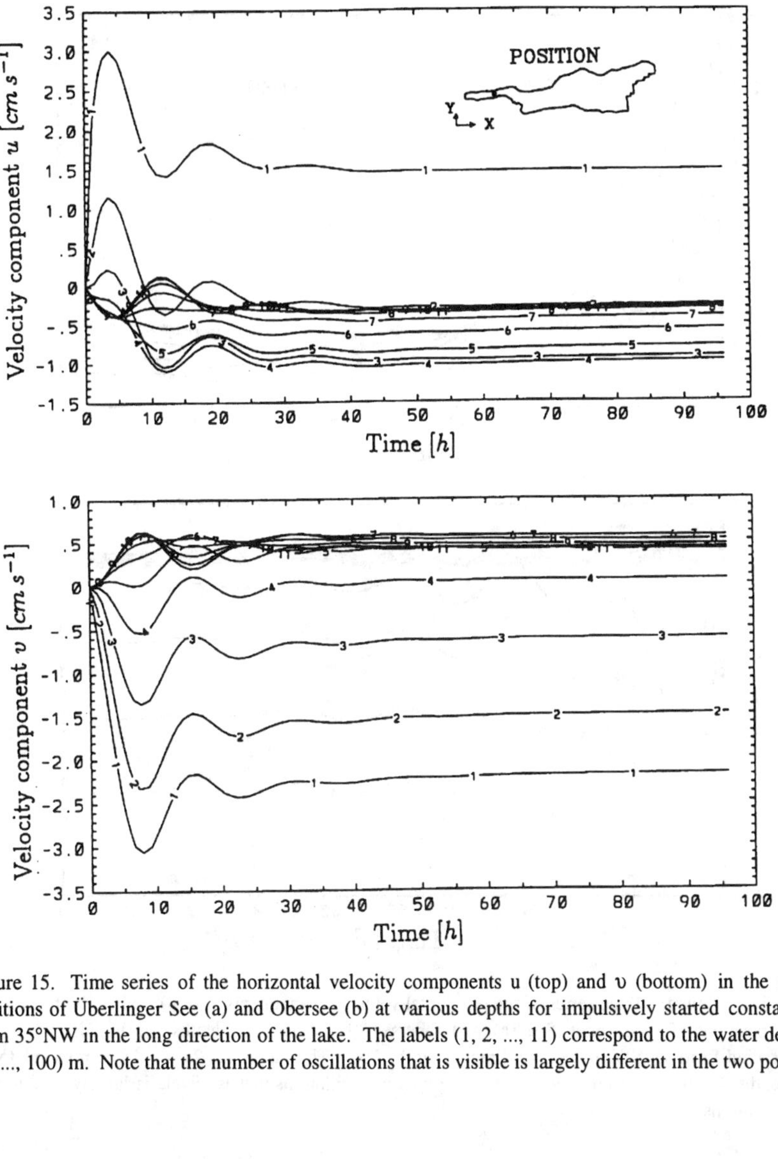

(a)

Figure 15. Time series of the horizontal velocity components u (top) and υ (bottom) in the midlake positions of Überlinger See (a) and Obersee (b) at various depths for impulsively started constant wind from 35°NW in the long direction of the lake. The labels (1, 2, ..., 11) correspond to the water depths (0, 10, ..., 100) m. Note that the number of oscillations that is visible is largely different in the two positions.

Baroclinic Response

Rectangular Basin with Constant Depth

It turned out that the role played by the vertical diffusivities is as crucial in the prediction of baroclinic processes as it is for the barotropic ones. Computations were performed for the following two sets of selections for the diffusivities of momentum and heat:

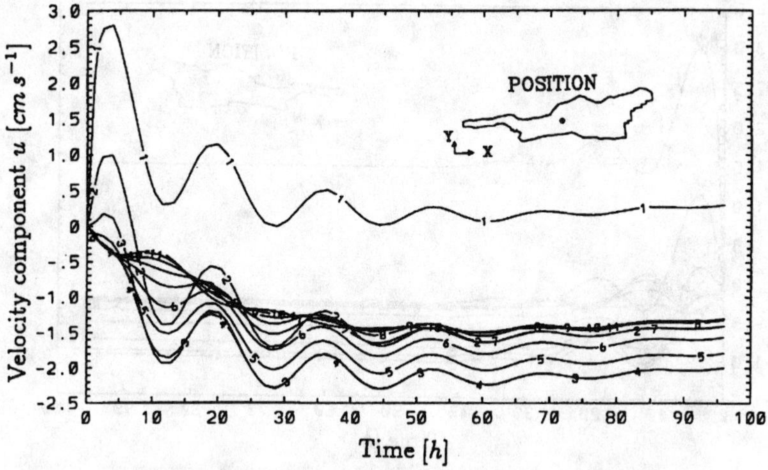

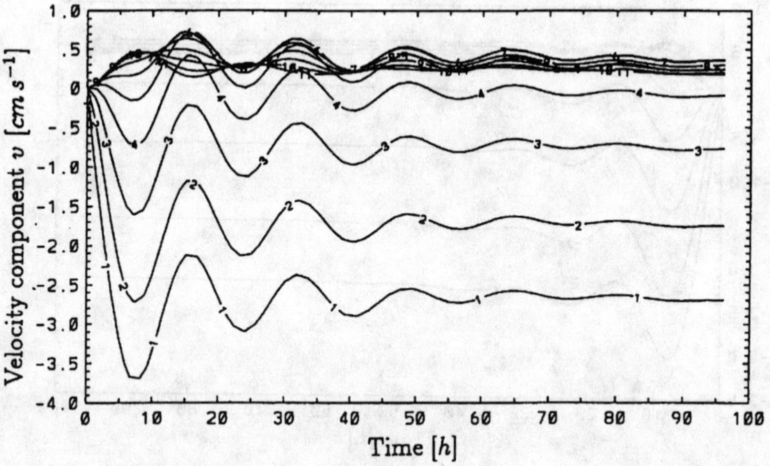

(b)

Figure 15 (continued). Time series of the horizontal velocity components u (top) and υ (bottom) in the midlake positions of Überlinger See (a) and Obersee (b) at various depths for impulsively started constant wind from 35°NW in the long direction of the lake. The labels (1, 2, ..., 11) correspond to the water depths (0, 10, ..., 100) m. Note that the number of oscillations that is visible is largely different in the two positions.

- Case (i):

$$\begin{aligned} \nu_H &= 1.0\,\mathrm{m}^2\,\mathrm{s}^{-1}, & D_H &= 1.0\,\mathrm{m}^2\,\mathrm{s}^{-1}, \\ \nu_V &= 0.02\,\mathrm{m}^2\,\mathrm{s}^{-1}, & D_V &= 0.0005\,\mathrm{m}^2\,\mathrm{s}^{-1}, \end{aligned} \tag{21}$$

- Case (ii):

$$\nu_V = \begin{cases} 0.04\ \mathrm{m}^2\,\mathrm{s}^{-1}, & z > -20\mathrm{m}, \\ 0.004\,\mathrm{m}^2\,\mathrm{s}^{-1}, & -20\mathrm{m} \geq z \geq -40\mathrm{m}, \\ 0.02\ \mathrm{m}^2\,\mathrm{s}^{-1}, & -40\mathrm{m} > z, \end{cases}$$

$$V_H = 1.0 \text{ m}^2 \text{ s}^{-1} \text{ (constant)}, \tag{22}$$

$$D_V = \begin{cases} 0.0005 \text{ m}^2\text{s}^{-1}, & z > -20\text{m}, \\ 0.00005 \ (0.0001) \text{ m}^2\text{s}^{-1}, & -20\text{m} \geq z \geq -40\text{m}, \\ 0.0001 \ (0.0002) \text{ m}^2\text{s}^{-1}, & -40\text{m} > z, \end{cases}$$

$$D_H = 1.0 \text{ m}^2 \text{ s}^{-1} \text{ (constant)},$$

(Note, the values in parentheses apply later for Lake Constance.)

Case (i) assumes constant diffusivities with values unrealistically high, however, they were needed for numerical stability when (only) 12 Chebyshev polynomials were used. Case (ii) is more realistic as it accounts for the fact that diffusivities (viscosities) are smaller in the metalimnion than in the epi- and hypolimnion. Compared with physically realistic values they are, however, still somewhat large. As before, computations were performed for impulsively applied constant wind in the long direction from a state of rest, and the lake is stratified by the vertical temperature profile

$$T(t = 0) = \begin{cases} 17 - 2 \times \exp(-(z+20)/5), & z \geq -20\text{m} \\ 5 + 10 \times \exp((z+20)/20), & z < -20\text{m} \end{cases} [°\text{C}] \tag{23}$$

which is shown as curve A in Figure 16. As thermal boundary conditions at the lake bottom and free surface vanishing heat flow is assumed.

Figures 16a, b show two snapshots each of the midlake temperature profile 4 days (curves B) and 8 days (curves C), respectively, after the wind set up. The situation after 8 days corresponds to near steady conditions. Results of panel a) were obtained with diffusivities (21), those of panel b) with diffusivities (22). In panel a) the epilimnion temperature is lowered after 4 days by 2°C and again by 1.5° during the subsequent 4 days. On the other hand, the hypolimnion temperatures are raised by only a few tenths of a degree. Of course, turbulent mixing paired with diffusion is responsible for this behavior.

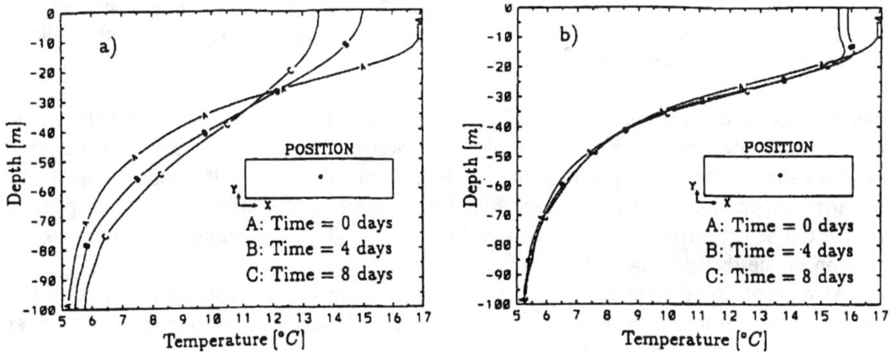

Figure 16. Vertical distribution of the temperature in the midpoint of a rectangular basin, as given by formula (23) and initially (curves A) as well as 4 days (curves B) and 8 days (curves C) after a constant uniform wind had blown from the West, a) as computed with viscosities (21), b) as obtained with the viscosities (22). 12 Chebyshev polynomials are used in the numerical implementation.

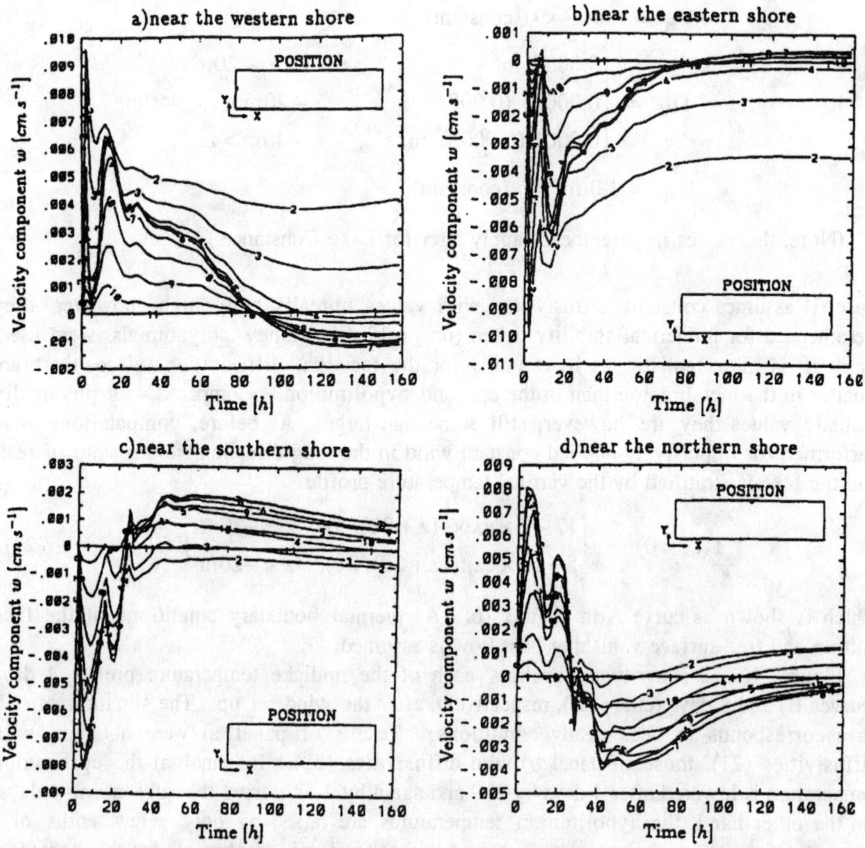

Figure 17. Time series of the vertical velocity component w at the four indicated near-shore midpoints at various depths in the homogeneous rectangular basin of constant depth subject to constant wind from W. The labels (1, 2, 3, ..., 11) correspond to the depths (0, 10, 20, ..., 100) m. Results were obtained with the constant diffusivities (21). Note that after wind secession the velocities essentially die out without any superimposed oscillations.

The difference to the results displayed in panel b) which are obtained with the diffusivities (22), shows how effective a reduction of the numerical values of the vertical diffusivities is. Now the maximum temperature drop in the epilimnion in 8 days is no more than 1.3°C, and the temperatures in the meta- and hypolimnion are hardly affected. The reason is of course the selection of the very small diffusivities in the metalimnion which block the erosion of the thermocline at larger depths.

As was the case for the barotropic processes, the smaller vertical diffusivities in (22) than in (21) let transient Kelvin-type and Poincaré-type waves be developed while they are largely damped away before they are fully developed when the diffusivities (21) are selected. Figures 17 and 18 compare time series of the vertical velocity component w at the four nearshore midpoints as obtained with the two sets of viscosities and for a constant wind in the long direction lasting 48 hours. Only in Figure 18 can we clearly identify two

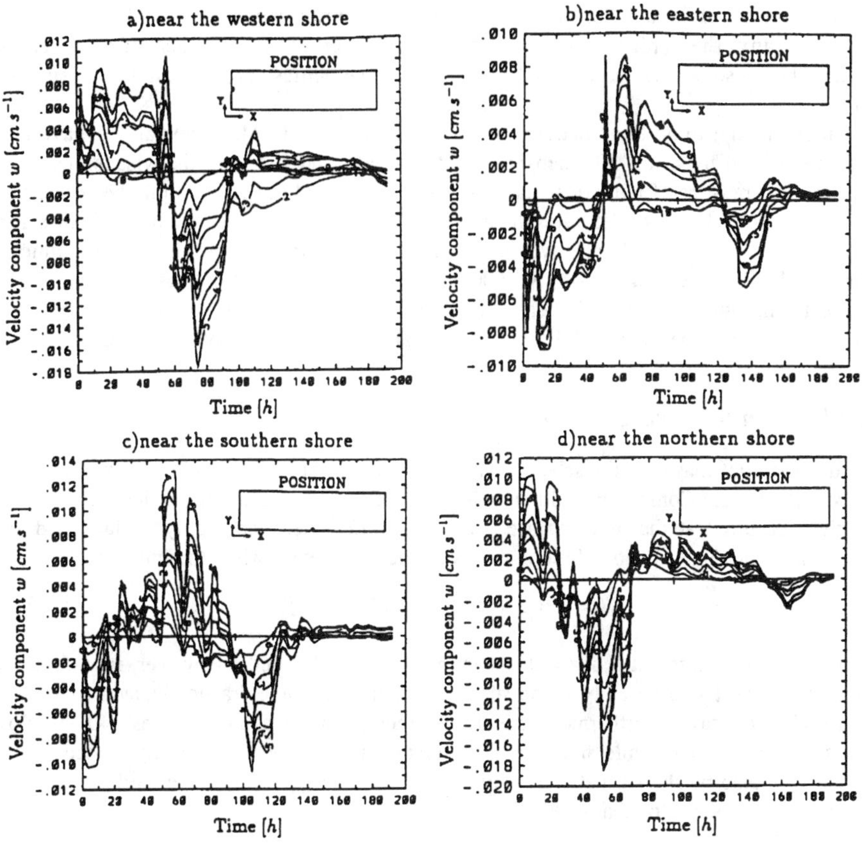

Figure 18. Same as Figure 17 but now as computed with the non-constant diffusivities (22). Note the downstroke, indicated by an arrow, that develops after wind secession, first arising in panel a) and propagating counterclockwise around the lake. It marks the Kelvin wave. The small periodic oscillation are Poincaré waves.

conspicuous components of oscillation. The longer periodic oscillation, barely visible, can be identified as an internal Kelvin-type wave, the shorter one as a Poincaré-type wave. Wang (1995) provides convincing details for this interpretation. A comparison of the two figure sets is particularly instructive. In panel a of Figure 17 there is no downstroke after wind secession (48 h) discernible, while this is conspicuously seen in Figure 18a at ~75 h. If we follow this downwelling signal around the lake then it is seen in panels (a, b, c, d) of Figure 18 at ~(75, 105, 135, 160) hours amounting to approximately 115-120 h travel time around the basin, corresponding to the internal Kelvin wave. For the computations with $\nu_v = 0.2\ m^2\ s^{-1}$ this does not even show up. Very clearly seen in Figure 18, but barely discernible in Figure 17 are the short periodic oscillations that are superimposed on the long term trend in the signals of Figure 18. These Poincaré waves persist for a very long time. Nevertheless, the Kelvin waves are still considerably attenuated suggesting that computations with a larger number of Chebyshev polynomials are needed. Finally we mention these Kelvin and Poincaré wave dynamics are not as well seen as in the vertical velocity if any other typical variable (isotherm depth, vertical displacement of particles) is used. This is somewhat unfortunate since vertical velocities are difficult to measure in-situ.

In an attempt to go to an extreme in the application of SPEM, computations have also been performed by using 30 Chebyshev polynomials and reducing the viscosities even further. For reasons of numerical stability, the values listed in (22) can only be reduced by approximately a factor of 2/5. Results are, naturally, improved as compared to Figure 16b, 18; in particular the Kelvin wave shows up more conspicuously and is less quickly attenuated. However, on an IBM RS 6000 Workstation, computations last about 1.5 times the real time duration. This indicates that realistic computations of wind induced lake circulations are expensive and a problem of main frame implementation and, possibly, parallel processing.

Application to Lake Constance

It turned out that with the selection (21) of the diffusivities no stable computations for the stratified Lake Constance could be achieved, unless the maximum vertical temperature gradient occurred at the free surface. So, with 12 Chebyshev polynomials and the temperature profile (23) no solutions were obtained; however, when the initial profile was

$$T(t=0) = 5 + 15 \times \exp(z/20) \qquad [°C] \tag{24}$$

whose maximum gradient is at z = 0, stable results were obtained. However, in a constant wind experiment (from 35° NW) the surface temperatures were drastically reduced to less than 8°C in 7.5 days. Furthermore, metalimnion temperatures as far down as 140 m below the surface increased, while hypolimnetic temperatures were left unchanged, see Figure 19a. It can also be shown that Kelvin-type and Poincaré-type waves are hardly excited in the transient process to steady state.

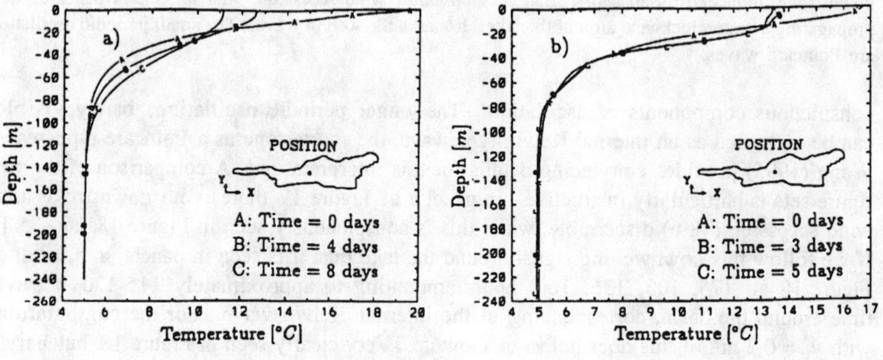

Figure 19. Vertical distribution of the temperature in the indicated midlake position of Lake Constance a) as obtained with the diffusivites (21) and the initial temperature profile (24) (curve A), after 4 days (curve B) and 7.5 days (curve C); b) as obtained with the diffusivites (22) and the initial temperature profile (25) curve A), and after 3 days (curve B) and 5 days (curve C), respectively.

Better conditions of numerical stability can be obtained if the region with large temperature gradients is better resolved, and this means that the number of Chebyshev polynomials must be increased. The choice of 18 polynomials and the selection of the diffusivities according to (22) and the initial temperature profile

$$T(t = 0) = \begin{cases} 17 - 2\times\exp(-(z+10)/5), & z \geq -10\text{m} \\ 5+10\times\exp((z+10)/20), & z < -10\text{m} \end{cases} \quad [^\circ\text{C}] \tag{25}$$

led to the evolution of the temperature profile as shown in Figure 19b. (Note: In formula (22) the numerical values in parentheses are applied for Lake Constance.) Fairly detailed information is obtained if the horizontal and vertical velocity components at typical positions are studied. Figure 20 shows time series for these at various depths of the positions shown in the insets; these are computed with the diffusivities (21) and the initial temperature (24). Somewhat simplified, the motion commences with an upwelling of approximately 80 h duration at the Western end of the Überlinger See that goes over into a downwelling, that relaxes after 160 h into a steady state with asymptotic vanishing vertical velocity. Correspondingly, the motion starts at the Eastern end with a downwelling which after 50 h goes into an upwelling, at least at the lower depths. In the Northern (Southern) mid boundary points there occurs first an upwelling (downwelling) and thereafter a downwelling (upwelling), even though only weakly, after which the flow approaches steady state with vanishing vertical velocity. This behavior is, clearly, due to the Coriolis force that causes a velocity drift to the right and therefore is responsible for the initial downwelling at the Southern shore. However, no indication of Kelvin waves being established are discernible, and the formation of Poincaré waves can only be recognized at early times in panels b) and c) of Figure 20.

Figure 21 shows how much more complex and richer the response is when (22) is employed for the diffusivities, (25) as the initial temperature profile and 18 Chebyshev polynomials are used. Grossly speaking, the behavior is similar but with more pronounced Poincaré wave activity. A conspicuous difference is the rebounding of the vertical velocities at the Western end in Lake Überlingen at t = 75 h which is much larger in Figure 21 than in Figure 20. With the exception of this position the dominant transient interplay between Kelvin-like and Poincaré-like waves is also better visible than in Figure 20. Wang (1995) reports many more details on the baroclinic response of Lake Constance.

Concluding Remarks

In this paper we have reported experiences gained by a nonlinear two-layered model and an implicit semi-spectral primitive equation model with the aid of which various computations were performed for the wind induced currents in barotropic and baroclinic processes of a rectangular lake as well as lake Constance. With the *two-layer model* we showed that non-linear modeling was crucial if the fine structure of the internal waves was to be analysed. The direct response to wind forcing and the oscillation after wind secession were reasonably well predicted; in particular the model showed a surprisingly long persistence of the existence of Kelvin-type waves. This indicates that a model, which is capable of predicting wind induced currents and the wave dynamics due to it, most likely ought to be formed with flexible layers.

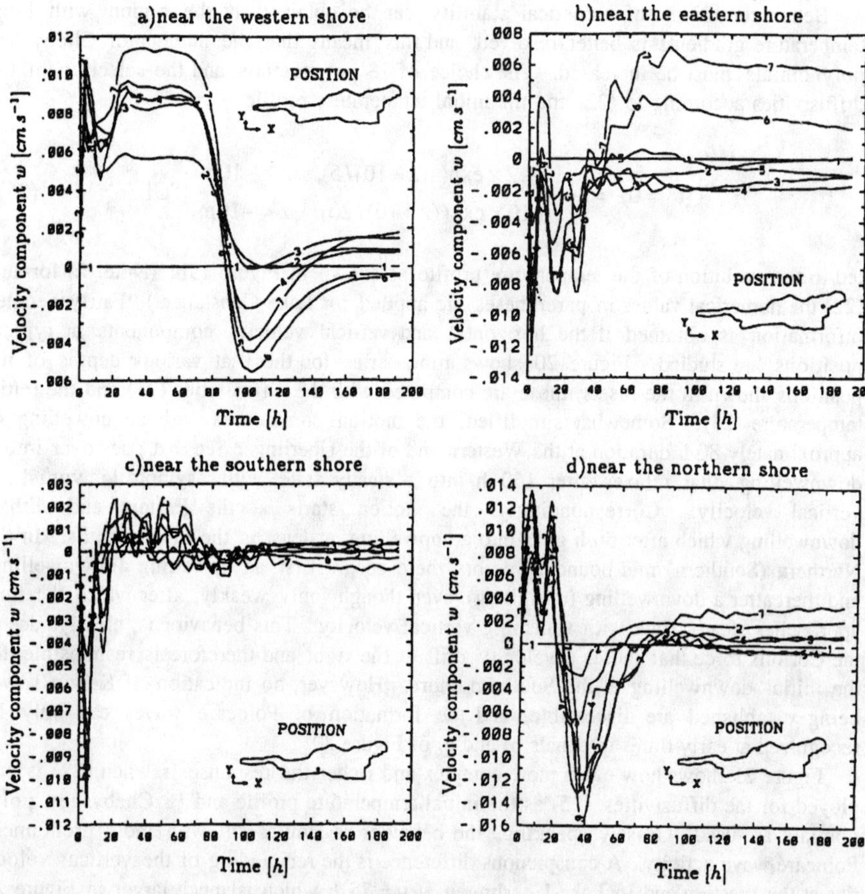

Figure 20. Time series of the vertical velocity component w at various depths in the indicated near shore positions in the stratified Lake Constance subject to constant wind from 35°NW lasting 48 hours. Diffusivities are as in (21) and initial temperature as in (24) and computations are done with 12 Chebyshev polynomials. the labels (1, 2, 3, ..., 7) correspond to the depths (0, 10, 20, ..., 60) m. In panels b), c) small initial signals of Poincaré waves are discernible.

The *semi-spectral model*, on the other hand, was built on equations incorporating the hydrostatic pressure assumption, that necessarily excludes the positive effects of amplitude dispersion. Computations in basins with homogeneous and stratified waters showed that the selection of the number of spectral functions, the choice of the horizontal and (more so) vertical diffusivities is very crucial; in fact the velocity distribution, both in the vertical and in the horizontal depend on the selection of the diffusivities. However, while our program allows to choose the values in the range that is thought to be physically acceptable, the vertical distribution cannot be chosen with sufficient assurance of physical reliability. This indicates that the Reynolds-closure conditions must be computed along with the balances of mass, momentum and energy. We have adopted the k-ε closure condition, but are not yet far enough to be able to report results.

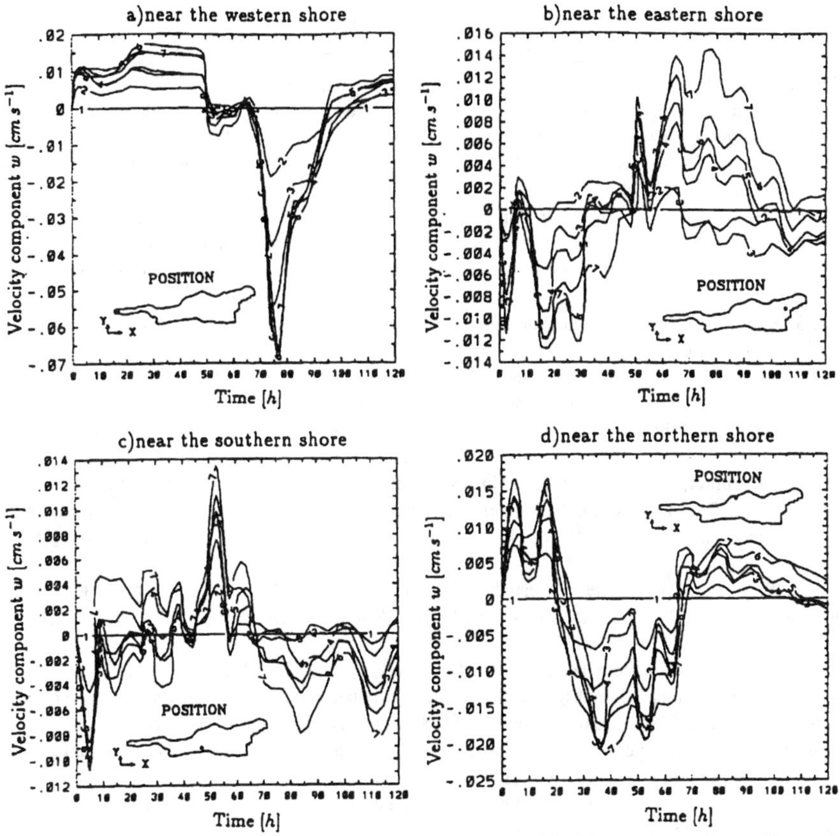

Figure 21. Same as Figure 20 but now computed with 18 Chebyshev polynomials and (22) as the chosen diffusivities and (25) as initial temperature distribution.

Future computational schemes will have to incorporate such closure conditions, apply these in semi-spectral-model techniques - perhaps by employing the spectral-element technique and incorporating sufficiently fine resolution to reduce the required numerical diffusion. In the horizontal, finite element techniques may well prove superior to the use of the Schwarz-Chrystoffel transformation in combination with finite-difference techniques.

Acknowledgments. We thank Erich Bäuerle for many interesting and stimulating discussions. The participation of Kolumban Hütter in the IUTAM-Symposium was made possible by support from the Deutsche Forschungsgemeinschaft and the A. v. Humboldt-Stiftung and Max Planck-Gesellschaft through the Max Planck-Preis, 1994. We thank Dale Haidvogel for allowing us to use SPEM.

Throughout this paper references have been kept to the absolute minimum. Historical preference has not necessarily be observed and if several references could have been stated, a recent one has been selected to serve as a source other references. We apologize to all those who are not mentioned but feel they have significantly contributed.

References

Arakawa, A. and V. R. Larm, Computational Design of the Basic Dynamical Processes of the UCLA General Circulation Model. *Methods of Computational Physics, 17*, 173-265, 1977.

Bauer, G., Windangeregte Strömungen im geschichteten Bodensee, Modellrechnungen und Feldbeobachtungen. Diplomarbeit, Technische Hochschule Darmstadt, 1993.

Bauer, G., S. Diebels, and K. Hutter, Nonlinear internal waves in ideal rotating basins. *Geophys. Astrophys. Fl. Dyn., 78*, 21-46, 1995.

Benjamin, T. B., Internal waves of finite amplitude and permanent form. *J. FluidMech., 25*, 241-270, 1966.

Beckers, Thesis No. 819, Ecole Polytechnique Federale de Lausanne (unpublished), 1989.

Csanady, G. T., *Circulation in the Coastal Ocean*, D. Reidel Publ. Comp., Dordrecht, Boston, Lancaster, 1984.

Diebels, S., B. Schuster, and K. Hutter, Nonlinear internal waves over variable topography. *Geophys. Astrophys. Fl. Dyn., 76*, 165-192, 1994.

Fennel, W., and H. -U. Lass, Analytical theory of forced oceanic waves. Akademie Verlag Berlin, 1989.

Haidvogel, D. B., J. L. Wilkin, R. Young, A semi-spectral primitive equation ocean circulation model using vertical sigma and orthogonal curvilinear horizontal coordinates. *J. Comput. Phys., 94*, 151-185, 1991.

Heaps, N., Vertical structure of current in homogeneous and stratified waters. In *Hydrodynamics of Lakes*, edited by K. Hutter, CISM Lecture Notes No. 286, Springer-Verlag, Wien-New York, 1984.

Hutter, K., Waves and oscillations in the ocean and in lakes. *In Continuum Mechanics in Environmental Sciences and Geophysics*, edited by K. Hutter, CISM Lecture Notes No. 337, Springer-Verlag, Wien-New York, 1993.

Hutter, K., Strömungsdynamische Untersuchungen im Zürich- und Luganersee. Ein Vergleich von Feldmessungen mit Resultaten theoretischer Modelle. *Schweizerische Zeitschrift fur Hydrologie, 45*, 101-144, 1983.

Hutter, K., Large scale water movements in lakes. *Aquat. Sci., 53*, 100-135, 1991.

Kielmann, J. and T. J. Simons, Some aspects of baroclinic circulation models. *In Hydrodynamics of Lakes*, edited by K. Hutter, CISM Lecture Notes No. 286, Springer-Verlag, Wien-New York, 1984.

Korteweg, D. J. and G. de Vries, On the change of form of long waves advancing in a rectangular channel, and on a new type of long stationary waves. *Phil. Mag., 39*, 422-443, 1895.

Lin, T. -Y., Windangeregte nichtlineare Wellen in zweigeschichteten Rechteckbecken. Diplomarbeit, Technische Hochschule Darmstadt, 1995.

Maurer, J., K. Hutter, and S. Diebels, Experiments on and computational analysis of viscous effects in internal waves of two-layered fluids with variable depth. *European J. Mech., B/Fluids* (in press), 1996.

Melville, W. K., G. G. Tomasson, and D. P. Renouard, On the stability of Kelvin-waves, *J. Fluid Mech., 206*, 1-23, 1989.

Miura, R. M., The Korteweg-de Vries equation: a survey of results. *SIAM Review, 18*, 412-459, 1976.

Mysak, L. A., Topographic waves in lakes. In: *Hydrodynamics of Lakes*, edited by K. Hutter, CISM Lecture Notes No. 286, Springer-Verlag, Wien-New York, 1984.

Ostrovsky, L. A. and Y. A. Stepanyants, Nonlinear surface and internal waves in rotating fluids, in *Nonlinear Waves 3*, edited by Gaponov-Gretch et al., Springer Verlag (Berlin, New York, Tokyo), 106-128, 1990.

Russell, J. S., Report on waves. Rep. Meet. Brit. Assoc. Adv. Sci., 14th, York, 311-390, John Murray, 1844.

Sander, J. and K. Hutter, On the development of the theory of waves - A historical essay. *Acta Mechanics, 86*, 111-152, 184, 1991.

Simons, T. J., Circulation models of lakes and inland seas. *Can. Bull. Fish. and Aquat. Sci. 203*, 1-145, Government of Canada, Fisheries and Oceans, 1980.

Vries, G., de, Bijdrage tot de kennis der lange golven. Doctoral Dissertation, Univ. of Amsterdam, 1894.

Wang, Y., Windgetriebene Strömungen in einem Rechteckbecken und im Bodensee. Pending Dissertation, Technische Hochschule Darmstadt, 1995.

11

Excitation of Internal Seiches by Periodic Forcing

E. Bäuerle

Abstract

A two-layer model of stratified basins and lakes is driven by tide-generating forces in order to verify the possibility of internal oscillations with tidal frequencies. The results show that medium-sized lakes are favourable to semi-diurnal tidal response as they exhibit large-scale Poincaré-wave like structures with periods far enough from the inertial period. Some observations from two alpine lakes (Lake Constance and Lake Zug) give some evidence for the theoretical findings of the model. Theoretically, small lakes should be tidally forced when the periods of low order internal seiches reach 24 h (or 12 h).

Introduction

The pre-alpine lakes (Lake of Geneva, Lake Constance, Lake Zug,...) are far too small to exhibit significant barotropic tidal oscillations. As was demonstrated by Hamblin et al. (1977), the tidal surface displacements on Lake Constance are no more than 0.8 mm with currents less than 0.01 mm s^{-1}. However, during seasons of stratification there is some evidence that a sporadic semi-diurnal constituent is present in thermistor-chain and current-meter measurements. The question arises as to whether this is caused by some semi-diurnal components in the wind-field or by favourable interplay of the permanently working tide generating forces with free internal modes of the lake.

Near-inertial oscillations are an almost ubiquitous phenomenon in stratified lakes. In the Great Lakes of North America their periods are very near (but always below) the local inertial period (≈ 16.9 h). The most comprehensive observations are presented by Mortimer (1971). It has become customary to call them Poincaré-wave like internal modes as in certain cases their periods and some characteristics of their structure (e.g., anti-cyclonic rotation of the current vectors) can easily be modelled by two oppositely propagating Poincaré waves in a channel with dimensions appropriate to the basin under consideration. However, (firstly) not all real lakes are suitable to be described by any channel and (secondly) it was shown by Rao (1977) that even with a flat-bottomed rectangular basin the structures of the super-inertial oscillations can be rather complicated if the influence of the Earth's rotation is not yet well established – a well known fact in smaller lakes. Schwab (1977) calculated the super-inertial internal oscillations in a flat-bottomed model of Lake Ontario resulting in a dense spectrum of oscillations near, but always above, the inertial frequency. In small lakes like Lake of Geneva or Lake Constance

Physical Processes in Lakes and Oceans
Coastal and Estuarine Studies Volume 54, Pages 167-178

– the "nearinertial" oscillations exhibit frequencies 20 % (and more) above the inertial frequency, and it was shown by Bäuerle (1985) that the model structures are rather complicated and only parts of the basin oscillate with anti-cyclonic rotating currents. Furthermore, the modes need not be basin wide (Bäuerle, 1994).

From a theoretical viewpoint there is no doubt that the periods of free internal oscillations match the tidal periods – at least for some instants – when stratification varies. The principle is sketched in Figure 1 by means of the forced oscillation of a damped harmonic oscillator with temporally varying eigenfrequency subject to periodic forcing. When the eigenfrequency passes through the driving frequency the response signal is amplified locally; when it stays at resonance, the response signal grows monotonically. In the same way it should be possible to accelerate baroclinic motions in stratified basins when the frequencies of tidal forcing and free internal oscillations coincide.

In the present paper we introduce a linear theory of baroclinic forced oscillations in two-layered basins with constant or non-constant bottom topography and present some observational results which may strengthen the proposed hypothesis of resonant forcing. Finally we discuss the lack in frequent observations of that quasi-permanent phenomenon.

Theoretical Considerations

Basic Equations

We shall consider a simple model for free and periodically forced shallow water waves in a rotating two-layered basin of variable topography with bottom friction. We neglect non-linear terms and assume that the solution is periodic in time with frequency ω of the forcing function. The equations then read in a right-handed cartesian coordinate system with z pointing downward for the upper layer ($0 \leq z \leq h_1$)

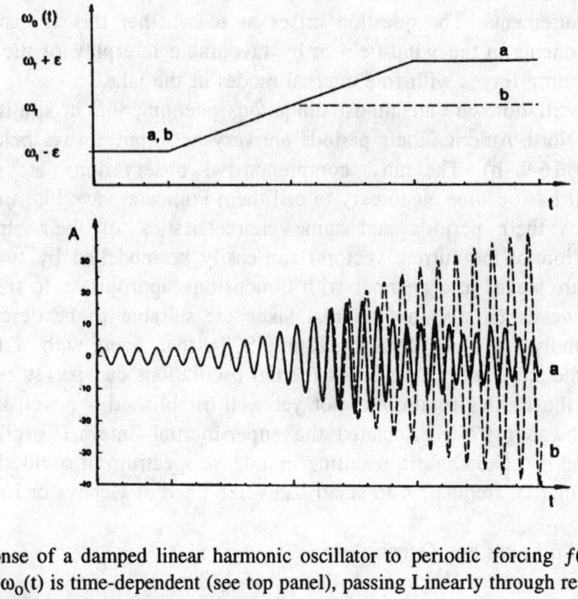

Figure 1. Response of a damped linear harmonic oscillator to periodic forcing $f(t) \sim \sin(\omega_f t)$ when eigenfrequency $\omega_0(t)$ is time-dependent (see top panel), passing Linearly through resonance (solid lines, a) or staying in resonance (dashed lines, b) .

$$i\omega U + fV_1 + gh_1 \frac{\partial Z_1}{\partial x} = \frac{1}{\rho_1} F_1^{(x)}, \tag{1}$$

$$i\omega V - fU_1 + gh_1 \frac{\partial Z_1}{\partial y} = \frac{1}{\rho_1} F_1^{(y)}, \tag{2}$$

$$\frac{\partial U_1}{\partial x} + \frac{\partial V_1}{\partial y} + i\omega Z_1 - i\omega Z_2 = 0, \tag{3}$$

and for the lower layer ($h_1 \leq z \leq h_1 + h_2(x, y)$)

$$i\omega U_2 + \rho_2 k_B U_2 + fV_2 + g\delta h_2 \frac{\partial Z_1}{\partial x} + g\varepsilon h_2 \frac{\partial Z_2}{\partial x} = \frac{1}{\rho_2} F_2^{(x)}, \tag{4}$$

$$i\omega V_2 + \rho_2 k_B V_2 - fU_2 + g\delta h_2 \frac{\partial Z_1}{\partial y} + g\varepsilon h_2 \frac{\partial Z_2}{\partial y} = \frac{1}{\rho_2} F_2^{(y)}, \tag{5}$$

$$\frac{\partial U_2}{\partial x} + \frac{\partial V_2}{\partial y} + i\omega Z_2 = 0. \tag{6}$$

The variables Z_i ($i = 1,2$) denote the amplitudes of the displacement of the surface ($i = 1$) and of the interface ($i = 2$), respectively, and U_i and V_i are the amplitudes of the longitudinal and transverse components of the volume transports in the upper ($i = 1$) and lower ($i = 2$) layers. g is the acceleration of gravity and δ and ε are dimensionless quantities defined as $\delta = \rho_1 / \rho_2$ and $\varepsilon = (\rho_2 - \rho_1)/\rho_2$, where ρ_1 and ρ_2 are the densities of the upper and lower layer, respectively. δ is about 1 and ε of order 10^{-4}. $g\varepsilon$ is the so-called "reduced gravity". h_1, the depth of the upper layer, is constant, whereas h_2 is variable. The shoreline of our model lake is an approximation of the intersection of the interface with the sloping bottom. Thus we ignore the effects which enter the problem from the shallow (one-layer) parts of the basin. Friction at the interface is very small and thus ignored. Bottom friction is assumed to be linearly dependent on the volume transport in the lower layer in order to keep the equations linear. The coefficient of bottom friction, k_B, controls the amount of amplification when resonance occurs.

The driving forces on the right-hand sides of (1), (2), (4) and (5) represent wind stress $\mathbf{F}_s = \rho C_D \mathbf{W}\mathbf{W}$, horizontal gradients of the atmospheric pressure ∇p_a, seismic accelerations $\mathbf{b} = -\omega^2 \mathbf{A}$ (where $\mathbf{A}$ is the horizontal displacement due to seismic waves with frequency ω), and horizontal components of the tide-generating forces $\mathbf{F}_t$, and may be written in the form

$$\mathbf{F}_1 = \mathbf{F}_s + h_1 \nabla p_a + h_1 \rho_1 \mathbf{b} + h_1 \mathbf{F}_t, \tag{7}$$

$$\mathbf{F}_2 = h_2 \nabla p_a + h_2 \rho_2 \mathbf{b} + h_2 \mathbf{F}_t. \tag{8}$$

In a two-layer model the contribution of the wind stress arises from an integration procedure over the upper layer, and so we may think it to be uniformly distributed in there. In the same way the contributions of the bottom friction are uniformly distributed over the lower layer. Contrary to this, the other forces are by nature of barotropic origin. Thus, with respect to baroclinic oscillations, wind forcing is in advantage since momentum which accelerates the motion in one layer will not at the same time decelerate the motion in the other (oppositely moving) layer.

In this paper, atmospheric pressure gradient and seismic forces are ruled out as they peak in a much higher spectral domain than long internal waves (spectacular examples of seismic and atmospheric forcing of external seiches are reported, e.g., by Donn, 1964, in connection with the Alaskan earthquake, and by Ballister et al., 1982, in connection with resonant bay modes on the Baleares excited by atmospheric Kelvin waves). The most effective agency of forcing, the wind stress will not be investigated explicitly, but it should be kept in mind that, whenever tidal forcing is possible, wind-forcing at the same frequency is as well and, in general, much more efficient.

We assume the forces to be horizontally uniformly distributed. The assumption of no variation of longitude for the tide-generating forces is reasonable because of the limited extent of the lakes under consideration. Even at Lake Erie the phase differences in the semi-diurnal constituents are no more than 0.3 h (see Platzman, 1966).

We only take the semi-diurnal components of the tidal forces into consideration as we are interested in the forcing of super-inertial internal oscillations. To remain as general as possible we assume that the M_2 and S_2 tides have horizontal components of the tidal generating forces which rotate in the clockwise sense with constant amplitude. The local magnitude and form of the tidal generating forces ($\mathbf{F}_t$) may be obtained from any standard reference on tides.

The Numerical Procedure

The solution of the equations (1) – (6) together with the boundary conditions of vanishing transport through the shoreline is sought by expressing the spatial derivatives by finite differences (as explained in Bäuerle, 1981). The resulting difference equations form a quasi-tridiagonal linear inhomogeneous set of algebraic equations, which is solved with a direct method proposed by Schechter, 1960.

Results of the Numerical Calculations

5× 1 Rectangular Basins With Flat Bottom

To demonstrate the great variety of possible super-inertial modes with varying stratification we calculate the free modes in a 5 × 1 rectangular basin with flat bottom. From Figure 2 it is apparent that in the ω-c_i-plane (or ω-ε-plane) the frequency-curves form two different families of curves with different slopes. Those with smaller slope correspond to modes with purely anti-cyclonic rotating amphidromic systems, those with steeper slopes correspond to modes where cyclonic amphidromic systems are possible (together with anti-cyclonic ones) and which become solely cyclonically rotating when the inertial frequency is passed from above. (There is a strange not yet explained regularity concerning the intersection of the curves: intersection points and "touching regions" alternate.)

We confine our interest to the modes above the line $\omega/f = 1$. The structure of the individual super-inertial modes varies considerably with increasing influence of the Earth's rotation. This fact is pointed out in Figure 3 by showing the current ellipses of some modes marked by capitals in Figure 2. The mode marked with A in the ω-c_i-diagram is a mixture of cyclonically and anti-cyclonically rotating amphidromic systems. The structures of the modes indicated with B and C, respectively, resemble the superposition of two oppositely propagating Poincaré waves with wavelength $\lambda_x = 2$ L and $\lambda_x =$ L, respectively, in a channel of width B = L/5; but it is obvious that this approximation is

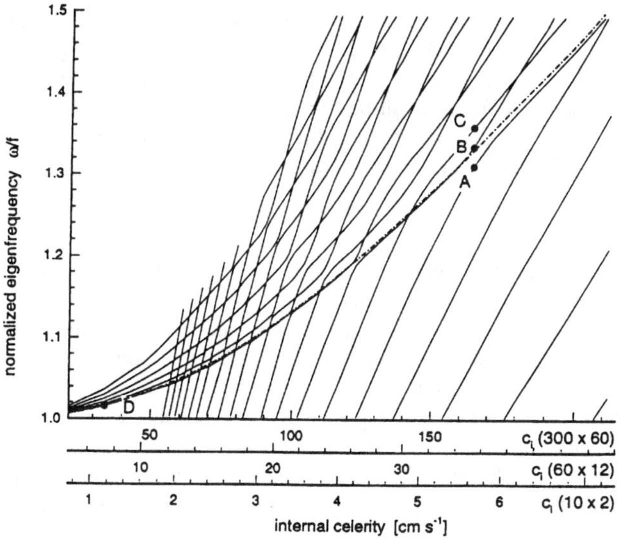

Figure 2. Dependence of the normalized eigenfrequencies of internal oscillations in stratified 5 × 1 rectangular basins (with flat bottom) on the phase velocity (celerity) of long internal waves. Different abscissa are introduced to allow interpretation for basins with different extent: upper scale for 300 km × 60 km (Lake Ontario), middle scale for 60 km × 12 km (Lake Constance), lower scale for 10 km × 2 km (Lake Zug). The ordinate is scaled by the respective local inertial frequencies. The meaning of the dashed-dotted curve is explained in the text.

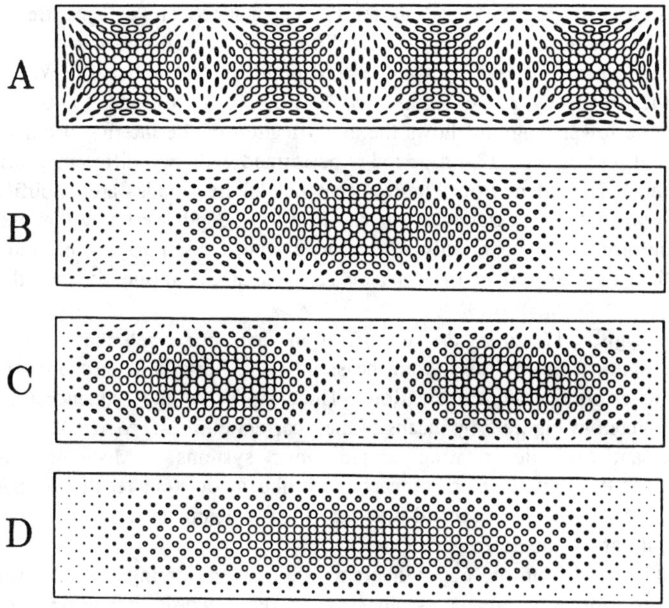

Figure 3. Current ellipses of some modes marked by capitals in Figure 2. Full ellipses indicate cyclonic rotation of the current vectors, empty ellipses indicate anti-cyclonic rotation.

not yet perfect as the stratification is relatively strong and the basin is small ($c_i \approx 35$ cm s^{-1} in a basin with 60 km × 12 km in extent). The approximation gets much better if stratification weakens or the basin becomes larger. The mode with indication D is representative for Lake Ontario conditions and the structure is nearly identical with the Poincaré wave approximation. The broken line in the diagram is representative of the dispersion relation of internal Poincaré waves

$$\omega^2 = f^2 + c_i^2(k^2 + l^2), \quad (9)$$

where c_i is the internal celerity and $k = \pi/L$ and $l = \pi/B$ are the lowest order along-channel and cross-channel wavenumbers, respectively.

Whereas the eigenfrequencies of modes A – C are much greater than the inertial frequency and may become semi-diurnal with small changes of stratification, the eigenfrequencies with $c_i = 35.2$ cm s^{-1} on the upper scale (like mode D) are near-inertial with scales similar to low order internal Poincaré waves. Otherwise they would possess very high mode numbers at semi-diurnal frequencies with very small-scaled horizontal structures. Thus we conclude from Figure 2, that the semi-diurnal free oscillations in large lakes are much too small-scaled in order to match with the (quasi-uniform) distribution of the tidal forces.

5 × 1 Basin With Parabolic Bottom

Next we consider semi-diurnal tidal forcing in a 5 × 1 basin with parabolic depth configuration. We take the bottom topography such that the mean depth stays equal to the flat-bottomed basin of the preceding section. The minimum depth of the basin is 50 m (along the boundaries), the greatest depth is 142 m (at the center). The upper diagram of Figure 4 is similar to Figure 2 and shows the dependence of the eigenfrequencies on stratification (here represented by the density difference between upper and lower layer, ε, scaled by $g\overline{h_e}$, where $\overline{h_e}$ is the mean reduced depth over the basin). We confine our interest to the frequency range around the S_2-tide (12.0 h) and M_2-tide (12.4 h), respectively. The lower diagram shows the amplification of the internal oscillations in the basin when tidal forcing with 12.4 h period is performed with two different coefficients of bottom friction. It is evident from Figure 4 that not all free modes are amplified by tidal forcing. From non-rotating rectangular basins of uniform depth we know that resonance to a spatially uniform force occurs only in odd-order modes. Similarly, some modes in the present rotating two-layered non-flat basin are not at all influenced by the field of anti-cyclonically rotating forces.

For values $\sqrt{g\overline{h_e}} < 17$ cm s^{-1} the amplification does not reach the levels explicitly shown in the figure. This follows from the small-scale horizontal structure of the semi-diurnal free modes with weak stratification. Three modes of the sequence (marked with arrows) are depicted in Figure 5 by means of their current ellipses. Mode A is a mixture of cyclonic and anti-cyclonic rotating amphidromic systems. Mode B exhibits the characteristics of the lowest order Poincaré wave like model (but recall: the basin is not flat-bottomed). Mode C shows second-order cross-basin features, nearly vanishing in the (deeper) central part of the basin.

As can be inferred from the upper diagram in Figure 4 forcing with S_2 would move resonance to considerably greater ε-values or, in other words, changing stratification could move M_2-resonance to S_2-resonance (and vice versa).

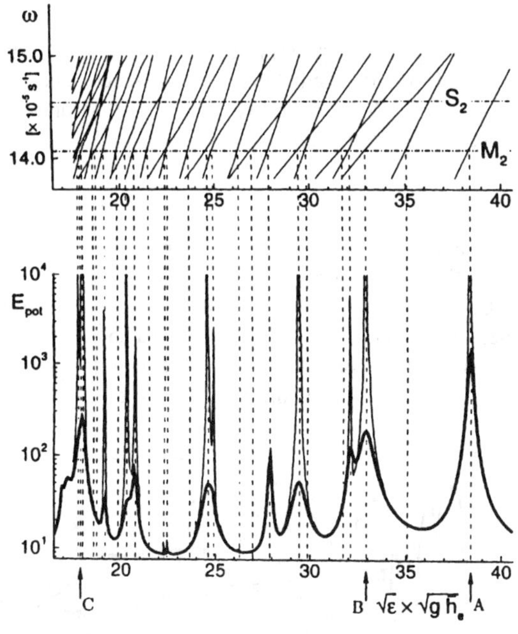

Figure 4. Above: Dependence of the eigenfrequencies of internal oscillations in a two-layer 5 × 1 rectangular basin of 60 km length with parabolic bottom topography on the relative density difference, ε, between upper and lower layer (mean depth as in Figure 2; for better comparison with Figure 2, the square root of ε scaled by $\sqrt{g h_e}$ is taken as abscissa). Frequencies of M_2 and S_2, respectively, are indicated by dash-pointed lines. Below: Amplification (represented by potential energy; arbitrary units) of the above basin when forced by the M_2 tide generating force with different coefficients of bottom friction $k_B = 5 \times 10^{-6}$ (solid); $k_B = 1 \times 10^{-8}$ (dashed).

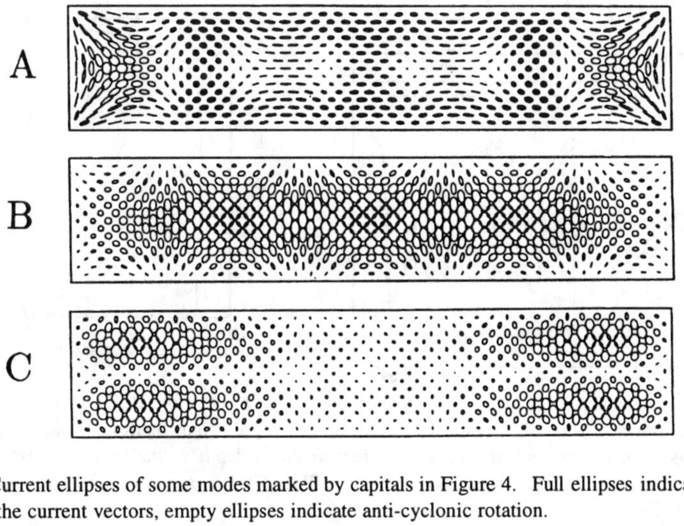

Figure 5. Current ellipses of some modes marked by capitals in Figure 4. Full ellipses indicate cyclonic rotation of the current vectors, empty ellipses indicate anti-cyclonic rotation.

Observations

In the remaining part of this paper we present some observations from two European alpine lakes, where semi-diurnal internal oscillations were observed. The semi-diurnal free modes of these lakes are computed with the same two layer model as presented above, and the calculations are compared with observations.

Lake Constance

The current speeds presented in Figure 6 (left side) originate from early spring of 1993 in 80 m depth at station MS. Stratification is very weak and it is even questionable whether the whole Lake Constance is uniformly stratified. Current speed and current direction exhibit five cycles of an oscillation with a period which is best fitted by 12.4 h (see inserted scale). The registration of current direction indicates the superposition of an anti-cyclonically rotating current with a mean current of nearly equal magnitude. From the

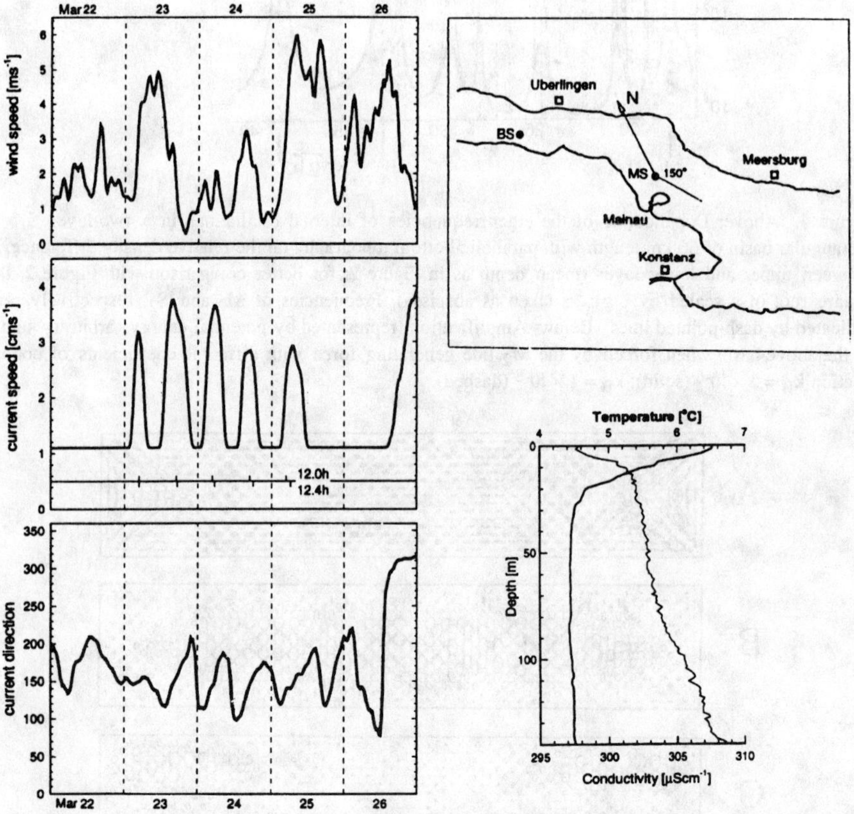

Figure 6. Lake Constance, 1993. Left row (above): Measurements from 03.22. 0.00 to 03.26. 24.00 of wind speed (in m s^{-1}) in 4.4 m height at station BS (see upper right figure), (middle and below): of current speed (in cm s^{-1}) and of current direction in 80 m depth at station MS. Right row (above): Location of the measuring sites; (below): vertical profiles of temperature and conductivity at station BS on 03.23. 09.25.

upper left figure it is evident that there was enough wind to excite the internal oscillation; therefore it is questionable whether it is reasonable that semi-diurnal tidal forces could have caused this oscillation. Nevertheless, it is worthwhile to investigate the structure of the free internal mode. With a two-layer finite difference approximation of the weakly stratified Lake Constance (see Figure 7) we obtain a free internal oscillation of very high mode number with largest amplitudes in the region under consideration (see lower picture of Figure 7). As stratification in early spring changes rapidly, conditions for resonant forcing only last for a few cycles of the oscillation.

Lake Zug

As the second example let us consider Lake Zug during winter conditions, when it is chemically - and persistent - stratified. Whereas the chemical stratification seems to be rather constant with time showing homogeneity in the upper 50 m, the temperature profile above 70 m undergoes changes which are typical for late winter and early spring situations. The pronounced density gradient at about 60 m is weakened and stratification develops from the heated-up surface.

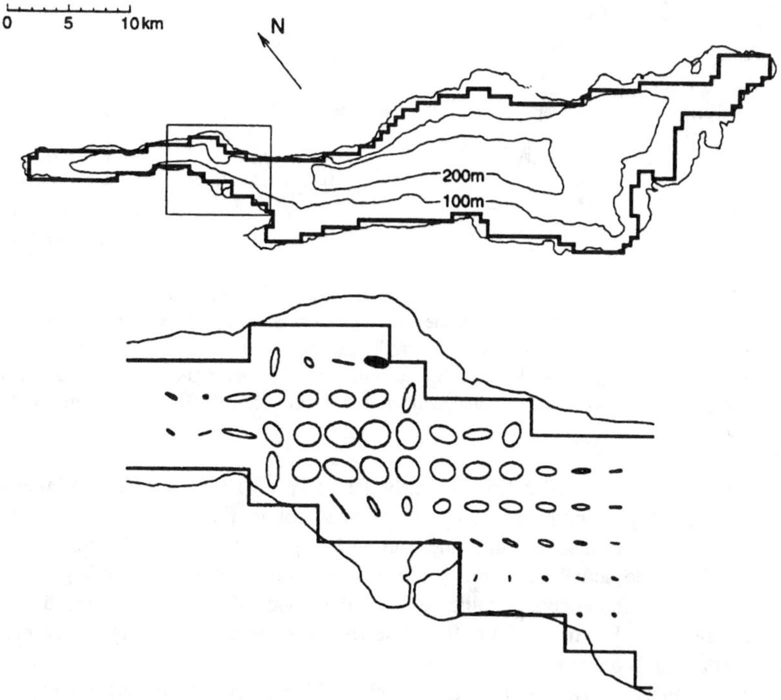

Figure 7. Above: Map of Lake Constance with depth contours (100 m and 200 m) and "shoreline" of the numerical rectangular grid approximating the intersection of the interface with the lake's bottom. Below: Ellipses of the volume transport vector of a high order mode (T, = 12.43h) in the part of :ake Constance near station MS (framed in the upper figure). Empty ellipses indicate anti-cyclonic rotation.

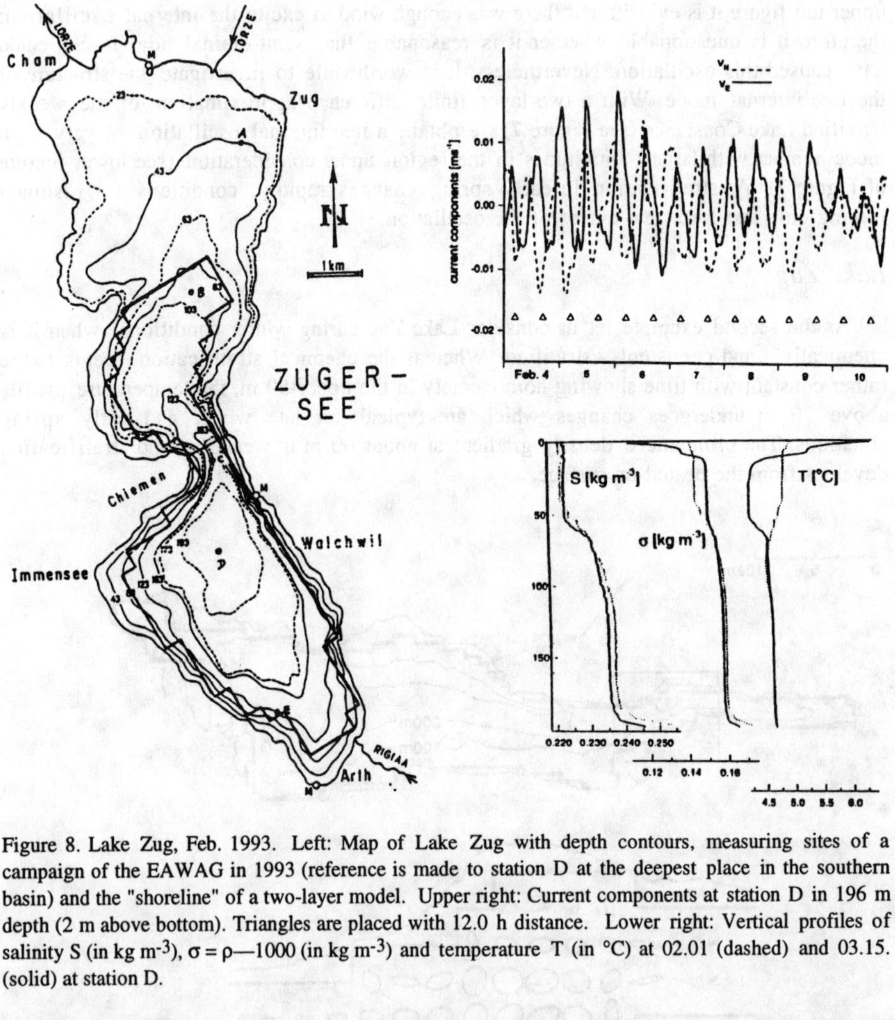

Figure 8. Lake Zug, Feb. 1993. Left: Map of Lake Zug with depth contours, measuring sites of a campaign of the EAWAG in 1993 (reference is made to station D at the deepest place in the southern basin) and the "shoreline" of a two-layer model. Upper right: Current components at station D in 196 m depth (2 m above bottom). Triangles are placed with 12.0 h distance. Lower right: Vertical profiles of salinity S (in kg m^{-3}), $\sigma = \rho - 1000$ (in kg m^{-3}) and temperature T (in °C) at 02.01 (dashed) and 03.15. (solid) at station D.

We modeled the density stratification in early February of 1993 with a two-layered basin of 70 m upper layer depth. From the requirement of vertical walls down to the intersection between density discontinuity and bottom topography we get a "shoreline" of the model far inside of the actual shoreline (at least in the northern basin; see Figure 8). With $\varepsilon = 1.25 \times 10^{-5}$ the fundamental Poincaré-wave like mode shows a 12 h period, and the modal structure is as shown in Figure 9. These findings are in surprising agreement with the actual situation (compare to Figure 8).

Although the temporal separation of the vertical profiles (Figure 8) is rather large we dare the conclusion that before the onset of thermal stratification the density stratification is constant enough to meet the preconditions of persistent resonant forcing. Furthermore we assert that friction is sufficiently small to allow the small but significant signal measured at station D (Figure 8) to be caused by the horizontal components of the semi-diurnal S_2-tide. Beyond that, the local winds during the episode were nearly vanishing (Wüest, pers. comm.). Consequently, the rare prerequisites for fairly undisturbed tidally forced internal oscillations likely were present at Lake Zug.

Discussion and Conclusions

From a theoretical viewpoint it is of no doubt that internal oscillations in stratified lakes can be excited by the tide-generating forces if the eigenfrequencies of the internal seiches agree with the periods of the tides. The most qualified candidates for such resonant forcing are the Poincaré-wave like modes whose eigenperiods during the stratified season regularly pass through the respective tidal periods. Although their modal structures may be of rather small scales (see Figure 7) their occurrence is of convincing observational evidence (see Figure 6). Getting in contest with wind-induced internal seiches the signals induced by tidal forces will ever be overtrumped. Consequently, in most cases (as, e.g., in the case of Lake Constance; see Figure 6) the observed internal oscillations owe their existence mainly to the driving forces at the surface of the lake. Nevertheless, in situations of faint winds or winds without energy in the tidal frequency domain the tidal signal may be observable in the periodic vertical displacements of the isopycnals and in periodically oscillating baroclinic horizontal currents. This situation most likely happened at Lake Zug (see Figure 8). However, the observations and the results of numerical modeling presented in this paper give further evidence of the overtone-rich response of ringing stratified lakes.

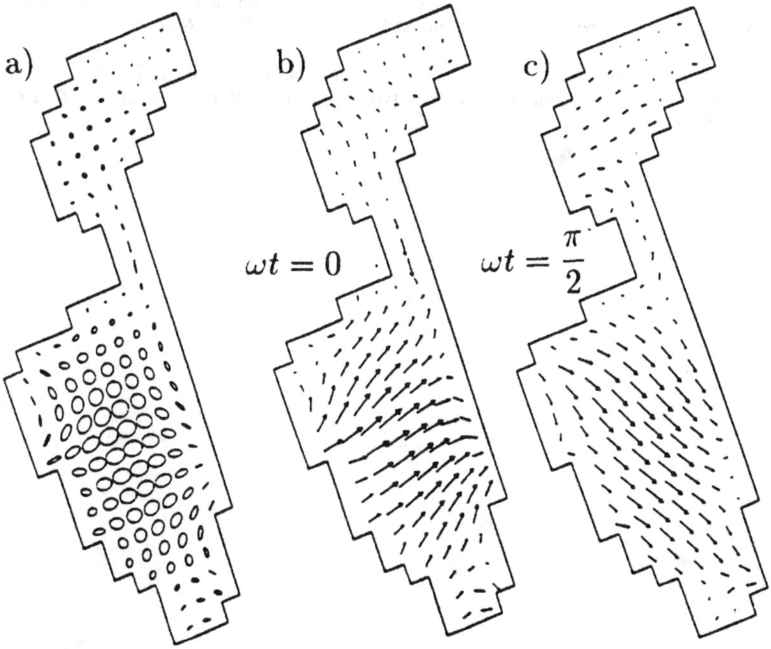

Figure 9. Horizontal structure of a high order mode (T_i = 12.0 h) with $\varepsilon = 1.25 \times 10\text{-}5$ and h_l = 70 m a) Ellipses of volume transport in the lower layer (empty ellipses indicate anti-cyclonic rotation of the vectors), b) volume transport at $\omega t = 0$, c) volume transport at $\omega t = \pi/2$.

Acknowledgments. This study was funded by the German Research Association (DFG) in the course of the special research project "Circulation of matter in Lake Constance". Special thanks go to A. Wüest who triggered the investigations with the data from Lake Zug and opened the view on this fascinating subject. My colleagues from the Institute of Environmental Physics Heidelberg provided the data from Lake Constance. K. Hutter corrected an earlier version of the manuscript. All this support is greatly appreciated.

References

Bäuerle, E., Die Eigenschwingungen abgeschlossener, zweigeschichteter Wasserbecken bei variabler Bodentopographie. *Berichte IfM Kiel, 85,* p 79, 1981.

Bäuerle, E., Internal free oscillations in the Lake of Geneva. *Annales Geophys, 3,* 199-206, 1985.

Bäuerle, E., Transverse baroclinic oscillations in Lake Uberlingen. *Aquatic Sci., 56,* 145-160, 1994.

Ballister, M., A. Jansa, and C. Ramis, Ondas cortas atmosfericas con interacciones aire-mar en el Mediterraneo. *Report of the Univ. of Palma de Mallorca,* 1982.

Donn, W. L., Alaskan earthquake of 27 March 1964: Remote seiches stimulation. *Science, 145,* 261-262, 1964.

Hamblin, P. F., R. Muhleisen, and U. Bosenberg, The astronomical tides of Lake Constance. *Dt. hydrogr. Z., 30,* 105-116, 1977.

Mortimer, C. H., Large-scale oscillatory motions and seasonal temperature changes in Lake Michigan and Lake Ontario. Part I, text, 111 p., Part II, illustrations, 106 p. *Ctr. Great lakes Stds., Univ. Wisconsin-Milwaukee, Spec. Rept. No. 12,* 1971.

Platzman, G. W., The daily variation of water level on Lake Erie. J. *Geophys. Res., 71,* 2471-2483, 1966.

Rao, D. B., Free internal oscillations in a narrow, rotating rectangular basin. In *Modelling of Transport Mechanisms in Oceans and Lakes,* edited by T.S. Murty, *Proc. Symp. CCIW, Burlington, Ontario,* 1977.

Schwab, D. J., Internal free oscillations in Lake Ontario. *Limnol. Oceanogr., 22,* 700-708, 1977.

Schechter, S., Quasi-tridiagonal matrices and type-insensitive difference equations. *Quart. App. Math., 3,* 285-295, 1960.

12

Thermohaline Transitions

E. C. Carmack, K. Aagaard, J. H. Swift, R. G. Perkin,
F. A. McLaughlin, R. W. Macdonald and E. P. Jones

Abstract

We here discuss the physical mechanism of transition from one stationary thermohaline structure to another. Such transitions may occur in both oceans and lakes, and may be either temporal, as in the case of climate variability, or spatial, as in the case of frontal regions separating thermohaline regimes. The example presented here uses data obtained during the Canada/U.S. 1994 Arctic Ocean Section to show that a major warming and ventilation of the mid-depth layers of the Arctic Ocean (200-1000 m) occured due to influx of anomalously warm waters from the Atlantic. This transition appears to have begun in the early 1990s; now, anomalously warm waters are found in the Nansen, Amundsen, and Makarov basins, with the largest temperature difference, as much as 1 °C, in the core of the Atlantic layer. In the Makarov Basin the core of the Atlantic layer is now shallower than before (200 versus 400 m), causing the displacement of cold, nutrient-rich Pacific waters, presumably into the North Atlantic and the convective gyres of the Greenland and Labrador Sea. The transition is occuring via the formation of multiple intrusions, 40-60 m thick, extending downward through the Atlantic and upper deep waters; these features appear to be laterally coherent across the Arctic Ocean (> 2000 km) and to persist over time-scales of several years. Once formed, the layers can support both diffusive and salt finger convection. Potential temperature (θ) versus salinity (S) correlation curves for each successive diffusive and salt finger regime are parallel, with higher values of the stability frequency (N) in the diffusive regime than in the salt finger regime.

Introduction

Thermohaline transitions are changes, either temporal or spatial, from one stationary thermohaline structure to another. The phenomenology generally involves one water mass either displacing or lying in juxtaposition to another, slightly different, water mass, with lateral exchanges occurring via the formation of intrusions along near-isopycnal surfaces. Spatial transitions are well known to exist in ocean settings such as frontal regimes or across mesoscale features (Federov, 1979; Ruddick, 1992). Similar intrusions can also occur in the hypolimnion waters of deep lakes; for example, where distinct water masses occur in basins separated by sills (cf. Wuest et al., 1988), or where differential

Physical Processes in Lakes and Oceans
Coastal and Estuarine Studies Volume 54, Pages 179-186

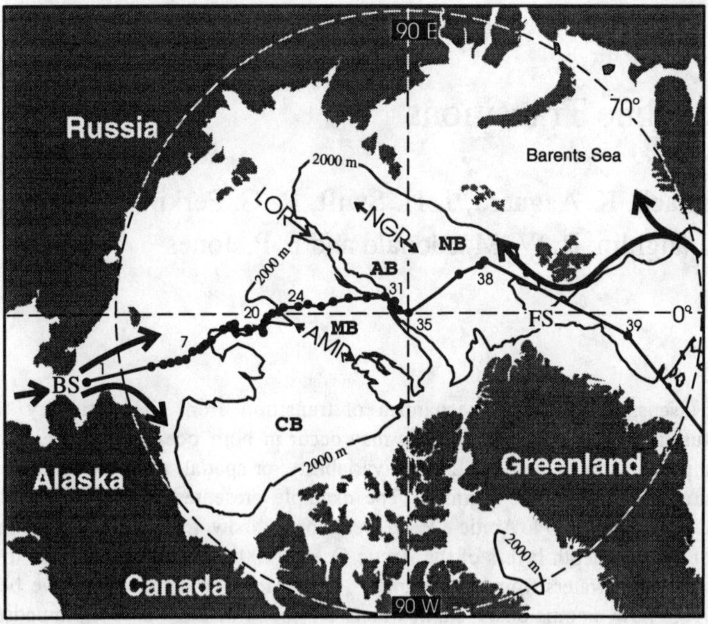

Figure 1. Map showing the Arctic Ocean and Arctic-94 station locations. Here, CB is Canada Basin, MB is Makarov Basin, AB is Amundsen Basin, NB is Nansen Basin, NGR is Nansen-Gackel Ridge, LOR is Lominosov Ridge, AMR is Alpha-Mendeleyev Ridge, BS is Bering Straight and FS is Fram Straight.

wind mixing or river inputs create lateral water mass boundaries (D. M. Imboden, pers. comm). We here focus on mechanisms associated with temporal transitions; that is, changes that take place when one water column laterally "pushes" on another. Such considerations may be important in the detection of climate change, as it may be important in both monitoring and modelling activities to distinguish between local and advective change.

In this paper effects confounding detection of climate variability in high-latitude oceans are examined using observations of recent changes in the thermohaline structure of the Arctic Ocean obtained aboard the *CCGS Louis S. St-Laurent* during the Canada/U.S 1994 Arctic Ocean Section (Arctic-94; Figure 1). Data were collected using a Neil Brown conductivity-temperature-depth (CTD) system (with accuracies estimated at T = ± 0.001 ˚C; S = ± 0.002; and D = ± 2m). Potential temperature (θ) and density (σ) were calculated from algorithms in Unesco (1983).

Thermohaline Circulation and Changes in the Arctic Ocean

Warm, relatively salty water of Atlantic origin enters the Arctic Ocean through Fram Strait and the Barents Sea, subducts beneath the cold and relatively fresh surface water, and flows counterclockwise along the continental slope and submarine ridges (Figure 1; Carmack et al., 1990; Jones et al., 1990). Fresher and less dense water of Pacific origin enters through Bering Strait and likewise flows in a counterclockwise direction (Aagaard, 1989). Both waters exit the Arctic into the North Atlantic via Fram Strait and the complex

Canadian Archipelago. Earlier views held that this flow maintains quasi-stationary water mass distributions in the interior Arctic Ocean, and that significant variability occurred only on the adjacent marginal seas and shelves. Data obtained in the Canada and Makarov basins prior to the 1990's show the Atlantic layer temperature maximum (θ_{max}) between 0.4 to 0.6°C (Carmack et al., 1995). Above this layer cold, nutrient-rich Pacific water was believed to form a stationary structure which extended across the Canada and Makarov basins to the Lomonosov Ridge (Kinney et al., 1970; Pounder, 1986; McLaughlin et al., 1996).

During the past five years, however, evidence has appeared of a major thermohaline transition event within the core of the Atlantic layer, first in 1990 in the Nansen Basin north of Svalbard (Quadfasel, et al., 1991) and then in 1993 in the Makarov Basin north of the East Siberian Sea (Carmack, et al., 1995; McLaughlin et al., 1996). Data from Arctic-94 show that this trend now extends across the Nansen, Amundsen, and Makarov basins. For example, the vertical section of θ from the Chukchi Sea to the North Pole (Figure 2a) shows cold (θ<0 °C) surface water in the upper 150-200 m and warm Atlantic water (θ>0 °C) at mid-depths. The multiple cores of water warmer than 0.8 °C between 200-350 m are associated with currents following the continental slope and submarine ridges crossed by the ship (cf. Aagaard, 1989). Figure 2b shows a section of θ anomaly, derived by subtract-

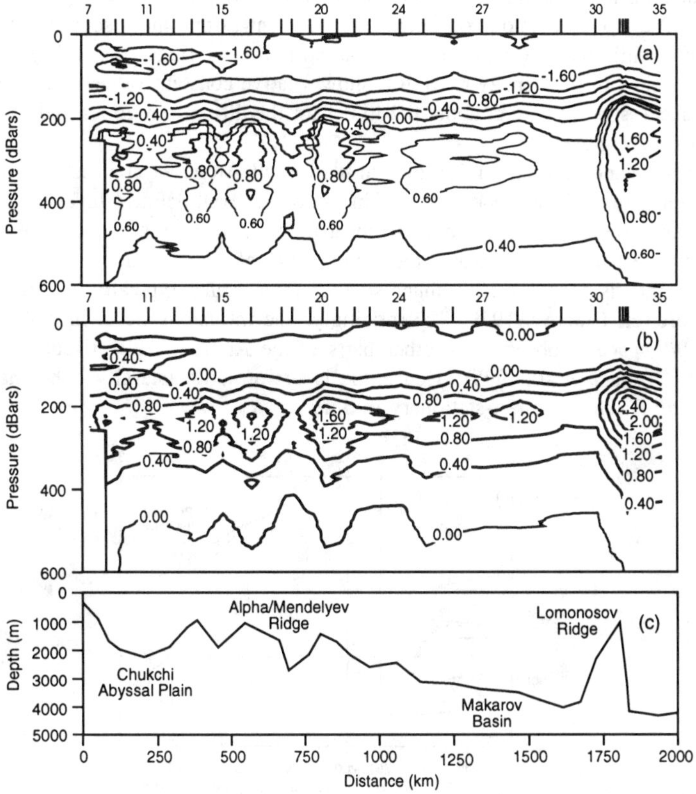

Figure 2. Sections across the Canadian Basin of (a) θ, (b) θ anomaly and (c) bathymetry.

ing the mean θ profile constructed from four North American ice camps prior to the 1990s (cf. Carmack et al., 1995) from each Arctic-94 station. The largest temperature change, over 2°C, occurs near 200 m since the Atlantic core is now both warmer and shallower. However, near-surface water (50-100 m) is now actually colder due to the replacement of waters of Pacific origin by surface waters from the Eurasian Basin. The missing Pacific water likely exited into the North Atlantic via Fram Strait or the Canadian Archipelago. If so, such exports of low salinity waters may have influenced convection in the gyres of the Greenland and Labrador seas. Cooling is also evident in the lower Atlantic layer, perhaps reflecting shelf inputs by sinking plumes of brine enriched water (cf. McLaughlin et al., 1996).

Thermohaline Structure

Profiles of θ and S versus depth (Figure 3) show the basic thermohaline structure of the Arctic Ocean. In the Atlantic layer above θ_{max} both temperature and salinity increase with depth, a condition supported of the diffusive instability (cf. Padman and Dillon, 1987), while below θ_{max} the water column is stabilized by both θ and S gradients. Superimposed on the large scale distributions, however, are multiple, finescale horizontal intrusions in which potential temperature inversions are stabilized by a density compensating salinity profile. The upper portion of each finescale intrusion may support the diffusive instability, while the lower portion may support the salt finger instability (cf. Schmitt (1994). Profiles of the density ratio, R_ρ, show relative constant values within each regime for each successive layer; e.g. about 0.2 to 0.4 for the diffusive regimes and 1.5 to 1.8 for the salt finger regimes.

Expanded scale correlation θ/S diagrams from all Arctic-94 stations (Figure 4) show that the intrusion of new water occurs in coherent, well-defined layers. Further, the intrusions are aligned in θ/S space, and appear immune to mixing and disruption by shear flow and internal waves across the full width of the section. The aspect ratio (length/height) is near 2 x 10^{-6}, much smaller than scales typically reported for ocean finestructure (cf. Federov, 1979). Surprisingly, the intrusions occur in exactly the same place on θ/S space as observed in other parts of the Arctic in 1991 (Rudels et al., 1994) and 1993 (Carmack et al., 1995). Hence, the layers are both laterally coherent across the entire Arctic Basin and quasi-stationary in time.

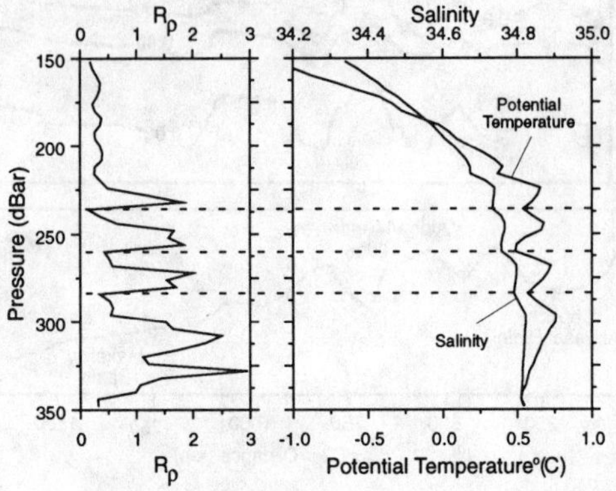

Figure 3. Vertical profiles of θ, S, and density ratio (R_ρ) at station 22 from Arctic-94.

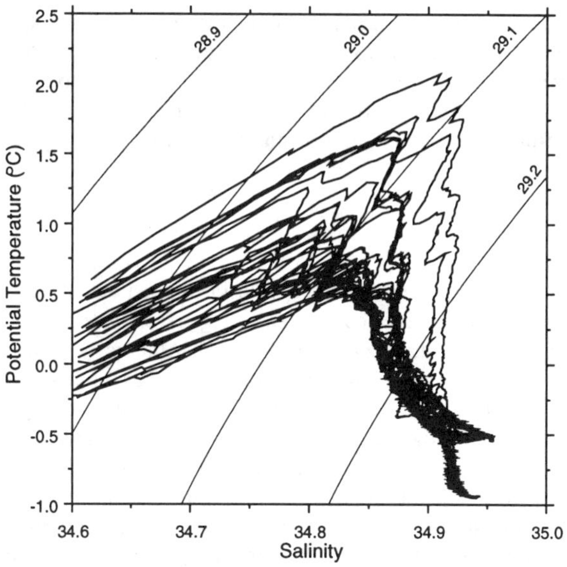

Figure 4. Correlation θ/S plots for Arctic-94 stations; also shown are lines of constant $\sigma_{0.25}$.

Figure 4 also shows that the slopes of θ/S for each successive layer in the vertical form a set of near-parallel lines for both the diffusive and salt finger regimes. The salt finger regimes cross isopycnals at a small angle (R_ρ about 1.2) suggesting that the transformation of layers as they spread laterally is not a purely isopycnal mixing process, but rather is such as to increase the salinity (and density) more rapidly than can be compensated for by the heat flux. The θ/S slopes associated with the diffusive regimes (about 4) are close to those giving maximum contraction on mixing, while the θ/S slopes associated with the salt finger regimes (about 20) are close that those yielding maximum stability to mixing (minimum potential energy) when the effects of contraction on mixing and differential compressibility are considered (Bennett, 1995). It is possible that the layers are, in fact, self-propelled by the conversion of potential to kinetic energy by salt-flux convergence, contraction on mixing, or both.

Because the salt finger regimes of each intrusion have constant θ/S slopes, it is useful to "rotate" the salinity axis by defining $S_{mix} = S - (\Delta S/\Delta\theta)\cdot\theta$, where $\Delta S/\Delta\theta = -0.056$, so as to vertically align the salt finger regimes, as is done in the θ/S_{mix} scatter diagram shown in Figure 5. Here, each θ/S_{mix} data point represents a layer of water 4 m thick, so that the density of points represents the volumetric distribution of water properties. Also shown in Figure 5 is the distribution of stability frequency, $N = ((g/\rho)(\delta\rho/\delta z))^{1/2}$.

Here, several features are evident. First, the water is volumetrically "banded" in θ/S_{mix} space; that is, there is a greater volume of water occurring along narrow S_{mix} lines aligned with the salt finger regime than elsewhere. Second, while N shows a general decrease with depth from values of about 4 hr^{-1} in the thermocline region to about 0.5 hr^{-1} in the deep water, the pattern is dominated by a series of alternating values associated with each intrusion: within each diffusive regime N is maximum; within each salt finger regime N is near-zero. Within each individual intrusion N (and potential vorticity) is constant across the full width of the Arctic Ocean.

The thermohaline transitions span not only the Atlantic layer, but in places extend into the deep water. Figure 6 is a θ/S_{mix} plot expanded in scale to show detail in the deep water. Here, layer structure (about 30 individul layers) is evident to depths exceeding 2000 m, the approximate sill depth of the Lomonosov Ridge which separates waters of the Makarov Basin from the Atlantic; R_ρ remains nearly constant despite major changes in ambient stratification with depth.

Conclusions

High-latitude oceans play an important role in the context of global warming. For example, the export of freshwater components (e.g. river water, sea ice, etc.) from the Arctic Ocean into the convective gyres of the North Atlantic is thought to influence the strength of the ocean's thermohaline circulation and thus global climate (cf. ACSYS, 1994). They are also presumed to be a bellwether of global warming (Walsh and Crane, 1992). The sensitivity of high-latitudes to change in such models, however, is due to albedo feedback effects. In contrast, Aagaard and Carmack (1989; 1994) argued that changes in ocean climate could also be effected by inter-basin advection of water masses, such as the export of fresh water from the Arctic to North Atlantic convective sites or the basin-wide modification of water mass structure due to changes in ocean circulation. Our data from the first oceanographic crossing of the Arctic Ocean documents what appears to be a recent influx of anomalously warm water from the Atlantic into the Arctic Ocean at mid-depth. This temperature increase, due to advection rather than albedo feedback effects, is sufficiently large to mask changes predicted by GCMs.

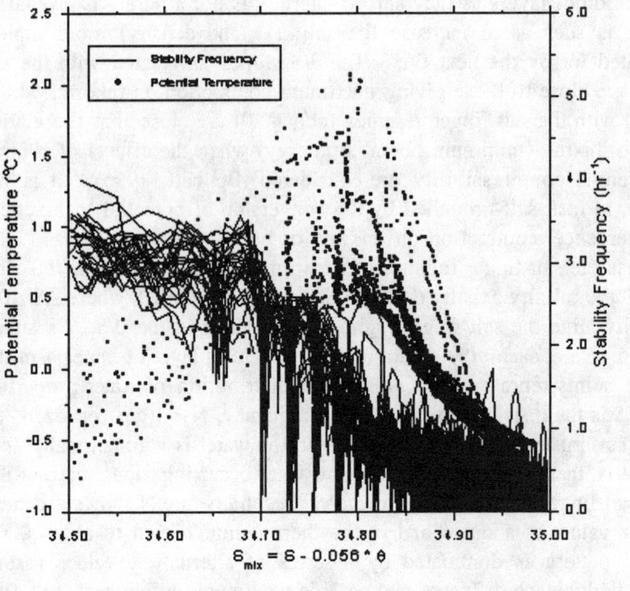

Figure 5. Correlation plots for Arctic-94 stations showing θ/S_{mix}, and N/S_{mix}, where S_{mix} is a parameter obtained by rotating the salinity axis to be parallel with the mixing bands, and N is the stability frequency.

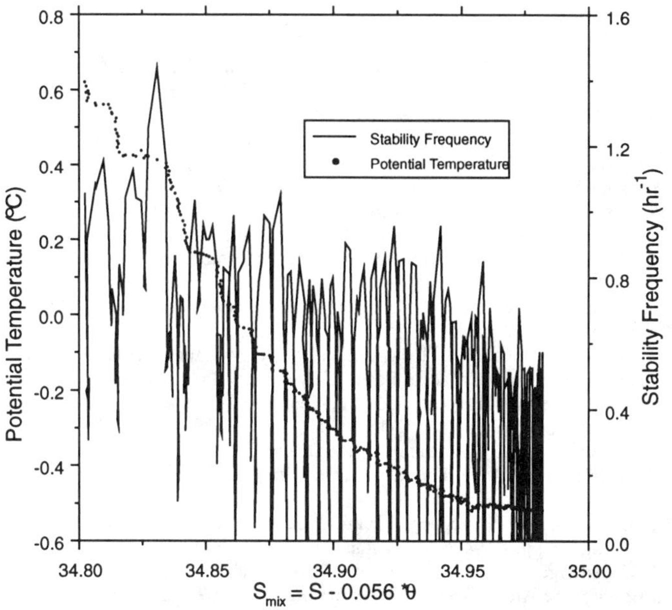

Figure 6. Plots from Station 23 showing correlations of θ/S_{mix} and N/S_{mix} in the range θ = -0.6 to 0.8 °C and S_{mix} = 34.8 to 35.0.

We interpret these data to signal a major change in the thermohaline structure of the Arctic Ocean driven by changes in the transport of Atlantic waters through Fram Strait and the Barents Sea, and an internal re-arrangement of water mass boundaries, rather than by local surface heat exchange. As such, this change could represent (a) a quasi-periodic interannual event akin to an ENSO event, (b) an episodic event analogous to the so-called great salinity anomaly that issued low salinity water from the Arctic to the North Atlantic (Dickson et al., 1989) or (c) a transition from one water mass structure to another (Bryan 1986). Clearly, this question must be resolved before models or observations (i.e. times series of temperature, ice thickness, etc.) can be interpreted as reliable indicators of global warming. However, the process of thermohaline transition via self-organized and self-propelled layers has not be simulated, parameterized, or accounted for in GCMs. Thus, if our data do indeed represent a transition from one stable state to another, and such a transition is a precursor to climate change, then GCM are not presently able to predict either the transition or quantify the change.

Acknowledgments. We deeply appreciate the efforts of Capt. P. Grandy, and the officers and crew of the *CCGS Louis S. St-Laurent*, in making Arctic-94 a successful mission. We are indebted to the extraordinarily competent technical support of all science personnel on Arctic-94; in particular: J. Elliott, D. Muus, R. Nelson, R. Pearson, D. Sieberg, and R. Williams.

References

Aagaard, K., A synthesis of the Arctic circulation, *Rapp. P.-v Reun. Cons. int Explor. Mer. 188*, 11-22, 1989.

Aagaard, K., and E. C. Carmack, On the role of sea-ice and other freshwater, *J. Geophys. Res. 94*, 14485-14498, 1989.

Aagaard, K., and E. C. Carmack, The Arctic and climate: A perspective, In *The Polar Oceans and Their Role in Shaping the Global Environment*, edited by O. M. Johannessen, R. D. Muench, and J. E. Overland, pp. 4-20, Geophys. Monogr. 85, Am. Geophys. Un., 1994.

Anderson, L. G., G. Bjork, O. Holby, E. P. Jones, G. Kattner, K. P. Koltermann, B. Liljeblad, R. Lindegren, B. Rudels, and J. Swift, Water masses and circulation in the Eurasian Basin: Results from the Oden-91 expedition, *J. Geophys. Res.*, *99*, 3273-3283, 1994.

ACSYS, Arctic Climate Systems Study, Initial Implementation Plan, WCRP-85, 66 p., 1994.

Bennett, E. B., Stationary Ocean Structures. *Prog. Oceanogr.*, Submitted, 1995.

Bryan, F., High-latitude salinity effects and interhemispheric thermohaline circulation, *Nature*, *323*, 301-304, 1986.

Carmack, E. C., Large-scale oceanography. In *Polar Oceanography*, Part A, edited by W. O. Smith, pp. 171-222, Academic, San Diego, 1990.

Carmack, E. C., R. W. Macdonald, R. G. Perkin, and F. A. McLaughlin., Evidence for warming of Atlantic Water in the southern Canadian Basin. *Geophys Res. Lett.* 22, 1061-1064, 1995.

Dickson, R. R., J. Meinke, S. A. Malmberg, and A. J. Lee, The "great salinity anomaly" in the northern North Atlantic, *Prog. Oceanogr.*, *20*, 103-151, 1988.

Federov, K. N., *The thermohaline finestructure of the Ocean*, 170 p., Pergamon, Oxford, 1979.

Gorshkov, S. G., *World Ocean Atlas, Vol. 3, Arctic Ocean* (in Russian) 189 p. Pergamon, New York, 1983.

Jones, E. P., D. M. Nelson, and P. Treguer, Chemical Oceanography. In *Polar Oceanography*, Part B, edited by W. O. Smith, pp. 407-476, Academic, San Diego, 1990.

Kinney, P., M. E. Arhelger, and D. C. Burrell, Chemical characteristics of water masses in the Amerasian Basin of the Arctic Ocean, *J. Geophys. Res.*, *75*, 4097-4104, 1970.

McLauglin, F. A., E. C. Carmack, R. W. Macdonald, and J. Bishop, The Atlantic/Pacific water mass boundary in the Southern Canadian Basin. *J. Geophys. Res.*, *101*, 1183-1197, 1996.

Padman, L., and T. M. Dillon, Vertical fluxes through the Beaufort Sea thermohaline staircase. *J. Geophys. Res.*, *93*, 10,799-10,806, 1987.

Perkin, R. G., and E. L. Lewis, Mixing in the West Spitsbergen Current, *J. Phys. Ocean.*, *14*, 1315-1325, 1984.

Pounder, E. R. Physical oceanography near the North Pole, *J. Geophys. Res.*, *91*, 11,763-11,773, 1986.

Quadfasel, D., A. Sy, D. Wells, and A. Tunik, Warming in the Arctic, *Nature*, *350*, 385, 1991.

Ruddick, B., Intrusive mixing in a Mediterranean salt lens - Intrusion slopes and dynamical mechanisms, *J. Phys. Oceanogr.*, *22*, 1274-1285, 1992.

Rudels, B., E. P. Jones, L. G. Anderson, and G. Kattner, On the intermediate depth waters of the Arctic Ocean, In *The Polar Oceans and Their Role in Shaping the Global Environment*, edited by O. M. Johannessen, R. D. Muench, and J. E. Overland, Geophys. Monogr. 85, Am. Geophys. Un., 1994.

Schmitt, R. W., Double Diffusion in Oceanography. *Annu. Rev. Fluid Mech.*, *26*, 255-285, 1994.

Unesco, Algorithms for computation of fundamental properties of seawater. *Tech. Pap. Mar. Sci.*, *44*, 53 pp., 1983.

Walsh, J. E., and R. G. Crane, A comparison of GCM simulations of arctic climate, *Geophys. Res. Lett.*, *19*, 29-32, 1992.

Wuest, A., D. M. Imboden, and M. Schurter, Origin and size of hypolimnic mixing in Urnersee, the southern basin of Vierwaldstattersee (Lake Lucerne), *Schweiz. Z. Hydrol.*, 50, 40-70, 1988.

13

Exchange Flows in Lakes

P. F. Hamblin

Abstract

Concern for transfer of contaminants between interconnected lakes, between lake basins and between the open lake and embayments or harbours has stimulated interest in exchange flows. In this review the term exchange flow refers to not only the flow but also the flux of material substances between two water bodies. Due to their intermittency and spatial complexity exchange flows in lakes have been mainly studied by field experiments on a site specific basis. Despite obvious differences in the nature of the forcing, much of their theoretical framework stems from oceanography. One reason that the theoretical understanding is still at an elementary stage in lakes is their complexity arising due to the need to consider both advective and diffusive effects. In general, the circulatory aspects of lakes have been studied more than their mixing characteristics. This is especially important for stratified exchange flows where vertical mixing within the exchange flow itself can lead to short-circuiting of the exchange of properties. A rapid improvement in the understanding of exchange flows is anticipated as their study becomes facilitated by the effective employment of advanced observational methods coupled with three-dimensional models. An example is presented of the use of mathematical models to obtain insight into the influence of water level fluctuations and wind driven circulation on the flushing between a harbour and a large lake.

Introduction

Owing to the increased use of lakes for recreational purposes, sources of water supply and as receiving waters for municipal and industrial discharges, gradients in water quality have developed within many lakes, between distinct basins and between nearshore and offshore areas. When the general lake circulation is constricted by shoreline configuration or by bathymetry, as is the case for entrances to some embayments or harbours, then the mixing and circulation of water masses between these basins is often termed exchange flow. It is important to stress that exchange flows do not simply refer to the local circulation in these restricted water bodies, but that the subject includes the combined effect of both advection and diffusion on the transport of water quality constituents. However, in many studies estimates of exchange flows have been based on measurements or modelling of flows alone when it was not possible to include diffusive effects. An outstanding characteristic of exchange flows is their complexity. Not only do they combine many of

Physical Processes in Lakes and Oceans
Coastal and Estuarine Studies Volume 54, Pages 187-198
Published in 1998 by the American Geophysical Union

the physical processes that occur in lakes with irregular topography, but due to the confinement of the geometry, flows tend to be concentrated to an extent seldom experienced in other regions of lakes. These flow concentrations result in unusually high Froude numbers and their gradients, which means that linear theory is seldom applicable. While much of our current understanding of exchange flow processes in lakes stems from oceanography, significant differences exist for lakes. The flushing of marine coastal embayments, as for example, the Western Australian marina studied by Schwartz and Imberger (1988), is dominated by the single and highly regular frequency of the astronomical tide. On the other hand, the forcing of exchange flows in lakes ranges from time scales of infragravity waves to meteorological disturbances of about 100 hours and lacks the regularity of tidal forcing. This episodic nature of exchange flow in lakes adds an element of complexity not generally found in the oceans.

In the present paper methods of observation of exchange flows are discussed, some examples from the limnological literature are cited illustrating the complexity and diversity of exchange flows, some theory is reviewed and finally an example of the use of numerical models in facilitating the understanding of exchange flow between a harbour and a lake is presented.

Methods of Observation of Exchange Flows in Lakes

The complex nature of exchange processes necessitates the collection of site specific field observations. In cases where gradients in properties exist between basins an observational approach that may provide useful information on exchange is the budget method whereby differences in the concentrations of a material substance such as dissolved solids between two water bodies may be exploited to determine the exchange. For example, the salinity of Hamilton Harbour which is connected to Lake Ontario (Figure 1), is about twice that of Lake Ontario. Klapwicjk and Snodgrass (1985) inferred the seasonal exchanges between Hamilton Harbour and Lake Ontario based on the salt balance of the harbour. Unfortunately, errors associated with the subtraction of large numbers common to budget methods meant that their method could not resolve variations in exchange having scales less than a month. Non conservative substances have been exploited too. Palmer et al. (1993) determined the exchange between Georgian Bay and a small bay favoured as a week-end anchorage by recreational boaters from the concentration of pathogens. The authors measured the effluent concentrations of bacterial contamination and attempted to relate them to physical conditions by means of a simple exchange model as the levels in the bay built up and dissipated over the course of a number of weekends. The principal advantage of budget methods is that they yield estimates of the combined effects of mixing and flow on the exchange. However, in many situations in lakes property gradients between basins are not sufficiently large or the inputs too uncertain to infer exchanges from the budget method.

In cases where the budget technique is unsuitable interbasin exchanges have been observed less directly by measurement of current based on moored current meters placed in the connecting passage. For example, Saylor and Sloss (1976) measured exchange between Lake Michigan and Lake Huron with an array of 11 recording current meters placed in the connecting passage and discovered what may be the largest lacustrine exchange ever recorded of 0.085 Sverdrup (1 Sverdrup = 10^6 m^3 s^{-1}) and associated current speeds of 1.1 m s^{-1}. Similarly, Bennett (1976) observed the long-term exchange between Lake Huron and Georgian Bay. Both studies found eastward flow in the upper layer and westward flow

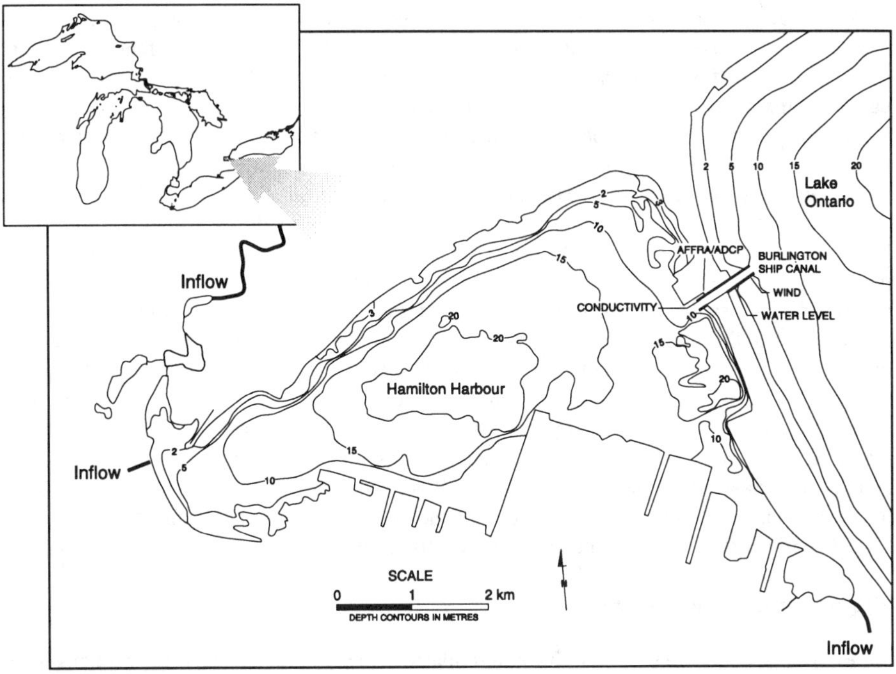

Figure 1. Instrument locations, inflows and bathymetry of Hamilton Harbour and the adjoining portion of western Lake Ontatio.

below during the stratified season. Interbasin exchanges have been extensively measured with current meters and thermistor arrays in Lake Erie by a considerable number of investigators and have been thoroughly reviewed by Bartish (1987). The difficulty with these current meter observations is that while they are capable of resolving all the relevant time scales except for the turbulent ones, they do not account for the effect of mixing on the exchange. In a subsequent section on model application the importance of the excursion distance of a flow intruding from one water body to another is discussed. For example, it is noteworthy in the case of Lake Erie that Chiocchio (1981) found that hypolimnetic exchanges between the eastern and central basins had frequent flow reversals of 2-6 days and that the excursion distance is typically 9 km, which is just sufficient to clear the connecting passage and enter the opposite basin. The net exchange in this case will depend critically on the hypolimnetic circulations in the two basins. Recording thermistor profilers have served as a valuable complement to current meters. Okamoto and Endoh (1995) have inferred net exchange volume associated with both internal co-oscillations and extreme upwelling events driven by typhoons from such data and the hypsometric curves for a large deep bay connected to Lake Biwa. Finally, the interesting case of exchange flow further complicated by the effects of the temperature of maximum density has been studied experimentally by Roy (1983) between Black Bay and Lake Superior.

In some instances the measurement of long-term exchanges between harbours and lakes presents special difficulties. In the case of frequent ship traffic in the connecting waterway conventional moored current meters are not possible due to hazards to unattended instrumentation from commercial shipping. Simons and Schertzer (1983) addressed this problem in the case of the Burlington Ship Canal by deploying a differential pressure gauge designed to measure the water level difference between the two ends of the ship canal. Unfortunately, they were unable to establish the calibration of this potentially useful device on account of the operational failure of an array of standard current meters moored during a rare occasion when the canal was closed to ships. Since 1989 devices which remotely sense currents have been commercially available. An evaluation of the field performances of two types of ultrasonic current meters in the Burlington Ship Canal has been reported by Hamblin (1991), one, a bottom mounted acoustic Doppler current meter (ADCP in Figure 1) and the other, a time-of-flight current meter (AFFRA in Figure 1) integrating across the channel at the level of the acoustic ray. Hamblin (1996) has provided a detailed analysis of the in-situ calibration of the latter instrument based on a method developed by Spigel (1989) employing a conventional profiling current meter. Hamblin's (1996) study indicated that this technique is limited to the unstratified season when refraction of the acoustic trajectories by the isopycnals is least. Furthermore, for the first time these measurements indicated on a monthly basis an inflow from Lake Ontario to Hamilton Harbour of 97 m^3 s^{-1}. Since this inflow is twenty times that of all other inflows to the harbour combined, the water quality of the harbour ought to be dominated by that of Lake Ontario if this inflow is mixed with the harbour waters.

Theoretical Considerations

Fischer et al. (1979) (p.264) have formulated a meaningful concept for exchanges in estuaries, the tidal exchange ratio, which is defined as the ratio of the volume of new ocean water actually exchanged by advection and diffusion to the total water entering an estuary over a flood cycle. Unfortunately, it is usually not possible to predict the exchange ratio from theory. In a situation where the exchange is dominated by the longitudinal seiche, Saylor and Miller (1987) have employed an analogous concept which they termed the tidal exchange prism length. They found that this length for exchanges between the western and central basins of Lake Erie ranged up to 5 km. With a prism length of 2 km, approximately 0.5 km^3 of water would be exchanged during each seiche cycle (14 hours), and at this rate complete flushing of the western basin by seiche activity alone would occur in 20-30 days. This example illustrates the shortcomings of such concepts as the tidal exchange prism length and much of the theory to follow compared to the tidal exchange ratio since only a fraction of the flow each cycle is actually exchanged by mixing processes. Without taking account of mixing actual flushing times will most likely be underestimated. The following theoretical considerations, while not leading to such useful quantities as the exchange ratio, at least shed some light on the pertinent temporal scales of interbasin exchange and the magnitude of exchanges excited by water level fluctuations and density gradients.

Resonant or Co-oscillations

Based on linear theory the equation of motion for the displacement, ζ_j, of a particle along the longitudinal axis, x, of a lake basin is, (Defant, 1961).

$$\frac{\partial^2 \zeta_j}{\partial t^2} + \beta_j \frac{\partial \zeta_j}{\partial t} - g h_j \frac{\partial^2 \zeta_j}{\partial x^2} = 0 \tag{1}$$

where g is the acceleration of gravity, h_j the depth and the damping coefficient, β_j may be expressed as μ/h_j^2 where μ is the vertical eddy viscosity and j refers to basin j. For a lake consisting of two basins, 1 and 3, connected by another, 2, the boundary conditions require zero displacement at the ends of 1 and 3 and continuity of mass and pressure between 1 and 2 as well as between 2 and 3. The free solution in basin j has the form $\zeta_j=A_j \sin(\alpha_j x)e^{i\sigma t} + B_j \cos(\alpha_j x)e^{i\sigma t}$ where i is the square root of minus one, σ is the complex frequency of oscillation and $\alpha_j = \surd\left((\sigma^2-i\beta_j\sigma)/gh_j.\right)$. From the boundary conditions it is possible to eliminate A_j and B_j in each basin, leading to the characteristic equation, where b_j and L_j are the breadth and length of basin j respectively,

$$b_2^2\ \alpha_1\ \alpha_3\ \tan\alpha_2 L_2 = b_1 b_3\ \alpha_2^2\ \tan\alpha_1 L_1 \tan\alpha_2 L_2 - b_2 b_3 \alpha_1 \alpha_2 \tan\alpha_3 L_3 - b_1 b_2 \alpha_2 \alpha_3 \tan\alpha_1 L_1 \tag{2}$$

By use of trigonometric identities it is straightforward to show that (2) reduces to the Merian formula when $b_1=b_2=b_3$, $h_1=h_2=h_3$ and all β_j are zero. Secondly, if it is assumed that the frequency of co-oscillation is less than the seiche frequency of any individual basin such that $\sigma < \surd(gh_j)/L_j$ and that there is no bottom friction then (2) reduces to

$$\sigma^2 = \frac{g b_2 h_2 A_L}{L_2 A_1 A_3} \tag{3}$$

where $A_L=A_1+A_2+A_3$ is the surface area of the lake and A_j is area of basin j.

The large spread between the fundamental periods of oscillation reported for Lake Biwa (240 min) and the next higher modes of 70 and 30 min (Imasato, 1971) suggests that the fundamental seiche is, in fact, a co-oscillation between the north and south basins. Based on the lake geometry published by Okuda and Kumagai (1995) and an assumed connecting channel length, L_2 of 4 km, breadth 1.4 km and mean depth of 3.5 m expression (3) yields a free period of 214 min. A similar application of this expression to the northern and southern basins of Lake Winnepeg results in inviscid periods of 31 and 45.5 hours when the extreme lengths of the connecting channel of 20 and 50 km respectively are assumed. These periods may be compared to that computed by means of a finite element model of 39.1 hours (Hamblin, 1976). In what is probably the longest exchange period found in any lake Saylor and Sloss (1976) observed a period of co-oscillation between Lakes Huron and Michigan from 44 to 76 hours which may be compared to that of (3) of 29 hours based on L_2 of 7 km and the geometrical data of Saylor and Sloss (1976). The poorer agreement in the last case likely due to the neglect of friction and to the difficulty of distinguishing between free and forced motions when the frequency approaches that of the wind forcing. The theory of equation (1) is extended is this study to five basins as is the case for the three coupled basins of the Lake of Lugano. Based on the basin geometry of Salvade and Zamboni (1987) an equation similar to equation (2) results in a co-oscillation period of 92 min which compares favourably to the observed period of 100 min.

Helmholtz Period

When one of the two basins, say 3, has a much larger area than the other then (3) reduces further to give the Helmholtz period, P, for an embayment,

$$P = 2\pi\sqrt{\frac{L_2 A_1}{g b_2 h_2}} \tag{4}$$

Applications of this flushing mode to three Great Lakes harbours have been provided by Freeman et al. (1974) as well as its extension to harbours with multiple inlets such as Toronto Harbour. These expressions in cases where the density field may be represented by two layers and the bathymetry not too complex may provide internal co-oscillations when density is replaced by reduced density, g'. The reduced gravitational acceleration $g'=g\ (\rho_2-\rho_1)/\rho_2$ and ranges from 10^{-2} - 10^{-4} m/s^2 in lakes. Okamoto and Endoh (1995) found that the summer flushing period of Lake Biwa's Shiozu Bay of 24h was often energized due to resonance with the daily wind. One practical implication of (4) in small harbours is that infilling reduces A_1, thereby decreasing the length of time for flushing or the excursion of the inflow.

Forced Oscillations

(a) Water level

The topic of tidally forced inflows into harbours and inlets is a well developed field with two excellent review articles providing sources of useful information to lakes as well, namely, Mehta and Joshi (1988) and van de Kreeke (1988). The approach is to predict the water level in the bay and the flow in the connecting channel forced by the ocean tide based on a one-dimensional momentum equation nearly identical to (1) except for the retention of full nonlinearity integrated spatially along the channel and a continuity equation equating the flux through the channel to the rise of water level in the bay. Both authors stress that in most cases it is important to retain the inertia term and various techniques have been developed to obtain analytical solutions for sinusoidally forced motion. In an early application of these equations (except that local acceleration of the flow was neglected) Dick and Marsalek (1973) found that inertia was significant in the case of exchange flow between Hamilton Harbour and Lake Ontario. Hamblin and He (1996) showed that integration of the complete one-dimensional equations in time can account for about 60% of the variance of 15-min observations of the channel flow over a period of a month. Despite the great success of these simple equations in simulating the flow they provide no information on the mixing of the exchange. This question will be examined in a subsequent section on the application of two higher dimensional models.

(b) Densimetric forcing

As a result of unequal heat capture between the shallower confined water body and the open lake as well as upwelling of colder bottom water in the lake there may be a contrast in density at the level of the connecting passage which drives a densimetric exchange flow. This concept of a summer exchange between Lake Ontario and Hamilton Harbour that has emerged from the studies of Dick and Marselek (1973), Klapwijk and Snodgrass (1985) and Spigel (1989) is illustrated schematically in Figure 2. Cold Lake Ontario water enters the Harbour along the bottom while warmer and lighter Harbour water exits as a surface layer. To model the two-layer densimetric exchange for flows where both friction and inertia are important, internal hydraulic theory may be applied as developed for an exchange problem in the Great Salt Lake by Holley and Waddell (1977). If it is assumed that exchange flows are sufficiently strong to develop internal hydraulic controls at or near the two ends of the joining passage then at these points the composite Froude number G2 is unity where $G2=F_1^2+F_2^2$ and the densimetric Froude numbers $F_i^2=u_i^2/g'h_i$, i=1,2. The subscript 1 refers to

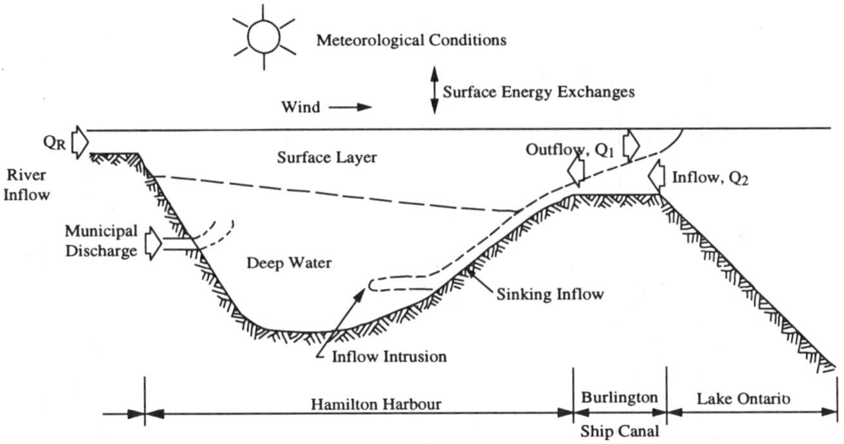

Figure 2. Schematic of the two-layer exchange flow between Hamilton Harbour and Lake Ontario, applicable on occasions from May to October.

the upper layer (outflow) while 2 denotes the lower layer (inflow). The thickness and average velocity of each layer are symbolized by h_i and u_i respectively. In accordance with internal hydraulic theory the slope of the interface along the channel, dh_1/dx, is given by the internal resistance equation, e.g. Rigter (1970),

$$\frac{dh_1}{dx} = -\frac{K_b F_2^2 + K_I \dfrac{(h_1 + h_2)\left(|u_1| + |u_2|\right)^2}{h_1 h_2 g'}}{1 - G^2} \tag{5}$$

Of course, this equation cannot apply at the control points where $G^2=1$. Apart from uncertainties in the specification of the frictional coefficients of K_b for bottom friction and K_I for internal friction the internal hydraulic equations must be supplemented by additional equations in order to determine all the variables. Hamblin (1989) and Hamblin and Lawrence (1990) have predicted the inflow based on the observed value of the outflow and compared the modelled inflows to Hamilton Harbour with sufficient correspondence to the field observations of Spigel (1989) to demonstrate that frictional effects are important. Further work on developing a time dependent model of the internal exchange flow has been sparked by this limitation of the internal hydraulic equations and by more recent observations of current profiles in the Burlington Ship Canal at the location marked as ADCP in Figure 1. The flows depicted in Figure 3 were measured with an acoustic current profiler. These data indicate a highly transient flow regime alternating between the vertically uniform barotropically forced exchange discussed in (a) above and the two-layer densimetrically dominated condition shown in Figure 2.

A related problem of the exchange due to the diurnal and seasonal cooling of sidearms has been recognized as mainly a theoretical problem since such exchange flows are usually masked by stronger wind forced circulations. Monismith et al. (1990) have studied such convective motions theoretically. Okubo (1995) found intermittent intrusions of more dense bottom water flowing into the northern basin of Lake Biwa formed in the shallow southern basin during episodes of intense autumnal cooling.

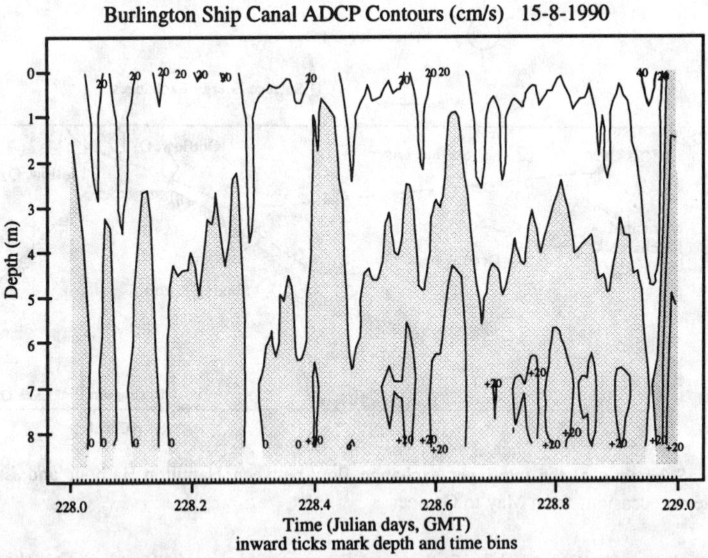

Figure 3. Isopleths of the along channel component of flow between Lake Ontario and Hamilton Harbour as a function of depth and time for July 15, 1990. Positive flow in cm s^{-1} is flow towards the lake. Grey shading is opposed.

Numerical and Physical Modelling

Physical limnologists have badly neglected laboratory and numerical experiments (Imberger, 1994). Similarly, few studies of physical modelling of exchange flows in lakes have been published. As the importance of exchange flows in lakes becomes more widely recognized this will hopefully change. G.A. Lawrence (pers. com.) is planning to conduct laboratory experiments on the exchange between Hamilton Harbour and Lake Ontario. Laboratory experiments could provide potentially valuable insight provided mixing processes can be accurately represented in the model.

Mathematical modelling of exchange processes is a particularly valuable tool for facilitating their understanding and providing practical input to other limnological studies. For example, ecologists modelling the ecosystems in lakes often require the specification of fluxes between physical boundaries of the cells of their models and turn to the physical limnologist for assistance. Recently, elaborate three-dimensional lake circulation models have been used to assist in the interpretation of exchange phenomena in lakes. For example, Oonishi (1995) has demonstrated by means of a three-dimensional model the blocking effect of the earth's rotation on the weak convectively driven exchange between the south and north basins of Lake Biwa. Moreover, he showed that with rotation average cooling rates were insufficient to drive a density current but at four times the average rate a strong intermittent exchange was evident. In the next section another example of the use of two-dimensional circulation and transport models to gain insight into exchange processes will be presented in some detail. On the subject of the exchange between the principal basins of Lake Biwa it is mentioned parenthetically that shortly our knowledge will be greatly enhanced as the results of the coordinated international field experiment conducted in 1993 appear in the literature.

Seiche and Wind Driven Flushing of Hamilton Harbour

Hamilton Harbour (Figure 1) has a surface area of 21.8 km^2, and is connected to the western extremity of Lake Ontario via a 828 m long canal of uniform depth (9m) and cross section (width 88 m). During the unstratified period a steady buildup of salt is observed which is attributed to the reduced exchange once the densimetric exchange (Figures 2 and 3) ceases. During this unstratified period exchanges are driven by water level fluctuations associated with storm surges and seiches of Lake Ontario, by local wind driven circulation and to a lessor extent by three tributaries. During the ice-free winter period wind stirring sufficiently homogenizes the water column to reasonably represent the harbour and adjoining region of the western end of Lake Ontario by a two-dimensional vertically uniform model.

Method

The resolution finite element model developed by Hamblin and He (1996) utilized an Euler-Lagrangian (ELM) or characteristic method for advection after the more standard wave equation solution of the governing shallow water equations failed to converge for the relatively high Froude number flow regime of the connecting passage. With a semi-implicit method of solution of free surface wave propagation the model is unconditionally stable. However, a time step of 2s was selected to maintain a Courant number of unity as the gravity wave accuracy degraded rapidly with higher Courant numbers. The graded finite element mesh shown in Figure 4 consisted of 4352 triangular elements and 7252 nodal points. A Chezy coefficient of 60 was specified for bottom friction but the model has no explicit diffusion of momentum. Water levels were specified at the outer boundary in Lake Ontario. Output from this model provided the advection field for a transport model also of

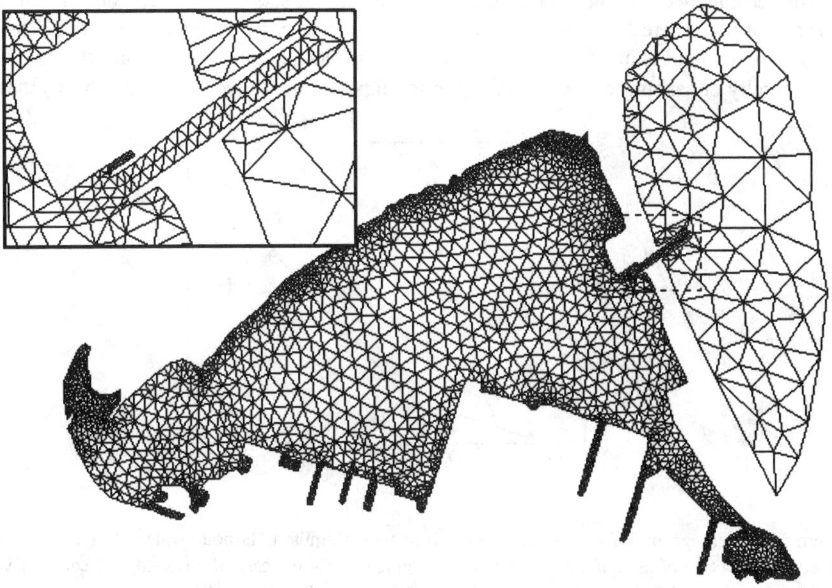

Figure 4. Triangular finite element mesh used in the modelling study.

finite element and ELM type, once a periodic solution had been obtained. Anisotropic diffusivity was represented in the advection-diffusion model with the major component parallel to the instantaneous streamlines. The magnitude of eddy diffusivity was assumed to be proportional to the local rate of dissipation and the scale of the element in the manner prescribed by Fischer et al. (1979) (p.76). The constant of proportionality was selected to maximize the portion of the observed variance accounted for by the simulation of a time series of 15-min salinity observations taken over a monthly period in the channel connecting the two water bodies (marked as conductivity in Figure 1).

In the series of flushing experiments to be presented subsequently the harbour water was set to a uniform salt concentration initially and then flushed by a differing lake concentration and run for four months. Since Hamilton Harbour demonstrates a resonant Helmholtz mode of 2.6 hours (Freeman et al., 1974) this period was chosen as a typical forcing period. Water level amplitudes ranging from 5 to 0.5 cm resulted in excursion distances of the order of the length of the connecting channel. Flushing response times were compared. In a second set of experiments steady wind forcing at 8 m s^{-1} directed along and normal to the channel connecting the two water bodies was added to one of the water level forced cases.

Results

The percentage flushing of harbour water by Lake Ontario after four months shows a definite sensitivity in Figure 5 to the ratio of the excursion distance to the connecting channel length or non dimensional excursion. It would appear that inflow excursions of about twice the channel length are required to significantly mix the two water bodies. In further numerical experiments with water level forcing combined with steady wind forcing for various directions it was found unexpectedly that winds along and across the canal did not further enhance mixing. Periodic water level forcing sets up stagnant zones at the ends of the channel where the concentrations are intermediate between the two water bodies Wind generated circulation would be expected to advect undiluted water into the exchange zone thereby enhancing exchange. Despite the input of undiluted water into the exchange.

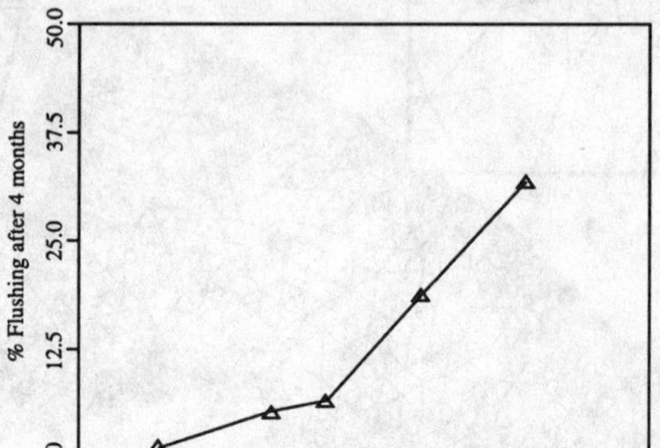

Figure 5. Percentage of Lake Ontario water mixed with Hamilton Harbour water after a four month period as a function of the non dimensional excursion length for the case of no wind and constant water level forcing at the Helmholtz period of 2.6 hours. Exc. is length of the exchange excursion and L is the length of the connecting channel.

zone by the wind generated circulation as evidenced by distinct cross channel concentration gradients this tendency was overwhelmed by the choking effect of the cross channel component of flow on the exchange and associated excursion length. It was evident from the numerical experiments that the pressure gradient along the canal and hence, the discharge in the connecting channel, was reduced by the velocity head associated with the transverse component of flow at the channel entrance. This effect is similar to the behaviour of outfalls in a cross flow and is explained by the theory of manifold hydraulics, Fischer et al. (1979) (p.413).

Discussion

In this section the effectiveness of models that combine mixing and circulation has been elaborated, albeit for two-dimensional models. For the first time in lakes the effects of both wind driven flow and water level forcing on the flushing efficiency of a confined water body have been examined. This study underscores the limitations of the more traditional one-dimensional approach unless the effects shown herein can be accurately parameterized.

Conclusions

With growing concern for the water quality of lakes, there is a clear need for increased understanding of exchange flows in lakes. Present knowledge has been hampered to a certain extent by the complexity of exchange processes in lakes and the difficulty of generalizing from one site to another. Promising news tools of investigation such as three-dimensional numerical models and remote sensing, when effectively combined, should lead to a rapid increase in understanding. Until now there has been a notable lack of attention devoted to mixing in exchange flows, both on the experimental and theoretical sides. Mixing processes are especially critical in stratified exchange problems since vertical mixing between the counterflowing layers acts as a short circuit for the transfer of substance through the exchange zone. This remains one of the most challenging problems confronting physical limnologists at present.

Acknowledgments. G. A. Lawrence, K. Hutter and a referee are thanked for their suggestions of useful references. I would like to acknowledge the help of C. He and R. Pieters in the preparation and display of the model output and field data.

References

Bartish, T., A review of exchange processes among the three basins of Lake Erie. *J. Great Lakes Res.*, *13*, 607-618, 1987.

Bennett, E. B., Aspects of the physical limnology of Georgian Bay, Ontario. *Unpublished Report.* Canada Centre Inland Waters. 1-33, 1976.

Chiocchio, F., Lake Erie hypolimnion and mesolimnion flow exchange between the central and eastern basins during 1978. *National Water Research Institute,* CCIW, Internal Report. APSD 009, 1981.

Defant, A., *Physical Oceanography.* Vol. II, Pergamon Press, New York, 598p, 1961.

Dick, T. M. and J. Marsalek, Exchange flow between Lake Ontario and Hamilton Harbour. *Unpublished Report* Sci. *Ser.* No. 36 Canada Centre for Inland Waters, Burlington, Canada, 1973.

Fischer, H. B., E. J. List, R. C. Y. Koh, J. Imberger and N. H. Brooks, Mixing in inland and coastal waters. *Academic Press,* 1979.

Freeman, N. G., P. F. Hamblin and T. S. Murthy, Helmholtz resonance in harbours in the Great Lakes. *Proc. 17th Conf. on Great Lakes Res.*, 399-411, 1974.

Hamblin, P. F., Remote flow sensing in the Burlington Ship Canal. *Manuscript submitted for publication,* 1996.

Hamblin, P. F., Acoustical remote sensing of exchange flows in lakes. *Proc. Hydrotechnical Conference, Can. Soc. Civil Eng., 1,* 11-15, 1991.

Hamblin, P. F., Notes on the hydraulics of Hamilton Harbour. *Unpublished Report, National Water Research Institute Contr., No. 89-36,* 1989.

Hamblin, P. F., Seiches, circulation and storm surges on an ice-free Lake Winnepeg. *J. Fish, Res. Bd. Can., 33,* 2377-2391, 1976.

Hamblin, P. F. and C. He, Unstratified harbour exchange hydraulics in lakes. *Manuscript in preparation,* 1996.

Hamblin, P. F. and G. A. Lawrence: Exchange flows between Hamilton Harbour and Lake Ontario. Proc. 1990 *An. Conf. Can. Soc. Civil Eng., 1,* 140-148, 1990.

Holley, E. R. and K. M. Waddell, Stratified flow in Great Salt Lake culvert. *J. Hydr. Div. ASCE, 102,* 969-985, 1976.

Klapwijk, A. and W. J. Snodgrass, Model for lake-bay exchange flow. *J. Great Lakes Res., 11,* 43-52, 1985.

Imasato, N., 1971: Study of seiches in Lake Biwa-ko (II) *Contribution Geophys. Inst., Kyoto Univ. 11,* 77-90.

Imberger, J. Transport processes in lakes: A review. In *Limnology Now: A Paradigm of Planetary Problems,* edited by R. Margalef, pp. 99-193, Elsevier Science, 1994.

Mehta, A. J. and P. B. Joshi, Tidal inlet hydraulics. *J. Hydr. Engr., 114,* 1321-1338, 1988.

Monismith, S., J. Imberger and M. L. Morison, Convective motions in the sidearm of a small reservoir. *J. Limnol. Oceangr., 35,* 1676-1702, 1990.

Okamoto, I. and S. Endoh, Water mass exchange between the main basin and Shiozu Bay. In *Physical Processes in a Large Lake: Lake Biwa, Japan,* edited by S. Okuda, J. Imberger, M. Kumagai, Coastal and Estuarine Studies, Vol. 48, pp. 31-42, American Geophysical Union, 1995.

Okubo, K., Field observations of the dense bottom water current between the north and south basins. In *Physical Processes in a Large Lake: Lake Biwa, Japan,* edited by S. Okuda, J. Imberger, M. Kumagai, Coastal and Estuarine Studies, Vol. 48, pp. 43-51, American Geophysical Union, 1995.

Okuda, S. and M. Kumagai, Introduction. In *Physical Processes in a Large Lake: Lake Biwa, Japan,* edited by S. Okuda, J. Imberger, M. Kumagai, Coastal and Estuarine Studies, Vol. 48, pp. 1-6, American Geophysical Union, 1995.

Oonishi, Y., Numerical simulation of density-induced currents between the north and south basins of Lake Biwa. In *Physical Processes in a Large Lake: Lake Biwa, Japan,* edited by S. Okuda, J. Imberger, M. Kumagai, Coastal and Estuarine Studies, Vol. 48, pp. 53-64, American Geophysical Union, 1995.

Palmer, M. D., M. F. Hollaran and M. J. Robert, The effect of indicator organisms in the waste water disposal on mooring embayments. *J. Great Lakes Res., 19,* 352-360, 1993.

Rigter, B. P., Density induced return currents in outlet channels. *ASCE J. Hydr. Div., 96* (HY2), 529-546, 1970.

Roy, F. E., Spring exchange flows Black Bay, Lake Superior. *Unpublished Manuscript, National Water Research Institute, Burlington,* 1983.

Salvade, G. and F. Zamboni, External gravity oscillations of the coupled basins of the Lake of Ligano. *Annales Geophysicae, 5b,* 247-254, 1987.

Saylor, J. H. and G. S. Miller, Studies of large scale currents of Lake Erie. *J. Great Lakes Res., 13,* 487-515, 1987.

Saylor, J. H. and P. W. Sloss, Water volume transport and oscillatory current flow through the Straits of Mackinac. J. *Phys. Oceanogr., 6,* 229-237, 1976.

Schwartz, R. A. and J. Imberger, Flushing behaviour of a coastal marina, *Proc. 21 Internat. Conf. on Coastal Engineering.* Torremolinas, Spain, 2626-2640, 1988.

Simons, T. J. and W. M. Schertzer, Analysis of simultaneous current and pressure observations in the Burlington Ship Canal. *Unpublished Report, National Water Research Inst.* Contr. No. 83-20, 1983.

Spigel, R. H., Some aspects on the physical limnology of Hamilton Harbour. *Unpublished Report, National Water Research Institute,* Contr. No. 89-08, 1989.

Van de Kreeke, J., Hydrodynamics of tidal inlets. In *Hydrodynamics and Sediment Dynamics of Tidal Inlets* Vol. 29 Lecture Notes on Coastal and Estuarine Studies, edited by M. J. Bowman, pp. 1-21. Springer-Verlag, New York, 1988.

14

Gyres Measured by ADCP in Lake Biwa

M. Kumagai, Y. Asada and S. Nakano

Abstract

ADCP measurements were performed in Lake Biwa from April to December, 1994, and the spatial distribution of gyres in the North Basin was obtained. It was only in August when we could see all three of the well-known gyres described by Suda et al. (1926). The biggest counter-clockwise gyre named "the first gyre" was formed in June and disappeared in November. The first gyre was initially located in the north part of the North Basin, and then moved to the south as temperature stratification in the upper layer became stronger. It returned to the north again as the stratification weakened. It is suggested that the movement of the first gyre can be related to the seasonal or annual difference in heating intensity. We also found that the second and the third gyres were not stable, and that these gyres vary in their position and size according to the wind field.

Introduction

Since Suda et al. (1926) first reported three gyres in Lake Biwa, various studies, such as field observations, laboratory tank experiments and numerical simulations, have been carried out to understand the generation, growth and dissipation of these gyres (see Endoh, 1995). Okamoto and Morikawa (1961a) showed that the first gyre in Lake Biwa is almost in geostrophic balance. The dynamic height of the first gyre was c.a. 3.0 at the center and c.a. 4.0 outside (Endoh et al., 1995). The gyres can be induced by both wind (Hidaka, 1927) and thermal convection (Oonishi, 1975). However, because past studies had spatial resolution deficiencies due to technical restrictions, definite distribution of the gyres in Lake Biwa has not yet been clarified precisely.

Since the late 1970's, indirect current measurements in the water by ADCP (Acoustic Doppler Current Profiler) have become popular (Row and Young, 1979; Pinkel, 1979). ADCP is based on the Doppler effect of ultrasonic scattering in the water and allows the three dimensional structure of the velocity field to be determined rapidly. The echo intensity of ADCP can also be used to estimate zooplankton abundance in the ocean (*e.g.* Charles and Smith, 1989) and in a lake (Trevorrow and Tanaka, 1997).

In 1993, the Lake Biwa Research Institute launched R/V *Hakken* equipped with an ADCP (RD-400300B) broad band and a differential GPS (Global Positioning System: 4000DL-II/IIR & 4000RL-II/IIR) which can provide high precision positioning data with a 1 m accuracy for the ADCP measurements.

Physical Processes in Lakes and Oceans
Coastal and Estuarine Studies Volume 54, Pages 199-208

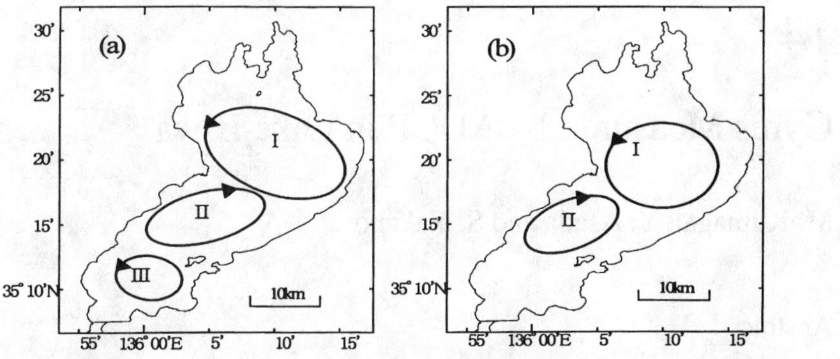

Figure 1. Gyres in Lake Biwa measured (a) by Suda et al.(1926) and (b) by Endoh and Okumura (1993).

Figure 1(a) shows the three gyres observed by Suda et al. (1926). The largest counter-clockwise gyre is called "the first gyre". However, as seen in Figure 1(b), the position of the first gyre measured by Endoh and Okumura (1993) moved southward from the original position shown by Suda et al. (1926). From the results of rotating tank experiment, Itakura and Kumagai (1988) showed the possibility that differences in heating intensity may change the position of the first gyre.

Although Suda et al. (1926) suggested that the second gyre was rotating clockwise and the third gyre was rotating counter-clockwise, as seen in Figure 1(a), Endoh and Okumura (1993) found that the second gyre was located to the south of the position reported by Suda et al. (1926), and they could not detect the third gyre.

The main aim of the present study was to determine the seasonal distribution of the gyres in Lake Biwa, and also to verify that the intensity of temperature stratification in the upper layer causes the gyres to shift.

Observations

The broad band ADCP on the R/V *Hakken* was used to measure the gyres in the North Basin of Lake Biwa from April to December, in 1994. The acoustic frequency of this ADCP is 300 kHz, and the depth cell size was fixed at 2 m. The averaging of the ADCP data over 15 seconds allowed us to obtain data at 60 m intervals along the ship track, because the vessel speed was kept at 8 knots during the ADCP measurement. Figure 2 shows eleven observation lines for the ADCP measurements and four profiling stations for the CTD measurements with the F-probe (Fine scale profiler).

As the vessel moved at 8 knots for the ADCP measurements, we could not cover the entire area within one day. Each observation was thus carried out for two days: the measurements from Line 1 to Line 6 were done on the first day and the rest were done on the second day. We could not see any serious discrepancy between the data taken at Line 6 and Line 7, although those were measured on successive days. It means that the first gyre has sufficient energy to resist disturbance by normal external forcing such as moderate wind. Only a strong typhoon may disturb the first gyre, but it quickly recovers within a few days.

At four stations (AS-1, AS-2, AS-3 and AS-4), we also lowered the F-probe to measure water temperature, pH, turbidity, chlorophyll *a* and dissolved oxygen for comparison with the ADCP data.

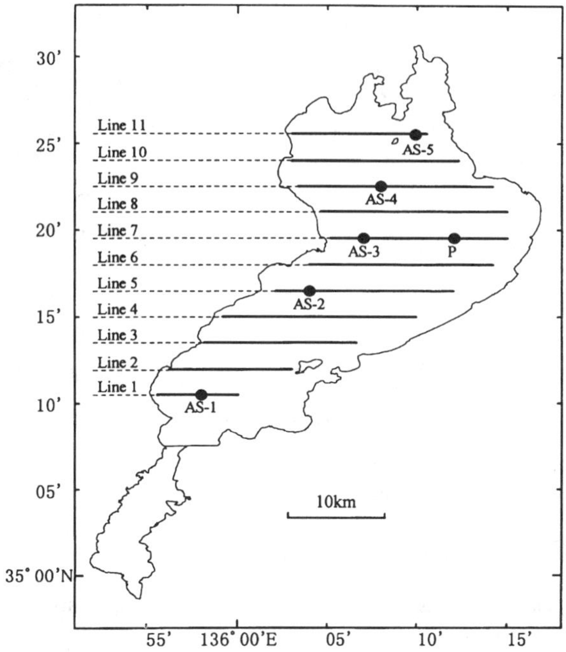

Figure 2. Observation lines of the ADCP measurements in 1994. The CTD profiling with F-probe were done at Stations AS-1, AS-2, AS-3 and AS-4. Station P is the place where vertical distribution of horizontal currents are depicted in Figure 6.

Observations started in April, and we performed the same measurements once a month. The weather conditions in May were not suitable for ADCP measurements because of strong winds over 8 m s^{-1}. We did not analyze the data taken in May. The weather conditions during observations from April to December 1994 are summarized in Table 1.

We experienced only weak wind during the ADCP measurements (Table 1), and were able to obtain excellent data on the gyres in the North Basin of Lake Biwa. The output data of the ADCP were calibrated with the data measured by an Alec Current Meter (ACM) moored at 29 m depth, where R/N *Hakken* was anchored near the ACM. Data from the ADCP and ACM were concordant as seen in Figure 3.

TABLE 1. Weather conditions during the ADCP measurements.

	Apr 19 / Apr 20	June 22/ June 23	July 26 / July 27	Aug 30/ Aug 31	Oct 6 / Oct 7	Oct 27 / Oct 28	Nov 14 / Nov 15	Dec 20 / Dec 21
Weather	fine / cloudy	fine / rainy	fine / cloudy	fine / fine	cloudy / fine	cloudy / rain	cloudy / fine	fine / rainy
Wind speed (m/sec)	2.1/ 1.7	2.1/ 1.1	2.6/ 1.6	1.9/ 2.0	1.7/ 2.2	1.0/ 0.8	4.4/ 4.2	1.6/ 2.2
Wind dir.	NW / ESE	NNW / W	SSE/ NW	NW / NW	NNW / NW	NNW / ESE	NW / WNW	S / WNW
Rainfall (mm)	1.5 / 0.0	0.0 / 2.5	0.0/ 0.0	0.0/ 0.0	0.0/ 0.0	0.0/ 1.5	0.0 / 0.0	0.0 / 0.0

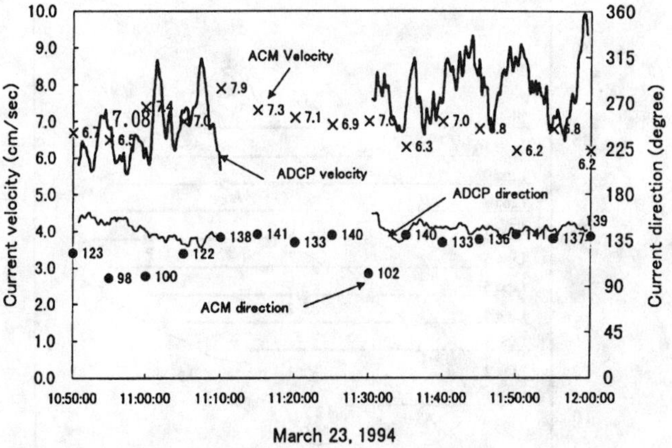

Figure 3. Comparison of field data from the ADCP and moored current meters (ACM) measured at 29 m depth. The numbers show the Julian days.

Results and Discussions

In Figure 4 seasonal changes of isothermal contours at Stations AS-1 to AS-4 are shown. We had a hot summer and less rainfall than normal in 1994. Strong stratification could be seen between 10 m and 20 m depth. July and August (Julian Days 211-241) were warmest in the epilimnion at every station. We had Typhoon 9426 on September 29 (Julian Day 272), and the strong stratification was relaxed after it had passed.

In Figure 5, horizontal current vectors in the epilimnion are shown. The original ADCP data were obtained at every 60 m along the observation lines and every 2 m, and were interpolated into 0.5 km × 0.5 km mesh points using the objective interpolation method with hyperbolic function. Averaged data from 7.5 m and 9.5 m depths in the epilimnion were plotted at every 1.5 km because of plotting complexity.

Figure 5(a) shows the data taken from April 19 to 20. A small clockwise gyre, which may be called the third gyre, was seen in the south part of the North Basin, but there were no remarkable gyres on the whole. Figure 5(b) shows the data taken from June 22 to 23. A large counter-clockwise gyre was formed in the north part of the North Basin. Also, northward jet-like currents were found along the west coast line, and turning to the east off the Ado river. The data taken from July 26 to 27 are shown in Figure 5(c), the counter-clockwise gyre had moved southward and had became wider. As seen in Figure 5(d), three gyres had been finally formed, and the positions of the first and second gyres were almost the same as those reported by Endoh and Okumura (1993). The jet going from the Ado River to the southeast was observed, a phenomenon also suggested by Itakura and Kumagai (1988).

After Typhoon 9426 passed near Lake Biwa on October 29, the horizontal currents in Figure 5(e) were completely different from the currents seen in Figure 5(d). Only the first gyre could be seen, but no other gyres were recorded. This may have been caused by strong mixing due to the typhoon. On October 27 and 28, the first and second gyres again observed, but the third gyre had disappeared in Figure 5(f). November, the first gyre had disappeared, and only the second gyre was found in Figure 5(g). By December, most of the distinguishable gyres had disappeared because overturning started to mix the surface lake water Figure 5(h).

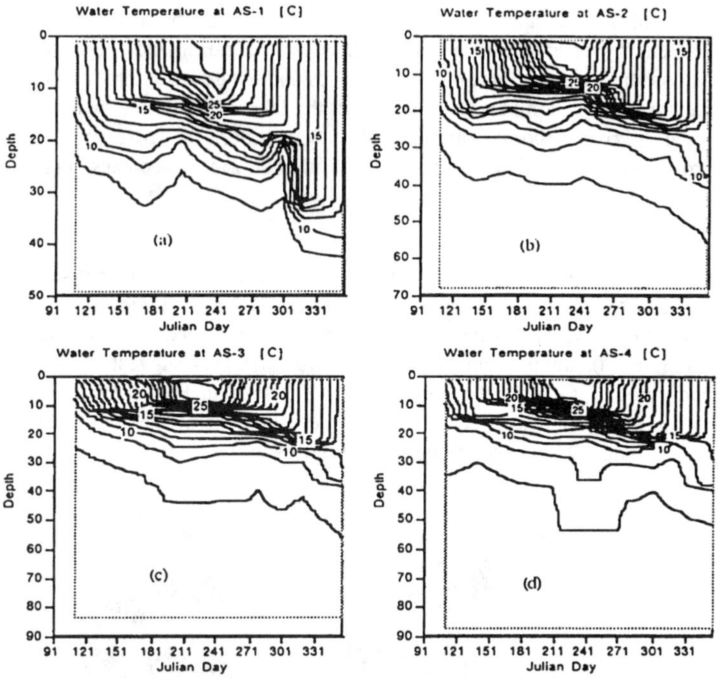

Figure 4. Isothermal changes at Stations (a) AS-1, (b) AS-2, (c) AS-3 and (d) AS-4.

Conclusion

First gyre

The first gyre was formed in the epilimnion above 12 m (Figure 6) and was therefore baroclinic. The isothermal depth of 15°C is showed in Figure 7, where the depth of 15°C near AS-3 is less than 14m, but the depths of 15°C near AS-2 and AS-5 are greater than 17m. This can suggest that the first gyre is basically geostrophic.

As seen in Figures 5(a) to (h), the number of gyres in the North Basin varied between seasons, and the position of the first gyre also varied between sampling dates. Apparently, the water temperature gradient becomes greater during summer time, and symmetrically the latitude of the first gyre moves southward (see Figure 8). This suggests that the position of the first gyre depended on the stratification in the upper water.

Figure 9 shows the temporal change of Rosbby internal deformation radius and the latitude of the first gyre's center in 1994. It can be said that the first gyre is born in the north part of the North basin of Lake Biwa, and it moves southward as stratification becomes stronger and Rossby internal deformation radius becomes smaller. After water surface cooling occurs, the first gyre goes back to northward. This would explain the difference of the first gyre's position as observed by Suda et al. (1926) and Endoh and Okumura (1993).

From these results, the following hypothesis can be proposed: eddies smaller than Rossby internal deformation radius move from the south to the north along the east shallow coast, and reach to the north end of Lake Biwa. These eddies cannot come back southward until the size approaches the deformation radius, because the west coast is too steep. The energy of each eddy is accumulated at the north and the first gyre becomes dominant. As the gyre becomes bigger, it moves southward and stops moving at the place where the gyre size is comparable in size with the lake width. When heating becomes weak and vertical mixing becomes stronger after September, the first gyre starts to move northward again, because Rossby internal deformation radius becomes small.

Second gyre

The second gyre was detected in June, August, October and November. This indicates that the second gyre is not stable and it may be very changeable due to wind, the intensity of the first gyre, or internal waves. The location of the second gyre was similar in Figure 5(b) and Figure l(a), and in Figure 5(d) and Figure l(b). This suggests that the position of the second gyre is located in the north initially as seen in Figure 5(b), and moves to the south as seen in Figure 5(d) in relation to the development of the first gyre. We could see the coastal jet and off-shore jet along the second gyre.

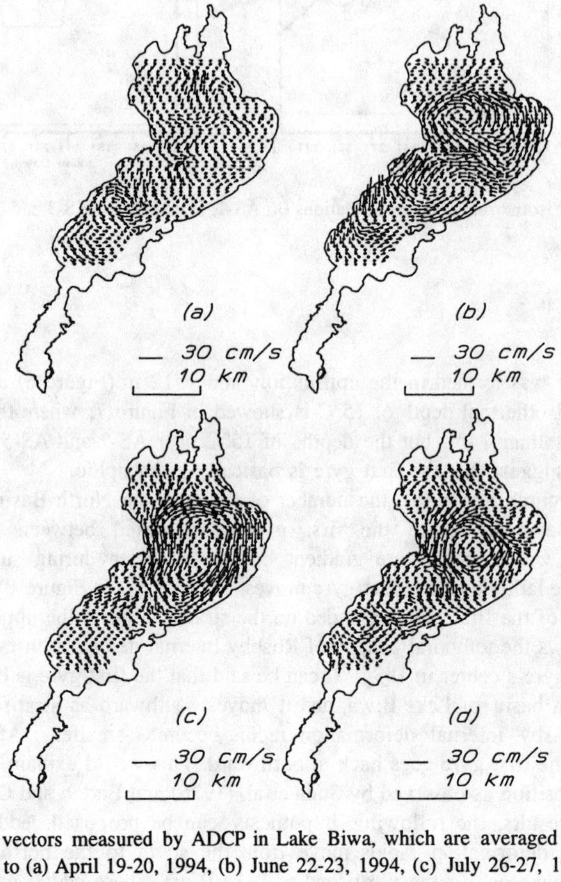

Figure 5. Current vectors measured by ADCP in Lake Biwa, which are averaged above 10 m depth. Panels correspond to (a) April 19-20, 1994, (b) June 22-23, 1994, (c) July 26-27, 1994 and (c) August 30-31, 1994, respectively.

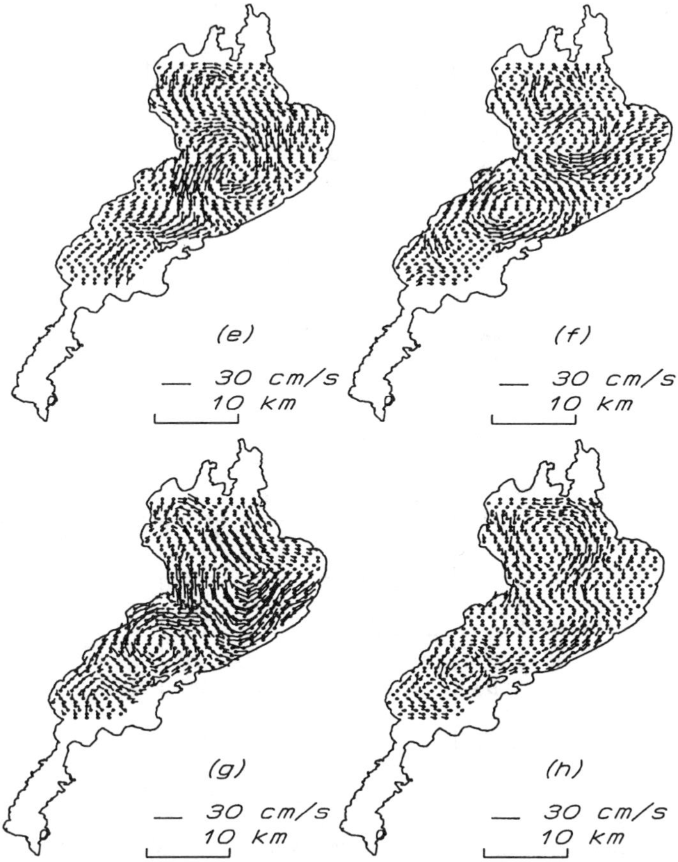

Figure 5. Continued. Panels correspond to (e) October 6-7, 1994, (f) October 27-28, 1994, (g) November 14-15, 1994 and (h) December 20-21, 1994, respectively.

Third gyre

The third gyre is more unstable and its changes were more complex. For example, the clockwise gyre is seen in Figure 5(a), but the counter-clockwise gyre is seen in Figures 5(d) and 5(h). In Figures 5(b), 5(c), 5(e), 5(f) and 5(g), we cannot see the third gyre. BITEX'93 (Biwako Transport Experiment) was carried out between August and September in 1993, and we found that the third gyre rotated clockwise with a period of 24 due to diurnal variations in wind strength (Kumagai et al., 1994). This clockwise gyre strongly depends on the geographical features around the lake. When wind stops in summer, a counter-clockwise gyre, such as in Figure 5(d), can be formed to keep the geostrophic balance due to a horizontal temperature difference (Ookubo et al., 1984). The counter-clockwise gyre seen in Figure 5(h) cannot be easily explained, but it might be formed by the south wind on December 30 and modified by the surrounding topography.

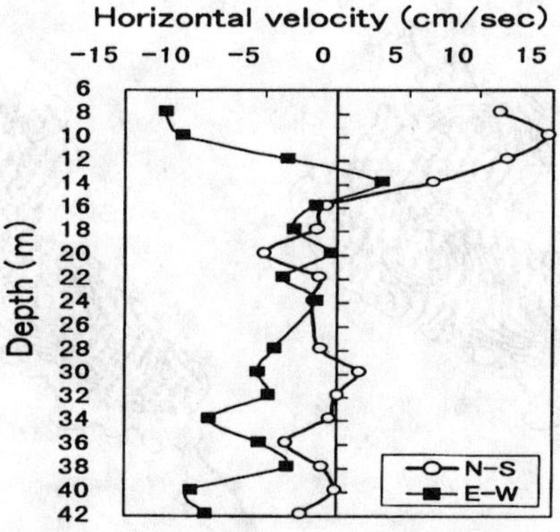

Figure 6. Vertical distribution of north-south components and east-west components of horizontal currents at Station P.

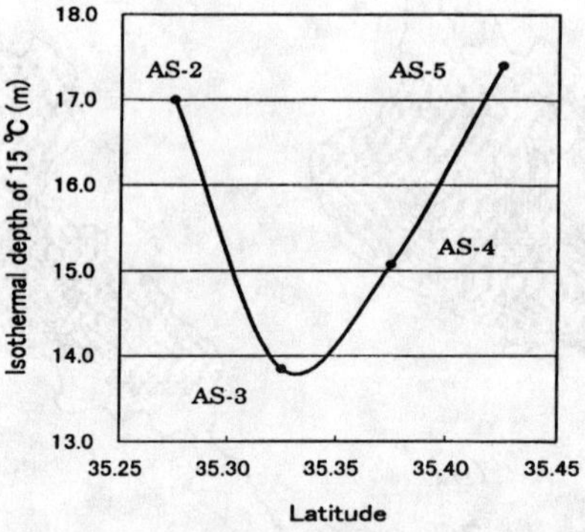

Figure 7. Isothermal depth of 15°C measured at Stations AS-2, AS-3, AS-4 and AS-5 on August 5 in 1994.

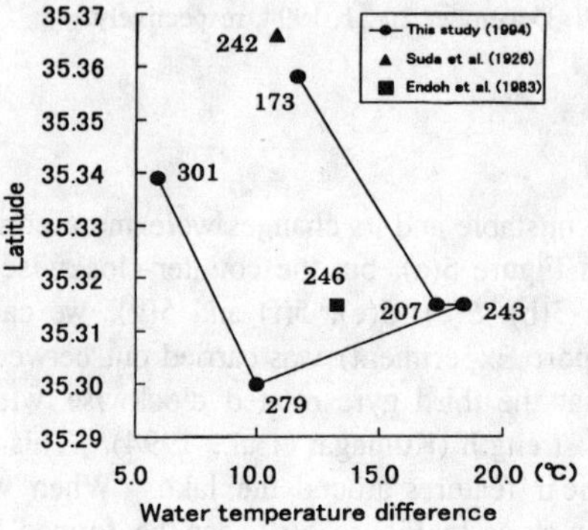

Figure 8. Relationship between stratification in the upper layer and the latitude of the first gyre. The solid circles show the results of this study, the solid triangle shows the results of Suda et al. (l926) and the solid square the results in 1983 after Endoh et al. (1987). The numbers are the Julian days of the observations.

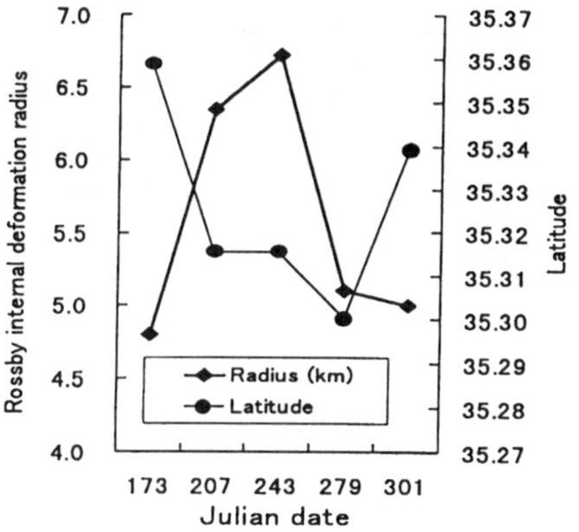

Figure 9. Changes of Rossby internal deformation radius and the latitude of the first gyre's center from the Julian date 173 to 301 in 1994.

Acknowledgments. The authors want to express their sincere thanks to Setsuo Okuda for his encouragement on this experiment, and to Richard Robarts for reviewing the manuscript. They also wish to thank Yuichi Hayami for his providing necessary information, Rieko Tanimoto and Louise Ferland and Michiko Nakagawa for typing and checking the manuscript, and the crews of R/V *Hakken* for their technical support in carrying out the field observations.

References

Charles, N. F. and S. L. Smith, On the use of the acoustic Doppler current profiler to measure zooplankton abundance, *Deep Sea Res., 36*, 455-474, 1989.

Endoh, S., Y. Review of geostrophic gyres, in *Physical Processes in a Large Lake: Lake Biwa, Japan*, edited by S. Okuda, J. Imberger and M. Kumagai, pp 7-13, Coastal and Estuarine Series, American Geophysical Union, Washington DC, USA, 1995.

Endoh, S., I. Okamoto, Y. Okumura, T. Tamura, K. Takano, Y. Hamai, T. Kodani, Y. Hayami, H. Asada, H. Kawamura and K. Iwane; Measurements of lake current by using a radar, *Mem. Fac. Educ., Shiga Univ.*, 37, 27-38, (in Japanese), 1987.

Endoh, S., and Y. Okumura, Gyre system in Lake Biwa derived from recent current measurements, *Jap. J. Limnol., 54*, 191-197, 1993.

Endoh, S., Okumura, and I. Okamoto, Field observation in the North Basin, *Physical Processes in a Large Lake: Lake Biwa, Japan*, edited by S. Okuda, J. Imberger and M. Kumagai, pp 15-29, Coastal and Estuarine Series, American Geophysical Union, Washington DC, USA, 1995.

Hidaka, K., Model experiment on the lake currents in the surface layer of Lake Biwa. *Bull. Kobe Marine Obs. 13*, 1-4, (in Japanese), 1927.

Itakura, Y. and M. Kumagai, Observation of currents in Lake Biwa using infrared remote-sensing, *Japanese Society of Electronics, LAV-888*, 29-38, (in Japanese), 1988.

Kumagai, M., Y. Azuma, S. Nakano, H. Hashitani, M. Okumura, Y. Seike, Y. Okumura, K. Kudo, C. Jiao, Y. Hayami and M. Nakanishi: Large scale transport - Water quality and phytoplankton related to currents in Lake Biwa, *BITEX symposium~workshop, Short Papers*, 53-64, 1994.

Ookubo, K., Y. Muramoto, Y. Oonishi and M. Kumagai, Laboratory experiments on thermal induced currents in Lake Biwa, *Bull. Disas. Prev. Res. Inst., Kyoto. Univ., 34*, 304, 19-54, 1984.

Oonishi,Y., Development of the current induced by the topographic heat accumulation (I) - the case of the axis-symmetric basin. *J. Oceanogr. Soc. Japan, 31*, 243-254, 1984.

Pinkel, R., Observation of nonlinear internal motion in the open sea using a range-gated Doppler sonar, *J. Phys. Oceano., 9*, 675-850, 1979.

Rowe, F. D., and J. W. Young, An ocean current profiler using Doppler sonar, *IEEE Proc, Oceans,* 292-297, 1979.

Suda, K., K. Seki, J. Ishii, S. Takaishi and S. Mizuuchi, The report of limnological observation in Lake Biwa (I), *Bull. Kobe Marine Obs. 8*, 104, (in Japanese), 1926.

Trevorrow, M. V. and Y. Tanaka, Acoustic and in situ measurements of freshwater amphipods (*Jesogammarus annandalei*) in Lake Biwa, Japan. *Limnol. Oceanogr., 42,* 121-132, 1997.

15

Circulation, Convection and Mixing in Rotating, Stratified Basins with Sloping Topography

P. B. Rhines

Abstract

Sloping sidewalls, or 'bowl' geometry of large lakes and ocean basins have a strong effect on their dynamics. Here we describe several related problems: the development of convective plumes and circulation driven by surface cooling, including the general problem of the building of a stable stratification; the role of boundary currents and topographic waves over the sloping bottom in filling out basin circulations; the great differences between bowls and basins with vertical-sidewalls, in which Kelvin-wave dynamics and nonlinear vorticity layers are important; the 'hypsometric' effect of the non-constant area of the basin, at different depths, on the overall circulation, and finally, brief discussion of the suppression of Ekman layer dynamics and spin-up/spin-down by sloping boundaries with density stratification.

Introduction

Sloping topography introduces so many new dynamical properties to a rotating, stratified fluid that it is remarkable to have the rectangular basin with side-walls still a paradigm of modelling. Numerical models of lake circulations do routinely incorporate *depth* variations, as do oceanic general circulation models, but the physics of flow with sloping boundaries is poorly represented in such models, when they are based on a rectangular Cartesian grid. Their redesign now often incorporates sigma-coordinate surfaces, which follow the bathymetry, or isopycnal (constant-potential-density) surfaces, and these seem to perform significantly better than horizontal grid levels.

The topography of real lakes and ocean basins affects dynamics both through its *slope* and through the its cumulative depth variation. Oftentimes buoyancy variations lead to sinking of surface waters, perhaps as a widely distributed convective overturning. With non-uniform depth this vertical motion will require considerable lateral circulation, as the densest water seeks out the greatest depth. Lateral circulation may also occur due to uneven flux of buoyancy at the surface, in its interaction with the atmosphere and with run-off or outflow sources and sinks at its margins. When lateral circulation becomes significant, we also find that what was an unstably stratified or neutrally stratified body of water (with uniform wintertime surface cooling, say) becomes *stably* stratified, as the denser waters

Physical Processes in Lakes and Oceans
Coastal and Estuarine Studies Volume 54, Pages 209-226

flow to the more weakly convecting regions of the basin, and recirculate warmer, more buoyant surface water toward the most strongly convecting region. The building of the thermocline is a central problem of both lakes and oceans, even those suffering active convection. One of the interesting elements of thermocline building is the advection/diffusion balance for temperature or a tracer, in which lateral circulation interacts with vertical mixing. Simple laboratory experiments show how long it takes to for such circulations bring the density field to a steady state. Numerical models of lake circulation (e.g., Bennett, 1977) have to incorporate many such effects, often in context with strong forcing by atmospheric wind and pressure fields.

Earth's rotation adds strong constraints, both restricting the size of convective plumes and the pathways of mean circulation. It provides a new set of wave modes, rotational topographic waves that are distant relations of both the Kelvin wave and the Rossby wave. It also provides a new set of turbulent/viscous boundary layers, notably the Ekman layer and the bottom boundary layer, which may either be an Ekman layer or an 'arrested' Ekman layer in the presence of a sloping bottom.

The Coriolis force may be significant even for very small bodies of water. Wherever the time characteristic of the general circulation is greater than f^{-1} (typically 4 hours), then this may be the case. Let L be the lateral scale of the basin, H its mean depth, N its buoyancy frequency, and U the typical horizontal velocity. When the Rossby number, U/fL, is smaller than unity, Coriolis effects are likely to be important unless for some reason turbulent friction is very strong. The relative strength of buoyancy forces and Coriolis forces is measured by the parameter NH/fL, known as the Prandtl ratio or Burger number. If a rotating, stably stratified fluid is excited by forcing from above, with lateral scale L, it will tend to penetrate a depth ~fL/N downward into the fluid; conversely, motions with imposed vertical length scale H will penetrate a distance ~NH/f laterally across the fluid. This gives, for example the off-shore e-folding scale of an internal Kelvin wave at a vertical coast. Rotation and stratification are demonstrably active in large lakes. Endoh (1995) describes the great geostrophic gyres in Lake Biwa, Japan, which is approximately 60km × 15km in size with an average depth of about 40m.

Convectively Forced Circulations: Laboratory Experiments

Convective Plumes: Length and Velocity Scales

When a rotating basin is cooled from the top, it responds with a lengthy life-cycle of convection and circulation. If Coriolis effects are negligible (for example, with episodes of convection lasting less than about 4 hrs, at middle latitudes) convective plumes tend to develop with widths similar to the depth of the convective layer. There are however exceptions; with very weak plumes which may be much taller than they are wide. Also, convection driven by imposed heat flux rather than an imposed surface temperature may develop horizontal scales much exceeding the fluid depth.

After about 4 hrs the Coriolis force causes a ring of converging fluid near a plume to spin, tending to conserve its absolute (= relative + planatary) angular momentum. The lateral scale of individual convective plumes is then limited on energetic grounds: the available potential energy embodied in a thin upper layer of unstably stratified fluid, with density perturbation $\delta\rho$, gives an upper limit to the kinetic energy of the motion. Large convective cells can develop only slowly, or not at all, in the presence of strong Coriolis forces. The relative angular momentum of the fluid converging on the sinking region corresponds to large kinetic energy density ($\sim\rho(fX)^2$ for fluid that has converged a distance X). For a fully developed plume, $X \sim L_h$, the horizontal scale of the plume. The potential energy anomaly ($^1/_2 g\ \delta\rho\ h^2$) supplies this kinetic energy, and thus

$$\frac{f^2 L_h^2}{gH\delta\rho/\rho_o}\left(\frac{H}{h}\right)^2 \sim 1 \qquad (1)$$

where h is the thickness of the cooled region, and H the depth of penetration of the convective plumes, which tends to be very much greater. Taking values corresponding to 0.2°C cooling of a 3-m surface layer, penetrating to depth H = 10m, with $f = 10^{-4}\ s^{-1}$ we find $L_h \sim 300$m. For oceanic deep convection h might be 100m and H 1000m, for which the plume width $L_h \sim 800$m. The Taylor-Proudman theorem in geophysical fluid dynamics describes the 'stiffness' that Coriolis effects add to a fluid: little vertical shear is allowed, unless it is balanced by horizontal density gradient, and hence the marvelous appearance of a forest of rotating tree-trunks, with fluid spiralling cyclonically inward and downward first, then anticyclonically (depending on the 'memory time' for potential vorticity ('pv')) outward at the lower half (approximately) of the convective plume. There is a tendency for the cyclones to remain small, and to dart downward from the surface, feeding surrounding anticyclonic regions that may be rather bigger in extent. The implied lateral scale of the plumes is so small that, likely, diffusion processes are important in the laboratory, and indeed figure in the marginal stability problem.

The above is an energetic form of Rossby-radius scaling, based on an unstable density perturbation, $\delta\rho$. The appearance of the ratio H/h emphasizes that the potential energy change due to a unit cooling will be the greater, the deeper the cooling signal is carried. The arguments of Fernando et al (1991) give the analogous scales for a constant surface buoyancy flux, $B_o = g\alpha\ Q/\rho C_p$ where Q is the net heat gain of the water in watts m^{-2}, α is the thermal expansion coefficient and C_p is the specific heat at constant pressure. B_o has dimensions L^2/T^3. The familiar result is the prediction for lateral length scale $(B_0/f^3)^{1/2}$ and velocity scales (both horizontal and vertical), $(B_0/f)^{1/2}$. If the depth of penetration (sometimes the full fluid depth) is H, then L_h/H is proportional to $Bo^{1/2}/f^{3/2}$ H, which is sometimes called a convective Rossby number Ro*. The implied scale L_h is < 1cm in the lab, and ~300m to 800m for intense wintertime cooling of the ocean surface. The famous 'spiral eddies' seen by astronauts on the sea surface in sun-glitter reflection are nearly always cyclonic, yet have a much larger lateral scale (5 to 10 km). Convection by nocturnal cooling is suggested, yet the large scale is unexplained, and may relate to the length scale of the air/sea buoyancy flux. Much smaller rotating convection plumes are indeed seen with SAR imagery at high latitude, and by implication, in vertical velocity time-series from moored and drifting deep instruments in the Mediterranean, Greenland and Labrador Seas. Aiming at subpolar or subtropical regions where the water temperature lies above 2°C, we omit nonlinear equation-of-state effects that are important in polar regions (e.g., Garwood, Isakari and Gallacher, 1994).

Rotating convection is described in the beautiful papers from Dave Fultz's lab at Chicago, e.g. Fultz et al. (1959), Nakagawa and Frenzen (1955), and developed by many groups, for example Boubnov and Golitsyn (1986, 1995), Fernando et al. (1991), Maxworthy and Narimousa (1994), Brickman and Kelly, (1993), Coates, Ivey and Taylor, (1995). Fultz et al. (1959) describe the earlier history of rotating convection including a 3m wide rotating room ('Karussell') built in Goettingen: "Prof. Prandtl and Hr. W. Mueller said that some dishpan-type experiments using a 1-m diameter pan were made in the

Karussell soon after it was built but that they were judged unsuccessful and were not continued because the patterns seemed too irregular and ill-defined".

In order to look at the 'basic physics' of convection we have carried out experiments to examine the horizontal and vertical velocities in simple convection without sea-floor topographic slopes. The Eulerian 'portrait' of laboratory rotating convection (much as one would see at a deep-sea mooring) is shown in Figure 1a, perhaps for the first time. It shows 'spikes' of both upward and downward vertical velocity, w (lower panel) and a horizontal (u-, upper panel) velocity field that is 'red-shifted' toward lower frequencies characteristic of mesoscale eddies. This experiment is driven by a uniform heat flux (actually in this case, heating from below), with 707 watts m^{-2} thermal forcing. The container depth is 15 cm, and the laser observations are made 15 cm vertically from the forced surface, and 2.5 cm from the vertical sidewall. With uniformly distributed cooling, the root-mean square w exceeds u by a factor of about 1.5, and the probability density function for w is skewed, favoring intense downward plumes with less concentrated updrafts. It is very likely that the two-dimensional turbulence cascade toward large horizontal scale is beginning to operate, with merging of vorticies of the same sign, although here the vertical velocities are still uniformly strong. The frequency spectra typically are flatter than $\omega^{-5/3}$ owing to the extreme high frequency energy in the plumes. The u- and w- variances show a crude agreement with the above Bo/f scaling (Figures 1b, 1c), although the scatter about this relationship is not small. The proportionality constant for w is 2 ± .5, which is within the estimates of Fernando et al. (1991). The work of Jones et al. (1993) and Maxworthy et al. (1994) emphasize that the Rossby number Ro* is of order 0.2 in intense oceanic convection. Ro* is essentially the predicted plume width divided by the depth of the convective zone, H. We concur, yet the small- Ro* range is also of geophysical interest.

The above picture of roughly equal horizontal and vertical velocities, u~w and $L \sim (Bo/f^3)^{1/2}$ does provide a guide for the purest case of convection driven by a large region

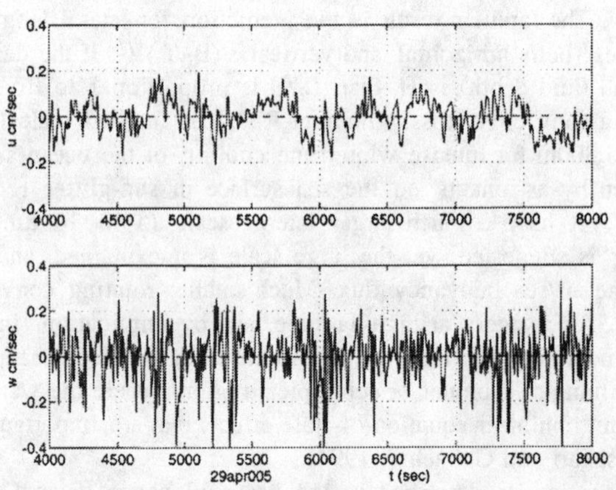

Figure 1a. Horizontal (u-) and vertical (w-) velocity timeseries at a fixed point in a uniformly forced rotating fluid ($2\Omega = 0.75\ s^{-1}$, 707 watt m^{-2} buoyancy forcing). The spikes of vertical velocity due to convective cyclonic plumes are evident (note, this case is heated from below rather than cooled from the top, yet we have plotted -w to mimick the cooled-surface case). The u-field has a longer timescale than w, characteristic of larger, mesoscale eddies. With uniform buoyancy forcing the variance of w exceeds that of u, but with nonuniform forcing, horizontal velocity typically is much greater than vertical velocity.

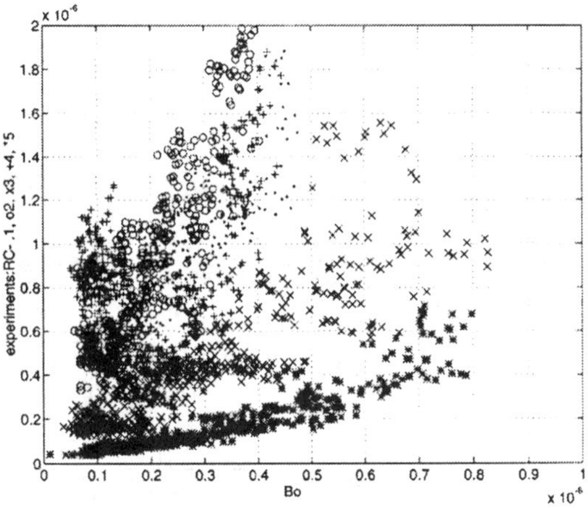

Figure 1b. Variance of w plotted against the imposed surface buoyancy flux, B_0, and (right) replotted against B_0/f. The B_0/f normalization greatly reduces the scatter. Variance in $m^2\ s^{-2}$, buoyancy flux in $m^2 s^{-3}$.

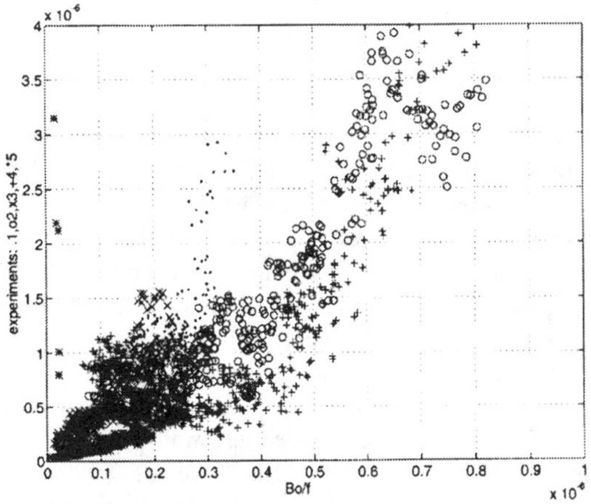

Figure 1c. As in Figure 1b but with variance of w plotted against Bo/f. The Bo/f normalization greatly reduces the scatter.

of uniform cooling, but it seriously underestimates the lateral scale of the horizontal velocity. Particularly in more realistic circumstances with topography, and non-uniform sea-surface buoyancy flux, the mesoscale eddy field becomes dominant (e.g., Brickman and Kelley, 1993). Then geostrophic turbulence deformations cause an expansion of the eddy diameters, and maintain a strong barotropic mode. In a mature convective field driven by non-uniform surface buoyancy flux, the vertical velocity dies very quickly, as one moves away from the most intense sites and times of convection, but the horizontal velocity is long-lived: this is the situation that is relevant to the oceans. We thus have a conversion of small scale convective motions into much larger eddies, with predominantly horizontal velocities. Such eddies are seen to populate thoroughly the high latitude oceans. With laterally varying cooling, eddies of larger scale are also generated by baroclinic instability of the lateral density gradient, which may have a sharp, frontal character.

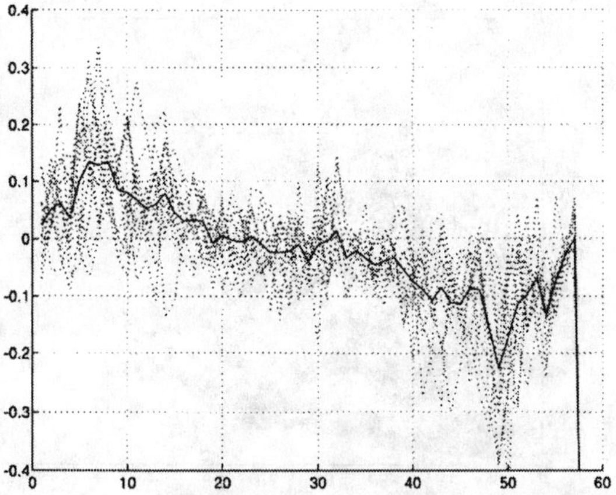

Figure 2a. Azimuthal horizontal velocity along a section cutting across the center of a uniformly cooled, rotating bowl. Individual profiles (dashed) and the average profile (full curve) show anticyclonic circulation. The basin is in the form of a spherical cap, diameter 0.9m, with maximum water depth of about 0.2m; see Condie and Rhines, 1994a.

Bowl-like Basins Driven by Surface Cooling

We have found that cooling the surface of a rotating basin generates a basin-scale mean circulation in the form of a gyre of anticyclonic circulation, if the cooling is uniform, Figure 2a (Condie and Rhines 1994a,b). One reason for the sense of this circulation is that uniform buoyancy flux will cause a greater temperature change in the shallow region near the shore, and this will promote an overturning circulation with downslope flow at depth. Outward radial flow occurs above this, and it is less subject to bottom friction. Hence the anticyclonic Coriolis driven circulation corresponding to this outflow is the winner. An apparent problem with this explanation is that energetic Rossby-radius arguments suggest that the radial/vertical overturning of the basin should have a small radial scale, similar to L_h above. Indeed, the anticylone has embedded jets in it, which seem to be organization of the fine-scale convection at the L_h scale. But there are great subtleties in the angular momentum balance of fluids driven in this way, which are seen when one does simple source-sink experiments in basins of rotating fluid. When this is done the axisymmetric response of the fluid involves overturning at the (large) scale of the basin, and angular momentum does what it must to obey. Dissipation of this swirling flow occurs in the bottom frictional layer, and in interior shear layers when there is density stratification.

Development of anticyclonic circulation is suggested by the variation in horizontal area of the basin, with depth. This 'hypsometry' causes rising fluid to diverge horizontally, flowing toward larger radius. The anticyclonic circulation is thus enhanced, in any situation where heavy fluid seeks out the deep center of the basin and displaces less-dense fluid upward (Rhines, 1993). We see the effect particularly strongly when a dense source flow is directed into the basin (and surface water is removed at the same rate), which is another problem relevant to lakes fed by cold, inflowing streams. Working in the other direction, in lakes, is the cyclonic tendency favored by wind-forcing and by stirring of potential vorticity.

With non-uniform cooling (a 'patch' of surface cooling), there develops cyclonic circulation in the upper ocean surrounding the region of surface cooling, and anticyclonic elsewhere. A velocity transect is shown in Figure 2b, (the section crosses the basin through its center, passing close to the region of cooling) and a streak photograph of the surface circulation in Figure 3. The velocity was measured by repeated passes of a laser velocimeter, with its 3 beams focused about 3 cm. beneath the water surface. A great tent-shaped chimney of cold water lies beneath the cooling region. This represents a 'sorting out' of clockwise and counterclockwise circulations by the continental slope topography of the 1-meter-diameter basin. Essentially, deep topographic waves carry the 'signal' of the circulation round the basin (counterclockwise in the Northern Hemisphere) and determine its shape and extent. With uniform cooling this flow is symmetric with respect to the rotation axis, and intensified near the rim of the basin: it is a natural mechanism for generating boundary currents. With a 'patch' of surface cooling that is much smaller than the basin, a cyclonic surface circulation surrounds the cooling zone, and the anticyclonic response is shifted to fill much of the rest of the basin. The repeat velocimeter passes show intense mesoscale eddy activity, and have been verified also with videos and photographs of tracers.

These events differ somewhat from experiments in the more familiar basins with uniform depth and vertical sidewalls. In those circumstances (e.g., Jones and Marshall 1993), a 'cold patch' at the surface creates a convected 'chimney' in the fluid below, with a cyclonic upper-level rim current that then generates mesoscale eddies by baroclinic instability. We find, by contrast, that the fine-scale convective plumes (of order 200m wide in the ocean) are much scarcer and mesoscale eddies populate the fluid in the cooled-bowl experiments. There is a cyclonic 'rim current' for the first few minutes of the experiment, but it propagates into a larger cyclonic gyre, as discussed below. This is particularly true when the buoyancy forcing region is smaller than the basin, and when Ro* is small. The plumes stand out at the early stages of convection, but quickly give way to much larger mesoscale eddies (80 - 150km diameter in the ocean, as found by Brickman et al., 1993). At larger Ro*, due to smaller f or larger buoyancy flux, the rim-current structure is significantly stronger, just as one finds in classic annulus experiments.

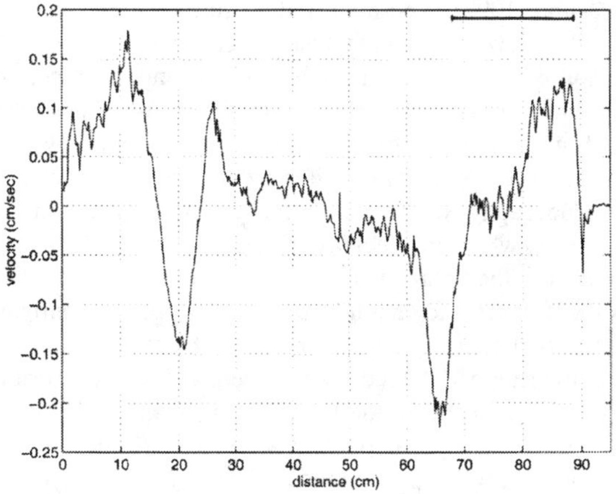

Figure 2b. As in Figure 2a, but for a rotating bowl-shaped basin driven by an off-center disk of surface cooling. The jets of mean circulation define a thin anticyclonic gyre (at left) and a cyclonic gyre surrounding the cooled region (at right). The transect crosses the basin in Figure 3 from '8' to '2 o'clock'.

Figure 3. Streak photograph of the surface circulation in the off-center cooled bowl. The cooling disk is seen at right. The intense rightward jet and eddies in the upper center of the basin is the boundary between the two gyres: cyclonic (surrounding the cooling region, extending anticlockwise round the basin) and anticyclonic (filling the rest of the basin). The jet originates at 'northwestern' boundary where boundary currents separate from shore.

Convectively Forced Circulations: Numerical Experiments

Numerical modelling of convection in three dimensions is a challenging task, but convincing experiments have been carried out by Jones and Marshall (1993) and Garwood, Paluskiewicz and Denbo (1994). Non-hydrostatic effects are crucial to plume convection, where w' ~ u'. But the view is emerging that the convective plumes efficiently mix the fluid inside the region of convection, and subsequent events at larger scale may be somewhat insensitive to the fine-scale detail. In this spirit, we show hydrostatic numerical simulations carried with a new isopycnal-layer model built by Robert Hallberg at University of Washington. The aim of this new circulation model is to deal accurately with the sloping topography of the ocean margin. It uses a conservative scheme of Arakawa and Hsu (1990) which is specially aimed at accurate behavior where constant-density surfaces intersect the continental slope.

Figure 4 shows a 10-layer integration in which the model is configured almost exactly like the laboratory experiments just described, in a bowl-shaped basin. The basin is a spherical cap 1m in diameter and 0.2m deep at its center. The equivalent of 57 w of cooling is applied to a disk 0.3m in diameter, located at the righthand side of the basin. In the layer model this is achieved by a diapycnal velocity w_o at the uppermost layer interface, chosen such that $g' w_o = B_o$, where $g' = g\delta\rho/\rho$ is the reduced gravity at the layer interface. It gives a buoyancy flux $B_o = 7.5 \times 10^{-7}\ m^2\ s^{-3}$, or a convective Rossby number Ro* of 0.57×10^{-2}. The section view (Figure 5) shows the model starting with homogeneous fluid, developing a 'cold dome' beneath the surface cooling, and gradually building a stable stratification throughout. The associated velocity field involves a cyclonic gyre surrounding the region of cooling, where fluid is converging and sinking, and a large anticyclone filling the rest of the basin (Figure 3). This flow is very dependent upon the sloping topography: companion experiments done with uniform fluid depth show extremely weak basin-scale circulation, dominated by a small cyclone round the cooling zone. The amplitude of the basin-scale circulation is strongly dependent on Ro*, and smaller f or larger B_0 make for stronger circulation. The parameters here were chosen for the annual forcing of the Labrador Sea, rather than for smaller basins or lakes.

Hallberg's model has been developed for basic dynamical studies (with a single density variable and linear equation of state): particularly those involving buoyancy driven general circulation in a much larger basin, in which the β–effect is important (Hallberg and Rhines, 1994). In this case, the guiding effect of continental-slope topography causes the high-latitude cooling to generate a deep western boundary current that encircles the basin. As in the laboratory model, there is a large region of anticyclonic flow in the upper layers. These circulations are very different from the classic Stommel-Arons circulation of an ocean whose bottom is flat. An extensive theoretical study shows the action of deep topographic waves in establishing the boundary currents, which contain flow both away from and toward the source.

Labrador Sea Convection

Convection and the generation of mesoscale eddies are evident in direct measurements from the Labrador Sea, Figure 6. The figure shows the temperature at fixed depths for one year (May 1994-May 1995). In this region the upper ocean has low salinity, owing to the

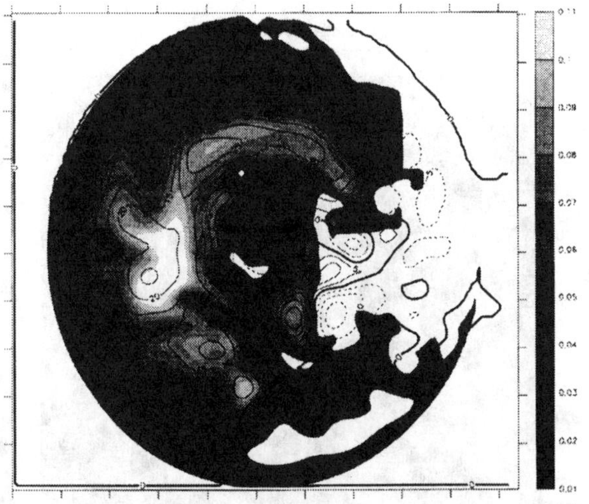

Figure 4. As in Figure 3, for a 10-layer isopycnal numerical model simulation of the same circulation, on a 100x100 grid. Plotted is the vertically integrated stream-function (curves) and in grey-tones the depth of the uppermost density interface. This constant density (or constant-temperature) surface intersects the upper fluid surface at the edge of the pure white region, which defines the cold cyclonic gyre. Elsewhere darker tones are colder temperatures or shallower isotherm depths. The jet-like boundary between circulation gyres is not as concentrated in the numerical model.

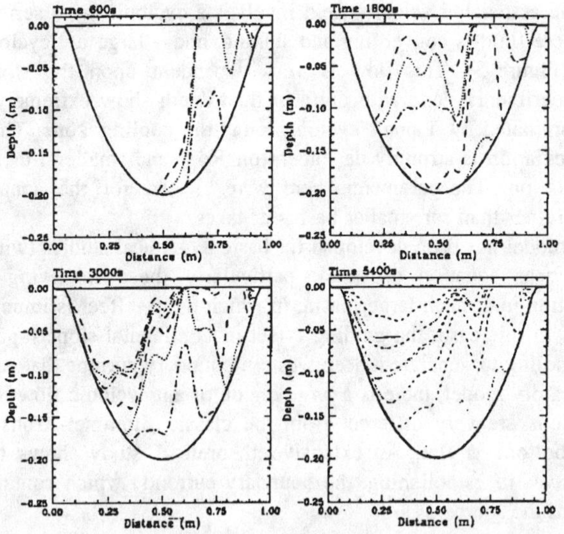

Figure 5. Cross sections of the developing temperature/density field for the cooled, rotating basin. The initial temperature is uniform. These 'hydrographic sections' show cold water forming beneath the circular patch of surface cooling (at the right-hand .3m of the 1.0 meter diameter basin). The cold chimney spreads and builds a stable stratification throughout the basin. The sloping isotherms are balanced by the thermal-wind vertical shear of the double gyre circulation. The 10 layers have equal increments of density.

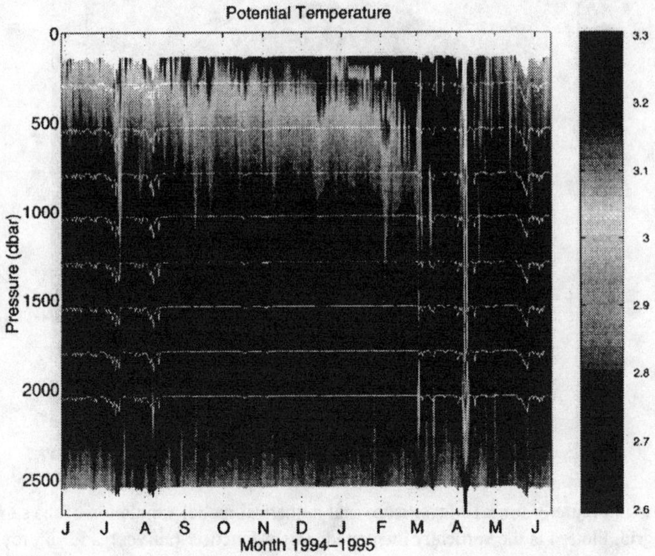

Figure 6. Temperature timeseries from a mooring site (the 'Bravo mooring') in the Labrador Sea, May 1994-May 1995. Here, intense wintertime cooling by air from the Canadian Arctic causes deep convection, reaching as deep as 2500m in the 3600m deep Sea.

invasion of fresh water from continental runoff, from the Arctic, and from locally melted sea-ice. Wintertime cooling must overcome this haline buoyancy, and convection occurs when the surface waters are about 0.5C colder than the deep water. High-frequency temperature oscillations occur, which are due primarily to horizontal advection of small-scale temperature anomalies past the mooring. The net result of this convection is the creation and maintenance of a 500km × 700km × 2.3 km deep Labrador Sea Water mass. Labrador Sea Water affects the world ocean, spreading as part of the Upper North Atlantic Deep Water, at about 1500m depth. A complete description of the Labrador Sea deep convection process is given by Lilly et al. (1998).

Building the Thermocline

Before leaving the subject we mention the basic problem of the creation of stable density stratification. If the buoyancy forcing varies spatially, or if there is geography introduced by depth variations, we quickly begin to see three important changes in the nature of the flow: an initially homogeneous fluid now develops a strongly stable stratification, the convection field becomes shifted toward larger, nearly geostrophic scales, and a global (basin-filling) circulation mode develops. Actually predicting the parametric dependence of these three properties on the forcing and geometry is an interesting challenge. Above, we have seen the development of the thermal stratification, started from an initially homogeneous density distribution. In the rotating cylinder driven by a cold patch at the surface, it is a gradual process requiring the baroclinic instability induced spreading of the chimney. The 'stride' of the lateral random walk of the spreading chimney is typically λ, the deformation radius, and this is small so long as the stratification is weak. In a bowl it happened more quickly, encouraged by the geometry to develop both rim currents and sinking toward the basin center.

There is also the fascinating, basic prototype of the two-dimensional, non-rotating Stommel-Rossby (Rossby, 1965) convection problem. Varying the buoyancy flux along the surface of the fluid seems generally to produce a time-averaged overturning circulation (which is probably more efficient than the small-scale convection it replaces) with small regions of sinking and very large regions of upwelling into the thermocline. We have found that when full non-hydrostatic dynamics is at work in the Stommel-Rossby problem, the idealized nature of the slow, diffusive, nearly hydrostatic numerical or laboratory model changes: turbulent entrainment into the sinking branch drives a strong overturning cell in which fluid recirculates many times before passing up through the thermocline, and warming up (Figure 7, Pierce and Rhines, 1996). There are important mid-depth recirculations in large-scale ocean circulation, possibly as a result, and very likely in small bodies of water as well. A recent example is the recirculation of Labrador Sea Water interacting through entrainment with the intense, sinking Denmark Strait Overflow Water. These convecting circulations have natural 'loop' oscillations as well, which seem to excite the climate community.

Transient behavior of this thermally forced system can last for a very long time, since the diffusive timescales are so long. For this reason, the appearance of density stratification in the deeper regions of natural fluids may often be a sign of unsteady forcing.

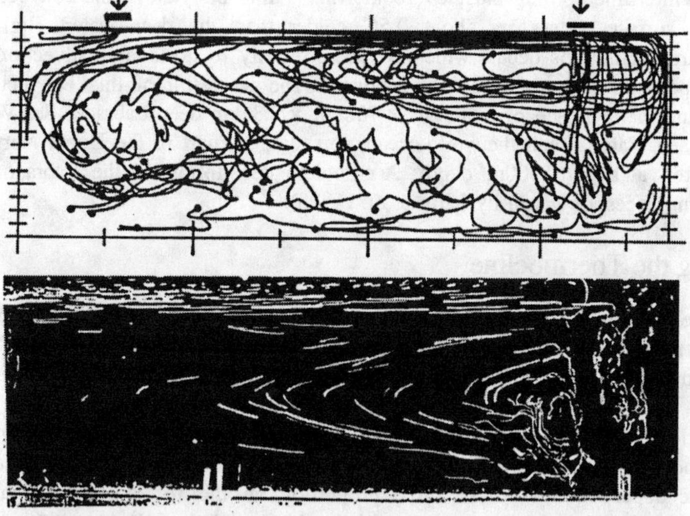

Figure 7. Numerical (above, 2-dimensional) and laboratory- (below, thin aspect ratio) experiments, with varying buoyancy forcing at the surface of the non-rotating fluid (Pierce and Rhines, 1996,1997). Unlike the linear variation of heating to cooling in the classic Stommel-Rossby problem, we have narrow regions of positive and negative buoyancy flux (indicated above the upper figure). The narrow sinking branch of the overturning cell entrains fluid, causing deep recirculation. The numerical version oscillates between two families of circulation, with the sinking region alternately lying at the end wall, and beneath the 'cooling'. There is intense recirculation of fluid beneath the thermocline, which has a long transient behavior. The flow at this stage is very unlike the single cell of overturning in the classic problem.

Sloping-bottom Boundary Waves and Currents

To look more deeply into mechanisms, we want to examine the propagators of time-dependent information. The free wave motions in a laterally unbounded stratified fluid are internal gravity waves, in combination with a surface mode. In simple geometry they are a complete linear set, describing the subsequent evolution from arbitrary (though not pathelogical) initial conditions, of small amplitude. Exciting weakly nonlinear motions, notably internal undular bores and solitons, exist and are very often seen in coastal and lake geometries. Sloping-bottom geometry introduces Ursell's edge waves, and their stratified generalization.

Adding Earth's rotation we find that the reflection laws for both surface waves and internal waves are altered. Consider an idealized situation with a single, plane, rigid wall lying at an arbitrary angle, bounding a rotating, stratified fluid. If we wanted to construct Green's function for a point disturbance, oscillatory in time, we might first try to use an image disturbance, in the wall, to satisfy the no-normal-flow boundary condition. This works if the boundary condition is the vanishing of the wave function, or its normal derivative, at the wall. Typically, however, when two 'orthogonal' (loose useage) restoring forces are present, say rotation and stratification or stratification and compressibility, the boundary condition becomes mixed rather than homogeneous:

$$a\phi + b\frac{\partial\phi}{\partial n} = 0 \tag{2}$$

Here ϕ is a wave function and n is a coordinate normal to a rigid, plane boundary, which may have any orientation relative to the vertical. a/b may be complex. This boundary condition introduces a phase shift between incident and reflected waves. The completeness relation (which is the expansion of a delta function in terms eigenfunctions) is now an integral over the freely propagating waves from the source and its image, plus an integral over edge modes with imaginary wavenumber in the n-direction (i.e., exponentially trapped). This bit of Cambridge mathematics helps to explain the appearance of Lamb waves (stratification+compressibility, horizontal boundary), Kelvin waves (rotation + free surface), internal Kelvin waves (rotation + stratification, vertical boundary), and their generalization at a sloping boundary, which limit on bottom-trapped topographic waves as the angle of the boundary from horizontal, α, becomes smaller than f/N (Rhines, 1970).

In the spirit introduced by James Lighthill many years ago (1968), we can use these wave modes in a sloping coastal- or continental slope geometry to express the 'forerunners' of steady circulation generated by buoyancy or external stresses. The distant signals due to the switch-on of a forcing effect involve waves of ever lower frequency, as time progresses. Eventually friction arrests the pattern, and the group velocity of the low (perhaps vanishing-) frequency waves says much about the eventual steady circulation (though without telling us about fundamentally nonlinear balances). Thus, internal gravity waves without rotation limit on nearly horizontally propagating 'tubular' disturbances that can represent wakes and upstream blocking, in the weak nearly linear sense. The group velocity remains substantial, $\sim NL_z$ as $\omega \to 0$, where L_z is the vertical scale of the motion. On the β-plane Rossby waves become 'β-plumes' in this same limit, extending westward from a forcing effect with group velocity $c_g \sim \beta L_y{}^2$ where L_y is the north-south length scale of the forcing effect.

In the present circumstance, the bottom slope provides a potential vorticity gradient (see below), which supports topographic waves, the same 'edge modes' described above. For horizontal scales smaller than λ, the internal Rossby deformation radius, they do not reach the top of the fluid, and their frequency, $\omega = N \sin\alpha \sin\phi$, is independent of f; α is the bottom slope and ϕ the angle between the wavevector and the depth gradient (the wavevector lies in the plane of the boundary). ω is just the component of N resolved along the inclined path of the oscillating fluid particles. The fluid moves in sheets lying parallel with the boundary in this limit. The exponential decay normal to the boundary has the Prandtl scale $H_s \sim fL/N$, and in general there is phase variation in that direction. Thus, $\omega \sim N\alpha \sim (f\alpha/H_s)L$, showing the dual nature of this mode as a 'buoyancy oscillation trapped by rotation, like the Kelvin wave' on the one hand, and a 'topographic Rossby wave in a container of virtual depth H_s ' on the other. The exact dispersion relation in the long-wave, barotropic, small-slope limit is just that of simple topographic waves,

$$\omega = -\frac{(f\alpha / H)k}{k^2 + l^2} \tag{3}$$

for a wave with wavevector (k,l) whose component l lies antiparallel with the depth gradient, and k along depth contours. These modes provide 'forerunner' group velocity $c_g \sim N\alpha L$ for short scales L (measured in the direction of the depth gradient) $< \lambda$, or $\sim f\alpha L^2/H$ for longer waves that become increasingly barotropic. These are relatively fast modes: for $\alpha =$

0.01, L = 25 km, N = 2 × 10^{-3} s^{-1} typical of the deep western boundary currents on the continental slope, c_g ~ 0.5 m s^{-1}. In small to moderate-size lakes with, say, N=0.2 s^{-1}, N/f~200, and the Prandtl vertical height scale L/200 is probably too small for these topographic modes to be important (though Coriolis effects may still be very significant to circulation). In a large lake with $\alpha = 0.05$, L = 5 km, N = 0.02 we find H_s ~ 25m and c_g ~ 2 m sec^{-1} (note for these estimates L is the rational length scale, ~ wavelength/2π. A key parameter here is just $\alpha L/H$, which determines whether the frequency is small or large compared with f.

The important analysis of these modes in wedge-shaped geometry appropriate to coasts, basins and lakes begins with the inspired work of Ou (1980), who realized that the bottom-trapped topographic wave problem in a wedge is isomorphic to Ursell's surface edge-wave problem. The complex free-surface boundary condition in Ursell's problem becomes the complex bottom boundary condition in the stratified, rotating wedge (now with rigid lid) and the solutions follow immediately. The sense of the modes is that, in deep water, they tend to be bottom trapped, like the short wave limit above, while coming into shallow water they become nearly barotropic. The analogous modes about a seamount were calculated by Brink (1989).

The idea of linearized plumes of current extending cyclonically around depth contours is attractive, but as soon as substantial advection of 'new' potential vorticity arrives from the convectively (or otherwise) forced region, the waveguide is altered, and one must look again. In Hermann et al. (1988) we discuss this effect in a prototype problem, the Rossby adjustment of a single layer of fluid, with a free surface, in a rotating channel. The linear problem treated by Gill (1976) demonstrates the action of Kelvin waves in just the manner described here, draining fluid along the channel walls in jets of width given by the external deformation radius. The vorticity of this boundary current is opposite in sign to that needed to accommodate the incoming fluid with its anomolous pv, and this leads to 'Rossby readjustment' in which the pv dynamics plays a more active role; the boundary current width changes, there is 'creep' of fluid in the 'anti-Kelvin wave' direction and, of greatest importance, separation from the wall and the creation of roundish eddies that fill the channel (Monismith and Maxworthy, 1989).

This particular problem provides a good lesson in the tools of oceanography: the 'Rossby stage' (slumping of the free surface and creation of a jet by Coriolis forces) lasts a fraction of a day, the 'Gill stage' (formation of the boundary jets) a few days, the 'Hermann et al., stage' (nonlinear modification of the boundary jets and creep of fluid anticyclonically round the channel) another few days, and then the Monismith et al., stage (the circulation breaking away from the boundary and filling the interior) for the rest of time. The first three stages were solved theoretically while the final stage, occupying much of time, required laboratory simulation.

Sloping Bottom as a Source of Large Potential Vorticity

The Ertel-Rossby potential vorticity equation is

$$\frac{\partial(\rho q)}{\partial t} + \nabla \bullet (\rho q \vec{u} - \Im(\vec{\zeta} + 2\vec{\Omega}) - \vec{F} \times \nabla\sigma) = 0$$

$$\frac{D\sigma}{Dt} = \Im \qquad (4)$$

$$q = (\vec{\zeta} + 2\vec{\Omega}) \bullet \nabla\sigma / \rho$$

q is the Ertel potential vorticity and σ the potential density. $\zeta = \nabla x\, \mathbf{u}$ is the relative vorticity vector, so $\zeta + 2\Omega$ is the absolute vorticity vector. In general σ must be some variable that is conserved in absence of forcing, following the fluid motion, and is a function of density and pressure only. The external forcing is given by **F**, the vector body force per unit mass due small scale turbulence and friction, or other body forces, and, $\mathfrak{I}$, the diabatic forcing of potential density. The righthandside terms are described by Haynes and McIntyre (1987). They remark that the transport of q, even in the presence of righthand terms, is directed parallel with σ–surfaces, which can readily be seen by integrating q between neighboring σ–surfaces: the pv integrated this way is just $(2\Omega + \zeta)_n$, the absolute vorticity in the direction normal to the surface. If this is also intergrated over an area of the surface out to regions where no rhs terms are non-zero, then the statement that 'integrated pv cannot be created or destroyed' or 'isentropic surfaces are impermeable to pv, despite mixing and diffusion' (Haynes et al., op. cit.) become true. This is really a form of Kelvin's circulation theorem, which can be derived for closed circuits of fluid lying on a potential density surface.

The problem with this potential vorticity is that it contains no inkling of the sloping bottom, which we use above to provide a form of Rossby-/bottom-/edge wave, geostrophic contours and other pv effects, when no stratification is present. I suggested (Rhines, 1979), following the idea of Bretherton for effects of sloping isopycnals in Eady's instability problem, that topographic effects can be incorporated in the quasi-geostrophic problem by adding to the definition of q, a delta-function sheet of pv along the bottom. What this does is to provide a potential vorticity whose vertical integral is just the familiar barotropic potential vorticity of the unstratified problem.

The suggestion I make here (it is not yet thoroughly examined) is that fluid circulations can strip off the sheets of high pv, which, when taken away from the topography, become sources of large cyclonic relative vorticity, and hence of circulation. Essentially, the wedge-shaped regions defined by the intersection of isopycnals with the sloping bottom represent the large pv, and can be advected out to sea. Notice that this opposes the sense of 'hypsometric' circulations induced in the far-field of convective overturning. In Hallberg's numerical model experiments, an isopycnal model that resolves well the sloping bottom, such stripping of high pv layers is evident. Where boundary current separation occurs, in particular, the high boundary pv is carried off to mid-ocean. On the one hand, it helps to express the high-pv of the outcropping density interfaces, found in open ocean, at the surface. In addition, the 'potential vorticity crisis', as Joe Pedlosky calls it, of β-plane fluid forced to leave the western boundary at an uncomfortably high latitude, is alleviated by the continuing strong control of the pv by the layer thickness. The vertical resolution of the Hallberg and Rhines (1996) model has been increased from 4 layers to 20, and the effect remains significant.

It is important to characterize the stability properties of deep boundary currents on slopes, as these can decide whether fluid can mix three-dimensionally with the interior. Reversal of the mean potential vorticity gradient, the criterion for baroclinic or barotropic instability, is discouraged by the bottom slope, yet can occur particularly on the deep side of cyclonically circulating boundary current.

Suppression of Ekman Dynamics on a Sloping Bottom

The reaction of a rotating stratified fluid to bottom friction forms the classic problem of stratified spin-up. The debate began in the remote solar interior, where the physicist

Robert Dicke wished to prove that the interior spin of the sun should not have reached that of the outer photosphere, so as to affect its gravational lens; the GFD community warned that Ekman driven spin-up is a fast process, compared to simple turbulent-viscous momentum diffusion. A triumphal scale analysis convinced us (Sakurai, 1969) that Ekman layers are so thin, $\partial^2u/\partial z^2$ so big, that stratification can be ignored within the layer. Ekman layers suck and blow, and it is up to the quasi-geostrophic interior to accommodate with secondary circulation and deceleration of the swirling flow. Penetration reaches a Prandtl height fL/N into the fluid, and accomplishes partial spin-down there; further evolution is left as a difficult problem in diffusion of momentum and temperature/salinity.

What is wrong with this tour-de-force of GFD is that the sloping bottom is left out. It was clear that sidewalls are important and difficult parts of the classic problem (leading to various misdirections in the literature); for a right circular cylinder with insulating boundaries, the Ekman flux cannot climb the wall; delta-function jets squirt from (or into) the corners, and then quickly broaden into a wider region of Ekman divergence.

An equally simple scale analysis tells us that buoyancy forces paralleling a sloping boundary become comparable with the Coriolis force when $g\rho'\sin\alpha \sim fu$ for typical density and horizontal velocity perturbations ρ',u' and bottom slope α. In an idealized spin-down experiment begun at t=0, ρ' is given by the net vertical displacement of a fluid particle, $\sim ut\sin\alpha$, and hence after a 'shut-down time', τ, given by

$$f\tau = \frac{f^2}{N^2\sin^2\alpha} \cong \frac{f^2}{N^2\alpha^2} \tag{5}$$

buoyancy forces are able strongly to resist the up- or down-slope motion (the second expression is for small slope, α). Once shut-down of this component of boundary-layer velocity has occurred, Ekman pumping all but ceases. Spin-down is suppressed: a sloping bottom is slippery. The phenomenon can be visualized in the laboratory by inserting a vertical dye sheet in the fluid when it is in solid-body rotation, and then changing the rotation rate. The spin-up of the fluid is seen by the winding of the dye into a helical spiral, which is then cut by a plane of light (Figure 8, MacCready and Rhines, 1991, 1993). The 'onion slice' of surfaces of equal angular rotation shows the directions of principal shear. In the figure they are aligned with the spherical-cap basin, indicating that rather simple shear normal to the boundary is diffusing momentum inward (the core fluid has rotated about 12 times relative to the container); in experiments where stratification has not 'shut off' the Ekman layers, these fluoresein surfaces bow upward near the bottom, indicating the lack of vertical shear there (the fluid is spun up through a Prandtl thickness fL/N).

The subsequent behavior is very interesting. Momentum must diffuse vertically but in doing so, its vertical shear will have to be accommodated by thermal-wind tilting of the constant-density surfaces: this in turn does require a meridional redistribution of mass, as in normal spin-up but much more slowly. The large velocity shear of the original boundary layer is gradually replaced by much weaker shear of a near-geostrophic thermal-wind balance. Stress on the bottom is much reduced. The scale analysis above is verified by one-dimensional (z,t) numerical solutions of the 'slow-diffusion equation' which is the result of noting that after a rotation period or so, time-rate-of-change is so small that one can neglect it in all but the density/temperature/salinity equations. The Coriolis force on the meridional (r,z) circulation argued above enhances or retards the diffusion of momentum...but just a bit. A somewhat similar retarded diffusion equation occurs an unbounded problem of 'slumping' of an gravitationally unbalanced flow; that is geostrophic adjustment with diffusion, Gill (1981), Garrett (1983).

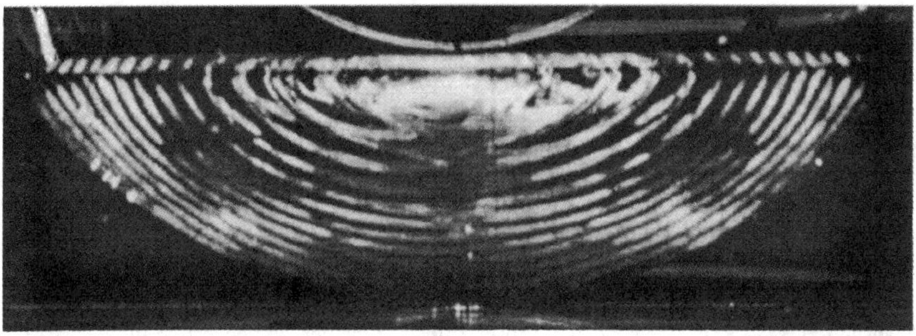

Figure 8. Fluorescein dye pattern in spin-up of a rotating stratified fluid in a 1m. diameter bowl (viewed from the side; MacCready and Rhines, 1991). The fluid is initially at rest in the rotating frame, and then a small change in rotation rate sets it into motion. The dye is initially in a vertical sheet extending across the basin, but the flow winds it into a helical spiral, which is then 'cut' by a vertical sheet of light. The central core of the fluid continues to rotate long after the fluid at the rigid boundary has come to rest, and by counting the 'rings' we can see just how many revolutions have occurred. The constant density surfaces are nearly horizontal, and the bowl-shape of the dye surfaces tells us that the principal shear is normal to the rigid boundary, and momentum is simply diffusing through the fluid. Ekman dynamics and rotating spin-up have been suppressed by the stratification/bottom slope.

These arguments come to grips with turbulent mixing, only insofar as a convective-adjustment parameterization is accurate in descrbing the mixed layer with heavy fluid over light. Down-slope boundary layer flow then creates very thick bottom mixed layers (or, in a ways not yet understood, decides to separate from the boundary and form an interior mixed layer). Upslope boundary layer flow, by contrast, is stable and the mixed layer tends to remain thin. This asymmetry between upwelling- and downwelling-favorable flows has many implications. It has been observed in the coastal ocean. A boundary current on a slope runs down slowly, and its potential energy provides a source for circulation long beyond the simple Ekman spindown time (if conditions are such that the Ekman flux has not shut down). However, it is worth noting that so long as Ekman flux does occur, it will drive secondary circulations that may be much 'taller' than the dense boundary current fluid. These secondary circulations, mismatched to the jet that created them, will tend to drive a counter current in the water above. Ideas of boundary mixing impinge on this problem, and the combined situation is reviewed by Garrett et al., 1993.

Acknowledgments. This work is sponsored by the US Office of Naval Research. I am grateful to Eric Lindahl, who is the research engineer in the GFD laboratory at University of Washington. The Hallberg Isopycnal Model is available from Robert Hallberg, Geophysical Fluid Dynamics Laboratory, Princeton University (rwh@gfdl.gov).

References

Arakawa, A. and Y. -J. G. Hsu, Energy conserving and potential-enstrophy dissipating schemes for the shallow water equations. *Mon. Wea. Rev. 118*, 1960-1969, 1990.

Bennett, J. R. Three-dimensional model of Lake Ontario's summer circulation, Pt. 1, Comparison with observations. *J. Phys. Oceanogr., 7*, 591-601, 1977.

Boubnov, B. M. and G. S. Golitsyn, Experimental study of convective structures in rotating fluids, *J. Fluid Mech. 167*, 503-531, 1986.

Boubnov, B. M. and G. S. Golitsyn, *Convection in Rotating Fluids*, Kluwer Academic Publishers, Dortrecht, 1995.

Brickman, D. and D. E. Kelly, Development of convection in a rotating fluid: scales and patterns of motion. *Dyn. Atmos. Ocean 19*, 389-405, 1993

Brink, K. The effect of stratification on seamount-trapped waves. *Deep-Sea Res., 36*, 825-844, 1989.

Coates, M. J., G. N. Ivey and J. R. Taylor, Unsteady, turbulent convecttion into a rotating, lineraly stratified fluid: Modeling deep ocean convection, *J. Phys. Oceanogr. 25*, 3032-3050, 1995.

Condie, S. A. and P. B. Rhines, Topographic Hadley cells, *J. Fluid Mech., 280*, 349-366, 1994.

Condie, S. A. and P. B. Rhines, Superadiabatic jets on Jupiter and Saturn, *Nature, 367*, 711-713, 1994.

de Madron, X. D. and G. W. Weatherly, Circulation, transport and bottom boundary layers of the deep currents in the Brazil Basin, *J. Mar. Res., 52*, 583-638, 1994.

Endoh, S., Review of geostrophic gyres, *in Physical Processes in a Large Lake: Lake Biwa, Japan*, edited by S. Okuda, J. Imberger and M. Kumagai., American Geophysical Union, Coastal and Estuarine Studies, 216pp, 1995.

Fernanando, H. J. S., R. R. Chen and D. L. Boyer, Effects of rotation on convective turbulence, *J. Fluid Mech., 228*, 513-547, 1991.

Fultz, D., R. R. Long, G. V. Owens, W. Bohan, R. Kaylor and J. Weil, Studies of thermal convection in a rotating cylinder with some implications for large-scale atmospheric motions, *Meteorol. Monographs, 4*, 104pp, American Met. Soc., Boston, 1959.

Garrett, C. J. R., On the spin-down of the ocean interior, *J. Phys. Oceanogr., 12*, 989-993, 1983.

Garrett, C. J. R., P. M. MacCready and P. B. Rhines, Boundary mixing and arrested Ekman layers: rotating stratified flow near a sloping boundary, *Ann. Revs. Fluid Mech. 25*, 291-323, 1993.

Garwood, R. W., R. Paluskiewicz and D. Denbo, Deep convection plumes in the ocean, *Oceanogr., 7*, 1994.

Gill, A. E., Adjustment under gravity in a rotating channel, *J. Fluid Mech. 77*, 603-621, 1976.

Gill, A. R., Homogeneous intrusions in a rotating stratified fluid, *J. Fluid Mech., 103*, 275-295, 1981.

Hallberg, R. H. and P. B. Rhines, Buoyancy driven circulation in an ocean basin with isopycnals intersecting the sloping boundary, *J. Phys. Oceanogr., 26*, 913-940, 1994.

Haynes, P. and M. E. McIntyre, On the evolution of vorticity and potential vorticity in the presence of diabatic heating and fricitional or other forces, *J. Atmos. Sci., 44*, 828-841, 1987.

Hermann, A., P. B. Rhines and E. R. Johnson, Nonlinear Rossby adjustment in a channel: beyond Kelvin waves. *J. Fluid Mech., 205*, 469-502, 1988.

Jones, H. and J. Marshall, Convection with rotation in a neutral ocean: a study of open-ocean deep convection, *J. Phys. Oceanogr., 23*, 1009-1029, 1993.

Lighthill, M. J., On the waves generated in dispersive systems by travelling forcing effects, with application to the dynamics of rotating fluids, *J. Fluid Mech., 27*, 725-752, 1967.

Lilly, J., P. B. Rhines, M. Visbeck, R. Davis, J. R. N. Lazier and D. Farmer, Deep convection in the Labrador Sea observed during winter 1994-1995, *J. Phys. Oceanogr., sub judice,* 1998.

MacCready, P. M. and P. B. Rhines, Buoyant inhibition of Ekman transport on a slope and its effect on stratified spin-up, *J. Fluid Mech. 223*, 631-661, 1991.

MacCready, P. M. and P. B. Rhines, Arrested Ekman layers on a slope, *J.Phys.Oceanogr.,23*, 5-22, 1993.

Maxworthy, T. and S. Narimousa, Unsteady turbulent convection into a homogeneous rotating fluid, with oceanographic applications, *J. Phys. Oceanogr. 24*, 865-887, 1994.

Monismith, S. and T. Maxworthy, Selective withdrawal and spiunup o a rotating stratified fluid, *J. Fluid Mech., 199*, 377-401, 1989.

Nakagawa, Y. and P. Frenzen, A theoretical and experimental study of cellular convection in rotating fluid, *Tellus 7*, 1-21, 1955.

Ou, H. W., On the propagation of free topographic Rossby waves near continental margins, 1. Analytical model in a wedge, *J. Phys. Oceanogr., 10*, 1051-1060, 1980.

Pierce, D. P. and P. B. Rhines, Convective building of a pycnocline - laboratory experiments, *J. Phys. Oceanogr., 26*, 176-190, 1996.

Rhines, P. B., Edge-, bottom- and Rossby waves in a rotating stratified fluid. *Geophys. Fluid Dyn, 1*, 273-302, 1970.

Rhines, P. B., Geostrophic turbulence. *Ann. Rev. Fluid Mech. 11*, 401-441, 1979.

Rhines, P. B., Oceanic general circulation: wave- and advection dynamics, in *Modelling Ocean Climate I Interactions*, edited by J. Willebrand and D. Anderson, NATO ASI Series, Springer, 67-149, 1993.

Rossby, H. T. On thermal convection driven by non-uniform heating from below; an experimental study. Deep-Sea Res., *12*, 9-16, 1965.

Sakurai, T., Spin-down problem of rotating, stratified fluid in a thermally insulating circular cylinder, *J. Fluid Mech., 37*, 689-699, 1969.

16

Internal Solitary Waves in Shallow Seas and Lakes

R. Grimshaw

Abstract

We give a brief account of the basic theory of internal solitary waves with the emphasis on applications to observations in shallow seas and lakes. Starting with the equations of motion for an inviscid incompressible fluid, we indicate how various asymptotic models for weakly nonlinear long waves are derived, including the well-known Korteweg-de Vries equation. We then describe the solitary wave solutions, and relate their properties to observations. Various generalisations are considered taking into account effects such as friction, variable bottom topography, and wave refraction.

Introduction

Solitary waves are nonlinear waves of permanent form which owe their existence to a balance between nonlinear wave-steepening effects and weak linear dispersion. Typically they consist of a single isolated wave of elevation, or depression, whose speed is generally an increasing function of the amplitude. For many years solitary waves were regarded as curiosities and largely ignored, but it is now recognized that they are ubiquitous in nature, and in particular internal solitary waves are a commonly occurring feature in coastal seas, fjords and lakes (Apel, 1980, 1995; and Ostrovsky and Stepanyants, 1989). Simultaneously with these observational developments has come the discovery that solitary waves can typically be described by certain generic canonical model equations, such as the now well-known Korteweg-de Vries equation, which although nonlinear, are exactly integrable and posses certain remarkable properties (see, for instance, the texts by Ablowitz and Segur, 1981, Dodd et al., 1992, Drazin and Johnson, 1989, or the review article by Miles, 1980).

In this brief review, we will demonstrate in the next section how the canonical model evolution equations such as the Korteweg-de Vries equation can be asymptotically derived from a fuller set of fluid dynamic equations, such as those appropriate to describe the motion of density stratified fluids. Then we will discuss the solitary wave solutions of these model evolution equations emphasising those properties which are robust and can be easily related to observations. Finally we will discuss various generalizations which can account for effects such as friction, variable bottom topography and wave refraction.

Physical Processes in Lakes and Oceans
Coastal and Estuarine Studies Volume 54, Pages 227-240

Derivation of Evolution Equations

Formulation

We begin by considering an inviscid incompressible fluid which is bounded above by a free surface and below by a rigid boundary. We shall use spatial co-ordinates (**x**, z) where **x** = (x, y) is the horizontal co-ordinate and z is the vertical co-ordinate, while t will be the time co-ordinate. In the state of rest the fluid has density $\rho_0(z)$ and the corresponding pressure $p_0(z)$ is obtained from the hydrostatic equilibrium so that $p_{0z} = -g\,\rho_0$. In standard notation, the equations of motion are

$$\rho_0 \mathbf{u}_t + \nabla p = -(\rho_0 + \rho)(\mathbf{u}\cdot\nabla \mathbf{u} + w\mathbf{u}_z) - \rho \mathbf{u}_t, \tag{1}$$

$$p_z + g\rho = -(\rho_0 + \rho)(w_t + \mathbf{u}\cdot\nabla w + w w_z), \tag{2}$$

$$g\rho_t - \rho_0 N^2 w = -g(\mathbf{u}\cdot\nabla\rho + w\rho_z), \tag{3}$$

$$\nabla\cdot\mathbf{u} + w_z = 0, \tag{4}$$

Here (**u**, w) is the fluid velocity, where **u** is the horizontal component and w is the vertical component, $p_0 + p$ is the pressure, $\rho_0 + \rho$ is the density and

$$\rho_0 N^2(z) = -g\rho_{0z}. \tag{5}$$

Thus N(z) is the buoyancy frequency and defines the equilibrium density stratification. The boundary conditions are

$$w = 0 \quad \text{at} \quad z = -h, \tag{6}$$

$$p_0 + p = 0 \quad \text{at} \quad z = \eta(\mathbf{x}, t), \tag{7}$$

and

$$\eta_t + \mathbf{u}\cdot\nabla\eta = w \quad \text{at} \quad z = \eta(\mathbf{x}, t) \tag{8}$$

Here the fluid has undisturbed depth h, and η is the displacement of the free surface from its undisturbed position, z = 0. Note that we are neglecting the effect of the earth's rotation here.

For the application to internal solitary waves in shallow coastal seas, fjords or lakes we can anticipate that we are seeking solutions whose horizontal length scales are much greater than h. We will also assume initially that the waves have small amplitude. Then the dominant balance is given by equating the terms on the left-hand side of equations (1) to (4) to zero, and these then describe linear long wave theory. To proceed it is useful to replace the vertical velocity w by the vertical isopycnal displacement ζ. These are related by noting that on a given isopycnal surface (i.e. $(\rho_0 + \rho)$ = constant), $w = \zeta_t + \mathbf{u}.\nabla\zeta$, where we recall that (3) states that the total density $(\rho_0 + \rho)$ is conserved on fluid particles. When this expression is converted to the vertical velocity evaluated at the Eulerian position z, we get

$$w = \zeta_t + \nabla\cdot(\mathbf{u}\zeta) + \ldots \tag{9}$$

where the omitted terms are of cubic order and will not concern us here. Note that the boundary condition (8) simply becomes $\zeta = \eta$ at $z = \eta$.

Linear long wave theory can now be developed in terms of the linear long wave modes, $\phi_n(z)$, which are defined by the eigenvalue problem,

$$\left(\rho_0 \phi_{nz}\right)_z + \frac{\rho_0 N^2}{c_n^2} \phi_n = 0, \quad \text{in} \quad -h < z < 0, \tag{10}$$

$$\phi_n = 0, \quad \text{at} \quad z = -h, \tag{11}$$

$$\phi_{nz} = \frac{g}{c_n^2} \phi_n, \quad \text{at} \quad z = 0. \tag{12}$$

Typically there is an infinite sequence of modes, n = 0, 1, 2, ... , and a corresponding infinite sequence of long wave speeds c_n. Here we label n = 0 as the surface gravity wave mode, and the remaining modes describe internal waves. It is often useful to make the Boussinesq approximation in which ρ_0 in (10) is regarded as a constant, and this is usually valid in the present applications since it requires that $|h\rho_{0z}| \ll \rho_0$, or. $|N^2 h/g| \ll 1$. Using this approximation, it is readily shown that the surface gravity wave mode is given by the familiar relations, $c_0 \approx \sqrt{gh}$, and $\phi_0(z) \approx (z + h)/h$. In contrast the internal modes have much slower phase speeds which scale with Nh/n, and their modal functions vanish at both boundaries but have n extrema in the interior of the fluid. Indeed, using the Boussinesq approximation, and assuming that N is a constant, it is readily shown that

$$\phi_n = \sin\left(\frac{n\pi z}{h}\right), \tag{13}$$

and

$$c_n^2 = \frac{N^2 h^2}{\pi^2 n^2}. \tag{14}$$

Another explicit result is the case of a two-layer fluid, which is represented by a fluid of constant density ρ_1 and basic depth h_1 overlying another fluid of constant density ρ_2 and basic depth h_2, where $h_1 + h_2 = h$. For this situation the density gradient ρ_{0z} is given by the δ-function $-(\rho_2 - \rho_1)\delta(z + h_1)$. There are now just two modes, a surface mode and an internal mode. Again using the Boussinesq approximation the internal mode is given by

$$\phi_1 = \left\{ \begin{matrix} -z/h_1 & \text{for} \quad 0 > z > -h_1, \\ (z+h_2)/h_2 & \text{for} \; -h_1 > z > h \end{matrix} \right\} \tag{15}$$

and

$$c_1^2 = \frac{g' h_1 h_2}{h_1 + h_2}, \quad \text{where} \quad g' = \frac{g(\rho_2 - \rho_1)}{\rho_2}. \tag{16}$$

Here g′ is the so-called reduced gravity.

The modes defined by the eigenvalue problem are mutually orthogonal, and it is convenient to express this in the following form,

$$\int_{-h}^{0} \rho_0 \phi_{nz} \, \phi_{mz} \, dz = \delta_{nm} \frac{I_n}{c_n}, \tag{17}$$

We can then proceed to solve the linear long wave equations in terms of the modal expansion,

$$\zeta = \sum_{0}^{\infty} A_n(\mathbf{x}, t)\, \phi_n(z), \tag{18}$$

$$\mathbf{u} = \sum_{0}^{\infty} B_n(\mathbf{x}, t)\, \phi_{nz}(z), \tag{19}$$

$$p = \rho_0 \sum_{0}^{\infty} C_n(\mathbf{x}, t)\, \phi_{nz}(z) \tag{20}$$

It can then be shown that $C_n = c_n^2$ An, and

$$A_{nt} + \nabla \cdot B_n = 0, \tag{21}$$

$$B_{nt} + c_n^2 \nabla A_n = 0. \tag{22}$$

Elimination of B_n shows that A_n satisfies the familiar linear wave equation. The general initial-value problem can now be readily solved, and the outcome is that any initially localized disturbance will break up into a family of modes, each propagating outward at their own speed c_n. A similar approach can be employed for internal waves generated by localized forcing, such as flow over an isolated topographic feature (Grimshaw and Smyth, 1986).

Asymptotic Analysis

The discussion of the previous section shows that any localized initial disturbance will develop into a family of outwardly propagating modes. In practice, for the internal wave components, this family is often dominated by the first and second modes. In any case, assuming that the speeds c_n are sufficiently distinct, it is enough for large times to examine just a single mode. Henceforth we shall omit the index "n". Also, for simplicity, we shall assume in the remainder of this sub-section that the mode is propagating uni-directionally in the x-direction. Then, according to the linear long wave theory described above, it is simply given by A(x - ct), which is just a wave of permanent form propagating with speed c. However, as time increases, we can expect nonlinear effects to come into play and cause wave steepening. But this in turn is opposed by weak linear dispersion which has been neglected in the linear long wave theory. Thus we expect a balance between these two effects to emerge as time evolves, and it is now well-known that the outcome is the Korteweg-de Vries (KdV) equation, or a related equation, for the wave amplitude.

The formal derivation of the evolution equation is best done through the introduction of the small parameters α and ε charaterizing the wave amplitude and the weak dispersion respectively (the length scale is proportional to ε^{-1}). The KdV equation requires the balance $\alpha = \varepsilon^2$ and the corresponding timescale is ε^{-3}. The asymptotic analysis required to derive the KdV equation is lengthy but now well-understood, and hence we shall not give details. Essentially the KdV equation arises as a condition for the validity of an asymptotic expansion in powers of α. The outcome is the KdV equation for the wave amplitude $A(\theta, t)$ (see, for instance, Benney, 1966; Lee and Beardsley, 1974; Ostrovsky, 1978; Maslowe and Redekopp, 1980; Grimshaw, 1981; Tung et al., 1981),

$$A_t + \mu A A_\theta + \delta A_{\theta\theta\theta} = 0. \tag{23}$$

Here $\theta = x - ct$ is a phase variable corresponding to the linear long wave speed c, while the coefficients μ and δ are given by,

$$I\mu = \frac{3}{2}\int_{-h}^{0} \rho_0 c^2 \phi_z^{\ 3} \, dz, \tag{24}$$

and

$$I\delta = \frac{1}{2}\int_{-h}^{0} \rho_0 c^2 \phi^2 \, dz, \tag{25}$$

and we recall that I is defined by (17). Note that the dispersion coefficient δ is always positive, but the nonlinear coefficient μ can take either sign. For the usual situation of a near-surface pycnocline, μ is negative for the first internal mode. Explicit evaluation of the coefficients μ and δ requires a knowledge of the modal function defined by (10) to (12), and hence normally they are evaluated numerically. However, for the two-layer fluid we can use the expressions (15) and (16) to find that (in the Boussinesq approximation)

$$\mu = \frac{3}{2}\frac{c}{h_1 h_2}(h_1 - h_2), \qquad \delta = c\frac{h_1 h_2}{6}. \tag{26}$$

Note that $\mu < 0 (>0)$ according as $h_1 < h_2$ ($h_1 > h_2$). Also, a special situation arises when $h_1 \approx h_2$ as then $\mu \approx 0$, and in this case the KdV equation (23) must be replaced by a modified KdV equation which retains nonlinear terms up to the cubic order, and is given by

$$A_t + \mu A A_\theta + \omega A^2 A_\theta + \delta A_{\theta\theta\theta} = 0. \tag{27}$$

The evaluation of the coefficient ω is given by Djordjevic and Redekopp (1978) and Kakutani and Yamasaki (1978) (see also Gear and Grimshaw, 1983), and it can be shown that $\omega < 0$ for a two-layer fluid. A similar situation arises for a fluid with constant stratification, when μ vanishes in the Boussinesq approximation. In this case also the KdV equation can be replaced by the modified KdV equation. Note that the validity of the modified KdV equation requires a different balance between the small parameters, namely $\alpha = \varepsilon$, where μ is $O(\alpha)$ and the timescale is again ε^{-3}.

The KdV equation (23) has been derived on the assumption that the waves are perturbations about a state of rest. Often in applications this is not the case and there is a significant basic shear flow. If we suppose that this is uni-directional and in the x-direction, given by $u_0(z)$, then the theory described above needs to be modified. First the modal equations (10) to (12) are replaced by

$$\left\{\rho_0 (c - u_0)^2 \phi_z\right\}_z + \rho_0 N^2 \phi = 0, \text{ in } -h < z < 0, \tag{28}$$

$$\phi = 0, \quad \text{at } z = -h, \tag{29}$$

$$(c - u_0)^2 \phi_z = g\phi, \quad \text{at } z = 0. \tag{30}$$

The modes are now no longer orthogonal, and may contain a critical layer where $c = u_0$. Here we shall assume that this is not the case for the mode of interest, but for the modifications necessary when a critical layer occurs, see Maslowe and Redekopp (1980) and Tung et al. (1981). It may now be shown that the amplitude $A(\theta, t)$ of a given mode again satisfies the KdV equation (23) but now the coefficients μ and δ are given by

$$I\mu = \frac{3}{2}\int_{-h}^{0} \rho_0\left(c_0 - u_0\right)^2 \phi_z^3 \, dz\,, \tag{31}$$

$$I\delta = \frac{1}{2}\int_{-h}^{0} \rho_0\left(c_0 - u_0\right)\phi^2 \, dz\,, \tag{32}$$

and

$$I = \int_{-h}^{0} \rho_0\left(c - u_0\right)\phi_z^2 \, dz\,. \tag{33}$$

In many oceanic applications, the ocean depth h is not necessarily small compared to the horizontal length scale of the solitary wave, but nevertheless the density stratification is effectively confined to a near-surface layer of depth h_1, which is much shorter than the horizontal length scale. It was pointed out by Benjamin (1967) and Davis and Acrivos (1967) that then a different theory is needed. To be specific, we shall suppose that $\rho_0(z)$ and $u_0(z)$ vary only over a near-surface layer of depth h_1, and that in the deep-fluid region $\rho_0(z) = \rho_\infty$ (a constant) and $u_0(z) = 0$, while the ocean bottom is now defined by $z = -H/\varepsilon$ (i.e. $H = \varepsilon h$). The modal function is again defined by (28) in the near-surface layer, with the upper boundary condition (30), but (29) is replaced by the condition that $\phi_z \to 0$ as $z \to -\infty$. To derive the evolution equation the appropriate parameter balance is now $\alpha = \varepsilon$, and the outcome is the intermediate long-wave (ILW) equation (Kubota et al., 1978; Maslowe and Redekopp, 1980; Grimshaw, 1981; Tung et al., 1981),

$$A_t + \mu A A_\theta + \gamma \mathcal{L}(A_\theta) = 0, \tag{34}$$

$$\mathcal{L}(A) = -\frac{1}{2\pi}\int_{-\infty}^{\infty} k \coth kH \, \exp(ik\theta)\, \mathcal{F}(A)dk, \tag{35}$$

and

$$\mathcal{F}(A) = \int_{-\infty}^{\infty} A \exp(-ik\theta)d\theta\,. \tag{36}$$

Here the coefficient μ is again given by (31) and (33) with -h now replaced by $-\infty$, while $I\gamma = (\rho_0 c^2 \phi^2)_{z\to-\infty}$. For the two-layer fluid, where a shallow layer of density ρ_1 and depth h_1 overlies a deep layer of density ρ_2 and depth H_2/ε (i.e. $H_2 = \varepsilon h_2$), we find that $c^2 = g' h_1$, $\mu = -3c/2h_1$ (the limits of $c_1{}^2$ (16) and μ (26) as $h_2 \to \infty$) while $\gamma = {}^1/_2 c h_1$, in the Boussinesq approximation. In the limit $H \to \infty$, $k \coth kH \to |k|$ in the integrand of (35), and (34) becomes the Benjamin-Ono (BO) equation. In the opposite limit $H \to 0$, (34) reduces to a KdV equation.

Solitary Waves

Structure

Each of the evolution equations described in the previous section has solitary wave solutions. That for the KdV equation (23) is the best known and is given by

$$A = a \operatorname{sech}^2 \beta(\theta - vt), \tag{37}$$

where

$$v = \frac{1}{3}\mu a = 4\delta\beta^2. \tag{38}$$

Note that the speed v is relative to the phase variable θ, and the total speed is c + v. Since the dispersion coefficient δ is positive for internal solitary waves, it follows from (38) that v is always positive and so solitary waves always propagate faster than linear long waves. Further, we see from (38) that μa is also always positive and so internal solitary waves are waves of elevation or depression according as μ is positive or negative. Note that the expression (38) shows that the width β^{-1} varies as $|a|^{-1/2}$, and hence the larger solitary waves are narrower.

For the modified KdV equation (27) the corresponding solitary wave solution is given by Kakutani and Yamasaki (1978) and Gear and Grimshaw (1983)

$$A = \frac{a}{b+(1-b)\cosh^2\beta(\theta - vt)}, \tag{39}$$

where

$$v = \frac{1}{3}a\left(\mu + \frac{\omega a}{2}\right) = 4\delta\beta^2, \tag{40}$$

and

$$b = \frac{-\omega a}{(2\mu + \omega a)}. \tag{41}$$

For the main case of interest when $\omega < 0$, it follows that $0 < b < 1$ and $\mu a > 0$. Further, as b increases from 0 to 1, the amplitude $|a|$ also increases from 0 to a maximum of $|\mu/\omega|$, while the speed also increases from 0 to a maximum of $-\mu^2/6\omega$. For the case when this modified KdV equation is relevant $|\mu| << |\omega|$, and hence it is of interest to examine further the structure of this solitary wave as $b \to 1$. In this limit the solution (39) describes a flat crest of constant amplitude $a \approx |\mu/\omega|$, terminated at each end by the bore-like solutions

$$A = \frac{1}{2}a_m\left\{1 \mp \tanh \beta_m(\theta - v_m t)\right\}, \tag{42}$$

where a_m is the maximum amplitude, $a_m = -\mu/\omega$, and $v_m = a_m - \mu^2 / 6\omega = 4\delta\beta_m{}^2$. In the limit $\mu \to 0$, there are no exact solitary wave solutions of (27) when $\omega < 0$, but instead there is a solution in the form of a travelling bore,

$$A = a \tanh \beta(\theta - vt), \tag{43}$$

where

$$v = \omega\frac{a^2}{3} = -2\delta\beta^2. \tag{44}$$

Indeed, this solution can be derived by observing that the transformation $A \to A - \mu/2\omega$ reduces the modified KdV equation with a quadratic nonlinear term into a "pure" modified KdV equation with no quadratic nonlinear term and only a cubic nonlinear term, and then utilising the solution (42). Note that the amplitude of the travelling bore (42) is a free parameter, and that the speed ω is negative.

For the case when $\omega > 0$, it follows that $b < 0$ and there are two families of solitary waves described by (39). One has $\mu a > 0$ and the other $\mu a < -2\mu^2/\omega$. In this case solitary waves exist if $\mu = 0$ and are given by

$$A = a \text{ sech } 2\beta(\theta - vt), \tag{45}$$

where

$$v = 4\delta\beta^2 = \frac{\omega a^2}{6} \tag{46}$$

Finally, for the ILW equation (34) the solitary wave solution was obtained by Joseph (1977), and is given by

$$A = \frac{a\,sin^2\Delta}{cosh^2\beta(\theta - vt) - cos^2\Delta}, \tag{47}$$

where $$v = 2\gamma\beta cot 2\Delta, \quad \tfrac{1}{4}\mu a = \gamma\beta cot\Delta, \quad \Delta = -\beta H. \tag{48}$$

In the limit $H \to \infty$, $\Delta \to 0$ and $\beta \to 0$ when (42) reduces to the BO solitary wave, which is distinguished by its algebraic structure,

$$A = \frac{a}{1 + \kappa^2(\theta - vt)^2}, \tag{49}$$

where $$v = \frac{1}{4}\mu a = \gamma\kappa. \tag{50}$$

In this limit, the solitary wave width (κ^{-1}) is inversely proportional to the amplitude, with the consequence that the mass carried by this solitary wave is independent of its amplitude. In the opposite limit $H \to 0$, $\Delta \to \pi/2$ and (47) reduces to the familiar KdV-solitary wave.

Evolution and Applications

Each of the nonlinear evolution equations (viz. the KdV equation (23), the modified KdV equation (29) and the ILW equation (34) are exactly integrable (see, for example, Ablowitz and Segur, 1981; Dodd et al., 1982), with the consequence that the initial-value problem with a localized initial condition is exactly solvable. From the point of view of the present application, the most important implication is that an arbitrary localized initial disturbance will evolve into a finite number (N) of solitary waves (usually called solitons in this context) and a dispersive, oscillatory, decaying tail. This, together with the stability of solitary waves, helps to explain why internal solitary waves are so commonly observed. Because these solitary waves have speeds increasing with amplitude, the N solitary waves are rank-ordered by amplitude as $t \to \infty$. Also, in order to produce solitary waves the initial disturbance must be such that μA is sufficiently positive. For instance, for the KdV equation, if $\int_{-\infty}^{\infty} A(\theta, 0)d\theta$ is positive, then $N \geq 1$.

In attempting to compare the theory with observations, we first recall that the initial condition for the nonlinear evolution equation is determined by solving an initial-value problem (or boundary-value problem) for the linear long wave equations, (21) and (22). The outcome as $t \to \infty$, for a given long wave mode, then determines, through asymptotic matching, the initial condition for the nonlinear evolution equation. Alternatively, of course, the initial data could be directly determined from observations. Then, with the initial condition known, the appropriate nonlinear evolution equation can be solved, analytically or numerically, to determine the number (N) and amplitudes of the evolving solitary waves. This information can then be compared with observations. However, although there are now a large number of observations of internal solitary waves, there have been relatively few attempts to carry out this complete comparison, notable exceptions being the study of Apel et al. (1985) and Liu et al. (1985) of internal solitary waves in the Sulu Sea, and the study by Holloway (1987) and Smyth and Holloway (1988) of the development of the internal tide into solitary waves on the North West Shelf of

Australia. Some of the reasons for this are that the initial data is often not known completely, and that the basic density $\rho_0(z)$ and shear flow $u_0(z)$ are also not known with sufficient accuracy to determine the coefficients μ(31) and δ(32) (or γ) in the nonlinear evolution equation.

One of the most striking features of the present theory is that the solitary waves will be rank-ordered by amplitude as $t \to \infty$, and this feature alone serves as a useful indication that a set of observed waves are indeed solitary waves. Indeed, this is commonly seen in observations of internal solitary waves in coastal seas, fjords and lakes (Apel, 1995; Ostrovsky and Stepanyants, 1989). Another feature which has proved useful in identifying solitary waves from observations is that they are characteristically waves of elevation when $\mu > 0$ (or depression when $\mu < 0$), and their speed increases with amplitude, although in practice it often turns out that the dominant component of the observed speed is just the linear long wave speed. However, attempts to use the inverse relation between solitary wave amplitude and width as an identifying feature of observations has not proved very successful. Finally, we note that many observed solitary waves have very large amplitudes suggesting that weakly nonlinear theory is not appropriate and at least it should be necessary to take into account higher-order amplitude terms. Nevertheless, the weakly nonlinear theory described here has proved to be surprisingly successful.

Extensions

Variable Background

Although the weakly nonlinear theory leading to the nonlinear evolution equations such as the KdV equation (23), or the other variants, forms the basis for studying internal solitary waves, it is necessary in practice to take into account other effects such as bottom friction, variable bottom topography and wave refraction. For simplicity, we shall describe these modifications in the context of the KdV equation (23) since the corresponding modifications to the other variants are completely analogous.

First, let us suppose that the background environment in which the wave propagates varies slowly in the wave propagation direction (i.e. the x-direction). This can occur, for instance, when the depth h varies, and also when the basic density ρ_0 and shear flow u_0 also vary spatially. In this situation, provided the lengthscale of the variation in the background is much greater than the solitary wave width, the KdV equation (23) is replaced by a variable-coefficient KdV equation given by (see, for example, Grimshaw (1981) or Zhou and Grimshaw (1989)).

$$A_s + \frac{1}{2}\frac{\sigma_s}{\sigma}A + \frac{\mu}{c}AA_\theta + \frac{\delta}{c^3}A_{\theta\theta\theta} = 0, \tag{51}$$

where

$$s = \int_0^x \frac{dx'}{c(x')}, \quad \theta = s - t, \quad \sigma = c^2 I. \tag{52}$$

Here the linear long wave speed c depends on x, as in general does ρ_0 and u_0. Hence the linear long wave modal function $\phi(z, x)$ also now depends parametrically on x, and thus the coefficients μ, δ and I defined by (31), (32) and (33) also likewise depend on x. Note that s is a time-like variable used to describe the evolution of the wave in the x-direction, and that σ, μ, δ and c can all be regarded as functions of s. The significance of the coefficient σ is that σA^2 is proportional to the wave action flux in the x-direction. When the background varies slowly, but in both spatial directions, the appropriate nonlinear evolution equation has the same form as (51) but now s is a time-like variable along the ray path determined by the linear long wave speed c (see Ostrovsky and Pelinovsky (1975) for surface waves and Grimshaw (1981) for internal waves).

Unlike the constant-coefficient KdV equation, the variable-coefficient equation (51) is not integrable in general, and hence must be solved either numerically or by approximate analytic methods. An instance of the latter is the case when the background environment varies slowly compared to the solitary wave width, which is relevant to many oceanic situations. Here, multi-scale perturbation techniques can be used (see, for instance, Grimshaw, 1979) in which the leading order term is the slowly varying solitary wave,

$$A \approx a\,\mathrm{sech}^2\beta\left(\theta - \int_0^s v\,ds\right), \qquad (53)$$

where

$$v = \frac{\mu a}{3c} = \frac{4\delta\beta^2}{c^3}. \qquad (54)$$

Note that this expression is analogous to the solitary wave solution, (37) and (38), of the constant-coefficient KdV equation, but now the speed v, and hence also a and β, are slowly varying functions of s. This variation is most easily determined by noting that (51) possesses the conservation law,

$$\frac{\partial}{\partial s}\int_{-\infty}^{\infty} \sigma A^2\,d\theta = 0 \qquad (55)$$

which expresses conservation of wave action flux. Substitution of (53) into (55) gives the law for the amplitude variation

$$\sigma a^2\beta^{-1} = \text{constant.} \qquad (56)$$

On using (54) this expression can be used to determine each of v, a and β in terms of the known coefficients σ, μ, δ and c. Because each of these depends in a complicated way on the depth h, the basic density ρ_0 and the shear flow u_0 through the modal equations (28) to (30), it is not possible in general to obtain simple explicit expressions for the amplitude a as a function of the depth h for instance. Nevertheless, the result (56) does provide a systematic method for determining how the solitary wave properties change in a variable background. We should also note here that while the slowly varying solitary wave conserves the wave action flux law (55), it cannot simultaneously conserve mass, and hence is accompanied by a trailing shelf of small amplitude but long length scale whose amplitude A_1 at the rear of the solitary wave is given by

$$A_1 = \frac{3c\beta_s}{\mu\beta^2}. \qquad (57)$$

From (56) and (57) we can see that when the wave action flux σa^2 increases (decreases) with s, then the trailing shelf amplitude A_1 has the same (opposite) polarity as the solitary wave.

Friction

In practice, internal solitary waves in shallow seas, lakes of fjords are subject to dissipation which can take a variety of forms. These include turbulent mixing in the interior of the fluid associated with local shear instability, scattering induced by bottom roughness, and the viscous decay due to the bottom boundary layer. These, and other forms of dissipation can be incorporated into the variable-coefficient KdV equation (51) by the addition of an extra terms, so that (51) now becomes

$$A_s + \frac{1}{2}\frac{\sigma_s}{\sigma}A + \frac{\mu}{c}AA_\theta + \frac{\delta}{c^3}A_{\theta\theta\theta} + \eta\mathcal{D}(A) = 0, \tag{58}$$

where η is a friction coefficient and $\mathcal{D}$ (A) is an operator representing the effects of friction. Here we shall consider only the case of a laminar bottom boundary layer when

$$\mathcal{D}(\mathcal{A}) = \frac{1}{2\pi}\int_{-\infty}^{\infty}(-ik)^{1/2}\exp(ik\theta)\ \mathcal{F}(A)dk, \tag{59}$$

where we recall that $\mathcal{F}(A)$ is the Fourier transform (37) of A, while $I\eta = {}^1/_2$ $(\rho\nu|c-u_0|^3/|c|)^{1/2}(\phi_z)^2$ evaluated at z = -h, where ν is the kinematic viscosity.

To obtain some information about the effect of dissipation we again use the slowly varying wave hypothesis, and assume that η is a small coefficient. Then to leading order the solitary wave is again given by (53) and (54) but now (56) is replaced by

$$\frac{\partial}{\partial s}\left(\frac{2}{3}\frac{\sigma a^2}{\beta}\right) = -\eta\sigma a^2\int_{-\infty}^{\infty}\operatorname{sech}^2\beta\theta\mathcal{D}\left(\operatorname{sech}^2\beta\theta\right)d\theta\,. \tag{60}$$

For the case of a laminar bottom boundary layer when $\mathcal{D}(A)$ is given by (59), the right-hand side of (60) can be evaluated to yield a differential equation for the amplitude a whose solution in the case when σ, μ, δ, η and c are all constants is

$$a = a_0\left[1 + 0.168\left(\frac{c^2\mu a_0}{9\delta}\right)^{\frac{1}{4}}\eta s\right]^{-4} \tag{61}$$

where α_0 is the amplitude at s = 0. The expression (61) has been found to provide a reasonably reliable estimate of the decay of a solitary wave due to dissipation in the bottom boundary layer.

Wave Refraction

Because the background environment generally has variability in both spatial directions it is sometimes necessary to take into account wave refraction effects in directions transverse to the propagation direction. When these refraction effects are weak, the appropriate evolution equation is the variable-coefficient Kadomtsev-Petviashvili (KP) equation,

$$\left(A_s + \frac{1}{2}\frac{\sigma_s}{\sigma}A + \frac{\mu}{c}AA_\theta + \frac{\delta}{c^3}A_{\theta\theta\theta}\right)_\theta + \frac{1}{2}c^2A_{yy} = 0. \tag{62}$$

Here y is a coordinate perpendicular to the propagation direction x, and s, θ are defined by (52). Clearly the KP equation (62) is a two-dimensional generalization of the variable-coefficient KdV equation (51). With s, θ given by (52), the background environment varies predominantly only in the x-direction. However, when the spatial variability occurs in both directions, an evolution equation similar in form to (62) can be derived (Grimshaw, 1981), in which s is a time-like variable along a ray path determined by the linear long wave speed c, and y is a co-ordinate transverse to that path. The coefficients μ, δ, σ as well as c are in general functions of s along the ray path. When these coefficients are all constants, the KP equation is exactly integrable.

To this point, the nonlinear evolution equations that we have considered here have been predicated on the assumption that the generation mechanism for these internal solitary waves is essentially one-dimensional analogous to a line source. Sometimes, however, it is more appropriate to consider a point source. The simplest evolution equation in this case can be obtained from (62) by noting that s is a time-like variable in the radial direction r, whose origin is at the point source. Omitting the transverse variation and considering the case when the background environment has no spatial variability, it follows that $s = r/c$ and $\sigma = r$. Letting $\theta = r - ct$, we see that (51) reduces to the cylindrical KdV equation,

$$A_r + \frac{1}{2r}A + \frac{\mu}{c}AA_\theta + \frac{\delta}{c}A_{\theta\theta\theta} = 0. \quad (63)$$

Like the KdV equation this is also exactly integrable. However, for our present purposes, it is more instructive to examine the effect of radial divergence using the slowly varying solitary wave solution of (63). Thus, from (56) with $\sigma = r$ we readily find that the amplitude varies as $r^{-2/3}$. This can be contrasted with the corresponding result for linear waves where the amplitude varies as $r^{-1/2}$.

References

Ablowtiz, M. J. and H. Segur, Solitons and the inverse scattering transform, SIAM, Philadelphia, 425pp. 1981.

Apel, J. R., Satellite sensing of ocean surface dynamics, *Ann. Rev. Earth Planet. Sci., 8*, 303-342, 1980.

Apel, J. R., J. R. Holbrook, A. K. Liu and J. J. Tsai, The Sulu Sea internal soliton experiment, *J. Phys. Oceanogr, 15*, 1625-1651, 1985.

Apel, J. R., Linear and nonlinear internal waves in coastal and marginal seas, in Oceanographic Applications of Remote Sensing, edited by M. Ikeda and F. Dobson, 512 pp., CRC Press, Boca Raton, Florida, 1995.

Benjamin, T. B., Internal waves of permanent form in fluids of great depth, *J. Fluid Mech., 29*, 559-592, 1967.

Benney, D. J., Long non-linear waves in fluid flows, *J. Math. Phys., 45*, 52-63, 1966.

Davis, R. E. and A. Acrivos, Solitary internal waves in deep water, *J. Fluid Mech., 29*, 593-607, 1967.

Djordjevic, V. D. and L. G. Redekopp, The fission and disintegration of internal solitary waves moving over two-dimensional topography, *J. Phys. Oceanogr., 8*, 1016-1024, 1978.

Dodd, R. K., J. C. Eilbeck, J. D. Gibbon and H. C. Morris, Solitons and Nonlinear Wave Equations, 630 pp., Academic, London, 1982.

Drazin, P. G. and R. S. Johnson, Solitons: An Introduction, 226 pp., CUP, Cambridge, 1989.

Gear, J. and R. Grimshaw, A second-order theory for solitary waves in shallow fluids, *Phys. Fluids, 26*, 14-29, 1983.

Grimshaw, R., Slowly varying solitary waves. I Korteweg-de Vries equation, *Proc. Roy. Soc. London, A368*, 359-375, 1979.

Grimshaw, R., Evolution equations for long nonlinear waves in stratified shear flows, *Stud. Appl. Maths., 65*, 159-188, 1981.

Grimshaw, R. and N. Smyth, Resonant flow of a stratified fluid over topography, *J. Fluid Mech., 169*, 429-464, 1986.

Holloway, P. E., Internal hydraulic jumps and solitons at a shelf break region on the Australian North West Shelf, *J. Geophys. Res., 92 (C5)*, 5405-5416, 1987.

Joseph, R. J., Solitary waves in a finite depth fluid, *J. Phys A: Math and Gen., 10*, L225-L227, 1977.

Kakutani, T. and N. Yamasaki, Solitary waves on a two-layer fluid, *J. Phys. Soc. Japan*, *45*, 674-679, 1978.

Kubota, T., D. R. S. Ko, and L. Dobbs, Weakly nonlinear long internal gravity waves in stratified fluid of finite depth, *AIAA J. Hydronautics*, *12*, 157-165, 1978.

Lee, C-Y. and R. C. Beardsley, The generation of long nonlinear internal waves in a weakly stratified shear flow:, *J. Geophys. Res.*, *79*, 453-462, 1974.

Liu, A. K., J. R. Holbrook and J. R. Apel, Nonlinear internal wave evolution in the Sulu Sea, *J. Phys. Oceanogr*, *15*, 1613-1624, 1985.

Maslowe, S. A. and L. G. Redekopp, Long nonlinear waves in stratified shear flows, *J. Fluid Mech.*, *101*, 321-348, 1980.

Miles, J. W., Solitary waves, *Ann. Rev. Fluid Mech.*, *12*, 11-43, 1980.

Ostrovsky, L. A., Nonlinear internal waves on a rotating ocean, *Oceanology*, *18*, 181-191, 1978.

Ostrovsky, L. A. and E. N. Pelinovsky, Refraction of nonlinear ocean waves in a beach zone, *Izv. Atmos. Oceanic Physics*, 11, 67-74, 1975.

Ostrovsky, L. A. and Yu. A. Stepanyants, Do internal solitons exist in the ocean?, *Rev. Geophysics.*, *27*, 293-310, 1989.

Smyth, N. F. and P. E. Holloway, Hydraulic jump and undular bore formation on a shelf break, *J. Phys. Oceanogr.*, *18*, 947-962, 1988.

Tung, K-K., D. R. S. Ko and J. J. Chang, Weakly nonlinear internal waves in shear, *Stud. Appl. Math*, *65*, 189-221, 1981.

Zhou, X. and R. Grimshaw, The effect of variable currents on internal solitary waves, *Dyn. Atmos. Oceans*, *14*, 17-39, 1989.

17

Two Intersecting Internal Wave Rays: A Comparison Between Numerical and Laboratory Results

A. Javam, S. G. Teoh, J. Imberger and G. N. Ivey

Abstract

The interaction between two downward propagating internal wave rays in a linearly stratified fluid was examined numerically and experimentally. The numerical simulations employed a SIMPLE scheme with a third order QUICK discretization for the advective terms and second order Crank-Nicholson scheme on a non-staggered grid to solve the full unsteady equations of motion in an open domain with boundary conditions based on the Sommerfield radiation condition. The laboratory experiments were performed in a glass-walled tank filled with a linearly stratified salt solution with two identical wave paddles used to generate two internal wave rays. The interaction between the two wave rays was visualized by bull eye's rainbow schlieren and shadowgraph techniques, in conjunction with velocity measurements made by particle image velocimetry. Good agreement was found between the observed and simulated interaction mechanisms. The nonlinear nonresonant interaction of two wave rays with identical properties, but opposite horizontal phase velocities, led initially to the formation of small spatial scales followed by the development of evanescent modes. The evanescent modes, with frequencies greater than the local buoyancy frequency, were trapped within the intersection region. The energy transferred to the trapped evanescent modes ultimately causing overturning of the density field in the intersection region.

Introduction

Internal waves play an important role in the distribution of the momentum and energy throughout stratified water bodies (Thorpe, 1975; Imberger, 1994) and provide a linkage between the energy-containing scales and the dissipative scales (Munk, 1981; Muller et al., 1991; Imberger, 1994). Therefore, an understanding and parameterization of these transfer, mixing and dissipation processes are essential for a description of the circulation and transport in stratified water bodies.

Resonant triad interaction is believed to be a significant mechanism in energy transfer to other modes. As a result, this mechanism has been widely investigated (e.g. Hasselmann, 1966; Bretherton, 1964; Phillips, 1960; Thorpe, 1966, 1987a, 1987b; Davis and Acrivos, 1967; Simmons, 1969; Martin et al., 1972; McComas and Bretherton, 1977;

Physical Processes in Lakes and Oceans
Coastal and Estuarine Studies Volume 54, Pages 241-250

Thorpe, 1987b). Triad interaction among internal waves redistributes energy and momentum among different wave components. Resonant interactions may also cause the amplitude of parasitic waves to grow once the amplitude of host waves exceeds a critical value (McEwan, 1971). Eventually, an instability occurs and the ambient stratified water is mixed (McEwan, 1971, 1983a, 1983b; Taylor, 1992).

Wave energy can be transferred towards frequencies lower than the forcing frequency (Martin et al., 1969, 1972; McEwan, 1971; McEwan and Robinson, 1975). Any resonant triad consisting of a finite amplitude wave and two infinitesimal components is unstable for the sum interaction and neutrally stable for the difference interaction (Hasselmann, 1967; McEwan, 1971). Note that each of the three waves participating in a resonant triad is an exact solution of the equation of motion (within the inviscid and Boussinesq approximations).

Nonlinear nonresonant interactions have received comparatively little attention, even though they also transfer energy across the spectrum. McEwan (1973) studied the nonlinear nonresonant interactions between two rays and showed that the Richardson, number was large before the occurrence of 'traumata'; isopycnal slopes were, however greater than 30°. Thorpe (1987a) illustrated that second order resonance between an incident and a reflected wave from a slope only occurs when the slope and the incident ray are less than 8.4° and 30°, respectively, to the horizontal. The evanescent modes can be excited by the effects of nonlinearity (Thorpe, 1987b; Ivey et al, 1995). The excited evanescent modes play a crucial role in producing density overturning. In this paper it will be shown both experimentally and numerically that two intersecting beams of internal waves do become unstable, not by resonant triad interaction, but rather by transferring energy to standing evanescent modes, ultimately leading to overturning of the density field.

Laboratory and Numerical Models

Laboratory Model

A schematic of the experimental configuration is shown in Figure 1. The experimental glass-walled tank of length 5900 mm, depth 540 mm and width 535 mm, was filled with linearly stratified salt solution, using the standard two-tank system. Two identical wave-paddles, each consisting of eight blades which could pivot independently about their long central axes, were used to generate internal wave rays. The wave paddles were separated, centre to centre, by 615 mm and located horizontally near the centre of the tank, 40 mm below the free surface. Only the central six blades were oscillated, forming either an M or W shape, and internal waves rays of width 1.5 times the wavelength were generated. The central amplitude of each paddle was twice that at both sides of the paddle to minimise the net volume flux induced by blade displacements.

Flow visualisation inside the intersection region was achieved by rainbow colour schlieren (Howes, 1984; Ivey & Nokes, 1989) and shadowgraph (Merzkirch, 1974). The evolution of the density field was complemented by time series of density and profiles obtained with conductivity and temperature measurements. Digital particle image velocimetry (DPIV) was used to complement the results from flow visualisation (Stevens & Coates, 1994). A detailed description of the experimental configuration is given in Teoh et al. (1997).

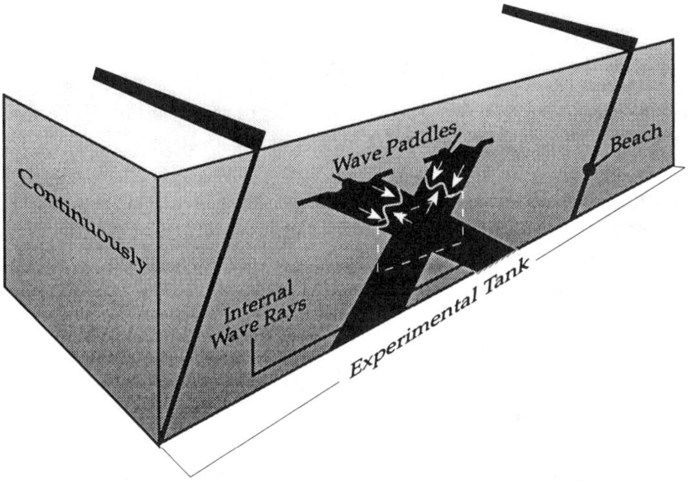

Figure 1. Schematic of the experimental tank.

Numerical Model

The conservation of momentum, mass and volume are expressed in non-dimensional form in Euclidean coordinates, (x, y) with corresponding velocity components (u, v) as follows,

$$\frac{\partial u}{\partial t}+\mathrm{Fr}\left[u\frac{\partial u}{\partial x}+v\frac{\partial u}{\partial y}\right]=-\frac{\partial p}{\partial x}+\mathrm{Re}^{-1}\nabla^2 u, \tag{1}$$

$$\frac{\partial v}{\partial t}+\mathrm{Fr}\left[u\frac{\partial v}{\partial x}+v\frac{\partial v}{\partial y}\right]=-\frac{\partial p}{\partial y}-\rho+\mathrm{Re}{-1}\,\nabla 2v+\sin(t)\big(f_1,(x,y)+f_2(x,y)\big), \tag{2}$$

$$\frac{\partial \rho}{\partial t}+\mathrm{Fr}\left[u\frac{\partial \rho}{\partial x}+v\frac{\partial \rho}{\partial y}\right]=\mathrm{Ri}\,v+\mathrm{Re}^{-1}\,\mathrm{Pr}^{-1}\,\nabla^2\rho, \tag{3}$$

$$\frac{\partial u}{\partial x}+\frac{\partial v}{\partial y}=0, \tag{4}$$

$$\mathrm{Ri}=\frac{(-g/\rho_0)(d\hat{\rho}/dy)}{\omega^2}=\left[\frac{N}{\omega}\right]^2, \tag{5}$$

$$\mathrm{Fr}=\frac{F}{L\omega^2}, \tag{6}$$

$$\mathrm{Pr}=\frac{\nu}{\kappa}, \tag{7}$$

$$\mathrm{Re}=\frac{L^2\omega}{\nu}, \tag{8}$$

$$\nabla^2 = \frac{\partial^2}{\partial x^2} + \frac{\partial^2}{\partial y^2},$$

and where $N = (g/\rho_0 (d\hat{\rho}/dy))^{1/2}$ is the buoyancy frequency, t is the time, ρ, $\hat{\rho}$ and ρ_0 are the fluctuating, background and reference densities, respectively, ν is the coefficient of kinematic viscosity, κ is the diffusivity, $f_1(x,y)$ sin t and f_2 (x, y) sin t are the momentum sources used to generate two internal wave beams, ω is the frequency of the momentum sources, $f_1(x,y)$ and $f_2(x,y)$ are the dimensionless localization functions, F is the amplitude of the momentum source, and L is the scale of the standing wavelike momentum source. The localization functions $f_i(x,y)$ are given by:

$$f_i(x,y) = \begin{cases} \cos(2\pi x_1)\exp(-300\,|\,y_1\,|^3) & \text{if } |\,x_1\,| \le 0.5 \\ \left[\cos(2\pi x_1)\exp(-300\,|\,y_1\,|^3)\right]/2 & \text{if } 0.5 < |\,x_1\,| \le 0.75 \\ 0.0 & \text{otherwise} \end{cases} \qquad (9)$$

in which x_l and y_l are the horizontal and vertical dimensionless distances from the location of the momentum source.

The equations of motion (1)–(4) were solved by the SIMPLE scheme (Patankar, 1980) with a QUICK correction for the convection terms (Leonard, 1979) on a non-staggered grid using a finite volume method (Armfield, 1994). In order to ensure that waves continue to propagate unchanged through the boundaries, an open boundary condition, based on the Sommerfield radiation condition, was formulated. A detailed description of the numerical method, the discrete equations and open boundary conditions is given in Javam et al. (1997a).

Comparison between Numerics and Experiments

To investigate the wave-wave interaction, the spatial domain was discretized with 240 × 160 uniform cells with the nondimensional mesh size of $\Delta x = \Delta y = 0.05$. The two momentum sources were located at $(0.35L_x, 0.50L_y)$ and $(0.65L_x, 0.50L_y)$ where L_x and L_y are the length and width of the computational domain, respectively (Figure 2). Velocity vector plots are shown in Figure 2 for the case: Fr = 0.1, Ri = 2.5, Re = 25,000 and Pr = 700, ten wave periods after start up from rest. The rays generated by the momentum sources travelled away from the sources and the wave energy was transported along the ray path or direction of the group velocity vector. The wave number vectors were normal to the group velocity vectors and were directed to the horizontal. The rays intersected in two regions called interaction regions (Figure 2); we will only present the results from the lower interaction region for comparison with the experiments.

In the laboratory experiment, the dimensional scales were: L = 191 mm, N = 1 r/s, ω = 0.4 r/s and salt was used to stratify the fluid. In the numerical model this choice resulted in Re = 15000, Ri = 6.25 and Pr = 700. In order to have a similar wave field between numerical and experimental study, we set the Froude number Fr to 0.75 which resulted in the same wave amplitude within the rays in both the experiments and the numerics.

Figure 3 shows a comparison between a schlieren image from the laboratory experiment and density gradient contours from the numerical model in the intersection region. The schlieren images were sensitive to the changes in the first derivative of the density field (Howes, 1984). Within the intersection, the wave induced density gradient fluctuations grew and eventually developed a regular pattern propagating vertically upwards as shown in Figure 3. The density gradient fluctuations in Figure 3 were symmetric about the vertical axis. The schlieren images from the experiment (Teoh et al., 97) show the appearance of a new mode and later a turbulent patch along the horizontal axis near the centre of the image. A black colour region was observed to surround this patch indicating a strong density gradient boundary. New rays generated by turbulent fluctuations were subsequently seen to radiate outwards from a point above the patch.

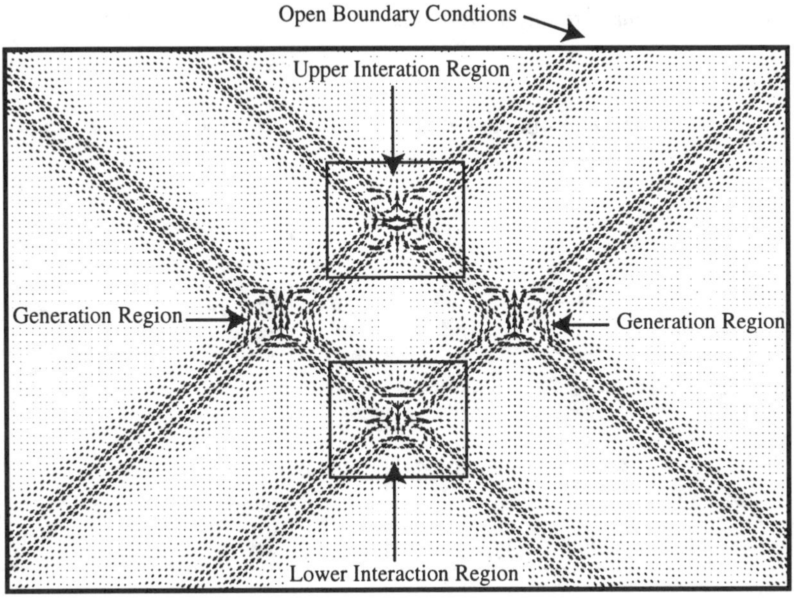

Figure 2. Velocity vector plots for the internal wave, Ri = 2.5, Fr = 0.1, Pr = 700 and Re = 25000; at time T=10. Two momentum sources are used to generate internal waves.

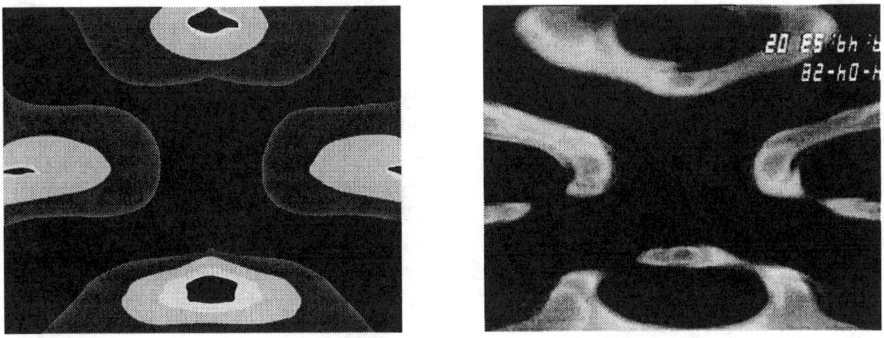

Figure 3. Rainbow colour Schlieren picture (on the right) and density gradient contours from the numerics.

The corresponding velocity field is shown in Figure 4. The velocity field exhibits stagnation points propagating vertically upwards within the intersection region. Horizontal compression and dilation straining are visible at the stagnation points, demonstrating that a fluid volume centred around a fixed point on the centreline of the interaction region is successively compressed and dilated.

Density profiles, at one wave period apart, illustrating the effect of this forcing in the progression towards gravitationally unstable overturning from T = 4.3 to T = 10.3 are shown in Figure 5a. The density profile at T = 8.3 in the numerical model shows the presence of the primary resultant wave and the locally produced new modes which had

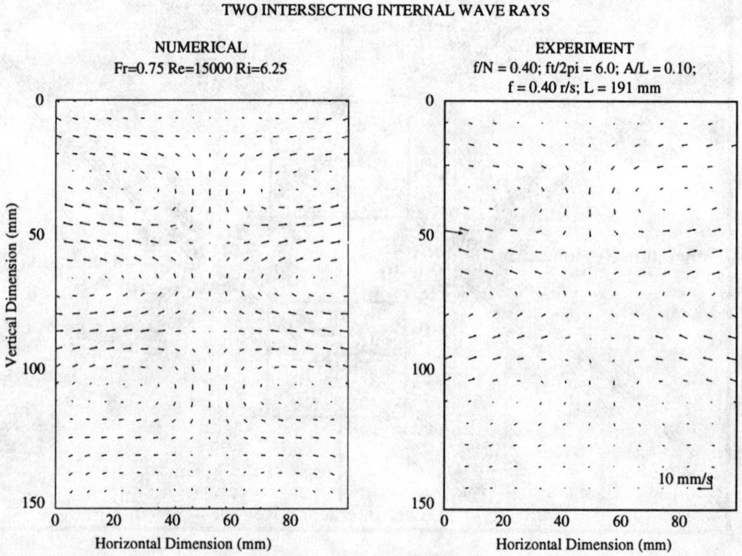

Figure 4. Velocity vector plots. Horizontal compressive and dilation straining visible at the stagnation points.

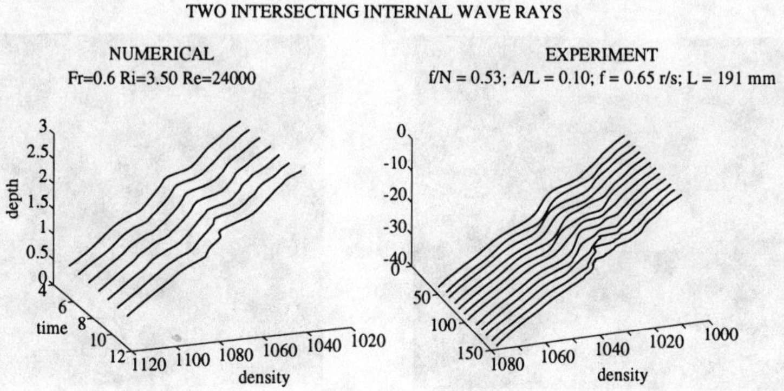

Figure 5. Vertical profiles of density for numerical and experimental results. New modes with the shorter vertical wavelengths are produced.

shorter vertical wavelengths. The result from the laboratory experiment shows a similar feature to the numerical model but shows much smaller scales which can not be resolved by the numerical model due to the restriction in mesh sizes.

The density fluctuation spectra within an incoming ray, an outgoing ray, and within the intersection region illustrate the production of new modes by nonlinear nonresonant interactions between the two rays (Figure 6). Comparison between spectra for incoming and out-going rays (Figures 6a, b) shows that most of the incoming energy was trapped within the interaction region and this trapped energy excited higher modes (Figures 6a,b and c). In the laboratory experiments the central amplitudes of the wave-paddles were twice those at both sides of the paddles; consequently, the wave amplitude varied across the rays. Because of the strong spatial variation in wave amplitude and the difficulty in precisely locating sensors, the result from the experiment can not be directly compared with the numerics (Figures 7a, b). The laboratory experiments were also run for six periods longer than the numerical model and therefore, a number of higher (evanescent) modes were developed which are clearly visible in Figure 7c.

Figures 6 and 7 also show that the level of the spectral energy inside the interaction region was much higher than that of the incoming ray, which implies that the energy was trapped in the interaction region. The density fluctuation spectra within the interaction region (Figures 6 and 7) indicate that the excited modes are multiples of the primary frequency ω. The new modes with frequencies greater than the local buoyancy frequency are thus evanescent modes and cannot propagate (LeBlond and Mysak, 1978). Therefore the energy transfer from ω to the evanescent modes provides a mechanism to accumulate energy locally and subsequently leading to overturning as seen in Figure 5b for longer times.

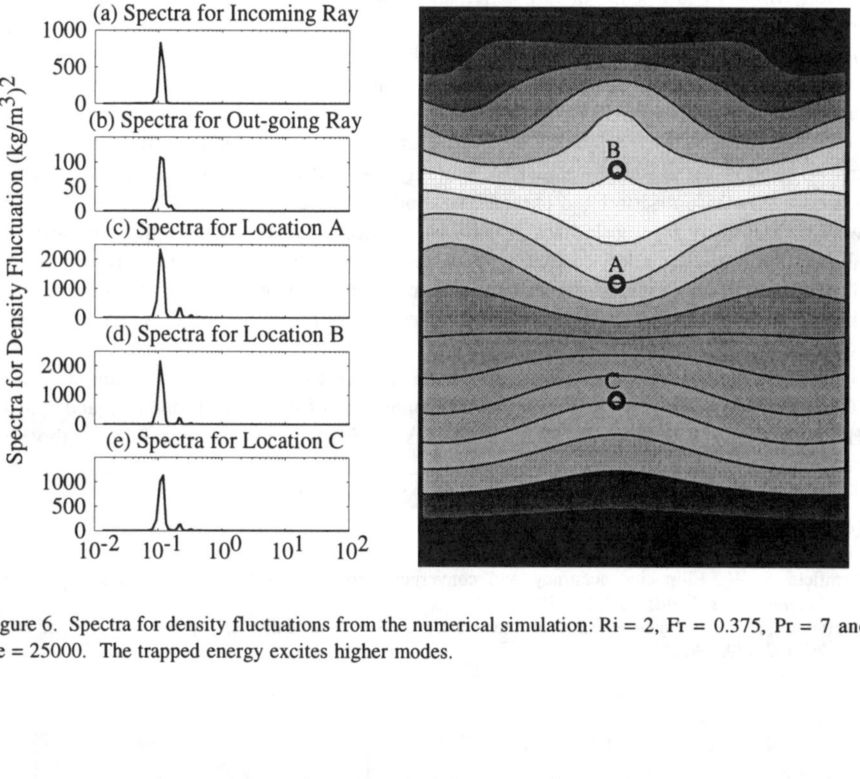

Figure 6. Spectra for density fluctuations from the numerical simulation: Ri = 2, Fr = 0.375, Pr = 7 and Re = 25000. The trapped energy excites higher modes.

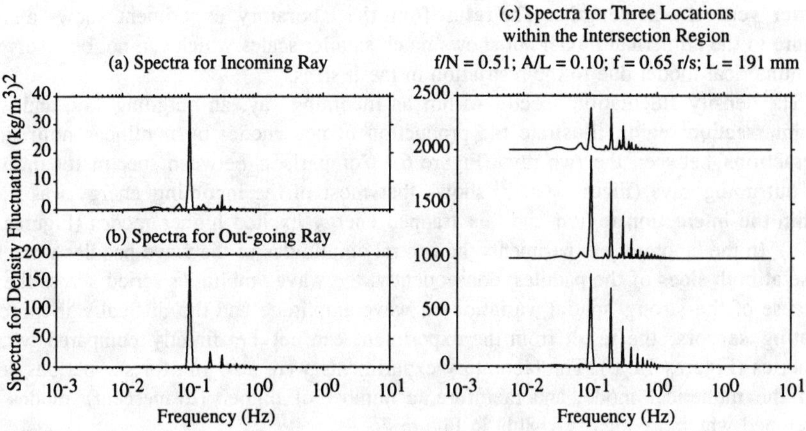

Figure 7. Spectra for density fluctuations from the laboratory experiments Higher modes are excited.

Conclusion

We have presented the results from a combined numerical and experimental study of the nonlinear interaction of internal waves. Both numerical and experimental results showed that nonlinear nonresonant wave-wave interactions transfer wave energy to higher harmonics. Higher harmonics with frequencies greater than buoyancy frequency cannot propagate out of the interaction region. This increases the local energy density within the region which in turn leads to turbulence and mixing. This mechanism of energy transfer towards higher frequencies is similar to either nonlinear interactions between an incident and a reflected wave above a sloping boundary (Thorpe, 1987a; DeSilva, et al, 1997; Javam et al., 1998d) or to interactions near a turning point of a ray propagating against a steady shear (Javam et al, 1997b).

In the numerical experiments the overall energetics and turbulence dynamics could be investigated in detail and the effects of key physical and nondimensional quantities studied via both flow visualisation and parameterization (see Javam et al, 1997b). However, they were restricted by the mesh sizes which can be used whereas the laboratory experiments allowed somewhat smaller scales to be documented as well as the assessment of any three dimensional motions. The above comparison clearly showed that the motions were two dimensional for all scales up to the point where the overturning led to active small scale turbulence.

Acknowledgments. We wish to thank the Centre for Environmental Fluid Dynamics (CEFD) for providing extensive computer resources and for financial support, and also the Hydraulics and Geophysical Fluid Dynamics (GFD) group for providing laboratory equipment.

References

Armfield, S. W., Ellipticity, accuracy and convergence of the discrete Navier-Stokes equations *J. Comput. Phys. Fluids, 114*, 176-184, 1994.

Bretherton, F. P., Resonant interactions between waves: The case of discrete oscillations. *J. Fluid Mech. 20*, 457-479, 1964.

Davis, R. E. and A. Acrivos, Solitary internal waves in deep water. *J. Fluid Mech. 29*, 593-607, 1967.

DeSilva, I. P. D., J. Imberger, and G. N. Ivey, Localised mixing due to a breaking internal wave ray at a sloping bed. *J. Fluid Mech., 350*. 1-27, 1997.

Hasselmann, K., Feyman diagrams and interaction rules of wave-wave scattering processes. *Rev. Geophys. 4*, 1-32, 1966.

Howes, W. L., Rainbow schlieren and its application. *Applied Optics 23*, 2449-2460, 1984.

Imberger, J., Transport processes in lakes: A review article, in *Limnology Now: A Paradigm Of Planetary Problems*, edited by R.. Margalef, pp 99-193, 1994.

Ivey, G. N. and R. I. Nokes, Vertical mixing due to the breaking of critical internal waves on sloping boundaries. *J. Fluid Mech. 204*, 479-500, 1989.

Ivey, G. N., I. D. P. DeSilva, and J. Imberger, Internal waves, bottom slopes and boundary mixing. In *Proceedings 'Aha Huliko' a Hawaiian Winter workshop*, University of Hawaii, pp 199-206, 1995.

Javam, A., J. Imberger, and S. W. Armfield, Numerical study of internal wave-wave interaction. *J. Fluid Mech.* submitted, 1997a.

Javam, A., J. Imberger, and S. W. Armfield, Numerical study of internal wave-shear interactions. *J. Fluid Mech.* submitted, 1997b.

Javam, A., J. Imberger, and S. W. Armfield, Numerical study of internal wave breaking on sloping boundaries. *J. Fluid Mech.* Submitted, 1998.

Martin, S., W. F. Simmons, and C. I. Wunsch, Resonant internal wave interactions. *Nature 224*, 1014-1016, 1969.

Martin, S., W. F. Simmons, and C. I. Wunsch, The excitation of resonant triads by single internal waves. *J. Fluid. Mech. 53*, 17-44, 1972.

McComas, C. H. and F. P. Bretherton, Resonant interaction of oceanic internal waves. *J. Geophys. Res. 82*, 1397-1412, 1977.

McEwan, A. D. and R. M. Robinson, Parametric instability of internal gravity waves. *J. Fluid. Mech. 67*, 667-687, 1975.

McEwan, A. D., Degeneration of resonantly-excited standing internal gravity waves. *J. Fluid. Mech. 50*, 431-448, 1971.

McEwan, A. D., The kinematics of stratified mixing through internal wavebreaking. *J. Fluid. Mech. 128*, 47-57, 1983a.

McEwan, A. D., Internal mixing in stratified fluids. *J. Fluid. Mech. 128*, 59-80, 1983b.

Merzkirch, W., *Flow Visualization.* Academic Press, 1974.

Moller, P., E. A. D'Asaro and G. Holloway, Internal gravity waves and mixing, *in Proceedings 'Aha Huliko' a Hawaiian winter workshop* edited by P. Muller and D. Robinson, pp 499-508, 1991.

Munk, W. H., Internal waves and small scale process. In *Evolution of Physical Oceanography*, edited by B. A. Warren and C. Wunsch, pp. 264-291 Mussachusetts Institute of Technology Press, Cambridge, 1980.

Phillips, O. M., On the dynamics of unsteady gravity waves of finite amplitude. Part I. *J. Fluid Mech. 9*, 193-217, 1960.

Simmons, W. F., A variational method for weak resonant wave interactions. *Proc. Roy. Soc. A. 309*, 551-575, 1969.

Taylor, J. R., The energetics of breaking events in a resonantly forced internal wave field. *J. Fluid Mech. 239*, 304-340, 1992.

Teoh, S. G., G. N. Ivey and J. Imberger, Experimental study of two intersecting internal waves. *J. Fluid Mech. 336*, 91-122, 1997.

Thorpe, S. A., On wave interactions in a stratified fluid. *J. Fluid Mech. 24*, 737-751, 1966.

Thorpe, S. A., The excitation, dissipation, and interaction of internal waves in the deep ocean. *J. Geophys. Res. 80*, 328 338, 1975.

Thorpe, S. A., On the reflection of a train of finite-amplitude internal waves from uniform slope. *J. Fluid Mech. 178*, 279-302, 1987a.

Thorpe, S. A., Transitional phenomena and the development of turbulence in stratified fluids: A review. *J. Geophys. Res. 92*, 5231-5248, 1987b.

18

Breaking Internal Waves and Fronts in Rotating Fluids

A. V. Fedorov and W. K. Melville

Abstract

We consider the evolution of nonlinear Kelvin waves using analytical and numerical methods. In the absence of dispersive (nonhydrostatic) effects, such waves may evolve to breaking. We find that one of the effects of rotation is to delay the onset of breaking in time by up to 60%, with respect to a comparable wave in the absence of rotation. The onset of breaking occurs almost simultaneously over a distance comparable to the Rossby radius of deformation. We also study three-dimensional hydraulic jumps travelling near boundaries in rotating fluids. The transverse structure of the wave field behind such jumps is similar to the transverse structure of Kelvin waves. We obtain the jump relations and derive an evolution equation for the jump as it moves along the boundary. The model is suitable for describing the propagation of oceanic and atmospheric fronts. We show that after some initial adjustment the Kelvin-type jump assumes a permanent form and travels with a constant velocity along the boundary or the coast. The jump is normal to the boundary, but within one Rossby radius it becomes oblique to the coastline.

Introduction

Significant progress has been achieved in nonlinear wave dynamics in recent years. The theory of solitons and studies of the instabilities of the surface waves are just a few examples (for recent reviews see Debnath (1994) and Brandt (1991)). Still there remain many questions and unresolved problems. One concerns wave breaking in rotating fluids. Here we will discuss one such problem, namely, the combined effect of nonlinearity and rotation on internal waves, including the influence of nonlinearity on internal Kelvin wave propagation and breaking.

It is well-known that internal waves play a significant role in the dynamics of oceans, lakes and the atmosphere. When travelling in the vicinity of the coast (or near mountain ridges in the atmosphere), they may be trapped, provided the Rossby radius of deformation is comparable to the wavelength. Thereby they acquire some properties of Kelvin waves. An example of such development can be seen in Figure 1, which displays oceanic internal waves travelling along the coast of Italy. Nonlinear internal Kelvin waves are a common feature in larger lakes and are discussed by a number of authors at this conference (e.g. Saggio and Imberger, 1998; Hutter et al., 1998). Dorman (1985) investigated the possibility of atmospheric Kelvin waves propagating in the low marine layer adjacent to

Physical Processes in Lakes and Oceans
Coastal and Estuarine Studies Volume 54, Pages 251-260

the coast of California, which is seen as overcast propagation. Similar phenomena have been observed along the coast of Australia (Baines, 1980) and southern Africa (Gill, 1977, Bannon, 1980).

In recent years a considerable amount of work has been done in both experimental and theoretical research on the nonlinear aspects of the Kelvin wave evolution. Most of the theoretical work has been within the framework of KdV-type (Grimshaw, 1985; Melville et al., 1989, 1990; Grimshaw and Melville, 1989) or Boussinesq-type equations (Tomasson and Melville, 1992). Regardless of the particular approach, an important assumption of these studies, as well as laboratory experiments (Renouard et al., 1992), was that the dispersion (nonhydrostatic effects) and nonlinearity were of the same order. This assumption led to an approximate balance between nonlinearity and dispersion. However, in many physical situations involving long waves the dispersion is weak and waves may behave in a typical nonlinear hyperbolic manner leading to wave breaking (Whitham, 1974; Boyd, 1980). For instance, an estimate of the parameters for a coastal internal Kelvin wave of about 5km wavelength and 1.5 hour period corresponding to a 10km Rossby radius of deformation shows that the dispersive effects may be four orders of magnitude weaker than the nonlinear effects (Fedorov and Melville, 1995). In turn, breaking of the internal waves may be important in mixing and momentum and energy transfer in the coastal oceans and lakes.

In this paper we consider two particular models accounting for wave breaking. In the first, a Kelvin wave evolves to breaking under the influence of weak nonlinearity. We show that breaking does occur, and is similar to long gravity waves breaking in the absence of rotation. The wave steepens, while travelling along the coast, then breaks almost simultaneously along the coast, with the breaking region extending over a distance comparable to the Rossby radius of deformation.

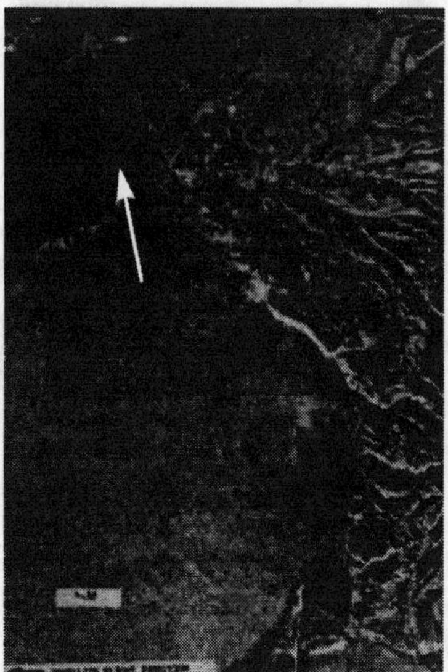

Figure 1. A satellite image of oceanic internal waves near the coast of Italy. (From Alpers and Salusti, 1983.)

Having established that Kelvin waves can break, we study the possibility of hydraulic jumps with a transverse structure behind the jump similar to that of a Kelvin wave. We refer to them as Kelvin jumps (also shocks or fronts). Such shocks can be induced by a rapid elevation of the isopycnals over a large area adjacent to the coast, which may result from either a storm crossing the coastline (Welander, 1961), or a flood increasing fresh water influx from a river plume (Garvine, 1987). A steady hydraulic jump may also be established when a coastal current is incident on topography (Pratt, 1983, 1987). Similar phenomena may occur in meteorology (Parret and Cullen, 1984; Dorman, 1987) and limnology.

Although hydraulic jumps in rivers and straits are well-studied (Lighthill, 1978), only during the last decade has work on three-dimensional rotating jumps progressed (Pratt, 1983, 1986; Nof, 1984, 1986). For the present study, the numerical and experimental work of Pratt (1983, 1986) is most relevant. In particular, in numerical work Pratt (1983) showed that the jump amplitude may decay offshore in a manner similar to a Kelvin wave. Also, an experimental study (Pratt, 1986) demonstrated that the jump became oblique to the coastline at some distance offshore. Our analysis yields the possibility of exponential decay of the wave amplitude and the prediction of the oblique angle offshore.

Evolution of Nonlinear Kelvin Waves: Formulation

To describe nonlinear long wave propagation we use coupled evolution equations valid for one- and two-layer fluids under assumptions of weak nonlinear, transverse and rotational effects

$$\eta_t + \eta_x + \alpha\eta\eta_x + \gamma(v_y - v) = 0, \tag{1}$$

$$v_t + \eta_y + \eta = 0. \tag{2}$$

Full derivations of such coupled evolution equations, including dispersive effects, are given in several papers (Macomb, 1986; Grimshaw and Melville, 1989; Tomasson, 1991). The details of the derivation of (1) and (2) based on the shallow-water equations are given in Fedorov and Melville (1995). Weak rotation is essential for their derivation. Also, it is shown that with certain limitations, this system is asymptotically equivalent to a non-dispersive modified KP equation (Grimshaw and Melville, 1989).

Here η is the elevation of the interface between two fluids (or the elevation of a free surface for a single-fluid model), v is the offshore velocity, x is the along-shore coordinate, and y is the coordinate normal to the coastline (see Figure 2). Parameters α and γ, responsible for nonlinear and rotational effects, respectively, are much smaller than unity and $\gamma = 0(\alpha)$. The system (1) – (2) appears to be the simplest model which incorporates the transverse velocity into the description of long nonlinear Kelvin waves. Note, that when α is zero, the system describes a linear Kelvin wave with zero transverse velocity. Nonlinearity leads to steepening of the wave front and ultimately wave breaking. Eventually the breaking will be affected by either dissipation or dispersion, but until then the system (1) – (2) will describe the phenomenon well.

We study the evolution of an initial localized disturbance in a semi-infinite ocean subject to a no-flux condition at the boundary:

$$\eta|_{t=0} = e^{-y}\,\Phi(x), \; v|_{t=0} = 0 \tag{3}$$

$$v|_{y=0} = 0\,, \tag{4}$$

The initial profile chosen is

$$\Phi(x)=\frac{1}{\cosh(2x)}, \tag{5}$$

with

$$|\Phi_x| \leq 1. \tag{6}$$

The numerical scheme used for solving the system (1) – (2) with (3) – (4) is simple and stable. It is an implicit staggered-mesh scheme, similar to that described by Melville et al. (1989) and Tomasson (1991), with centered differences for spatial derivatives. Similar staggered-mesh grids for linear Kelvin waves were presented by Hsieh et al. (1983) and for nonlinear forced disturbances by Hua and Thomasset (1983). The consistency and reliability of the scheme was tested by switching off one of the parameters, either α or γ. In both cases comparison with available analytical solutions confirmed the accuracy of the numerical approach.

The main question which arose in the numerical treatment was how to define the moment of wave breaking. As long as we do not introduce any additional terms in the system, the only possible breaking would be numerical, as profiles develop gradients beyond the stability limits of the numerical scheme. If the spatial resolution of the numerical scheme is sufficiently high, we expect that numerical breaking will be a good indicator of physical breaking.

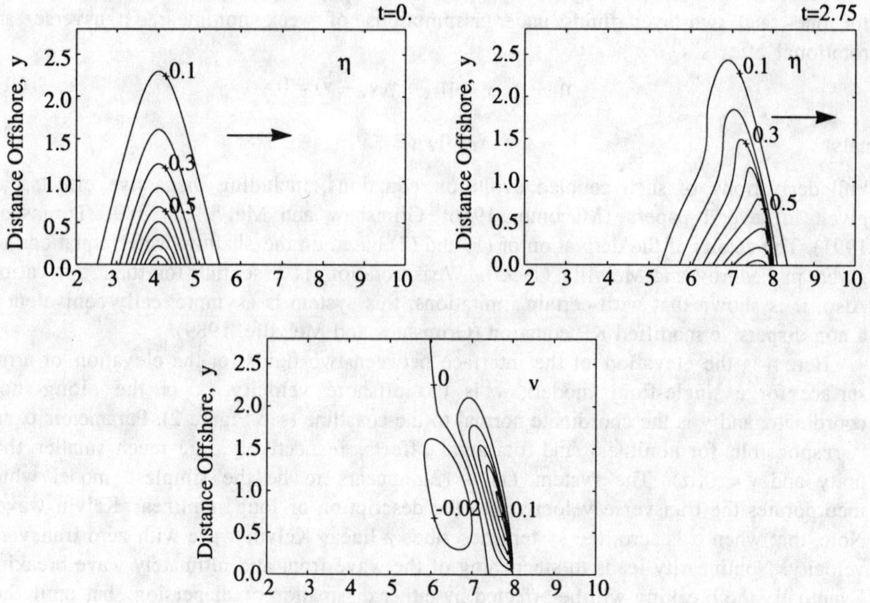

Figure 2. The propagation of a Kelvin wave along the coast: Contour map of the elevation of the interface η at the initial time, t=0, and direct numerical solution giving the elevation of the interface η and transverse velocity v at a time close to breaking, t=2.75, (α = 0.5, γ = 0 5). Breaking occurs at t=3.0. The concentration of isolines in the forward face of the wave indicates the steepening front of the wave and imminent breaking.

To understand the dynamics involved in the problem one may average equations (1) – (2) with respect to the transverse coordinate y. If one assumes that to leading order the wave is still a Kelvin wave, and utilizes an expression similar to (3), then the system reduces to

$$\bar{\eta}_t + \bar{\eta}_x + \frac{\alpha}{2}\bar{\eta}\bar{\eta}_x = 0(\alpha^2), \tag{7}$$

where

$$\bar{\eta} = \int_0^\infty \eta dy. \tag{8}$$

Neglecting the higher order terms, we obtain an implicit solution for (7):

$$\bar{\eta} = \Phi\left(x - t - \frac{\alpha}{2}t\bar{\eta}\right) \tag{9}$$

It follows from (9) that the onset of breaking may begin at $t < 2/\alpha$ (the x-derivative of $\bar{\eta}$ becomes infinite, or the solution develops a multi-valued profile, at $t - 2/\alpha$). That is, the Kelvin wave may propagate without breaking for up to twice the time of a corresponding nonlinear wave in the absence of rotation (c.f. (10)).

Evolution of Nonlinear Kelvin Waves to Breaking: Numerical Results

Our direct numerical calculations show that the breaking is delayed by up to 60%. As an example, Figure 3 demonstrates the evolution of the wave amplitude at the coast for $(\alpha, \gamma) = (0.5, 0.5)$. For comparison we also present the graph for the evolution of the amplitude in the absence of rotation (dashed line), which is obtained from the same numerical scheme but without rotational terms.

In the absence of rotation the time to breaking may be calculated from a well-known analytical result

$$t_{breaking} = 1/\alpha \tag{10}$$

see Whitham, 1974), which is in complete agreement with the numerical result. In Figure 3 one can see that the rotational effect is to delay the onset of breaking by 50% (from t=2 to t=3) and to slightly decrease the wave amplitude as well. To demonstrate the global character of the breaking we also show the graphs for the evolution of other wave parameters, including the wave amplitude at y=0.5 and the maximum transverse velocity in the wave. Similar results are obtained for different combinations of parameters, with (smaller delay for larger α and smaller γ.

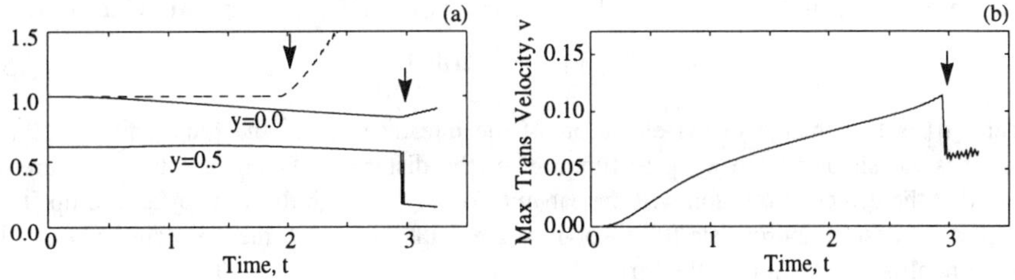

Figure 3 (a) The amplitude of a Kelvin wave vs. time (solid line) at the coastline (y=0) and offshore (y=0.5) from numerical calculations. The onset of breaking is shown by arrows. The amplitude of the wave in the absence of rotation is given by the dashed line. $\alpha = 0.5$, $\gamma = 0.5$. (b) The evolution of the maximum value of the transverse velocity in the wave.

The breaking occurs almost simultaneously over a zone of uniform phase that is normal to the coast and extends over a distance comparable to the Rossby radius of deformation. In other words, the process of breaking embraces the most energetic area of the wave. Figure 2 shows the wave approaching breaking, with the concentration of isolines on the forward face of the wave indicating imminent breaking.

These and other features of the numerical solution can be accounted for by an analytical model which is based on slowly-varying averaged variables. It may be shown that the wave phase can be described by an inhomogeneous Klein-Gordon equation

$$\chi_{tt} - \gamma\chi_{yy} + \sigma^2\chi = i\alpha k e^{i\frac{k}{2}t-2y}, \tag{11}$$

with certain initial and boundary conditions. For details see Fedorov and Melville, 1995.

Hydraulic Jumps at Boundaries in Rotating Fluids

Next, we study internal hydraulic jumps (or shocks), with a transverse structure similar to that of a Kelvin wave. Hydraulic jumps are usually associated with regions of intense wave breaking separating areas with a sharp change in their characteristics. We consider a semi-infinite ocean, and again use the assumption of long waves, so that we can neglect dispersion, which is important for coastally-trapped waves on shorter scales (Tomasson and Melville, 1990, 1992). A coastal current entering still water gives rise to a Kelvin jump. The jump is the boundary between the still water and the following current (see Figure 4). The equation of motion for this boundary can be derived from the nondimensionalized shallow-water equations (Pedlosky, 1987, p. 88)

$$u_t+\alpha(uu_x + vu_y) + \eta_x - v = 0 \tag{12}$$

$$v_t + \alpha(uv_x + vv_y) + \eta_y + u = 0 \tag{13}$$

$$\eta_t + u_x + v_y + \alpha(u\eta)_x + \alpha(v\eta)_y = 0\,. \tag{14}$$

Here, x is the along-shore coordinate, y is the coordinate normal to the coastline and η, u and v are, respectively, the elevation of a free surface, and the alongshore and offshore velocities. The Rossby radius of deformation is equal to unity, while the nonlinearity parameter α, i.e. Rossby number, is relatively small (α, used in (12) – (14), differs from that in (1) – (2) by a factor of 2/3). We follow the conventional procedure and apply the conservation laws across the jump, integrating equations (12) - (14). Neglecting higher order terms and making several transformations gives (see Fedorov and Melville, 1996)

$$-2r_t + (r_y)^2 - \frac{3}{2}\alpha[\eta] = 0, \tag{15}$$

where $[\eta]$ is the change in the elevation of the interface across the jump and r = r(y, t) describes the shape of the jump as function of the distance offshore and time (Figure 4). Note that the effect of rotation will appear in (15) only through the value of the jump $[\eta]$.

If we choose η appropriate for a Kelvin wave, with $[\eta] = -e^{-y}$, the problem reduces to a forced nonlinear equation of the form

$$\rho_t + \rho\rho_y = \frac{3}{4}\alpha e^{-y}, \tag{16}$$

with

$$\rho = -r_y \tag{17}$$

and the boundary condition

$$\rho|_{y=0} = 0, \tag{18}$$

corresponding to the no-flow condition at the boundary.

We can easily find steady wave solutions of (16) – (18), which describe a hydraulic jump simply translating along the coast. Letting $r_t = s$ allows us to integrate (16) (note that the speed of the jump relative to the coast is actually $1 + s$). The value of s is determined immediately from the boundary condition giving

$$s = \frac{3}{4}\alpha, \tag{19}$$

which leads to

$$r = st - \sqrt{\frac{3\alpha}{2}}\left(2q + \log\left(\frac{1-q}{1+q}\right)\right), \tag{20}$$

with

$$q = \sqrt{1-e^{-y}}. \tag{21}$$

Such a jump of permanent form is shown in Figure 4. One of the characteristic features of this solution is that offshore the jump line and the normal to the coast form an oblique angle φ, where

$$\varphi \approx \sqrt{3\alpha/2} \tag{22}$$

A similar oblique angle was observed in experiments by Pratt (1987), which is in qualitative agreement with our result.

Evolution of Hydraulic Jumps: The Initial Value Problem

Now we can solve the initial value problem for (16) – (18), which will give us the evolution of coastal fronts. Since (16) is hyperbolic we can use the method of characteristics. As the basic system for the analysis we choose

$$\left(\frac{\rho^2}{2} + e^{-y}\right)_t + \rho\left(\frac{\rho^2}{2} + e^{-y}\right)_y = 0, \tag{23}$$

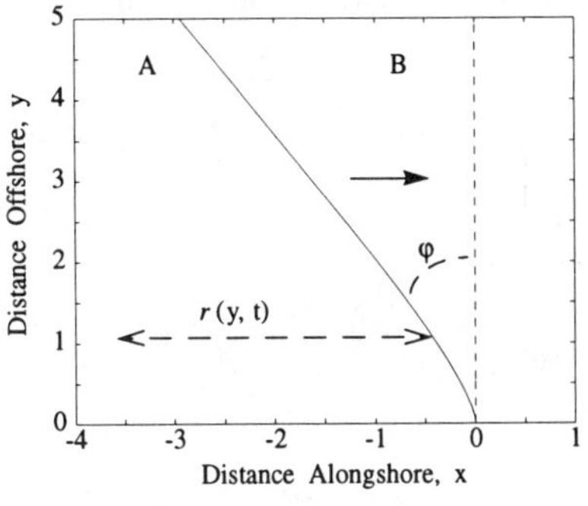

Figure 4. The shape of a stationary Kelvin jump travelling to the right along the coast, plan view, for α = 1 /3, $\varphi = \sqrt{3\alpha/2}$. Variable $r = r(y, t)$ denotes the x-coordinate of the jump. There is an alongshore current in area A, but no motion in area B.

with the boundary and initial conditions

$$\rho|_{y=0} = 0, \tag{24}$$

$$\rho|_{t=0} = F(y), \tag{25}$$

We have rewritten (16) as (23), so that it becomes apparent that

$$I \equiv \frac{\rho^2}{2} + e^{-y} \tag{26}$$

is conserved along the trajectory, described by

$$\frac{dy}{dt} = \rho. \tag{27}$$

Using this fact one can construct two different family of characteristics, corresponding to initial and boundary conditions. Their interaction yields the solution. The results of this method are presented in Figure 5.

As the initial condition we choose a jump which is a straight line normal to the coastline (F(y) = 0). One can see how an initially straight line gradually bends. With time it develops a kink with a jump in the derivative, which corresponds to the boundary between the characteristic families. From the coastline to the kink our solution is simply the steady translating shock described in the previous section. With time the kink propagates offshore. This means that after some adjustment time the Kelvin shock acquires a permanent form over any finite distance from the shore. In general, a more fundamental result for (23) – (25) is valid: With weak restrictions on the initial profile a Kelvin shock evolves into a steady shock, determined by expressions (19) – (20). The shape of the wave of permanent form depends only on the shock amplitude.

Summary

From the above discussion, we conclude that nonlinear and nondispersive Kelvin waves may evolve to breaking. The breaking is similar to the breaking of long gravity waves in the absence of rotation. That is, to the extent that the numerical solutions resolve it, the approach to breaking is typical of long nonlinear waves in nonrotating hyperbolic systems. As in the corresponding linear problem, the influence of the coastal boundary is felt to distances comparable to the Rossby radius, but is now extended to

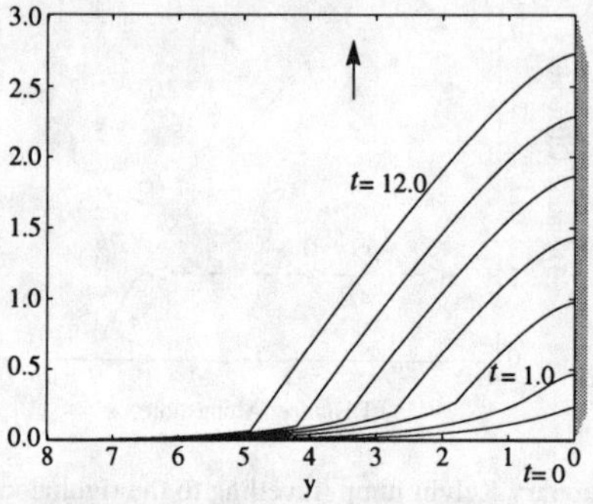

Figure 5. The solution of the initial value problem showing the shape of the jump at different times (r = r(y, t)), for α = l/3 . Plan view. Initially the jump is a straight line normal to the coast.

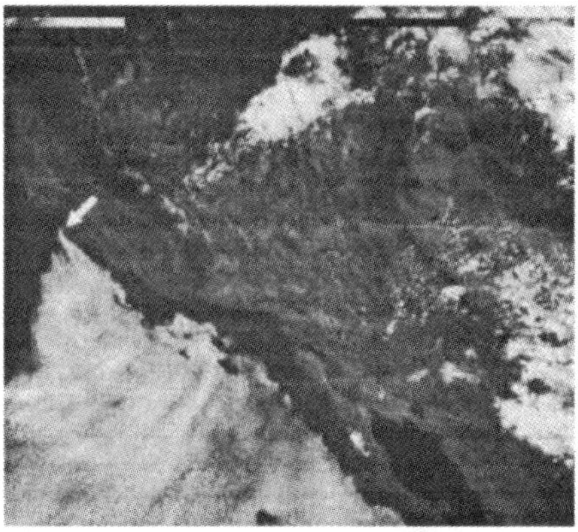

Figure 6. A satellite photograph on July 15, 1982 of the southwestern United States. Overcast stratus moves along the California coast. The arrow points to the leading edge of the stratus. (from Dorman, 1987). Corresponding Rossby radius is approximately 150km.

include the nonlinear phenomenon of breaking. The delay in breaking as compared to a nonrotating case is related to the transverse flow, which tends to suppress the steepening of the wave profile. The onset of breaking is determined by the evolution of the wave phase and appears to be independent of the local wave amplitude, which varies with distance from the boundary.

In the absence of dissipation we can provide several estimates of the time and distance to breaking. For an internal Kelvin wave of 10m amplitude, 5km length and characteristic period 1.5 hours, for a Rossby radius of 10km we expect the distance and time to breaking to be about 80km and 21 hours, respectively. In another example, a Kelvin wave of 50m amplitude, 50km length, propagating in the marine atmospheric layer adjacent to the Californian coast (height 500m, Rossby radius of 150 km - from Hermann et al., 1990) will have a characteristic period of about 1 hour. This gives the distance and time to breaking as 750 km and 14 hours.

Another form of breaking Kelvin waves is Kelvin hydraulic jumps. The properties of the Kelvin jumps are as follows:

1) In the lee of the jump, the wave field decays exponentially offshore in a manner similar to that of a Kelvin wave.

2) After initial adjustment the jump travels with a constant speed and maintains a permanent shape, which depends only on the jump strength.

3) The jump curves back from the normal orientation at the coast to a straight oblique line offshore. The shape of the resulting jump solution, especially for larger amplitudes, has a strong resemblance to satellite imagery of atmospheric disturbances propagating to the north along the western coast of North America (Mass and Albright, 1987 - Figure 6).

Acknowledgment. This research is supported by grants from the Office of Naval Research (Coastal Sciences) and the National Science Foundation (Physical Oceanography).

References

Alpers, W. and Salusti, E., Scylla and Charybdis observed from space. *J. Geophys. Res. 88*, 1800-1808, 1983.

Baines, P. G. The dynamics of the southerly buster. *Aust. Meteorol. Mag. 28*, 175-200, 1980.

Bannon, P. R. synoptic scale forcing of coastal flows: forced double Kelvin waves in the atmosphere, *Quart. J. Roy. Met. Soc. 107*, 313-327, 1981.

Boyd, J. P., The nonlinear equatorial Kelvin wave. *J. Phys. Oceanogr. 10*, 1-11, 1980.

Brandt, A., Ramberg, S. E. and Shlesinger, M. F. (Eds), *Nonlinear Dynamics of Ocean Waves*. World Scientific. 220pp., 1991.

Debnath, L., *Nonlinear Water Waves*. Academic Press. 544pp, 1994.

Dorman, C. E., Evidence of Kelvin waves in california's marine layer and related eddy generation. *Month. Weather Rev. 113*, 827-839, 1985.

Dorman, C. E., Possible role of gravity currents in northern California's summer wind reversals. *J. Geophys. Res. 92*, 1497-1505, 1987.

Fedorov, A. V. and W. K. Melville, On the propagation and breaking of nonlinear Kelvin waves. *J. Phys. Oceanogr. 25*, 2518-2531, 1995.

Fedorov, A. V. and W. K. Melville, Hydraulic jumps at boundaries in rotating fluids. *J. Fluid Mech. 324*, 55-82, 1996.

Garvine, R. W. Estuary plumes and fronts in shelf waters: A layer model. *J. Phys. Oceanogr. 17*, 1877-1896, 1987.

Gill, A. E. Coastally trapped waves in the atmosphere. *Quart. J. Roy. Met. Soc. 103*, 431-440, 1977.

Grimshaw, R., Evolution equations for weakly nonlinear, long internal waves in a rotating fluid. *Stud. Appl. Math. 73*, 1-33, 1985.

Grimshaw, R. and W. K. Melville, On the derivahon of the modified Kadomtsev-Petviashvili equation. *Shld. Appl. Math. 80*, 183-202, 1989.

Hsieh,W. W., M. K. Daley, and R. C. Wajsowitcz, The free Kelvin wave in finite-difference numerical models. *J. Phys. Oceanogr. 13*, 1383–1397, 1983.

Hutter, K., G. Bauer, P. Guting, and Y. Wang, Forced motion response in enclosed lakes, in *Physical Processes in Lakes and Oceans*, edited by J. Imberger, Coastal and Estuarine Studies, pp. 132-161, this volume. AGU, Washington, 1998.

Lighthill, M. J., *Waves in Fluids*, Cambridge Univ. Press, 1978.

Macomb, E. S. The interaction of nonlinear waves and currents with coastal topography. M. S. thesis, Dept. of Civil Engineering, MIT., 1986.

Mass, C. F. and M. D. Albright, Coastal southerlies and alongshore surges of the west coast of North America. *Month. Weather Rev. 115*, 1707-1738, 1987.

Melville, W. K., D. P. Renouard and X. Zhang, On the generation of nonlinear internal Kelvin waves in a rotating channel. *J. Geophys. Res. 95*, 18,247-18,254, 1990.

Melville, W. K., G. H. Tomasson, and D. P. Renouard, On the stability of Kelvin waves. *J. Fluid Mech. 206*, 1-23, 1989.

Nof, D., Shock waves in currents and outflows. *J. Phys. Oceanogr. 14*, 1683-1702, 1984.

Nof, D., Geostrophic shock waves. *J. Phys. Oceanogr. 16*, 886-901, 1986.

Parret, C. A. and M. J. P. Cullen, Simulation of hydraulic jumps in the presence of rotation and mountains. *Quart. J. Roy. Meteor. Soc. 110*, 147-165, 1984.

Pedlosky, J., *Geophysical Fluid Dynamics*, Springer-Verlag, 1987.

Pratt, L. J., On inertial flow over topography. Part 1. *J. Fluid Mech. 131*, 195-218, 1983.

Pratt, L. J., Rotating shocks in a separated laboratory channel flow. *J. Phys. Oceanogr. 17*, 483-491, 1987.

Renouard, D. P., G. Chabert D'Hires, and X Zhang, An experimental study of strongly nonlinear waves in rotating system, *J. Fluid Mech. 177*, 381-394, 1987.

Renouard, D. P., G. G. Tomasson, and W. K. Melville, An experimental and numerical study of nonlinear internal waves. *Phys. Fluids A,5*, 1401-1411, 1992.

Saggio, A. and J. Imberger, J., Internal wave weather in a stratified lake. *Limnol. Oceanogr.* (in press), 1998.

Tomasson, G. G. and W. K. Melville, Geostrophic adjustment in a channel: Nonlinear and dispersive effects. *J. Fluid Mech. 241*, 23-57, 1992.

Tomasson, G. G., Nonlinear waves in a channel: three-dimensional and rotational effects. Ph. D. thesis, Dept. Of Civil Engineering, MIT, 1991.

Tomasson, G. G. and W. K. Melville, Nonlinear and dispersive effects in Kelvin waves. *Phys. Fluids A 2(2)*, 189-193, 1990.

Welander, P., Numerical prediction of storm surges. *Advan. Geophys. 8*, 316-379, 1961.

Whitham, G. B., Linear and nonlinear waves. John Wiley & Sons, 1974.

19

A Laboratory Demonstration of a Mechanism for the Production of Secondary, Internal Gravity-Waves in a Stratified Fluid

T. Maxworthy, J. Imberger and A. Saggio

Introduction to the Proposed Production Mechanism

One of the major aims, if not the ultimate one, of physical limnology is to construct models that can predict the physical and biological state of a natural aquatic system at any given moment. This is clearly an ambitious programme but one that has taken enormous strides in the past several years. It is now possible to generate computer models that can simulate the largest scales of motion in any given physical system. The major constraint on such models is the need to model the energy transfers and ultimately the dissipation at the smaller, non-resolved scales in an accurate and realistic way. This is a major problem even for the conceptually simple case of homogeneous turbulence, but one that is magnified enormously when the basic state is a stratified one. Under these circumstances a large number of new phenomena present themselves, some of which are reviewed in Thorpe (1987) and Imberger (1994) among others. Here we consider a nonlinear energy-transfer and dissipation scenario that has not been considered before, as far as we are aware, although the individual pieces of the sequence are well documented.

Field Observations

The original motivation came from observations of temperature variations in Lake Biwa, Japan, during the BITEX campaign of 1992 and 1993 (Imberger, 1994). Certain temperature signatures were seen repeatedly in the thermistor-chain data-sets in the Southern part of the main basin of the lake at the site marked BN 50 on Figure 1.

At this site, five fast response, high resolution, thermistor chains sampling at 1.5 second intervals to an accuracy of 0.01°C with a resolution of 0.001°C operate over a period of 10 days in 1992 (Julian Day 253.8 to 263.3). The longer temperature-time trace shows conclusive evidence of eight mode-one waves, two of which are shown in detail in one day section in Figure 2 (Julian Day 254.4 to 255.5). The average frequency of the whole eight-wave train is approximately 1.2×10^{-5} Hz, while that of the two waves shown in Figure 2 is 2.8×10^{-5} Hz, as indicated in spectrum of the 16°C isotherm displacement (Figure 3). On the Figure 4 is shown the vertical temperature distribution taken by

Physical Processes in Lakes and Oceans
Coastal and Estuarine Studies Volume 54, Pages 261-270

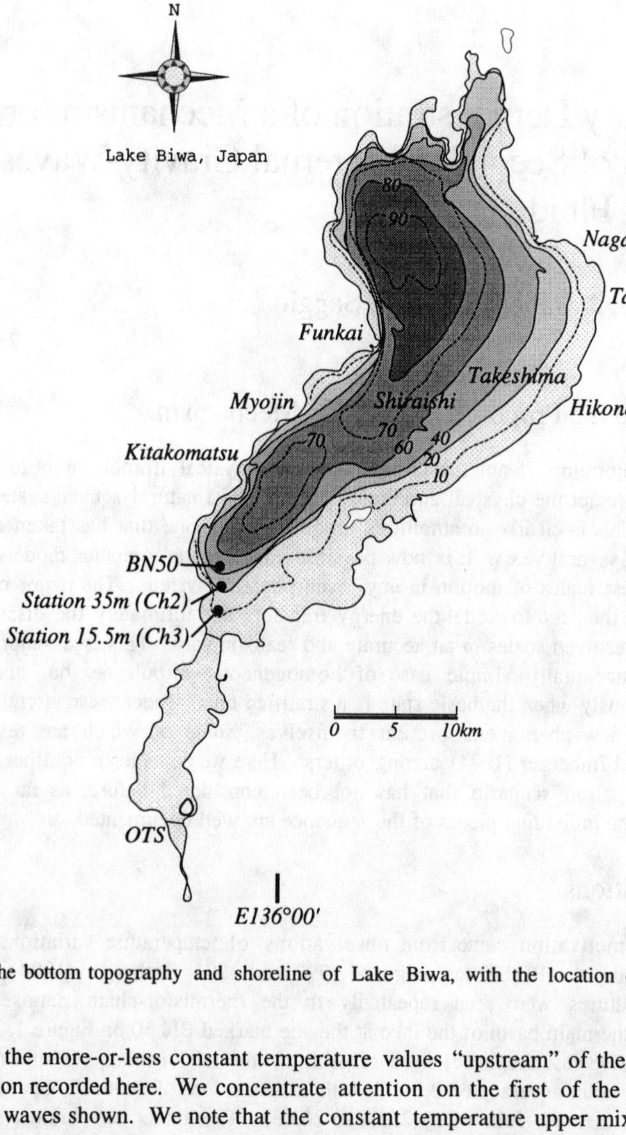

Figure 1. The bottom topography and shoreline of Lake Biwa, with the location of the site BN 50 marked.

averaging the more-or-less constant temperature values "upstream" of the large amplitude wave motion recorded here. We concentrate attention on the first of the large amplitude mode-one waves shown. We note that the constant temperature upper mixed layer, which is approximately 5-7 m deep upstream, is depressed to a depth of 19 m at the peak of the wave, to give, in summary, a total wave amplitude of some 14 m on a 5-7 m mixed layer in a 50 m deep water column. If we estimate the temporal wave width to the $1/e$ amplitude of the 26° C isotherm (Figure 5) we obtain a value of 0.34 days or 2.9×10^4 s, i.e., a frequency of 3.4×10^{-5} Hz. From the upstream temperature distribution given in Figure 4 we estimate a linear, mode-one, long wave-speed of 45 cm/sec, to give a lower value for the wavelength of 11.7 km, assuming that no mean convective motion exists at this scale. This is clearly much longer than any equilibrium solitary wave that can be supported on this observed density distribution. We calculate that the internal Rossby Radius of

Deformation of this wave is some 5.4 km which is considerably smaller than the width of the lake at this location. Thus we interpret the wave as an evolving, long Kelvin wave, presumably generated by the response of the stratification to the variable wind-stress over the lake surface. We note, however, that although the wave amplitude is large the wave slope is very small because of the very long horizontal extent of this wave. We anticipate that the nonlinear correction to the linear wave speed will be small under these circumstances, a prediction that is borne out by direct measurement of the phase-speed of these waves (Saggio and Imberger, 1998).

Next we need to determine the likelihood that the flow generated under or near the wave crest is unstable in the Kelvin-Helmholtz sense (see for example Turner, 1973). In Figure 4 we plot also the temperature distribution under the wave crest and from this we estimate an unstable layer thickness (ΔH) of some 9.5 m with a temperature difference (ΔT) of approximately 9° C. From arguments similar to those given in Bogucki & Garrett (1993) we estimate a velocity difference across this layer (ΔV) of approximately 45 cm/sec (see Figure 5) assuming that the velocity profile in each layer is uniform. Thus a crude estimate of the bulk Richardson number (Ri) across this layer is:

$$\mathrm{Ri} \approx \frac{g\alpha\Delta T\Delta H}{(\Delta V)^2} \approx 0.8,$$

where α is the volumetric coefficient of thermal expansion of water and g is the acceleration due to gravity.

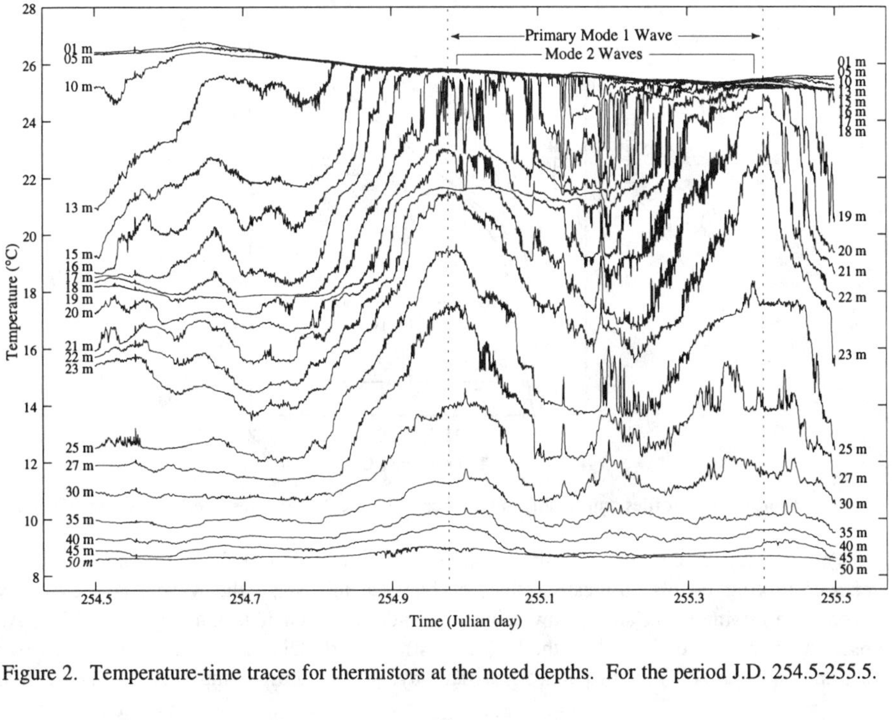

Figure 2. Temperature-time traces for thermistors at the noted depths. For the period J.D. 254.5-255.5.

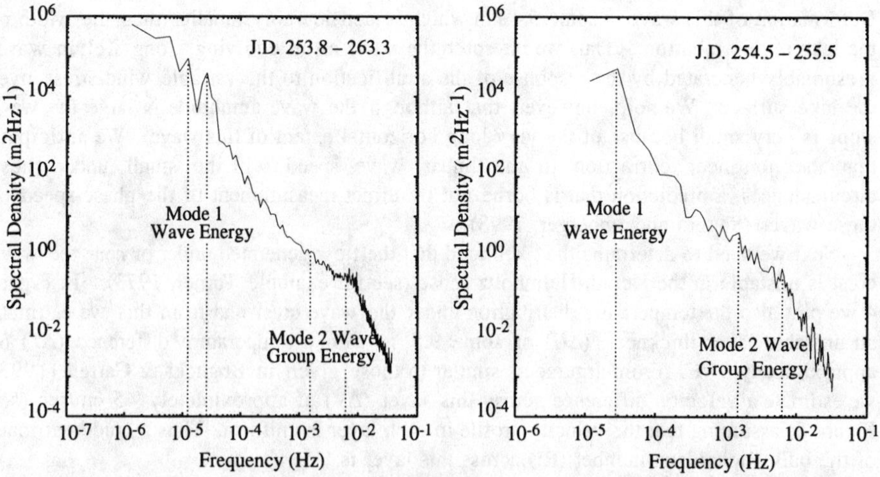

Figure 3. Long and short time spectra for isotherm 16°C, with the frequencies of the long primary waves and the short mode two waves noted.

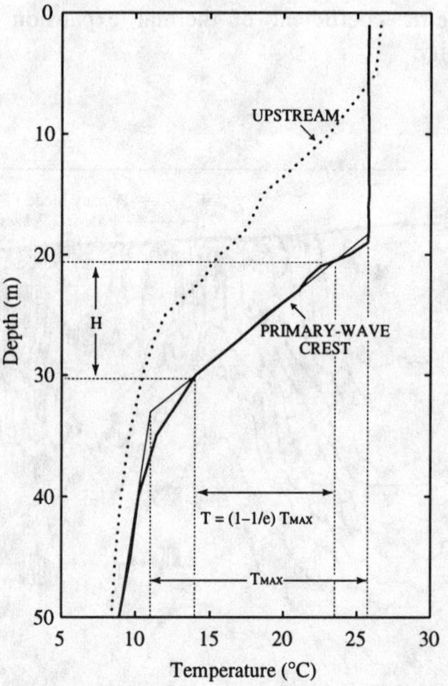

Figure 4. Temperature profiles "upstream" of the trace of Figure 2 at its crest, i.e. at time J.D. 254.97.

This value is at the upper limit of the range over which such layers are known to generate mixing events suggesting that, perhaps, at this stage the wave has previously generated a mixing event and is now at an amplitude at which it is marginally stable. An observation that is confirmed by the lack of a strong turbulent signal is the traces shown here and on Figure 6. The turbulence generated by this instability must have done several things, initially. Firstly it generated some dissipation directly, secondly it stirred and partially mixed the fluid in the thermocline region to generate a mean density profile that had an excess of potential energy over the initial, undisturbed surrounding state. It is well known that when such mixed regions evolve they do so to reduce this potential energy excess and are capable of generating internal solitary wave trains (e.g., Amen & Maxworthy (1950); Maxworthy (1983); Maxworthy and Monismith (1986) among several examples). The number and amplitude of these waves depends on the spatial extent of the mixed region both vertically and horizontally, and the potential energy excess in the region (see Whitham (1971) p. 595 for some simple examples). In the present case with mixing confined to the thermocline region we anticipate that mode two internal solitary waves can form and we interpret the short wavelength disturbances seen downstream of the primary wave in this way. We have marked the more obvious examples of mode-two waves on Figure 2 and note that they have the vertical structure typical of such waves with their central, flat isopycnal at approximately 19 m depth and with a depressed temperature above this depth and an elevated one below (see the sketch on Figure 5). On Figure 6 we show an expanded version of the signal between Julian days 255.1 and 255.4 which indicates even more structure and a sequence of about 50 mode two waves of very short period, of the order of 140 s, i.e., a frequency of 7×10^{-3} Hz. Figure 7 shows the isotherms for this period and the phase between isotherms after the trough of the long mode-one wave (J.D. 254.97). Further consequences of this energy exchange can be seen in the spectra of Figure 3 where we show for the isotherm 16°C both the long term (J.D. 254-263) and short term spectra (J.D. 255.5-255.5). On them we show the frequencies corresponding to the 8 and 2 wave trains calculated earlier and the dominant mode 2 wave frequency calculated above. Also in the longer, 10 day record the secondary peak at approximately 7-8 $\times 10^{-4}$ Hz is enhanced by wave generation at later times (J.D. 256-258).

This observation, plus its laboratory interpretation discussed in the next section, indicates what could possibly be a direct energy transfer from the long waves to short waves with a two order-of-magnitude increase in frequency with no intermediate energy cascade or nonlinear triad interactions. The suggestion is that the mode two waves evolve from the very non-linear mixing event that has occurred during the propagation of the

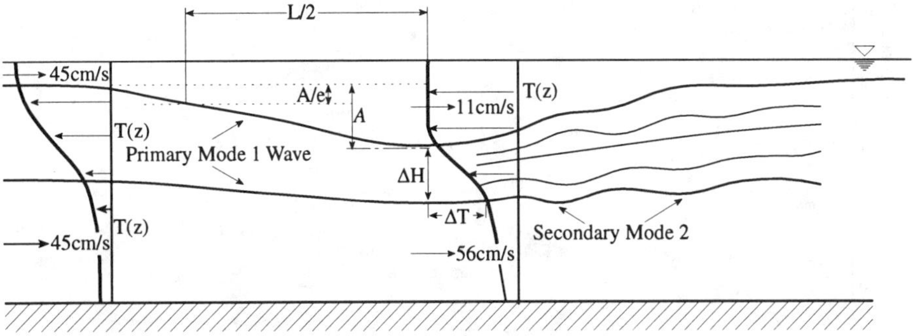

Figure 5. Sketch of the flow in the Primary Mode-1 wave in a frame of reference moving with the wave-speed. The horizontal scale is greatly distorted. The structure of the Secondary Mode-2 waves is indicated as they propagate on the density profile distorted by the Primary wave. The definition of the wave-length (L) is indicated as are the depth (ΔH) and the temperature difference (ΔT) of the shear layer at the wave crest.

primary wave. This is presumably true for all of the intermediate wave numbers, there being an interaction between them only in so far as they compete to fit into the available space within the thermocline. Unfortunately it is only possible to estimate one of the more interesting properties of the mode 2 wave, that is their amplitudes, because we do not know the speed at which they are being swept past the thermistor chain. We can estimate a wave speed but since they are also being convected by the flow generated by the mode 1 wave, their overall speed is, at best, a very crude estimate and of limited interest for our present purposes. Further detailed analysis of the total data set will enable us to refine this process considerably.

To reiterate we propose the following sequence of events, some of which are well known but which are combined in such a way as to give a reasonable explanation for existence of the observed mode-two waves. The temporally varying wind stress on the lake surface generates a thermocline set-up that evolves into a sequence of long, solitary Kelvin waves propagating around the lake in a counter-clockwise sense. As these waves enter the shallowing region near site BN 50 they grow to an amplitude that can generate strong Kelvin-Helmholtz instability near their crests. The stirred and mixed fluid thus formed evolves to a new state in which some of the potential energy that has been gained is transformed to mode two and possible mode three wave internal-wave energy depending on the distributions of mixing within the thermocline. Since the mode 2 waves have such a low wave speed it seems unlikely that they could have originated too far away from BN50, and this argues for the more local source, as described above.

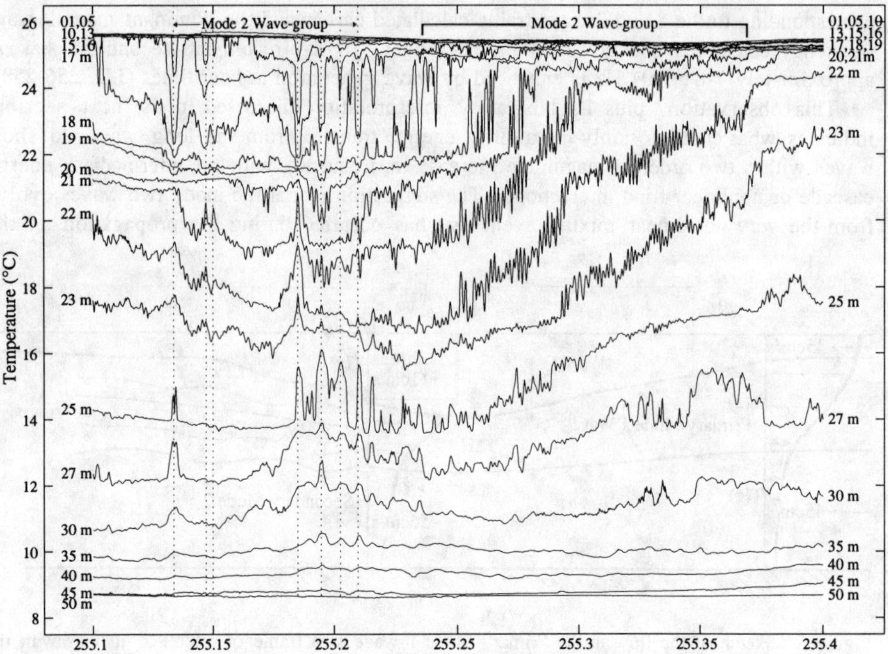

Figure 6. An expanded temperature-time trace for the period J.D. 255.2-255.4 showing details of the large number (≈80) of mode-two waves present after the primary wave crest.

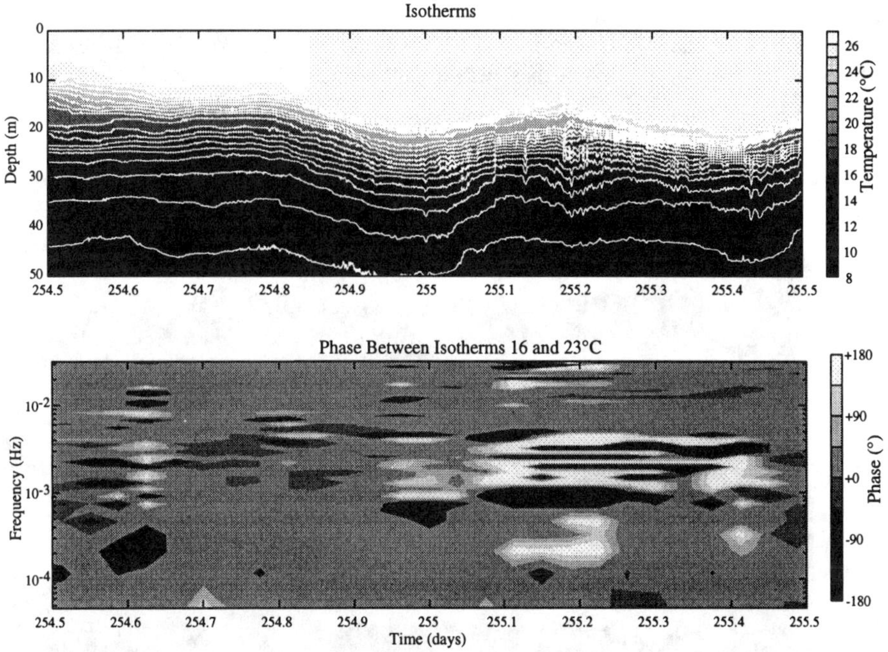

Figure 7. Long and short time spectra from the isotherm 16°C, with the frequencies of the short mode two waves between 10^{-3} and 10^{-2} Hz approximately, clearly evident.

A Laboratory Demonstration

There are numerous ways to produce primary waves of the type discussed in the previous section. We have tried three of them and found identical results in all cases. In what follows we concentrate on one of these techniques, the use of a tilting duct to simulate the evolution of the wind set-up of the thermocline after the forcing ceases. The apparatus consisted of a perspex duct 3 m long with an internal cross-section of 10 cm horizontally and 30 cm vertically. The flow was viewed through the 3 m × 30 cm sides and was back-lit. The tube was placed on a central pivot so that it could be tilted through 90°. An experiment was started by filling the tube in a vertical position with a two-layer, salt stratified system. The filling procedure generated an interface that was 1-2 cm thick. The tube was then slowly tilted to slightly below the horizontal so that the heavy, bottom layer was thicker at one end of the tube than the other (see Figure 8a). Due to the increase in length of the interface during this tilting process the final interface thickness was reduced to the order of 1-2 mm. To initiate an experiment the tube was suddenly raised to the horizontal position and the interface evolution photographed at approximately 2 frames per second with a 35 mm camera. As set-up here, with the heavier, thinner layer on the bottom, the flow was inverted from the natural flow being modelled. However, the dynamics are not affected by this change. We show a sequence of photographs (Figure 8) of the evolution for a density ratio of 1.023. The slumping of the initial triangular shaped region evolves to a very long wave which, in Figures. 8j-m, exhibits mixing at its crest and is longer than the equilibrium solitary wave that conforms in this situation. The wave reflected from the end wall is bore-like and is evolving towards a solitary wave sequence. Mixing at the crest of the first of these waves is pronounced.

(a)
(b)
(c)
(d)
(e)
(f)
(g)
(h)
(i)
(j)
(k)
(l)
(m)

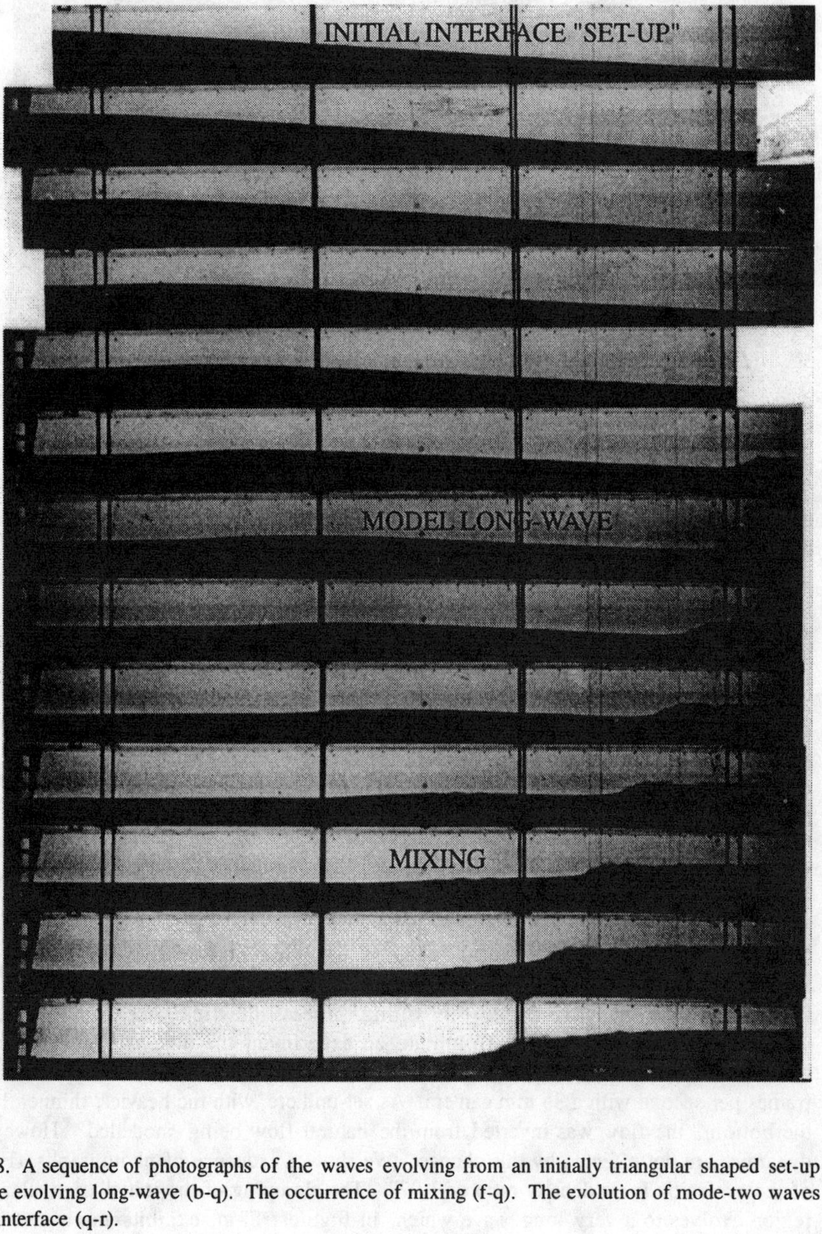

Figure 8. A sequence of photographs of the waves evolving from an initially triangular shaped set-up (a). The evolving long-wave (b-q). The occurrence of mixing (f-q). The evolution of mode-two waves on the interface (q-r).

(n)

(o)

(p)

(q)

(r)

(s)

(t)

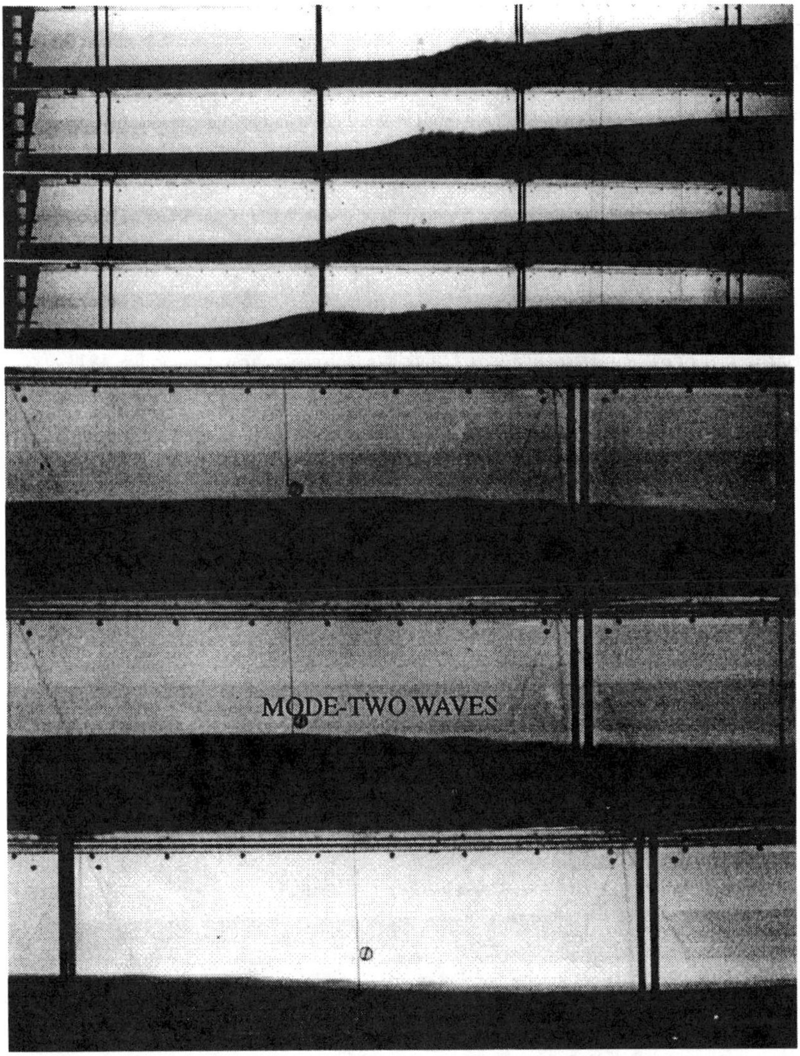

Figure 8 (Continued). A sequence of photographs of the waves evolving from an initially triangular shaped set-up (a). The evolving long-wave (b-q). The occurrence of mixing (f-q). The evolution of mode-two waves on the interface (q-t).

The last three photographs of this sequence show the region behind the moving mixing region with a clear indication of 6-7 mode-two waves on the thickened interface between the two layers. Thus the energy in the original distorted interface has been transferred to a sequence of large amplitude mode-one waves, a thickened interface, some dissipation and a sequence of weaker mode-two waves, as outlined in the introduction to this note.

References

Amen, R., and T. Maxworthy, The gravitational collapse of a mixed region into a linearly-stratified fluid. *J. Fluid Mech., 93*, 47-63, 1980.

Bogucki, D. J. and C. Garrett, A simple model for the shear-induced decay of an internal solitary wave. *J. Phys. Oceanogr.,* 28, 1-10, 1993.

Imberger, J. Transport processes in lakes: A review article. *In Limnology Now: A Paradigm of Planetary Problems*, edited by R. Margalef, 99-193, Elsevier Science, 1994.

Maxworthy, T., Experiments on solitary, internal Kelvin waves. *J. Fluid Mech., 129*, 259-282, 1983.

Maxworthy, T. and S. Monismith, Differential mixing in a stratified fluid. *J. Fluid Mech., 189*, 571-598, 1986.

Saggio, A. and J. Imberger, Internal wave weather in stratified lakes. *Limnol. Oceanogr.* (in press) 1998.

Thorpe, S., Transitional phenomena and the development of turbulence in stratified flow: A review, 1987.

Turner, J. S. *Buoyancy Effects in Fluids*, Cambridge University Press, Cambridge, 1973.

20

Direct Numerical Simulation of Wave-Mean Flow and Wave-Wave Interactions: A Brief Perspective

C-L. Lin, J. R. Koseff, J. H. Ferziger, and S. G. Monismith

Abstract

In this paper we examine the interaction of internal waves with the critical layer and the role of the parametric subharmonic instability (PSI) in the energy transfer process when multiple triad interactions are generated by one strong primary mode and background white noise or a Garrett-Munk (GM79) model spectrum. We performed direct numerical simulations of the incompressible Navier-Stokes equations with the Boussinesq approximation, using the pseudo-spectral method (see Rogallo, 1981, Holt et al., 1992). For the critical layer a spanwise Rayleigh-Taylor instability should be the most unstable mode if the Richardson number is less than zero locally. The mixing efficiency associated with this process still remains an open question. For the triad interactions the PSI energy transfer mechanism is feasible if local sum resonance is suppressed (i.e. under certain geometric constraints), but it is not clear if the mechanism is resonant. When examining triads with a background GM79 spectrum we found that even when the energy level of the spectrum was reduced by a factor of 100 the induced diffusion triads (non-resonant) were the strongest, indicating the dominance of wave-mean flow interactions. In this instance triads of the PSI type were still not found, suggesting that they may not be significant in the distribution of wave energy in the ocean.

Introduction

Breaking internal waves are thought to be the dominant source of turbulence in the interior of the ocean (Gregg et al., 1993). However, a great deal of uncertainty remains as to the details (Thorpe, 1994). The interaction of internal waves with critical layers is complicated and includes wave steepening, wave-mean flow interaction, wave transmission and reflection, and instability (Munk, 1981). This problem has received considerable attention in the past with studies ranging from theoretical (Booker and Bretherton, 1967, for example) to experimental (Thorpe, 1981, Koop and McGee, 1986) to numerical (Fritts, 1982, Winters and d'Asaro, 1989, 1994, Lin et al., 1993). [For a more complete review see Winters and d'Asaro (1994) and Lin et al. (1993).] The outcome of the numerical studies by Winters and d'Asaro and Lin et al. is that the "wave breakdown" process appears to be strongly three-dimensional. However, some significant questions remain as to the exact

Physical Processes in Lakes and Oceans
Coastal and Estuarine Studies Volume 54, Pages 271-284

form of the breakdown - is it true "wave-breaking" due to steepening or simply a gravitational instability, what is the scale at which this breakdown occurs, and what is the mixing efficiency associated with the breakdown?

Even putting these uncertainties aside, the distribution of the turbulence in the thermocline region of the ocean cannot be understood or calculated without understanding how internal wave energy is redistributed (Gregg et al., 1993); much work has focused on this subject (Levine, 1983). Although Levine deems this calculation to be "straightforward" because of lack of dependence on sources and sinks, opposing viewpoints have developed over the years. Most studies of nonlinear transfer among internal waves have used weak-interaction theory, which assumes that the time-scale of the interaction is large compared with the period of the wave. This approach was pioneered by Hasselman (1966), Olbers (1976), McComas and Bretherton (1977), and Phillips (1977), among others. Perhaps the most significant result emerging from this work was the identification of three special triad interactions that dominate the energy transfer process: induced diffusion (ID), elastic scattering (ES), and parametric subharmonic instability (PSI). Other significant work in this area includes that of Pomphrey et al (1980) using Langevin methods, Henyey and Pomphrey (1983), Flatte et al. (1985) using eikonal techniques, and Martin et al. (1972) using experimental techniques. An excellent review is provided by Muller et al. (1986).

There is some disagreement, however, as to the applicability of weak-interaction theory to the ocean. According to the energy transfer model of McComas and Muller (1981), energy is generated at low vertical wave numbers, and dissipated at high wavenumbers with the ID and PSI mechanisms transferring energy to the high wavenumbers. Holloway (1980,1982), however, estimated that the oceanic wave energy level is about 100 times the limit at which weak interaction theory should apply, and claimed that the nonlinear transfer time is not large (compared with the period of the wave) as required by weak interaction theory. Furthermore numerical experiments by Orlanski and Cerasoli (1981) of a 2D random internal wave field suggest strong interactions may be important in the transfer of energy. As a result Lin et al. (1995, 1996), following Riley et al. (1981), Shen and Holloway (1986), Lelong and Riley (1991), Ramsden and Holloway (1992), and Siegel and Domaradzki (1994), studied single and multiple triad interactions, as well triad interactions using a Garrett-Munk energy spectrum to represent the linearly-stratified ocean. This work will be discussed below.

We will focus on two of the issues for which there is not universal agreement, and on the role of direct numerical simulations in resolving these issues. Most of the work presented here is Lin's but we will refer to other simulations. The first issue is the interaction of internal waves with the critical layer, and the second is the role of the parametric subharmonic instability in the energy transfer process. First we briefly describe the numerical approaches we have used in this study and discuss their strengths and weaknesses.

Numerical Techniques

Approach

The incompressible Navier-Stokes equations with the Boussinesq approximation, the continuity equation, and the scalar transport equation are solved numerically. For computational efficiency, the pseudo-spectral method (see Rogallo, 1981, Holt et al., 1992) is adopted for the spatial derivatives. The Fourier representation requires periodic boundary conditions, and the computational domain is normalized to a $2\pi \times 2\pi$ box. A second-order Runge-Kutta method is used to advance the solutions in time, and the inverse Brunt-Väisälä (BV) frequency is used to scale time. In both sets of simulations a hyper-viscosity (applied at a cut-off wavenumber) was used to prevent accumulation of energy at the highest wavenumbers. This model acts like a dissipative wall and does not affect the energetic modes whose wavenumbers are much less than the cut-off wavenumber.

The results presented in this paper are primarily for a planar flow solution in which two components of the velocity field are resolved. For convenience we shall refer to these calculations as two-dimensional or 2D. Lin (1993) did perform three-dimensional (3D) simulations as well which are referred to in the text.

Initial Conditions

Critical layer

The unperturbed flow has a Gaussian profile (see Figure 1) and linear stratification. The wave packet propagates from near the upper boundary into the critical layer. A more detailed discussion of the initial conditions is given by Winters and d'Asaro (1989), whose configuration we have used, and by Lin et al. (1993). The initial Reynolds number based on the width of the jet (vertical distance AB in Figure 1) and maximum velocity is about 10^7, the length of the model domain is 250m, and the Richardson number, based on the initial mean flow and initial background stratification near the critical level is 14.8 for the case discussed below. For further details and other cases, see Lin et al. (1993). The resolution is 512×512 points unless otherwise noted in the text.

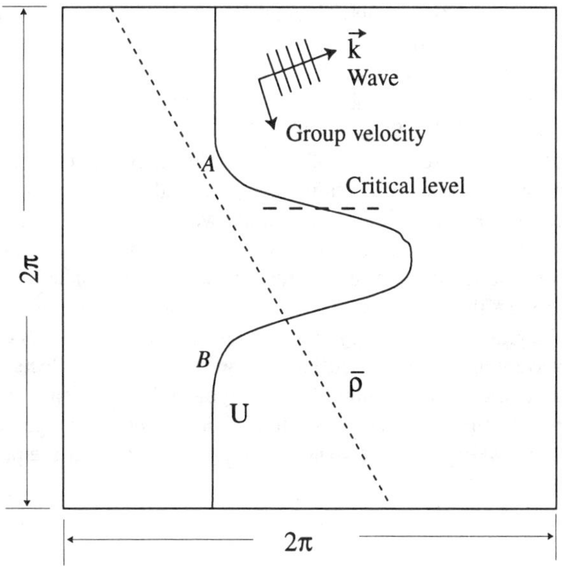

Figure 1. Schematic of initial configuration for wave-critical layer interaction simulation. (From Lin et al., 1993.)

Triads

In the 2D cases multiple triad interactions generated by one strong primary mode and background white noise were investigated. The white noise has a much smaller amplitude than the energetic mode in the triad. Linear internal wave solutions were used for initial fluctuating quantities, so the kinetic and potential energy are initially equi-partioned at all wavenumbers. For the GM79 initial conditions the GM spectrum was used with a factor to control the total energy level. The wavelengths resolved in the simulations are as follows: the minimum horizontal wavelength is 61m and the maximum is 5.2km; the minimum vertical wavelength is 30.5m and the maximum is 2.6km.

The Role of Direct Numerical Simulations

The primary advantages of using direct numerical simulation are (i) the ability to examine strong interactions, and (ii) accurate resolution. Lin et al. (1993, 1996) simulated both infinitesimal as well as finite amplitude waves, and examined the usefulness of resonant interaction coefficients and the Manley-Rowe relations for looking at weak interactions, as well as the limitations of weak interaction theory. One feature of this approach is the ability to "tune" the resolution of the code by using stretching factors.

We would like to be able to perform all DNS in three dimensions. However, cost prohibits this approach so most of the simulations performed by Lin et al. (1993, 1995, 1996), Ramsden and Holloway (1992), Bouruet-Aubertot et al. (1995), and Javam et al. (1995) are two-dimensional. Both Lin et al. (1995) and Ramsden and Holloway (1992) found that the results from 2D simulations compared very favorably with those from 3D simulations and that extension to 3D was, therefore, not always necessary. Lin (1997) compared the energy transfer mechanisms for multiple triad interactions and found that 3D triadic energy transfer is similar to its 2D counterpart, indicating that 2D simulations can provide qualitative properties of resonant interactions. 2D simulations do not, however, allow wave-vortex interactions or include the effects of rotation.

Wave-vortex interactions are believed to be significant in energy transfer (Lelong and Riley, 1991, and Muller et al., 1986). In his 3D simulations of multiple triad interactions using background white noise, Lin (1993) found that (a) if the white noise consisted of internal waves the wave-vortex interaction was weak, and (b) if the white noise is characterized by turbulence the wave-vortex interaction was stronger. In both cases, however, the interactions were weak compared with wave-wave resonant interactions; only 12% of energy transfer was due to wave-vortex interaction in the case studied by Lin (1993). Lin (1993) also examined the effects of rotation using 3D simulations and found that in general rotation increases total energy transfer rates, and tends to push the sum resonant triads to higher wavenumbers. Our justification for ignoring the effects of rotation below is that the Earth' rotation rate is about 1% of a typical BV frequency in the ocean. The inertial wave period in this instance is about 33 hours which is longer than a 2D typical simulation with the GM79 spectrum. Therefore, ignoring rotation seems reasonable.

Discussion

Our discussion will focus only on the 2D direct numerical simulations of critical layer processes and triad interactions. Where appropriate the results will be compared and contrasted with other 2D simulations, 3D simulations, and experimental work.

Critical Layers

Summary of 2D Results

The two-dimensional critical layer process occurs in two stages. In the first stage the wave packet approaches the critical level, part of the wave energy is transferred to the mean flow and the remainder is confined to increasingly thin regions. Subsequently the vertical gradients of horizontal velocity and density variations are increased, and the strong shear and unstable density gradients lead to instability. The simulations show that a shear instability occurs even though unstable density profiles occur earlier.

The instability is illustrated in Figure 2 where vorticity and Richardson (Ri) number (based on local density and velocity profiles) contours are plotted at a non-dimensional time of $t^* = 39.5$ for the case described above. For this case density overturns occur at a $t^* = 20$: a Rayleigh-Taylor instability is possible but not a shear instability because the regions with the strongest shear all have local Ri > 0.25. Figure 2(a) shows vorticity contours at $t^* = 39.5$. The wave-like motions appear to be an instability: an enlargement of region (b) is shown in Figure 2(b), and vorticity and Ri contours for region (b) are shown in Figure 2(c). Figure 2(c) shows that the region with 0 < Ri < 0.25 is located exactly where the shear is strongest and the resemblance to a Kelvin-Helmholtz instability is strong; the unstable fluid motions are confined to very thin layers.

To further examine this instability Lin et al. (1993) applied a local linear stability analysis to the flow: this approach is justified because the numerical simulations show that the local velocity and density profiles change very slowly in a convected frame. This analysis showed that there are two principal modes of instability: a shear-dominated mixed instability and one dominated by the unstable density gradient. The analysis was extended by looking at the sensitivity of the instability to oblique modes: this is done by reducing the shear. Because oblique modes are found to be more unstable, a spanwise Rayleigh-Taylor instability should be the most unstable mode if Ri < 0 locally: a similar conclusion was reported by Winters and Riley (1992). The three-dimensional simulations by Winters and d'Asaro (1994) confirm that the "breakdown" is three-dimensional and leads to the formation of "quasi horizontal vortical structures".

The resolution of the fine scales should be addressed. Figure 2(d) shows vorticity contours for a lower resolution simulation (64 × 256) of the case discussed above. Some kind of "breakdown" is occurring, but the exact process is unclear. When the resolution is increased to 512 × 512 (Figure 2(b)) the scale at which the "breakdown" occurs is smaller, and a Kelvin-Helmholtz-like picture emerges. The resolution problem arises because the horizontal wavelength of the instability is related to the vertical wavelength of the internal wave near the critical layer. Once the instability becomes non-linear and significant energy is transferred to horizontal scales (comparable in size to the small vertical scales in the critical layer), the structures cannot be resolved on the lower resolution grid and numerical "noise" is generated.

The resolution problem has direct implications for calculation of quantities such as mixing efficiency which depend on the fine-scale physics: insufficient resolution of the flow tends to artificially increase the mixing efficiency at the smallest scales, hence resulting in higher efficiencies. If DNS is to be useful in resolving the differences in mixing efficiency for stratified flows of various types highly resolved, high (turbulent) Reynolds number, three-dimensional simulations of such flows are necessary. The mixing efficiency in a breaking internal wave may be higher than that in a stratified shear flow because of the structure of the flow (coherence) and the distribution of local Ri within the flow. A combination of DNS and the technique of Winters et al. (1995) for computing mixing efficiencies may be extremely useful in resolving this issue.

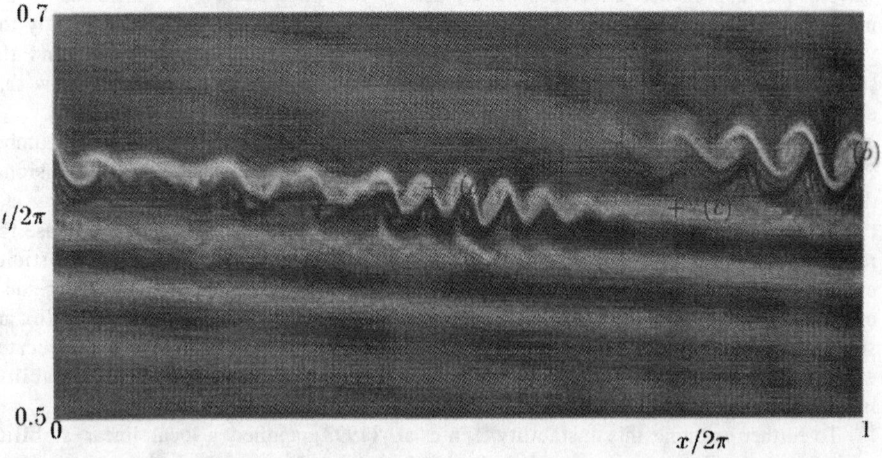

Figure 2(a). Vorticity contours at t* = 39.5: black, negative vorticity; white, positive vorticity.

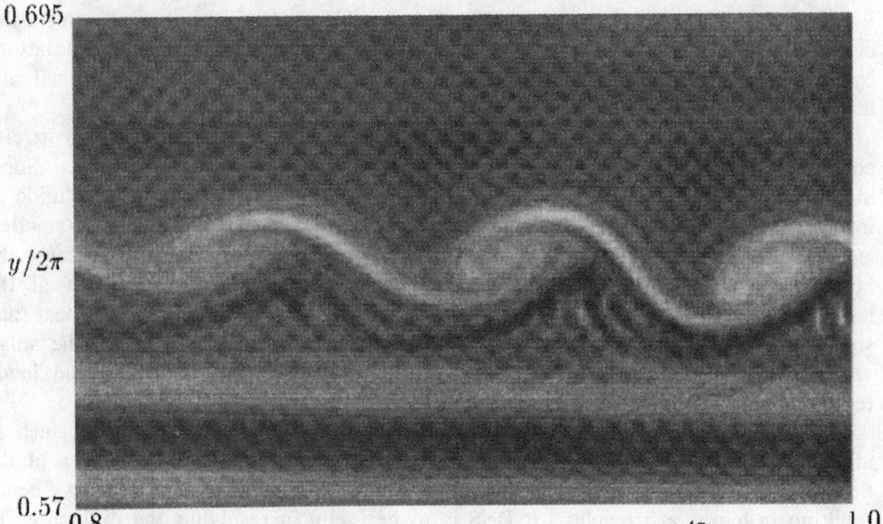

Figure 2(b). Enlargement of region (b) shown in Figure 2(a). (From Lin et al., 1993).

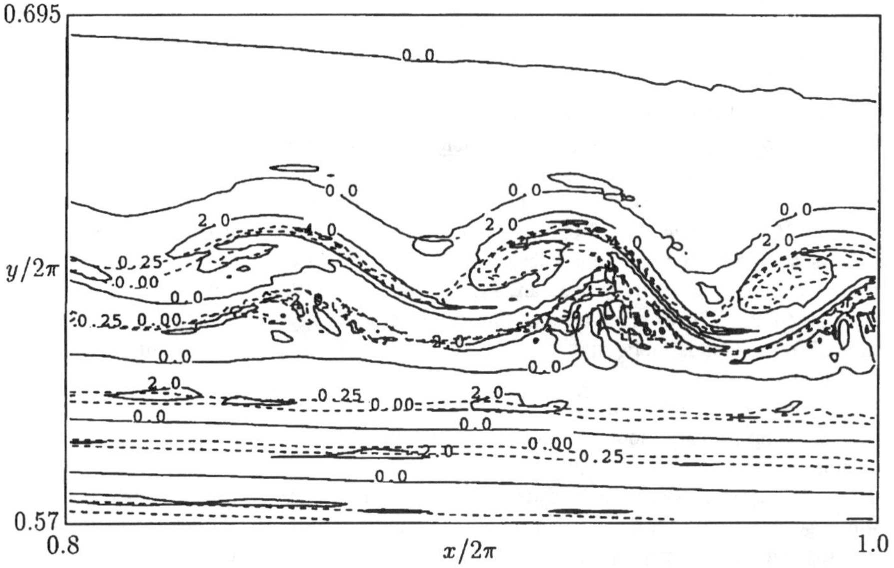

Figure 2(c). Vorticity and Richardson number contours at t* = 39.5: ----------- , vorticity (label format with one digit beyond decimal); - - - - - , Ri (label formats with two digits beyond decimal).

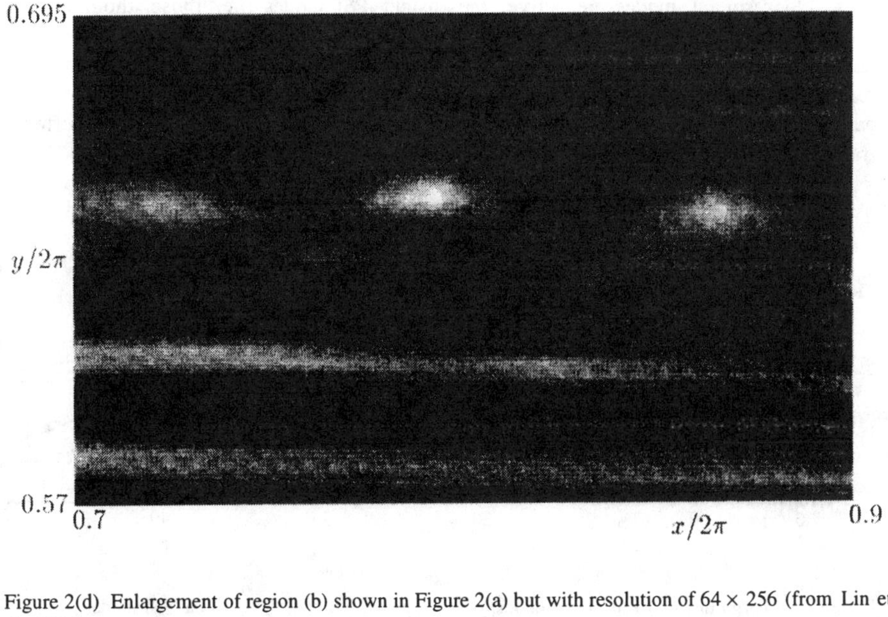

Figure 2(d) Enlargement of region (b) shown in Figure 2(a) but with resolution of 64 × 256 (from Lin et al., 1993).

Parametric Subharmonic Instability

General Findings

Lin et al. (1997) used the concept of resonant interaction coefficients (RIC) together with solutions of the Manley-Rowe relation and integration of the triadic Navier-Stokes equations to examine single triad interactions with infinitesimal and finite amplitude waves. The RIC revealed two energy transfer patterns: a cyclic one typical of local sum resonant triads, and another, found in PSI triads, with strong interactions between the two small-scale modes but small net energy transfer. Solutions of the Navier-Stokes equations for finite amplitude waves showed an abrupt decrease of both interaction period and energy transfer on the PSI resonant traces, confirming that energy transfer to very small scales through resonance is ineffective, while transfer through local sum resonant triads is most effective.

This study was extended by Lin et al. (1997) to multiple triad interactions, generated by one strong primary mode and background white noise, with the purpose of determining whether resonant triads are active in such an environment. As in single triad interactions the triad energy transfer pattern is determined by triad geometry. He concluded that local sum resonant triads are the most important, that triads of ID type gain energy rapidly initially but less energy later, and that the interaction strength of triads of the ES type depends on the inclination angle of the primary mode (which tends to be large in the ocean), but are always weak compared with local sum resonant modes.

Lin et al. (1995, 1997) also examined the role of the third special triad, Parametric Subharmonic Instability (PSI), which is thought to play an important role in oceanic mixing and in energy transfer from large to small scale modes. He performed a simulation with a primary mode with an inclination angle of 30°. The energy spectra show that while local PSI resonant modes are active, the distant PSI modes (i.e. those thought to be important in energy transfer) are inactive. Furthermore, as shown in Figure 3 (a plot of the energy history for selected modes) the PSI mode is also much weaker than the local sum resonant (LS) triad: the LS triad is always at least 5 times stronger than the PSI triad. Other cases were run to confirm that viscous dissipation was not the cause of the ineffective transfer of distant PSI resonant modes.

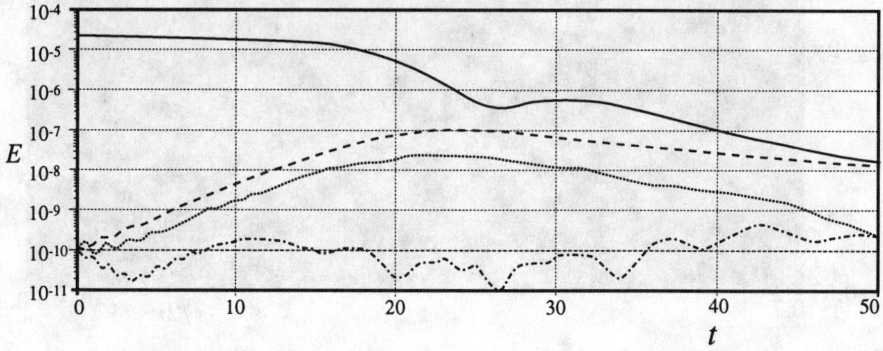

Figure 3. Energy history of resonant modes: ---------, primary mode; ·······, a local PSI mode; - - - - - , a local sum mode; -- · -- · --, a difference mode (from Lin, 1993).

Comparison with Other Studies

Bouruet-Aubertot et al. (1995) proposed that the strong instability leading to wave breaking is preceded by a slow transfer of energy to secondary waves by resonant interactions, primarily of the PSI type. McEwan (1983) hypothesized that these resonant triads transfer energy to new triads (a secondary instability) and the resulting cascade transfers energy to ever smaller scales producing steeper and steeper waves until wavebreaking occurs. The results of Lin et al. (1995), therefore, seem to differ from the conclusions of McEwan and Robinson (1975), Thorpe (1994), and Bouruet-Aubertot et al. (1995) who all concluded that PSI type energy transfer leads to the observed wave-breaking. The primary differences between these studies and ours are (a) their geometries do not permit local sum resonance modes (due to finite tank size or large-scale primary wave), and (b) their forcing is continuous.

Furthermore, the influence of the propagation angle of the primary mode needs to be accounted for as well in any comparison. The simulations of Lin et al. (1995, 1997) show local sum resonant modes to be dominant, especially if the inclination angle of the primary mode is large (see Figure 3) and there are no geometric constraints. This implies that if the primary mode is forced and wave breakdown is to occur, the local modes should be responsible and not the PSI modes. This result is consistent with that of Bouruet-Aubertot et al. (1995) who show in their Figure 11 that as the inclination angle of the primary mode gets steeper the local sum modes become more dominant.

Geometric Constraints and Forcing Mechanism

We have not yet investigated any differences due to the type of forcing used but Lin (1993, 1997) did investigate the effects of geometry by performing a simulation where the primary wavenumber is small thus constraining local resonance. Energy spectra from this simulation show that energy *is* indeed transferred to very small scale modes on PSI resonant traces when the LS triad is inhibited. However, the phase correlation and energy transfer rates of the higher wave number modes suggest they are not resonant. The phase correlations are irregular, implying non-resonance, and strong net energy transfer between the two small-scale modes occurs, which is not a feature of resonant interactions. Thus the PSI energy transfer mechanism is active if local sum resonance is inhibited (i.e. under certain geometric constraints), but it appears to be non-resonant.

The role of resonant and non-resonant PSI triads in leading to wave-breaking was also investigated by Bouruet-Aubertot et al. (1995). Figure 14 of their paper shows that the simulation results agree well with the growth rate predicted by the resonant theory. However, there is some other evidence in their paper which suggests that non-resonant PSI triads may be active as well. For example, in Figure 7 (b) of this paper the resonance criterion is satisfied only for about 10 BV periods for run V9, but Figure 13 shows that it takes about 80 BV before waves break, indicating that most of the time the triads are not resonant. The presence of these off-resonant triads, associated with non-linear interactions, which may finally lead to wave-breaking was also confirmed through a private communication by one of the authors (Staquet). Staquet also makes the point that at the time of publication of Aubertot et al. (1995) they had not computed the energy transfer by the triads and so could not confirm the exact process leading to wave breaking.

We can offer a conjecture about the role of the forcing mechanism. In our studies we have focused on a system where the energy of the strong mode is not constant. In such a system resonant interactions are stable and the energy of the weak resonant modes tends to be transferred back to the strong mode once it reaches a maximum. However, if the forcing is constant the energy of the weak resonant modes is unlikely to be transferred back to the strong mode, and will cascade down to higher wavenumbers leading to small-scale instabilities and breakdown. This distinction could explain the differences between our simulations and the experiments of McEwan and Thorpe: even if their experiments were modified to allow local sum resonance the PSI mechanism may still be active simply due to continuous forcing. Clearly the final breakdown and the associated mixing processes still need further investigation. Examination of the wavenumber, i.e. the scale, at which "breaking" occurs may also explain the differences between the experimental and numerical results.

Relevance to the Ocean

Finally, because PSI is thought to be important in energy transfer in ocean Lin et al. (1995) looked at triads in a GM79 spectrum. Figure 4 shows time-averaged bispectra (time averaging is allowed due to the slow variation of the flow with time) (a) for a mode which has an inclination angle typical of strong modes in the ocean, and (b) for the same mode but with the mean-flow eliminated. The loci of the modes satisfying resonant conditions (Phillips, 1977), known as the resonant traces, are superimposed: if resonant interactions dominate the strongly interacting modes should lie on the resonant traces. In Figure 4(a) the strongly interacting modes do not lie on the resonant traces but rather on two vertical lines, one at k_1=0 corresponding to modification of the mean flow, and the other at the horizontal wavenumber of the chosen mode due to strong wave-mean flow interaction. The data shown in Figure 4(a) therefore suggest that the energy transfer is mainly of non-resonant induced diffusion type. In this energy transfer mechanism, the mean flow does not gain or lose much energy, but rather its strength determines the energy transfer rate between the two small scale modes. The high wavenumber modes are weak enough for resonance, but the strong mean current increases nonlinearity and resonance is destroyed.

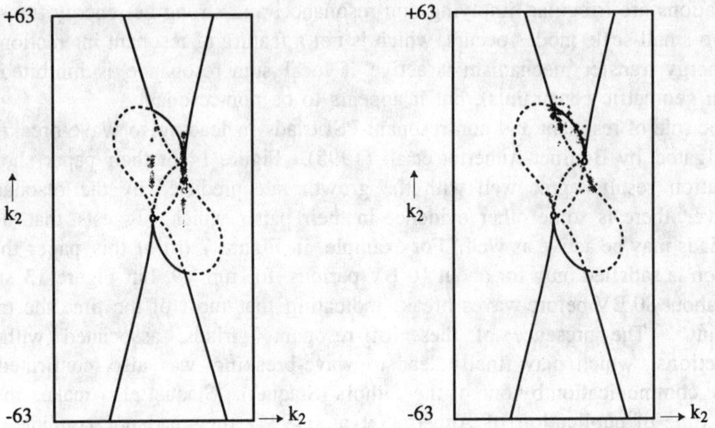

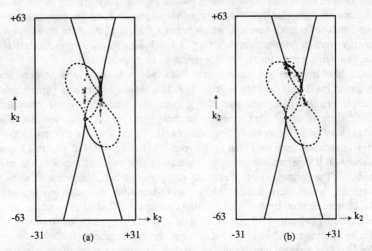

Figure 4. Time averaged bispectra from t* = 0.1 to 10; (a) with mean flow; (b) with mean-flow eliminated. Resonant traces are indicated with lines (from Lin, 1993, 1997).

When the mean flow is eliminated (Figure 4(b)) the distribution of the strongly interacting modes approximately matches the resonant traces, implying that resonant interactions can exist in the absence of wave-mean flow interaction. These interactions are of the local sum type. PSI triads do not play a significant role due to weak interaction coefficients, decreasing energy with increasing wavenumber, and the large inclination angle of the strong modes. Even when the energy level of the spectrum was reduced by a factor of 100, to examine Holloway's (1980) conclusion that oceanic waves were too energetic to be treated by weak interaction theory, non-resonant induced diffusion triads were the strongest indicating the dominance of wave-mean flow interactions. In this instance triads of the PSI type were still not found, suggesting that they may not be significant in the ocean.

A further interesting point about the role of PSI in the ocean stems from both the work of Lin et al. (1995) and that of Bouruet-Aubertot et al. (1995). Several researchers attribute the inertial peak of the internal wave spectrum to the PSI, i.e., energy is transferred from waves of frequency $2f$ to waves of frequency f. Using the linear dispersion relationship and f/N=0.01 (where N is BV frequency), we obtain a propagation angle of 89 degree for which tan (89)=57.3 for a wave of frequency $2f$, and a wave number of (1, 58). Looking at the growth rate of the C branch (local sum mode) in Figure 11 (f) of Bouruet-Aubertot et al. (1995) we see that the C branch has a higher growth rate for a wave number (1,5) than the A and B branches (PSI modes). We can, therefore, imagine that at the steeper propagation angle represented by a (1,58) wave the C branch will become much more important than the PSI branches A and B. This result also suggests that the role of PSI in oceanic mixing may be overemphasized.

Conclusions

1. For internal wave-critical layer interaction a spanwise Rayleigh-Taylor instability should be the most unstable process if the Richardson number is less than zero locally.

2. For the triad interactions the PSI energy transfer mechanism is feasible if local sum resonance is suppressed (i.e. under certain geometric constraints), but the mechanism *may* be non-resonant.

3. For triads with a background GM79 spectrum even when the energy level of the spectrum is reduced by a factor of 100 the induced diffusion triads (non-resonant) were the strongest indicating the dominance of wave-mean flow interactions. Triads of the PSI type were still not found, suggesting that they may not be significant in the ocean.

4. A number of issues still remain to be resolved including: (a) the mixing efficiency associated with wave breakdown for both wave-critical layer and wave-wave interactions; (b) the final breakdown process in both cases; and (c) the role of forcing in the case of wave-wave interactions.

Acknowledgments. We would like to thank Jörg Imberger, Sanjiva Lele, Chris Rehmann, and Chantal Staquet for valuable discussions, and the reviewers of a previous draft of this paper for their excellent suggestions. This work was supported by the Office of Naval Research under Grant N00014-89-J-1909 with the able guidance of Alan Brandt and Lou Goodman. The computations were performed on the Cray Y-MP at the NAVOCEANO Supercomputer Center.

References

Booker, J. R., and F. P. Bretherton, The critical layer for internal gravity waves in a shear flow, *J. Fluid Mech., 27*, 513-539, 1967.

Bouruet-Aubertot, P., J. Sommeria, and C. Staquet, Breaking of standing internal gravity waves through two-dimensional instabilities, *J. Fluid Mech., 285*, 265-301, 1995.

Flatte, S. M., F. S. Henyey, and J. A. Wright, Eikonal calculations of short wavelength internal wave spectra. *J. Geophys. Res., 90(C4)*, 7265-7272, 1985.

Fritts, D. C., The transient critical-level interaction in a Boussinesq fluid, *J. Geophys. Res., 87*, 7997-8016, 1982.

Garrett, C. and W. H. Munk, Internal waves in the ocean, *Ann. Rev. Fluid Mech., 11*, 339-369, 1979.

Gregg, M. C., D. P. Winkel, and T. B. Sanford, Varieties of fully resolved spectra of vertical shear, *J. Phys. Ocean.*, 23, 124-141, 1993.

Hasselman, K., Feynman diagrams and interaction rules of wave-wave scattering processes. *Rev. Geophys., 4*, 1-32, 1966.

Henyey. F. S, and Pomphrey, N. Eikonal description of internal wave interactions: a nondiffusive picture of "induced diffusion". *Dyn. Atmos. Oceans, 7*, 189-219. 1983.

Holloway, G., Oceanic internal waves are not weak waves, *J. Phys. Ocean., 10*, 906-914, 1980.

Holloway, G., On interacting time scales of oceanic internal waves, *J. Phys. Ocean., 12*, 293-296, 1982.

Holt, S. E., J. R. Koseff, and J. H. Ferziger, A numerical study of the evolution and structure of homogeneous stably stratified sheared turbulence, *J. Fluid Mech., 237*, 499-539, 1992.

Itsweire, E. C., J. R. Koseff, D. A. Briggs, and J. H. Ferziger, Turbulence in stratified shear flows: Implications for interpreting shear-induced mixing in the ocean, *J. Phys. Ocean., 23*, 1508-1522. 1993.

Ivey, G. N., and J. Imberger, On the nature of turbulence in a stratified fluid. Part I: The energetics of mixing, *J. Phys. Ocean., 21*, 650-658, 1991.

Javam, A., J. Imberger, and S. W. Armfield, Numerical study of internal wave-wave interactions in a stratified fluid, submitted to *J. Fluid Mech*, 1997.

Koop, C. G., and B. McGee, Measurements of internal gravity waves in a continuously stratified shear flow, *J. Fluid Mech.*, 172, 453-480, 1986.

Lelong, M. -P., and J. J. Riley, Internal wave-vortical mode interactions in strongly stratified flows. *J. Fluid Mech., 232*, 1-19, 1991.

Levine, M. D., Internal waves in the ocean: A review, *Rev. Geophysics and Space Physics, 21, 5*, 1206-1216, 1983,

Lin, C. -L., J. H. Ferziger, J. R. Koseff, and S. G. Monismith, Simulation and stability of two-dimensional internal gravity waves in a stratified shear flow, *Dyn. Atmos. Oceans, 19*, 325-366, 1993.

Lin, C. -L., Numerical study of wave-mean flow and wave-wave interactions in linearly stratified flows, Ph. D Thesis, Dept of Mechanical Engineering, Stanford University, Stanford, CA, 315p., 1993.

Lin, C-L., J. R. Koseff, and J. H. Ferziger, On triad interactions in a linearly stratified ocean, *J. Phys. Ocean., 25*, 1,153-1,167, 1995.

Lin, C-L., J. H. Ferziger, and J. R. Koseff, A study of triad interactions in a linearly stratified fluid using direct numerical simulations, in preparation, 1997.

Martin, S., W. Simmons, and C. Wunsch, The excitation of resonant triads by single internal waves. *J. Fluid Mech., 53*, 17-44, 1972.

McComas, C. H., and F. P. Bretherton, Resonant interaction of oceanic internal waves. *J. Geophys. Res., 82(9)*, 1397-1411, 1977.

McComas, C. H., and P. Muller, Time scales of resonant interactions among oceanic internal waves. *J. Phys. Ocean., 11*, 139-147, 1981.

McEwan, A. D. and R. M. Robinson, Parametric instability of internal gravity waves, *J. Fluid Mech., 67*, 667-687, 1975.

McEwan, A. D., Internal mixing in stratified fluids, *J. Fluid Mech., 128*, 59-80, 1983.

Muller, P., G. Holloway, F. S. Henyey, and N. Pomphrey, Non-linear interactions among internal gravity waves. *Rev. Geophys., 24(3)*, 493-536, 1986.

Munk, W. H., Internal waves and small-scale processes. In *Evolution of Physical Oceanography*, edited by B. A. Warren and C. Wunsch, pp. 264-291, The MIT Press, 1981.

Olbers, D. J., Nonlinear energy transfer and the energy balance of the internal wave field in the deep ocean, *J. Fluid Mech., 74(2)*, 375-399, 1976.

Orlanski, I., and C. P. Cerasoli, Energy transfer among internal gravity modes: weak and strong interactions, *J. Geophys. Res.*, *86*, 4103-4124, 1981.

Phillips, O. M., in *The Dynamics of the Upper Ocean*, Cambridge University Press, 1977.

Phillips, O. M., Wave interactions-the evolution of an idea, *J. Fluid Mech.*, 106, 215-227, 1981.

Pomphrey, N., J. D. Meiss, and K. M. Watson, Description of nonlinear internal wave interactions using Langevin methods, *J. Geophys. Res.*, *85(C2)*, 1085-1094, 1980.

Ramsden, D., and G. Holloway, Energy transfers across an internal wave-vortical mode spectrum, *J. Geophys. Res.*, *97*, 3659-3668, 1992.

Riley, J. J., R. W. Metcalfe, and M. A. Weissman, Direct numerical simulations of homogeneous turbulence in density-stratified fluids. In *Nonlinear Properties of Internal Waves,* edited by B. J. West, *76*, 79-112, 1981.

Rogallo, R. S., Numerical Experiments in Homogeneous Turbulence, *NASA, Tech. Memo*. 81315.

Shen, C. Y. and G. Holloway, 1986: A numerical study of the frequency and the energetics of nonlinear internal gravity waves, *J. Geophys. Res.*, *91(C1)*, 953-973, 1981.

Siegel, D. A., and J. A. Domaradzki, Large-eddy simulation of decaying stably stratified turbulence, *J. Phys. Ocean.*, *24*, 11,2353-11,2386, 1994.

Thorpe, S. A., An experimental study of critical layers, *J. Fluid Mech.*, *103*, 321-344, 1981.

Thorpe, S. A., Observations of parametric instability and breaking waves in an oscillating tilted tube, *J. Fluid Mech.*, *260*, 333-350, 1994.

Winters, K. B., and E. A. d'Asaro, Two-dimensional instability of finite amplitude internal gravity wave packets near a critical level, *J. Geophys. Res.*, *94(C9)*, 12709-12719, 1989.

Winters, K. B., and E. A. d'Asaro, Three-dimensional wave instability near a critical level, *J. Fluid Mech.*, *272*, 255-284, 1994.

Winters, K. B., and, J. J. Riley, Instability of internal waves near a critical level, *Dyn. Atmos. Oceans*, *16*, 249-278, 1992.

Winters, K. B., P. N. Lombard, J. J. Riley, and E. d'Asaro, Available potential energy and mixing in density-stratified fluids, *J. Fluid Mech.*, *289*, 115-128, 1995.

Thorpe, S. A., [illegible], *J. Fluid Mech.*, *260*, 333–350, 1994.

21

Momentum Exchange Due to Internal Waves and Wakes Generated by Flow Past Topography in the Atmosphere and Lakes

P. G. Baines

Abstract

Topographic features on the earth's surface with length scales of 5–100 km have a significant effect on the general circulation of the atmosphere. This is accomplished by the drag that they exert on the atmosphere through the generation of internal gravity waves that may propagate to great heights, through hydraulic transitions in low level air-streams across mountain ranges, and through the formation of wakes. The satisfactory representation of these processes in general circulation models is important for both weather and climate forecasting. This may be accomplished by characterising the small-scale topography by suitable statistical parameters, and (conceptually) replacing it by simpler topography that has the same values of these parameters, and for which the dynamics have been studied and are reasonably well known. The drag of this topography on the atmospheric flow may then be estimated for modelling purposes.

The extension of these concepts to flow past topography in lakes is discussed. As a beginning, this is done by examining the properties of stratified flow past obstacles on an inclined plane surface, simulating part of the bottom of a typical lake. The three dynamical processes listed above for the atmosphere are again relevant, with conspicuous asymmetries present because of the sloping terrain.

Introduction

In recent years it has become apparent that the drag of small-scale orographic features (i.e. on scales of 100 km or less) can have a significant effect on the general circulation of the atmosphere (Palmer et al., 1986). This implies that the effects of unresolved sub-grid-scale (sgs) topographic features should be represented or parametrised in global atmospheric circulation models with some care. The physical processes involved may be divided into two parts: determining the drag of the atmospheric flow on a given mountain or region of topography, and secondly, determining where the corresponding opposing drag on the atmosphere is felt.

One approach to this problem is to observe the properties of the flow past a typical obstacle, and then to represent the effects of the observed flow patterns in the numerical

Physical Processes in Lakes and Oceans
Coastal and Estuarine Studies Volume 54, Pages 285-294

models in a parametrised manner. This approach has been adopted for atmospheric models at the European Weather Centre (ECMWF) and at CSIRO Atmospheric Research. In principle, the same approach can be applied to flow in lakes. In this paper we first describe the salient features of stratified flow past obstacles, and their implications for momentum transfer. The application of this to large-scale numerical models is then summarised. Some observations of flow patterns more relevant to conditions in lakes are then described and some implications for parametrisation of sgs effects in lakes are discussed. Similar dynamics should apply to the upper regions of the oceanic continental slope, although the Coriolis force is omitted in the present discussion.

The Nature of Stratified Flow Past an Isolated Obstacle

The principal parameters are the maximum height of the obstacle, h, the buoyancy frequency N, and the mean fluid speed U. The flow is then characterised by the dimensionless number Nh/U, and its properties are described in great detail in Baines (1995). If Nh/U is small, the flow pattern is essentially that described by linear perturbation theory, with lee waves, low-level flow splitting and divergence (and if the obstacle is sufficiently steep, lee-side separation). Drag here is primarily manifested in vertically propagating lee waves, plus lee-side separation if present. As Nh/U increases to values above unity, additional features become apparent as shown in Figure 1: lee-wave overturning and upper-level stagnation, low-level hydraulic transition, upstream flow splitting with two stagnation points, and lee-side separation with a separated wake. The drag here is primarily manifested in lee waves, hydraulic transition of the low-level stream, and the separated wake. Observations show that for obstacles with sufficiently small slopes, flow from upstream heights z in the range $0 < z < z_b = h - U/N$ splits and flows around the obstacle, whereas flow from upstream heights $z > z_b$ passes over it.

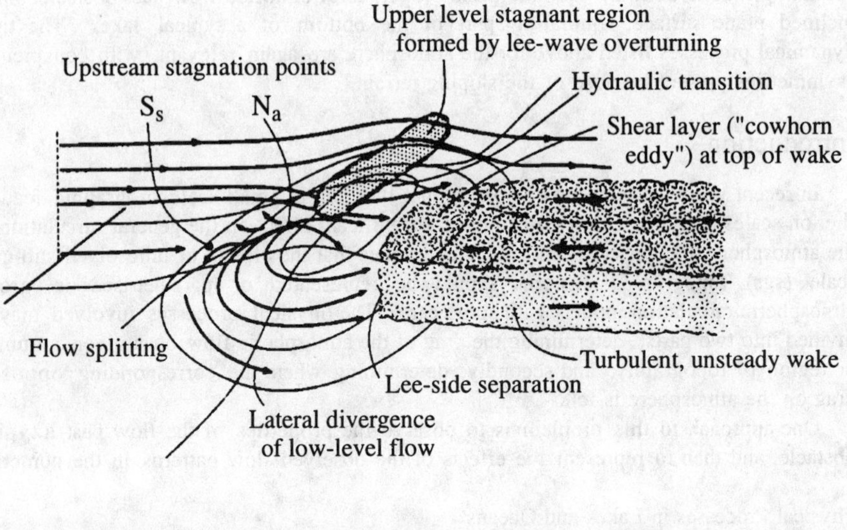

Figure 1. Schematic diagram of the flow past a symmetrical obstacle where Nh/U>>1. Lines denote streamlines on the surface of the obstacle, and in the central plane of symmetry. (From Baines, 1995).

These properties imply that the net drag on the obstacle is felt by the fluid in different ways and places, as follows. Firstly, the flow over the obstacle generates vertically propagating gravity waves that carry momentum (with the flux represented by $\overline{w'u'}$) to heights where they are dissipated or break. The momentum so carried is then deposited at these heights, in proportion to the divergence of the momentum flux. For long obstacles where the flow is hydrostatic, analytic expressions may be obtained for the surface drag, and this drag is transmitted vertically by the waves to locations where the momentum flux diverges.

Secondly, the low-level stream undergoing an hydraulic transition exerts a drag on the obstacle, because a thick, dense lower layer on the upstream side changes to a thinner, faster one on the downstream side, and undergoes a loss of momentum flux in the process. Thirdly, the flow splitting beneath the hydraulic stream results in a relatively stagnant wake (with lee-side vortices etc.) that also causes a net drag on the low-level flow. This drag may be represented by use of a drag coefficient, as for homogeneous flow past a cylinder, for example.

Application to Parametrisation of SGS Orographic Effects in Numerical Models of the Atmosphere

Representation of the Topography

Most numerical models of the atmosphere are represented by a grid that fills the fluid in question, so that a region of the earth's surface is represented by values at one grid point, termed a grid-point region or gpr. This region is assumed to be locally flat, with horizontal coordinates x & y. The topography within this region may be represented by five parameters. The first two of these are the mean height $\bar{h}$ and the variance μ^2, defined by

$$\mu^2 = \frac{1}{S}\iint (h - \bar{h})^2 dS, \tag{1}$$

where S is the area of the gpr. The other three parameters stem from the topographic gradient correlation tensor

$$H_{ij} = \overline{\frac{\partial h}{\partial x_i}\frac{\partial h}{\partial x_j}}, \tag{2}$$

where $x_1 = x$, $x_2 = y$, and the overbar denotes the mean over the whole area of the gpr. For a given gpr, the components of this tensor may be calculated from a digitised data set of sufficient resolution. Once found, it may be diagonalised to find the principal axes and the degree of anisotropy of the topographic variations within the gpr. If the principal axis is oriented at an angle θ to the x-axis, and new coordinates (x', y') are defined with this as the x' -axis, the corresponding information about the topography is contained in the three parameters θ, σ and γ, where σ and γ are defined by

$$\sigma^2 = \overline{\left(\frac{\partial h}{\partial x'}\right)^2}, \gamma^2 = \overline{\left(\frac{\partial h}{\partial y'}\right)^2} \Big/ \overline{\left(\frac{\partial h}{\partial x'}\right)^2}. \tag{3}$$

As an illustration, if the topography has the form

$$h(x',y') = h_o(1 + \cos kx' \cos my'), \tag{4}$$

where necessarily $k \geq m$, then

$$\overline{h} = h_o, \quad \mu = h_o/2 \quad \sigma = kh_o/2, \quad \gamma = m/k. \tag{5}$$

One may therefore obtain the same values of these parameters by replacing the existing topography by a doubly periodic topography with a representative shape. Hence, if we know what the drag is for this particular shape, we may represent the total drag on the gpr.

The nature of the drag and its effect on the fluid is primarily dependent on $2N\mu/U$, since the effective height of the topography relative to the mean is 2μ. If $2N\mu/U < 1$, for gently sloping obstacles without flow separation, most of the drag on the fluid is carried upwards by internal waves, whereas for $2N\mu/U > 1$, there are contributions from hydraulic transition and low-level blocked flow. These three types of drag may be represented as follows.

Distribution of Stress in the Atmosphere

We first consider the case where $2N\mu/U < 1$. From linear perturbation theory it may be shown that the mean stress on the surface of the gpr is approximately (Phillips, 1984, Baines, 1995)

$$\tau = 0.1\rho NU\mu\sigma\,\hat{\tau} \tag{6}$$

where ρ is the fluid density and the direction of the unit vector $\hat{\tau}$ is somewhere between the directions of the principal axis and the incident wind. This gives the magnitude and direction of the stress carried vertically by the internal waves, and the effect on the mean flow is given by the divergence of this stress:

$$\frac{\partial \overline{u}}{\partial t} = -\frac{\partial \overline{u'w'}}{\partial z}, \tag{7}$$

where here $\overline{u}$ denotes the mean velocity component (as represented by the numerical model) in the direction of τ, and u', w' denote the velocity components in the waves. The simplest representation of a wave field with this momentum flux is a single monochromatic wave, propagating in this plane, and we use this wave as a surrogate for the whole wave field. This is justified by the relative simplicity of hydrostatic wave fields. One may then calculate the Richardson number Ri for the combined mean flow plus this wave. As the waves propagate vertically, their amplitude may increase due to changing mean flow and decreasing density of the air. For the next step, one appeals to the saturation hypothesis of Lindzen (Lindzen, 1981, Baines, 1995, Kim and Arakawa, 1995), which states that the amplitude of internal gravity waves in a random field is limited to a maximum value controlled by local wave breaking (in a manner that parallels the equilibrium spectrum of ocean surface waves). We assume that this can be approximately described by a limiting value of the Richardson number - say $R_i > 1/4$, and that overturning and mixing occurs to maintain this condition. If the value of this Richardson number falls below 1/4, therefore, the momentum flux is reduced to a value that increases it. The result is a nett vertical stress gradient that drives the mean flow.

When $2N\mu/U > 1$, the same equations are applicable except that the relevant height of the topography is now U/N rather than 2μ. For the other two forms of drag, the hydraulic component may be represented by using results from stratified hydraulics (Baines, 1995), but in the atmosphere this effect appears to be somewhat smaller than the third type, which is the "blocked flow" or aerodynamic drag (Lott and Miller, 1999). The latter applies for heights $0 < z < z_b = 2\mu - U/N$. Laboratory experiments with flow past complex terrain with comparable maximum height show that in this depth range the flow between the obstacles is greatly reduced. This may be approximated by assuming that the fluid is locally stagnant, or by invoking drag coefficients for these nearly horizontally two-dimensional flows around the obstacles.

Extension of These Ideas to Lakes

The numerical modelling of flow in lakes has not yet reached the level of sophistication of that for the atmosphere, partly because of the variety of complex and diverse physical processes involved (Imberger and Patterson, 1990). These include the effects of topography that have scales that are two small to be resolved by the model.

When one considers the possible application of the above framework to lakes, a number of salient differences present themselves. Firstly, the lake has a free surface (normally with an associated upper mixed layer), imposing a limited vertical scale on the motion. Secondly, the bottom of a lake is not normally horizontal, and thirdly, the stratification in lakes is normally closer to two-layer than uniform stratification. The presence of the free surface and the stratification structure with mixed layers above and below implies that internal waves generated may be formed into horizontally propagating modes.

There does not appear to be a vast literature on small-scale topography in lakes, but some of the prominent features must include ridges and valleys due to the continuation below the surface of the lake of the topographic features evident above it. The flow of low-frequency currents and seiches driven primarily by the wind past such features may then produce some of the phenomena described above. If we take a nominal buoyancy frequency N of 10^{-2} s^{-1}, topographic wavelengths of 100 to 1000 m and current speeds $U \sim 0.05$-0.1 ms^{-1}, and we assume that these mean flows are approximately steady for a period of several hours, the space and time scales involved are short enough to justify the omission of the Coriolis force.

We may investigate the effects of localised bottom topography in a lake under these circumstances by examining the dynamics of idealised situations. Here we examine the properties of uniformly stratified horizontal flow past a simple obstacle on a sloping plane surface, inclined at angle θ to the horizontal, and consider the resulting changes to the three dynamical features described above. We again take axes with z vertical and mean velocity U directed in the x-direction, with the y-axis directed horizontally toward shallow water, to the left when facing downstream. The equation for the topography is then

$$z = \tan\theta .\ y + h(x,y) \qquad (8)$$

where h(x,y) is the perturbation due to topography. We also use coordinates z^* normal to the plane and y^* upslope, with topography given by $z^* = h^*(x,y^*)$. The properties of internal waves produced by flow past such topography with sufficiently small amplitude may be readily calculated. For periodic topography of the form

$$h = h_o e^{i(kx + my)}, \qquad (9)$$

the solution for the pressure field assuming infinite depth is

$$p' = iA\rho_o U h_o e^{i(kx+m^*y+nz)}, \tag{10}$$

where

$$m^* = m - n \tan\theta,$$

$$A = \frac{(N^2 - k^2U^2)k^2U}{(nk^2U^2 + m^*\tan\theta.(N^2 - k^2U^2))}, \quad n^2 = (k^2 + m^{*2})\left(\frac{N^2}{k^2U^2} - 1\right). \tag{11}$$

The drag per unit surface area of the plane is then obtained by integrating the pressure over the surface of the topography (e.g. Baines, 1995, Section 5.2), to obtain

$$\mathbf{F}_D = \tfrac{1}{2}\rho_o AUh_0^2\,(k\hat{\mathbf{x}}, m^*\hat{\mathbf{y}}^*). \tag{12}$$

directed normal to the ridges and valleys of (9). This drag implies an oppositely directed momentum flux carried away by the three-dimensional internal wave field, upslope if $m > 0$, downslope if $m < 0$. The effects of this momentum flux on the mean flow are felt where the waves are dissipated, or are absorbed by a critical layer. Waves propagating into deeper water propagate along "rays" in the direction of the group velocity (e.g., Baines, 1971a, b), which are inclined to the vertical. If the bottom slope decreases so that the depth tends toward a constant value, these waves on the upward and downward rays (reflected from both the surface and the bottom) may be described by a set of horizontally propagating modes. Whether this happens or not, these waves may encounter critical layers due to horizontal shear, and the resulting interactions would appear to have received little attention.

When the denominator in A (equation 11) vanishes, the group velocity of the forced plane wave lies in the plane of the topography, and vice versa. This is the condition for resonance, and hence for given N, U and θ, topographic Fourier components that satisfy these conditions should generate a large response. This may partially explain some of the observations reported by Lemmin et al. (1998), since it implies that a large response is expected in general when flow past topographic variations generates waves with group velocity in the plane of the sloping topography.

Hydraulic effects causing drag on the mean flow are very likely with near-two-layer flow if the topography is sufficiently large, and/or the local Froude number is close to unity (*i.e.* the local wave or mode speed is close to zero relative to the topography). Such effects have yet to be studied with three-dimensional geometries.

Finally, flow past sufficiently high three-dimensional obstacles ($Nh/U > 1$) will result in flow-splitting and consequent formation of a wake with lee-side vortices, but on a sloping surface the structure of this wake will be asymmetric. To investigate the properties of this system, a set of observations of the properties of wakes of towed obstacles on horizontal surfaces, and on inclined surfaces set at 5° to the horizontal has been made. The results to date are largely qualitative, and the salient points are as follows. (i) For small, steep obstacles with h = 2A = 2B (where A and B are the half-widths in the along-and across-flow directions respectively), a stratified vortex street wake develops from two initially symmetric eddies, which is largely coherent in the vertical, and in which the flow at lower levels is more advanced in time (in terms of its non-linear evolution) than at upper levels (see Figure 2). Observations showing similar wake structures have been described by Boyer et al. (1987) for stratified flow past obstacles of conical and cosine-squared shape, under much the same conditions.

(ii) For large obstacles with h = 2A << 2B, on the other hand, the observed wake is again coherent in the vertical, but consists of two symmetric eddies that are elongated in the downstream direction. An example is shown in Figure 3. Here the development of the pattern is limited by the finite towing distance and the length of the tank, and it is possible that a periodic vortex-street type of wake would eventually develop in a much longer tank.

The above experiments have been repeated by towing the same two obstacles through stratified fluid but with the tank tilted at an angle of 5° to the horizontal about an axis along the towing direction. Flow visualisation is again due to dye released in upstream rakes at fixed distances above the floor of the tank; however, the density of the dyed fluid from each rake is approximately uniform, and quickly settles to a uniform mean height when it emerges from the tubes. In these experiments the wake is asymmetric and more complex than before, as shown by the examples in Figures 4, 5 and 6 and the salient points observed so far are as follows:

Figure 2. Plan view of a "vortex street" wake formed by stratified flow past a ridge of finite length and uniform triangular cross-section, with the ridge line lying perpendicular to the flow with vertical ends, towed at constant speed U from right to left. (Hence the relative flow impinges from the left in all these figures). Here Nh/U = 9.31, and the obstacle has h = 2A = 2B = 9 cm, being 9 cm in height, breadth and width. Flow is visualised with dye released from two rakes upstream, at heights 3 cm (less distinct pattern, with upstream dye lines deflected to the left) and 6.5 cm (distinct pattern, with conspicuous eddies). Reynolds number $R_e \equiv Uh/\nu = 864$, where ν is kinematic viscosity.

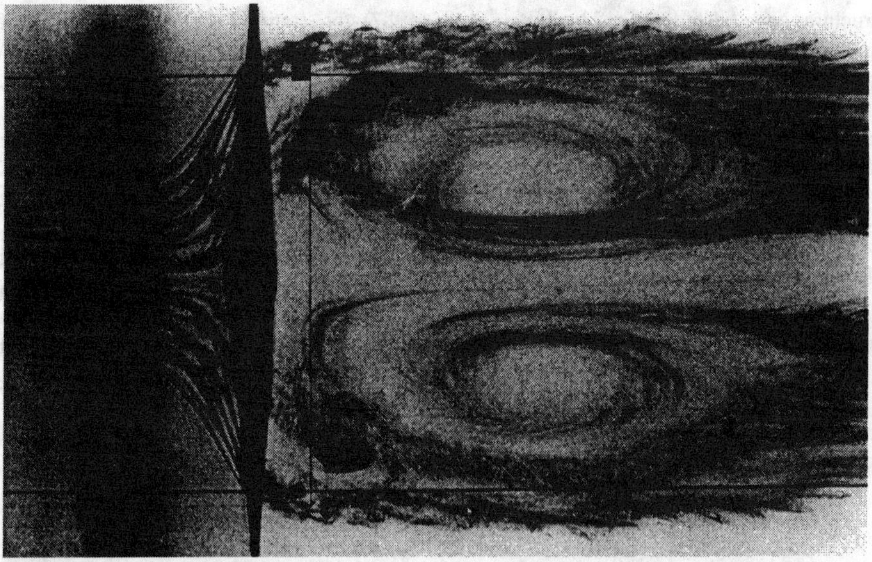

Figure 3. As for Figure 2, but for an obstacle with a single peak and tapered triangular cross-section, decreasing linearly to zero at each end. Nh/U = 9.2, with h = 9.2 cm, 2A = 114 cm, 2B = 10.8 cm. Here the dye from the lower rake (height 3 cm) is spread more broadly on the upstream side, and that for the upper rake (6.5 cm) less broadly, and these streams may be traced downstream. R_e = 891. The observations of flow past symmetric obstacles with triangular cross sections on a horizontal surface with Nh/U >>1 and large Reynolds number show the following.

(i) Fluid flowing around the obstacle tends to pass it on the deep side.
(ii) With the smaller obstacle, vortex street wakes still occur as shown in Figure 4, but the vortices on each side are now different: those shed on the deep side are generally larger and rounder, and the vortices shed on the shallow side are smaller and tend to be elongated in the direction of the flow. As before, the flow at different heights shows similarities but the features are often not coincident, and the lower pattern generally has larger amplitude and is more advanced in its evolution with time.
(iii) For the larger obstacles (Figures 5 and 6), elongated eddies are again seen but they are definitely not symmetric about the centreline and resemble "figure of eight" patterns, although cross-overs do not occur. The deviations are of larger amplitude at lower levels.

Despite the complexity of these effects, it is possible that they may be parameterised as sub-grid-scale features in numerical models of lakes in a similar manner to that for the atmospheric models, provided that the topography is known to sufficient detail.

References

Baines, P. G., The reflexion of internal/inertial waves from bumpy surfaces. *J. Fluid Mech. 46*, 273-291, 1971a.

Baines, P. G., The reflexion of internal/inertial waves from bumpy surfaces. Part 2. Split reflexion and diffraction. *J. Fluid Mech.*, *49*, 113-131, 1971b.

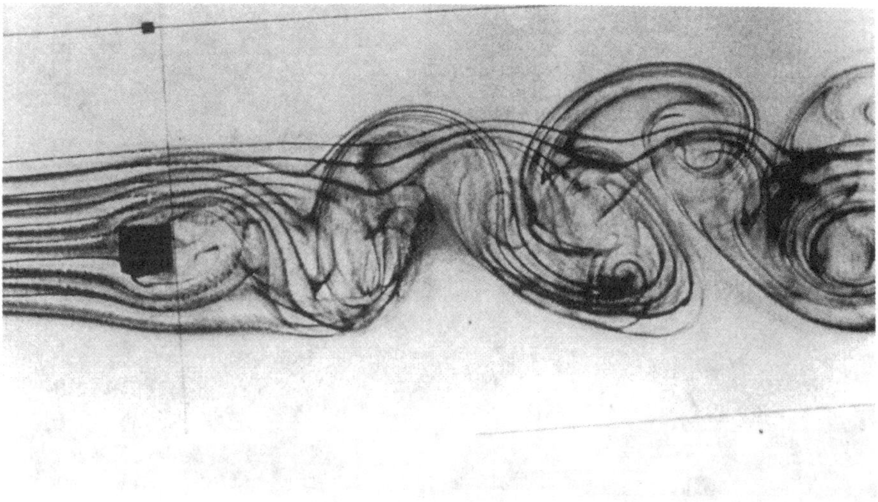

Figure 4. As for Figure 2, but now the tank has been tilted to an angle of 5° to the horizontal, with the deeper side at the bottom of the picture. Here Nh/U = 9.4, with h measured perpendicular to the bottom of the tank. The dye rakes are again parallel to the bottom, but after emerging from the rake the streams of dye rapidly adjust vertically to lie in an approximately horizontal plane. The dye rakes have been offset towards the shallow side to better reveal the flow, and the lower dye is released at 3 cm (with the more widely spread, coherent pattern), the upper at 6.5 cm as for Figure 3. $R_e = 855$.

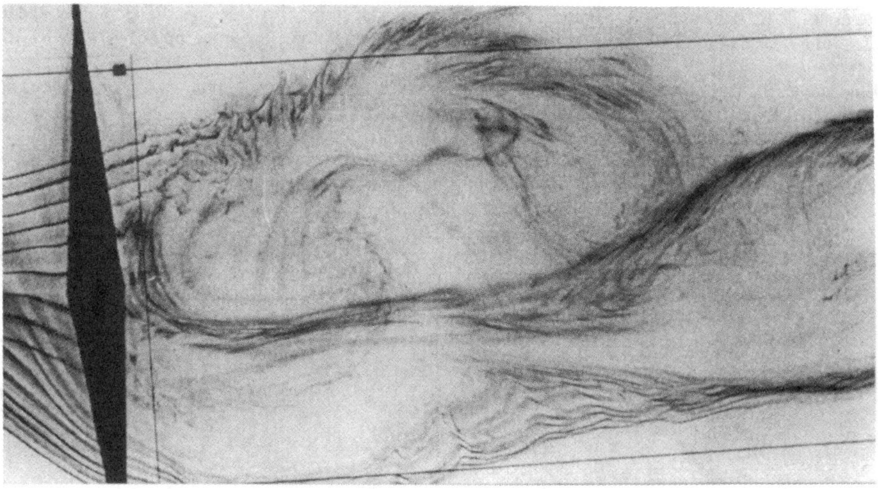

Figure 5. As for Figure 4 but with the obstacle of Figure 3, and Nh/U = 12.5, $R_e = 657$. The dye from the lower rake is deflected to the right on the upstream side.

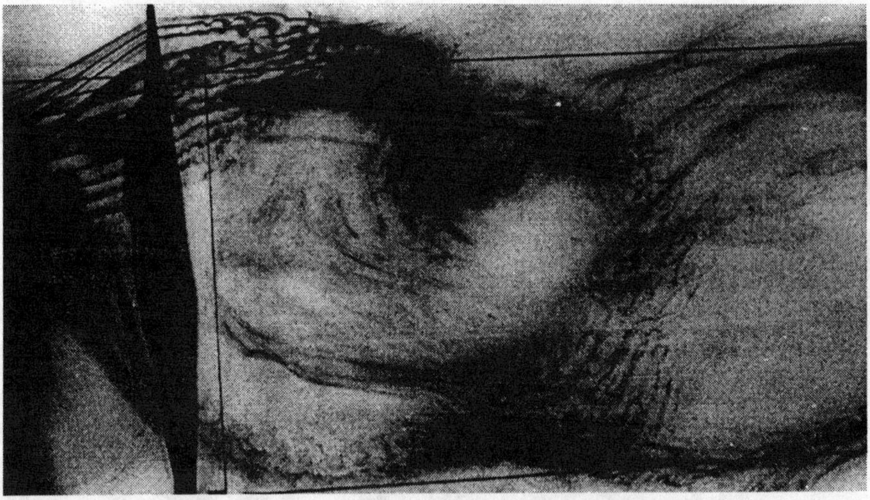

Figure 6. As for Figure 5, but with the dye rake further offset toward the shallow (upper) side of the tank. Nh/U = 7.6, R_e = 1080. Dye from the lower rake spreads and flows around both sides, whereas dye from the upper flows over the left-hand side only, with much less deflection.

Baines, P. G., *Topographic Effects in Stratified Flows*. Cambridge University Press, 482pp, 1995.

Boyer, D. L., P. A. Davies, W. R. Holland, F. Biolley and H. Honji, Stratified rotating flow over and around isolated three-dimensional topography. *Phil. Trans. Roy. Soc.* A, 322, 213-241, 1987.

Imberger, J. and J. C. Patterson, Physical Limnology. In *Advances in Applied Mechanics*, Academic Press, Cambridge, 303-475, 1990.

Kim, Y-J. and A. Arakawa, Improvement of orographic gravity wave drag parametrisation using a mesoscale gravity wave model. *J. Atmos. Sci. 52*, 1875-1902, 1995.

Lemmin, U. R. Jiang and S. A. Thorpe, Finescale dynamics of stratified waters near a sloping boundary of a lake, in Physical Processes in Lakes and Oceans, edited by J. Imberger, pp 461-474, AGU, this volume, 1998.

Lindzen, R. S., Turbulence and stress due to gravity wave and tidal breakdown, *J. Geophys. Res., 86*, 9707-9714, 1981.

Lott, M. and M. Miller, A new sub-grid-scale orographic drag parametrisation: its formulation and testing. *Quart, J. Roy. Met. Soc., 123*, 101-127, 1997.

Palmer, T. N., G. J. Shutts and R. Swinbank, Alleviation of a systematic westerly bias in general circulation and numerical prediction models through an orographic gravity wave drag parametrisation, *Quart. J. Roy. Met. Soc., 112*, 1011-1039, 1986.

Phillips, D. S., Analytical surface pressure and drag for linear hydrostatic flow over three-dimensional elliptical mountains. *J. Atmos. Sci. 41*, 1073-1084, 1984.

22

In Search of Holmboe's Instability

G. A. Lawrence, S. P. Haigh, and Z. Zhu

Abstract

The stability of stratified shear flows with a density interface much thinner than the velocity interface has been studied theoretically, numerically and experimentally. When the interfaces are centered about the same level linear stability analysis and numerical simulations predict the Holmboe instability. Unfortunately attempts to reproduce this instability in mixing layer facilities have not been successful, because it is difficult to avoid a displacement between the two interfaces. Here we report on exchange flow experiments where the interfaces are centered about each other and Holmboe instabilities are observed. The results are compared with numerical simulations. We also compare numerical and laboratory experiments where there is a displacement between the two interfaces

Introduction

The study of the hydrodynamic stability of stratified shear flows with a thin density interface relative to the shear layer thickness dates back to the linear stability analysis of Holmboe (1962). For small bulk Richardson numbers he predicted a Kelvin-Helmholtz type instability travelling with a phase speed equal to the mean velocity; and at higher Richardson numbers a second mode of instability (now known as the Holmboe instability) consisting of two trains of interfacial waves travelling at the same speed, but in opposite directions, with respect to the mean flow. Although there has been considerable interest in the Holmboe instability, there is little experimental verification of its existence and behaviour. Pouliquen et al. (1994) have made brief (3 - 4 s) observations of Holmboe instabilities by spatially forcing the flow of immiscible fluids in a tilting tube. Browand and Winant (1973), Koop (1976) and Lawrence et al. (1991) have performed mixing layer experiments, under conditions that might have been expected to yield Holmboe instabilities, without conclusive results. Their flows were "one-sided" in that disturbances and mixing were confined to one side of the density interface (Figure 1).

In mixing layer facilities the density interface is generally displaced with respect to the shear layer. This asymmetry is caused by the fact that to create the shear layer two layers of water flowing at different velocities, and with different boundary layer thicknesses, are initially separated by a splitter plate. After they merge at the trailing edge of the splitter plate the point of maximum vorticity is found on the high speed side of

Physical Processes in Lakes and Oceans
Coastal and Estuarine Studies Volume 54, Pages 295-304

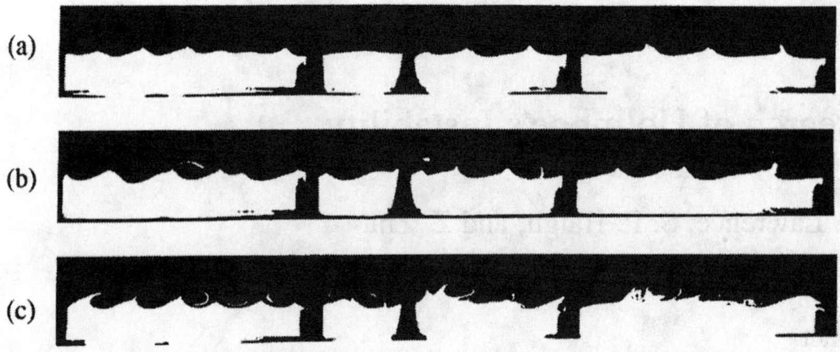

Figure 1. Side views of experiments performed in a mixing layer facility by Lawrence et al. (1991) showing the effect of decreasing Richardson number, J: in (a) J = 0.12, in (b) J = 0.09 and in (c) J = 0.06. Flow is from left to right. The horizontal field of view of each photograph is approximately 25 cm. The three photographs for each experiment are not taken simultaneously. The left hand sides of each of the photographs are 20, 40 and 60 cm downstream of the splitter plate, respectively.

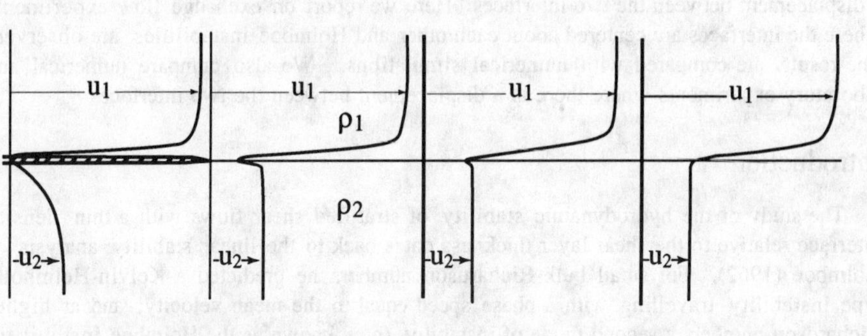

Figure 2. Evolution of an asymmetric flow in a mixing layer facility (adapted from Koop 1976).

the density interface (Figure 2). The profile asymmetry can be varied, but not eliminated, by repositioning fine mesh screens placed at, or near the trailing edge of the splitter plate, since they partially remove the boundary layers generated by the splitter plate. Far downstream of the splitter plate the profile asymmetry approaches zero and Holmboe instabilities appear to form; however, this phenomena has not been well documented. Note that far downstream of the splitter plate the growth of the bottom and sidewall boundary layers, as well as the shear layer, results in a very different flow than just downstream of the splitter plate. Thus mixing layer experiments do not appear to be an effective means of studying Holmboe instabilities.

Linear Stability Analysis

Piecewise Linear Profiles

Following Taylor (1931), Goldstein (1931), Holmboe (1962), and others we start by examining the stability of a sheared density interface using piecewise linear approximations to the velocity and density profiles. We choose the profiles given in Figure 3. We

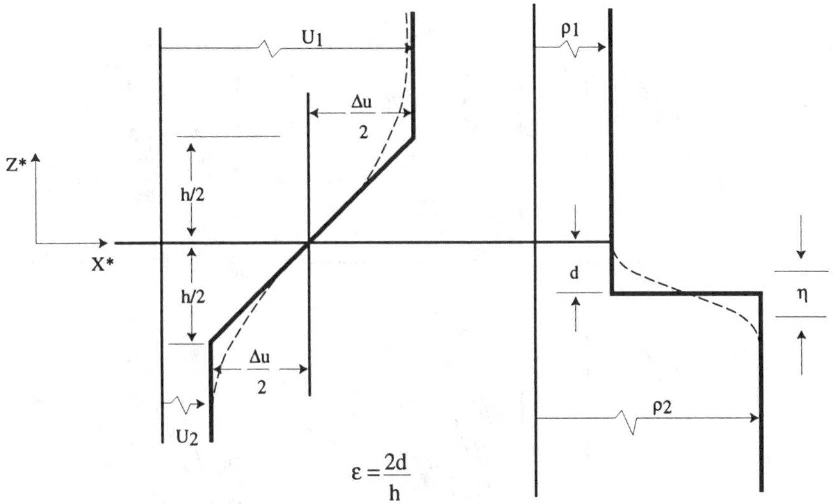

Figure 3. Definition diagram

define the shear layer thickness, $h = \Delta U/(dU/dz)_{max}$, and the thickness of the density interface, $\eta = \Delta\rho/(d\rho/dz)_{max}$, where the velocity difference, $\Delta U = |U_2 - U_1|$, and the density difference, $\Delta\rho = \rho_2 - \rho_1$, and the densities and velocities of the upper and lower layers are ρ_1, U_1 and ρ_2, U_2 respectively. To model flows with a displacement, d, between the two interfaces, and where the thickness of the density interface, $\eta \ll h$, the density is assumed to vary discontinuously at $z^* = -d$, where z^* is the distance above the center of the shear layer.

For unbounded flows Lawrence et al. (1991) obtained a quartic for the dimensionless complex wavenumber, $c = 2(c^* - \bar{U})/\Delta U$, where the dimensional complex wave speed $c^* = c_r^* + i\, c_i^*$, and the mean velocity $\bar{U} = (U_1 + U_2)/2$. Contours of the growth rate, αc_i (where the non-dimensional wave number, $\alpha = kh$, $k = 2\pi/\lambda$, and λ is the wavelength), are plotted in Figure 4a as a function of α and the bulk Richardson number, $J = g'h/\Delta U^2$ (where $g' = g\,\Delta\rho/\rho_2$), for different values of the profile asymmetry, $\varepsilon = 2d/h$. Haigh (1995) has investigated the effects of boundaries at $z = \pm H/2$, where $z = z^*/h$ (Figures 4b, c, and d). The results for $H = 10$ are little different than those for $H = \infty$. For $\varepsilon > 0$ the stability diagram bifurcates into two limbs: the limb closer to the J axis represents positive waves ($c_r^+ > 0$) that protrude into the upper layer (where we adopt the convention that the upper layer is the faster moving layer); whereas, the limb closer to the α axis represents negative waves ($c_r^- < -\varepsilon$) that protrude into the lower layer. For a given J the maximum growth rate for positive waves is greater than for negative waves. Yonemitsu et al. (1996) have shown that viscosity strengthens this effect, since it preferentially dampens the higher wavenumber negative waves. Thus, linear stability analysis is consistent with experimental observations in mixing layer channels with $H = O(10)$, where for $\varepsilon > 0$ the flow is "one-sided" with the positive wave dominating. As H is decreased positive waves are stabilized more quickly by the presence of boundaries, since they occur at lower wave numbers than negative waves. For sufficiently small values of H the fastest growing wave is a negative wave (Haigh, 1995; Haigh and Lawrence, 1998).

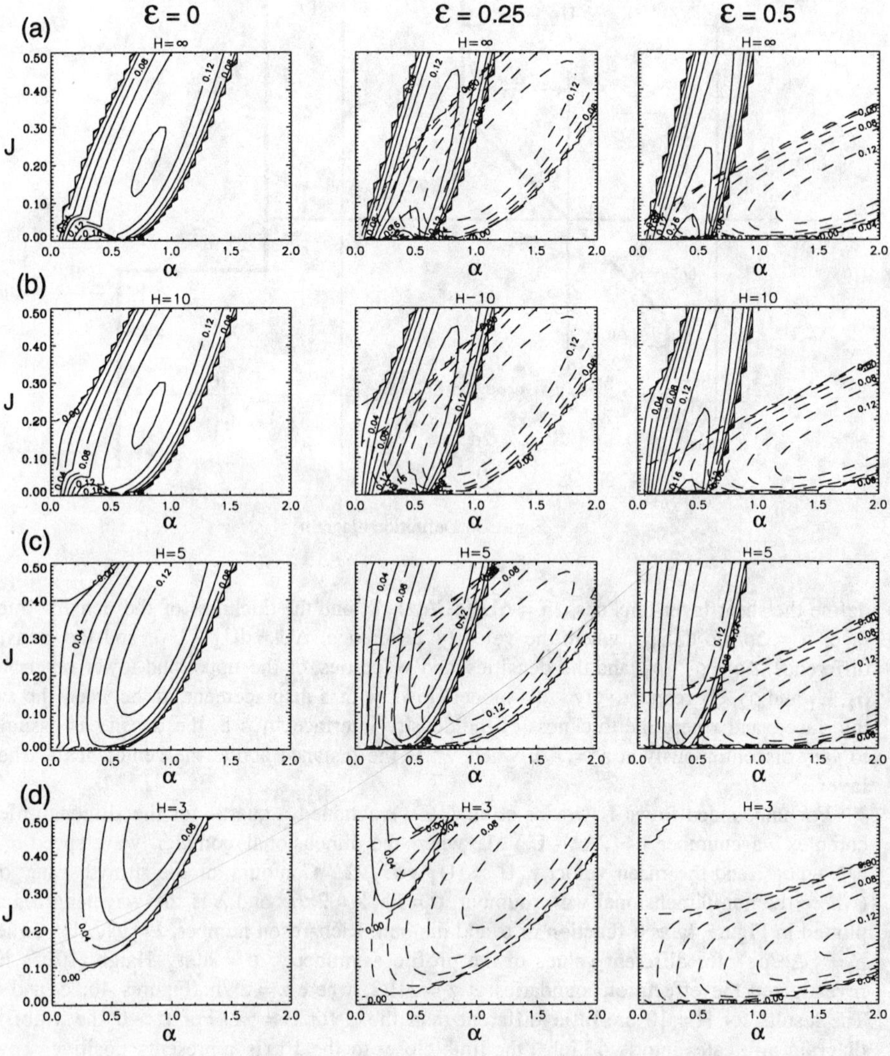

Figure 4. Linear stability diagrams for piecewise linear profiles with $\varepsilon = 0$, 0.25, and 0.5 for $H = \infty$, 10, 5, and 3. The contours are of constant growth rate αc_i spaced at an interval of 0.02 starting at 0.0. For $\varepsilon > 0$, solid contours correspond to positive waves and dashed contours to negative waves. Adapted from Haigh (1995).

Smoothed Profiles

The above results were extended by Haigh (1995) who assumed that the nondimensional mean velocity, U, and the non-dimensional mean density, ρ_a, can be described using hyperbolic tangent functions:

$$U(z) = \bar{U} + \delta U \tanh(z) \tag{1a}$$

$$\rho_a(z) = \bar{\rho} - \delta\rho \tanh R(z+\varepsilon) \tag{1b}$$

where $\delta U = \Delta U/2$, $\delta\rho = \Delta\rho/2$, $\rho = (\rho_1 + \rho_2)/2$, and $R = h/\eta$. Haigh also included viscosity and diffusivity yielding two additional parameters: the Reynolds number, $Re = \delta U\, h/\nu$; and the Prandtl number, $Pr = \nu/\kappa$, where ν is the kinematic viscosity, and κ is the diffusivity of the stratifying agent. The stability diagrams for $\varepsilon = 0$, 0.25 and 0.5, and $R = 3$ and 8, with $Re = 300$, and $H = 10$ are presented in Figure 5. We chose $H = 10$, since, as shown above, the boundaries only have a significant effect when $H < 10$. We chose $Pr = R^2$, since Smyth et al. (1988) have shown that diffusion drives the flow to a state in which $R = \sqrt{Pr}$. Results are presented for $R = 3$, since $Pr = 9$ for heat in water at 12 °C. We have also computed solutions for $R = 8$, since we would like to have data for as large a value of R as possible, but were limited by the computer resources needed to maintain resolution across the density interface.

Despite the smoothing of the profiles, and the inclusion of viscosity, diffusivity, and a finite thickness density interface, the stability diagrams for $R = 8$ have many similarities with the piecewise linear solutions (Figure 4). For $\varepsilon = 0$ there are, for large J, two modes with equal growth rates and equal phase speeds, but travelling in opposite directions to the mean flow; for small J the phase speed vanishes as in the Kelvin-Helmholtz instability.

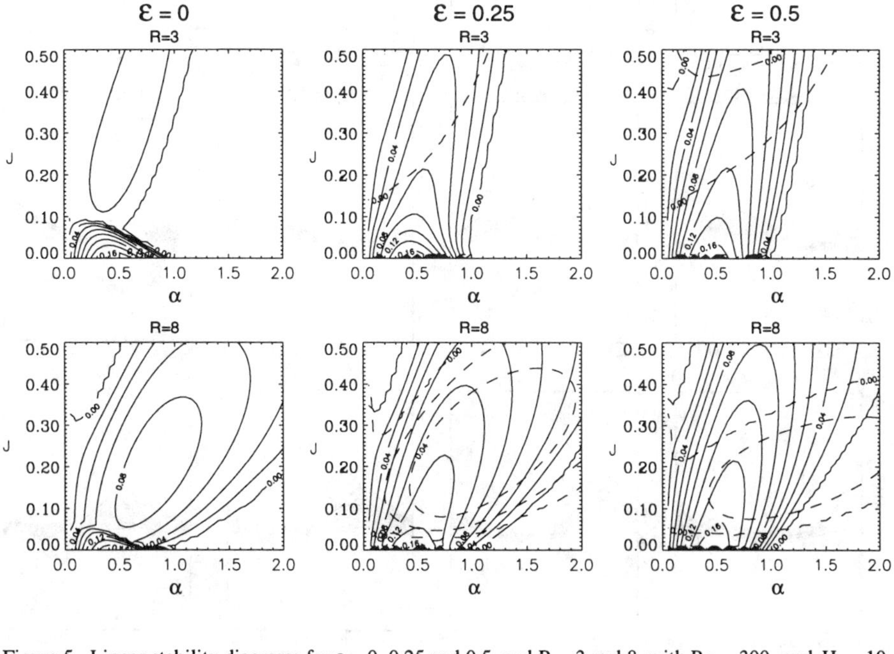

Figure 5. Linear stability diagrams for $\varepsilon = 0$, 0.25 and 0.5, and $R = 3$ and 8, with $Re = 300$, and $H = 10$. The contours are of constant growth rate αc_i spaced at interval of 0.02 starting at 0.0. For $\varepsilon > 0$, solid contours correspond to positive waves and dashed contours to negative waves. Adapted from Haigh (1995).

For $\varepsilon > 0$ the stability diagram bifurcates as before and the positive mode becomes increasingly more dominant as ε increases. The major difference is that the growth rates of both modes generally decrease with decreasing R. This continues until $R \approx 2.4$ when the flow becomes stable to both modes (Smyth and Peltier, 1989).

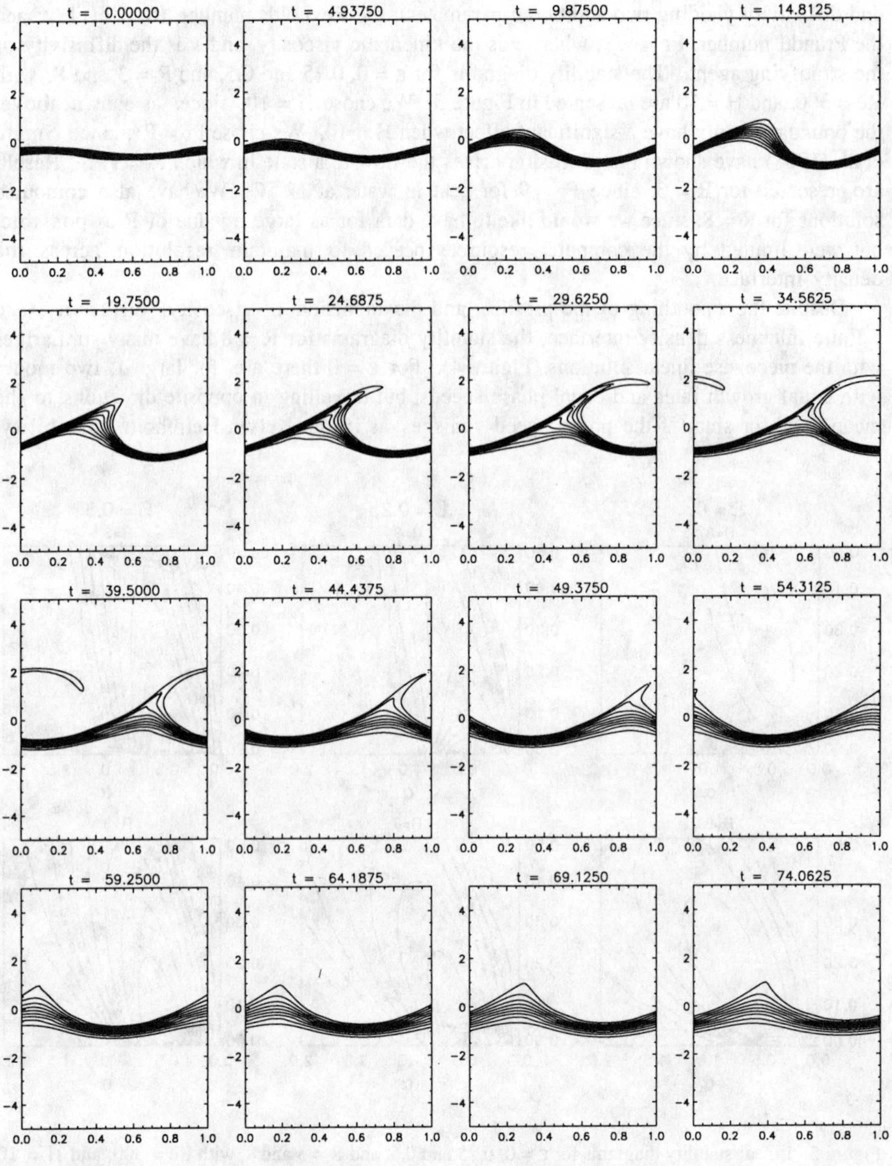

Figure 6. Density contours of results from numerical simulations with J = 0.06, Re = 25, R = 6, Pr = 36, ε = 0.5, with α = 0.45. Contours are from -0.72 to 0.72 in equal increments of 0.18, x has been scaled with respect to its period $2\pi/\alpha$. From Haigh (1995).

Non-Linear Simulations

We have numerically computed the non-linear evolution of small perturbations with wave numbers determined from linear stability analysis using parameters that match as closely as practicable flows that have been observed in the laboratory. For details of the numerical procedure see Haigh (1995). A simulation made for conditions similar to those of the laboratory experiment shown in Figure 1c is presented in Figure 6. In both cases the positive mode dominates and thin billows of fluid are drawn from these positive waves. The major difference is that the billows appear thinner in the laboratory experiments than in the numerical simulation. This is because the laboratory experiments were salt stratified with higher values of Pr and R than we were able to model numerically. Haigh (1995) has shown, however, that as R and Pr are increased the billows do become thinner, suggesting that if we were able to use higher values of these parameters the comparison would be improved.

Unfortunately, for the reasons discussed above, mixing layer experiments have not provided good realisations of Holmboe instabilities; however, in the laboratory experiments of Zhu and Lawrence (1996), performed to study exchange flow over an underwater sill (Figure 7), Holmboe instabilities unexpectedly developed. Kelvin-Helmholtz instabilities develop downstream (with respect to the denser layer) of the sill crest, where the shear is strong due to the thinning of the lower layer. Holmboe instabilities develop upstream, where the shear is not as strong. At first only waves cusping into the upper layer are observed. They appear to be generated near the sill crest by the disturbances caused by the Kelvin-Helmholtz instabilities and move upstream. Eventually negative waves, which move to the left and cusp into the lower layer, are observed. Numerical simulations (Haigh 1995) indicate that the negative waves are generated by the propagation of the positive waves.

A sequence of close-up photographs of the flow upstream of the obstacle is presented in figure 8a. In this portion of the flow $J \approx 0.3$, and the mean velocity is about 0.4 cms^{-1}. The speed of waves moving to the right is about 1.0 cms^{-1}, and the speed of waves moving to the left is about -0.2 cms^{-1}. Thus both waves are travelling at about the same speed with respect to the mean flow, satisfying the requirements for a Holmboe instability. Estimates of the wavelengths and wave speeds have been obtained from measurements of the variations in interface elevation by Zhu (1995), and are consistent with the predictions of linear theory. A numerical simulation under a similar set of parameters exhibits similar

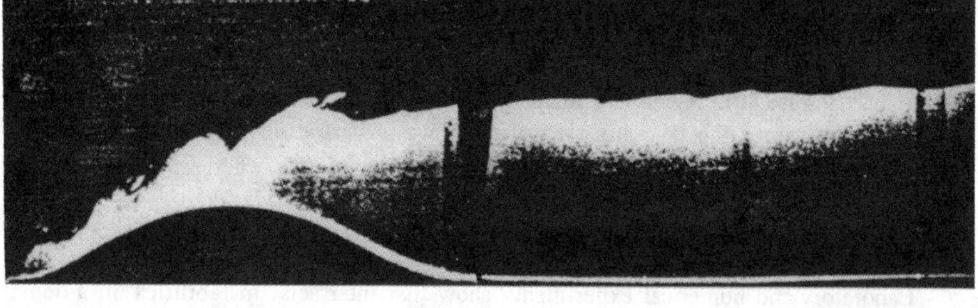

Figure 7. Photograph of exchange flow over a sill. The lower, dense layer is moving to the right. Holmboe instabilities form to the right of the sill. From Zhu (1996).

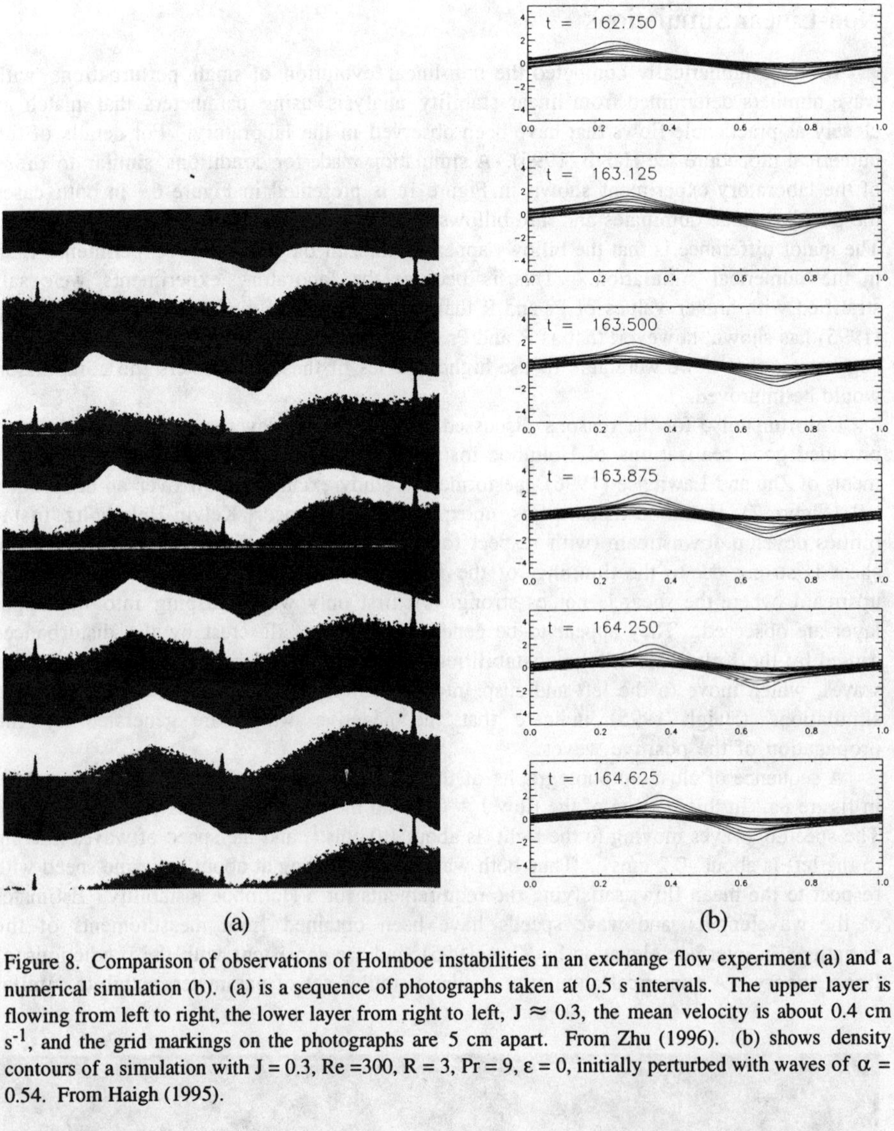

Figure 8. Comparison of observations of Holmboe instabilities in an exchange flow experiment (a) and a numerical simulation (b). (a) is a sequence of photographs taken at 0.5 s intervals. The upper layer is flowing from left to right, the lower layer from right to left, $J \approx 0.3$, the mean velocity is about 0.4 cm s^{-1}, and the grid markings on the photographs are 5 cm apart. From Zhu (1996). (b) shows density contours of a simulation with $J = 0.3$, $Re = 300$, $R = 3$, $Pr = 9$, $\varepsilon = 0$, initially perturbed with waves of $\alpha = 0.54$. From Haigh (1995).

behaviour (Figure 8b). Again, the major limitation of the numerical simulation is that R and Pr are low since the laboratory experiment was salt stratified, this may explain why the waves are not as steep in the numerical simulation.

Discussion and Conclusions

Laboratory and numerical experiments show that interfacial instabilities in a density stratified flow with a thin density interface are dependent on the profile asymmetry, e, as well as the bulk Richardson number J. The effects of these two parameters on the nature of

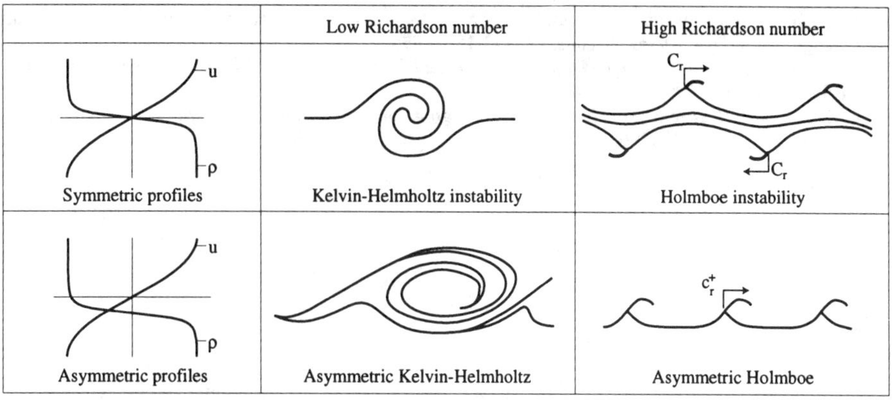

Figure 9. Sketches of the nature of interfacial instability as a function of the asymmetry of the density and velocity profiles and the Richardson number.

the instabilities are sketched in Figure 9. Given that the asymmetry affects the nature of the instability, then it must also affect the shear stresses and the amount of vertical mixing that occurs, and is worthy of further study. Pure Holmboe instabilities have not been observed in mixing layer facilities, but they have been observed in an exchange flow experiment. With increases in computing power the opportunities for comparison of numerical simulations with laboratory experiments will increase. At present we are limited to simulating flows with relatively low values of R, Pr, and Re.

Acknowledgments. The authors gratefully acknowledge the advice and assistance of Noboru Yonemitsu and Craig Stevens, funding from the Canadian Natural Sciences and Engineering Research Council, and support from the Centre for Water Research while preparing this paper.

References

Browand, F. K. and C. D. Winant, Laboratory observations of shear-layer instability in a stratified fluid, *Boundary Layer Met., 5*, 67-77, 1973.

Goldstein, S., On the stability of superposed streams of fluid of different densities, *Proc. Roy. Soc. A, 132*, 524-548, 1931.

Haigh, S. P., *Non-symmetric Holmboe waves*, PhD Thesis, Dept. Mathematics, Un. British Columbia, 1995.

Haigh, S. P. and G. A. Lawrence, Symmetric and non-symmetric Holmboe instabilities in an inviscid flow. Submitted to *Physics of Fluids.*, 1998.

Holmboe, J., On the behaviour of symmetric waves in stratified shear layers, *Geophys. Publ., 24*, 67-113, 1962.

Koop, C. G., Instability and turbulence in a stratified shear layer, *Tech. Rep. USCAE 134,* Dept. Aerospace Eng., Univ. Southern Calif., 1976

Lawrence, G. A., L. G. Redekopp, and F. K. Browand, The stability of a sheared density interface, *Phys. Fluids A, 3*, 2360-2370, 1991.

Pouliquen, O., J. M. Chomaz, and P. Huerre, Propagating Holmboe waves at the interface between two immiscible fluids, *J. Fluid Mech., 266*, 277-302, 1994.

Smyth, W. D. and W. R. Peltier, The transition between Kelvin-Helmholtz and Holmboe instability: An investigation of the overreflection hypothesis, *J. Atmos. Sci., 46*, 3698-3720, 1989.

Smyth, W. D., G. P. Klaassen and W. R. Peltier, Finite amplitude Holmboe waves, *Geophys. Astrophys. Fluid Dyn*, 52, 181-222, 1988.

Taylor, G. I., Effect of variation in density on the stability of superposed streams of fluid, *Proc. Roy. Soc. A, 132*, 499-523, 1931.

Yonemitsu, Y., G. E. Swaters, N. Rajaratnam, and G. A. Lawrence, Shear instabilities in arrested salt-wedge flows. *Dyn. Atmos. Oceans*, *24*, 173-182, 1996.

Zhu, Z., *Exchange flow through a channel with and underwater sill*, PhD Thesis, Dept. Civil Eng., Univ. British Columbia, 1996.

Zhu, Z. and G. A. Lawrence, Exchange flow through a channel with and underwater sill. *Dyn. Atmos. Oceans*, *24*, 153-161, 1996.

23

Estimation and Geography of Diapycnal Mixing in the Stratified Ocean

M. C. Gregg

Abstract

Microstructure measurements to determine diapycnal mixing in the ocean have been made for the past 25 years. Although covering only a negligible fraction of the ocean, these observations form consistent patterns showing extensive background areas of the Pacific where mixing is produced by internal waves close to the background state modeled by Garrett and Munk (1972). This diffusivity is approximately 5×10^{-6} m^2 s^{-1}, much less than obtained by one-dimensional models of large-scale properties. Minimum levels in the Atlantic appear to be several times larger, (1-2) $\times$ 10^{-5} m^2 s^{-1}. Mixing increases near fronts and in warm-core rings to 10^{-5} m^2 s^{-1} to about 10^{-4} m^2 s^{-1}. The most intense mixing, however, occurs where strong currents flow over irregular topography, e.g., in the Denmark and Gibraltar Straits. The observations responsible for these patterns are reviewed, along with the sensors and sampling involved. Also reviewed is recent theoretical work examining the Osborn and Cox (1972) model and how diffusivities from temperature microstructure can be related to the large-scale circulation.

1. Introduction

Ocean models are usually formulated in vertical and horizontal coordinates. Fluxes, however, are oriented more closely along or across density surfaces and thus expressed most naturally as isopycnal (along) or diapycnal (perpendicular) components. Isopycnal fluxes are usually much larger than diapycnal ones, e.g., turbulent diffusivities are usually considered to be about 10^8 times larger than diapycnal diffusivities. Because isopycnals have small slopes, e.g., 10^{-4} – 10^{-5}, diapycnal fluxes are very nearly vertical. Owing to their much larger magnitudes, however, isopycnal fluxes can have vertical components as large or larger than diapycnal fluxes.

For the past 25 years microstructure measurements have been made to estimate the magnitude of diapycnal fluxes, which result from small-scale mixing. The measurements are usually expressed as diapycnal diffusivities and have been collected in enough places to permit a first-order description of the levels and patterns of mixing. These observations are reviewed in section 5. The data summary is preceded by discussions of recent analyses of the methods of inferring diapycnal diffusivities from microstructure (section 2),

Physical Processes in Lakes and Oceans
Coastal and Estuarine Studies Volume 54, Pages 305-338

limitations of microstructure sensors (section 3), and the adequacy of microstructure sampling (section 4).

2. Diapycnal Fluxes and Microstructure Measurements

2.1. *The Distinction Between Diascalar and Advective Turbulent Fluxes*

Fluxes produced by molecular diffusion of a scalar are inherently perpendicular to isoscalar surfaces. Winters and D'Asaro (1996) term these fluxes diascalar. For heat and salt the diascalar fluxes are

$$\begin{aligned} J_{Q,\mathrm{dia}} &\equiv \int_{\mathrm{isotherm}} \rho c_p \kappa_T \nabla\theta \cdot d\hat{n} \qquad [\mathrm{W\ m^{-2}}], \\ J_{S,\mathrm{dia}} &\equiv \int_{\mathrm{isohaline}} \rho \kappa_S \nabla s \cdot d\hat{n} \qquad [\mathrm{kg_{ss}\ m^{-2}\ s^{-1}}] \end{aligned} \tag{1}$$

where ρ is density, s is salinity in concentration units, c_p is the specific heat of seawater at constant pressure, and κ_T and κ_S are the molecular diffusivities of heat and salt. The $d\hat{n}$ are unit vectors normal to isothermal and isohaline surfaces and directed toward increasing temperature or salinity.

Winters and D'Asaro demonstrate that when scalars obey a simple diffusion equation, e.g., $d\theta/dt = \kappa_T \nabla^2\theta$ where κ_T is the molecular diffusivity of heat, then as a direct mathematical consequence

$$J_{Q,\mathrm{dia}} = -\rho c_p \kappa_T \frac{\left\langle |\nabla\theta|^2 \right\rangle_{x_3^*}}{d\theta / dx_3^*} \qquad [\mathrm{W\ m^{-2}}] \tag{2}$$

The new vertical coordinate x_3* is computed by keying the original vertical coordinate, x_3, to θ as θ is resorted to increase monotonically with depth. Resorting temperature in this manner was begun by Thorpe (1977) to estimate overturning scales. Taking x_3* as the resorted x_3, x_3* has a unique value everywhere on isothermal surface θ, and $|\Delta x_3^*| \equiv |z_3^*(\theta + \Delta\theta) - z_3^*(\theta)|$ equals the volume of fluid between isotherms $\theta + \Delta\theta$ and θ divided by the cross-sectional area of the volume. The symbol $\langle\rangle x_3^*$ represents averaging on the isothermal surface defined by constant x_3^*. That is, the temperature gradient variance is computed after θ is sorted to be monotonic.

Equation 2 is similar to the Osborn and Cox (1972) expression

$$J_Q = -\rho c_p K_T \frac{\left\langle |\nabla\theta'|^2 \right\rangle}{\langle d\theta / dx_3 \rangle} \qquad \left[\mathrm{W\,m^{-2}}\right] \tag{3}$$

Osborn and Cox obtained (3) from a turbulence length-scale analysis of the θ'^2 balance equation by arguing that the balance reduces to local production and dissipation. The differences are that (3) contains θ', temperature fluctuations relative to a mean profile, instead of temperature, and that $\langle|\nabla\theta'|^2\rangle$ and $\langle d\theta/dx_3\rangle$ are calculated without sorting θ to be monotonic.

Obtaining mean gradients from monotonic profiles is already being done as an ad hoc means of estimating 'background' gradients in large overturns (Dillon and Park, 1987; Wesson and Gregg, 1994). Computing variances from sorted profiles, however, differs considerably from present procedures and raises practical issues about removing instrumental effects from the recorded data. At present, spectra of the recorded data are divided by transfer functions that represent the changes in amplitude and phase introduced by the probes and electronics. The corrected spectra are then scaled to represent gradients in wavenumber space and integrated to form χ_T. The Winters-D'Asaro procedure requires that these corrections be applied in the time domain so the corrected gradient records can be resorted using the same displacements applied to temperature records. This rearrangement must be done before computing $\langle \chi_T \rangle$.

The diascalar heat flux in (1) differs from the usual turbulent heat flux,

$$J_{Q,ad} \equiv \rho c_p \langle u_3' \theta' \rangle_{\langle\theta\rangle} \qquad \left[\mathrm{W\,m^{-2}}\right] \tag{4}$$

Because diffusion is not directly involved, Winters and D'Asaro (1996) refer to $\rho C_p \langle u_3' \theta' \rangle_{\langle\theta\rangle}$ as the advective heat flux through the $\langle\theta\rangle$ surface. Averaging is done on $\langle\theta\rangle$, and fluctuations θ' and u_3' are relative to averages on $\langle\theta\rangle$.

Figure 1 illustrates the difference between the diascalar and advective fluxes in a shear instability. The continuous line in the upper panels is an isotherm wrapped up in a billow, and the arrows show the instantaneous diffusive heat flux across the isotherm from hot to cold. At any instant the diascalar flux (1) is the total diffusive flux across the isotherm regardless of its orientation. After the billow collapses, the θ surface is no longer continuous but is multiply connected with pieces in many small blobs. The diascalar flux is always negative, representing the molecular transport of heat from warm to cool across the instantaneous θ surface. At any instant the advective flux is the average of $u_3'\theta'$ on the mean isothermal surface $\langle\theta\rangle$. It is the net advective transport through the surface and changes sign as the instability evolves. The advective flux is negative during rollup, owing to warm parcels moving down ($u_3' < 0$, $\theta' > 0$) and cool parcels moving up ($u_3' > 0$, $\theta' < 0$). After collapse the advective flux is positive as the parcels restratify.

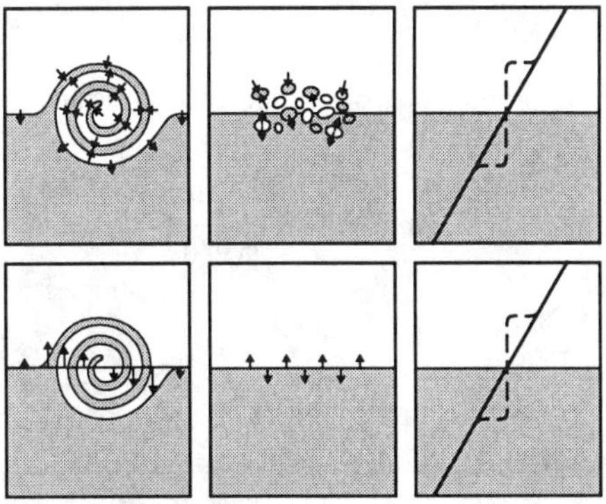

Figure 1. Schematic of the difference between diascalar and advective heat fluxes in a shear instability. The upper panels illustrate the diascalar flux across the instantaneous isothermal surface shown by the solid line. This surface is continuous when wrapped up in the billow (upper left panel), is multiply connected after collapse (upper middle panel), and continuous and horizontal after the event ends (upper right panel). Also in the right panel are the original (solid) and final (dashed) temperature profiles. Arrows in the upper panels show instantaneous diffusive heat fluxes across the instantaneous position of the isotherm. The lower panels illustrate the advective flux across the mean position of the same isotherm. The mean position is shown by the horizontal solid line in each panel, and arrows show the instantaneous velocities on it.

Advective and diascalar fluxes differ in principle and often in magnitude. The diascalar flux depends directly on diffusion and occurs mostly at the smallest scales of temperature variability. The advective flux depends directly on transport and occurs mostly at the largest turbulent scales. It, however, depends indirectly on diffusion, as seen by considering a shear instability in an ideal fluid. The diascalar flux would always be zero in the absence of diffusion. The instantaneous advective flux, however, would not be zero as long as turbulent motions persisted. However, the mean advective flux would be zero if the averaging interval were sufficiently long and included all of the restratification that returns the profile to its initial configuration. In diffusive fluids the restratification (positive) flux component is smaller than the rollup (negative) flux component to the degree that diffusion alters density and the water column does not return completely to its initial state.

In some cases, the volume being analyzed must be tailored to fit the situation. For instance, complex oceanic fronts often have several salt-stabilized temperature inversions that contain common isotherms, e.g., several intrusions may have temperatures of 8–9°C. Because resorting to produce a monotonic temperature profile scrambles the identities of the intrusions, applying (1) to the entire column gives the total diffusive heat flux across the isotherms involved, of interest for water mass formation, but does not provide heat fluxes through individual intrusions. Fluxes applicable to the dynamics of the intrusions are obtained by restricting the analysis volume, such as by placing boundaries at θ minima and maxima.

2.2. *Applying the Osborn-Cox Method to the General Circulation*

To relate the advective flux to temperature microstructure, Osborn and Cox (1972) analyzed the $\langle \theta'^2 \rangle$ balance using turbulent length scales and obtained

$$2\langle u_3'\theta' \rangle \left\langle \frac{\partial \theta}{\partial x_3} \right\rangle = -\chi_T \qquad \left[K^2\, s^{-1} \right] \tag{5}$$

Parameterizing turbulent production of temperature variance, $\langle \theta'^2 \rangle$, with an eddy coefficient,

$$\left\langle u_3'\theta' \right\rangle = -K_T \left\langle \frac{d\theta}{dx_3} \right\rangle \tag{6}$$

gives the eddy coefficient as

$$K_T = \frac{\chi_T}{2\left\langle \frac{d\theta}{dx_3} \right\rangle^2} \qquad \left[m^2\ s^{-1} \right] \tag{7}$$

The heat flux follows as

$$J_Q = -\rho c_p K_T \left\langle \frac{d\theta}{dx_3} \right\rangle = -\frac{\rho c_p \chi_T}{2\langle d\theta / dx_3 \rangle} \qquad \left[W\ m^{-2} \right] \tag{8}$$

where c_p is the specific heat at constant pressure. These expressions are applied to temperature microstructure data to obtain turbulent fluxes and diffusivities for comparison with large-scale, long-time balances, such as those in Abyssal Recipes by (Munk, 1966).

Davis (1994a) points out that applying fluxes estimated by the Osborn-Cox method to the general circulation is inconsistent because Osborn and Cox defined turbulence to include only microstructure, thereby assigning variability from internal waves, mesoscale circulation, and interannual variability to the mean field. Analyses such as Abyssal Recipes, on the other hand, are based on averages over several years and include microstructure, internal waves, mesoscale circulation, and interannual variability in the eddy field.

Davis (1994a) and Davis (1994b) follow the later approach in re-analyzing the $\langle \theta'^2 \rangle$ balance with eddy fluxes defined relative to average density surfaces. This approach avoids using turbulent length scales, consistent with observations subsequent to Osborn and Cox (1972) showing that there is no spectral gap separating turbulence from internal waves. The balance equation for temperature variance is

$$\left\langle \partial_t \rho \theta'^2 \right\rangle + \nabla \cdot \left[\langle m \rangle \left\langle \theta'^2 \right\rangle + \left\langle m' \theta'^2 \right\rangle \right] + 2 \left\langle m' \theta' \right\rangle \cdot \nabla \langle \theta \rangle \approx -\langle \rho \rangle \left\langle \chi_T \right\rangle \tag{9}$$

where $m \equiv \rho u$, and $\langle \cdot \rangle$ is an average in fixed spatial coordinates over a time interval appropriate to the general circulation, considered as a slowly evolving regime characteristic of a particular climate. The Osborn-Cox method is to neglect the time derivative and divergence terms on the left and assume production balances dissipation. The difference from Osborn and Cox (1972) is that now m' and θ' include fluctuations produced by internal waves and mesoscale eddies as well as by small-scale turbulence. Moreover, if the averaging interval is several years, slow interannual accumulations of θ'^2 must also be considered before neglecting the time derivative and divergence terms.

Davis (1994a) uses data compiled by Levitus (1982a) to compare rates of heating by diapycnal turbulence to temperature changes observed in the North Atlantic. Levitus (1982b) reports statistically significant changes of 0.1–0.5°C in averages over 5 years on potential density surfaces at intermediate depths. These changes correspond to time derivatives of $(0.2\text{–}1) \times 10^{-9}$ °C s^{-1} and are significantly larger than the temperature changes resulting from diapycnal turbulence equivalent to $K_T = 10^{-5}$ m^2 s^{-1} (Figure 2).

Consequently, diapycnal fluxes produced by $K_T = 10^{-5}$ m^2 s^{-1} would be nearly undetectable in a heat budget that neglected temporal variations. Moreover, inferences from large-scale budgets could easily obtain $K_T = 10^{-4}$ m^2 s^{-1} by applying a steady state model to profiles that are changing in time.

To compare diapycnal and lateral fluxes, Davis uses the commonly accepted lateral diffusivity of $K_H = 10^3$ m^2 s^{-1} and lateral gradients computed from the Levitus data to estimate heating rates from lateral flux divergences. Comparison with heating rates from diapycnal fluxes shows that lateral fluxes usually change temperature much faster than do divergences of vertical turbulent fluxes (Figure 2). Therefore, lateral fluxes must be known very accurately before diapycnal fluxes of 10^{-5} m^2 s^{-1} can be inferred as residuals from local balances.

Taking the same approach to the density field, Davis expresses the conservation equation for mean density as

$$\langle \partial_t \rho \rangle + \langle \mathbf{u} \rangle \cdot \nabla \langle \rho \rangle + \nabla \cdot \langle \rho' \mathbf{u}' \rangle = -\langle \rho \rangle \langle \nabla \cdot \mathbf{u} \rangle \tag{10}$$

Following the mean flow, changes result from divergence of the advective flux, $\nabla \cdot \langle u' \rho' \rangle$, and from the production term $\langle \nabla \cdot u \rangle$, which expresses the non-Boussinesq nature of the flow. Production is

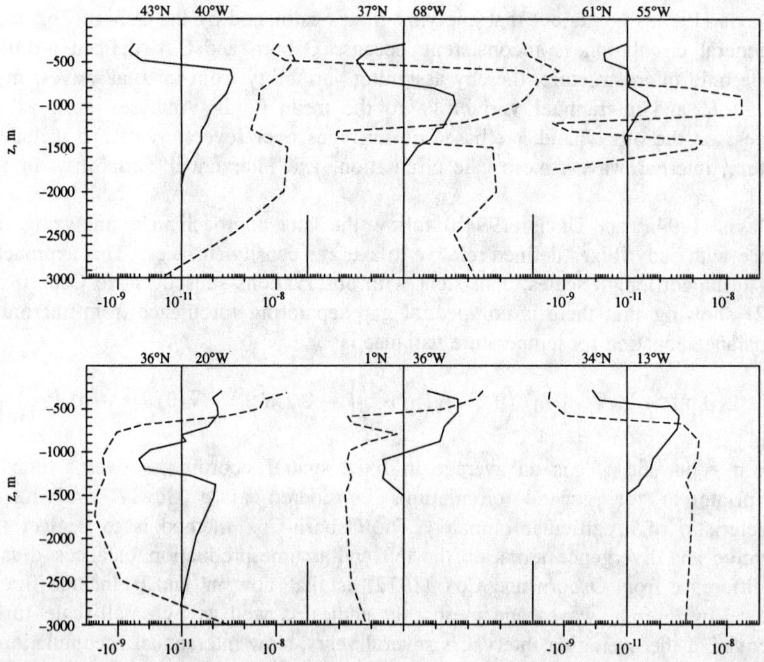

Figure 2. Comparisons of rates of heating rates (K/s) in the Atlantic (Davis, 1994a). Solid lines are estimated from vertical divergences of diapycnal turbulent heat fluxes for diapycnal diffusivities of $K_T = 10^{-5}$ m^2 s^{-1}. Dashed lines are estimated from horizontal divergences of lateral heat fluxes for isopycnal diffusivities of $K_H = 10^3$ m^2 s^{-1}.

$$-\langle \nabla \cdot u \rangle \approx \qquad - \underbrace{g\Gamma_p \langle w \rangle}_{\text{adiabatic compression}}$$

$$-\sum_{\phi=s,T} \left[\underbrace{\frac{\kappa_\phi \rho_\phi}{\rho} \left\langle \nabla^2 \phi \right\rangle}_{\text{diffusion}} + \underbrace{g \frac{\partial \Gamma_p}{\partial \phi} \langle w' \phi' \rangle}_{\text{thermobaricity}} + \underbrace{\frac{\kappa_T}{\rho} \frac{\partial^2 \rho}{\partial T \partial \phi} \langle \nabla \phi' \cdot \nabla T' \rangle}_{\text{cabelling}} \right] \tag{11}$$

where κ_T and κ_ϕ are the molecular diffusivities of heat and salt, Γ_p the adiabatic compressibility, and T and s subscripts indicate partial differentiation with respect to temperature and salt concentration. Compressibility changes are reversible, but the other terms represent irreversible changes. The first of these is molecular diffusion. The remaining terms result from nonlinearities in the equation of state. Changes in compressibility with temperature are referred to as thermobaricity (McDougall, 1987), and the last term represents the thermal and haline components of cabelling, in which mixing of waters of the same density produces water with greater density.

After neglecting molecular diffusion, which is a negligible density source, Davis (1994a) compares the magnitudes of the remaining irreversible terms with the density changes resulting from the vertical divergence of the diapycnal flux, $\partial\langle \rho' u_3' \rangle / \partial x_3$. These comparisons are expressed by the ratios

$$P_H = \frac{\text{production by lateral fluxes}}{\text{divergence of the diapycnal turbulent flux}}$$

$$P_V = \frac{\text{production by diapycnal fluxes}}{\text{divergence of the diapycnal turbulent flux}} \qquad (12)$$

$$P_\Gamma = \frac{\text{production by } \Gamma \text{ fluctuations}}{\text{divergence of the diapycnal turbulent flux}}$$

Each production term is estimated with horizontal and vertical eddy coefficients, and Davis assumed $K_H = 10^3$ m^2 s^{-1} and $K_V = 10^{-5}$ m^2 s^{-1}, giving a ratio of horizontal to vertical diffusivity of $K_H/K_V = 10^8$. Figure 3 displays these ratios for representative Atlantic sites.

Considering Figure 3 as well as the Pacific ratios shown by Davis, in no case is $P_V \gg 1$, demonstrating that density production by advective turbulent fluxes never dominates density changes resulting from vertical divergences of the diapycnal fluxes. P_Γ is generally even smaller and exceeds P_V only where neither is important. Lateral production, however, is often dominant; P_H is as large as 10 in the Gulf Stream recirculation and in the Antarctic Circumpolar Current. Owing to the importance of cabelling and the dominance of lateral fluxes in local balances of the mean field, Davis concludes that inferences of diffusivities are doubtful when they are based on one-dimensional balances.

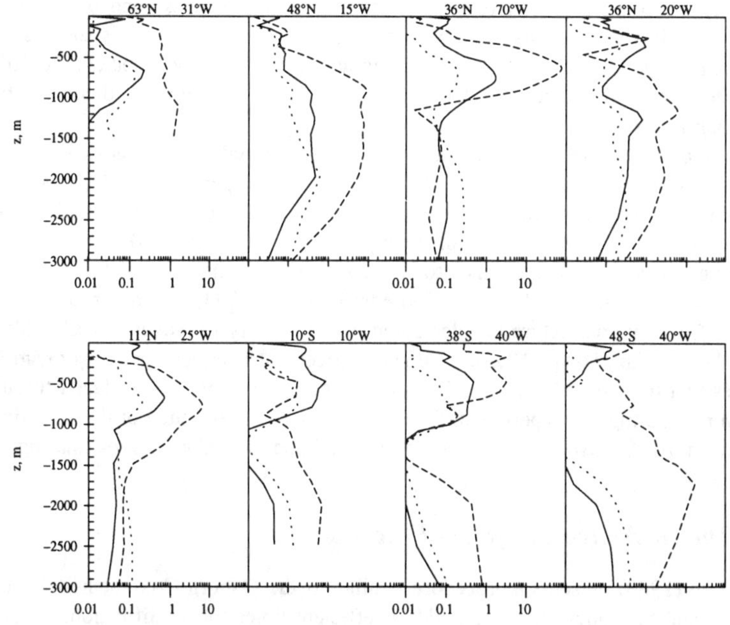

Figure 3. Rates of density change resulting from divergences of diapycnal fluxes divided by the rates of density change produced by density production accompanying the diapycnal fluxes estimated by Davis (1994a). Solid lines are for densification accompanying diapycnal fluxes (P_V) and do not depend on the eddy diffusivity. Long dashes are for densification accompanying isopycnal fluxes (P_H) and are computed for $K_H/K_V = 10^8$. Short dashes are for thermobaricity (P_Γ), which depends weakly on $K_V/K_H = 10^8$.

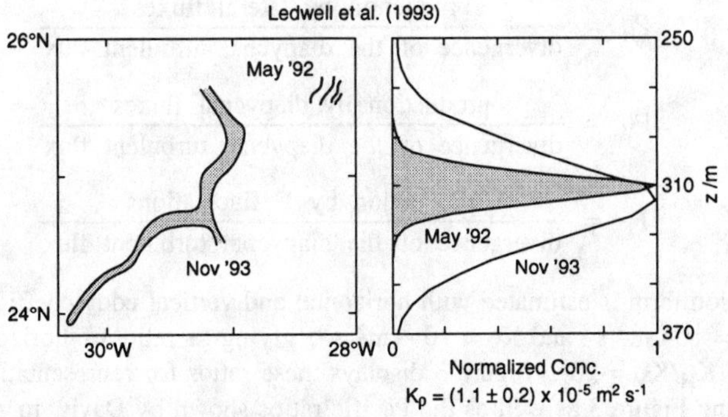

Figure 4. The plan view, left, compares the SF_6 injected in May 1992 with the continuous tracer cloud found in November 1993. (Figure adapted from Ledwell et al., 1993.) Plots on the right show the vertical thickening.

Davis (1994b) finds that using multi-year time averages to separate eddy and mean fields removes worries about the lack of a spectral gap arising when the fields are separated by spatial scales. In examining sources of error in the Osborn-Cox method, he finds that temporal changes of variance are usually not important except in upwelling areas. However, the dominance of lateral production at high latitudes, and other places where stratification is weak, results in significant overestimation of K_T, as is the case in frontal areas such as the California Current. Davis also finds that the turbulent flux of variance cannot be dismissed and concludes that the Osborn-Cox method is a theory to be verified rather than a deduction.

In response to uncertainty about diffusivities inferred from large-scale balances and from the Osborn-Cox method, the World Circulation Experiment (WOCE) sponsored the North Atlantic Tracer Release Experiment (NATRE). Streaks of an artificial tracer, SF_6, were injected into the thermocline of the eastern Atlantic (Figure 4).

Fitting a simple diffusion curve to the rate of thickening during the first five months gives $K_\rho = (1.1 \pm 0.2) \times 10^{-5}\ m^2\ s^{-1}$ (Ledwell et al., 1993). Similar diffusivities were obtained from microstructure profiles taken in the vicinity for several weeks (Toole et al., 1994; Sherman and Davis, 1995). The good agreement is either fortuitous or, more likely, indicates that the internal wave field was nearly steady in time and produced turbulence at a uniform rate during the experiment. This success has allayed much of the uneasiness about application of the Osborn-Cox method as an ad hoc procedure for estimating diapycnal fluxes.

2.3. *The Advective Buoyancy Flux*

Osborn (1980) presents a procedure similar to the Osborn-Cox method for estimating the diapycnal buoyancy flux as an eddy coefficient times the stratification,

$$J_B \equiv -\frac{g}{\rho} \langle \rho' u_3' \rangle = -K_\rho N^2 \qquad \left[\mathrm{W\ kg^{-1}} \right] \tag{13}$$

Again neglecting time derivatives and divergences, he reduces the turbulent kinetic equation to a balance between local production and dissipation. This balance involves three terms,

$$\langle u_1' u_3' \rangle \frac{\partial \langle u_1 \rangle}{\partial x_3} = J_B - \varepsilon \qquad \left[W\ kg^{-1}\right] \tag{14}$$

Using (13) and the flux Richardson number, $R_f = J_B / \langle u_1' u_3' \rangle \, \partial \langle u_1 \rangle / \partial x_3$, results in

$$K_\rho = \frac{R_f}{1 - R_f} \frac{\varepsilon}{N^2} \qquad \left[m^2 s^{-1}\right] \tag{15}$$

Arguing that $R_f \leq 0.15$ in stratified turbulence, Osborn considers the estimate an upper bound, resulting in

$$K_\rho \leq 0.2\varepsilon / N^2 \qquad \left[m^2 s^{-1}\right] \tag{16}$$

This expression is used for the K_ρ values in Tables 1–3, but must also be considered an ad hoc expression.

A different expression for K_ρ is obtained when double diffusion is the principal source of mixing, such as in thermohaline staircases. McDougall (1988) expresses the diapycnal diffusivities of heat and salt in these staircases as

$$K_{T,dd} = \frac{\gamma_f}{R_\rho} \frac{R_\rho - 1}{1 - \gamma_f} \frac{\varepsilon}{N^2}, \quad K_{S,dd} = \frac{R_\rho - 1}{1 - \gamma_f} \frac{\varepsilon}{N^2} \qquad [m^2 s^{-1}] \tag{17}$$

where $\gamma_f \equiv (\alpha/c_p) J_Q / \beta J_S$ is the buoyancy flux ratio for salt fingering, and $R_\rho \equiv (\alpha d\theta/dx_3)/(\beta ds/dx_3)$ is the ratio of thermal to salinity stratification. Using $\gamma_f = 0.7$ and $R_\rho = 1.6$ for the large salt fingering staircase east of Barbados gives $K_{T,dd} = 4.4 \times K_\rho$ and $K_{S,dd} = 10 \times K_\rho$ with K_ρ from (16).

Buoyancy fluxes resulting from double diffusion are positive, in contrast to the negative buoyancy fluxes produced by turbulence in stratified profiles. Combining the heat and salt fluxes for double diffusion in

$$J_B = \frac{g}{\rho}\left(\frac{\alpha}{c_p} J_Q - \beta J_S\right) \qquad \left[W\,kg^{-1}\right] \tag{18}$$

gives $J_{B,dd} = -5 \times K_\rho$ for the Barbados staircase.

3. Spatial Resolution and Signal Strength

Estimating K_T and K_ρ accurately is difficult, even in open ocean background areas, because the small-scale gradients contributing to χ_T and ε vary greatly in wavenumber content as well as in amplitude. Part of the variability is proportional to the mean stratification, $\langle N \rangle$, which decreases from 10^{-3} s^{-1} in the seasonal thermocline to 10^{-4} s^{-1} at abyssal depths. Mean temperature gradients vary somewhat more, 5×10^{-4} K $m^{-1} \leq (\partial\theta/\partial x_3) \leq 3 \times 10^{-2}$ K m^{-1}. To obtain K_ρ and K_T of 5×10^{-6} to 10^{-4} m^2 s^{-1} in profiles with these mean gradients, it is necessary to measure average dissipation rates of $\langle \varepsilon \rangle = 2.5 \times 10^{-11}$ to 5×10^{-8} W kg^{-1} and $\chi_T = 2.5 \times 10^{-12}$ to 2×10^{-7} K^2 s^{-1}.

The variances in ε and χ_T are obtained by integrating spectra of shear and temperature gradients over short sections, typically 0.5 to 5 m. Probability distributions of these estimates are highly skewed and much closer to lognormal than to normal probability densities (Gregg, 1975b; Osborn, 1978). If random variable x is lognormal, then ln x is

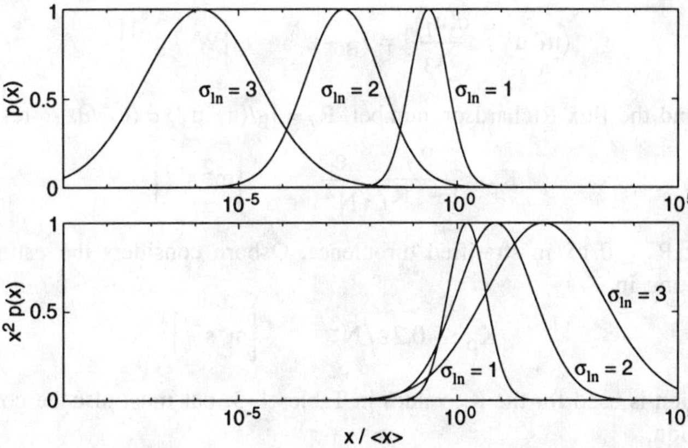

Figure 5. Lognormal probability density functions having the same mean but different widths, $\sigma_{ln} = 1$, 2, 3. Distributions in the upper panel appear normal because they are plotted on logarithmic axes. In the lower panel the same probability densities are plotted in area-preserving form. Comparing the two panels shows that as σ_{ln} increases: 1) most of the samples have values much smaller than the mean, and 2) those samples that contribute to the mean are increasingly larger than the mean.

normal and the width of the distribution can be characterized by σ_{ln}. For ε and χ_T, σ_{ln} is usually between 1 and 3, the range shown in Figure 5. The upper panel shows three lognormal probability densities having the same mean, $\langle x \rangle$, but differing widths, $\sigma_{ln} = 1$, 2, and 3. Modes of lognormal distributions, i.e., the most probable values, become much smaller than the means as σ_{ln} increases. For instance, when $\sigma_{ln} = 3$ the mode is about $1.5 \times 10^{-6} \langle x \rangle$ and only a tiny fraction of the samples exceeds the mean.

The tiny fraction larger than the mean, however, makes large contributions to $\langle x \rangle$. These contributions are shown in the lower panel of Figure 5, where the three lognormal probability densities are replotted in area-preserving form, characterized by equal contributions to $\langle x \rangle$ from equal areas under the curve. This property, achieved by plotting $x^2p(x)$ on a linear y-axis versus a logarithmic x-axis, is a consequence of $\langle x \rangle \equiv \int xp\, dx = \int x^2p\, d \ln x$. The curves for $\sigma_{ln} = 1$ show significant contributions to $\langle x \rangle$ from $10^{-1} \langle x \rangle \leq x \leq 10^{1} \langle x \rangle$. The greater width for $\sigma_{ln} = 3$ expands the range of contributions to $10^{-1} \langle x \rangle \leq x \leq 10^{5} \langle x \rangle$.

Because the lower bound for variance contributions remains near $0.1 \langle x \rangle$, the equivalent thresholds for detection are $\varepsilon = 2.5 \times 10^{-12}$ W kg^{-1} and $\chi_T = 2.5 \times 10^{-13}$ K^2 s^{-1}. Upper bounds of 10^5 times the mean are $\varepsilon = 5 \times 10^{-3}$ W kg^{-1} and $\chi_T = 2 \times 10^{-2}$ K^2 s^{-1}. Because ε determines the wavenumber bandwidth of turbulence, these estimates establish the requirements for spatial resolution.

3.1. Spatial Resolution

Dissipation rates are estimated most accurately by integrating spectra that fully resolve the temperature and velocity gradients. The area-preserving spectra in Figure 6 compare wavenumber bandwidths that must be resolved with dynamic response functions of the probes used most commonly. For a given ε, the principal contributions to ε come from a 40-fold range of wavenumbers, and those to χ_T from a 20-fold range.

Most ε estimates are made with airfoils (Osborn, 1974) mounted on profilers falling at 0.5 to 1 m s^{-1}. Oakey (1982) estimates the amplitude-squared response of airfoils as $H_{af}^2(f) = 1/(1 + (\lambda_c f/w)^2)$, where $\lambda_c = 0.02$ m, f is frequency in Hz, and w is the speed in m s^{-1}. Ninnis (1984) measured the response by comparing probe outputs with a laser Doppler velocimeter. His response is similar to Oakey's at scales less than the probe cutoff, which is similar to a sinc-squared function. Because we do not observe the signature of this cutoff in measurements of intense turbulence, we use Oakey's estimate. Superposition of Oakey's response on the variance-preserving shear spectra in Figure 6 reveals little attenuation for $\varepsilon < 10^{-9}$ W kg^{-1} but large loss of variance for $\varepsilon > 10^{-7}$ W kg^{-1}, which is the upper bound for resolving spectra with present airfoils and requires some correction for probe attenuation. In view of the uncertainty of the probe response, spectra are usually not corrected beyond 100 cpm. Consequently, samples with $\varepsilon > 10^{-7}$ W kg^{-1} must be estimated from low-wavenumber portions of their spectra by assuming that the shape is universal. After integrating to 100 cpm, the correction factor for missed variance is 1.055 for $\varepsilon = 10^{-6}$ and 3.85 for $\varepsilon = 10^{-3}$ W kg^{-1} (Wesson and Gregg, 1994).

Most χ_T measurements are made with thermistor beads coated with glass to insulate them electrically from seawater. Temperature fluctuations in the water change the temperature of the bead after diffusing through a boundary layer around the probe in the water and through the glass. Owing to the boundary layer, the response varies with speed, w. Gregg

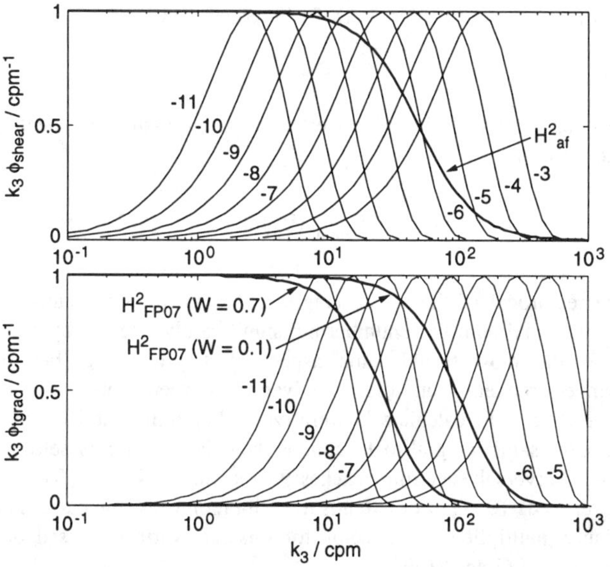

Figure 6. Variance-preserving universal turbulence spectra of shear (upper) and strain (lower) for the ε range in most of the open ocean overlaid. The spectra are overlaid with dynamic transfer functions of airfoil probes (upper) and FastTip thermistors (lower). Spectra are computed using the forms of Panchev and Kesich (1969) and Batchelor (1959) and are normalized by their maximum amplitudes. Each spectrum is labelled with $\log_{10}\varepsilon$. The spatial transfer function for the airfoil is estimated by Oakey (1982), and that for FP07 FastTip thermistors was measured by Gregg and Meagher (1980). The airfoil response is independent of speed, but thermistors are limited by diffusion and attenuate signals more strongly at faster speeds. The two thermistor transfer functions are for fall rates, w, of 0.1 m s^{-1} and 0.7 m s^{-1}.

and Meagher (1980) give the response of glass-rod thermistors as $H^2(f) = (1 + (2\pi f\tau)^2)^{-2}$, where the form is that of a double-pole filter and $\tau = \tau_1 w^{-03.2}$ is the response time. When expressed in terms of wavenumber, $H^2(k) = (1 + (2\pi k r_1\ w^{0.68})^2)^{-2}$. Consequently, at a given wavenumber, k, the attenuation increases with speed. Lueck et al. (1977) measure similar responses but find better fits to single-pole filters, in accord with their theoretical model. Subsequent to these response measurements, glass-rod thermistors have been supplanted by FastTips, a trademark of Thermometrics Inc., which have nominal responses of $\tau_1 = 0.005$ s. Probes vary significantly as a result of manufacturing imperfections, but response tests are too costly to permit testing individual units.

Area-preserving spectra of temperature gradients are narrower than those of shear and have peak wavenumbers nearly 3 times larger (Figure 6, lower panel). Because the temperature gradient variance occurs at higher wavenumbers and the signals are more attenuated, χ_T is resolved more poorly than ε. Cox's original Microstructure Recorder (MSR) fell at 0.05–0.1 m s^{-1} to optimize the measurement of χ_T. At that speed, FastTip thermistors can resolve temperature gradients produced by $\varepsilon < 10^{-9}$ W kg^{-1} with little attenuation, but thermal gradients are poorly resolved where ε is larger. To optimize ε measurements and to increase the rate of collecting samples, most recent microstructure profiling is done at speeds of 0.5–1 m s^{-1}. At these speeds, χ_T is so poorly resolved that it is usually not reported, even though it is measured. The only recent temperature microstructure measurements at comparable speeds are those by the Long-Term Autonomous Microstructure Profiler (LAMP) (Sherman and Davis, 1995).

3.2. *Signal Strength*

Shear measured from airfoils is converted to ε by assuming spectral isotropy and integrating the spectra,

$$\varepsilon = 7.5\nu \int_{k_0}^{k_c} \Phi_{shear}(k_3)dk_3 \qquad [\mathrm{W\,kg^{-1}}] \qquad (19)$$

where k_0 is the reciprocal of the record length and k_c is the upper cutoff chosen for each spectrum to be the maximum wavenumber not significantly affected by noise or vibration. Choosing k_c is the most troublesome aspect of ε processing because sensors are sometimes changed frequently and the noise varies between probes and within individual records. Our practice is to calculate k_c iteratively, beginning at 10 cpm. The universal spectrum having the same integral to 10 cpm as the data spectrum is selected. The next k_c is set where that universal spectrum reaches 90% of its variance. Then these steps are repeated. If k_c exceeds 100 cpm or extends to a vibration peak, the cutoff is reduced and the measured variance multiplied by a factor for missing variance based on the universal spectrum (Wesson and Gregg, 1994).

Figure 7 shows examples of cutoffs for two averaged spectra obtained with the MSP, which falls at 0.35 m s^{-1} and is usually free of vibrations in the variance band. The spectrum with $\varepsilon = 4.4 \times 10^{-10}$ W kg^{-1} is well resolved, and k_c captures most of the variance. Because the spectrum with $\varepsilon = 3.5 \times 10^{-11}$ W kg^{-1} does not follow the universal shape, we set 10^{-10} W kg^{-1} as the noise level, although individual estimates go below 10^{-11} W kg^{-1}. Owing to the very high impedance of airfoils, $10^{13}\ \Omega$, measuring noise spectra in laboratories has not been very successful, but rising f^{-1} noise at low frequencies is a likely contributor.

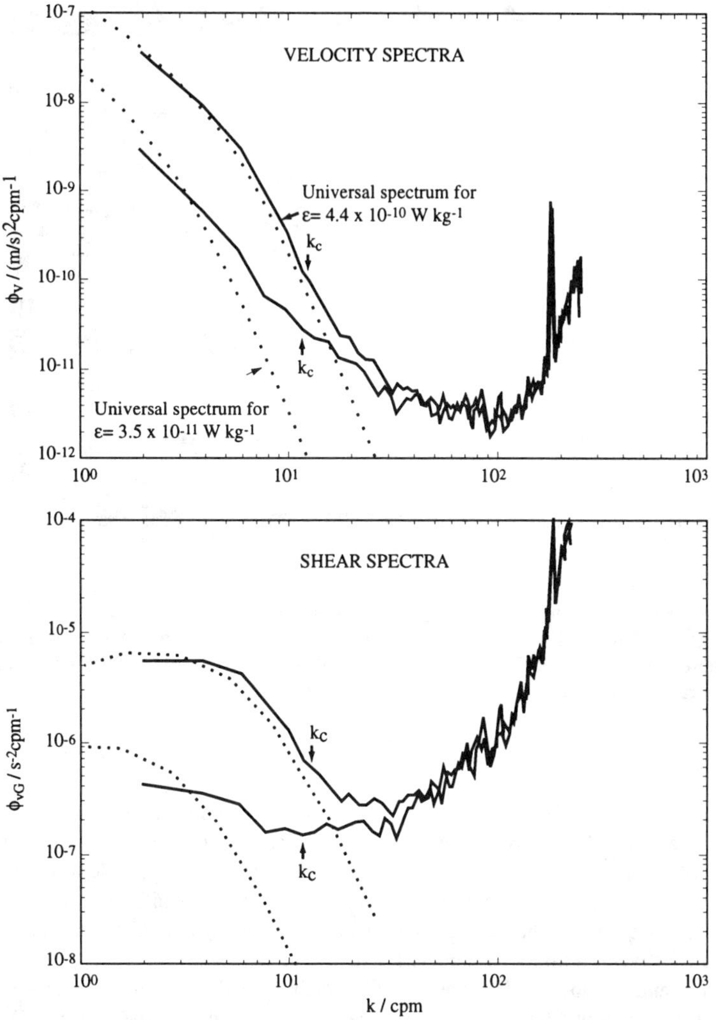

Figure 7. Velocity and shear spectra from the MSR during PATCHEX compared with universal turbulent shapes for the same ε (Gregg and Sanford, 1988). Each spectrum is an average of 20 spectra computed over 0.5 m. k_c is the upper bound used in integrating the spectrum.

Understanding the low-frequency noise of airfoils is central to accurate estimation of K_ρ below 10 MPa, where most dissipation rates are less than 10^{-10} W kg^{-1}. Figure 8 shows how the fraction of samples with $\varepsilon < 10^{-10}$ W kg^{-1} increases with pressure when the internal wave field is in its background state modeled by Garrett and Munk (1972, 1975); Munk (1981). At 9.5 MPa $\langle N^2 \rangle = 3 \times 10^{-3}$ s^{-2}, $\langle \varepsilon \rangle = 1.7 \times 10^{-10}$ W kg^{-1}, and 80% of the samples are less than 10^{-10} W kg^{-1}. Even for lognormal distributions, this large a fraction of noisy samples affects the mean. For instance, setting to zero all samples less than 10^{-10} reduces $\langle \varepsilon \rangle$ at 9.5 MPa by half, and the change would be much greater in the abyss.

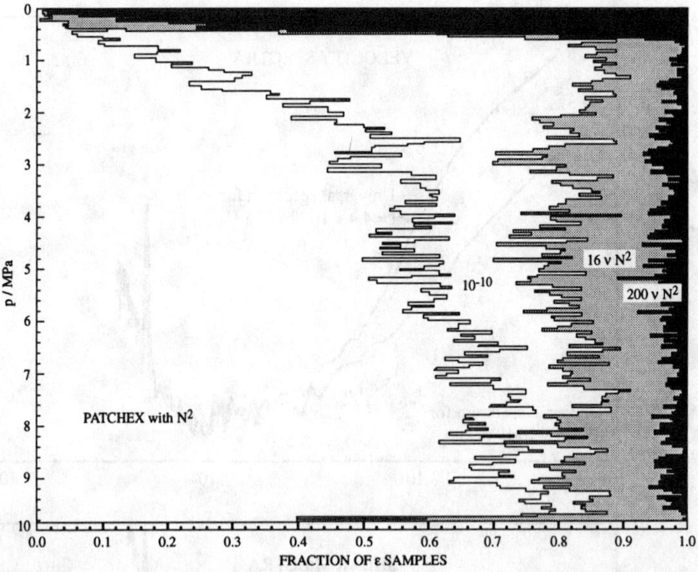

Figure 8. Fractions of 0.5 m ε's in activity classes when internal waves are at the Garrett and Munk level (Gregg and Sanford, 1988). The activity classes represent the approximate noise level, 10^{-10} W kg^{-1}, the threshold for a net buoyancy flux, $16\nu N^2$ (Rohr et al., 1987), and $200\nu N^2$ for isotropy in the dissipation range (Gargett et al., 1984).

4. Sampling

Statistical confidence limits are computed from probability distributions of ensembles of independent samples. Some of the ensembles whose means are summarized in Tables 1–3 have not been tested for independence and probably are not independent. Their distributions, however, are close to lognormal and sometimes pass significance tests for lognormality. Consequently, the nature and origins of lognormal distributions have been examined by several authors as models for turbulent samples.

Gurvich and Yaglom (1993) predict lognormality for dissipation rates in steady, fully-developed, homogenous turbulence as a consequence of applying the theory of breakage to the turbulent cascade. Yamazaki and Lueck (1990) review breakage theory and examine the probability distributions of point samples of ε, which they obtain by filtering raw data to correct for instrumental effects. The point samples are not lognormal, but short vertical averages of samples within turbulent patches rapidly approach lognormality.

Without assuming a turbulent cascade, Davis (1996) demonstrates that $|\nabla\theta|$ is likely to be locally lognormal in the thermocline. Starting from the diffusion and hydrodynamic equations, Davis obtains

$$\ln|\nabla\theta| = -\int \frac{\partial u_\theta}{\partial x_\theta}\, dt + \int\left(\frac{\kappa}{|\nabla\theta|}\,\frac{\partial \nabla^2\theta}{\partial x_\theta}\right) dt \qquad (20)$$

where the coordinate system is always aligned with the local $\nabla\theta$, and u_θ and x_θ are parallel to the temperature gradient. As a consequence of (20), $\ln|\nabla\theta|$ of a fluid parcel is the sum of the cumulative rate of strain and a diffusion term. Davis argues that ln $|\nabla\theta|$ will tend to be lognormal because the strain term usually dominates, and strain is the sum of contributions from many different spatial scales and times. These contributions are not fully correlated, and many strains result from internal waves which tend to be normally distributed. For even a relatively small number of independent strain components, applying the Central Limit Theorem shows that local distributions of ln $|\nabla\theta|$ will be normally distributed.

Strict lognormality, however, will not occur when statistics of the strain field vary across the data ensemble. Moreover, Davis points out that most observed χ_T distributions are based on $|\partial\theta/\partial x_3|$, which cannot be lognormal if $|\nabla\theta|$ is lognormal and the temperature field is isotropic, as usually assumed. Therefore, strict lognormality of χ_T seems unlikely, but there are strong tendencies for the distributions to be nearly lognormal.

A tendency to lognormality for ε cannot be shown simply from the hydrodynamic equations (Davis, 1996), but in most parts of the thermocline $\langle \varepsilon \rangle$ is proportional to moments of internal wave shear and consequently highly skewed (Gregg et al., 1993b). Observations confirm that internal wave shears over vertical scales of 10–15 m are independent and normally distributed, as assumed by Garrett and Munk (1972). Moments of normal distributions become increasingly skewed with the power of the moment. Theoretical calculations by McComas and Muller (1981) and Henyey et al. (1986) find that wave-wave interactions in a Garrett and Munk internal wave field produce a net flux of energy toward dissipation scales. The magnitude of this flux is proportional to E^2, were E is the energy level of the field. For internal waves close to the Garrett and Munk spectrum, E^2 is equivalent to Shear^4, which has a highly skewed distribution with its mode less than its mean. Scalings of dissipation rates confirm that $\langle \varepsilon \rangle \propto \text{Shear}^4$ (Gregg, 1989; Polzin et al., 1995), demonstrating that the source function for dissipation in the thermocline is strongly skewed and closer to lognormal than normal.

Ensembles of raw ε estimates are computed over distances of 0.5–5 m and do not contain independent samples, at least when the ensembles are obtained in the thermocline. Averaging over 10 m is required to obtain independent samples when internal waves are at the Garrett and Munk level (PATCHEX) and over 15 m when the internal wave shear spectrum is several times larger (PATCHEX north) (Gregg et al., 1993a). These averaging scales correspond to the scales at which the shear spectrum rolls off from being flat, k_3^0, to sloping downward as k_3^{-1}. Correlation lengths of 0.5 m samples of χ_T are much shorter; Gregg (1975a) obtained 1.1 m and speculated that this was a consequence of the short correlation lengths of $d\theta/dx_3$.

Although strictly applicable only when the distributions are demonstrated to pass significance tests, lognormal distributions are useful for approximating confidence limits and for estimating sampling requirements. Figure 9 shows 95% confidence limits for $\sigma_{\ln}$ = 0.5 to 3. Ensembles of independent ε samples are unlikely to have $\sigma_{\ln}$ as large as 3; independent samples from PATCHEX (10 m) and PATCHEX north (15 m) have $\sigma_{\ln} = 1.03$ and 1.30. Davis (1996) considers $\sigma_{\ln}$ for χ_T and obtains $\sigma_{\ln} = 2$ theoretically when $\nabla\theta$ has constant magnitude but random direction. This is the theoretical lower bound for χ_T distributions, and Davis notes that observed values are not much smaller. Nor are they much larger.

The lognormal confidence limits are multipliers of the expected value. For example, when $\sigma_{\ln} = 2$, an ensemble of 70 independent samples has 95% confidence limits of 0.5 and 3 times the expected value. When internal waves match the Garrett and Munk spectrum, limits for ε are considerably tighter than this and thus require less sampling. With $\sigma_{ln} = 1.1$ only 10 independent samples are needed for a lower limit of 0.5, and 30 samples give an upper limit of 2. Obtaining these limits requires only 300 m of sampling. Tightening the limits, however, drastically increases the required sampling owing to the asymptotic behavior of the limits. For $\sigma_{ln} = 1.5$ confidence limits of ±10% require about 2,000 independent samples, or 3 km of data.

5. Observations

To compare K_T and K_ρ in the stratified ocean, four regimes are considered: background areas, mesoscale features, on the equator, and near topography. Background areas are those where mixing is believed to result from breaking internal waves unaffected by interactions with mesoscale features or topography. Mesoscale features include the Gulf Stream, rings, and fronts. The equator is considered separately because some of its unique features have been sampled more extensively than the other regimes.

Figure 10 shows where the observations were made. To facilitate comparisons, most observations are reduced to single numbers in Tables 1–4. Doing so is justified for many profiles because diapycnal diffusivities are observed to be constant with depth when internal wave intensities increase with the gradual decrease in stratification (Gregg and Sanford, 1988; Gregg, 1989; Polzin et al., 1995).

5.1. Background Areas

Figure 11 shows the continuous profiles available to me from background areas, and Table 1 summarizes eddy coefficients from these profiles as well as from other locations and earlier times.

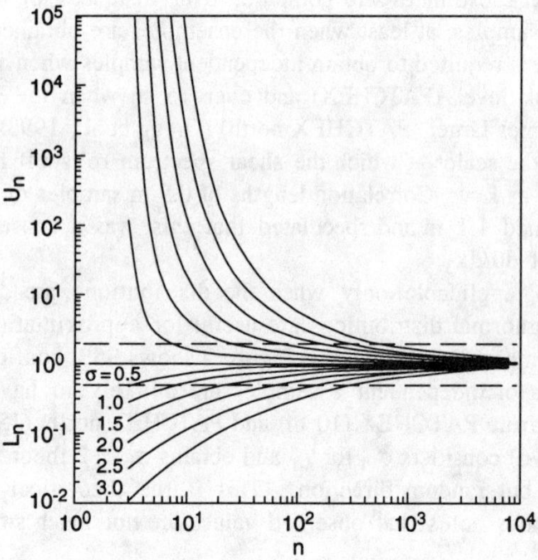

Figure 9. Upper (U_n) and lower (L_n) 95% confidence limits for averages of lognormal distributions (Gregg et al., 1993a). The confidence limits depend on the number of independent samples, n, and on σ_{ln}, the standard deviation of ln x.

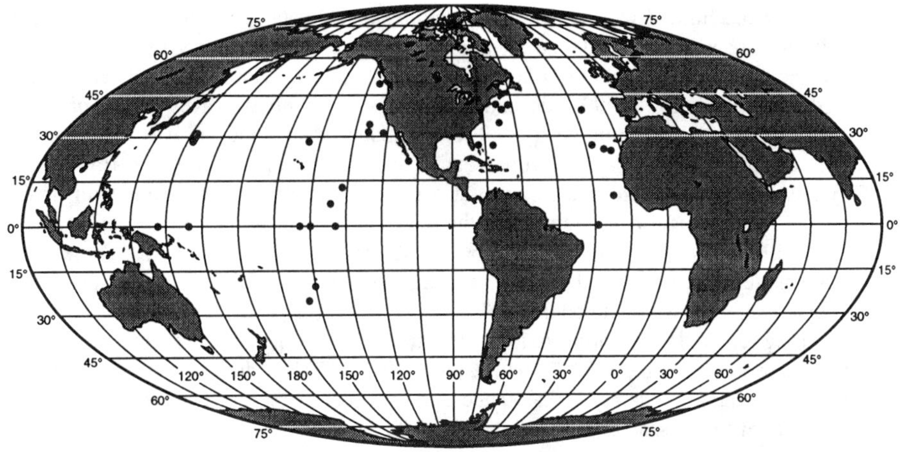

Figure 10. Locations of K_T and K_ρ estimates.

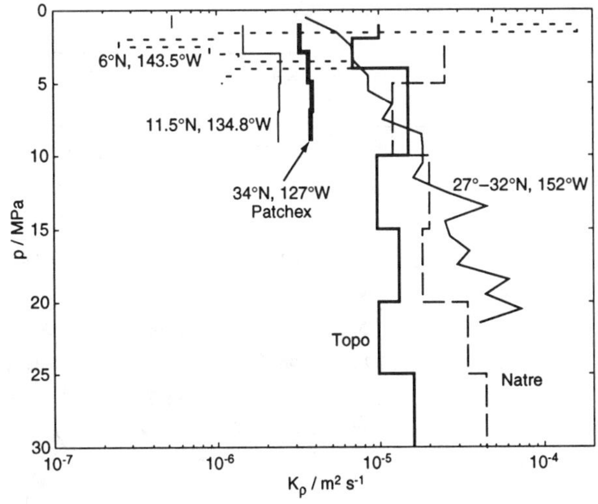

Figure 11. Profiles of K_ρ where mixing is not known to be influenced by topography or mesoscale features. The thick line was taken when internal waves were very close to the background state modelled by Garrett and Munk (1972).

Observations to test Abyssal Recipes began in September 1971 when Chip Cox and I took his Microstructure Recorder (MSR) to the subtropical gyre north of Hawaii. Munk estimated $K_T \approx 10^{-4}$ m^2 s^{-1} by assuming that the main thermocline between 10 and 40 MPa (1 to 4 km) is maintained by the downward turbulent heat flux balancing the flux of upwelling cold bottom water. Six MSR profiles were obtained between 2.2 and 22 MPa. Each profile was limited to pressure intervals of 1–2 MPa by the slow fall rate, w = 0.08 m

s^{-1}, and by the limited capacity of frequency-modulated data recording. Only temperature microstructure was measured, and these profiles averaged $K_T = 2 \times 10^{-7}$ m^2 s^{-1}, less than twice the molecular diffusivity of heat (Gregg et al., 1973). The records totaled 0.9 km in length, corresponding to 90 independent samples for internal waves at the background state. Conservatively assuming $\sigma_{ln} = 3$ for the independent samples gives 95% confidence limits of approximately (0.3,8) times the mean. The upper bound is one fiftieth of the *Abyssal Recipes* estimate and demonstrates that mixing was much weaker than predicted by Munk. Subsequent measurements at the same site found $K_T = 1.4 \times 10^{-6}$ m^2 s^{-1} in June 1973 and $K_T = 6.4 \times 10^{-6}$ m^2 s^{-1} in February-March 1974 (Gregg, 1977). These are also much smaller than 10^{-4} m^2 s^{-1}, but their differences indicate significant variability due to a seasonal cycle or mesoscale eddies.

During June 1972, observations at similar subtropical sites in the southern hemisphere averaged $K_T = 6.1 \times 10^{-6}$ (Gregg, 1977). Taken during southern winter, this average is nearly the same as that found during winter at the northern site. One of the profiles went to abyssal depths, 35.0–36.5 MPa, and found $K_T = 4.2 \times 10^{-6}$ m^2 s^{-1}, not significantly different than in the main thermocline.

TABLE 1. Average diapycnal diffusivities in background areas. Latitudes are positive in the northern hemisphere, and longitudes in the western hemisphere. L is the total length of data used for the estimate. a) K_ρ increased continuously with depth. These averages are over 0.4–1.0 and 1.0–1.4 km. b) Larger and smaller values are above and below 1.5 MPa.

Site	Lat/Lon	Date	K_T m^2 s^{-1}	K_ρ m^2 s^{-1}	p MPa	L km	Source
Sub-trop. N.Pac	28/155	8/71	2e-7		2.2-21.5	0.9	GCH73
Sub-trop. S.Pac	–19/–25,155	7/72	6.1e-6		1.0-36.4	0.7	G77
Sub-trop. N.Pac	28/155	6/73	1.4e-6		3.5-11	1.3	G77
Sub-trop. N.Pac	28/155	2/74	6.4e-6		7.5-12.0	1.6	G77
Trop. N.Atl.	9/22.6	8/74	5.3e-7		0.3-1.2	12.0	E080
near Azores	37/26	3/75		2e-4	2.0-7.0	1.5	O78
Sargasso Sea	35/66.5	10/75	5.9e-6		1.4-12.2	0.8	GS80
Sargasso Sea	35/66.5	10/75		3e-5	3.0-7.0	2.8	O80
near Bermuda	33/64	10/75	9.6e-6		8-11	1.0	GS80
near Bermuda	33/64	10/75		3e-5	1-8	2.1	O80
Vancouver Is.	49/128	5/80		(0.7–1.1)e-5	1-11	13	L83
Kuroshio Ext.	27–32/–152	5/82		(2.1,4.6)e-5[a]	4.0-10.0	3.0	MO86
Calif. Current	31/121	10/82	3.2e-6	3.5e-6	1.4-2.0	5.8	GDSL86
Calif. Current	34/127	10/86		3.7e-6	1.0-9.0	22.4	GS88
Trop. N.Pac.	11.5/134.8	4/87		2.2e-6	1.0-9.0	4.0	GWSP93
Trop. N.Pac	6/143.5	4/87		(1.0–0.018)e-4[b]	0.5-5.0	2.0	GWSP93
Calif. Current	32.5/127.7	3/91	2.1e-5	1.8e-5	25.0-30.0	2.5	TPS94
Sub-trop. N.Atl.	24.5/20.3	3/92		1.1e-5	30.0-39.0	2.7	TPS94
Sub-trop. N.Atl.	25.5/23.9	3/92		2.6e-5	2.5-30.0	25.0	PTS95
Sub-trop. N.Atl.	26/28	5/92	1.4e-5		2.0-3.5	11.4	SD95

In August 1974 Elliott and Oakey dropped their loosely tethered profiler Octuprobe twice a day for a month while at 9°N during GATE. This series plus several sequences of rapidly repeated profiles produced a much larger data set than previously acquired at one site. They found very weak mixing in the seasonal thermocline, $K_T = 5.3 \times 10^{-7}$ m^2 s^{-1} between 30 m and 120 m (Elliott and Oakey, 1980). Considering that estimates are independent only when averaged over 10 m, their ensemble contained about 1200 independent samples and had 95% confidence limits close to (09,1.1).

Osborn (1978) deployed his profiler Camel to obtain the first ε profiles in the open ocean while working in the Atlantic near the Azores during March 1975. When he reported K_ρ values for these data, the profiles 80 km from the nearest island were considered free of topographic influence (Osborn, 1980) and averaged $K_\rho = 2 \times 10^{-4}$ m^2 s^{-1} between 2 and 7 MPa. Based on simultaneous measurements of ε and finescale shear, levels this high require internal wave energy levels about 7 times the Garrett and Munk background. Winds reached 28 m s^{-1} two days before the start of the measurements, so these levels may have resulted from a storm, or possibly the site was in an island wake.

During the Fine and Microstructure Experiment (FAME) in October 1975, Camel and MSR were dropped close to each other and to Sanford's Electromagnetic Velocity Profiler (EMVP) at three sites in the north Atlantic. Gargett and Osborn (1981) describe the Camel data, and Osborn (1980) reports their K_ρ values, which averaged 3×10^{-5} m^2 s^{-1} in the Sargasso Sea, significantly higher than $K_T = 5.9 \times 10^{-6}$ m^2 s^{-1} obtained by Gregg and Sanford (1980). A set of profiles taken 48 km from Bermuda and considered beyond the island's influence had a smaller discrepancy: $K_T = 9.6 \times 10^{-6}$ m^2 s^{-1} and $K_\rho = 3 \times 10^{-5}$ m^2 s^{-1}. Both Atlantic K_T values are close to the wintertime observations north of Hawaii. The discrepancy between them and K_ρ could not be explained because temperature and velocity were measured on different vehicles and could not be compared directly. Osborn, however, notes that K_ρ in the main thermocline would only be 1/3 as large if averaging were done after setting noisy ε estimates to zero.

In May 1980 a line of Camel profiles 90 km west of Vancouver Island in the North Pacific averaged $K_\rho = (0.7\text{–}1.1) \times 10^{-5}$ m^2 s^{-1} between 1 and 11 MPa (Lueck et al., 1983). These are somewhat larger than the K_T values from mid-gyre in the Pacific, but they are significantly smaller than K_ρ in the Atlantic. There was no evidence of the observations being directly affected by topography. Camel observed much larger levels in May 1982 when it was dropped along 152°W in the Kuroshio Extension (Moum and Osborn, 1986): K_ρ increased from 4×10^{-6} m^2 s^{-1} at shallow depths to 6×10^{-5} m^2 s^{-1} at 22 MPa (Figure 11). The line of profiles passed through a cold-core ring, but the same K_ρ increase with decreasing stratification is shown by the average of 11 profiles south of the ring. This is the only report of a systematic increase of K_ρ with decreasing stratification and caused Moum and Osborn to consider the effects of instrumental noise, which Moum and Lueck (1985) report as $\varepsilon = 3 \times 10^{-10}$ W kg^{-1} for Camel. Noise does not appear to have been a major factor; many deep mixing patches had dissipation rates well above the noise. For instance, a 30-m-thick patch at 20.3 MPa had $\overline{\epsilon} = 3 \times 10^{-9}$ W kg^{-1}. The most likely explanation is that the internal wave intensity increased with depth at that location and time.

To examine mixing following the local flow, the Advanced Microstructure Profiler (AMP) was dropped alongside a drogue in the California Current for a week in October 1982. The profiles were free of the thermohaline intrusions between 1.4 and 2.0 MPa, where the eddy coefficients for heat and density were very close: $K_T = 3.2 \times 10^{-6}$ m^2 s^{-1} and $K_\rho = 3.5 \times 10^{-6}$ m^2 s^{-1} (Gregg et al., 1986). Expendable Velocity Profilers (XCPs) revealed shear levels slightly below the Garrett and Munk level.

Nearly the same results were found four years later during the Patches Experiment (PATCHEX) several hundred kilometers farther north in the California Current. Internal waves between 1 and 9 MPa were very close to the Garrett and Munk level, and $K_\rho = 3.7 \times 10^{-6}$ m^2 s^{-1}, nearly the same as during Drifter (Gregg and Sanford, 1988). These were the first measurements with finescale shear and microstructure reported from the same vehicle, the Multi-Scale Profiler (MSP), and established the level of mixing produced by internal waves at the Garrett and Munk level. They also demonstrated that K_ρ is constant with depth when internal wave shear follows WKB scaling. No K_T values were reported owing to numerous thermohaline intrusions.

While passing to and from the equator during Tropic Heat I in April 1987, the MSP was dropped at low latitudes in the central Pacific. At 11.5°N between 1 and 9 MPa, $K_\rho = 2.2 \times 10^{-6}$ m^2 s^{-1}; at 6°N between 1.5 and 5 MPa, $K_\rho = 1.8 \times 10^{-6}$ m^2 s^{-1}, but in a strongly sheared flow between 0.5 and 1.5 MPa, $K_\rho = 1 \times 10^{-4}$ m^2 s^{-1} (Gregg et al., 1993b). Unlike the flat Garrett-Munk spectrum, these low-latitude internal waves have relatively strong shear concentrated at scales of 30–50 m and rolloff at higher wavenumbers with slopes of −1.3 instead of −1. Similar shapes were also found in the western Pacific during TOGA/COARE (Gregg et al., 1996). At low latitudes, dissipation is weaker than predicted using scalings developed at mid-latitudes, consistent with internal wave interaction rates being proportional to f, the Coriolis parameter.

During the Topography Experiment at Fieberling Guyot in March 1991, Toole et al. (1994) sampled abyssal depths with the High Resolution Profiler (HRP), which also carries an acoustic current meter to measure finescale shear. A set of profiles 18–37 km from the summit were thought free of influence from the seamount and had $K_T = 2.1 \times 10^{-5}$ m^2 s^{-1} and $K_\rho = 1.8 \times 10^{-5}$ m^2 s^{-1}. These are the largest eddy coefficients reported from Pacific background areas except for the shallow measurements at 6°N.

HRP observations during NATRE in March 1992 found $K_\rho = 2.6 \times 10^{-5}$ m^2 s^{-1} between 2.5 and 30.0 MPa (Polzin et al., 1995). Another set of profiles on the African continental rise obtained $K_\rho = 1.1 \times 10^{-5}$ m^2 s^{-1} between 30 and 39 MPa (Toole et al., 1994). The deeper profiles came close to the bottom and demonstrate that eddy coefficients do not always increase in the abyss or near topography.

Also during NATRE, Sherman and Davis (1995) made a test deployment of the Long-Term Autonomous Microstructure Profiler (LAMP). Executing repeated profiles by changing its ballast, LAMP measures temperature microstructure while descending at 0.05–0.08 m s^{-1}. It stores the raw data on a hard disk and must be recovered so the data can be processed ashore. The LAMP took a profile every 3 hours and obtained 76 profiles in ten days while Ledwell was laying the SF_6 streaks. The average is $K_T = 1.4 \times 10^{-5}$ m^2 s^{-1} between 2.0 and 3.5 MPa. This was obtained after reducing χ_T by about 1/3 to account for departures in isotropy inferred by comparing signals from two microstructure probes.

5.2. *Mesoscale Features*

Mesoscale features such as fronts and rings are likely places for elevated shear to increase turbulence above background. Table 2 summarizes eddy coefficients from mesoscale features, and the available average profiles are included with other hot spots in Figure 12.

Some K_T averages are difficult to interpret near fronts because the dominant contributions to χ_T often come from the boundaries of thermohaline intrusions. When this occurs, dividing χ_T by the mean gradients within the intrusion usually yields moderate

K_T, but using the much weaker large-scale gradients gives artificially large eddy coefficients. As an example, a profile containing prominent intrusions observed near Cabo San Lucas, Mexico, has $K_T = 8.0 \times 10^{-4}$ m^2 s^{-1} for the entire record, but $K_T = 4.1 \times 10^{-5}$ m^2 s^{-1} in the most active feature when the local mean gradient is used (Gregg, 1975b).

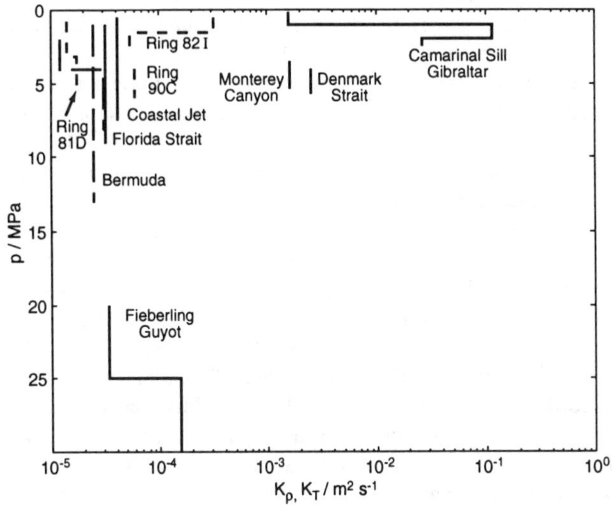

Figure 12. Average K_ρ profiles in suspected or confirmed mixing hot spots, including fronts, rings, islands, seamounts, straits, and canyons.

TABLE 2. Average diapycnal diffusivities in mesoscale features. a) Values are in monotonic and intrusive profiles, respectively. b) Because the influence of thermohaline intrusions was not considered, the larger K_T are probably overestimates. c) Higher value is from sections with thermohaline intrusions; lower value is from sections lacking obvious lateral influences. c) Data averaged over transitional and subducted mixed layers. d) Average is from two different sections through the ring.

Site	Lat/Lon	Date	K_T m^2 s^{-1}	K_ρ m^2 s^{-1}	p MPa	L km	Source
C. San Lucas	25/110	3/71	(0.017,8.0)e-4[a]		0.2-2	0.2	G75
Gulf Stream	40/64	10/72	(3.8-4.9)e-6[b]		1-10	10.0	OE77
Gulf Stream	40/64	3/74	(0.069-6.0)e-4[b]		1-10	8.0	OE77
Gulf Stream	38/69	10/75	(0.59-8.4)e-5[c]		0.5-5	0.5	GS81
Gulf Stream	38/69	10/75		(0.7-3.0)e-5			O80
ST Front N.Pac.	30/154	1/80		(1.5-0.54)e-3	< 1.5	0.24,0.08	L88
Ring 81D	40/64	9/81		1.6e-5[d]	0.8-5.5	21.0	LO86
Ring 82I	41/66	1/83		1.8e-4	0.5-2.5	6.0	G89
Coastal Jet	42/126	10/86		3.1e-5	1-9	22.4	G89
ST Front N.Atl	28/70	1/86	1.8e-5	2.5e-5	1.3-2.5	10,6.8	POTS95
Ring 90C	40.5/63.5	4/90		5.9e-5	4-6	2.0	POTS95

A profile only a few kilometers away had no intrusions and weaker mixing, $K_T = 1.7 \times 10^{-6}$ $m^2 s^{-1}$, similar to background areas in the California Current. Owing to the uncertainty about interpretation, K_T is often not reported from fronts in recent measurements, as evident in Table 2.

In October 1972 and March 1974, Oakey and Elliott (1977) took two sections of temperature microstructure across the Gulf Stream. During the first crossing, profiles between 1 and 10 MPa averaged $K_T = (3.8 - 4.9) \times 10^{-6}$ $m^2 s^{-1}$. Considering that no attempt was made to isolate contributions from intrusions, these are surprisingly low and close to background. In their second section, Oakey and Elliott found a wider range, $K_T = (0.069 - 6.0) \times 10^{-4}$ $m^2 s^{-1}$, but we do not have an average. Whether the higher values resulted from more frequent intrusions cannot be determined, but this seems likely in view of measurements made in the Gulf Stream with MSR and Camel while returning from FAME in November 1975; K_T varied from 6.9×10^{-6} $m^2 s^{-1}$ to 6.0×10^{-4} $m^2 s^{-1}$ (Gregg and Sanford, 1980), with the larger values resulting from intrusions having temperature inversions up to 5°C. Small staircases indicative of both double diffusive regimes were observed on the boundaries of the intrusions, even in the peak shear zones of the stream. K_ρ was more moderate, $(0.7 - 3.0) \times 10^{-5}$ $m^2 s^{-1}$ (Osborn, 1980).

Contours of K_ρ across the Florida Strait in June 1990 found levels in the Gulf Stream no larger than elsewhere in the strait (Figure 13). K_ρ is less than 10^{-5} $m^2 s^{-1}$ near the velocity maximum, and it is 10^{-5} to 10^{-4} $m^2 s^{-1}$ in the high-shear zones. The average across the strait excepting surface and bottom boundary layers is 4.1×10^{-5} $m^2 s^{-1}$. Therefore, the evidence to date is that K_ρ in the Gulf Stream is similar to the north Atlantic background.

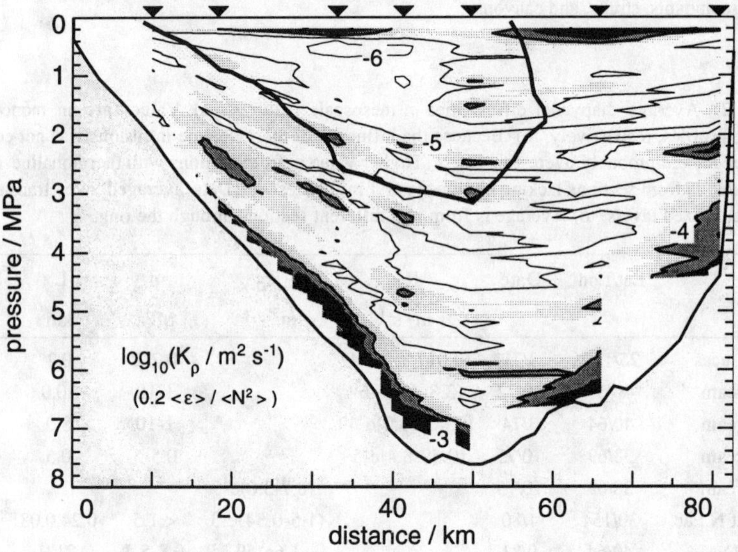

Figure 13. K_ρ contoured from average profiles at 7 stations across the Florida Strait at 27°N (Winkel et al., 1992). Shaded contours are labelled with $\log_{10} K_\rho$; unshaded areas are less than -5 and light shading is for values between -5 and -4. the thick line is the isotach for a mean northerly current of 1 m s^{-1}. About 10 profiles were taken at each station.

The first observations in a warm-core ring were two radial sections taken with Camel II through Ring 81D in September 1981. Their average is $K_\rho = 1.6 \times 10^{-5}$ m^2 s^{-1} (Lueck and Osborn, 1986). Use of AMP in Ring 82I in January 1983 found levels nearly 10 times higher, $K_\rho = 1.8 \times 10^{-4}$ m^2 s^{-1} (Gregg, 1989). The stronger mixing occurred beneath a surface mixed layer that was deepening rapidly in response to intense convection during a cold air outbreak. Elevated shear, measured with XCPs, was also found. The third average was obtained with HRP in Ring 90C during April 1990 and found levels midway between the others, $K_\rho = 5.9 \times 10^{-5}$ m^2 s^{-1} (Polzin et al., 1995). These three observations in warm-core rings demonstrate substantial variability in which the weaker mixing is comparable to the background outside rings.

In January 1980, Camel II was used in FRONTS80, an intensive study of the Pacific subtropical front north of Hawaii (Lueck, 1988). The microstructure profiles were taken during a hydrographic survey along a grid with stations 37 km apart. The separations were too large for observing detailed correlations, but well-mixed layers were shown to cross the front and appeared to have been produced by surface mixed layers sinking below more dense water while spreading laterally. Lueck averaged K_ρ in what he defined as transitional and submerged mixed layers. The transitional layers were well stratified and averaged $K_\rho = 1.5 \times 10^{-3}$ m^2 s^{-1}. The submerged mixed layers were more weakly stratified and averaged $K_\rho = 5.4 \times 10^{-4}$ m^2 s^{-1}. These averages include only the most active features and are likely to overestimate average mixing at the front. This is borne out by measurements obtained with Epsonde and HRP in the subtropical front of the North Atlantic during FASINEX in January 1986. Averages of all data give $K_T = 1.8 \times 10^{-5}$ m^2 s^{-1} and $K_\rho = 2.5 \times 10^{-5}$ m^2 s^{-1} (Polzin et al., 1996). These averages are similar to those in Atlantic background areas.

Observations from 1 to 9 MPa beneath a shallow coastal jet in October 1986 during PATCHEX north averaged $K_\rho = 3.1 \times 10^{-5}$ m^2 s^{-1} (Gregg and Sanford, 1988). This is 10 times the PATCHEX average taken farther offshore in the California Current and is consistent with the elevated internal wave shear observed with the MSP. These constitute a hot spot relative to other measurements in the eastern North Pacific, but do not stand out compared to Atlantic levels. This is characteristic of the overall pattern seen in comparing Figures 11 and 12. Most of the mesoscale hot spots are not much above the Atlantic background, which is considered background only because there were no obvious mesoscale features identified. This implies that background internal wave levels in the North Atlantic may be several times higher than background in the eastern Pacific. Often mesoscale features do not increase K_ρ above the Atlantic background.

5.3. *On the Equator*

Microstructure measurements in the Pacific and Atlantic have consistently found the shallow equatorial thermocline to be a major mixing hot spot. Intensive measurements have been made in the central and western Pacific owing to the importance of the turbulent fluxes to heat budgets of the surface waters. The turbulent momentum fluxes are also important and are significant components of dynamical balances of the equatorial undercurrent.

In the central and eastern Pacific the high-shear zone above the equatorial undercurrent extends into the surface layer and often produces low mean Richardson numbers in stratified water below the surface layer. Initially, the undercurrent was thought to be a turbulent jet, mixing most intensely at its center, an interpretation supported by towed measurements showing high dissipation rates near the velocity maximum. This view was contradicted by the first microstructure profiles, which showed strong mixing in the shear zones above and below the velocity maximum but only low, background mixing in the low-shear zone at the core of the undercurrent (Gregg, 1976; Osborn, 1980; Osborn and Bilodeau, 1980). Accordingly, except as noted, Table 3 gives average eddy coefficients in stratified profiles above, in, and below the velocity maximum. Although subsequent measurements reveal significant variability with time and place, the initial values of (4–5) $\times$ 10^{-4} m^2 s^{-1} above the core and (0.6 to 1.5) $\times$ 10^{-5} m^2 s^{-1} in the core remain representative.

5.3.1. Central Pacific

Three sets of measurements have intensively sampled turbulence at 0°N, 140°W, near the western end of the tongue of cold water extending along the equator from the Americas. Figure 14 compares average background conditions and K_ρ for these measurements.

Tropic Heat I was the first program to focus on turbulence in the undercurrent and found strong shear and turbulence in November and December 1984. The undercurrent was at full strength; the trade winds were well developed; and the tropical instability waves were at peak amplitude in their yearly cycle. In the shear zone above the velocity maximum, Richardson numbers of the mean flow were between 0.25 and 0.5, and K_ρ increased nearly exponentially from 10^{-6} m^2 s^{-1} at the velocity maximum to 10^{-3} m^2 s^{-1} at the base of the surface mixed layer (Gregg et al., 1985; Moum and Caldwell, 1985). Eddy coefficients increased below the core, but not as much as above.

The second intensive measurements were made in April 1987 during Tropic Heat II and found the undercurrent at half speed and the stratification stronger than during Tropic Heat I as the central Pacific recovered from an El Niño. In addition, the tropical instability waves were at their annual minimum, further reducing the total shear. K_ρ was less than during Tropic Heat I above the velocity maximum and would have been even weaker had not an equatorial wave been trapped above the maximum (Peters et al., 1991).

TABLE 3. Microstructure-based K_T and K_ρ in the equatorial undercurrent. Triplets of values are from the high-shear zone above velocity maximum, the velocity maximum, and the high-shear zone below the maximum. a) Average above velocity maximum, which was near 150 m. b) Averages are above and below the N^2 maximum at 0.8 MPa.

Site	Lon	Date	$10^4\ K_T$ $m^2\ s^{-1}$	$10^4\ K_\rho$ $m^2\ s^{-1}$	p MPa	L km	Source
Central Pac.	155	7/72	4.6,0.06,0.14,0.03		0.2-5.0	1.3	G76
East Atl.	28	7/74	3.8,0.06,0.4-0.008	4.0,0.015,3-0.3	0-3.5	1.7	OB80a,O80
Central Pac.	150	1/79		8.5[a]	0.2-1.6	0.6	C82
Central Pac.	140	11/84		5.3,0.014,0.11			GPWOS85,PGT88
Central Pac.	140	4/87		2.5,0.012,0.07			PGS91
West Pac.	-147	2/90		2.5,0.090[b]	0.3-2.7	165	BG96
Central Pac.	140	10-12/91		6.5,0.063,0.064	0.5-2.0	900	LCGM95
West Pac.	-156	2/93		0.54,0.035,0.094	0.8-3.2	7.9	GWSP95

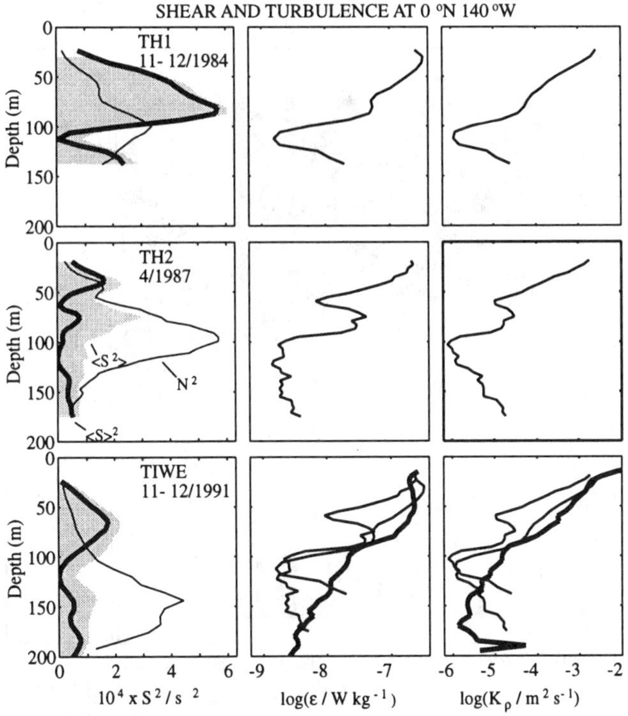

Figure 14. Representative background conditions at 0°N, 140°W during Tropic Heat I (upper), Tropic Heat II (middle), and TIWE (lower). R. C. Lien kindly prepared this figure.

The highest K_ρ at 140°W were found in November–December 1991 during the Tropical Instability Waves Experiment (TIWE). Two ships successively occupied the station and profiled continuously for 38 days to obtain 5590 profiles to 2 MPa (Lien et al., 1995). While the measurements were being made, an El Niño was beginning in the central Pacific, resulting in the instability waves being absent and the upper thermocline being strained and displaced by large Kelvin waves. For reasons not understood, $K_\rho = 6.5 \times 10^{-4}$ m^2 s^{-1} in the upper shear zone, somewhat larger than the 5.3×10^{-4} m^2 s^{-1} found during Tropic Heat I when the instability waves were present. Owing to the weaker mean gradients, however, the diapycnal heat flux produced by the turbulence was less.

At 140°W mixing is stronger above the undercurrent core than below it as a result of the 'deep daily cycle,' a term describing the extension of the daily convective cycle into stratified water far below the bottom of the surface mixed layer. The mixed layer is usually no deeper than 0.3 MPa, and K_ρ between the base of the layer and about 0.8 MPa fluctuates by factors of 10 to 100 in approximate unison with turbulence in the surface layer (Figure 15). Several mechanisms have been proposed to explain the deep cycle. The issue is not resolved, but essential background conditions appear to include strong mean shear and the daily convection cycle driven by surface forcing. Daily convective cycles occur at many places, but the mean shear is usually weak and the turbulence extends only a few meters below the bottom of the surface layer.

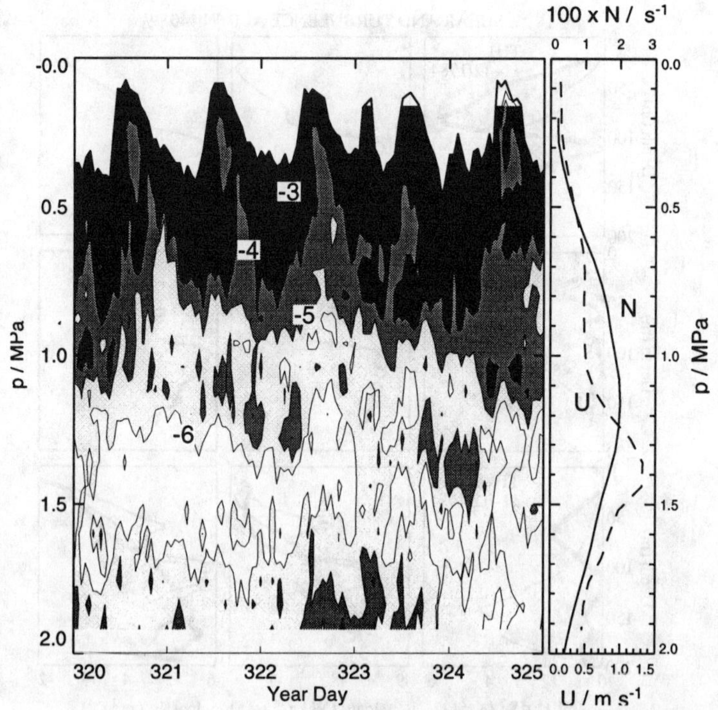

Figure 15. Typical deep daily mixing cycle on the equator at 140°W during TIWE, shown by contours of $\log_{10} K_\rho$, with shading beginning at the base of the surface mixed layer. Profiles on the right are average stratification and zonal velocity for the times shown. The velocity maximum of the undercurrent is at 1.4 MPa, deeper than usual owing to a Kelvin wave. R. C. Lien kindly made this figure.

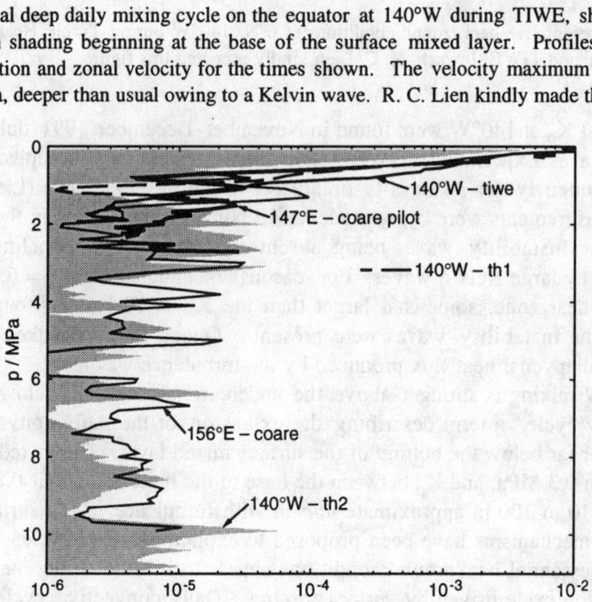

Figure 16. K_ρ profiles in the central and western Pacific.

Mixing below the undercurrent is highly variable with depth (Figure 16). The K_ρ profile for 140°W is an average of 68 records and varies from 10^{-6} to 4×10^{-5} m^2 s^{-1} below 2 MPa. The high values occur in the shear zones of deep jets beneath the undercurrent. Between these shear zones little mixing is detectable. This contrast is partially smeared out in the average plotted in Figure 16.

5.3.2. Western Pacific

Intensive measurements were made at 147°E in February 1990 during the COARE pilot cruise (Brainerd and Gregg, 1996). The undercurrent was centered at 1.6 MPa and played no role in mixing above 1 MPa, but wave-like meridonal shear produced other shallow mixing zones, and K_ρ immediately below the mixed layer was 2.5×10^{-4} m^2 s^{-1}, the same as found at 140°W during Tropic Heat II (Figure 16). Owing to the meridional shear, the velocity maximum of the undercurrent was not a minimum in total shear and therefore was not a minimum in K_ρ either.

Mixing was even weaker at 156°E in February 1993 during leg 3 of COARE (Gregg et al., 1996). The undercurrent had a broad velocity maximum between 1.5 and 2.5 MPa and produced shear extending upward to the base of the surface layer. K_ρ, however, averaged only 5.4×10^{-5} m^2 s^{-1}, the lowest of the equatorial averages. Strong mixing was found during legs 1 and 2 when bursts of winds from the west generated near-inertial oscillations that elevated shear for days. Except during those times, the prevailing winds were lighter, contributing to higher Richardson numbers than typical of the central Pacific.

A deep daily cycle was not found during the intensive profiling at 147°E. Nor was one evident in time series collected slightly off the equator at 156°E during legs 1 and 2 of TOGA/COARE. Its absence is puzzling, as the surface layers are forced by a daily convective cycle and shears are strong below the surface layers. Probable factors are larger mean Richardson numbers in the western Pacific and more steplike stratification. Mixing below the undercurrent at 156°E has magnitudes similar to that at 140°W and is equally intermittent (Figure 16).

5.4. Near Topography

Observations from sites affected by topography and those where topographic influence is suspected are included with the hots spots in Figure 12 and Table 4. This selection includes not only the most intense hot spots but also sites where mixing is not much above background levels.

By averaging profiles taken about 24 km from Santa Maria Island in the Azores in March 1975 by Osborn (1978), Osborn (1980) obtains $K_\rho = 2.7 \times 10^{-4}$ m^2 s^{-1} and 2.0×10^{-5} m^2 s^{-1} for pressures of 1.0 to 1.5 MPa and 1.5 to 7.7 MPa, respectively. The elevated levels over the shallower pressure range may reflect the passage of a storm several days before the measurements rather than island influence. On the other hand, the deeper values are comparable to background and give no evidence of island influence.

In August 1973 Oakey and Elliott (1980) used Octuprobe to take several sections of temperature and χ_T across the Denmark Strait, where cold water from the Icelandic Sea flows over a sill into the North Atlantic. Contours of temperature are complicated and reveal several thick intrusions extending across the strait. Because local temperature gradients were not used in estimating K_T, diapycnal fluxes cannot be assessed in the intrusions and only data said to be free of intrusions have been used. These come from deeper sections of the profiles toward the Icelandic side of the strait. The average, $K_T = 2.5 \times 10^{-3}$ m^2 s^{-1}, is 100 times the typical Atlantic background and 6 times the average K_ρ across the Florida Strait. Therefore, flow through the strait appears very turbulent.

During FAME, hot spots in thermal finestructure were found at several sites around Bermuda by Hogg et al. (1978), who attributed the elevated levels to the effect of mesoscale eddies impinging on the island and generating strong flows across the bottom. Microstructure was very intermittent in the finestructure hot spots, with 1 MPa averages of K_T varying by factors of 2–10. The elevated levels in the hot spots, however, were not sufficiently numerous or intense to raise the overall average of data near the island above background; the overall averages are $K_T = 2.5 \times 10^{-5}$ m^2 s^{-1} and $K_\rho = (1.2-3) \times 10^{-5}$ m^2 s^{-1} (Gregg and Sanford, 1980; Osborn, 1980).

Nine drops with Expendable Dissipation Profilers (XDPs) on the south flank of Monterey Canyon found $K_\rho = 1.6 \times 10^{-3}$ m^2 s^{-1} averaged over a total length of 0.2 km in active sections close to the bottom (Lueck and Osborn, 1985). These are the only microstructure data reported from a submarine canyon in spite of several observations of elevated internal waves in canyons (Wunsch and Webb, 1979; Hotchkiss and Wunsch, 1982). The average, however, is based on only 0.2 km of data and does not demonstrate that canyon-wide averages are comparable.

The average for Camarinal Sill in the Strait of Gibraltar, $K_\rho = 5.5 \times 10^{-2}$ m^2 s^{-1}, includes all data near the sill taken during the Gibraltar Experiment in October 1985 and May 1986 (Wesson and Gregg, 1994). Profiles west of the sill average $K_\rho \approx 0.1$ m^2 s^{-1}, owing to the inclusion of samples from the 50–75 m overturns produced when tidal currents carry water out of the Mediterranean. The flow becomes supercritical over the sill, and instabilities grow rapidly downstream on the interface separating Atlantic and Mediterranean waters. These billows collapse on the west flank of the sill, approximately where the flow becomes subcritical again in an internal hydraulic jump. Dissipation rates in these billows are much too high to be resolved by airfoil probes and were estimated from low wavenumber portions of the spectra by using the universal turbulent spectrum to estimate the unresolved variance. Turbulence in the overlying Atlantic water is much weaker but still well above background levels in the open ocean. The average includes measurements

TABLE 4. Near coasts and straits. a) These observations were taken in close proximity during the Fine and Microstructure Experiment (FAME). b) ε collection reported by Gargett and Osborn (1981) but K_ρ,s are given by Osborn (1980). c) Values are for seasonal and main thermocline. d) Average of author's category A, deep data without intrusions. e) First value is for 0.1–0.5 km; second is for deeper data. f) Values are from east, of over, and west of Camarinal Sill in the strait.

Site	Lat/Lon	Date	K_T m^2 s^{-1}	K_ρ m^2 s^{-1}	p MPa	L km	Source
Azores	37/26	3/75		(2.7,0.2)e-4	1.0-1.5,1.5-7.7	0.5	O78
Denmark St.	67/25	8/73	2.5e-3[d]		4.0-5.7	0.8	OE80
Bermuda[a]	32/65	10/75	2.5e-5		1-13	1.3	GS80
Bermuda[a]	32/65	10/75		(1.2,3)e-5[c]	1-8	2.1	O80[b]
Monterey Canyon	37/122	12/82		1.6e-3	3.5-5.4	0.2	LO85
Gibraltar St.[f]	36/5.8	5/86		5.5e-2	0.3-2.0	29.6	WG95
Florida St.	27/79.6	6/90		4.1e-5	0.5-7.5	29.7	WGS92
Fieberling Guyot	32.5/127.7	3/91	3.0e-4	2.2e-4	25-30		TPS94

from flooding as well as ebbing tides and thus is considerably less than an average including only the large overturns.

6. Summary and Discussion

6.1. Summary

The distinction between diascalar and advective turbulent fluxes by Winters and D'Asaro (1996) clarifies much of the current confusion about which flux is more fundamental and hence more desirable to measure. Neither is more fundamental; both are important measures of mixing. Diascalar fluxes are diffusive fluxes across instantaneous isoscalar surfaces and occur at the smallest turbulent scales. Advective fluxes are the net correlations of turbulent transport across the mean positions of isoscalar surfaces and occur at the largest turbulent scales. Advective fluxes depend indirectly on diffusion, as the turbulent motions produce no net correlation in the absence of diffusion. Sometimes diascalar and advective fluxes are equal numerically.

Analyses of large-scale budgets by Davis (1994a) and of the Osborn-Cox procedure by Davis (1994b) provide a common framework for relating χ_T measurements to the general circulation. This is not the end of the issue, however, because Davis concludes that inferences of diapycnal diffusivity from one-dimensional balances are not reliable for diffusivities of 10^{-5} m^2 s^{-1}, the zeroth-order estimate of background mixing. Nor can Osborn-Cox estimates of K_T be applied to the general circulation in particular locations without careful analysis of the local temperature variance budget. In many situations Osborn-Cox appears a useful ad hoc procedure, but major problems exist in weakly-stratified profiles at high latitudes and near fronts.

K_T is being omitted from most recent microstructure reports owing to inadequate spatial resolution of the temperature gradient variance at the speeds of the profilers. Thermistors have adequate sensitivity in most places, but their time response is too slow when moving 0.5 m s^{-1} or faster. Cold films respond more quickly but lack the required sensitivity when used below the seasonal thermocline.

The estimation of K_ρ from ε (Osborn, 1980) has not been investigated as carefully as Davis's analysis of Osborn-Cox and should also be considered an ad hoc procedure. However, the agreement between K_ρ and the diffusivity obtained from a tracer during NATRE indicates that it is a useful ad hoc procedure.

The quality of K_ρ estimates depends on the adequacy of the ε measurements. Present airfoils resolve the shear variance adequately in much of the thermocline but not near topography, where turbulence is most intense. Even when the variance is well-resolved, procedures for cutting off spectral integration at high wavenumbers are ad hoc and differ among investigators. The estimation of K_ρ at abyssal depths raises another issue, the ability of airfoils to resolve very weak turbulence, e.g., $\varepsilon < 10^{-10}$ W kg^{-1}. Here the limitation is not spatial resolution but the low-frequency noise characteristics of the probes. Low-frequency noise has not been investigated seriously, but f^{-1} noise is a concern owing to the very high impedance of the airfoils, 10^{13} Ω.

Statistical confidence limits for estimates of K_T and K_ρ are lesser issues than spatial resolution, determination of cutoff wavenumbers for spectral integration, and noise in abyssal measurements. Nevertheless, they must be considered carefully owing to the tendency of χ_T and ε distributions to lognormality. Davis (1996) demonstrates a strong tendency toward lognormality for $|\nabla\theta|$. Because $|\nabla\theta|$ and $|\partial\theta/\partial x_3|$ cannot both be lognormal, the near-lognormality of $|\nabla\theta|$ means that the vertical gradient magnitude is not strictly lognormal. Therefore, lognormal distributions should not be assumed a priori, and Davis recommends arithmetic averaging to obtain $\langle\chi_T\rangle$ rather than fitting lognormal curves to observed distributions.

Summarizing internal wave and microstructure measurements available at the time as the 'zeroth-order' view, Garrett (1983) gives the diapycnal diffusivity as 10^{-5} m^2 s^{-1} when internal waves are at the background state parameterized by Garrett and Munk (1972). Incorporating more recent observations gives us the outlines of a first-order view:

- Substantial areas of the Pacific have internal waves at or slightly below the Garrett and Munk level. The corresponding diapycnal diffusivities are about 5×10^{-6} m^2 s^{-1}. Minimum levels observed in the North Atlantic are higher: internal wave energies are 1.5–2 times Garrett and Munk and diapycnal diffusivities are (1-2) $\times 10^{-5}$ m^2 s^{-1}. In both the Atlantic and Pacific, the background varies seasonally and at lower frequencies.
- Most mixing hot spots in the open ocean thermocline have average diapycnal diffusivities of 10^{-5} m^2 s^{-1} to 10^{-4} m^2 s^{-1}. These include warm-core rings, fronts, and some regions close to islands and seamounts. The shear zone above the equatorial undercurrent in the central Pacific is an exception and sometimes has diapycnal diffusivities exceeding 10^{-3} m^2 s^{-1}.
- The most intense mixing in stratified water occurs where strong currents flow over irregular topography. The Denmark Strait and the Strait of Gibraltar are prime examples, and submarine canyons may be equally intense.
- After being normalized for changes in stratification, internal wave energy levels do not appear to increase systematically with depth. Consequently, neither do diapycnal diffusivities. Examples of increases have been found but so have cases with uniform levels. Evidence for abyssal behavior is scanty, but the pattern seems similar to that in the thermocline, i.e., large areas are at background and there are localized hot spots.

6.2. Discussion

In view of the evidence that diapycnal diffusivities over large areas are closer to 10^{-5} m^2 s^{-1} than to 10^{-4} m^2 s^{-1}, future sampling should be guided by answers to the question: 'How accurately must diapycnal diffusivities be known and what are the priorities for sampling different oceanic regimes?'

One way of addressing the question is to consider the length, ℓ, and time, τ, scales involved in mixing by using the turbulent scaling for diffusivity, $K_\rho = \ell^2/\tau$. Table 5 evaluates ℓ and τ for $K_\rho = 5 \times 10^{-6}$ m^2 s^{-1}, obtained when internal waves are at background, and for $K_\rho = 10^{-4}$ m^2 s^{-1}, obtained by one-dimensional balances (Munk, 1966). For comparison, Gerdes (1989) quotes vertical grid spacings of Δx_3 of 50–900 m for coarse resolution models of the general circulation. Time steps, $\Delta\tau$, for temperature and salinity are typically one day, and the models are integrated for 1,500–2,700 years. Fine resolution models make relatively minor changes, decreasing the minimum vertical spacing to 20 m and time steps to 6 hours. For $K_\rho = 5 \times 10^{-6}$ m^2 s^{-1}, ℓ matches the smaller grid spacings, used in the thermocline, when $\tau \approx 100$ years, and remains much smaller than the large grid spacings, used in the abyss, even when $\tau = 1{,}000$ years. For $K_\rho = 10^{-4}$ m^2 s^{-1}, ℓ matches the shorter grid scales in only a few years but still is appreciably smaller than the larger grid scales after 1,000 years.

The time and length scale estimates in Table 5 are upper bounds because the $K_\rho = \ell^2/\tau$ scaling is based on first-order changes, e.g., the temperature difference over distance I

TABLE 5. Vertical diffusive length scales as a function of K_ρ and time interval τ.

	τ / years			
K_ρ/m^2 s^{-1}	1	10	100	1000
5×10^{-6}	3 m	8 m	26 m	81 m
1×10^{-4}	12 m	36 m	115 m	360 m

changing by a large fraction of its magnitude. Smaller changes than these can affect model results qualitatively as well as quantitatively. For example, Gargett et al. (1989) show that, when $K_\rho \approx 10^{-5}$ m^2 s^{-1}, meridional heat flux, stream function, and flow direction in the abyss are relatively insensitive to the deep diffusivity. Applying $K_\rho \propto N^{-1}$, however, changes the deep dynamics.

To define the role of mixing in the ocean, other parameters than K_ρ need to be considered, such as the turbulent heat flux. Large diffusivities are not necessarily important when they occur in very weak mean gradients, such as those in the abyss. As an example, Figure 17 shows the turbulent heat flux corresponding to Abyssal Recipes, i.e., for $K_T = 1 \times 10^{-4}$ m^2 s^{-1}. The flux decreases from – 1 Wm^{-2} at 1 km to less than –0.1 W m^{-2} at depths exceeding 4 km. By comparison, turbulent heat fluxes in the shear zone above the equatorial undercurrent are usually tens of Watts per square meter in the central Pacific.

Large-scale averages of turbulent parameters cannot be obtained by intensive ship-based measurements as done for the past 25 years. Shipboard measurements are far too expensive and cover too little space and time to yield large-scale averages. The most promising technique for the upper thermocline is to deploy large numbers of LAMPs and retrieve the data by satellite when cellular phone service is expanded to cover the globe. To prepare for this capability, expected within a few years, a project is presently funded to add electromagnetic velocity profiling to LAMP (T. Sanford, personal communication, 1995). By relating the mixing directly to the internal wave field, these measurements will provide a basis for modeling the global internal wave field. To the degree that diapycnal mixing is important and varies in space and time, mixing must be incorporated into models of the general circulation via the internal wave field producing the mixing.

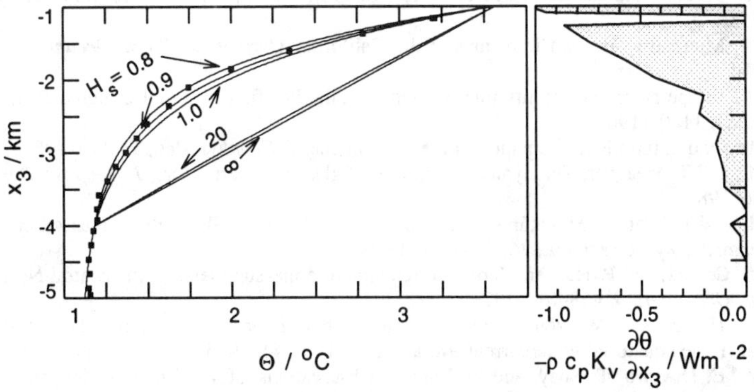

Figure 17. The left panel shows a deep potential temperature profile and Abyssal Recipes fits of exponential curves with length scales, H_s, in kilometers. The right panel is the diapycnal heat flux for the observed temperature profile and $K_T = 1 \times 10^{-4}$ m^{-1} s^{-1}.

Acknowledgments. This review was prepared for the IUTAM conference on Physical Limnology organized by Jörg Imberger and held in Broome, Australia, from l0 to 14 September 1995. The U.S. Office of Naval Research funded the writing, and Kraig Winters, Eric D'Asaro, and Johny Wüest provided valuable discussions and comments. Contribution 2193 of the School of Oceanography.

References

Batchelor, G. K., Smallscale variation of convected quantities like temperature in turbulent fluid, *J. Fluid Mech., 5,* 113-139, 1959.

Brainerd, K., and M. Gregg, Turbulence and mixed layer dynamics in the Western Pacific Warm Pool—the TOGA-COARE microstructure pilot cruise, J. *Geophys. Res.,* in press, 1996.

Davis, R., Diapycnal mixing in the ocean: Equations for large-scale budgets, *J. Phys. Oceanogr., 24,* 777-800, 1994a.

Davis, R., Diapycnal mixing in the ocean: The Osborn-Cox model, *J. Phys. Oceanogr., 24,* 2560-2576, 1994b.

Davis, R., Sampling turbulent dissipation, *J. Phys. Oceanogr., 25,* in press, 1996.

Dillon, T. M., and M. M. Park, The available potential energy of overturns as an indicator of mixing in the seasonal thermocline, *J. Geophvs. Res., 92,* 5345-5353, 1987.

Elliott, J. A., and N. S. Oakey, Average microstructure levels and vertical diffusion for Phase III, GATE, *Deep-Sea Res., 26,* 273-294, 1980.

Gargett, A., P. Cummins, and G. Holloway, Effects of variable diffusivity in the GFDL model, in *Parameterization of Small-Scale Processes,* Proceedings, Hawaii Winter Workshop, Jan. 17-20, edited by P. Müller and D. Henderson, pp. 11-20, University of Hawaii at Manoa, Honolulu, HI, 1989.

Gargett, A. E., and T. R. Osborn, Small-scale shear measurements during the Fine and Microstructure Experiment (FAME), *J. Geophys. Res., 86,* 1929-1944, 1981.

Gargett, A. E., T. R. Osborn, and P. W. Nasmyth, Local isotropy and the decay of turbulence in a stratified fluid, *J. Fluid Mech., 144,* 231-280, 1984.

Garrett, C., and W. Munk, Space-time scales of internal waves, *Geophys. Fluid Dyn., 2,* 225-264, 1972.

Garrett, C. J. R., Diapycnal mixing in the ocean interior, in *Proceedings of the Joint Oceanographic Assembly 1982 General Assembly,* pp. 19-21, Halifax, Nova Scotia, Dalhousie University, 1983.

Garrett, C. J. R., and W. H. Munk, Space-time scales of internal waves: A progress report, *J. Geophys. Res., 80,* 291-297, 1975.

Gerdes, R., The role of numerical advection schemes in general circulation models, in *Parameterization of Small-Scale Processes,* Proceedings, Hawaii Winter Workshop, Jan. 17-20, edited by P. Müller and D. Henderson, pp. 21-57, University of Hawaii at Manoa, Honolulu, HI, 1989.

Gregg, M., Oceanic fine and microstructure, *Rev. Geophys. and Space Phys., 13,* 586-591 and 635-636, 1975a.

Gregg, M., Microstructure and intrusions in the California Current, *J. Phys. Oceanogr., 5,* 253-278, 1975b.

Gregg, M., Temperature and salinity microstructure in the Pacific Equatorial Undercurrent, *J. Geophys. Res., 81,* 1180-1196, 1976.

Gregg, M., Scaling turbulent dissipation in the thermocline, *J. Geophys. Res., 94,* 9686-9698, 1989.

Gregg, M., and T. Meagher, The dynamic response of glass rod thermistors, *J. Geophys. Res., 85,* 2779-2786, 1980.

Gregg, M., and T. Sanford, Signatures of mixing from the Bermuda Slope, the Sargasso Sea and the Gulf Stream, *J. Phys. Oceanogr., 10,* 105-127, 1980.

Gregg, M., C. Cox, and P. Hacker, Vertical microstructure measurements in the central North Pacific, *J. Phys. Oceanogr., 3,* 458-469, 1973.

Gregg, M., H. Peters, J. Wesson, N. Oakey, and T. Shay, Intensive measurements of turbulence and shear in the equatorial undercurrent, *Nature, 318,* 140-144, 1985.

Gregg, M., E. D'Asaro, T. Shay, and N. Larson, Observations of persistent mixing and near-inertial internal waves, *J. Phys. Oceanogr., 16,* 856-885, 1986.

Gregg, M., H. Seim, and D. Percival, Statistics of shear and turbulent dissipation profiles in random internal wave fields, *J. Phys. Oceanogr., 23,* 1777- 1799, 1993a.

Gregg, M., D. Winkel, and T. Sanford, Varieties of fully resolved spectra of vertical shear, *J. Phys. Oceanogr.*, 23, 124-141, 1993b.
Gregg, M. C., Variations in the intensity of small-scale mixing in the main thermocline, *J. Phys. Oceanogr.*, *7*, 436-454, 1977.
Gregg, M. C., and T. B. Sanford, The dependence of turbulent dissipation on stratification in a diffusively stable thermocline, *J. Geophys. Res.*, *93*, 12,381-12,392, 1988.
Gregg, M. C., D. P. Winkel, T. B. Sanford, and H. Peters, Turbulence produced by internal waves in the oceanic thermocline at mid and low latitudes, *Dyn. Atmos. Oceans*, 24, 1-14, 1996.
Gurvich, A. S., and A. M. Yaglom, Breakdown of eddies and probability distributions for small scale turbulence, *Phys. Fluids*, *10*, 59-65, 1993.
Henyey, F. S., J. Wright, and S. M. Flatté, Energy and action flow through the internal wave field, *J. Geophys. Res.*, *91*, 8487-8495, 1986.
Hogg, N., E. J. Katz, and T. B. Sanford, Eddies, islands, and mixing, *J. Geophys. Res.*, *83*, 2921-2938, 1978.
Hotchkiss, F. S., and C. Wunsch, Internal waves in Hudson Canyon with possible geological implications, *Deep-Sea Res.*, 29, 415-422, 1982.
Ledwell, J. R., A. J. Watson, and C. S. Law, Evidence for slow mixing across the pycnocline from an open-ocean tracer-release experiment, *Nature*, *364*, 701-703, 1993.
Levitus, S., Climatological atlas of the world ocean, Professional Report 13, Geophysical Fluid Dynamics Laboratory, Princeton, N. J., 1982a.
Levitus, S., Interpentadal variability of temperature and salinity at intermediate depths of the North Atlantic, *J. Geophys. Res.*, 94, 6091-6131, 1982b.
Lien, R. C., D. R. Caldwell, M. C. Gregg, and J. N. Moum, Turbulence variability at the equator in the central Pacific at the beginning of the 1991–1993 El Niño, *J. Geophys. Res.*, *100*, 6881-6898, 1995.
Lueck, R. G., Turbulent mixing at the Pacific Subtropical Front, *J. Phys. Oceanogr.*, *18*, 1761-1774, 1988.
Lueck, R. G., and T. R. Osborn, Turbulence measurements in a submarine canyon, *Deep-Sea Res.*, 4, 681-698, 1985.
Lueck, R. G., and T. R. Osborn, The dissipation of kinetic energy in a warm-core ring, *J. Geophys. Res.*, *91*, 803-818, 1986.
Lueck, R. G., O. Hertzman, and T. R. Osborn, The spectral response of thermistors, *Deep-Sea Res.*, *24*, 951-970, 1977.
Lueck, R. G., W. R. Crawford, and T. R. Osborn, Turbulent dissipation over the continental slope off Vancouver Island, J. *Phys. Oceanogr.*, *13*, 1809-1818, 1983.
McComas, C. H., and P. Müller, The dynamic balance of internal waves, *J. Phys. Oceanogr.*, *11*, 970-986, 1981.
McDougall, T., Neutral surfaces, *J. Phys. Oceanogr.*, *17*, 1950-1964, 1987,
McDougall, T. J., Some implications of oceanic mixing for ocean modelling, in *Small-Scale Turbulence and Mixing in the Ocean*, edited by J. C. J. Nihoul and B. M. Jamart, pp. 21-36, Elsevier, Amsterdam, 1988.
Moum, J. N., and D. R. Caldwell, Local influences on shear-flow turbulence in the equatorial ocean, *Science*, *230*, 315-316, 1985.
Moum, J. N., and R. G. Lueck, Causes and implications of noise in oceanic dissipation measurements, *Deep-Sea Res.*, *32*, 379-390, 1985.
Moum, J. N., and T. R. Osborn, Mixing in the main thermocline, *J. Phys. Oceanogr.*, *16*, 1250-1259, 1986.
Munk, W. H., Abyssal recipes, *Deep-Sea Res.*, *13*, 707-730, 1966.
Munk, W. H., Internal waves and small-scale processes, in *Evolution of Physical Oceanography*, edited by B. A. Warren and C. Wunsch, pp. 264-291, MIT Press, Cambridge, MA, 1981.
Ninnis, R., The effects of spatial averaging on air-foil probe measurements of oceanic velocity microstructure, Ph.D. thesis, Univ. of British Columbia, Vancouver, Canada, 1984.
Oakey, N. S., Determination of the rate of dissipation of turbulent energy from simultaneous temperature and velocity shear microstructure measurements, *J. Phys. Oceanogr.*, 12, 256-271, 1982.
Oakey, N. S., and J. A. Elliott, Vertical temperature gradient structure across the Gulf Stream, *J. Geophys. Res.*, 82, 1369-1380, 1977.
Oakey, N. S., and J. A. Elliott, The variability of temperature gradient microstructure observed in the Denmark Strait, *J. Geophys. Res.*, *85*, 1933-1944, 1980.
Osborn, T. R., Vertical profiling of velocity microstructure, *J. Phys. Oceanogr.*, *4*, 109-115, 1974.

Osborn, T. R., Measurements of energy dissipation adjacent to an island, *J. Geophys. Res., 83*, 2939-2957, 1978.

Osborn, T. R., Estimates of the local rate of vertical diffusion from dissipation measurements, *J. Phys. Oceanogr., 10,* 83-89, 1980.

Osborn, T. R., and L. E. Bilodeau, Temperature microstructure in the equatorial Atlantic, *J. Phys. Oceanogr., 10,* 66-82, 1980.

Osborn, T. R., and C. S. Cox, Oceanic fine structure, *Geophys. Fluid Dyn., 3,* 321-345, 1972.

Panchev, S., and D. Kesich, Energy spectrum of isotropic turbulence at large wavenumbers, *Comptes rendus de l'Académie bulgare des Sciences,* 22, 627-630, 1969.

Peters, H., M. Gregg, and T. Sanford, Equatorial and off-equatorial fine-scale and large-scale shear variability at 140° W, *J. Geophys. Res., 96,* 16,913-16,928, 1991.

Polzin, K., J. M. Toole, and R. W. Schmitt, Finescale parameterization of turbulent dissipation, *J. Phys. Oceanogr.*, 25, 306-328, 1995.

Polzin, K., N. S. Oakey, J. M. Toole, and R. W. Schmitt, Fine and microstructure characteristics across the northwest Atlantic Subtropical Front, *J. Geophys. Res.*, 101, 14111-14121, 1996.

Rohr, J. J., K. N. Helland, E. C. Itsweire, and C. W. V. Atta, Turbulence in a stably stratified shear flow: A progress report, in *Turbulent Shear Flows,* vol. 5, Springer-Verlag, Berlin, 1987.

Sherman, J. T., and R. E. Davis, Observations of temperature microstructure in NATRE, *J. Phys. Oceanogr., 25,* 1913-1929, 1995.

Thorpe, S. A., Turbulence and mixing in a Scottish loch, *Phil. Trans. Roy. Soc. London, 286,* 125-181, 1977.

Toole, J. M., K. L. Polzin, and R. W. Schmitt, Estimates of diapycnal mixing in the abyssal ocean, *Science, 264,* 1120-1123, 1994.

Wesson, J., and M. Gregg, Mixing at Camarinal Sill in the Strait of Gibraltar, *J. Geophys. Res.*, 99, 9847-9878, 1994.

Winkel, D., M. Gregg, and T. Sanford, Simultaneous observations of shear and turbulence in the Florida Current, in *AMS 10th Symposium on Turbulence and Diffusion, September 29-October 2, 1992, Portland, Oregon,* pp. (J5)101-(J5)104, 1992.

Winters, K., and E. A. D'Asaro, Diascalar flux and the rate of fluid mixing, *J. Fluid Mech.*, submitted, 1996.

Wunsch, C., and S. Webb, The climatology of deep ocean internal waves, *J. Phys. Oceanogr.*, 9, 235-243, 1979.

Yamazaki, H., and R. Lueck, Why oceanic dissipation rates are not lognormal, *J. Phys. Oceanogr., 20,* 1907-1918, 1990.

24

Special Closure for Stratified Turbulence

J. Weinstock

Abstract

A spectral (closure) equation for stably stratified turbulence is derived from the Navier-Stokes equation. For simplicity, the spectral equation is limited to quasi-stationarity and homogeneity of the velocity field. The closure is then solved for $E_H(k_1)$, the horizontal kinetic energy spectrum at horizontal wavenumber k_1, and compared with measurements in the atmosphere.

Derivation of the spectral equation is obtained by first dividing the spectral stress equation into two components, horizontal and vertical, and correspondingly, the kinetic energy spectra are divided into horizontal and vertical components $E_H(\underset{\sim}{k})$ and $E_\perp(\underset{\sim}{k})$ at wavenumbers $\underset{\sim}{k}$. The (nonlinear) spectral transfer term is closed by an EDQNM (eddy-damped quasi-normal Markovian) approximation, but extended here to strongly stratified turbulence. The pressure-velocity correlation is evaluated with account taken for stable buoyancy. The overall closure is then solved for $E_H(k_1)$.

Introduction

Atmospheric measurements show that horizontal wavenumber spectra of wind fluctuations behave differently than do vertical wavenumber spectra in at least one quantitative aspect — the transition wavenumber at which the spectral power law changes from -5/3 to about -3 (e.g., Nastrom and Gage, 1985; Lilly and Peterson, 1983; Sidi et al., 1981). For horizontal wavenumber spectra, this transition occurs at about 200 km, greatly exceeding the buoyancy length L_R. That length is defined by $L_R = 2\pi\ (\varepsilon_0/N^3)^{1/2}$, where N is the Brunt-Väisälä frequency and ε_0 is the dissipation rate of turbulence kinetic energy density, and is less than 1 km in the lower atmosphere. For vertical wavenumber spectra the transition occurs close to the buoyancy length. Related observations have been made in oceans (e.g., Gregg, 1977; Gargett et al., 1981; Garrett and Munk, 1979). A possible explanation of these observations has been suggested by Lilly (1981) and Riley et al. (1981) but a quantitative spectral theory remains to be given.

Physical Processes in Lakes and Oceans
Coastal and Estuarine Studies Volume 54, Pages 339-349

Published in 1998 by the American Geophysical Union

The purpose of our paper is to derive a spectral closure of stably stratified turbulence. The motivation is to see to what extent such a theory can account for observed spectra. What complicates a theoretical treatment of stratified turbulence is the large anisotropy of kinetic energy spectra at large wavenumber — the horizontal component being much larger than the vertical. Perhaps it is because of this complication that an analytical spectral closure including triple correlations is not yet available, in so far as we are aware. To obtain such a closure is our goal. A spectral theory was derived many years ago by Deissler (1962), but triple velocity correlations were neglected thereby limiting the results to Reynolds number too small to establish an inertial range as found in the atmosphere.

To do so, an evolution equation for the stress spectrum $R_{ij}(\underline{k})$ is first derived from the Navier-Stokes in terms of the nonlinear energy transfer term T_{ij}, a triple velocity covariance, as well as in terms of the other familiar correlations that arise (in the spectral equations). These other terms include the buoyancy term coming from the pressure when incompressibility is applied to the Navier-Stokes equation. This particular buoyancy term influences the spectrum markedly at small horizontal wavenumber. [This term is included implicitly, but not explicitly, in the work of Riley et al. (1981).] The spectral equation, with buoyancy terms, is then divided into horizontal and vertical parts to thereby obtain two coupled equations for $E_H(\underline{k})$ and $E_{22}(\underline{k})$, the horizontal and vertical components of velocity (stress) spectra at wavevector $\underline{k}$. Closure of these coupled equations is affected by a direct interaction approximation type of method (Kraichnan, 1959 — to be referred to as DIA) but simplified by approximation to obtain an analytical solution. For our purpose it was convenient to use the approximation referred as the eddy-damped quasi-normal Markovian closure (also referred to as EDQNM — Orszag, 1970; Leith, 1971; Kraichnan, 1971; Herring and Kraichnan, 1972; Pouqet et al., 1975; Cambon et al., 1981). This closure is of the DIA class introduced by Kraichnan (1959). Our intention is to obtain an analytical, although approximate expression for $T_H(k1)$, the horizontal component of the energy transfer function. Such an expression makes possible an analytical solution of (17) and permits additional insight into the physics of the problem — but at the cost of some uncertainty in the values of numerical coefficients.

The closure determines the spectra at any directional orientation of wavevector $\underline{k}$. Here, the closure is solved for $E_H(k_1)$, the kinetic energy spectrum at horizontal wavenumber k_1 — a widely measured spectrum — but restricted by time stationarity and spatial homogeneity. The spectrum is thereby determined by a balance between nonlinear energy transfer and buoyancy loss. Because of this balance the modes making up the spectrum are not purely horizontal. They have a small but significant vertical velocity component induced by a buoyancy contribution, (which originated in the pressure-strain rate term) at sufficiently small k_1.

Derivation of Spectrum

The evolution equation for the stress spectrum in a stratified flow can be straightforwardly derived from the Navier-Stokes equation. The fluctuation part of the latter equation for a stratified flow is

$$\frac{\partial u_i}{\partial t} + \underset{\sim}{u}\cdot\underset{\sim}{\nabla} u_i = -\frac{\partial p}{\rho_o \partial x_i} + \frac{g\theta\delta_{i2}}{\theta_o} + \nu\nabla^2 u_i, \tag{1}$$

where $\underset{\sim}{u} \equiv \underset{\sim}{u}(\underset{\sim}{x},t)$ is the fluctuation part of the fluid vector velocity at position $\underset{\sim}{x}$ at time t, the indices i denote a Cartesian coordinate direction 1, 2, or 3 [e.g.,, $\underset{\sim}{u} \equiv \{u_1, u_2, u_3\}$], $p \equiv p(\underset{\sim}{x},t)$ denotes the fluctuation pressure, $\rho_o \equiv \rho_o(x_2,t)$ is the mean particle density at height x_2, $\theta_o \equiv \theta_o(x_2,t)$ is the mean potential temperature, $\theta \equiv \theta(\underset{\sim}{x},t)$ is the fluctuation of potential temperature, g is the acceleration of gravity, and ν is the molecular viscosity.

To obtain the Fourier transform of (1) — required to derive the stress spectral equation — account must be taken of the variation of ρ_0 and θ_0 with height x_2. Again, for the sake of simplicity, we have restricted ourselves to a spatial and temporal quasi-stationarity of the velocity field. Spectral homogeneity implies that $\langle p^2\rangle/\rho_0^2$ and $\langle\theta^2\rangle/\theta_0^2$ are spatially constant (do not grow with height or depth). It is further assumed that although ρ_0 and θ_0 grow with height x_2, their scale heights $H \equiv \rho_0 (\partial\rho_0/\partial x_2)^{-1}$ and $H_\theta \equiv \theta_\theta(\partial\theta_0/\partial x_2)^{-1}$ are sufficiently large that they can be ignored in comparison with the vertical wavelengths of the velocity fluctuations — the situation prevalent in oceans. A side effect of this assumption is that it eliminates the growth of gravity waves with height. In that idealized case, the Fourier transform of (1) is

$$\frac{\partial u_i(\underset{\sim}{k})}{\partial t} + (\underset{\sim}{u}\cdot\underset{\sim}{\nabla} u_i)_{\underset{\sim}{k}} = -\frac{ik_i p(\underset{\sim}{k})}{r_o} + \frac{g\theta(\underset{\sim}{k})\delta_{i2}}{q_o} + \nu k^2 u_i(\underset{\sim}{k}), \tag{2}$$

where $2\Pi/k_2 \lll H$ and H_0, $u_i(\underset{\sim}{k}) \equiv u_i(\underset{\sim}{k},t)$, $p(\underset{\sim}{k}) \equiv p(\underset{\sim}{k},t)$ and $\theta(\underset{\sim}{k}) \equiv \theta(\underset{\sim}{k},t)$ denote the Fourier transforms of u_i, p, and θ, respectively [with a relatively slow x_2 dependence for $p(\underset{\sim}{k})$ and $q(\underset{\sim}{k})$ since ρ_0 and θ_0 vary slowly with x_2] and δ_{i2} denotes the Kronecker delta (i.e., $\delta_{i2} = 0$ for $i \neq 2$ and $\delta_{22} = 1$). The quantity $(\underset{\sim}{u}\cdot\underset{\sim}{\nabla} u_i)_{\underset{\sim}{k}}$ denotes the Fourier transform of $(\underset{\sim}{u}\cdot\underset{\sim}{\nabla} u_i)$ and is given by

$$(\underset{\sim}{u}\cdot\underset{\sim}{\nabla} u_i)_{\underset{\sim}{k}} \equiv \frac{i}{(2\Pi)^3}\int d\underset{\sim}{q}\, k_m u_m(q) u_i(\underset{\sim}{r}) \tag{3}$$

$$\underset{\sim}{r} = -(\underset{\sim}{q} + \underset{\sim}{k}), \tag{4}$$

where repeated indices are summed (e.g., $k_m u_m \equiv \underset{\sim}{k}\cdot\underset{\sim}{u}$), and the wavevectors $\underset{\sim}{k}$, $\underset{\sim}{q}$, and $\underset{\sim}{r}$ satisfy (4) corresponding to momentum conservation among the Fourier modes $\underset{\sim}{u}(\underset{\sim}{k})$, $\underset{\sim}{u}(\underset{\sim}{q})$, and $\underset{\sim}{u}(\underset{\sim}{r})$.

The evolution equation for the stress spectrum:

$$R_{ij}(\underset{\sim}{k}) \equiv (2\Pi)^{-3} V^{-1}\langle u_i(\underset{\sim}{k}) u_j^*(\underset{\sim}{k})\rangle,$$

where V denotes the volume of the system and the asterisk denotes the complex conjugate, is obtained by postmultiplying (1) with $u_j^*(\underset{\sim}{k})$ and then premultiplying its complex conjugate with $u_i(\underset{\sim}{k})$. Addition of the ensemble average of these two equations results in

$$\frac{\partial R_{ij}(\underset{\sim}{k})}{\partial t} = -\frac{(2\Pi)^{-3}}{V}\left(1+\Gamma^{*}\right)\left[\left\langle(\underset{\sim}{u}\cdot\underset{\sim}{\nabla} u_i)_{\underset{\sim}{k}} u_j^{*}(\underset{\sim}{k})\right\rangle + \frac{\langle ik_i p(\underset{\sim}{k}) u_j^{*}(\underset{\sim}{k})\rangle}{\rho_o}\right] \quad (5)$$

$$+ (2\Pi)^{-3}\left(1+\Gamma^{*}\right) g\delta_{i2}\frac{\langle u_j^{*}(\underset{\sim}{k})\theta(\underset{\sim}{k})\rangle}{\theta_o} - 2\nu k^2 R_{ij}(\underset{\sim}{k}),$$

where Γ^{*} denotes the complex conjugate of the transpose of a matrix element (e.g., $\Gamma^{*} a_i b_j^{*} = b_j a_i^{*}$).

For an incompressible flow, $k_i u_i(\underset{\sim}{k}) = 0$, the pressure is determined from (2) as

$$\frac{p(\underset{\sim}{k})}{\rho_o} = -\frac{igk_2\theta(\underset{\sim}{k})}{k^2\theta_o} - \frac{k^{-2}}{(2\Pi)^3}\int d\underset{\sim}{q}\, k_m k_s u_m(\underset{\sim}{q}) u_s(\underset{\sim}{r}), \quad (6)$$

using $(\underset{\sim}{u}\cdot\underset{\sim}{\nabla} u_i)_{\underset{\sim}{k}} = 0$, as mentioned after (5). The second term on the right is the usual velocity-velocity term for an incompressible turbulence while the first term is the contribution of buoyancy to the pressure. This contribution causes transfer of some vertical buoyancy forcing into the horizontal direction via the pressure-strain rate term in (5). It is the principal influence of stratification on horizontal motion in the absence of mean shear. Otherwise shear production would be important. Substitution of (6) in (5) gives the spectral equation as

$$\frac{\partial R_{ij}(\underset{\sim}{k})}{\partial t} = T_{ij}(\underset{\sim}{k}) + \left(\delta_{i2} - \frac{k_i k_2}{k^2}\right) B_j(\underset{\sim}{k}) - \varepsilon_{ij}(\underset{\sim}{k}), \quad (7)$$

NON-LINEAR ENERGY TRANSFER — BUOYANCY — DISSIPATION

$$T_{ij}(\underset{\sim}{k}) \equiv -V(2\Pi)^{-6}\int d\underset{\sim}{q}\, P_{ims}\langle u_m(\underset{\sim}{q}) u_s(\underset{\sim}{r}) u_j^{*}(\underset{\sim}{k})\rangle, \quad (8)$$

$$B_j(\underset{\sim}{k}) \equiv -V^{-1}(2\Pi)^{-3} 2\frac{g}{\theta_o}\langle\theta(\underset{\sim}{k}) u_j^{*}(\underset{\sim}{k})\rangle, \quad (9)$$

$$\varepsilon_{ij}(\underset{\sim}{k}) \equiv 2\nu k^2 R_{ij}(\underset{\sim}{k}), \quad (10)$$

$$P_{ims}(\underset{\sim}{k}) \equiv k_m P_{is}(\underset{\sim}{k}) + k_s P_{im}(\underset{\sim}{k}), \quad (11)$$

$$P_{is}(\underset{\sim}{k}) \equiv \delta_{is} - k_i k_s / k^2, \quad (12)$$

where $T_{ij}(\underset{\sim}{k})$ is the usual energy transfer spectrum describing three-wave interactions, $B_j(\underset{\sim}{k})$ is the buoyancy spectrum flux in direction j and $\varepsilon_{ij}(\underset{\sim}{k})$ is the dissipation. Equation (7) is simply the Karman-Howarth equation (e.g., Hinze, 1975; Deissler, 1962) in a (homogeneous) stratified fluid. For stable stratification $\langle\theta(\underset{\sim}{k}) u_2^{*}(\underset{\sim}{k})\rangle$ is negative so that

(7) directly causes energy loss in the vertical direction 2, some of which (loss) is transferred to the horizontal directions 1 and 3 by virtue of the pressure-strain rate in (5) from which arises the $k_i k_2/k_2$ quantity that occurs in the (total) buoyancy term of (7).

Since the turbulence is anisotropic, it will be useful to divide (7) into horizontal and vertical parts, as follows:

$$\frac{\partial R_H(\underline{k})}{\partial t} = T_H(\underline{k}) + \left(\frac{k_2^2}{k^2}\right) B_2(\underline{k}) - \varepsilon_H(\underline{k}) \tag{13}$$

$$\frac{\partial R_{22}(\underline{k})}{\partial t} = T_{22}(\underline{k}) + \left(1 - \frac{k_2^2}{k^2}\right) B_2(\underline{k}) - \varepsilon_{22}(\underline{k}) \tag{14}$$

where the horizontal components $R_H(\underline{k}), T_H(\underline{k}), \varepsilon_H(\underline{k})$ simply denote the sum of horizontal components $[R_{11}(\underline{k}) + R_{33}(\underline{k})]$, $[T_{11}(\underline{k}) + T_{33}(\underline{k})]$, $[\varepsilon_{11}(\underline{k}) + \varepsilon_{33}(\underline{k})]$, respectively. Equation (13) differs from that of Riley et al. (1981) and Lilly (1983) in that the buoyancy term $B_2(\underline{k})$ which originates in the pressure term is accounted for explicitly. The form of the second term on the right side of (13) was obtained by use of the incompressibility condition $k_1 u_1(\underline{k}) + k_3 u_3(\underline{k}) = -k_2 u_2(\underline{k})$ to arrive at

$$k_1 k_2 B_1(\underline{k}) + k_3 k_2 B_3(\underline{k}) = -k_2^2 B_2(\underline{k}) . \tag{15}$$

To complete the closure, there remains to express $T_H(\underline{k})$ and $T_{22}(\underline{k})$ in terms of velocity covariances or spectra. This is done next, but in the meantime, only for one-dimensional spectra defined by

$$E_H(k_1) \equiv \int dk_2 dk_3 R_H(\underline{k}) . \tag{16}$$

To determine such spectra, one integrates (13) over k_2 and k_3 to obtain

$$\frac{\partial E_H(k_1)}{\partial t} = \bar{T}_H(k_1) + \bar{B}_2(k_1) - \varepsilon_H(k_1) , \tag{17}$$

where the other 1-D spectra are defined by

$$\{\bar{T}_H(k_1), \bar{B}_2(k_1), \varepsilon_H(k_1)\} \equiv \int dk_2 \int dk_3 \left\{T_H(\underline{k}), \frac{k_2^2}{k^2} B_2(\underline{k}), \varepsilon_H(\underline{k})\right\} .$$

For our case of (quasi-) stationary spectra, we have $\partial E_H(k_2)/\partial t = 0$, while for k_1 smaller than the viscous dissipation range of wavenumbers one has $\varepsilon_H(k_1) << T_H(k_1)$ so that (17) reduces to

$$\bar{T}_H(k_1) = -\bar{B}_2(k_1) . \tag{18}$$

This equation determines k_1 spectra in a stationary, stably stratified flow. It (the equation) is seen to be a balance between nonlinear energy transfer and buoyancy.

To solve (18) for $E_H(k_1)$ requires that $T_H(k_1)$ be derived in terms of the second moments $R_{ij}(\underset{\sim}{k})$ or $E_H(k_1)$, a derivation which constitutes the classic problem of turbulence closure theory. Fortunately, closures are available.

We have derived such an expression for $T_H(k_1)$, although approximately, from the EDQNM. It is given by

$$T_H(k_1) \approx -\frac{\partial}{\partial k_1}\left[\left(\frac{L_H}{L_v}\right) k_1^{5/2} E_H(k_1)^{3/2}\right], \qquad k_1 << k_v \tag{19}$$

where L_H^{-1} and L_v^{-1} are integral scales for horizontal and vertical wavenumber spectra of horizontal winds (or currents) respectively. $k_v \equiv (\nu^3/\varepsilon_0)^{-1/4}$ is the Kolmogorov wavenumber and ε_0 is the dissipation rate of kinetic energy. The proof of (19) is a separate subject in itself and is given elsewhere (Weinstock, 1995). Suffice it to say that (19) reduces to the form of $T_{ii}(k)$ used by Leith (1967), Besnard et al. (1991), and Weinstock (1993) in the isotropic limit where $L_v \approx L_H$, expect that in those references, the scalar wavenumber k occurs instead of k_1 and the scalar energy spectrum $E(k) \equiv [E_{11}(k) + E_{22}(k) + E_{33}(k)]$ instead of the horizontal component spectrum $E_H(k) \equiv [E_{11}(k_1) + E_{22}(k_1)]$.

The other quantity we need to solve (18) is the buoyancy flux spectrum $B_2(k_1)$. A calculation of that spectrum has been given, for small isotropy [eqn. (23) of Weinstock, 1978]. For large anisotropy, a similar derivation shows that

$$\bar{B}_2(k_1) \approx -(0.3)N^2 \frac{E_H(k_1)^{1/2}}{k_1^{3/2}}, \tag{20}$$

where $N \equiv (g\theta_0^{-1}\,\partial\theta_0/\partial x_2)^{1/2}$ denotes the Brunt-Väisälä frequency. This expression is more general than, but reduces to the inertial range expression of Lumley (1964).

Finally, substitution of (19) and (20) into (18) we have the closure equation for $E_H(k_1)$:

$$\frac{\partial}{\partial k_1}\left[k_1^{5/2} E_H(k_1)^{3/2}\right] = -\left(\frac{0.3L_v}{L_H}\right) N^2 \frac{E_H(k_1)^{1/2}}{k_1^{3/2}}. \tag{21}$$

This equation determines the vertical wavenumber spectrum $E_H(k_1)$ in a (quasi) stationary stratified flow. Solution of this equation, subject to the boundary condition $E_H(k_1) \sim (2\alpha/3)\,\varepsilon_0^{2/3} k_1^{-5/3}$ for $k_1/k_R >> 1$, finally yields the desired result

$$E_H(k_1) = \frac{2\alpha}{3}\varepsilon_0^{2/3} k_1^{-5/3}\left[1 + \left(\frac{0.15L_v}{L_H}\right)\frac{3}{2\alpha}\left(\frac{k_1}{k_R}\right)^{-4/3}\right]. \tag{22}$$

where $k_R \equiv (N^3/\varepsilon_0)^{1/2}$ is the buoyancy wavenumber, and it is assumed that k_1 exceeds an energy-containing wavenumber which we denote as $k_1{}^*$.

It can be seen that the spectrum approaches the $k_1^{-5/3}$ form, as imposed by the boundary condition, when $k_1 > k_R$, whereas for $k_1 < C_o k_R$,

$$C_o \equiv \left(\frac{0.45}{2\alpha}\frac{L_v}{L_H}\right)^{3/4}, \tag{23}$$

the spectrum approaches k_1^{-3}, a so-called "buoyancy" subrange — a range where buoyancy decreases the spectral power law below -5/3 under conditions of stationarity. These two limiting forms are given by

$$E_H(k_1) = \frac{2\alpha}{3}\varepsilon_o^{2/3} k_1^{-5/3}, \qquad k_1 > C_o k_R \tag{24}$$

$$E_H(k_1) = \left(\frac{0.15L_v}{L_H}\right) N^2 k_1^{-3}, \qquad k_1 < C_o k_R \tag{25}$$

Both these behaviors have been observed in the lower atmosphere (e.g., Nostrum and Gage, 1985; Lilly and Peterson, 1983).

Discussion

What we believe is new, and to be emphasized, about (22) is that, since $C_0 << 1$ when anisotropy is strong, the spectrum behaves as $k_1^{-5/3}$ for wavenumbers much smaller then k_R, wavenumbers for which the spectra are still strongly anisotropic.

Considerations of isotropy and anisotropy are better explained by comparison of (22) with $E_H(k_2)$, the vertical wavenumber spectrum of horizontal kinetic energy density. The latter is approximately given as (e.g., Gargett et al., 1981; Weinstock, 1978)

$$E_H(k_2) = \frac{2\alpha}{3}\varepsilon_o^{2/3} k_2^{-5/3}\left[1 + \left(\frac{k_2}{k_R}\right)^{-4/3}\right]. \tag{26}$$

For wavenumbers k_1 and k_2 exceeding k_R, (22) and (26) reduce to

$$\left.\begin{aligned} E_H(k_1) &= \frac{2\alpha}{3}\varepsilon_o^{2/3} k_1^{-5/3} \\ E_H(k_2) &= \frac{2\alpha}{3}\varepsilon_o^{2/3} k_2^{-5/3} \end{aligned}\right\} \qquad \begin{aligned} k_1 &>> k_R \\ k_2 &>> k_R \end{aligned} \tag{27}$$

In this case the turbulence is nearly isotropic and three-dimensional — the Kolmogorov inertial range.

On the other hand, for wavenumbers k_1, k_2 in an intermediate range, the range $C_o k_r <$ $k_1, k_2 < k_R$, (22) and (26) become

$$\left.\begin{aligned} E_H(k_1) &= \frac{2\alpha}{3}\varepsilon_o^{2/3} k_1^{-5/3} \\ E_H(k_2) &= \frac{2\alpha}{3} N^2 k_2^{-3} \end{aligned}\right\} \qquad C_o k_R << \begin{matrix} k_1 \\ k_2 \end{matrix} << k_R \tag{28}$$

Here, it is seen that $E_H(k_1)$ still behaves as $k_1^{-5/3}$ whereas $E_H(k_2)$ behaves as k_3^{-3}, and the spectrum is no longer isotropic. For example, $E_H(k_1) << E_H(k_2)$ when $k_1 = k_2$. In this interval, buoyancy (flux) is strong enough to deform the k_2 spectrum away from inertial but not strong enough to deform the k_1 spectrum away from inertial — at stationarity. The k_1 spectrum resists such deformation simply because $T_H(k_1)$, energy transfer along the horizontal, is enhanced by stratification — the factor L_H/L_v in (19). It is noted that although $E_H(k_1)$ behaves as $k_1^{-5/3}$ buoyancy never-the-less influences the fluctuations, causing them to be elongated along the horizontal. Such fluctuations are gravity waves or vortical modes. It is this elongation which enhances $T_H(k_1)$. ($T_H(k_2)$ is not enhanced.)

Finally, there is the k_1 range $k_1^* < k_R << C_0\, k_R$, where, we recall, k_1^* is an energy-containing(energy production)wavenumber. For such a range, (22) reduces to

$$E_H(k_1) = \frac{2\alpha}{3} C_o N^2 k_1^{-3}, \qquad k_1^* < k_R << C_o k_R \tag{29}$$

In this case, buoyancy deforms the power law of the k_1 spectrum as well as that of the k_2 spectrum. We hasten to add that although $E_H(k_1)$ has the small coefficient C_0 whereas $E_H(k_2)$ in (26) does not, (22) will give the same mean square horizontal kinetic energy as does (26) because $k_1^* << k_2^*$ where k_2^* is an energy-containing wavenumber of the k_2 spectrum. In fact, we estimate $k_1^*/k_2^* \sim C_0^{1/2}$.

In sum, there are three ranges of k_1 and k_2; $k_1, k_2 > k_R$; $C_0 k_R < k_1, k_2 < k_R$ and $k_1^* < k_1 < C_0 k_R$, $k_2^* < k_2 < C_0 k_R$. The spectra of these ranges are, respectively: inertial for both $E_H(k_1)$ and $E_H(k_2)$; inertial for $E_H(k_1)$ with buoyancy deformed power law (k_2^{-3} form) for $E_H(k_2)$; and buoyancy deformed power law for both $E_H(k_1)$ and $E_H(k_2)$ — provided that $k_1 > k_1^*$ and $k_2 > k_2^*$. The fluctuations are near isotropic in the first range, corresponding to a Kolmogorov range, and become more and more anisotropic and horizontally elongated as k_1 and k_2 decrease below k_R. The k_1 spectrum is viewed as having three ranges although the power law does not change in the first two ranges. What does change, (in going from the first to the second range) is the character of the fluctuations being isotropic in one range and elongated (anisotropic) owing to buoyancy in the other.

Comparison with an Experiment

An experiment that helped motivate our article are measurements of atmospheric k_1 spectra made as part of the Global Atmospheric Sampling Program (e.g., Nastrom and Gage, 1985 — referred to as GASP). Other measurements are given by Lilly and Peterson (1983). The GASP measurements were abundant, and satisfied the theoretical conditions of quasi-stationarity and large Reynolds number — a large Reynolds number being required to ensure a distinct inertial range for $E_H(k_1)$. The results of these measurements is given in Figure 1, in which is graphed observed $E_{11}(k_1)$, zonal wind spectra versus wavenumber. For comparison, the theoretical relation (22) is also plotted. For this plot, we use the observed values $L_H \sim 10^4$ km, $L_v \sim 1$ km (e.g., Sidi et al., 1988). The (average) value of ε_0 is chosen to be $\sim 10^{-4}\ m^2/s^2$. This value is taken from the $k_1^{-5/3}$ slope of the GASP data (Nastrom and Gage, 1985) which ensures (22) agree asymptotically with the observed

$E_{11}(k_1)$ at large k_1. This value is also independently found to be an average for tropopause heights (10 to 14 km) under quiet conditions of the free atmosphere — the usual situation at those heights (e.g., Pinus, 1974; Vinnichenko and Dutton, 1969). With this ε_0, and with $N \approx 0.01\ s^{-1}$, the buoyancy wavenumber is $k_R \equiv (N^2/\varepsilon_0)^{1/2} \sim 0.1\ m^{-1}$. Substitution of these L_H, Lv, ε_0 and k_R into (22) with $\alpha \approx 1.5$ determines $E_H(k_1) \approx 2E_{11}(k_1)$ which is plotted in Figure 1 — the large circles. It can be seen in this figure that the theory follows the general slope of the observed $E_H(k_1)$. The feature to be emphasized is the scale predicted for the transition from $k_1^{-5/3}$ to k_1^{-3} behavior is about 200 km. This scale greatly exceeds the buoyancy length, as does the observed transition scale.

With regard to the quantitative agreement between theory and experiment, this must be taken with a grain of salt since agreement is forced at large k_1 by choice of ε_0, and, additionally, the theoretical values depend on estimates of the length scales L_v and L_H.

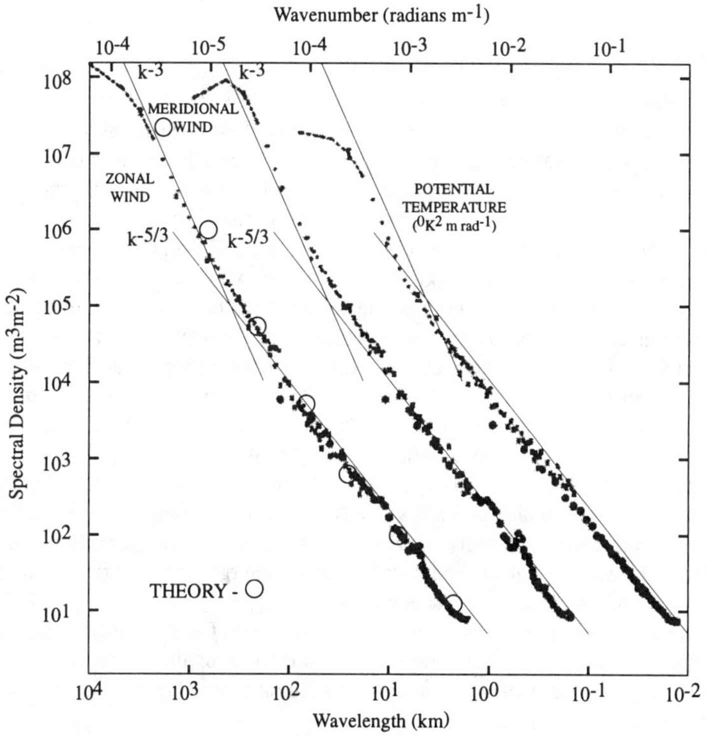

Figure 1. Variance power spectra of wind and potential temperature near the tropopause from GASP aircraft data. The spectra for meridional wind and temperature are shifted one and two decades to the right, respectively; lines with slopes - 3 and - $^5/_3$ are entered at the same relative coordinates for each variable for comparison. (Taken from Nastrom and Gage, 1985.). The theory, equation (22) is given by large circles, O.

Previously, the $k_1^{-5/3}$ spectrum for the (mesoscale) range $C_0k_R < k_1 < k_R$ was attributed, as a possibility, to an upscale (reverse) cascade of vortical modes, corresponding to a two-dimensional enstrophy cascade (Lilly, 1983; Riley et al., 1981). The energy source of the upscale cascade was speculated upon in a qualitative way. In our work the turbulence (fluctuations) is three-dimensional although very anisotropic (elongated) and the cascade is downscale from large to smaller scales. The source is the downward cascade of synoptic scales — a widely accepted source of energy input at such large scales. As the synoptic scale cascade to smaller scales they become less two-dimensional, although still very anisotropic, and more influenced by buoyancy than by rotation, until they finally cascade down to isotropic turbulence. The downward cascade has been suggested in the past (e.g., Vinnichenko and Dutton, 1969), but a quantitative theory had not been given.

Summary

As spectral closure for stably stratified turbulence is derived from the Navier-Stokes equation. For simplicity, the closure is restricted to an idealized case of spatial homogeneity and (quasi) stationarity of the velocity field. Spectral energy transfer is represented by an EDQNM expression. A key feature of the closure is a calculation of the buoyancy dependent pressure-strain rate term. It is this term which converts vertical into horizontal momentum flux, and drives the horizontal fluctuations at large scales. With this pressure term, the closure can be solved for spectrum $E_{11}(k_1)$. It is given by (22).

What is emphasized is that C_0k_R, the transition wavenumber that separates $k_1^{-5/3}$ behavior from k_1^{-3} behavior, can be much less than the buoyancy wavenumber k_R. Such a small wavenumber transition is observed in the free atmosphere. It (the transition at C_0k_R) is attributed to an enhancement of $T_H(k_1)$ spectral energy transfer rate along the horizontal, when the stratification is strong. This transition contrasts with that of the vertical wavenumber spectrum $E_H(k_2)$. For that spectrum, the transition occurs at about k_R (e.g., Gargett et al., 1981), there being no enhancement of energy transfer along the vertical. In view of these transitions, $E_H(k_1)$ is viewed as having three ranges: a nearly isotropic Kolmogorov range for $k_1 > k_R$; a (non-isotropic) range where the fluctuations are elongated, indicative of gravity waves or vortical modes (but for which the spectrum remains inertial owing to enhancement of horizontal energy transfer) for $C_0k_R < k_1 < k_R$; and a range of buoyancy deformed spectral power law, varying as k_1^{-3}, for $k_1^* < k_1 \ll C_0k_R$, where k_1^* is an energy-containing wavenumber. On the other hand, for the k_2 spectrum there are the two familiar ranges — elongation of the fluctuations occurring at the same wavenumber as does the transition of spectral power law.

References

Besnard, D. C., F. H. Harlow, R. M. Rauenzahn, and C. Zemach, Spectral transport model for turbulence. Los Alomos Report No. LA-UR-91-1930, Los Alomos Lab, Los Alomos, NM, 1991.

Cambon, C., D. Jeandal, and J. Mathieu, Spectral modeling of homogeneous non-isotropic turbulence. *J. Fluid Mech., 104*, 247-262, 1981.

Deissler, R. G., Turbulence in the presence of a vertical body force and temperature gradient. *J. Geophys. Res. 67*, 3049-3061, 1979.

Gargett, A. E., P. J. Hendricks, T. B. Sanford, T. R. Osborn, and A. J. Williams, III, A composite spectrum of vertical shear in the upper ocean. *J. Phys. Oceon., 11*, 1258-1271, 1981.

Garrett, C. and W. Munk, Internal waves in the ocean. *Ann. Rev. Fluid Mech. 11*, 339-369, 1979.

Gregg, M. C., Variations in the intensity of small-scale mixing in the main thermocline. *J. Phys. Ocean., 7*, 436-454, 1977.

Fritts, D. C., T. Tsuda, T. Sato, S. Fukau, and S. Kato, Observational evidence of a saturated gravity wave spectrum in the troposphere and lower stratosphere. *J. Atmos. Sci., 45*, 1741-1759, 1988.

Hinze, J. O., Turbulence, Mc-Graw Hill, New York , 1975.

Holloway, G., A conjecture relating oceanic internal waves and small-scale processes. *Atmos. Ocean., 21*, 107-122 , 1983.

Herring, J. R. and R. H. Kraichnan, Comparison of some approximations of isotropic turbulence, in: *Statistical Models and Turbulence*. Springer, p. 148, 1972.

Kraichnan, R. H., The structure of turbulence at very high Reynolds numbers. *J. Fluid Mech., 5*, 497-543, 1959.

Kraichnan, R. H., An almost-Markovian Galilean-invariant turbulence model. *J. Fluid Mech., 47*, 513-524, 1971.

Leith, C. E., Diffusion approximation to inertial energy transfer in isotropic turbulence. *J. Fluid Mech., 10*, 1409-1416, 1967.

Leith, C. E., Atmospheric predictability and two-dimensional turbulence. *J. Atmos. Sci., 28*, 145-161, 1971.

Lilly, D. K., Stratified turbulence and the mesoscale variability of the atmosphere. *J. Atmos. Sci., 40*, 749-761, 1983.

Lumley, J. L., The spectrum of nearly inertial turbulence in a stably stratified flow. *J. Atmos. Sci., 21*, 99-102, 1964.

Orszag, S. A., Analytic theories of turbulence. *J. Fluid Mech., 44*, 363-386, 1970.

Pouquet, A. M. Lesieur, J. C. Andre, and C. Basdevant, Evolution of high Reynolds number two-dimensional turbulence. *J. Fluid Mech., 72*, 305-319, 1975.

Nastrom, G. D. and K. Gage, A climatology of atmospheric wavenumber spectrum of wind and temperature observed by commercial aircraft. *J. Atm. Sci., 42*, 950-960, 1985.

Riley, J. J., R. W. Metcalfe, and M. A. Weissman, Direct numerical simulations of homogeneous turbulence in density stratified fluids, in *Nonlinear Properties of Internal Waves*, edited by B. J. West, Amer. Inst. of Phys., 80-112, 1981.

Senft, D. C. and C. S. Gardner, Seasonal variability of gravity wave activity in the upper mesosphere region over Urbana. *J. Geophys. Res., 96*, 17,229-17,264, 1991.

Sidi, C., J. Lefrere, F. Dalaudier, and J. Barat, An improved atmospheric buoyancy wave spectrum model. *J. Geophys. Res., 93*, 774-790, 1988.

Vinnichenko, N. K. and J. A. Dutton, Empirical studies of atmospheric spectra in the free atmosphere. *Radio Science, 4*, 1115-1126, 1969.

Weinstock, J., On the theory of turbulence in the buoyancy subrange of stably stratified flows. *J. Atmos. Sci., 35*, 634-649, 1978.

Weinstock, J., Spectral transport and turbulence cascade in two-point closure theories. To be submitted to *J. Fluid Mech.*, 1997.

Wilson, R., M. L. Chanin, and A. Hauchecorne, Gravity waves in the middle atmosphere observed by Rayleigh lidar 2. Climatology. *J. Geophys. Res., 96*, 5169-5183, 1991.

25

Turbulent Mixing in Stably Stratified Flows: Limitations and Adaptations of the Eddy Diffusivity Approach

H. J. S. Fernando, J. C. R. Hunt, E. J. Strang, A. L. Berestov and I. D. Lozovatsky

Abstract

We review how turbulence in stably stratified flows differs from that in unstratified flows, especially as regards the Lagrangian motion of fluid particles and mixing, so that the vertical fluxes of heat, matter and momentum in environmental flows are controlled both by macro- and micromixing, the former being related to large scale features of the flow and the latter to small-scale turbulence and molecular diffusion. The variation of these fluxes determines the distribution of heat and different species through the turbulence layer. These fluxes lead to density profiles being modified by the presence of turbulence; in some situations so as to be effectively eliminated, but in others so as to be formed into thin stable layers separating regions of turbulence or regions of turbulent and non-turbulent flow. The prevailing evolutionary scenarios are governed by the dependence of fluxes on local conditions or in mathematical terms on the local eddy diffusivity.

It is shown that although fluxes are most commonly modelled using the concept of eddy diffusivity, there are serious defects with this approach. Other models have been developed for research studies. However, since diffusivity parameterizations are the basis of most practical numerical models, some proposals are made here for parameterising eddy diffusivities based on mixing mechanisms and their mathematical representations.

Introduction

Layers of varying density (i.e., stratification) are ubiquitous features of natural water bodies. The changes in temperature and density of small volumes of fluid caused by mixing in such flows are key elements that determine the physical state of such flows as well as their ability to maintain ecological balances. These water bodies mainly receive energy through solar radiation. The distribution of thermal energy and other species at various layers and their conversion to alternative forms, in quantities adequate to maintain ecosystem balances, depend on vertical and horizontal mixing. Another consequence of solar insolation on the Earth is the generation and destruction of stratification, but in the

Physical Processes in Lakes and Oceans
Coastal and Estuarine Studies Volume 54, Pages 351-362

overall sense the response of natural fluids to solar insolation is to generate stratification variations in density cause in environmental fluids. Because the buoyancy forces and thence fluid motions, any mixing in natural water bodies, whether in horizontal or vertical directions, there is a strong coupling between the mixing and the dynamics of these flows (Fernando and Hunt, 1996). Therefore, understanding the interaction between density stratification and turbulence remains a key aspect in studies of environmental flows.

The aim of environmental forecasting models is to predict the state of natural flows, such as atmosphere and oceans with a reasonable degree of accuracy. To achieve this objective, forecasting models are based on equations of motion and transport of species such as heat and salt in oceans and heat and moisture in the atmosphere. The complexity of these models demands numerical solutions that utilize intense computer resources. Because of the wide variety of scales involved, all scales of motions cannot be resolved in these models and hence often small-scale turbulent quantities are expressed in terms of bulk mean quantities such as mean velocity and concentration fields. Of course, the scales resolvable in models depend on the largest scale of the motion being modelled, grid size and Reynolds number of the flow. For example, the mean flow patterns of global scales are computed using General Calculation Models (GCM); with currently available computer resources, the atmospheric GCM's can allow vertical and horizontal grid sizes of 100 m $\times$ 300 km (in the GCM used at the UK Meteorological Office).

The mean equations for velocity and scalar transports in turbulent fluids contain an extra unknown term – either in the form of the Reynolds stress or the flux transports, respectively, – thus requiring certain closure hypotheses to be invoked in solving them. A large number of different closures have been developed in this regard (e.g., Stull, 1988). In forecasting models, however, for calculational simplicity it is customary to use simple first order closures (e.g., Pacanowski and Philander, 1981) based on eddy diffusivity defined as

$$F_i = -K_i^c \frac{\partial \overline{C}}{\partial x_i} \tag{1}$$

where $\overline{C}$ is the mean concentration of the scalar quantity S, K_i^c is the eddy diffusivity which in general is determined both by local statistics of the turbulent velocity field u_i and by molecular properties of S. In equation (1), the value of K_i^c is independent of the form of the concentration profile only if certain conditions are satisfied, chief of which is that the lengthscale of the velocity fluctuations is smaller than that of the mean concentration field (Hunt, 1982).

Although it is quite common in engineering practice, the value of K_i^c derived in one type of stratified turbulent flow should not necessarily be applied in other types; this practice is based on the assumption that stratified turbulence has universal characteristics that can be defined by a few non-dimensional numbers (e.g., Richardson number Ri and the integral scale and variance of turbulence). It is generally necessary to consider other aspects of the flow such as whether the flow is in local equilibrium or is developing (Fernando and Hunt, 1996). A particularly important point in stratified flows is that the values of K_i^c are generally different in the same flow for momentum, for buoyant scalars (such as temperature and density) and for passive scalars (such as pollutants), because of the differences in their molecular diffusivities (see Schumann, 1996; Pearson et al., 1983; Hanazaki and Hunt, 1996; Komori and Nagata, 1996).

In GCM's, the vertical eddy diffusivities K_i^c are either estimated as having typical values or are related to variables defining the large-scale features of the flow (or 'parameterized'). Using global considerations, Munk (1966) estimated that the vertical scalar and momentum diffusivities K_s and K_V for the deep oceans are of the order 10^{-4} m^2 s^{-1}. However, available (sparse) measurements from oceans have consistently yielded K_vs values of the order 10^{-5} m^2 s^{-1} (Ledwell et al., 1994; Toole et al., 1994), with the exception of near the rough bottom and side boundaries of oceans (Caldwell and Moum, 1995; Polzin et al., 1997). Another case that may yield a large localized eddy diffusivities is the deep ocean convection (or chimneys), where convection effectively brings down cold surface water deep into the ocean interior. In oceans, such active spots with large diffusivities appear to dominate the overall eddy diffusivities, and hence the use of uniform diffusivities across entire oceans is highly questionable.

In parameterizing turbulent mixing in atmospheric, oceanic and limnological situations, it is necessary to understand mixing mechanisms, the temporal and spatial frequency of their causations, the mixing efficiency associated with each mechanism and the local conditions under which they make appearance. If all such information were available, the simplest scheme of parameterization of overall mixing would involve the appropriate use of different eddy diffusivities associated with different mechanisms, at a given resolution and location of the flow. (For a review of different 'canonical' stratified turbulent flows see Fernando and Hunt, 1996). However, in reality, these different mechanisms interact, thus modifying the flux laws. Such complications arise due to simultaneous existence of different types of turbulent diffusion processes have not been adequately assessed in previous studies (e.g., see the approach used by Large et al., 1994).

The purpose of this correspondence is to discuss some issues related to eddy diffusivity parameterizations in stratified turbulent flows. In Section 2, we will discuss some general concepts related to mixing in homogeneous stratified turbulent flows and illustrate that, unlike in non-stratified fluids, the scalar fluxes at a given point in stratified turbulence are mainly determined by molecular-scale (micro) mixing. In Section 3, issues related to mixing in inhomogeneous stratified turbulence, in particular mixing across their interfaces, will be discussed. The sensitivity of eddy-diffusivity parameterizations on the predictions of numerical models is illustrated in Section 4.

Some General Concepts of Mixing

Mixing involves the migration of fluid particles of a given species concentration to locations of different concentrations, thus changing average and fluctuating concentrations at various measurement points. These concentration changes can occur by two entirely different processes known as "macromixing" and "micromixing." In macromixing, the fluid particles themselves may not change their concentration C, as they wander around by the motion field; the mixing simply increases the probability of finding particles of concentration C at a measurement point. In the limiting case of $dC/dt \approx 0$, the process becomes formally reversible.

Some examples of macromixing are the increase of the width of a turbulent plume by the transport of fluid particles of higher concentration to its boundaries and the breakdown of a stable density gradient by the turbulence via the movement of denser particles upward and vice versa (sometimes this is called "stirring"). Stirring increases the fluctuations of temperature, because particles of widely different concentrations pass through a given measurement point. Small-scale mixing and molecular diffusion between particles of different concentrations tend to erase these concentration fluctuations; this irreversible process is called micromixing.

These concepts can be illustrated through the basic equations and by considering an idealized problem. Consider the introduction of a quantity Q_0 of a substance into a confined, turbulent, incompressible fluid of volume V_0. The distribution of the concentration $C(\underline{x},t)$ of this substance is governed by

$$\frac{\partial C}{\partial t} = D_S \nabla^2 C - \nabla \cdot (C\underline{u}) \tag{2a}$$

or in Langrangian terms following a fluid particle $C = C(t;\underline{a})$

$$dC/dt = D_s \nabla^2 C, \tag{2b}$$

where D_s is the molecular diffusivity of S, $\underline{u}$ = is the Eulerian velocity field and $\underline{a}$ is the position of the fluid particle at time t=0. Integration of (2) over the fluid volume shows that the volume average (designated by $\langle\ \rangle$) satisfies

$$\langle C \rangle = \frac{1}{V_o} \int C dV; \quad \frac{\partial \langle C \rangle}{\partial t} = 0 \quad \text{and} \quad \langle C \rangle V_o = Q_o . \tag{3a}$$

Taking the ensemble mean of (2a) for the case $\underline{u}$ = 0 shows that the local mean concentration $\overline{C}$ is related to the mean scalar flux F_i in the ith direction by the divergence relation

$$\frac{\partial \overline{C}}{\partial t} = -\frac{\partial F_i}{\partial x_i}, \tag{3b}$$

where the flux has contributions from the turbulence and the molecular diffusion, viz.,

$$F_i = \overline{u_i C} - D_s \frac{\partial \overline{C}}{\partial x_i}. \tag{3c}$$

Micromixing causes changes in the concentration fluctuations defined by the quantity $(C - \langle C \rangle)^2$. Multiplying (2a) by $(C - \langle C \rangle)^2$ and integrating over the volume leads to

$$\frac{\partial \langle C^2 \rangle}{\partial t} = \frac{\partial}{\partial t} \langle (C - \langle C \rangle)^2 \rangle = \frac{\partial}{\partial t} \left[\langle C^2 \rangle - \langle C \rangle^2 \right] = -\frac{2D_s}{V_0} \int \nabla C . \nabla C dV, \tag{4}$$

indicating that the destruction of fluctuations or micromixing is associated with the smearing of small-scale density gradients by the molecular diffusion. Equation (4) can be generalized to the common situation of two scalars, e.g. A, B.

Alternatively, dispersion and mixing can be analyzed in Lagrangian terms by analyzing the movements and mixing of fluid particles. In a general turbulent flow, where the fluid particles' velocity is $w_L(t)$, they are displaced a vertical distance

$$Z(t) = \int_0^t w_L(\tau) dt. \tag{5a}$$

Following G.I. Taylor (1921), the mean square value of Z(t) for homogeneous and stationary turbulence tends to the value

$$\overline{Z^2} \sim 2(\sigma_{wL})^2 \left[T_L t - \int_0^\infty \tau R_{wL}(\tau) d\tau \right] \quad (t >> T_L), \tag{5b}$$

where σ_{wL} is a Lagrangian vertical r.m.s. velocity (which is equal to that of the Eulerian velocity $\sigma_w = \sqrt{u_3{}^2}$ (for homogeneous turbulence), T_L is the Lagrangian integral time-scale and $R_{wL}(\tau)$ is the Lagrangian auto-correlation function. Equation (5b) shows that, if the particles will disperse in the turbulent fluid without an upper bound, and at large times

$$\frac{\partial}{\partial t}\overline{Z^2} = 2\sigma_{wL}^2 T_L \quad \text{and} \quad \overline{Z^2} = 2\sigma_{wL}^2 T_L t. \tag{6}$$

However, in applying this analysis to stratified fluids, the *dynamical bound* introduced by the buoyancy forces should be considered (e.g., Csanady, 1964). For a given finite amount of kinetic energy, if the particles' density does not change as they are displaced, they cannot gain an infinite amount of potential energy and therefore can only be dispersed within a vertical distance of the order σ_w/N, where $N = [-g(\partial\bar{\rho}/\partial z)/\rho_0]^{1/2}$ is the buoyancy frequency. It also follows that $T_L = (2\sigma_{wL}^2)^{-1}\, d\,\overline{Z^2}/dt \rightarrow 0$.

However, if the particles' density can change due to *micromixing*, the equilibrium density level of each particle Z_e changes and $d\overline{Z^2}/dt \neq 0$. If we write $Z = Z_e + Z'$ where Z' is the oscillatory perturbation displacement (for that particle) relative to Z_e (Hunt, 1985), then

$$\frac{d\overline{Z^2}}{dt} = \frac{d\overline{Z_e^2}}{dt} + \frac{d\overline{Z'^2}}{dt} \approx \frac{d\overline{Z_e^2}}{dt} \tag{7}$$

In order to relate particle displacements to scalar flux, the latter can be expressed in Lagrangian terms essentially by integrating equation (2b). In a uniform vertical gradient of any scalar with concentration C, the mean vertical flux is

$$F_c = \overline{w_L\left[-Z d\bar{C}/dz + \Delta C\right]}.$$

Thus, when the scalar is the density ρ, the density flux is

$$F_\rho = \overline{w_L\left[(-d\bar{\rho}/dz)Z + \Delta\rho\right]} = -\frac{1}{2}\frac{d\bar{\rho}}{dz}\frac{d\overline{Z^2}}{dt} + \overline{\Delta\rho(t) w_L}, \tag{8a}$$

where w_L is the Lagrangian fluctuating velocity, $d\bar{\rho}/dz$ is the mean vertical density gradient, and ΔC, $\Delta\rho$ are the concentration or density changes of the particle due to micromixing caused by the diffusive term in (2a, 2b).

In neutrally stratified flows where $d\overline{Z^2}/dt$ is not confined by buoyancy forces and the micromixing is small, $F\rho \cong -K_v^\rho\, \partial\bar{\rho}/\partial z$ where $K_v^\rho = 1/2\; d\overline{Z^2}/dt$.

However, in stably stratified flows, following (7), the fluxes of a passive scalar S and the dynamic scalar ρ both have the same form,

$$F_c = -\frac{1}{2}\frac{d\overline{Z_e^2}}{dt}\frac{d\bar{C}}{dz} + \overline{\Delta C(t) w_L}, \tag{8b}$$

$$F_\rho = -\frac{1}{2}\frac{d\overline{Z_e^2}}{dt}\frac{d\rho}{dz} + \overline{\Delta\rho(t) w_L}. \tag{8c}$$

If one assumes that the density change during the up and down motions of particles can be modelled by $\gamma(\sigma_w/N)(d\bar{\rho}/dz)$, where the mixing parameter $\gamma \ll 1$ (Pearson et al., 1983), the first and second terms of (8c) become the order of $\gamma^2\; (\sigma_w^2/N)(d\bar{\rho}/dz)$ and $\overline{\Delta\rho(t) w_L} \sim \gamma(\sigma_w^2/N)(d\bar{\rho}/dz)$ respectively, showing that the second term dominates! The same argument applies to the passive scalar. Thus, passive scalar and density fluxes in stratified turbulent flows are dominated by fluid elements exchanging species with each other (micromixing). In this case the eddy diffusivity is not proportional to the 'macromixing' term $d\overline{Z^2}/dt$, i.e.,

$$F_\rho / (\partial\rho/\partial z) >> \frac{1}{2} d\overline{Z^2}/dt. \qquad (8d)$$

If a material is introduced uniformly over a horizontal plane at height $z = z_s$ and is allowed to diffused in a turbulent fluid, the vertical depth l_s of the patch where its concentration C is significant can be defined by the moments of $\overline{C}$ as

$$l_s^2 = \int_{-\infty}^{\infty} (z')^2 \overline{C} dz' / \int \overline{C} dz', \qquad (9)$$

where $z' = z - z_s$. If the vertical profiles of the mean concentration C(z) are Guassian, then it follows from the solutions to (2a), (3a) that the flux can be expressed in terms of an eddy diffusivity, e.g., where $d\overline{C}/dt = -\partial F/\partial z$ and

$$F = K_3^c \partial\overline{C}/dz, \qquad (10a)$$

If the turbulence is homogeneous and stationary, and since it is observed that at large times from the release time t = 0, $C \propto t^{-1/2}$, it follows that $K_3{}^c$ is uniform and constant, and

$$K_3^c = dl_s^2 / dt \qquad (10b)$$

In neutrally stratified flows at high Reynolds numbers, where (away from boundaries) micromixing has a negligible effect on the mean concentration $\overline{C}(t)$, if the turbulence is homogeneous and stationary, and (as is usual) the large scale turbulent velocity field is Guassian, $\overline{C}$ also has a Guassian distribution, and

$$l_s^2 = \overline{Z^2} \text{ and } K_3^c = \frac{1}{2} d\overline{Z^2}/dt. \qquad (11)$$

However, in turbulence (with length scale L) which is significantly affected by stable stratification, such that $\sigma_w/NL \leq 1$, it follows from (8b) (and the above discussion) that $F_c >> -1/2 \; d\overline{Z^2}/dt \; \partial\overline{C}/\partial z$ and therefore $K_3{}^c >> 1/2 \; d\overline{Z^2}/dt$. However, this result should not be misinterpreted as the diffusivity of the stratified case is larger than the unstratified case.

However, as Pearson et al. (1983) and Komori et al. (1996) have postulated, $K_3{}^c$ may be different from $K_3{}^\rho$ in part because the concentration distribution $\overline{C}(z)$ of the passive substance released into the flow is different from $\overline{\rho}(z)$, because of different micromixing processes. Typically $K_3{}^c \leq K_3{}^\rho$ and in both cases the fluxes are-down gradient so that $K_3{}^c$ is positive (see Hunt, 1985).

It should be noted that micromixing [i.e., the second term on the right hand side of (8)] can also be locally dominant in unstable and neutral flows near rigid or free surfaces, because in these regions also the vertical motions are inhibited, i.e. $d\overline{Z^2}/dt \rightarrow 0$.

Representation of Macromixing in Inhomogeneous Turbulent Flows

In the geophysical context, there are numerous processes that can lead to vertical mixing, which can be broadly classified into the following: (i) Mixing due to surface processes, which is typically confined to the upper boundary layer (e.g., wind-stirring induced surface wave breaking, convective cooling, surface turbulence due to shear, stirring by coherent structures such as Langmuir circulation or circulation cells aligned across the wind, evaporative and precipitation efforts); (ii) Mixing due to lateral and bottom boundary-induced processes (see the review by Garrett et al., 1993); (iii) Mixing due to internal processes (i.e., those due to local instability mechanisms in the interior not directly related to (i) and (ii), but may be driven explicitly by boundary and surface processes). The latter include: Kelvin-Helmholtz (K-H) billows in the deep oceans and continental shelf regions due to current shear or near inertial-wave shear; turbulent patches generated during the breaking of internal waves in the thermocline or inversion layers via resonant interactions, interaction with mean shear or critical-layer absorption or due to pure overturning and double-diffusive convection. In addition, mixing by artificial means is common in limnological situations, for example, the use of bubble plumes or artificial jets to improve the water quality in lakes and reservoirs. Owing to the space limitations, it is not possible here to discuss the above modes of mixing in detail, but the reader is referred to the appropriate sources of references for further information (Gregg, 1987; Fernando, 1991; Gargett, 1994; Wyngaard, 1992). Below, parameterization of a few simple mixing processes will be discussed, and the importance of identifying the mechanisms of mixing in such parameterizations is illustrated.

Consider the Eulerian computation of mixing using equations (1) and (3b), but discretized with grid boxes of volume $V = \Delta_x \Delta_y \Delta_z$ and have horizontal and vertical scales $\Delta_x \Delta_y$ and Δ_z. Then the ensemble averaged vertical buoyancy flux, further averaged over each box centered at $\underline{x} = \underline{x}_g$ is defined as

$$F_3^{(b,\Delta)} = \left\langle \left(\overline{bw}\right)_\Delta \right\rangle = (1/V)\int \overline{bw}\,(\underline{x}')\,\underline{dx}' \quad \text{where} \quad \underline{x}' = \underline{x}' - \underline{x}_g . \tag{12}$$

This average flux can be expressed in terms of the mean buoyancy gradient and a 'box' vertical eddy diffusivity of buoyancy by

$$F_3^{(b,\Delta)} = -K_3^{(b,\Delta)}\left(\frac{\partial \overline{b}}{\partial z}\right), \tag{13a}$$

and (3b) becomes

$$\frac{\partial \overline{b}^{(\Delta)}}{\partial t} = -\frac{\partial F_3^{(b,\Delta)}}{\partial z} . \tag{13b}$$

This method of eddy-diffusivity representation poses a series of questions. Some of them are: (i) How do we know what mixing mechanisms occur in scales smaller than and greater than that of the box; (ii) Can we develop a criterion based on the grid resolution which accurately models the appropriate mechanism within the grid box?; (iii) Is the buoyancy flux generation within the box aufficiently homogeneous that the eddy coefficients are scale independent?; and, (iv) How should we take into account the simultaneous presence of different mechanisms? All of these are important components of a complete model that describes the sub-grid scales processes. Lacking extensive knowledge on turbulent mixing in stratified fluid layers of complex and multi-component stratification, the current models often resort to simple semi-empirical parameterizations that have little physical and fluid dynamical grounds.

To illustrate the importance of understanding mixing mechanisms and possible scale dependence of mixing parameterization, consider the case of a shear-free density interface of thickness δ embedded in an otherwise homogeneous turbulent layer specified by r.m.s. velocity u_H and lengthscale L_H (see Figure 1). The governing parameters for the problem are u_H, L_H, Δb and δ, and hence the relevant non-dimensional parameters are the bulk Richardson number, $Ri = \Delta b L_H / u_H^2$ and the internal Richardson number $Ri_I = N^2\delta^2/u_H^2$ where $N^2 = \Delta b/\delta$. Carruthers and Hunt (1997) have shown that Ri_I is the key to the determination of the internal wave modes in the stratified layer and hence plays a major role in determining the fluxes. Accordingly, if $Ri_I < \pi^2$ (thin interfaces), the lowest mode of internal waves will be dominant and if $Ri_I < n^2\pi^2$ the first (n) modes will be present. Most atmospheric and oceanic layers have $h/L_H < 0.1$, and the case of a thin interface can be considered as generic to such flows.

The next question is: What should be the mixing mechanisms that can prevail at such interfaces? Laboratory experiments have provided some useful answers to this question (Fernando, 1991) in that: at low Ri, the eddies impinge on the interface and splash thin sheets of fluid from the opposite layer into the turbulent layer (Figure 2a); at moderate Ri ($2 < Ri < 10$), there is a possibility of generating K-H billows (Figure 2b) and at larger Ri ($20 < Ri < 100$) the interface is dominated by breaking of interfacial waves (Figure 2c). A plausible scheme of calculating the mixing rate for case (a) is to assume that the rate of supply of energy to the interface is compensated by the dissipation (ε) and the buoyancy flux (q) due to mixing. If the energy supply rate, as specified by the energy flux divergence $-\partial M_z/\partial_z$, can be modelled as w_i^3/L_ε, where w_i and L_ε are the relevant velocity and lengthscales, the quasi-stationary energy balance at the interface.

$$0 = -\frac{\partial M_z}{\partial z} - q - \varepsilon, \tag{14}$$

becomes $q/\eta \sim w_i^3/L_\varepsilon$, where $\eta = q/(q+\varepsilon)$ can be considered as the "mixing efficiency." The vertical 'box' eddy diffusivity can now be defined as $q = K_3^{(b,\Delta)}\,\Delta b/L_V$, or

$$K_3^{(b,\Delta)} \sim \eta u_H L_H (w_i/u_H)^3 (L_V/L_\varepsilon) Ri^{-1}, \tag{15}$$

where $L_V = \Delta_z$ is the vertical resolution of the box. In general, $w_i \sim u_H Ri^{-\alpha}$, where the parameter $\alpha = \alpha(Ri)$, and for the case (a) $\alpha \approx 0$ and $L_\varepsilon \sim L_H$ are good approximations (Linden, 1975). Furthermore, $\eta = \eta\,(Ri)$ but for continuously forced flows η = constant (McEwan, 1983). Thus, (15) can be represented as $K_3^{(b,\Delta)} \sim (L_V/L_H)\, u_H L_H Ri^{-1}$.

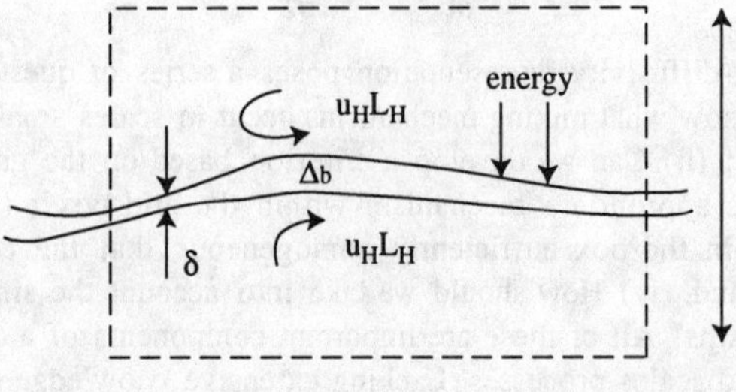

Figure 1. A density interface subjected to background shear-free turbulence. The vertical resolution of the computation grid is shown by the dashed lines.

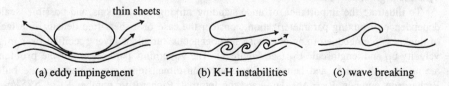

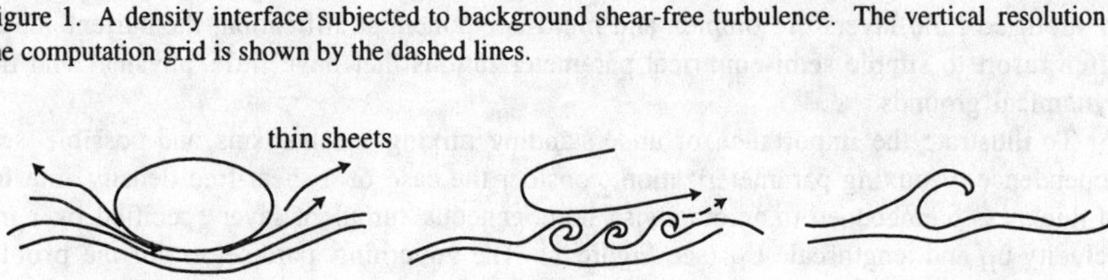

Figure 2. Possible mixing mechanisms in shear-free turbulence, active at different Richardson number ranges.

On the other hand, Mory (1991) presented a parameterization for the case 2(b) dominated by K- H billows. If the eddies of integral scales slosh smaller eddies of wave number k on the interface, the induced strain can be written as $\omega \sim k u_H$, and hence K-H billows are possible if $(\Delta b/\delta)/(k u_H)^2 < 1$ or $kL_H > Ri^{1/2} (L_H/\delta)^{1/2} [=k_c L_H]$. Now the energy absorption into the interface can take place over the range $k_c < k < \infty$, and hence the energy flux divergence can be parameterized by

$$-\frac{\partial M_z}{\partial z} \propto \int_{k_c}^{\infty} \frac{1}{T(k)} E(k)dk, \tag{16}$$

where T(k) is a time scale pertinent to the absorption of energy into the interface at wave number k and E(k) is the energy spectrum. If k_c is in the inertial subrange, a reasonable choice would be $T(k) \sim [E(k)k^3]^{1/2}$, where $E(k) \sim \varepsilon^{2/3} k^{-5/3}$. Using arguments analogous to those used in deriving (15), it is then possible to obtain $K_3{}^{b,\Delta} \sim (L_v/L_H) u_H L_H Ri^{-3/2}$.

Calculations pertinent to the energy absorption into a linear wave field of a thin density interface have been performed by Fernando and Hunt (1997). They found that the low frequency internal waves within the interface should break by the resonance induced by the pressure fluctuations of contiguous turbulent layers. Their results lead to an eddy diffusivity of $K_3{}^{b,\Delta} \sim (L_v/L_H) u_H L_H Ri^{-5/3}$.

Based on the above calculations for the case of shear-free interfaces, a suitable parameterization (that satisfies $K_3{}^{b,\Delta} \sim u_H L_H$ when $Ri \rightarrow 0$) should be $K_3{}^{b,\Delta} = c_1/(c_2 + Ri^n)$ where c_1 and c_2 are constants that may depend on grid resolution and n = 1 for case (a), n = 3/2 for case (b) and n = 5/3 for case (c). This simple example clearly illustrates that in representing turbulent mixing in inhomogeneous turbulent flows by using eddy diffusivity concepts, the results can be quite sensitive to the mixing mechanisms prevailing in the flows as well as resolution employed for the calculations. In the next section, we will demonstrate how the results of a numerical mixed-layer model can exhibit widely differing phenomena in response to varying eddy diffusivity parameterizations.

Sensitivity of Mixed-Layer Models to Mixing Parameterizations

Here we consider a simple one-dimensional mixed-layer model that describes erosion of an initially stable linear density gradient $\rho(0,z) = \rho_0(1 + \alpha z)$ by the action of a surface wind stress (Figure 3a). The equations of motion in the horizontal direction become

$$\frac{\partial \bar{u}}{\partial t} = \frac{\partial}{\partial z}\left(k_m \frac{\partial \bar{u}}{\partial z}\right) + f\bar{v} \tag{17a}$$

$$\frac{\partial \bar{v}}{\partial t} = \frac{\partial}{\partial z}\left(k_m \frac{\partial \bar{v}}{\partial z}\right) - f\bar{u}, \tag{17b}$$

where f is the Coriolis parameter ($f = 10^{-4}$ s^{-1}), k_m is the momentum eddy diffusivity and $\bar{u}$ and $\bar{v}$ are the horizontally averaged mean velocity in x and y directions. The mean buoyancy $\bar{b}$ equation becomes

$$\frac{\partial \bar{b}}{\partial t} = \frac{\partial}{\partial z}\left(k_b \frac{\partial b}{\partial z}\right), \tag{18}$$

where k_b is the scalar diffusivity. The energy equation takes the form

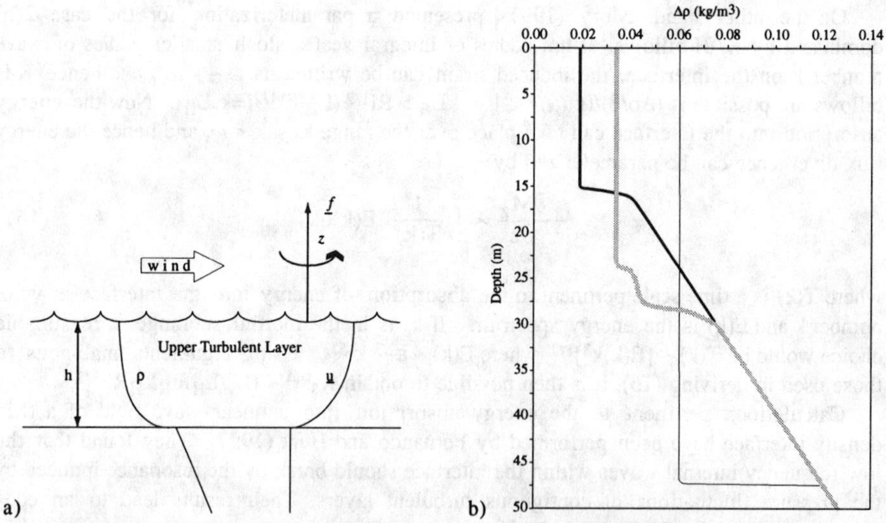

Figure 3. Simulation of mixed-layer deepening in a linearly stratified fluid subjected to a surface wind stress and background rotation. Three different parameterizations were used for eddy diffusivities of momentum, buoyancy and turbulent kinetic energy. Figure 3a illustrates the flow configuration considered and 3b shows the results of simulations. All microstructure features showed in the figure are well resolved in the simulation. Δρ represents the difference in densities at a given depth at time t and the surface at t=0.

$$\frac{\partial e}{\partial t} = \frac{\partial}{\partial z}(k_e \frac{\partial e}{\partial z}) + k_m S^2 - k_b N^2 - \varepsilon, \qquad (19)$$

where k_e is the diffusivity of turbulent kinetic energy e, $S^2 = (\partial \bar{u}/\partial z)^2 + (\partial \bar{v}/\partial z)^2$ is the vertical shear and ε is the rate of turbulent kinetic energy dissipation. The initial conditions are

$$\bar{u}(z,0) = \bar{v}(z,0) = 0 \quad \text{and} \quad e(z,0) = e_f \qquad (20)$$

where e_f is the initial background turbulence level. For the present calculations, $e_f \approx 5 \times 10^{-8}\,m^2\,s^{-2}$ was assumed. The surface (at z = 0) flux boundary conditions are taken as

$$k_m \frac{\partial \bar{v}}{\partial z} = -\frac{\tau_x}{\rho_0},\; k_m \frac{\partial \bar{v}}{\partial z} = -\frac{\tau_y}{\rho_0},\; k_b \frac{\partial \bar{b}}{\partial z} = F_b \text{ and } k_e \frac{\partial e}{\partial z} = F_e\,, \qquad (21)$$

where $\bar{\tau} = (\tau_x, \tau_y)$ is the surface wind stress, ρ_o is the reference density and F_b and F_e are surface buoyancy and energy fluxes, respectively. For the present calculations, $F_b = 0$, $F_e = 0.1|\bar{\tau}/\rho_0|^{3/2}$ were taken. Deep in the fluid layer, at an undisturbed depth of H, $\bar{u}$ (H,t) = $\bar{v}$ (H,t) = 0 and e (H,t) = e_f.

The momentum eddy diffusivities were parameterized using the following form proposed by Lozovatsky et al. (1993)

$$k_m = c_k \frac{e}{S\phi(Ri_g, Ri_{cr})}, \qquad (22)$$

where $c_k = 1.3$, ϕ is a function of the local gradient Richardson number $Ri_g = N^2/S^2$ and Ri_{cr} is 0.1. The form of $\phi(Ri_g, Ri_{cr})$ proposed by Lozovatsky et al. (1993) is

$$\phi = \sqrt{1 + Ri_g / Ri_{cr}}\,. \tag{23a}$$

An alternative form for ϕ that has the same asymptotic behaviour as (23a) for $Ri_g >> Ri_{cr}$ and $Ri \to 0$ was also used in the present study, viz.,

$$\phi^{mod} = 1 + \sqrt{Ri_g / Ri_{cr}}\,. \tag{23b}$$

Figure 3(b) shows the mixed-layer depth at time t=36 hrs. into the simulation, for, $\tau_x = 2.4 \times 10^{-4}$ kg/ms^2, $\tau_y = 0$, $\rho_0 = 10^3$kg/m^3, $\alpha = -2.5 \times 10^{-6}$ m^{-1} calculated using the following combinations and the assumption $k_e = 0.73k_m$:

$$k_m = \frac{c_k e}{S\phi},\ k_b = k_m, \qquad \text{(thin line)} \tag{24a}$$

$$k_m = \frac{c_k e}{S\phi},\ k_b = \frac{k_m}{1 + Ri/Ri_{cr}}, \qquad \text{(dark line)} \tag{24b}$$

$$k_m = \frac{c_k e}{S\phi^{mod}},\ k_b = \frac{k_m}{1 + Ri/Ri_{cr}}. \qquad \text{(thick line)} \tag{24c}$$

It is clear that the depth of the mixed layer as well as the density structure are sensitively dependent on the nature of the eddy diffusivity parameterization used (note that the parameterizations employed here has certain similarities in asymptotic limits described above). The form (24b) showed an interesting feature, in that it produced step-like density structures at the base of the mixed layer which was absent in the other two cases. This calculation clearly exemplifies the sensitivity of model results to the parameterizations and hence stresses the need for continuous improvements in eddy difffisivity parameterizations used for environmental forecasting models.

Closure

In foregoing sections, we have reviewed certain aspects of turbulent mixing in stably stratified flows. Using a Lagrangian formulation of fluid parcel displacements and dynamical constraints imposed by energetics, it was argued that the density fluxes in stably stratified homogeneous turbulent flows occur not due to displacement of fluid lumps, but due to the changes of their densities during the displacements. Thus the foundations of eddy diffusivities for the stratified case should fundamentally differ from the non-stratified case. The case of mixing across a thin shear-free interface was analyzed as an example of an inhomogeneous turbulent flow. The problems associated with using eddy difffisivities were illustrated by using a simple mixed-layer model which yielded widely different results, as a result of minor adjustments to the diffusivity parameterizations.

Acknowledgments. Stratified and rotating flow research at Arizona State University is supported by the Office of Naval Research (Physical Oceanography and High Latitude Processes Programs), the National Science Foundation (Fluid Mechanics and Hydraulics Program) and the Army Research Office (Geosciences Directorate). The authors wish to gratefully acknowledge this support.

References

Caldwell, D. R. and J. N. Moum, Turbulence and mixing in the ocean. *Rev. Geophys.*, *33*(2), 1385-1394, 1995.

Carruthers, D. J. and J. C. R. Hunt, Waves, turbulence and entrainment near an inversion layer. *J. Fluid Mech.*, submitted, 1998.

Csanady, G. T., Turbulent difffision in a stratified fluid. *J. Atmos. Sci. 21*, 439447, 1964.

Fernando, H. J. S., Turbulent mixing in stratified fluids. *Ann. Rev. Fluid Mech.*, *23*, 455- 493. 1991.

Fernando, H. J. S. and J. C. R. Hunt, Some aspects of turbulence and mixing in stably stratified layers. *Dyn. Atmos. Oceans.*, *23*, 35-62, 1996.

Fernando, H. J. S. and J. C. R. Hunt, Turbulence, waves and mixing at shear-free density interfaces. Part 1: A theoretical model. *J. Fluid Mech.*, *347*, 197-234, 1997.

Gargett, A. E., Ocean turbulence. *Ann. Rev. Fluid Mech.*, *21*, 419-451, 1994.

Garrett, C., P. MacCready, and P. Rhines, Boundary mixing and arrested Ekman layers. *Ann. Rev. Fluid Mech.*, *25*, 291-323, 1993.

Gregg, M. C., Diapycnal mixing in the thermocline: A review. *J. Geophys. Res.*, *92*, 5249-5286, 1987.

Hanazaki, H. and J. C. R. Hunt, Linear processes in unsteady stably stratified turbulence. *J. Fluid Mech.*, *318*, 303-337, 1996.

Hunt, J. C. R., Diffusion in the stable boundary layer. *in Atmospheric Turbulence and Air Pollution Modelling*, edited by F. T. M, Nieuwstadt and H. vanDop, Riedel, 231-274, 1982.

Hunt, J. C. R., Diffusion in the stably stratified atmospheric boundary layer. *J. Climate and Appl. Meteor.*, *24*, 1187-1195, 1985.

Komori, S., K. Nagata, and Y. Murakami, Heat and mass transfer in a stable thermally stratified flow. *Dyn. Atmos. Oceans.*, *23*, 235-245, 1996.

Large, W. G., J. C. McWilliams, and S. Doney, Oceanic vertical mixing: A review and a model with a non-local boundary layer parameterization. *Rev. Geophys.*, *23*, 363-403, 1994.

Ledwell, J. R., A. J. Watson, and C. S. Law, Evidence for slow mixing across the pycnocline from an open-ocean tracer-release experiment, *Nature*, *364*, 701-703, 1994.

Linden, P. F., The deepening of a mixed layer in a linearly stratified fluid. *J. Fluid Mech.*, *71*, 385-405, 1975.

Lozovatsky, I. D., A. S. Ksenofontov, A. Y. Erofeev and C. H. Gibson, Modeling of the evolution of vertical structure in the upper ocean by atmospheric forcing and intermittent turbulence in the pycnocline. *J. Mar. Sys.*, *4*, 263-273, 1993.

McEwan, A. D., The kinematics of stratified mixing through internal wave breaking. *J. Fluid Mech.*, *128*, 47-58, 1983.

Mory, M., A model of turbulent mixing across a density interface including the effect of rotation. *J. Fluid Mech.*, *223*, 193-207, 1991.

Munk, W. H., Abyssal Recipes. *Deep-Sea Res.*, *13*, 707-730, 1966.

Pacanowski, R. C. and S. G. H. Philander, Parameterization of vertical mixing in numerical models of tropical oceans, *J. Phys. Oceanog.*, *11*, 1443-1451, 1981.

Pearson, H. J., J. S. Puttock, and J. C. R. Hunt, A statistical model of fluid element motions and vertical diffusion in a homogeneous turbulent flow. *J. Fluid Mech. 129*, 219-249, 1983.

Polzin, K. L., J. M. Toole, J. R. Ledwell, and R. W. Schmitt, Spatial variability of turbulent mixing in the abyssal ocean. *Science*, *276*, 93-96, 1997.

Schumann, U., Direct and large eddy simulations of stratified homogeneous shear flow. *Dyn. Atmos. Oceans.*, *23*, 81-98, 1996.

Stull, R. B., *An Introduction to Boundary-Layer Meteorology*. Kluwer Academic, 1988.

Taylor, G. I. Diffusion by continuous movements. *Proc. Lond. Math. Soc.* Ser. 2, *20*, 196-212, 1921.

Toole, J. M., K. L. Polzin, and R. W. Schmitt, New estimates of diapycnal mixing in the abyssal ocean. *Science*, *264*, 1120-1123, 1994.

Wyngaard, J. C., Atmospheric turbulence. *Ann. Rev. Fluid Mech.*, *24*, 205-233, 1992.

26

Intermittency of Internal Wave Shear and Turbulence Dissipation

C. H. Gibson

Abstract

It is crucial to understand the extreme intermittency of ocean and lake turbulence and turbulent mixing in order to estimate vertical fluxes of momentum, heat and mass by Osborn-Cox flux-dissipation methods. Vast undersampling errors occur by this method when intermittency is not taken into account, Gibson (1990a, 1991ab). Turbulence dissipation and internal wave shear in the ocean are closely coupled. Often, oceanic turbulence is assumed to be caused by breaking internal waves. However, the extreme intermittency observed by Gregg et al. (1993), with intermittency factor $I_S{}^4 \approx 6$ for 10 m internal wave shears S, strongly suggests that wave breaking is not the cause of turbulence but an effect. The usual assumption about the wave-turbulence cause-effect relationship should be reversed. Wave motions alone reduce intermittency. Oceanic turbulence increases intermittency as the result of a self-similar nonlinear cascade covering a wide range of scales, mostly horizontal, but no such nonlinear cascade exists for internal waves. Extremely large I_ε and I_χ measured for oceanic turbulence are in the range 3-7, Baker and Gibson (1987). These values are consistent with the third universal similarity hypothesis for turbulence of Kolmogorov (1962) and a length scale range over 3-7 decades from viscous or diffusive to buoyancy or Coriolis force domination, where the measured universal intermittency constant $\mu = 0.44$, Gibson (1991a), is a result of singularities in multifractal turbulence dissipation networks and their degeneration, Bershadskii and Gibson (1994). The extreme intermittency of small scale internal wave shears in the ocean is a fossil turbulence remnant of the extreme intermittency of ocean turbulence, and the waves themselves may be fossil turbulence. Evidence that composite shear spectra and schematic temperature gradient spectra in the ocean reflect fossil turbulence effects, Gibson (1986), is provided by the decaying tidal sill temperature spectra of Rodrigues-Sero and Hendershott (1997).

Introduction

Davis (1994) has reviewed the flux-dissipation method, referred to as the Osborn-Cox method when applied to the ocean, to estimate vertical diffusivites K_V from the expression $K_V \approx DC$, where D is the molecular diffusivity of a conserved property like species

Physical Processes in Lakes and Oceans
Coastal and Estuarine Studies Volume 54, Pages 363-376

concentration or temperature T (internal energy) and C is the Cox number $C = \langle(\nabla T)^2\rangle/\langle\nabla T\rangle^2$, and points out that large discrepancies exist between such microstructure values and much larger values found by matching large-scale observations to models or budgets; e.g., Munk (1966). Several possible sources of small errors in K_V are considered, but not intermittency in space-time versus control volumes as emphasized by Gibson (1990a) in another review of the Osborn-Cox method. Davis (1996) considers some of the possible effects of intermittency on K_V estimates, but criticizes the suggestion of Baker and Gibson (1987) that since most microstructure data sets for ε, χ and C in identifiable ocean layers have sample distribution functions imperceptibly different from intermittent lognormal distributions, that maximum likelihood estimators of mean values based on lognormality should provide the best available estimators for K_V.

Davis (1996) suggests $|\nabla T|$ is more likely to be lognormal than χ and that if $|\nabla T|$ is lognormal then χ and averages of χ over various length scales cannot be lognormal, contrary to the proposal by Gibson (1981) that χ and averages of χ should be lognormal for oceanic mixing over a wide range of length scales. The Gibson (1981) analysis simply extends to turbulent mixing of conserved scalar fluid properties the very robust Gurvich and Yaglom (1967) breakdown coefficient analysis that provides the physical basis for Kolmogorov's third universal similarity hypothesis of intermittent lognormality of turbulence dissipation rates. This suggestion of Davis (1996) has no physical basis and is contrary to observations of χ averaged over scales $L_{ave.}$ from 1 to 1024 meters in the seasonal thermocline off the coast of California, Gibson (1991a), that were precisely lognormal with $I_\chi \approx 0.44 \ln (L_O/L_{ave.})$ for all $L_{ave.}$.

The Davis (1996) suggestion that fitting χ observations to lognormal distributions can give quite erroneous results is inconsistent with Gibson (1991b) estimates of K_V values from the 150 m average C values reported by Gregg (1977). These agree well with lognormality for various ocean layers sampled, and confirm the Munk (1966) estimates of K_V for vertical heat flux in the deep main thermocline, contrary to the Gregg (1989) correlation that underestimates the deep heat flux by a factor of about thirty. No statistically significant difference exists between K_V values inferred by the Munk (1966) abyssal recipe and the flux-dissipation methods (Osborn-Cox) based on the Gregg (1977) data, as shown by Gibson (1991a) taking the lognormal intermittency of C into account.

Some physical processes of nature are intermittent and some are not, where the degree of intermittency of a random variable X is measured by statistical parameters showing departures of its distribution function from that of a normal distribution; for example, skewness, kurtosis, and the intermittency factor I_X, where

$$I_X \equiv \sigma^2_{\ln X},\ X \text{ lognormal}, \tag{1}$$

is defined by Baker and Gibson (1987). What is the difference? Examples of non-intermittent processes in the ocean include the temperature, pressure, salinity, most chemical species distributions, and surface and internal wave displacements. Examples of intermittent oceanic processes include the viscous dissipation rate ε of turbulence, the thermal diffusive dissipation rate of temperature variance χ, the Cox number $C = \langle(\nabla T)^2\rangle/\langle\nabla T\rangle^2$ for temperature, the shear rate S of small scale internal waves in the ocean, the rate of biological productivity, and the vertical fluxes of heat, mass of chemical species, and momentum through the thermocline, Gibson (1990a). Why are these various examples different with respect to their intermittency?

Part of the difference has to do with the differences in intermittency of the sources of the various physical parameters. Temperature differences in the ocean are primarily determined by smooth large-scale variations in radiation transport on scales of the planet, and many of the non-intermittent chemical species distributions, like oxygen, also have uniform very large-scale sources and sinks. Chemical species with intermittent sources in space and time are more intermittent than those that have non-intermittent sources and sinks. As soon as the intermittent source is removed, the intermittency of such species concentrations begin to disappear as their maxima are damped and dispersed by diffusion and convection. Conservation laws for such random variables are linear in the variable and will tend to have normal probability distribution functions unless forced away from Gaussianity by non-Gaussian, intermittent, source terms. For example, tritium concentration in the ocean today is less intermittent than it was in the days of intermittent nuclear weapons testing. Biological growth rates are extremely intermittent because they depend on a coincidence of several random variables (such as nutrient concentrations, sunlight, turbulence, and even turbulence intermittency, Gibson and Thomas (1995)) having their proper value ranges simultaneously. Deviations from intermittent lognormality of oceanic dissipation rates occur when data sets include segments from different layers with different intermittency parameters, Yamazaki and Lueck (1990).

According to the intermittency hypothesis of Kolmogorov (1962), random variables of turbulence in the ocean such as ε are highly intermittent because the equations of turbulent momentum and vorticity are highly nonlinear and the range of scales of the turbulence cascade is enormous. Universal intermittency parameters of turbulence have been investigated in terms of singularities on dissipation networks, or caustics, and their degeneration, by Bershadskii and Gibson (1994), using multifractal asymptotics. The hallmark of intermittent random variables is a self-similar nonlinear cascade over a wide range of scales. Classic examples are turbulence, turbulent mixing, personal income, concentrations of precious metals, and the density of astrophysical fluids undergoing self gravitational condensation. Distributions of these random variables are close to intermittent lognormals. Each is subject to severe undersampling errors.

The intermittency parameter kurtosis K is defined as the fourth moment of X about its mean divided by its second moment, where

$$K = \exp(I_X) \qquad (2)$$

for a lognormal random variable X. For a normal random variable, K = 3. Therefore, from (2), I_X is 1.1 for normal random variables. Most data sets from particular oceanic layers show that the probability distributions of the viscous and temperature dissipation rates ε and χ are indistinguishable from intermittent lognormal distributions, Baker and Gibson (1987), and have intermittent I_ε and I_χ values in the range 3–7, much larger than 1.1. Such large departures from Gaussian behavior can produce severe problems for any experimental attempt to sample mean values of ε and χ because the mean to mode ratio G increases exponentially with I_X

$$G = \text{Mean/Mode} = \exp(3I_X/2). \qquad (3)$$

The ratio G is the expected undersampling error for intermittent random variables using sparse samples, and might be termed the Gurvich number. G equals 90 for $I_X = 3$, and 4×10^4 for $I_X = 7$. Thus, the expected undersampling errors for single samples of oceanic ε or χ are underestimates by two to four orders of magnitude. The first claim that the Munk (1966) abyssal recipe was incorrect was that of Osborn and Cox (1972), based on a single

25 m long microstructure record. Confirmation of this erroneous claim was provided by Gregg, Cox and Hacker (1973) with similar measurements of small C values (1.3, 2.2, 1.3, 2.0, 1.7 and 1.4) in the North Pacific Central Gyre, and 286 in the San Diego Trough neglecting G corrections.

The possibility of large undersampling errors demands extreme care in estimating mean dissipation rates from sparse data sets. Large numbers of samples are needed, but the samples must be independent and representative of the full range of space and time scales in the space-time control volume of interest. As mentioned above, Gibson (1991b) examines relatively independent 150 m average Cox numbers C, where $\langle C \rangle$ can be interpreted as the ratio of the average turbulent to molecular vertical diffusivities, measured by Gregg (1977) in three layers of the upper ocean of the mid-Pacific for three seasons of three years, and finds C to be an intermittent lognormal with I_C values in the range 3-6 for all layers. No statistically significant difference was found between the maximum likelihood estimator of the mean $\langle C \rangle$ and the value $\langle C \rangle \approx 700$ predicted by Munk (1966) for deep vertical diffusivity in the main thermocline.

Mode values of C are smaller than $\langle C \rangle \approx 700$ by factors of 1-2 orders of magnitude according to numerous investigators using microstructure dropsondes, leading to a "dark mixing" paradox (a term invented by Tom Dillon) analogous to the "dark matter" paradox of cosmology. No "dark mixing" paradox exists for $\langle C \rangle$ in the ocean if its extreme intermittency is taken into account. The Gregg (1995) conclusions that $\langle \varepsilon \rangle$, $\langle \chi \rangle$, and $\langle C \rangle$ are less near the equator than at mid-latitudes may actually reflect maximum undersampling errors (maximum Gurvich numbers) in dropsonde microstructure methodology near the equator where the intermittency factors for these quantities are maximum, due to the minimum in Coriolis forces and the huge range of horizontal turbulence scales up to hundreds of kilometers, Gibson (1983).

Any hydrophysical field of the ocean that is generated by an intermittent source will itself be intermittent, at least initially. Random variables like dissipation rates ε and χ are intrinsically intermittent because turbulence is intermittent, with I_ε and I_χ values increasing $\propto \ln L_O$. Random variables like temperature, internal wave shear, and chemical species concentration can disperse widely, and will be intermittent only as long as they are forced to be so by intermittent sources. Clearly if X is an intermittent random variable, then a random variable $Y \propto X$ will have the same intermittency parameters as X because Y has the same probability density function as X. Proportionality constants c cancel for the variance of logarithms, so that $I_X = I_x$ if $X = cx$. Other random variables $Z = f(X)$ will be more or less intermittent than X depending on the function f(X). Intermittency factors I_χ and I_ε can easily be measured because $I_\chi = I_{(\partial T/\partial x)^2}$ and $I_\varepsilon = I_{(\partial u/\partial x)^2}$ since $\chi \propto (\partial T/\partial x)^2$ and $\varepsilon \propto (\partial u/\partial x)^2$.

Measured intermittency factors I_χ not only reflect measured large intermittency factors I_ε but have the same values, Gibson (1991a). A model predicting this equality is given by Gibson (1981) based on the Gurvich and Yaglom (1967) model for the Kolmogorov (1962) intermittency hypothesis. Small scale internal wave shear S, S^2, S^4, ... , should have small I_{S^4} if the shear is dominated by linear wave mechanics and large I_{S^4} if small scale internal waves are caused by turbulence. Intermittency factors are good indicators of whether the processes that produce a particular random variable are linear or nonlinear. If I_{S^4} is large, it does not follow that turbulence is caused by breaking waves. Instead, it follows that the waves are caused by turbulence because the intermittency of the wave shear reflects the intermittency of the turbulence. If the turbulence were caused by waves I_ε values would be near 1.1, not 3-7.

Gregg et al. (1993) show that the internal wave shear S on scales of about 10 m is quite intermittent, and that S^4 is lognormal with $I_{S^4} \approx 6$. Because $\varepsilon \propto S^4$ according to Gregg (1989), Gregg et al. (1993) suggest the intermittency of ε may be explained as due to the intermittency of S^4 assuming turbulence is caused by breaking internal waves, but this conclusion is precisely reversed. Internal wave shears cannot be intermittent unless the waves are forced by nonlinear processes, and the most likely nonlinear forcing process for the ocean interior is turbulence. Waves break when they are forced to do so by the surface, beaches, or bottom topography (or by the turbulence triggered by these entities). Internal wave shears on 10 m scales are fossils of the rare turbulence events that intermittently reach such large vertical scales, Gibson (1986).

The Gregg et al. (1993) observation that $I_{S^4} \approx 6$ can be compared to a measured intermittency factor $I_\varepsilon \approx 5$, Washburn and Gibson (1984). $S^4 \propto \varepsilon^{4/3}L^{-8/3}$ from Kolmogorov's second hypothesis, since $\Delta V \propto (\varepsilon L)^{1/3}$ and $S \propto (\Delta V)/L$. Therefore,

$$I_{S^4} \approx (4/3)^2 I_\varepsilon. \tag{4}$$

from the definition of I_X. According to Kolmogorov's third hypothesis, $I_\varepsilon \approx 0.44 \ln (L_O/cr)$, where L_O is the energy, or Obukhov, scale of the turbulent cascade, r is the averaging scale for ε, and c is a constant of order ten. Taking cr = 0.3 m from Washburn and Gibson (1984) gives $L_O \approx 26$ km from $I_\varepsilon \approx 5$, and I_ε only 2.4 for cr = 100 m from (4), compared to $I_\varepsilon = 6/(4/3)^2 = 3.5$ from Gregg et al. (1993). This suggests a larger L_O value for Gregg et al. (1993).

Active turbulence in the ocean interior has never been observed simultaneously with internal waves with the extremely large shear rates S that must exist to produce the extreme intermittency factors $I_{S^4} \approx 6$ reported by Gregg et al. (1993). Fossil turbulence remnants of previous powerful mixing events, with vertical overturns up to 8 m, have been identified by Gibson (1982) from the Gregg (1977) profiles with largest C values, but S values were not measured. Wave spectra far downstream of powerful turbulence on a tidal sill are given below.

Temperature and density overturns on 20 m scales in the strong shear layer of the equatorial undercurrent were detected by Hebert et al. (1992) in a large patch of partially fossilized turbulence precisely in the center of a packet of internal waves with maximum amplitude of 20 m. Their claim that the strong simultaneous (but partially fossilized) turbulence observed was caused by the breaking of the 20 m amplitude internal wave packet they measured is questioned by Gibson (1991b) on the grounds that turbulence overturns caused by breaking waves should be on scales smaller than the amplitude of the wave and not identical to it. Identical maximum amplitudes for overturning scales and internal waves is a signature of internal waves caused by turbulence, Gibson (1980, 1986). Internal waves in the ocean interior typically have amplitudes of tens of meters, but turbulence overturn scales are ten meters or less. Recent measurements of Rodrigues-Sero and Hendershott (1997) showing evidence of saturated internal waves and persistent 60 m vertical amplitude fossil turbulence temperature overturns in the Sea of Cortez are described below. These internal waves are known to have been produced by powerful turbulence formed by tides flowing over sills, with overturn scales identical to the amplitude of the internal waves produced.

Causes of Intermittency in Physical Systems

The most important source of intermittency in physical systems is nonlinearity of the conservation equations describing the random variable. For the case of turbulence, turbulence dissipation rate ε, small scale internal waves, S^4 etc., the random variable in the ocean determining these random variables is the vorticity ω, that is described by the equation

$$\frac{\partial \omega}{\partial t}+(v\cdot\nabla)\omega=\omega\cdot e+2\Omega\cdot E+\frac{\nabla\rho\times\nabla_P}{\rho^2}-\frac{\nabla\rho\times\nu\nabla^2 v}{\rho}+\nu\nabla^2\omega \qquad (5)$$

where e is the rate of strain tensor with components $(\partial v_i/\partial x_j + \partial v_j/\partial x_i)/2$, Ω is the rotation vector of the coordinate system, ρ is density, p is pressure, and ν is the kinematic viscosity. The left hand side of (5) is the rate of change of the vorticity of a fluid particle. The vorticity becomes intermittent because of the nonlinearity of the first term on the right hand side of (5), representing the rate of increase due to vortex stretching. The term is nonlinear in velocity gradients, and causes rapid increases in vorticity magnitude because regions with large vorticity acquire large strain rates in the stretching direction when the vorticity aligns with the stretching axis of the strain rate tensor as it is amplified by it. The second (Coriolis) term on the right is generally smaller than the first. The third term shows that vorticity is produced on tilted density interfaces such as occur in small scale internal waves. However, the vorticity produced tends to disappear and reverse for internal waves with small amplitude that do not form turbulence.

When the density gradient tilt is steady, as on a front, the shear can build up until a turbulence patch forms. Such persistent density gradient tilts are characteristic of fronts formed at boundaries of horizontal turbulence eddies stirring horizontal density gradients, as shown in Figure 1. The relationship of equation (4) shown in Figure 1 between I_{S^4} and I_ε assumes equal averaging scales are used for both the shear S and the viscous dissipation rate ε. From Kolmogorov's third hypothesis $I_\varepsilon = 0.44 \ln (L_O/cr)$ and the relation $I_{S^4} \approx (4/3)^2 I_\varepsilon$ it is clear that such large values of shear intermittency factors as $I_{S^4} \approx 6$ on scales of 10 m require such a large range of scales in the turbulence cascade, to tens of kilometers, that most of the turbulence energy must be in the horizontal rather than in the vertical.

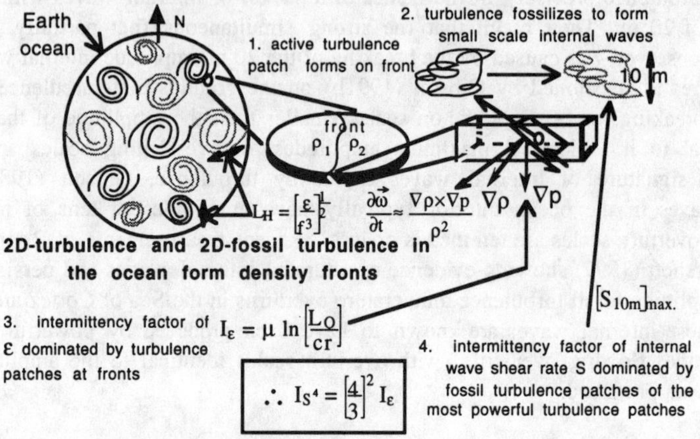

Figure 1. Four step model showing how the intermittency of ε causes the intermittency of internal wave shear rates S^4 if the waves are fossil vorticity turbulence.

As shown in Figure 1, horizontal scales of turbulence in the ocean are limited either by the time of application of the source as turbulence cascades from small scales to large, or by the Hopfinger scale $L_H = (\varepsilon/f^3)^{1/2}$, where f is the Coriolis parameter $2\Omega\sin\theta$ and θ is latitude. Horizontal eddy sizes up to hundreds of kilometers are observed at equatorial latitudes, compared to tens of kilometers near the poles, and intermittency factors estimated by Baker and Gibson (1987) show a similar correlation. It will be interesting to see whether values of the shear intermittency factor $I_S{}^4$ are also correlated with latitude. This would be further evidence that the dominant small scale internal wave shear is caused by the dominant turbulence events (those that control the vertical fluxes), rather than such turbulence being caused by internal wave breaking.

As mentioned previously, Gurvich and Yaglom (1967) provide a clear physical model for the intermittent, lognormal, distribution function for the dissipation rate of very high Reynolds number turbulence postulated by Kolmogorov (1962) in his third universal similarity hypothesis for turbulence. This model and Kolmogorov's hypothesis have been extended to passive scalar fields like temperature mixed by turbulence, Gibson (1981). Both models note that the mechanisms of turbulence and turbulent mixing are identical for every stage of the nonlinear cascades from small scales to large, and that every stage is independent of much larger or much smaller stages. If ε is averaged over a region of size r it should be independent of ε averaged over a larger region of size nr for some value of n>1. The ratio of ε_r to ε_{nr} and the ratio ε_{nr} to ε_{n^2r} should be independent and identically distributed. Therefore, the ratio ε_r to ε_{Lo} can be written as the product of a large number of independent, identically distributed random variables with the form $\varepsilon_{n^k r} / \varepsilon_{n^{k+1} r}$ for k an integer >1 if there is a very large range of scales between r and the energy, or Obukhov, scale L_O, Gibson (1990b).

Taking the logarithm of the product and applying the central limit theorem shows that ε_r must be lognormally distributed with intermittency factor increasing logarithmically with the Reynolds number, as assumed by the Kolmogorov (1962) third universal similarity hypothesis that the viscous dissipation rate averaged over scale r should be lognormal, with variance given by

$$I_\varepsilon = \sigma^2_{\ln\varepsilon r} = \mu \ln(L_O / cr) \tag{6}$$

where μ is a universal constant and c is a constant depending on the geometry of the energy scales. Measurements in the atmospheric boundary layer over the open ocean gave values of $\mu = 0.5 \pm 0.1$, Gibson, Stegen and McConnell (1970) from slopes of spectra of the dissipation $E_\varepsilon \approx k^{-1+\mu}$, and $\mu = 0.47 \pm 0.03$ from direct tests of (6) using different averaging intervals r, Gibson and Masiello (1971). Measurements of temperature dissipation rate χ in the seasonal thermocline of the California upwelling region in Monterey Bay by oceanographers of the former Soviet Union on the AKADEMIK KURCHATOV, Iosif Lozovatsky Chief Scientist, reduced by Mark Baker, gave values for the temperature equivalent of equation (6) of $\mu = 0.44 \pm 0.01$, with very precise agreement of χ averaged over length scales at each octave from a meter to a kilometer with lognormal distributions, Gibson (1991a). Falgarone and Phillips (1990) inferred $\mu = 0.4$-0.5 from measurements of line broadening in a dense molecular cloud of the galaxy attributed to intermittent turbulence, with ε values comparable to those in the upper ocean, Gibson (1991a). Structure functions were computed for separation lengths of 10^{15} to 10^{19} meters within this large non-star-forming gas cloud.

Estimation of the Kolmogorov Intermittency Index μ for the Ocean

Several authors have criticized the Kolmogorov (1962) hypothesis of lognormal intermittency for ε and the Gurvich and Yaglom (1967) physical justification; for example, Meneveau and Sreenivasan (1993). Exponential distributions have been predicted rather than lognormal, and claims of multifractal behavior of turbulence with no universal intermittency coefficient μ.

Bershadskii and Gibson (1994) show that $\mu = 0.5$ is a limiting value to be expected at very high Reynolds numbers Re, and that the lognormality of ε at such high Re follows from multifractal asymptotics. Turbulence dissipation at low Re is concentrated on caustic networks that appear due to vortex sheet instability in three dimensional space, leading to an effective fractal dimension D of 5/3 of the network backbone without caustic singularities and a turbulence intermittency exponent $\mu = 1/6$. As Re increases stable singularities form on the caustics, so that D decreases and μ increases. Four stable types of caustic singularities are identified: cusp ridge, swallowtails, pyramid, and purse points, Arnold (1990, 1986). For sheet-like caustics, cusp ridge (ordinary) singularities have $D = 3/2$ and $\mu = 1/4$; swallowtails $D = 1.4$ and $\mu = 0.3$; and pyramid and purse points, $D = 4/3$ and $\mu = 1/3$. Further increases of Re cause degeneration of the caustic networks into smooth vortex filaments with $D = 1$ and $\mu = 1/2$. The conclusion is that low Re numerical simulations of turbulence intermittency are likely to indicate μ values of 0.2, whereas field measurements are likely to indicate μ values near 0.5, as observed for atmospheric, oceanic and astrophysical turbulence with very long cascade ranges.

Figure 2 shows the Obukhov, or energy, scale L_O calculated from (6) as a function of I_ε for $\mu = 0.44$ and 0.2 for cr = 0.3 m. The energy scale L_O indicated for a given I_ε is very sensitive to the value selected for μ. Gurvich numbers G range from 20 to 3.3×10^6, from (3).

Smaller values of $\mu = 0.25$ and less have been inferred by other authors, Sreenivasan and Kailasnath (1993), usually from data with much lower Reynolds number, or from statistical estimates based on length scales outside the universal range (which is on much smaller scales for ε than it is for the universal range of velocity u, by a factor of about 7). Wang et al. (1996) find μ values in the range 0.20 to 0.28 using a massively parallel connection machine with Taylor microscale Reynolds number of about 200. The dashed lines show the range of energy scales implied by the Baker and Gibson (1987) intermittency

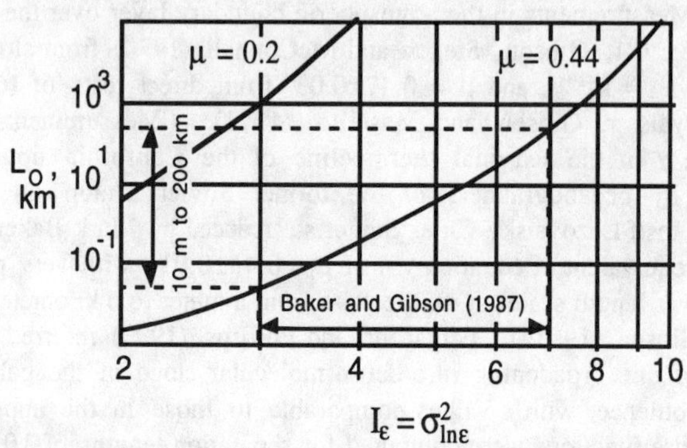

Figure 2. Energy scale L_O versus intermittency factor I_ε for $\mu = 0.44$ and 0.2 with cr = 30 cm, from (6). A reasonable range of L_O is found for $\mu = 0.44$, but not for $\mu = 0.2$.

factor estimates from ocean measurements are in the range 10 m to 200 km for $\mu = 0.44$, Gibson (1991a), which is reasonable. For $\mu = 0.2$ the range is from 1000 km to 4×10^{11} km, which is not reasonable.

Since some in the oceanographic community may not have accepted the fact that intermittency has crucial effects on the proper sampling of oceanic turbulence and turbulent mixing dissipation rates needed to apply the Osborn-Cox method, it might be instructive to review other self-similar nonlinear cascade processes that lead to intermittent random variables (and undersamplng errors) with different efficiencies. For example, the more money you have the more rapidly you can get more, especially if you have lots. Personal income rates for the not-so-rich (the lower 97%) is lognormal, but with $I_\$ = \sigma^2_{\ln\$}$ of only 0.31 compared to $I_\$ = 4.6$ for the super-rich (the upper 3%), Montroll and Shlesinger (1982). Tax collectors in Sultanates will clearly fail if they neglect to tax the Sultan and his richest friends.

Self-gravitational condensation of density in the universe is an extreme example, with variations of density covering a range of more than 50 decades, from intergalactic supervoid densities of 10^{-35} kg m^{-3} to neutron star densities of 10^{15} kg m^{-3}. Hubble (1934) showed that the density of galaxies is lognormal by counting over 44,000 "extra-galactic nebulae" on 1283 plates taken of three quarters of the sky north of -30 degrees declination. Galaxy densities measured by Hubble were smoothed over 639 fields and had intermittency factor I_ρ of only about 1.0, although intermittency factors I_ρ for modern red-shift sorted catalogues and cold-dark-matter calculations are much larger, up to 9, Columbi (1994), with an intermittency exponent μ about 2.5 compared to about 0.5 for turbulence. Estimating the average density of the universe from a small number of random samples is even more futile than estimating ε or χ in the ocean by this method, except that the Gurvich number for ρ is likely to be 10^9, underestimating the true value of 10^{-26} kg m^{-3} by a billion, rather than too small by only factors of 10^1 to 10^4 typical of oceanic undersampling errors for average C, ε and χ, Gibson (1996a). Estimation of the density of "massive compact halo objects" (MACHOs) in the galaxy halo by microlensing of nearby star fields gives smaller values than densities estimated by quasar microlensing which averages over larger halo volumes, when the extreme intermittency of MACHO density is not taken into account, Gibson (1996a). Part of the dark matter paradox is undersampling error.

Intermittency of Internal Waves and Turbulence in the Ocean

"*Because breaking internal waves produces most of the turbulence in the thermocline, the statistics of ε, the rate of turbulent dissipation, cannot be understood apart from the statistics of internal wave shear*" is the leading sentence of the Abstract of Gregg, Seim and Percival (1993). No proof is offered for the assumption that most of the turbulence in the thermocline is produced by breaking internal waves, but this is a crucial step in their goal of comparing the statistics of small scale internal wave shear and the statistics of ε. Later in the Abstract the authors state, "*It is hypothesized that the approximate lognormality of bulk ensembles of ε results from generation of turbulence in proportion to S^4*", where S is the internal wave shear measured over a 10 m vertical interval. This hypothesis depends on the previous assumption that turbulence is caused by waves, and is equally questionable. Certainly in some circumstances the reverse relationship will occur, where internal waves are caused by turbulence. Why should internal wave shear be intermittent without turbulence?

Internal waves, like surface waves, are generally governed by linear equations at the largest scales. Vertical displacements can be represented as the sum of a large number of independent random variables, and have normal probability distributions as expected from the central limit theorem. Wave energy disperses over larger and larger volumes and surface areas, reducing any nonlinear tendencies. Only at the smallest scales does nonlinearity develop. At small scales internal waves are described by the same equations as turbulence, and merge with turbulence at that scale. If large scale internal waves are forced to large amplitudes they develop thinner and thinner density steps with stronger and stronger density gradients that inhibit the formation of turbulence, Phillips (1969). The shear for a given patch of water subjected to internal wave motions increases and decreases reversibly. Concentrations of energy in a wave field are propagated to larger fluid volumes by well known linear mechanisms.

Clearly turbulence can generate internal waves in a stratified fluid. Mechanisms include the formation of fossil vorticity turbulence on scales smaller than those of the most active turbulence, Gibson (1980, 1986), and on larger scales external to the turbulent fluid, Townsend (1965, 1966), Carruthers and Hunt (1986), Uittenbogaard and Baron (1989). Internal wave breaking, like surface wave breaking, is most likely in the vicinity of the source of the waves and decreases as the wave energy is propagated to larger volumes of fluid where the wave mechanics become increasingly linear and wave shear rates increasingly non-intermittent, Phillips (1957), Munk (1981).

Evidence of Saturated Internal Wave Motions Produced by Turbulence

Gargett et al. (1981) have published a composite spectrum of vertical shear in the upper ocean that shows a flat subrange out to wave lengths of about 10 meters, followed by a k^{-1} subrange they attribute to saturated internal waves, and a bump they attribute to turbulence. Gibson (1986) has reinterpreted the k^{-1} spectrum as subsaturated internal waves—decayed fossil vorticity turbulence waves that were produced by previous powerful turbulence events on scales up to 10 meter vertical amplitudes—and the bump as fossil vorticity turbulence. Thus, the smallest scale internal waves of the Gargett et al. (1981) composite spectra, and also the intermittency of their shear rate S (and S^4), are caused by turbulence, *not vice versa!* Gregg (1977) published a corresponding "schematic" spectrum of temperature gradients with a similar flat subrange at large scales, a k^{-1} subrange attributed to saturated internal waves, and a bump at high wave numbers attributed to turbulence. This spectrum was also reinterpreted by Gibson (1986) in a similar way. The k^{-1} subrange was far below the universal temperature gradient spectrum of the Gibson (1980) fossil turbulence theory, by a factor of ten, and the bump at the end was classified as fossil or active-fossil temperature turbulence.

Rodrigues-Sero and Hendershott (1997) have measured temperature gradient spectra at several stations downstream of tidal sills in the Gulf of California. Near the sills the spectra have the form of universal temperature spectra mixed by turbulence, but tens of kilometers downstream and days later the spectra take the Gregg (1977) form, except with ten times higher k^{-1} subrange levels corresponding to the Gibson (1980, 1986) universal saturated internal wave spectrum rather than the lower Gregg schematic form. Clearly these saturated internal waves, and their universal saturated internal wave velocity and temperature spectra, are caused by turbulence. Their signatures appear to be decaying to the same subsaturated forms as those observed in the interior of the ocean by Gargett et al. (1981) and Gregg (1977), as shown in Figure 3.

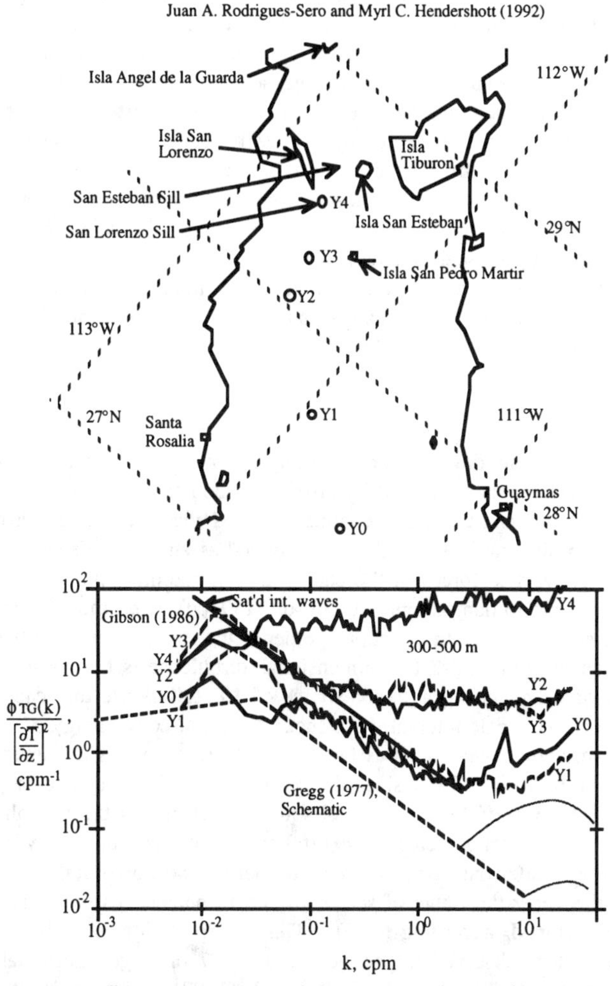

Figure 3. Temperature gradient spectra measured downstream of powerful turbulence events, with overturn scales up to 60 m, produced by flow over tidal sills. The spectra approach the universal form for saturated internal waves predicted by Gibson (1986), and are much higher than the schematic spectrum of Gregg (1977) which apparently represents subsaturated internal waves of the mid-Pacific thermocline. Measurements are from Rodrigues-Sero and Hendershott (1997).

Only if a persistent force maintains a tilted density interface long enough for significant shear to develop will nonlinearities and turbulence appear. When they do, the result is first the development of turbulence at the smallest scale possible for turbulence, the Kolmogorov scale, breaking the sheets of density gradient that produced the vorticity on tilting. Once the density concentration at the sheet has been broken, turbulence will cascade to larger scales and absorb all the kinetic energy that has accumulated on the tilted density surface while it has been tilted. If the tilting has been due to a passing internal wave, the patch of turbulence will be small because the maximum time period for kinetic energy accumulation is the internal wave period. If the tilting has gone on for time periods much longer than an internal wave period, as at a front, the concentration of density will be larger and the corresponding concentration of kinetic energy will be larger than the concentrations possible from large scale internal wave breaking. When the turbulence patch developed on a front reaches its maximum vertical size its kinetic energy is converted to the internal wave energy of fossil vorticity turbulence with persistence times much larger than the period of turbulence activity. Thus, the most extreme turbulence concentrations, and the most extreme values of small scale internal wave shears, will occur in association with fronts, and are caused by whatever causes the fronts. By this model, the small scale internal wave shear S, and the intermittency of S^4, are caused by turbulence, not the other way around. The small scale internal waves are fossil vorticity turbulence patches in various stages of decay from their initial saturated state and observed spectra, Gibson (1980, 1986).

Conclusions

Fronts in the ocean develop for a variety of reasons: including two-dimensional turbulence; a persistent source of density difference such as an island accumulating rain or a river emerging from a coastline; or a continental sized source of density difference like the polar caps. Two dimensional turbulence is defined as an eddy-like state of fluid motion where the inertial-vortex forces of the eddies in a [horizontal] plane are larger than any other forces that tend to damp them out, Gibson (1991cd), even though turbulence may be constrained at smaller scales in directions perpendicular to the plane. In the ocean interior far from islands or continents, two dimensional turbulence is the most likely source of fronts, and is therefore the most likely source of 3-D turbulence that causes the observed intermittency of small scale internal wave shear. The source of energy of two-dimensional turbulence is horizontal shear, such as boundaries between opposite currents in the ocean interior. The limitation of the size of two dimensional turbulence is presumably the Hopfinger scale $L_H = (\varepsilon/f^3)^{1/2}$ from Coriolis forces, where f is the Coriolis parameter, or the closely related Fernando scale $L_F = (q/f^3)^{1/2}$, where q is a buoyancy flux driving the turbulence. The smaller the Coriolis forces, the larger the horizontal scales of turbulence can become, the wider the range of scales in the turbulence cascade, and the larger the intermittency factors I_ε according to (6). This interpretation explains why the largest intermittency factors observed in the ocean, up to 7 or larger from Baker and Gibson (1987), are from equatorial data sets. Gregg et al. (1993) report intermittency factors for S^4 of over 6. If the Gregg (1989) correlation between ε and S^4 is functionally correct (even though the estimated proportionality constant is not) then the extreme intermittency of S^4 observed can be attributed to the extreme intermittency of ε resulting from a horizontal two-dimensional turbulence cascade.

The usual assumption that turbulence is caused by internal waves is questionable. Turbulence is caused by shear instability, and the largest shear instabilities in the ocean interior are those associated with fronts from two-dimensional turbulence with scales that can grow to tens of kilometers at midlatitudes and hundreds of kilometers at equatorial latitudes. The large intermittency factors of ε and S^4 observed in the ocean can both be attributed to the large range of scales of two dimensional turbulence following the Kolmogorov third hypothesis and (6). Horizontal measurements of turbulence and internal wave statistics are needed to test this model of oceanic intermittency, preferably at various latitudes.

If the connection between turbulence intermittency and internal wave shear intermittency, and the correct cause-effect relationship, can be established, it may be possible to use the internal wave shear intermittency to infer turbulence intermittency factors that are needed in estimating turbulence transport processes in the ocean interior, Gibson (1987, 1990a, 1991bc, 1996abc).

Acknowledgments. The author is grateful to Myrl Hendershott and Jim Brasseur for preprint versions of their very interesting papers, and to both of the referees for a number of helpful comments.

References

Arnold, V. I., Singularities of Caustics and Wave Fronts, Kluwer, 1990.

Arnold, V. I., *Catastrophe Theory*, Springer-Verlag, 1986.

Baker, M. A. and C. H. Gibson, Sampling turbulence in the stratified ocean, statistical consequences of strong intermittency. *J. Phys. Oceanogr.*, *17*, 1817-1837, 1987.

Bershadskii, A., and C. H. Gibson, Singularities in multifractal turbulence dissipation networks and their degeneration. *Physica, A 212*, 251-260, 1994.

Colombi, S., A "skewed" lognormal approximation to the probability distribution function of the large-scale density field. *Astrophysical J.*, *435*, 536-539, 1994.

Davis, R. E., Diapycnal mixing in the ocean, the Osborn-Cox model. *J. Phys. Oceanogr.*, *24*, 2560-2576, 1994.

Davis, R. E., Sampling turbulent dissipation. *J. Phys. Oceanogr.*, 26, 341-358, 1996.

Falgarone, E. and T. G. Phillips, A signature of the intermittency of interstellar turbulence, the wings of molecular line profiles, *Astrophysical J.*, *359*, 344-354, 1990.

Gargett, A. E., P. J. Hendricks, T. B. Sanford, T. R. Osborn, and A. J. Williams III, A composite spectrum of vertical shear in the upper ocean. *J. Phys. Oceanogr.*, *11*, 1258-1271, 1981.

Gibson, C. H., G. R. Stegen and S. McConnell, Measurements of the universal constant in Kolmogoroff's third hypothesis for high Reynolds number turbulence, *Phys. Fluids 13*, 2448-2451, 1970.

Gibson, C. H. and P. J. Masiello, Observations of the variability of dissipation rates of turbulent velocity and temperature fields, in *Lecture Notes in Physics, Vol. 12, pp.* 427-448, Springer-Verlag, 1972.

Gibson, C. H., Fossil temperature, salinity, and vorticity turbulence in the ocean, in *Marine Turbulence*, edited by J. C. H. Nihoul, pp 221-257, Elsevier Oceanography Series, Elsevier Publishing Co., Amsterdam, 1980.

Gibson, C. H., Buoyancy effects in turbulent mixing, sampling turbulence in the stratified ocean. *AIAA J.*, *19*, 1394-1400, 1981.

Gibson, C. H., Alternative interpretations for microstructure patches in the thermocline, *J. Phys. Oceanogr. 12*, 374-383, 1982.

Gibson, C. H., Turbulence in the equatorial undercurrent core, in *Hydrodynamics of the Equatorial Ocean*, edited by J. C. H. Nihoul, pp 131-154, Elsevier Oceanogr. Ser., Elsevier Pub. Co., Amsterdam, 1983.

Gibson, C. H., Internal waves, fossil turbulence, and composite ocean microstructure spectra. *J. Fluid Mech. 168*, 89-117, 1986.

Gibson, C. H., Fossil turbulence and intermittency in sampling oceanic mixing processes. *J. Geophys. Res. 92*, 5383-5404, 1987.

Gibson, C. H., Chapter 18. Turbulence, mixing, and microstructure, in *The Sea, Volume 9, Ocean Engineering Science*, edited by D. Hanes and B. LeMehaute, pp 631-659, John Wiley and Sons, New York, 1990a.

Gibson, C. H., On the definition of turbulence, *Proc. Intnl. Workshop, Anisotropy of fluid flows in external force fields, and geophysical, technological, ecological applications,* in honor of Professor A. M. Oboukhov, Jurmala, USSR, September 10-13, 1990, 1990b.

Gibson, C. H., Kolmogorov similarity hypotheses for scalar fields, sampling intermittent turbulent mixing in the ocean and galaxy. turbulence and stochastic processes, Kolmogorov's ideas 50 years on, *Proc. Roy. Soc. London, Ser. A, V434* (N1890), 149-164, 1991a.

Gibson, C. H., Turbulence, mixing, and heat flux in the ocean main thermocline, *J. Geophys. Res.*, 96, 20,403-20,420, 1991b.

Gibson, C. H., Laboratory, numerical, and oceanic fossil turbulence in rotating and stratified flows. *J. Geophys. Res.*, *96*, 12,549-12,566, 1991c.

Gibson, C. H., Fossil two-dimensional turbulence in the ocean, in *Turbulent Shear Flows, 7*, edited by F. Durst and W. C. Reynolds, pp 63–78, Springer-Verlag, Berlin, , 1991d.

Gibson, C. H. and W. H. Thomas, effects of turbulence intermittency on growth inhibition of a red tide Dinoflagellate, Gonyaulax polyedra Stein, *J. Geophys. Res. 100*, 24,841-24,846, 1995.

Gibson, C. H., Turbulence in the ocean, atmosphere, galaxy, and universe. *Applied Mech. Rev. 49*, 299-316, 1996a.

Gibson, C. H., Introduction to turbulent flow and mixing, Section 1.5. In *Handbook of Fluid Mechanics and Machinery*, edited by J. Schetz and A. E. Fuhs, pp. 83-90, John Wiley & Sons, 1996b.

Gibson, C. H., Stratified flow, Section 13.7. In *Handbook of Fluid Mechanics and Machinery*, edited by J. Schetz and A. E. Fuhs, pp. 832-841, John Wiley & Sons, 1996c.

Gregg, M. C., C. S. Cox and P. W. Hacker, Vertical microstructure measurements in the Central North Pacific, *J. Phys. Oceanogr.*, *3*, 458-469, 1973.

Gregg, M. C., A comparison of finestructure spectra from the main thermocline, *J. Phys. Oceanogr.*, *7*, 33-40, 1977.

Gregg, M. C., Scaling turbulent dissipation in the thermocline. *J. Geophys. Res. 94*, 9686-9698, 1989.

Gregg, M. C., H. E. Seim, and D. B. Percival, Statistics of shear and turbulent dissipation profiles in random internal wave fields. *J. Phys. Oceanogr.*, *23*, 1777-1799, 1993.

Gregg, M. C., The geography of diapycnal mixing, Symposium PS-09 on Turbulent Dispersion, Steve Thorpe, Convenor, IAPSO XXI General Assembly in Honolulu, Abstracts, 176, 1995.

Gurvich, A. S. and A. M. Yaglom, Breakdown of eddies and probability distributions for small-scale turbulence, *Phys. Fluids Supplement*, S59-S65, 1967.

Hebert, D., J. N. Moum, C. A. Paulson, D. R. Caldwell, Turbulence from internal waves at the equator, Part II, Details of a single event. *J. Phys. Oceanogr.*, *22*, 1346-1356, 1992.

Hubble, E., The distribution of extra-galactic nebulae. *Contributions from the Mount Wilson Observatory*, Carnegie Institution of Washington, No. 485, 1934.

Kolmogorov, A. N., A refinement of previous hypotheses concerning the local structure of turbulence in a viscous incompressible fluid at high Reynolds number. *J. Fluid Mech.*, *13*, 82-85, 1962.

Meneveau, C. and K. R. Sreenivasan, The multifractal nature of turbulent energy dissipation, *J. Fluid Mech. 224*, 429-484, 1991.

Montroll, E. W., and M. F. Shlesinger, On 1/f noise and other distributions with long tails, *Proc. Natl. Acad. Sci. U. S. A. Appl. Math. Sci. 79*, 3380-3383, 1982.

Munk, W., Abyssal recipes. *Deep Sea Res. 13*, 707-730, 1966.

Munk, W., Internal waves and small-scale processes, in *Evolution of Physical Oceanography, Scientific Surveys in honor of Henry Stommel*, edited by B. A. Warren and C. Wunsch, pp 264-291, MIT Press, 1981.

Osborn, T. and C. Cox, Oceanic fine structure, *Geophys. Fluid Dyn.*, *3*, 321-345, 1972.

Phillips, O. M., On the generation of waves by turbulent wind. *J. Fluid Mech.* 2, 417-445, 1957.

Phillips, O. M., *The Dynamics of the Upper Ocean*, Cambridge University Press, Cambridge, UK, 1969.

Rodrigues-Sero, J. A, and Myrl C. Hendershott, Vertical temperature gradient finestructure spectra in the Gulf of California, *J. Phys. Oceanogr.*, to be published, 1997.

Sreenivasan, K. R. and P. Kailasnath, An update on the intermittency exponent in turbulence, *Phys. Fluids A* 5, 512-514, 1993.

Uittenbogaard, R. E. and F. Baron, A proposal, extension of the K-ε model for stably stratified flows with transport of internal wave energy, in *7th Turbulent Shear Flows Symp.*, pp 21-23, Stanford, 1989.

Wang, Lian-Ping, Shiyi Chen, J. G. Brasseur, J. C. Wyngaard, Examination of hypotheses in the Kolmogorov refined turbulence theory through high-resolution simulations, Part 1. Velocity field, *J. Fluid Mech. 309*, 113-156, 1995.

Washburn, L. and C. H. Gibson, Horizontal variability of temperature microstructure in the seasonal thermocline during MILE, *J. Geophys. Res.*, *89*, 3507-3522, 1984.

Yamazaki, H. and R. Lueck, Why oceanic dissipation rates are not lognormal, *J. Phys. Oceanogr.*, *20*, 1907-18, 1990.

27

Buoyancy Fluxes in a Stratified Fluid

G. N. Ivey, J. Imberger and J. R. Koseff

Abstract

Direct numerical simulations of the time evolution of homogeneous stably stratified shear flows have been performed for mean flow Richardson numbers in the range from 0.075 to 1.00 and for Prandtl numbers ranging from 0.1 to 2. The local or instantaneous results indicate that when the turbulent Froude number $Fr_T = 1$ the peak value of mixing efficiency is $R_f \approx 0.25$, a result independent of the Prandtl number Pr and thus inconsistent with previous laboratory measurements at values of the Pr of 0.7 and 700. The results are consistent, however, with previous laboratory observations in demonstrating that the mixing efficiency R_f decreases rapidly away from this peak value at $Fr_T = 1$. Using data from both numerical simulations and laboratory experiments with both shear and grid generated turbulence, simple empirical relationships are developed to estimate R_f, and hence buoyancy flux b, for the entire range extending from active turbulence down to the limit where the effects of viscosity and/or density stratification suppress the vertical buoyancy flux.

Introduction

Specification of the turbulent buoyancy flux in a fluid with a stable density gradient is essential to our understanding of mixing processes in a wide variety of geophysical applications. Quantifying the rate of vertical mixing in an ocean or lake is central to parameterizing the transport of heat, salt, and passive tracers. For flows with turbulence driven by a mean shear, for example, this has led to the development of numerous closure models which express eddy diffusivities in terms of the instantaneous mean flow Richardson number Ri (e.g., Pacanowski and Philander, 1981). Such closure schemes rely on data obtained from laboratory and field measurements, although measurements are obtained in fundamentally different ways in the two cases. In the laboratory, data is typically derived from long time averages in steady mean flows, whereas in the field the flow is locally sampled by either horizontal or, more commonly, vertical profiling instruments with a limited number of casts available to obtain ensemble averaged statistics of the turbulence properties. These differences between laboratory and field procedures raise many fundamental questions in regard to parameterizing the turbulence, such as the relationship between mixing rates and mean flow properties and how to correctly account for time variability. If we use local or instantaneous measurements, how

Physical Processes in Lakes and Oceans
Coastal and Estuarine Studies Volume 54, Pages 377-388

are quantities such as the overturning lengthscale L_C, the dissipation rate ε and the buoyancy flux b related? If we only have single or limited numbers of realizations for L_C and ε in the field or laboratory, can we use these to predict b, and with what accuracy? From an energetics point of view, what is the efficiency of conversion of available turbulent kinetic energy to buoyancy flux b?

Ivey and Imberger (1991) examined the efficiency of this energy conversion by introducing a generalized definition of the flux Richardson number R_f, based on the full turbulent kinetic energy equation. Using data from laboratory studies where turbulence was generated by grids (Stillinger et al., 1983; Itsweire et al., 1986; Lienhard et al., 1990) and in studies where turbulence was generated by both a grid and a mean shear (Rohr et al., 1988), they argued that two dimensionless numbers were necessary and sufficient to characterize the mixing (cf Gibson, 1980, 1991), and suggested the use of the turbulent Froude number Fr_T and turbulent Reynolds number Re_T. Their results demonstrated that for fluids with a Prandtl number of 700 the peak value of R_f was 0.20, while for fluids with Prandtl number of 0.7 the peak value of R_f was 0.15. The data also demonstrated a very strong dependence of R_f on the turbulent Froude number Fr_T, with peak values of R_f observed near $Fr_T = 1$ and a rapid decrease of R_f for values of Fr_T either above or below this value. Imberger and Ivey (1991) further demonstrated the value of Fr_T varied greatly in field measurements. When combined with the results of the laboratory observations described above, this has significant implications when computing the buoyancy flux or eddy diffusivity from microstructure measurements of dissipation in the field.

Holt, Koseff and Ferziger (1992) and Itsweire, Koseff, Ferziger and Briggs (1993) examined a sheared stratified flow using direct numerical simulation (DNS) to study the temporal evolution of turbulence in a flow with constant mean density and velocity gradients. Using the pseudospectral method, they solved the Boussinesq form of the Navier Stokes equations for the three dimensional velocity and density fields on a 128^3 grid. The calculations were initialized by specifying pulse initial energy spectra for the turbulent velocity fluctuations (initially isotropic), while the initial fluctuating potential energy was set to zero. The advection imposed by the mean field was removed, leading to equations for the fluctuating quantities, although the imposed *mean* density and velocity gradients were held constant at each time step. Using the code described by Holt et al (1992), in the present study we have undertaken a series of numerical runs at values of the mean Richardson number $Ri = N^2/S^2$ in the range from 0.075 to 1.0, where N is the buoyancy frequency and S the mean shear rate, and with Prandtl numbers Pr ranging from 0.1 to 2 (computational restrictions limit the calculations to small Pr). Below we use the ensemble averaged measurements over the 128^3 point domain from the numerical results at each time step, compare these results with the earlier laboratory results described by Ivey and Imberger (1991), and develop expressions for mixing efficiency R_f and buoyancy flux b in terms of local or instantaneous quantities.

Governing parameters

The turbulent kinetic energy equation may be written as

$$m = b + \varepsilon \tag{1}$$

where ε is the rate of dissipation of turbulent kinetic energy, b denotes the buoyancy flux and m refers to the net mechanical energy required (or available) to sustain the turbulent motions from all possible sources, including unsteady terms. Following Ivey and Imberger (1991), the mixing efficiency is quantified by the generalized flux Richardson number

$$R_f = \frac{b}{m} = \frac{1}{1+\varepsilon/b} \tag{2}$$

The small scale or turbulent Froude number is defined as

$$Fr_T = \left(\frac{\varepsilon}{N^3 L_C^2}\right)^{1/3} = \left(\frac{L_O}{L_C}\right)^{2/3} \tag{3}$$

where L_C is the centred displacement scale, $L_O = (\varepsilon/N^3)^{1/2}$ is the Ozmidov scale, and N is the buoyancy frequency characterizing the stable background stratification. The turbulent Reynolds number Re_T and the small scale Froude number Fr_γ are defined by

$$Re_T = \left(\frac{\varepsilon L_C^4}{\nu^3}\right)^{1/3} = \left(\frac{L_C}{L_K}\right)^{4/3} \tag{4}$$

$$Fr_\gamma = \left(\frac{\varepsilon}{\nu N^2}\right)^{1/2} = \left(\frac{L_O}{L_K}\right)^{2/3} \tag{5}$$

where $L_K = (\nu^3/\varepsilon)^{1/4}$ is the Kolmogorov scale. The dimensionless numbers in equations (3), (4) and (5) are related by

$$Fr_T = \frac{Fr_\gamma}{Re_T^{1/2}} \tag{6}$$

One final length scale used in discussing the laboratory and numerical results below is the Ellison scale L_E defined as

$$L_E = -\frac{\left(\overline{\rho'^2}\right)^{1/2}}{\frac{\partial\bar{\rho}}{\partial z}} \tag{7}$$

where ρ is the mean density and ρ' the fluctuating density. Itsweire et al. (1993) demonstrated that, with the exception of flows with very high mean Richardson numbers when the turbulence is not active, L_C and L_E can be taken as the same scale for practical purposes, as we do below.

As the turbulence associated with the initial pulse spectra used to initiate the DNS calculations becomes self-adjusted for dimensionless times St > 2 (Holt et al., 1992), all results quoted here were computed by ensemble-averaging the data from the 128^3 computational domain for St > 2. In general the sources of turbulence for driving the mixing will come from both turbulence associated with the initial conditions and that generated by the mean shear, but both these possible contributions are captured by computing R_f according to the definition in (2). Note that while the turbulence therefore evolved from the initial values as the calculations proceeded, by construction the mean flow Richardson number Ri remained constant for all times.

Mixing Efficiency

The buoyancy flux b is given by

$$b = \frac{g}{\rho_0}\overline{\rho' w'} = \frac{g}{\rho_0} R_{\rho w}\,\overline{\rho'}\,\overline{w'} \tag{8}$$

where $R_{\rho w}$ is the correlation coefficient and $\overline{\rho'}$ and $\overline{w'}$ are the rms values of the density and vertical velocity fluctuations, respectively. Now writing

$$\overline{\rho'} = f_1 \frac{\rho_0}{g} N^2 L_C \tag{9}$$

$$\overline{w'} = f_2\, \varepsilon^{1/3} L_C^{1/3} \tag{10}$$

where f_1 and f_2 are unknown functions which could potentially be dependent on Ri and t as well as the method of computation of $\overline{\rho'}$ and $\overline{w'}$ (see below), then substituting (8), (9) and (10) into the definition of R_f in (2) yields

$$R_f = \frac{1}{1+\dfrac{Fr_T^2}{R_{\rho w} f_1 f_2}} \tag{11}$$

As discussed above, rather than use L_C we use the length scale L_E in (9) and (10) and thus, by the definition in (7), the corresponding function $f_1 = 1$. In order to determine f_2, in Figure 1a we plot $\overline{w'}$ against $(\varepsilon L_E)^{1/3}$ for the current DNS results for all Pr, while in Figure 1b we show the corresponding results from the earlier laboratory studies discussed by Ivey and Imberger (1991). Data in both figures cover a wide range of Fr_T, from the energetic turbulence case with $Fr_T > 1$ to the buoyancy controlled regime with $Fr_T < 1$. For comparison, in Figure 1a we include the best fit line of $\overline{w'} = 0.82\,(\varepsilon L_E)^{1/3}$ and in Figure 1b the best fit line of $\overline{w'} = 0.87\,(\varepsilon L_E)^{1/3}$. Included in the laboratory data in Figure 1b are also data points where the buoyancy flux is zero (these data were not included in the best fit). It is interesting to note that these points always lie *above* the best fit straight line, implying that there is likely some contamination of the measured velocities by internal waves (cf. Itsweire and Helland, 1989). Briggs et al. (1998) have recently argued that the contributions of internal wave-induced velocities is likely only felt when Ri is large - and certainly larger than for the runs plotted in Figure 1a which includes only active mixing cases with Ri equal to either 0.075 or 0.21.

If the linear correlations in Figures 1a and 1b hold in general, then it is consistent with the observations reported by Yamazaki (1990) of good agreement between his observed velocity spectrum and the Nasmyth spectrum, and hence good estimates of the dissipation, even in the limiting cases where the buoyancy flux was zero. In summary, providing that the turbulence is active, which in the present context we take as the requirement that $Re_T = (L_C/L_K)^{4/3} > 1$ (i.e. there exists a range of overturning velocity scales from L_K out to L_C), the data in Figure 1 suggests that if we use the length scale L_E then we may take $f_2 \approx 1$. Note, however, that for field data obtained from vertical casts, for example, where only L_C is available the value of the parameters f_1 and f_2 must be determined in order to use (11).

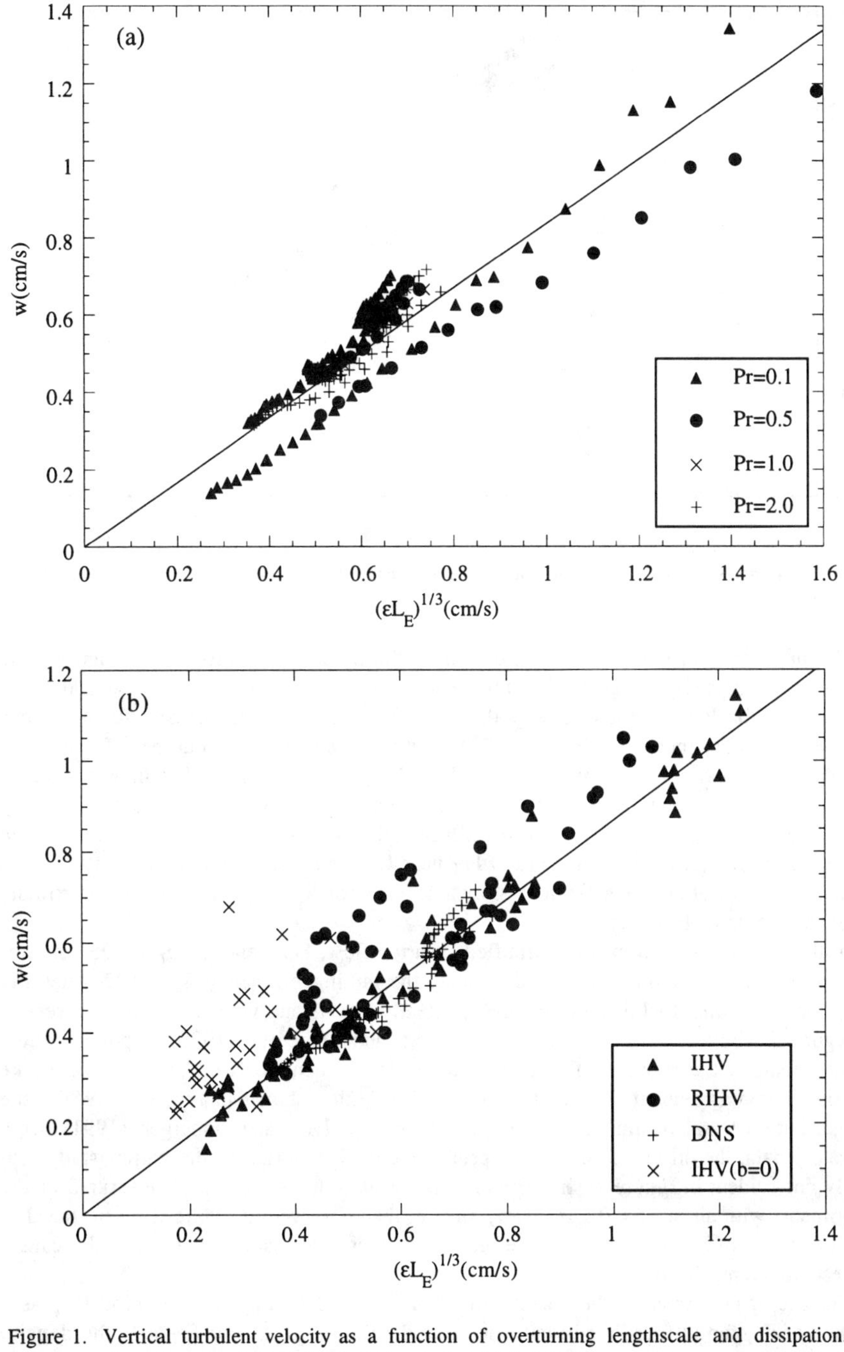

Figure 1. Vertical turbulent velocity as a function of overturning lengthscale and dissipation. Data sources are Itsweire et al. (1986) (IHV), Rohr et al. (1988) (RIHV), current numerical results (DNS). (a) DNS data (b) laboratory data

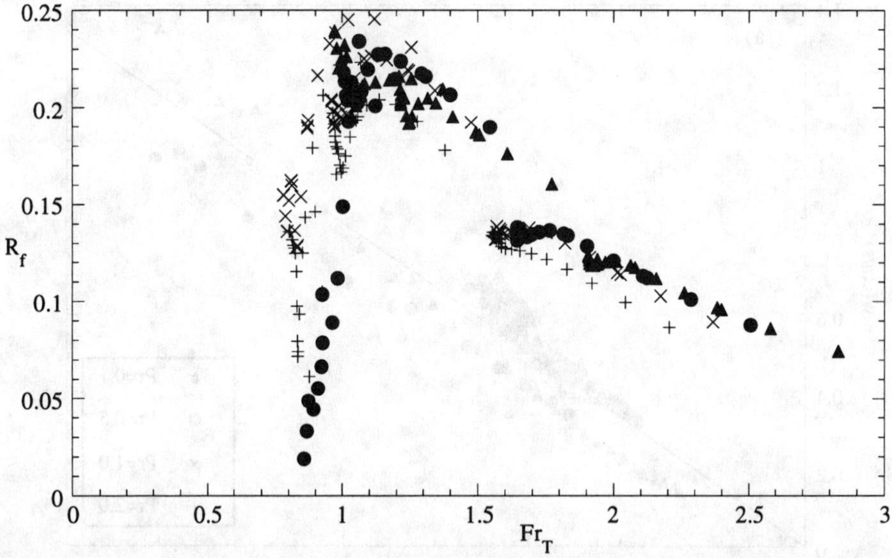

Figure 2. Mixing efficiency as a function of Froude number for the DNS results. Symbols are defined in Figure 1a.

The prediction in (11) can be tested against the numerical results and in Figure 2 we plot mixing efficiency R_f against Fr_T for the DNS runs with Pr = 0.1, 0.5, 1.0 and 2, all included. In the lower limit of $Fr_T \Rightarrow 0$, buoyancy completely suppresses all turbulence and $R_f = 0$. As Fr_T increases, R_f increases rapidly to reach a peak value near $Fr_T \approx 1$ and then decreases again with increasing Fr_T. Very large Fr_T is associated with vanishingly small N and this absence of a density gradient again causes R_f to become vanishingly small. Peak mixing efficiency of $R_f \approx 0.25$ for all runs, irrespective of Pr, is achieved between these two extremes at a value of Fr_T near 1. It is interesting to note from the form of equation (11) that, as the maximum allowable value of $R_{\rho w}$ is 1, the implied maximum achievable value of R_f is 0.5.

In their field observations of a stratified shearing flow, Seim and Gregg (1995) did not measure buoyancy flux directly but did estimate the mixing efficiency $R_f = 0.22$ when Fr_T lay in the range 0.85 to 1.3 (their Figure 3), clearly consistent with the numerical results in Figure 2 above. More recently, Lemckert and Imberger (1998) reported direct measurements of the buoyancy flux and R_f in the sloping benthic boundary layer of a lake and found good agreement with the form of the distribution shown in Figure 2. While the R_f distribution is also similar to the form found by Ivey and Imberger (1991) from laboratory data, the difference lies in the peak value of R_f reported in the earlier study, but clearly not evident in Figure 2. In particular, the peak value of $R_f = 0.15$ measured in the experiments with air at Pr = 0.7 is clearly smaller than observed in Figure 2 where the data span this value of Pr. Clearly the issue requires further investigation to resolve the causes of these apparent differences.

The use of (11) requires the specification of $R_{\rho w}$ and in Figure 3 we plot $R_{\rho w}$ as a function of Fr_T for all four Prandtl numbers from the DNS results. For $Fr_T > 1$, the density and vertical velocity fluctuations are strongly correlated with $R_{\rho w}$ in the range 0.2 - 0.6, independent of Fr_T. Ivey and Imberger (1991) suggested $R_{\rho w} = 0.3$ was a best estimate for the laboratory data and Weinstock (1992) has argued on the basis of a theoretical model that $R_{\rho w} = 0.26$. Taking all these observations into account, we thus take $R_{\rho w} = 0.4$ as the asymptotic value and for the range $Fr_T > 1$ write (11) as

$$R_f = \frac{1}{1+2.5\ Fr_T^2} \tag{12}$$

which is in good agreement with the observations in Figure 2. Note that in the limit of $Fr_T = 1$ equation (12) predicts $R_f = 0.28$. This value is consistent with the results shown in Figure 2 and while apparently larger than some laboratory estimates, we adopt (12) here as it is conservative in predicting an upper bound for R_f and hence for buoyancy flux.

The expression in (12) cannot always hold since as the turbulence weakens viscosity must start to become important at some point. To investigate this, we show in Figure 4 the dependence of $R_{\rho w}$ on $\varepsilon/\nu N^2$. Ivey et al. (1993) discussed some earlier DNS results for $Pr < 1$ which indicated that both the value of $R_{\rho w}$ and the transition value of the parameter $\varepsilon/\nu N^2$ which sustained a positive buoyancy flux decreased with decreasing Pr. With the application to limnology and oceanography in mind, we discuss here only the case for $Pr > 1$. In Figure 4 we show the results for the DNS runs with $Pr = 2$ and, for comparison, also the laboratory data from Itsweire et al. (1986) and Rohr et al. (1988) in salt stratified fluids with $Pr = 700$. The data sets indicate that $\varepsilon/\nu N^2 = 15$ is a good estimate of the lower limit which sustains a positive buoyancy flux. Note also that $R_{\rho w}$ is constant only for $\varepsilon/\nu N^2 > 75$. Thus given the relation in (6), along the transition line where $Fr_T = 1$ the value of R_f is given by (12) for $Re_T \geq 75$, but in the range $15 < Re_T < 75$ must decrease with reducing values of Re_T to the limit of $R_f = 0$ when $Re_T \leq 15$.

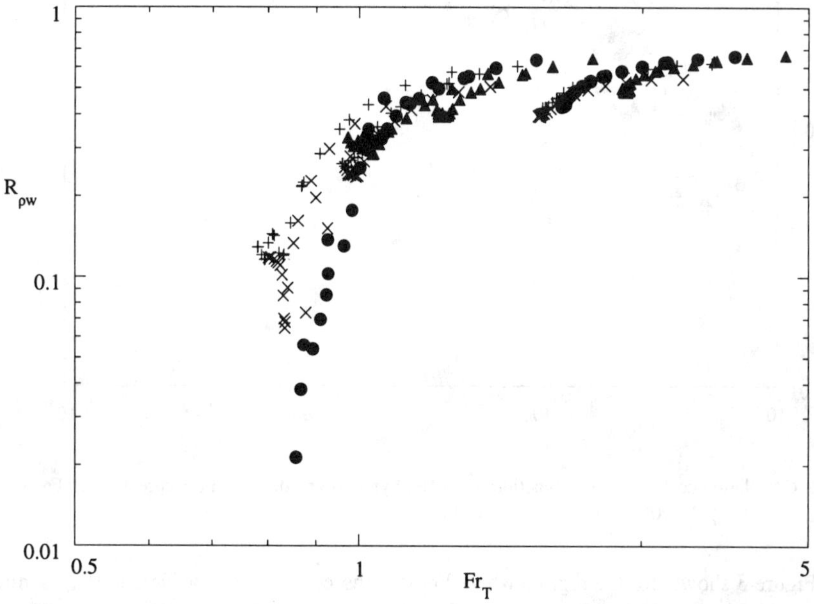

Figure 3. Correlation coefficient as a function of Froude number. Symbols are defined in Figure 1a.

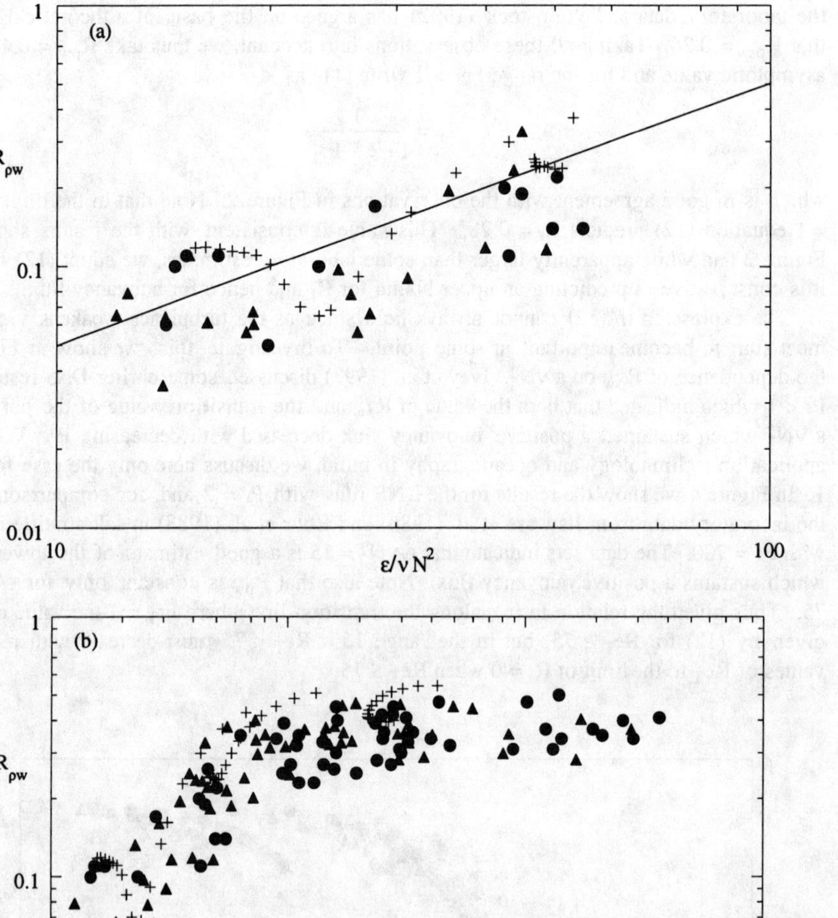

Figure 4. Correlation coefficient as a function of $\varepsilon/\nu N^2$. Symbols are defined in Figure 1a. (a) $Fr_T < 1$ (best fit shown: $R_{\rho w} = 0.0059\ (\varepsilon/\nu N^2)^{0.91}$ (b) All Fr_T.

As Figure 3 shows, for the regime where $Fr_T < 1$, the correlation coefficient $R_{\rho w}$ is not a constant and in fact rolls off very strongly with decreasing Fr_T. For these less active regimes, even combining the data from the numerical and laboratory results the data sets are clearly limited and more data, most likely obtainable in the field, is needed. However, we do know the behaviour of the mixing efficiency R_f in the two limits. Firstly, we know that the buoyancy flux and hence $R_f = 0$ when $Fr_\gamma = (\varepsilon/\nu N^2)^{1/2} = 15^{1/2}$. Secondly, in the upper limit when $Re_T \geq 75$ and $Fr_T = 1$ we have from (12) that $R_f = 0.28$. These two limiting conditions are satisfied by the interpolation given by

$$R_f = 0.28\left(\frac{Fr_\gamma - \sqrt{15}}{\sqrt{Re_T} - \sqrt{15}}\right) \tag{13}$$

Discussion

The arguments above imply that a number of regimes in Fr_T - Re_T parameter space are possible. In Figure 5 and Table 1, we summarise the results for the various regimes described above for the case of $Pr > 1$. In regime 3 where $Fr_T > 1$ and $Re_T > 75$, R_f is given by (12). In regime 2, we modify this by taking a simple linear interpolation between the

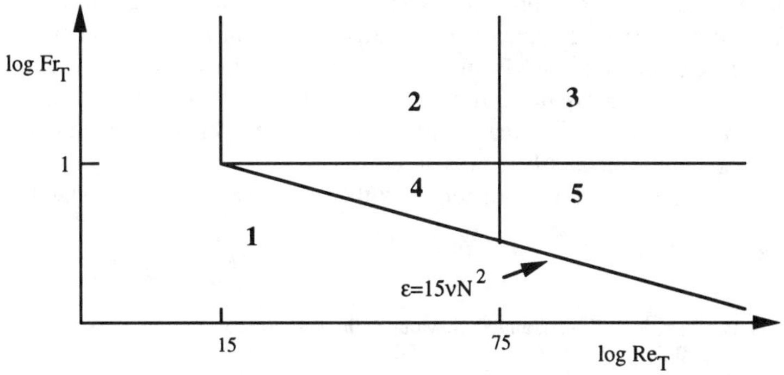

Figure 5. Froude number versus Reynolds number diagram with regimes marked.

TABLE 1. Regimes for Pr>1.

Regime	R_f
1	0
2	$R_f = \left(\frac{Re_T - 15}{60}\right)\frac{1}{1+2.5\ Fr_T^2}$
3	$R_f = \frac{1}{1+2.5\ Fr_T^2}$
4	$R_f = 0.0047\ (Re_T - 15)\left(\frac{Fr_\gamma - \sqrt{15}}{\sqrt{Re_T} - \sqrt{15}}\right)$
5	$R_f = 0.28\left(\frac{Fr_\gamma - \sqrt{15}}{\sqrt{Re_T} - \sqrt{15}}\right)$

upper limit given by (12) applied at $Re_T = 75$ and the lower limit of $R_f = 0$ when $Re_T = 15$. In regime 5 where $Fr_T < 1$ and $Re_T > 75$, R_f is given by (13). In regime 4, where $Fr_T < 1$ and $15 \leq Re_T \leq 75$, we again use a linear interpolation between (13) evaluated at the right hand boundary where $Re_T = 75$, and the requirement that $R_f = 0$ along the lower boundary where $\varepsilon/\nu N^2 = 15$ (the diagonal line in Figure 5). In regime 1 the value of $R_f = 0$ everywhere. For each regime, the expressions for R_f are summarized in Table 1, and the buoyancy flux can then be obtained directly from the definition in (2) which can be rearranged to yield $b = \varepsilon R_f/(1-R_f)$.

The model regimes depicted in Figure 5 and described in Table 1 are intended to provide simple yet complete formulae for the computation of R_f and buoyancy flux b which are as consistent as possible with available data and are conservative in the sense of predicting upper bounds on quantities. The underlying model extends or differs from previous work in several aspects. Unlike other studies (e.g., Schumann and Gerz, 1995), the model is applicable to both mechanically generated and shear generated turbulence. Unlike other studies (e.g., Weinstock, 1992) it makes no assumptions about the spectral forms or growth rates of the turbulence. Finally, the present model explicitly incorporates the effect of viscosity in defining a series of regimes which describe the mixing efficiency R_f, and hence buoyancy flux b, down to and including the limit where the combined action of buoyancy and viscosity suppress the vertical buoyancy fluxes. Such strongly buoyancy controlled situations occur both in the stratified ocean (e.g., Yamazaki, 1990) and in stratified lakes (e.g., Imberger and Ivey, 1991).

Acknowledgment. This work was supported by the Australian Research Council.

References

Briggs, D. A., J. H. Ferziger, J. R. Koseff, and S. G. Monismith, Turbulent mixing in a shear-free stably stratified two-layer fluid, *J. Fluid Mech., 354,* 175-208, 1998.

Gibson, C. H., Fossil temperature, salinity and vorticity in the ocean. *Marine turbulence*, edited by J. Nihoul, Elsevier, 221-258, 1980.

Gibson, C. H., Kolmogorov similarity hypotheses for scalar fields: sampling intermittent turbulent mixing in the ocean and galaxy, in *Turbulence and Stochastic Processes: Kolmogorov's ideas 50 years on,* edited by J. C. R. Hunt, O. M. Phillips and D. Williams, pp. 149-164. The Royal Society, London, 1991.

Gregg, M. C., Diapycnal mixing in the thermocline: A review. *J. Geophys. Res.*, *92*, 5249-5286, 1987.

Holt, S. E., J. R. Koseff, and J. H. Ferziger, A numerical study of the evolution and structure of homogeneous stably stratified shear turbulence. *J. Fluid Mech., 237*, 499-540, 1992.

Imberger, J. and G. N. Ivey, On the nature of turbulence in a stratified fluid. Part II: Application to lakes. *J. Phys. Oceanogr., 21*, 659-680, 1991.

Itsweire, E. C., K. N. Helland and C. W. Van Atta, The evolution of grid-generated turbulence in a stratified fluid. *J. Fluid Mech., 126*, 299-338, 1986.

Itsweire, E. C. and K. N. Helland, Spectra and energy transfer in stably stratified turbulence. *J. Fluid Mech., 207*, 419-439, 1989.

Itsweire, E. C., J. R. Koseff, D. A. Briggs, and J. H. Ferziger, Turbulence in stratified shear flows: implications for interpreting shear-induced mixing in the ocean. *J. Phys Oceanogr.*, *23*, 1508-1522, 1993.

Ivey, G. N. and J. Imberger, On the nature of turbulence in a stratified fluid. Part I: The energetics of mixing, *J. Phys. Oceanogr.*, *21*, 650-658, 1991.

Ivey, G. N., J. Koseff, D. Briggs, and J. H. Ferziger, Mixing in a stratified shear flow: energetics and sampling. *Annual Research Briefs - 1992, Centre for Turbulence Research*, Stanford University, 335-344, 1993.

Lemckert, C. and J. Imberger, The benthic boundary layer, in *Physical Processes in Lakes and Oceans*, edited by J. Imberger, Coastal and Estuarine Studies, pp. 485-498, this volume, 1988.

Lienhard, J. H. and C. W. Van Atta, The decay of turbulence in thermally stratified flow. *J. Fluid Mech., 210*, 57-112, 1990.

Pacanowski, R. C. and S. G. H. Philander, Parameterization of vertical mixing in numerical models of the tropical ocean. *J. Phys. Oceanogr., 11*, 1443-1451, 1981.

Rohr, J. J., E. C. Itsweire and C. W. Van Atta, Mixing efficiency in stably-stratifed decaying turbulence, *Geophys. Astrophys. Fluid Dyn., 29*, 2211-236, 1984.

Rohr, J. J., E. C. Itsweire, K. N. Helland and C. W. Van Atta, An investigation of the growth of turbulence in a uniform-mean-shear flow. *J. Fluid Mech., 187*, 1-33, 1988.

Schumann, U. and T. Gerz, Turbulent mixing in stably stratified shear flows. *J. Appl. Meteorology, 34,* 33-48, 1995.

Seim, H. E. and M. C. Gregg, Energetics of a naturally occurring shear instability. *J. Geophys. Res., 100*, 4943-4958, 1995.

Stillinger, D. C., K. N. Helland and C. W. Van Atta, Experiments on the transition of homogeneous turbulence to internal waves in a stratified fluid. *J. Fluid Mech., 131*, 91-122, 1983.

Yamazaki, H. Stratified turbulence near a critical dissipation rate. *J. Phys Oceanogr., 20*, 1584-1598, 1990.

Weinstock, J. Vertical diffusivity and overturning length in stably stratified turbulence. *J. Geophys. Res. 8*, 12653-12658, 1992.

Lienhard, J. H. and C. W. Van Atta, The decay of turbulence in thermally stratified flow, *J. Fluid Mech.*, 210, 57-112, 1990.

Pacanowski, R. C. and S. G. H. Philander, Parameterization of vertical mixing in numerical models of tropical oceans, *J. Phys. Oceanogr.*, 11, 1443-1451, 1981.

Rohr, J. J., E. C. Itsweire and C. W. Van Atta, Mixing efficiency in stably-stratified decaying turbulence, *Geophys. Astrophys. Fluid Dyn.*, 29, 221-236, 1984.

Rohr, J. J., E. C. Itsweire, K. N. Helland and C. W. Van Atta, An investigation of the growth of turbulence in a uniform-mean-shear flow, *J. Fluid Mech.*, 187, 1-33, 1988.

Schumann, U. and T. Gerz, Turbulent mixing in stably stratified shear flows, *J. Appl. Meteorology*, 34, 33-48, 1995.

Smyth, W. D. and [illegible], [illegible] secondary [illegible] shear instability, *J. Geophys. Res.*, [illegible], 1995.

Stillinger, D. C., K. N. Helland and C. W. Van Atta, Experiments on the transition of homogeneous turbulence to internal waves in a stratified fluid, *J. Fluid Mech.*, 131, 91-122, 1983.

Yamazaki, H., Stratified turbulence near a critical dissipation rate, *J. Phys. Oceanogr.*, 20, 1583-1598, 1990.

[illegible] ratios and overturning length in stably stratified turbulence, *J. Geophys. Res.*, [illegible], 1993.

28

Mixing Processes in a Highly Stratified River

S. Yoshida, M. Ohtani, S. Nishida and P. F. Linden

Abstract

Several of the mixing processes at density interfaces in highly stratified flow at the mouths of rivers flowing into the ocean bear close resemblance to mixing processes at the density interface seen in freshwater bodies, and can provide valuable insights into them. This paper addresses the mixing phenomena at the interface in the very highly stratified flow seen at the mouth of the Ishikari River during the dry season. It is shown that the chief factor influencing mixing across the interface is the wind, while factors disturbing the steadiness of the flow such as tide have little influence. Furthermore, when the entrainment coefficient is examined as a function of wind speed, it is found that the surface salinity at any given river station can be accurately predicted. Mixing mechanisms are also discussed.

Introduction

Commonality Between Limnology and Flows in Highly Stratified Flows at River Mouths

The physical processes of stratification and mixing take place between sea water and fresh water at the mouth of a river. When the volume flow rate is extremely high, as in the case of the Amazon River, or when the level of the river at the mouth is always higher than the level of the highest tide, the salt wedge never enters the river, and the processes all take place in the ocean; these are outside the concerns of limnology. However, the levels of many rivers are below the mean sea level at their mouths and their flow rates are not always high enough to flush the denser seawater from the riverbeds. As a result, sea water is sometimes included in the water drawn from rivers or lagoons for agricultural uses. In highly stratified flows, which are particularly common when the tidal range is small, it is not at all rare to observe salt intrusions of several tens of kilometers inland. Intrusions of 200 km have been documented in the Mississippi, which flows through an extremely flat plain in its lower reaches. Thus, the above processes may be observed between sea water and fresh water even at locations distant from the sea, whether in rivers or in wetlands, and relate directly to limnology. The processes of mixing which takes place in highly stratified flow at river mouths are qualitatively identical to processes of limnological stratified flows; for example, underflows, flow at the base of the surface layer and flow at the top of the benthic boundary layer. This study addresses several aspects of highly stratified flow at the mouth of a river which flows into a body of

Physical Processes in Lakes and Estuaries
Coastal and Estuarine Studies volume 54, Pages 389-400

comparitively low tidal range with a view to the insights they provide into current topics in limnology.

Review of Research on Mixing of River Water and Sea Water in Estuaries and River Mouths

Invasion of ocean water into rivers or estuaries has long posed problems in humankind's use of water resources. The Dutch complain of corrosion by sea water in records dating from the 1300's. The same set of problems beset man all over the world, and countries of Europe and North America have been tackling the practical challenge of salt invasion for over one hundred years (Proc. ICN, 1953). Therefore, research in the West has a long history, with river mouth flows generally classed into cases of estuary flows. The most significant findings are presented by Prichard (1963), Ippen (1966), Dyer (1972) and Officer (1976). Prichard classified river mouth flows into highly stratified (salt wedge), partially mixed and homogeneous (well mixed) flows. Because of the actual cases most often seen by Western researchers, their writings have focussed mainly on partially mixed and well mixed flows, though of course many of their results have significance with respect to highly stratified flows. Still, none of the reports from the West have been of any direct usefulness for solving the problems of highly stratified flows mentioned above. They do not apply to rivers like the Ishikari or Teshio, in which mixing is sometimes so low that they could perhaps be classed as 'very highly stratified' or even 'two-layered' flows.

In contrast, in Japan research into this field was only recently deemed necessary since nearly all the major Japanese rivers have steep gradients and there had been virtually no problems with salt damage inland. Japanese work in this field began from a purely academc viewpoint with Hisao Fukushima's observations of the Ishikari River in 1939 and 1940 (Fukushima, 1942). Those studies were limited in scope to the characteristics of the invading salt water body, or as it is usually called, the salt wedge. After World War II, however, authorities scaled up improvement of river mouths as part of broadened flood control projects. These works brought about sharp alterations in the structure of the salt wedge affecting fisheries and raising the salinity of water withdrawn for agricultural use beyond acceptable levels. Authorities thus became interested in the nature of salt water-fresh water mixing at river mouths; most of the development in this field has come since then. Research in Japan has however, focussed almost exclusively on highly stratified flows.

The most important result in this field has been the derivation of the time average coefficient of friction of the salt wedge. This is found from the following relation, which has been confirmed in nearly all rivers with highly stratified flows:

$$f_i = C(F_i^2 R_e)^n = C\Psi^n . \qquad (1)$$

Here, f_i, F_i and R_e, are, respectively, the coefficient of interfacial friction; the densimetric Froud number, which is calculated using the time mean velocity in the upper (fresh water) layer and the thickness of the upper layer; and the Reynolds number, which is calculated using the velocity and thickness values. Ψ is called the Iwasaki number. The value of coefficient C and n are about 0.25 and -0.5, respectively (Yoshida, 1990).

The physical significance of equation (1) remains unfounded. Little is yet known about the tidal oscillations of the salt wedge level and the process of diffusion of salt into river water. The range of interface oscillation is actually greater than the tidal range, by as much as a factor of seven. This amplitude decreases with distance up the river and is undetectable at the tip of the salt wedge in the Ishikari River, 30 km above the mouth. Therefore, though the condition holds only in the vicinity of the river mouth, the speed of the oscillatory flow of the salt wedge is great enough that it must be taken into account in considerations of speed of the local upper-layer flow. Time-average values of salt wedge velocity do not allow meaningful results. The instantaneous tidal flow velocity is also an essential consideration for stability of the interface.

Research on Mixing Mechanisims at an Interface

The stability of shear flows in the vicinity of an interface and the structure of turbulence which carries salt water into the fresh water layer have proven to be the most important factors influencing mixing across an interface. Theoretical studies of stability in statically stable two-layer flows, show that the instabilities listed in Table 1-A will occur for some forms of the velocity and density distributions. These are the principal instability flows that are known to occur in two-layer flow and have been observed by the researchers listed in Table1-B and others. However, as a result of diffusion and mixing in real stratified flows such as those at river mouths, there often develops a third, long-lived layer of median density between the two main layers. This structure shows instabilities distinct in nature from those of two-layer flows. They are described theoretically by the authors listed in Table 1-C and they have been confirmed experimentally (Caulfield et al., 1994). Yoshida (1977) used the fact that the salt wedge gradually deepens over the distance from its tip to the river mouth to demonstrate the simultaneous existence of K-H instabilities, Holmboe instabilities and Keulegan's 'one-sidedness'. Yoshida (1980) also found instabilities just below the interface which distorted it into a cycloidal-shape. As these instabilities remain even under flow conditions where Holmboe

TABLE 1. A list of selected references on instabilities in layered stably stratified flows.

A	Theoretically Described Shear Flow Instability in Two-Layer Flows
Kelvin-Helmholtz Flow	An explanation of the largest-amplitude instabilities in the interface
Holmboe Flow (1962)	An explanation of the largest-amplitude instabilities occurring simultaneously above and below the interface
Lawrence et al. (1991)	A description of so-called 'one-sidedness'
B	**Experimentally Observed**
Keulegan, G.H. (1949	Observed instabilities just above the interface (correspond to one-sidedness)
Thorpe, S.A. (1968)	Observed Kelvin-Helmholtz instabilities
Browand, F.K. and Wang, Y.H. (1972)	Observed Holmboe instabilities at the shear layer when the ratio of velocity to density length scales was 15
Yoshida, S. (1977)	Observed instabilities occurring independently of each other just above and just the interface when the ratio of velocity to density length scales was much larger than 10
C	**Theoretically Described Shear Flow Instability in Three-Layer Flows**
Taylor, G.I. (1931)	An explanation of the largest-amplitude instabilities within a middle layer of finite thickness
Caulfield, C.P. (1994)	An explanation of instabilities differing from the Holmboe mode which occur just within the upper or lower layers when there is a comparatively thin middle layer

instabilities disappear, it seems implausible to call them a one-sided phenomenon. The mechanism of development can be explained by stability theory on the assumption of a middle layer, however, and some now classify them as one kind of Taylor instability.

Occurrence of these instabilities is to some extent a function of Reynolds number (Nishida and Yoshida, 1983), but at the large scales seen in the field, the velocity and density distributions in the vicinity of the interface can be thought of as dominated by the Richardson number, when properly defined. Determining the Richardson number is the first task when investigating initiation of interfacial instability, the first stage of the transfer of salt across the interface. As mentioned earlier, the flow structure varies continuously from the tip of the salt wedge to the river mouth; thus, the Richardson number varies as a function of location. It is also a function of time in actual flows that are subject to the time-dependent forcing. To give an example, even if the river flowrate is constant the flow is influenced by the tide and so the interface position and shear vary with time. A single measurement of the Richardson number cannot be used to estimate the time average stability of the interface. The Richardson number must be observed over at least one tidal cycle. We emphasize here that even if interfacial instabilities are intense enough to form an intermediate layer, they are not themselves responsible for transport of salt up through the fresh water layer. Transport is brought about by instabilities in the upper layer, or turbulence. Thus, though we refer to this phenomenon as 'transport across the interface', we must go beyond instabilities at the interface to considerations of the turbulence structure, and its time dependencies. In the footsteps of Ellison and Turner (1959), many researchers have examined salt entrainment through turbulence at the interface in the vicinity of the interface; Tamai (1986) and Fernando (1991) have reviewed the most significant recent results. However, all the above researchers focussed on steady turbulent flows, while real flows into the ocean are never steady; it is essential first to identify what effect the factor of unsteadiness has on the structure of very highly stratified flows at river mouths.

Results of Observations in the Ishikari River

The outside factors responsible for time and location dependencies of the Richardson number and time dependencies of upper layer turbulence are not limited to discharge rate and tide. Several other factors such as atmospheric pressure, topography, wind and others play a role and only recently have research efforts been directed into clarifying which were the most influential (Yoshida et al., 1995). Figure 1 shows a map of the lower reaches of the Ishikari River and the points at which measurements were taken in the 1992-1994 Survey. Table 2 provides the distances of each point from the river mouth and the measurements performed there. Figure 2 shows a selection of the long-term results during July and August 1994. The discharge is taken from the river surface level read at St. 15 in the river and a discharge-surface level graph. The discharge was prepared using the relation

$$Q = 79.1\gamma (H + 0.11)^2 \text{ for } H = 0.00\text{m} \sim 1.98\text{m and}$$

$$Q = 34.2\,\gamma (H + 1.20)^2 \text{ for } H = 1.99\text{m} \sim 6.00\text{m} \qquad (2)$$

where Q is the discharge and H is the level. γ (1.32) is a correction factor to include the influences of the tributaries downstream of St. 15. The overall observation period for the observations listed in the figure was from July 1 to Aug 20, 1994, during the dry period. Two sets of measurements are not listed in Table 2; these were a one-week observation of the velocity at the surface and a 24-hour continuous observation of the height of surface waves,

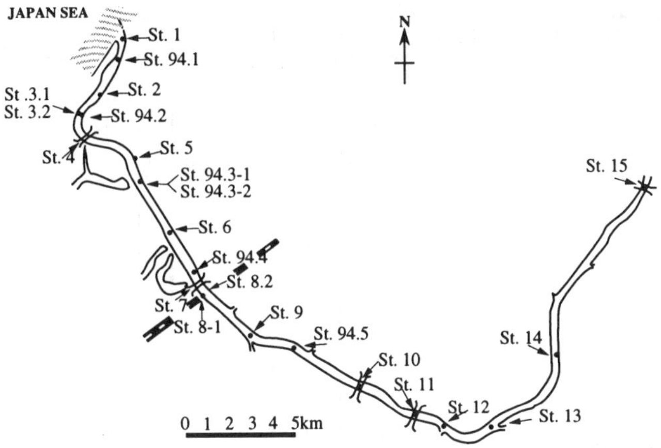

Figure 1. Ishikari River and observation stations.

turbulence and other parameters Aug. 7-8, 1994. Details will be provided during the following discussion as space permits.

Variations in River Surface Level and Salt Transport

The total river depth, one of the most important boundary conditions, is determined by the sea level and the flow rate. Sea level varies with tide and with atmospheric pressure. In order to estimate the latter relation, we used actual sea level data and compared them with the ideal levels predicted by the Meteorological Agency; the deviation is the second time series in Figure 2. This series shows an extremely consistent relation with the atmospheric pressure time series above it in the figure. The quantitative relation obtained is

$$h = a\,(1010 - P) \tag{3}$$

TABLE 2. Location of observations and measured variables on the Ishikari River during 1992-1994.

St.	Dist. from mouth (km)	Variable	St.	Dist. from mouth (km)	Variable
1	0	Interface observation (sonar)	7	14.6	-
94.1	1	Water temp. & Conduct.	8-1	15	Surface level
2	2.9	Water level	8-2	15	Interface level (Step Sensor)
94.2	4.2	Water temp. & Conduct.	9	17.9	-
3-1	4.2	Interface level	94.5	20	Water temp. & Conduct.
3-2	4.2	Surface level	10	23.4	-
4	5.4	-	11	26.6	Surface level
5	8.2	Meteorological data	12	27.9	-
94.3.1	9.3	Assorted data (Platform)	13	30.7	-
94.3.2	9.3	Water temp. & Conduct.	14	35.4	-
6	11.7	Assorted data (Platform)	15	44.6	Surface level
94.4	14	Water temp. & Conduct.			-

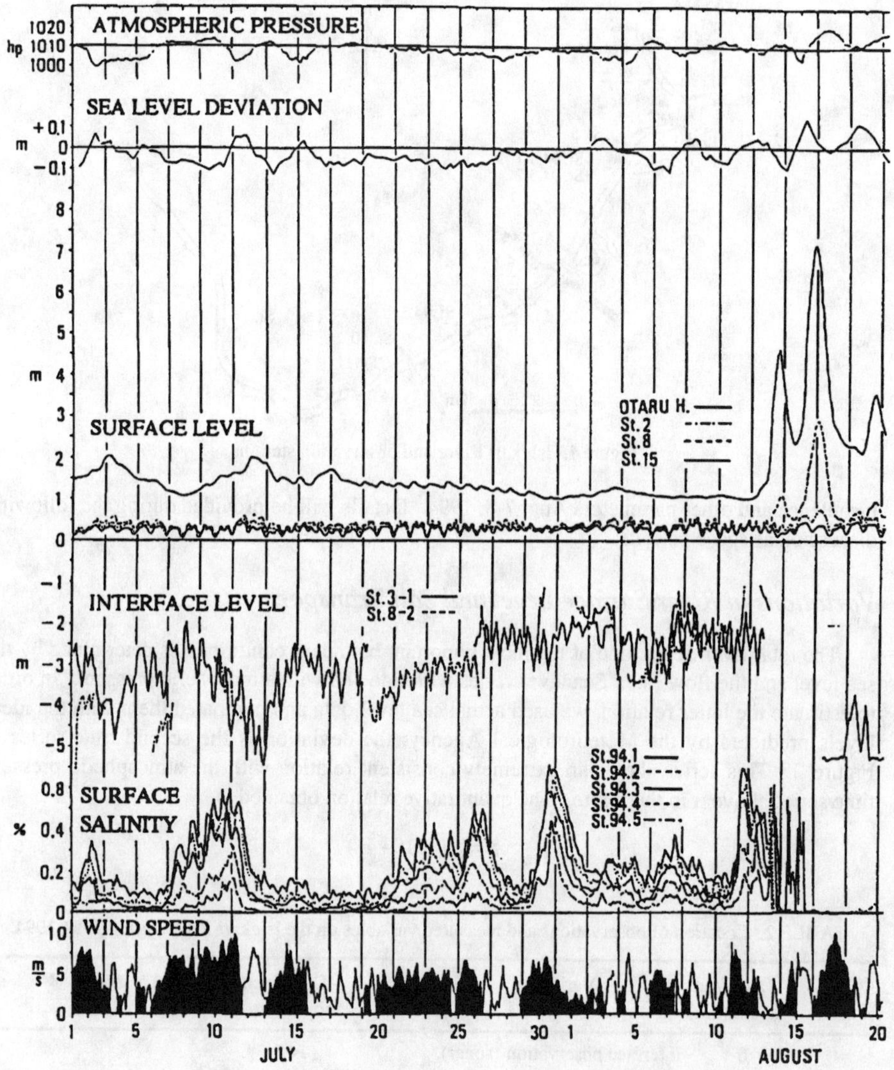

Figure 2. Data during July 1- August 20, 1994 periods

where h is the deviation in cm, a is a value between 1.4 and 1.5 for the Ishikari and P is the atmospheric pressure at the river mouth in hectopascals. Usually, h is a maximum of 20 cm, of the same order as the tidal range in the Japan Sea. Thus, it is necessary to take both tidal range and atmospheric pressure into account. However, the period of the cycle of air pressure change is several times that of the tidal cycle, and atmospheric pressure actually has quite a small effect on river velocity. Even if it acts to raise or lower the river surface, it is a quite minor factor in bringing about unsteady river velocity or salt transport. In comparison, the tide varies on a half-day period and, as mentioned earlier, has a vigorous influence on the two-layer structure and flow velocity, especially during the dry period. The record shows that there was indeed some influence in surface salinity (see Figure 2 record of surface salinity at St. 94.1, 94.2) from the tide, but that effect was limited to the vicinity of the river mouth and to times when the oscillation of the interface was of nearly the same order as the depth of the upper layer; it was virtually never detected at stations upstream of 94.5. From these data, it was clear that other factors than tide were governing the surface salinity.

Wind and Salt Transport

Let us examine the relation between wind speed and surface salinity as revealed in the record. The colored areas in Figure 2 indicate when the wind direction was directly downstream in the straight section of the river channel (from St. 12 to St. 5), and the uncolored areas represent when the wind was in the opposite direction. From these results, we observe that the short-term variations in wind speed make virtually no contribution to variations in the surface salinity, but there is a clear correlation between long-period salinity variations and periods of continued wind. When the data are examined in further detail, the times when the wind began to blow are observed match with the times the surface salinities at all observation points began to increase, but even after the wind velocity fell, the salinity did not fall immediately to the lower, corresponding value, but 'tailed off'. The delay in response increased with distance from the salt wedge tip and was over 1 day at the river mouth. It is impossible to discuss the effect of wind direction since the wind blew almost exclusively in the above two flowing directions in the vicinity of the Ishikari River mouth. However, it is unlikely that winds directions across the river course would bring about higher diffusion rates than winds parallelling the river. The power spectra of the wind speed and salinity support the above results and hypotheses, but cannot be shown here because of space considerations.

We now examine the process of salt transport in order to account for this correlation of wind and surface salinity. As the flow system under consideration is incompressible, both volume and mass are conserved and the following differential equations are assumed to apply.

$$\frac{d(Au)}{dx} = Bv_w \tag{4}$$

$$\frac{d(Au\rho_1)}{dx} = Bv_w\rho_2 \tag{5}$$

Here, A is the cross sectional area of the upper layer of the river, B is the river width, u is the cross-sectional mean velocity of the upper layer, v_w is the velocity of wind-caused entrainment of salt water into the river, x is the coordinate of position measured from the tip of the salt wedge downstream, and ρ_1 is the upper layer density.

The velocity of the lower layer is much smaller than u in this two-layer flow, and B is assumed uniform. Introducing h = A/B and Keulegan's entrainment coefficient $E_w = v_w/u$, (4) and (5) provide the following solution.

$$\frac{(\rho_2-\rho_1)|_{x=x2}}{(\rho_2-\rho_1)|_{x=x1}} = e^{-\int_{x1}^{x2}\frac{E_w}{h}dx} \tag{6}$$

Here ρ is the lower layer density and X_1 and X_2 are the positions of the upstream and downstream end of the river element, respectively. In the Ishikari River the density of the salt wedge has uniform value ρ_2 and generally, salinity S is proportional to the density. Therefore the left hand side of (6) is reduced to

$$\frac{\rho_2-(\rho_1)\big|_{x=x2}}{\rho_2-(\rho_1)\big|_{x=x1}} \cong \frac{S_2-(S_1)\big|_{x=x2}}{S_2-(S_1)\big|_{x=x1}} \quad . \tag{7}$$

The Keulegan coefficient was originally defined for the case where entrainment was considered as dependent only on the difference in the two layers' average velocities; Keulegan had not thought in terms of any effect of the wind. Thus, this differs somewhat in value from the original Keulegan coefficient, which would apply when the wind speed is zero. The average entrainment coefficients in the region between the tip of the salt wedge and St. 94.2 were obtained from the observations as follows:

$$E_w = 1.92 \times 10^{-6}\, e^{0.405\overline{W}} \quad \text{for} \quad 118 m^3/sec \quad \text{and} \quad 2.5 m/sec \le \overline{W} \le 10 m/sec \tag{8}$$

$$E_w = 2.24 \times 10^{-6}\, e^{0.281\overline{W}} \quad \text{for} \quad 142 m^3/sec \quad \text{and} \quad 2.5 m/sec \le \overline{W} \le 10 m/sec \tag{9}$$

$\overline{W}$ is the average wind speed during the time necessary for the river water to move from the tip of the salt wedge to St. 94.2 at the cross-sectional average speed. This time would be shorter in case (9) than in (8), but as the difference between the times was small, the average wind speed over a 27-hour period was used. The reason the exponent is higher in (8) is, the flowrate is lower, and the free surface, which is a source of turbulence, is closer to the interface.

Estimate of the Surface Salinity

In light of (8) and (9), we suggest that E_w will be accurately estimated by the following formula for any flow rate and $\overline{W}$:

$$E_w = E(Q)e^{k(Q)\overline{W}} \tag{10}$$

Here, E is the Keulegan entrainment coefficient estimated from the E_w for zero wind velocity, Q is the flow rate, and k is the function of flow rate. E is related to R_i, which is calculated using the depth of the upper layer as the length scale and the difference in velocities between the layers, as

$$kE = CR_i^{-3/2} \tag{11}$$

However, *C* shows much scatter and it is difficult to choose a credible value. Asaeda and Tamai (1985) explained this scatter by incorporating the Peclet number, which is an index of molecular diffusion. The Peclet number is meaningful in laboratory observations, where wind is not a factor, but in real river mouth flows the effect of wind is large, as shown in (8) and (9), and it is possible that the scatter in *C* in real flows is due largely to wind.

It is possible to predict the surface salinity of a river at any given location and time from wind data once the variables in (10) are found for the river in question. Figure 3 shows the calculated salinities for the points St. 3-1. The solid line in the graph shows the values found in the survey of 1994, and the dashed line shows the estimates based on the use of (9). To simplify the calculation, the depth of the upper layer and the width of the river were taken to be 3.1 m and 200 m, respectively. The estimated values show acceptable accuracy except for

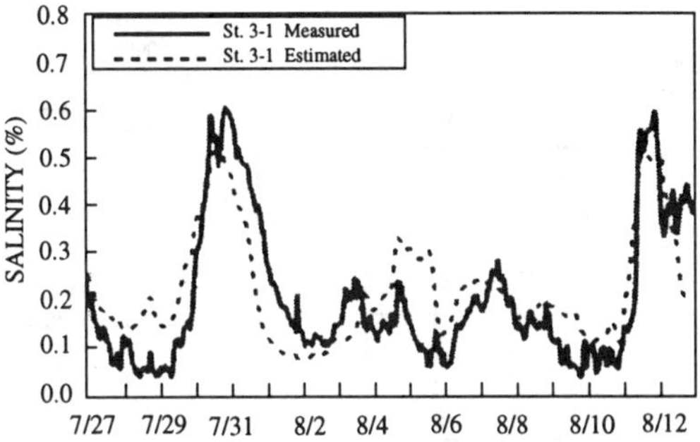

Figure 3. Estimates of surface salinity

the high-frequency variations. The reason that these calculations were of such high accuracy was that both the area of the actual upper layer cross section and the low-frequency component of wind speed were very uniform throughout this 29-km region of the river. As a result, E_w was uniform throughout the region. Accuracy would be improved by incorporating the actual cross sections, local E_w and tidal oscillations of the interface.

Mixing Mechanisms at the Interface

The overturn shown in Figure 4 can always be observed in the dry season just at the outlet of the rivers. The flow velocity and density distribution at the start position of overturn are shown in Figure 5. Because of the deep upper layer, the overturn mentioned above does not occur at the upstream interface of the Ishikari River in any wind conditions. If the wind speed is low, the shear flow is stable and no unstable waves appear at all the interface except at the outlet interface. Figure 6 depicts the distributions of flow speed and density in this low wind speed case where shear reaches a maximum during a tidal cycle. If the distributions of velocity and density Figure 5 are estimated according to the flow model proposed by G.A. Lawrence et al (1991) we obtain Table 3. The length of the instability waves observed in the Ishikari (see Figure 4) conforms closely to the length of the waves predicted by the theory. These results confirmed the validity of stability theory for analysis of real river flows and they have also shown that, rather than Kelvin-Helmholtz waves, one-sided instability waves occur just above the interface. The points of inflection of both curves in Figure 6 almost exactly coincide. Here, the Taylor-Goldstein model, which has linear velocity change and effectively linear density variation, is then used to estimate the Richardson number of this flow. The result (see the right column in Table 3) far exceeds the minimum 0.25 value necessary for stability.

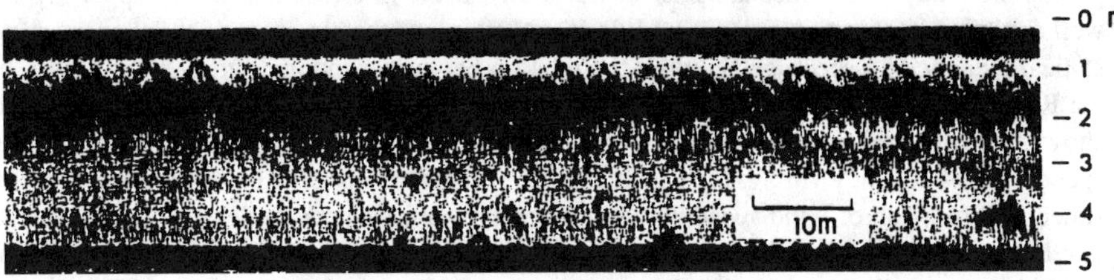

Figure 4. One-sided overturn at the outlet of the river

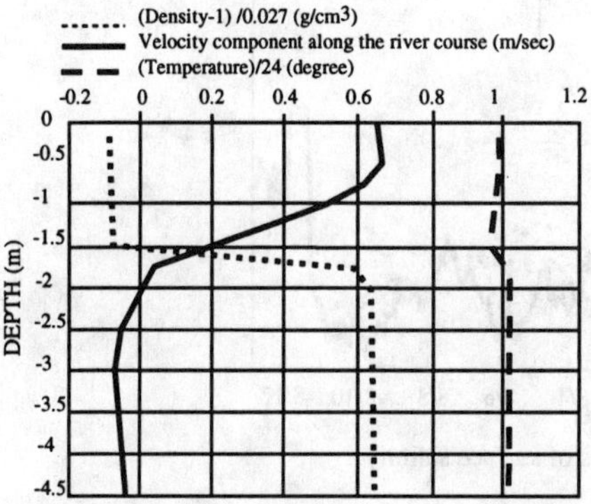

Figure 5. Velocity, density and temperature profile at St. 1 on Aug. 23, 1994

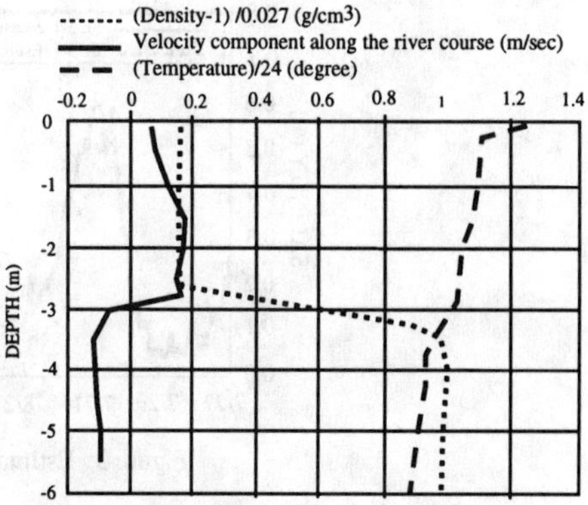

Figure 6. Velocity, density and temperature profile at St. 6 on August 7, 1994

TABLE 3. Measured and calculated values of various quantities relating to shear instability at the mouth of the Ishikari River.

Various Quantities	Measured (St. 1)	Calculated (St. 1)	Measured (St. 1)	Calculated (St. 1)
h: shear depth	1.2m	1.2m	0.73m	0.73m
velocity difference ΔU	0.75m/s	0.75m/s	0.29m/s	0.29m/s
((ΔU)	0.56	0.56	0.08	0.08
wave number k	1.27	1.33	no wave	no wave
α=kh	1.52	1.6	no wave	no wave
λ=2π/k	4.96m	4.71m	no wave	no wave
wave speed C	0.4m/s	0.42m/s	no wave	no wave
J: Richardson number	0.56	1.56	2.3	2.3
d: difference between infraction points of U and ρ	0.3m	0.3m	0	0
ε=2d/h	0.5	0.5	0	0

Nonetheless, the above considerations apply only when the effect of wind can be ignored. When the wind blows, gravity waves often arise to destroy the interface. The details of the complicated mechanisms of this transformation remain undescribed, however, as do those responsible for transport of salt to the surface once the interface is gone. It is necessary to pursue the study of the mechanisms for instability at the interface and in the upper layer as well as the structure of turbulence. Little is yet known of the structure of turbulence. Much statical evidence has been gathered on the relation between wind and turbulent intensity, but the Reynolds stress in real cases is too low for any correlation to be credible. Figures 7 and 8 show the horizontal and vertical components of the turbulent intensity with respect to wind speed. The buoyancy flux, which includes the vertical component of turbulence, probably stated as a function of wind speed, but the presently available data on density fluctuations are unreliable. Measurements at two vertical locations have reliably shown, however, that as wind speed increases, the vertical density gradient decreases to uniform salinity. This is clear evidence that salt transport out of the lower layer increases with wind speed (Yoshida, 1995).

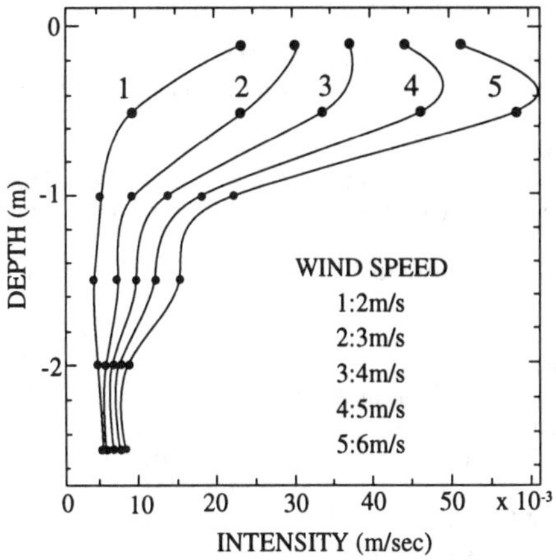

Figure 7. Horizontal turbulence intensity at St. 6.

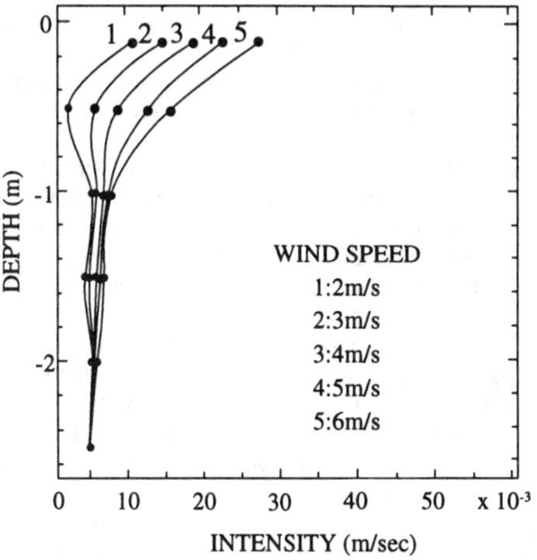

Figure 8. Vertical turbulence intensity at St. 6

Conclusions

This paper describes measurements on a salt wedge in a strongly stratified river. Stability of the flow in the Ishikari River can be determined using stability parameters from the Taylor Goldstein model. The entrainment of salt into the upper layer is determined by both the wind speed and shear velocity between the two layers.

Acknowledgments. This research was make possible by the support of the Research Project of a Grant-In-Aid for Scientific Research (B), The Ministry of Education, Science and Culture, Japan, under Grant No. 04452232 during 1992-1995 to S.Y. Hokkaido Aqua Culture Development Authority, Fukuda Hydrologic Center Co. Ltd. and M&C Co. Ltd.

References

Asaeda, T. and N. Tamai, Entrainment at a stable density interface at law Reynolds number, *J. Fac. Eng. Univ. Tokyo A-23*, 96-97, 1985.

Browand, F. K. and Y. H. Wang, An experiment on the growth of small disturbances at the interface between two streams of different densities and velocities, *Proc. Intnl Symp. on Stratified Flows*, 491-498, 1972.

Caulfield, C. P., Multiple linear instability of layered stratified shearflow, *J. Fluid Mech., 258*, 255-285, 1994.

Caulfield, C. P., S. Yoshida, W. R. Peltier and M. Ohtani, Multiple instability in a laboratory stratified shear layer, *Proc. 4 th Intnl Symp. on Stratified Flows, 1* (A1-139), 1994.

Dyer, K. R., *Estuaries. A Physical Introduction*, John Wiley & Sons, 1973.

Ellison, T. H. and J. S. Turner, Turbulent entrainment in stratified flows, *J. Fluid Mech., 6*, 423-448, 1959.

Fernando, H. J. S., Turbulent mixing in stratified fluids, *Ann. Rev. Fluid Mech., 23*, 455-493, 1991.

Fukusima, H., Stratified flow at a river mouth. *J. Oceanogr. Soc. of Japan, 1* (1), 57-73, 1942.

Holmboe, J., On the behaviour of symmetric waves in stratified shear layers. *Geophys. Publ. 24*, 67-113, 1962.

ICN, Proc. Intnl Conf. on Navigation, Rome, 1953.

Ippen, A. T., *Estuary and Coastline Hydrodynamics*, McGraw-Hill, New York, 1966

Keulegan, G. H., Interfacial instability and mixing in stratified flows, *J. Research, Nat., Bureau Stands., 43*, 487-500, 1949.

Lawrence, G. A., F. K. Browand, and L. G. Redekopp, The stability of a sheared density interface, *Phys. Fluids, A 3*, 2360-2370, 1991.
Nishida, S. and S. Yoshida. Stability of a two-layer shear flow. *Theoretical Appl. Mech., 32*, 35-45, 1983.
Officer, C. B., *Physical Oceanography of Estuaries (and Associated Coastal Waters)*, John Wiley & Sons, 1976
Prichard, D. W., Estuarine circulation patterns. *Proc. ASCE, 81 (717)*, 1-11, 1955.
Tamai, N., Unification of entrainment concepts and its application in buoyancy-associated flows. Report of Research project of a Grant in Aid Scientific Research (A), The Ministry of Science and Culture, Japan, Under Grant No. 59350034, 1986.
Taylor, G. I., Effect of variation in density on the stability of superposed streams of fluid. *Proc. Roy. Soc., A, 132*, 499-523, 1931.
Thorpe, S. A., A method of producing a shear flows. immiscible fluids, *J. Fluid Mech., 32*, 693-704, 1968.
Yoshida, S., M. Ohtani and S. Nishida. Unsteady factors affecting salt diffusion in the stratified flow system at a river mouth, Report of Research project of a Grant in Aid Scientific Research (B), The Ministry of Science and Culture, Japan, Under Grant No. 04452231, 1995
Yoshida, S., Time-space variations of hydraulic parameters and mixing processes at the density interface of river mouth flows. Report of Research project of a Grant in Aid Scientific Research (B), The Ministry of Science and Culture, Japan, Under Grant No. 62460158, 1990
Yoshida, S., Mixing mechanisms of density current system at a river mouth. *Proc. Second Intnl Symp. on Stratified Flows*, 2, 1062-1073, 1980.
Yoshida, S., On a mechanism for breaking of interfacial waves. *Coastal Engng in Japan, 20*, 7-15, 1977

29

Stratified Turbulence: Field, Laboratory and DNS Data

H. Yamazaki and D. Ramsden

Abstract

Direct Numerical Simulations (DNS) of stratified turbulence are performed in order to compare dynamic conditions of the flow with oceanic data. In addition to geophysical data, a laboratory data set obtained from a salt-stratified open channel facility is examined. The DNS data can achieve $\varepsilon/\nu N^2$ up to about 160. This is an important dynamic range for turbulence in a seasonal thermocline. It is concluded that DNS is a useful tool to study the nature of stratified turbulence and the results are applicable to the corresponding geophysical data.

Introduction

Stratified turbulence requires many independent parameters to characterize the dynamics of energy cascade and mixing. In general, it is assumed that turbulence in the ocean is created from a large scale motion or a large scale boundary process. Therefore, the associated Reynolds numbers are substantially larger than turbulence in a laboratory tank. It is therefore difficult to extrapolate the laboratory data to the generation scales of oceanic turbulence. However, turbulence in a seasonal thermocline can reveal the dynamic condition of the laboratory experiment (Yamazaki, 1990). Here, another experimental approach to study the dynamics of stratified turbulence is added. The Direct Numerical Simulation (DNS) codes of Gerz et al. (1989) are examined as a third data source for comparison to oceanic turbulence and the controlled experiments in the laboratory.

Riley et al. (1981) performed the first DNS of stably stratified turbulence and observed the oscillating behavior of the kinetic and the potential energy reservoir, partly as a consequence of the initialization procedure. They observed that the vertical kinetic energy is always less than the half of the horizontal kinetic energy showing that buoyancy suppresses the vertical kinetic energy. Gerz et al. (1989) developed a DNS code to study stably stratified turbulence with background shear. When the molecular Prandtl number was set to five, they observed counter-gradient heat flux resulting from the imbalance of dissipation of kinetic and potential energy. This simulation is consistent with a laboratory experiment of Komori et al. (1983). Holt et al. (1992) independently developed a pseudo spectral DNS code for sheared stratified turbulence flows. They compared the

Physical Processes in Lakes and Oceans
Coastal and Estuarine Studies Volume 54, Pages 401-416

numerical results to a comparable laboratory experiment reported by Rohr et al. (1988), and showed the turbulence statistics obtained from both methods agreed well.

Yamazaki (1990) compared geophysical turbulence data sets to the laboratory experiment of Itsweire et al. (1986). When $\varepsilon/\nu N^2$ is less than 100, the dynamic condition of the laboratory experiment is confirmed as identical to the turbulence in the seasonal thermocline. Itsweire et al. (1993) compared DNS for stratified turbulence to the laboratory experiment of Rohr et al. (1988) conducted in a salt-stratified shear flow tank. The DNS data confirmed the length scale evolution observed in laboratory experiments. No attempt has been made to compare the three different turbulence data, namely field, laboratory and DNS data. The objective of this study is to faithfully compare these different data, and to investigate to what extent the DNS approach is applicable to oceanic turbulence.

Data Sources

Oceanic Data

Two sources of oceanic turbulence data are used in this study. One is observed from the submarine *Dolphin* in a seasonal thermocline (Osborn and Lueck, 1985; Yamazaki et al., 1990), and the other is obtained from the submersible *Pisces IV* in a fjord (Gargett et al., 1984). The details of these data sets can be consulted in Yamazaki (1990). Briefly, two orthogonal velocity components, u_2 and u_3, relative to the moving platforms are measured from the air-foil probe (Osborn and Crawford, 1980). The coordinate system is moving and is fixed to the platform; the direction of the travel is taken as x_1. The velocity components, u_i, follow conventional notation.

The *Dolphin* experiment was conducted off the coast of San Diego, California, where the bottom depths of the experiment site exceeded 800 meters. The submarine was operated in a depth range between 50 and 120 meters, at which the influence of the surface mixed layer was absent. The turbulence data from the *Dolphin* represent a typical oceanic turbulence in a seasonal thermocline. On the other hand, the *Pisces IV* was operated in a much energetic mixing region in Knight Inlet on the western coast of Canada. Due to a strong current in the experimental site, a stationary lee wave was observed behind a sill. The turbulence observed during this experiment is an extremely energetic condition in comparison with the *Dolphin* data.

Laboratory Data

Itsweire et al. (1986) conducted a turbulence experiment in a salt-stratified channel. A ten-layer diffuser was used to develop a stratified fluid with a cross section of 0.25 m (width) × 0.4 m (height). The turbulence was generated by vertical grids at the inlet of the channel. The direction of mean flow in the channel is coincide with x_1. They measured u_1 and u_3 using hot film probes.

Direct Numerical Simulation (DNS) Data

The three-dimensional incompressible Navier-Stokes and temperature equation for perturbation velocities (u_1, u_2, u_3) and temperature T is integrated numerically in time in a cubic domain using the Boussinesq approximation. The original code is described in Gerz et al. (1989). The code allows for a sheared and an unsheared simulation. Mean shear, dU(z)/dz, is chosen in x_1 coordinate and the value is a linear function of x_3. The background reference temperature, $T_R(z)$, is also a linear function of the vertical coordinate. A spatial resolution of 64^3 grid points is used for the simulations.

Making use of the reference density ρ_o for density and l_o, v_o and $l_o\, dT_R/dz$ for initial values of integral length scale, rms velocity and temperature fluctuation respectively, the governing equations are nondimensionalized. The relevant nondimensional values become

$$Re = l_o\, v_o/\nu \quad (1)$$

$$Fr = v_o / (N\ l_o) \quad (2)$$

$$Pr = \nu / \gamma \quad (3)$$

$$Ri = (Sh\ Fr)^{-2} \quad (4)$$

where is ν kinematic viscosity, γ thermal conductivity, N buoyancy frequency, and Sh the shear number defined as $(dU/dz)(l_o/v_o)$. Three sets of simulation for sheared and unsheared condition reported in Gerz and Schumann (1991) have been computed:

Rms velocity	v_o	0.8226
Rms temperature	T_o	0
Prandtl number	Pr	1
Reynolds number	Re	42.7
Shear number	Sh	0 (unsheared)
		3 (sheared)
Froude number	Fr	1.42, 0.92, 0.29 (unsheared)
Richardson number	Ri	0.055, 0.13, 1.32 (sheared)

For a sheared experiment, Ri is specified, while Fr is specified when the flow condition is unsheared. Hence, six simulations were conducted in total following the same simulation procedure described in Gerz and Schumann (1991).

Scaling

In order to compare three different data sets, several scales of the motion are defined. Following the same argument stated in Yamazaki (1990) we define the energy containing eddy scale, L, as follows:

$$L = q^3/\varepsilon \quad (5)$$

where ε is the dissipation of the kinetic energy, and the rms turbulence velocity, q, is defined as:

$$q^2 = (\langle u_1^2\rangle + \langle u_2^2\rangle + \langle u_3^2\rangle)/3. \quad (6)$$

For the laboratory and the oceanic data sets, horizontal homogeneity is assumed for the velocity scales, namely $\langle u_1^2\rangle = \langle u_2^2\rangle$. The energy containing length scale is compared to several other length scales;

1) the Kolmogorov scale: $l_K = (\nu^3/\varepsilon)^{1/4}$, (7)

2) the Ozmidov scale: $l_{oz} = (\varepsilon/N^3)^{1/2}$, (8)

3) a buoyancy scale: $l_b = \langle u_3^2 \rangle^{1/2}/N$, (9)

4) a buoyancy-viscous length scale $\delta = (\nu/N)^{1/2}$. (10)

The Reynolds number, Re_l, and the Froude number, Fr_l, based on L are dynamically important, and are defined as follows:

$$Re_l = qL/\nu\,, \tag{11}$$

$$Fr_l = q/LN \tag{12}$$

This Reynolds number can be interpreted as a ratio of the inertial and viscous forces for the entire kinetic energy, and the corresponding Froude number is similarly considered a ratio of the inertial and gravitational forces. A Reynolds number, Re_v, and a Froude number, Fr_v, associated with the vertical scale of stratified turbulence are defined as follows:

$$Re_v = wh/\,\nu, \tag{13}$$

$$Fr_v = w/hN, \tag{14}$$

where $w = \langle u_3^2 \rangle^{1/2}$ is the vertical velocity scale, and h is a vertical length scale of turbulence. As Gargett (1988) suggested $h \approx l_{oz}$ when $Re_v >> 1$, Yamazaki (1990) assumed $h = l_{oz}$. In order to compare the DNS data with Yamazaki (1990) we used the same assumption. These parameters will be investigated in terms of $\varepsilon/\nu N^2$.

Following Yamazaki (1990), two approaches to the computation of the velocity scale of turbulence have been applied: a fixed method, and a variable method. Itsweire et al. (1986) estimated the rms velocity integrating the entire wavenumber range; they assumed that the noise due to internal waves was negligible. Using the fixed method, the velocity spectra for the DNS data were integrated over the entire wavenumbers, and the velocity spectra for the oceanic data were integrated between $k_L = 0.42$ cmp and the Kolmogorov wavenumber. Another set of the DNS and the oceanic data was created from the variable method; namely, the velocity spectra were integrated between the Kolmogorov and half the Ozmidov wavenumber as this is considered the dynamic range of the turbulence.

The fixed method may overestimate the rms velocity if turbulence is weak, as much of the power is due to internal waves. On the other hand it may also underestimate the rms velocity if the largest scale of overturn exceeds k_L^{-1}. The fixed lower bound crosses the lower limit of the variable method at approximately $\varepsilon/\nu N^2 = 150$; therefore, the fixed method may be overestimating the rms velocity when $\varepsilon/\nu N^2 < 150$.

The variable method may provide a good estimate of rms velocity, if l_{oz}^{-1} scales the upper limit of turbulent velocity Fourier components. This upper limit is a function of ε, so that the integration algorithm forces the rms velocity to be a function of ε. Considering the fact that the intensity of turbulent velocity must be physically related to the dissipation of its kinetic energy, this integration limit reflects the physical consequence of turbulent energy cascade. As turbulence becomes weak, the Kolmogorov scale increases, and as a result, the upper integration limit decreases in the wavenumber space. Conversely, the buoyancy scale decreases as the intensity of turbulence lessens, therefore, the lower integration limit increases in the wavenumber space. Both limits have roughly a linear relationship with $\varepsilon/\nu N^2$ in the log-log scale. Note that these limits do not collapse until $\varepsilon/\nu N^2$ becomes unity (Yamazaki, 1990). Therefore, no data exist for the variable method when $\varepsilon/\nu N^2 < 1$.

The ratio of $<u_3^2>$ to $<u_2^2>$ is shown in Figure 1. The DNS data for the fixed method extends below $\varepsilon/\nu N^2 = 1$ (Figure 1.a). The general agreement among the three data sets is good considering the different origin of the data sets. Therefore, the kinetic energy divides into the vertical and the horizontal components at an equal rate in terms of $\varepsilon/\nu N^2$ for all three data sets. Since the laboratory data are obtained from grid generated decaying turbulence experiments, these are equivalent to the unsheared case of the DNS. However, the ratios of $<u_3^2>$ and $<u_2^2>$ for the sheared DNS data are in close agreement with the laboratory data. For the variable method, the sheared DNS data follows the *Dolphin* data well. The unsheared data have smaller vertical velocity components than the horizontal components in comparison to the *Dolphin* data.

As Yamazaki (1990) noted for laboratory and field data, the fixed method for L/l_K of the *Dolphin* data show roughly constant values, and the laboratory data appeared in the same range with the *Dolphin* data (Figure 2.a). This tendency is due to the fact that as $\varepsilon/\nu N^2$ becomes smaller, the fixed method includes internal wave components in the estimation of L. The DNS data appeared quite different from the other data sources. For the sheared case, the ratios show roughly the same tendency as the other data sources, but the unsheared case showed much smaller values for L/l_K. These results suggest that a low Reynolds number experiment of the DNS for unsheared case does not generate internal waves. Consequently, the estimate of L declines sharply.

On the other hand, the variable method shows that both the field and the DNS data follow the same trend (Figure 2b). It is interesting point to notice that the field data appear in the mid range between the sheared and the unsheared DNS simulations. This may indicate that the geophysical data are somewhat mixtures of sheared and unsheared conditions.

When the energy containing eddy scale approaches that of the Kolmogorov scale, the turbulence should be dominated by viscosity, so it should be quickly approaching laminar flow. The geophysical data cross $L/l_K = 1$ at roughly $\varepsilon/\nu N^2 = 8$. For the DNS data of sheared case, L/l_K is always bigger than unity until the upper and the lower integration limits collapse. On the other hand, for the unsheared data $L/l_K = 1$ occurs roughly at $\varepsilon/\nu N^2 = 20$.

Both the fixed and the variable method for $l_{b/K}$ show a consistent feature (Figure 3). The reason is that l_b scales quite close to l_{oz} and l_{oz} is a function of $\varepsilon/\nu N^2$. Thus, the vertical length scale l_b is a strong function of $\varepsilon/\nu N^2$, and it does not depend on the data source. Therefore, the dynamical significance of this length scale is analogous to $\varepsilon/\nu N^2$.

For the Reynolds number, Re_l, the same feature as with the L/l_K diagram is roughly observed. In fact, L/l_K is identical to $(Re_l,)^{3/4}$. Again, the fixed method leads to the erroneous conclusion that the DNS and the field data do not show a consistent feature (Figure 4.a). When the Reynolds number for the variable method is bigger than about 50, the DNS data exhibit about the same range of $\varepsilon/\nu N^2$ for both data sets (Figure 4.b). This diagram clearly supports the contention that a locally defined Reynolds number for the field data can be in the same range of the DNS data.

For the Froude number, the sheared data from the fixed method shows a similar feature with the field data, but the unsheared data are quite different from both of them (Figure 5.a). It is difficult to interpret the dynamical significance of these different features. On the other hand, the variable method shows a consistent trend in the field and the DNS data. The Froude numbers for the sheared simulations are lower than unity, although there is a tendency to increase as the corresponding $\varepsilon/\nu N^2$ decreases. However, the unsheared data rapidly increase as $\varepsilon/\nu N^2$ decreases.

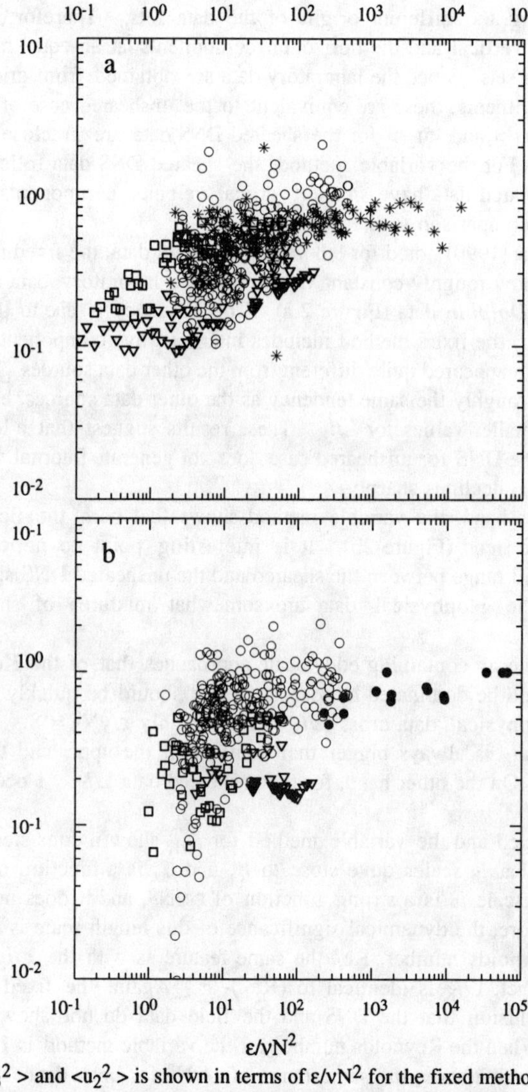

Figure 1. A ratio of $<u_3^2>$ and $<u_2^2>$ is shown in terms of $\varepsilon/\nu N^2$ for the fixed method (a: upper panel) and the variable method (b: lower panel). The thermocline data are O; Gargett et al. (1984) data are ●; and Itsweire et al. (1986) data are ∗. The DNS data with unsheared case are ∇, and the DNS data with sheared case are □.

Since the Froude number, Fr_l, of the fixed method for the DNS is quite different from the other data sources, the corresponding Re_l versus Fr_l diagram is difficult to interpret from a dynamical point of view (Figure 6.a). However, the variable method shows a consistent feature in the Re_l versus Fr_l diagram (Figure 6.b). When Re_l becomes less than about 10, Fr_l number for the *Dolphin* data increases. The same feature is observed from the DNS data

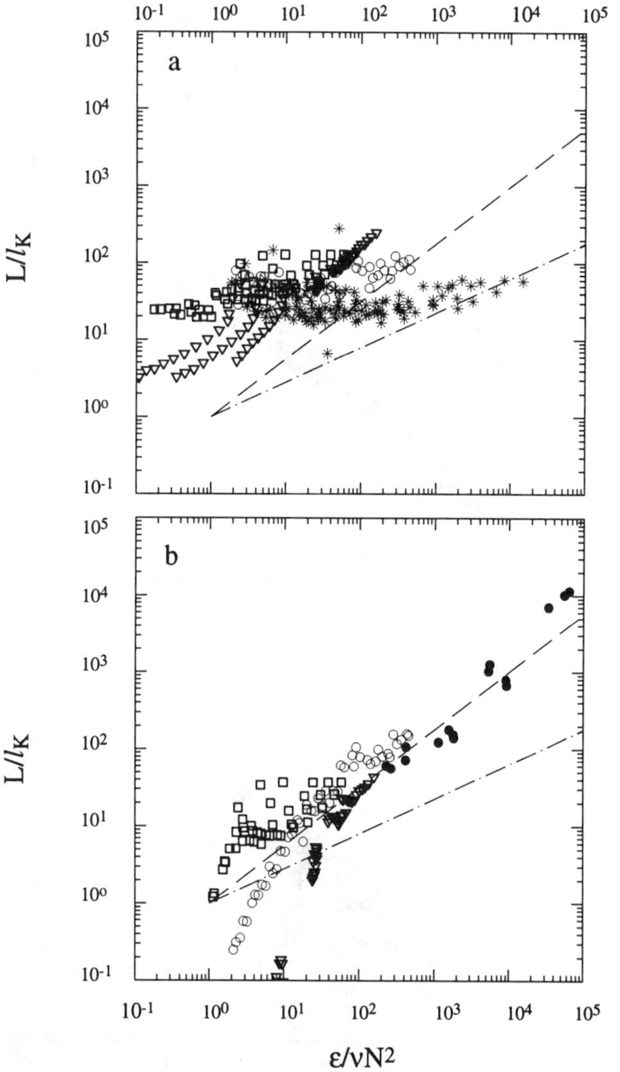

Figure 2. A ratio of the energy containing eddy size, L, and the Kolmogorov scale, l_K, is shown in terms of $\varepsilon/\nu N^2$ for the fixed method (a: upper panel) and the variable method (b: lower panel). The thermocline data are O; Gargett et al. (1984) data are ●; and Itsweire et al. (1986) data are ✳. The DNS data with unsheared case are ∇, and the DNS data with sheared case are □. A dash line is l_{oz}/l_K, and a chain dot line is δ/l_K.

for unsheared case, but much less similar tendency appeared in the sheared case. Yamazaki (1990) interprets that high Fr_l in low Re_l is due to horizontally dominating flow pattern, which does not require a work against gravity. If the same reasoning is applied to the DNS data, Fr_l with sheared experiments should increase as Re_l decreases. However, the results are opposite from the expectation. In order to consider the unexpected results we make use of the definition of q and L for Fr_l, namely $Fr_l = q/LN = \varepsilon/(q^2 N)$. When Fr_l is greater than 1, ε exceeds q^2 N. In other words, the entire kinetic energy dissipates on a shorter time

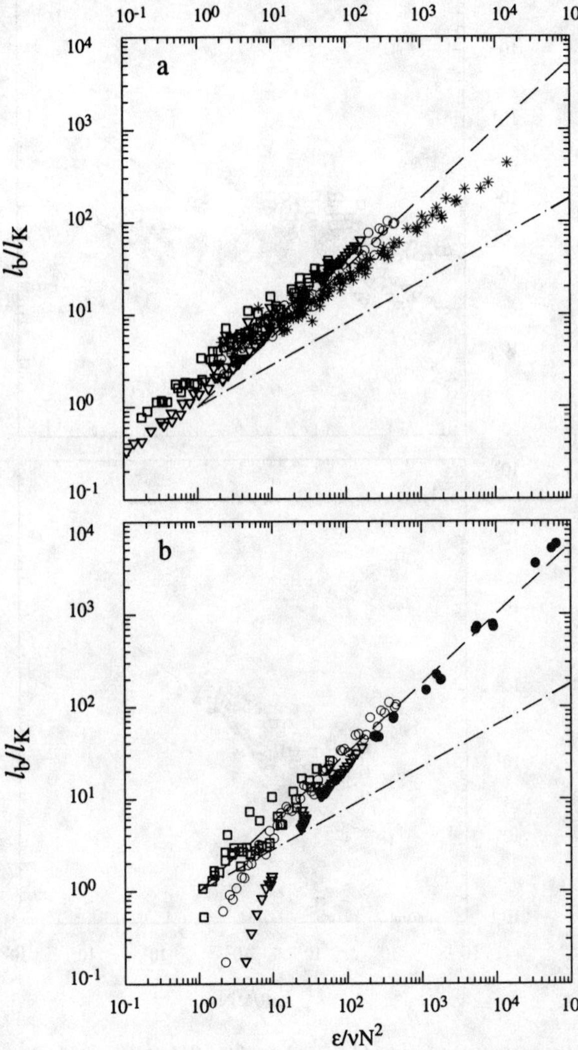

Figure 3. A ratio of the buoyancy scale, l_b, and the Kolmogorov scale, l_K, is shown in terms of $\varepsilon/\nu N^2$ for the fixed method (a: upper panel) and the variable method (b: lower panel). The thermocline data are O; Gargett et al. (1984) data are ●; and Itsweire et al. (1986) data are ✳. The DNS data with unsheared case are ∇, and the DNS data with sheared case are □. A dash line is l_{oz}/l_K, and a chain dot line is δ/l_K.

scale than the back ground buoyancy time scale. This is also the case for the unsheared experiment. On the other hand, when Fr_l is less than 1, the kinetic energy dissipation time scale is larger than the buoyancy time scale. Since the kinetic energy reservoir receives additional energy from shear, the dissipating time scale can be longer than ε. Therefore, it may not be correct to consider the interpretation for high Fr_l in low Re_l made in Yamazaki (1990).

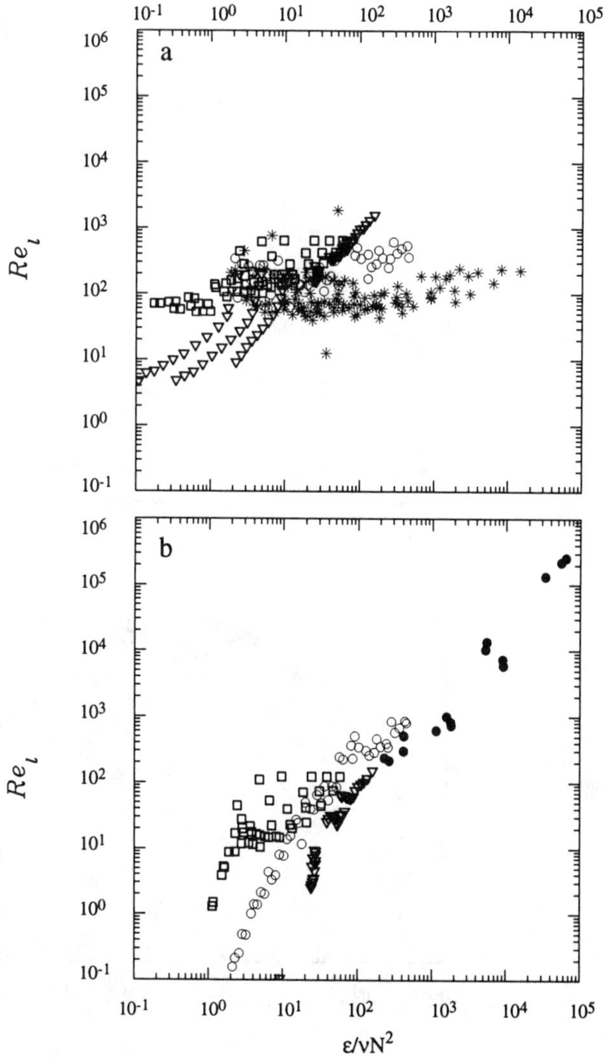

Figure 4. The Reynolds number, Re_l, is shown in terms of $\varepsilon/\nu N^2$ for the fixed method (a: upper panel) and the variable method (b: lower panel). The thermocline data are O; Gargett et al. (1984) data are ●; and Itsweire et al. (1986) data are ✳. The DNS data with unsheared case are ∇, and the DNS data with sheared case are □.

As Yamazaki (1990) noted, the Reynolds number Re_v exhibited the least dependence on the rms algorithm (Figure 7). Both methods show strong positive correlation of Rev against $\varepsilon/\nu N^2$. Note that this feature is quite similar with the l_b/l_K diagram. On the other hand, the Froude number Fr_v appeared quite different for the two algorithms, as Yamazaki (1990) reported (Figure 8). The DNS data followed the same trend with the *Dolphin* data. The Fr_v from the fixed method increased as $\varepsilon/\nu N^2$ decreases. This is physically

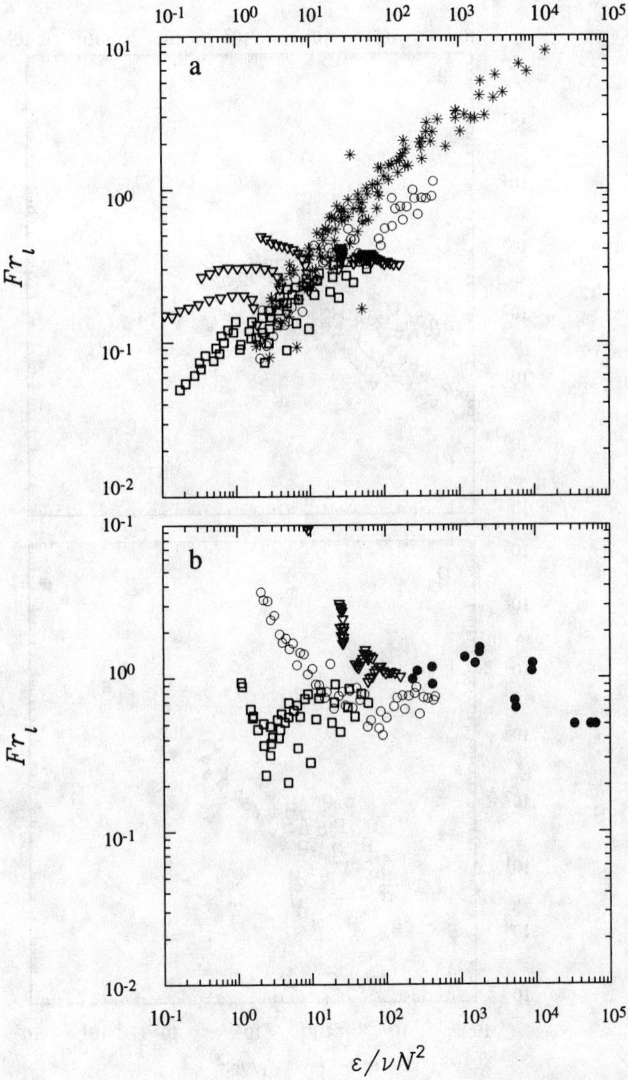

Figure 5. The Froude number, Fr_l, is shown in terms of $\varepsilon/\nu N^2$ for the fixed method (a: upper panel) and the variable method (b: lower panel). The thermocline data are O; Gargett et al. (1984) data are ●; and Itsweire et al. (1986) data are ✳. The DNS data with unsheared case are ∇, and the DNS data with sheared case are □.

inconsistent with the decaying process of turbulence. Whereas the variable method shows a dynamically consistent feature, namely, as $\varepsilon/\nu N^2$ decreases Fr_b decreases. Interestingly, Fr_v from the unsheared DNS never exceed one, but the sheared case values often exceed unity, even when $\varepsilon/\nu N^2$ is less than 10. This is a bit surprising result, considering the fact that low $\varepsilon/\nu N^2$ is a highly viscous regime, so the effects from shear are expected to be minimal. It is clear from Figure 2 to 8 that the DNS can be quite useful to study the oceanic turbulent data in the seasonal thermocline, but the dynamic range of the DNS is not

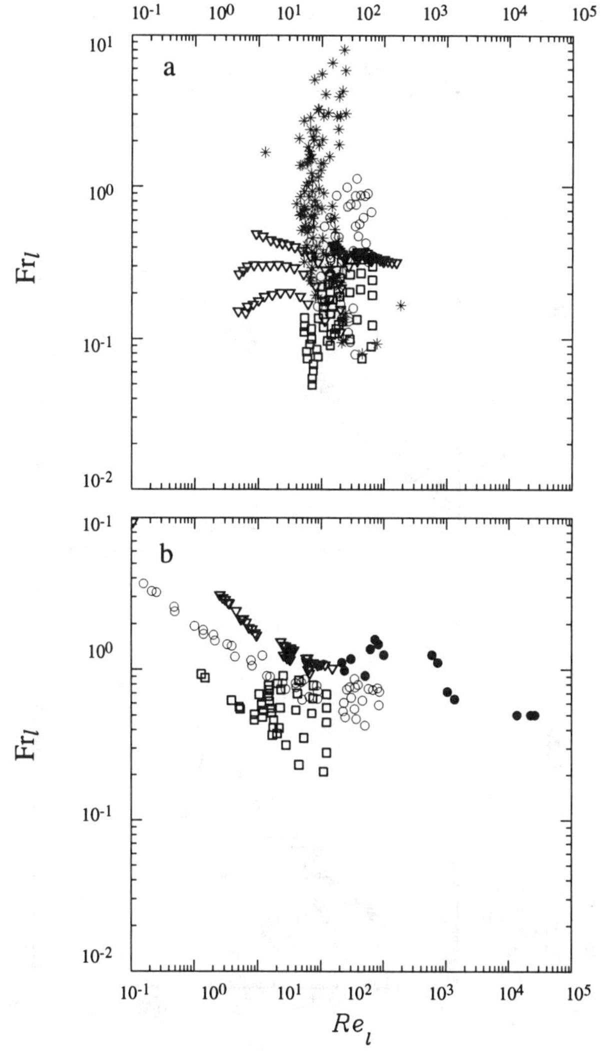

Figure 6. The Reynolds number, Re_l, is plotted against the Froude number, Fr_l, for the fixed method (a: upper panel) and the variable method (b: lower panel). The thermocline data are O; Gargett et al. (1984) data are ●; and Itsweire et al. (1986) data are ✳. The DNS data with unsheared case are ∇, and the DNS data with sheared case are □.

sufficient to compare the data from Gargett et al. (1984). It should be noted, however, that the Gargett et al (1984) fjord data is probably not representative of the ocean as a whole given that the turbulence is mechanically generated by tidal flow over a sill. In particular, the high values of $\varepsilon/\nu N^2$ observed do not necessarily preclude applying the DNS to oceanic regimes as these high values are probably limited to such locales.

One velocity spectrum from each sheared and unsheared case have been chosen to compare the DNS to the oceanic data. Figure 9 is obtained from an unsheared simulation at the early stage of the decaying process. Since the code excites the entire wavenumber band at the beginning of the simulation, all three components of velocity spectra show the

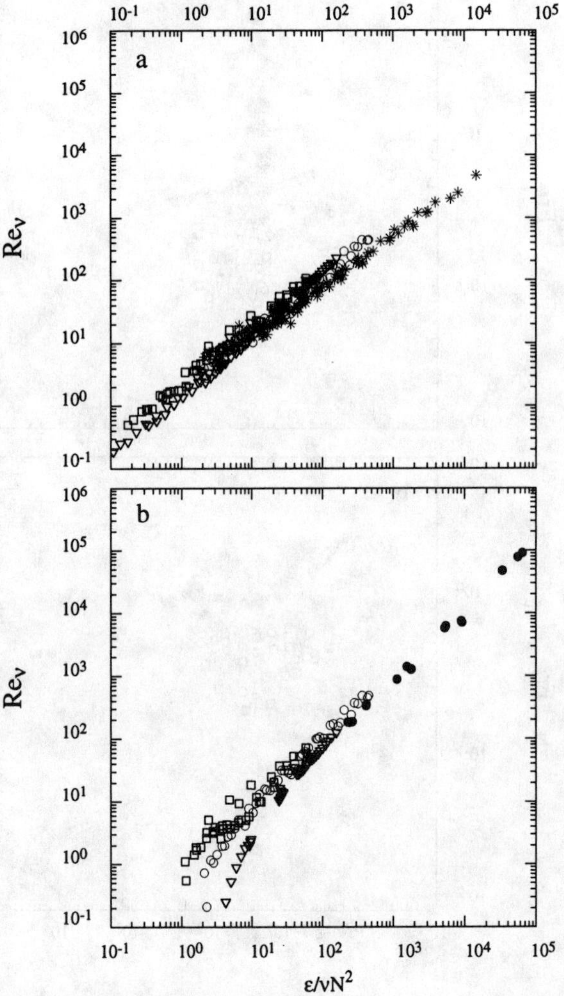

Figure 7. The Reynolds number, Re_ν, is shown in terms of $\varepsilon/\nu N^2$ for the fixed method (a: upper panel) and the variable method (b: lower panel). The thermocline data are O; Gargett et al. (1984) data are ●; and Itsweire et al. (1986) data are ✳. The DNS data with unsheared case are ∇, and the DNS data with sheared case are □.

same level. The corresponding three dimensional spectrum is compared to an empirical turbulent velocity spectrum of Nasmyth (1970). The agreement is generally good, but in the high end of the wavenumber band, the DNS spectrum is at a slightly higher level than the empirical spectrum. Gerz and Schumann (1991) reported the resolution of the viscous scale is sufficient for the present simulations. If the difference between the DNS and the field data is real, the difference may be due to Prandtl number dependence.

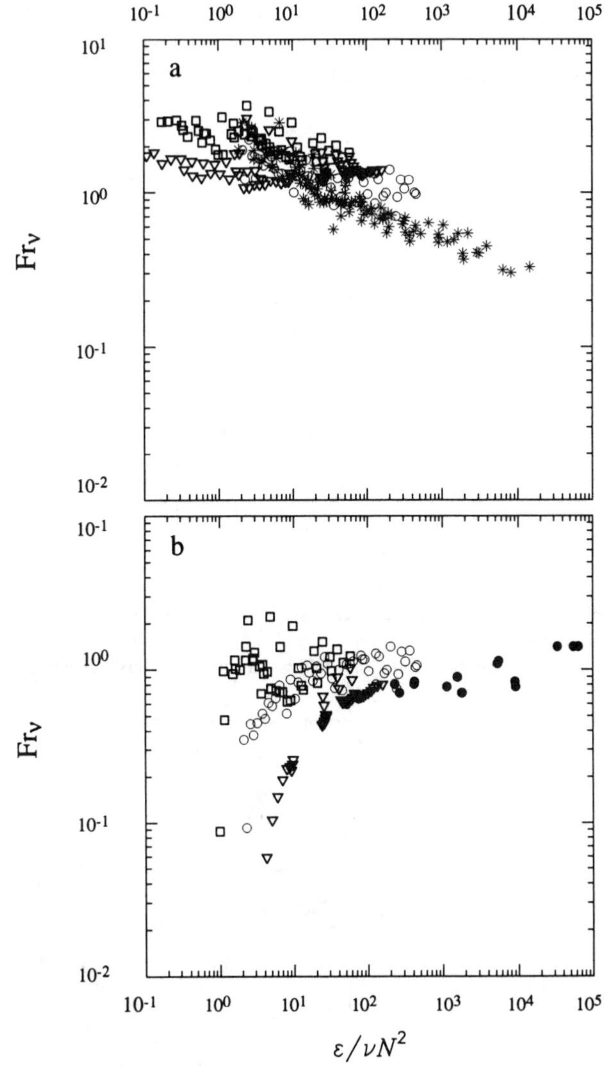

Figure 8. The Froude number, Fr_v, is shown in terms of $\varepsilon/\nu N^2$ for the fixed method (a: upper panel) and the variable method (b: lower panel). The thermocline data are O; Gargett et al. (1984) data are ●; and Itsweire et al. (1986) data are ✳. The DNS data with unsheared case are ∇, and the DNS data with sheared case are □.

For the sheared data spectra (Figure 10), we observe the vertical velocity level is lower than the horizontal velocity in the high wavenumber band. Hence, even at the viscous dominating wavenumber the partition of the kinetic energy is not identical to each other in the three components. This is quite different from a rundown experiment of the unsheared simulation, in which we observed all three components at the viscous dominating scale are comparable to each other. In Yamazaki (1990), the data were chosen so that all three velocity components are comparable at the high end of the spectra. This may suggest that Yamazaki (1990) selected turbulence without shear for low values of $\varepsilon/\nu N^2$. For high values of $\varepsilon/\nu N^2$, the velocity spectra should show a local isotropy as Gargett et al. (1984) demonstrated, so all three velocity spectra collapse at these high end of the wavenumbers.

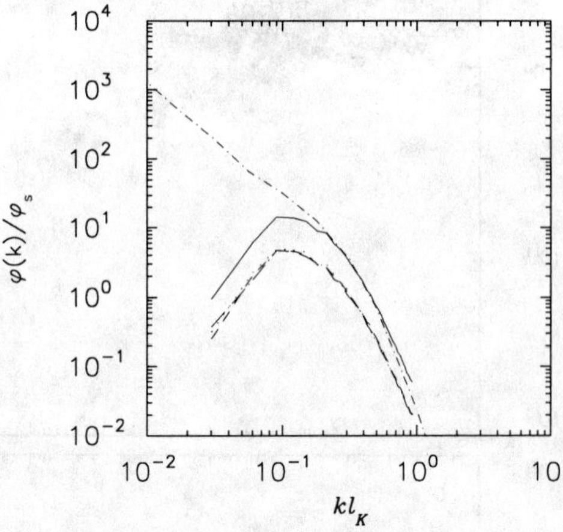

Figure 9. One dimensional velocity power spectra for u_1 (doted line), u_2 (dashed line) and u_3 (chain-dot line) are normalized by the Kolmogorov scale, $(\varepsilon\nu^5)^{1/4}$, for an unsheared case having $\varepsilon/\nu N^2 = 47$. The three dimensional power spectra (solid line) follow Nasmyth's universal spectrum (double chain dot line).

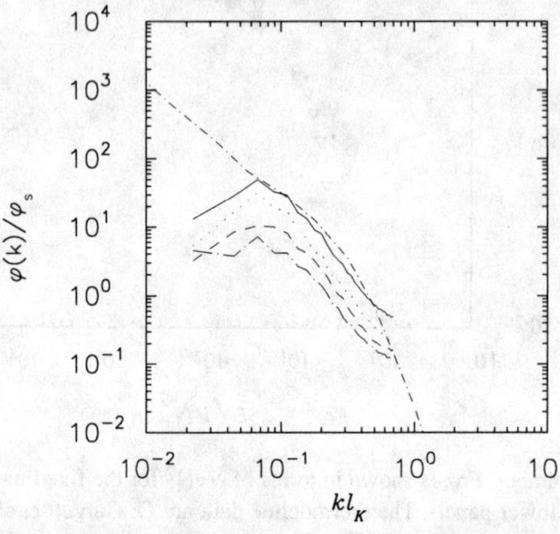

Figure 10. Same as Figure 9 for a sheared case having $\varepsilon/\nu N^2 = 67$.

Concluding Remarks

It has been demonstrated that DNS are a useful way to study stratified turbulence. In fact, oceanic turbulence can be at equivalent Reynolds and Froude numbers with the DNS, so the dynamics of both turbulence should be similar, if the Prandtl number dependence is minimal. If the Prandtl number dependence is significant, the difference between the DNS and the oceanic turbulence spectra at the viscous dominating wavenumbers should be investigated.

In order to faithfully compare the DNS data with the oceanic data the variable method must be applied to estimate turbulence parameters. Among all the non-dimensional plots shown in this paper, the Re_l - Fr_l diagram is the most important one because it can readily reveal the dynamics of turbulence.

In any event, the reasonable behavior exhibited by the DNS despite it's admittedly limited range of scales tends to suggest that oceanic turbulence is driven by local small-scale (up to a few decades larger than the Kolmogorov scale) shear. It is tempting to speculate that this small scale shear is generated by some weakly nonlinear larger scale process, where upon turbulence 'begins' and quickly dissipates the energy, leaving the region again slightly between the sheared and unsheared states.

Acknowledgment: This study was funded by the ONR.

References

Gargett, A. E., T. R. Osborn and P. W. Nasmyth, Local isotropy and the decay of turbulence in a stratified fluid. *J. Fluid Mech. 144*, 231-280, 1984.

Gerz, T., U. Schumann, and S. Elgobashi, Direct numerical simulation of stratified homogenous turbulent shear flows. *J. Fluid Mech., 200*, 563-594, 1989.

Gerz, T., and U. Schumann, Direct simulation of homogenous turbulence and gravity waves in sheared and unsheared stratified flows. In *Turbulent Shear Flow* 7, edited by W. C. Reynolds, pp. 27-45, Springer, 1991.

Holt, S. E., J. R. Koseff, and J. H. Ferziger, A numerical study of the evolution and structure of homogenous stably stratified sheared turbulence. *J. Fluid Mech., 237*, 499-539, 1992.

Itsweire, E. C., K. N. Helland and C. W. Van Atta, The evolution of grid-generated turbulence in a stably stratified fluid. *J. Fluid Mech., 155*, 299-338, 1986.

Itsweire, E. C., J. R. Koseff, D. A. Briggs, and J. H. Ferziger, Turbulence in stratified shear flows: implications for interpreting shear-induced mixing in the ocean. *J. Phys. Oceanogr., 23*, 1508-1522, 1993.

Komori, S., H. Ueda, F. Ogino and T. Mizushina, Turbulence structures in stably stratified open channel flow. *J. Fluid Mech., 130*, 13-26, 1983.

Nasmyth, P. W., Ocean turbulence, Ph. D. thesis, Department of Oceanography, University of British Columbia, Vancouver, Canada, 1970.

Osborn, T. R. and W. R. Crawford, An airfoil probe for measuring turbulent velocity fluctuation in water. In *Air-Sea Interaction: Instruments and Methods,* edited by F. Dobson, L. Hase and R. Davis, Chap. 19. Plenum, 1980.

Osborn, T. R. and R. G. Lueck, Turbulence measurement with a submarine. *J. Phys. Oceanogr., 15*, 1502-1520, 1985.

Riley, J. J., R. W. Metcalfe, and M. A. Weissman, Direct numerical simulations of homogenous turbulence in density stratified fluids. In *Nonlinear Properties of Internal Waves* edited by B. J. West, pp. 79-112, AIP Conf. 76, 1981.

Rohr, J. J., E. C. Itsweire, K. N. Helland and C. W. Van Atta, Growth and decay of turbulence in a stably stratified shear flow. *J. Fluid. Mech. 195*, 77-111, 1988.

Yamazaki, H., Stratified turbulence near a critical dissipation rate. *J. Phys. Oceanogr., 20*, 1583-1598, 1990.

Yamazaki, H., R. G. Lueck and T. R. Osborn, A comparison of turbulence data from a submarine and a vertical profiler. *J. Phys. Oceanogr., 20*, 1778-1786, 1990.

30

Waves, Mixing, and Transports over Sloping Boundaries

C. C. Eriksen

Abstract

Sloping boundaries in rotating, stratified fluids influence flow in their vicinity by supporting both trapped and reflected waves. Internal wave breaking caused by reflection at a sloping boundary produces microscale mixing. Spatial gradients of this mixing in turn may drive mean circulation near boundaries, both along and across isobaths, and may account for significant features of circulation. Trapped free wave and forced evanescent modes are possible at both subinertial and superinertial frequencies. These motions can account for much of the variance observed over topographic features in the ocean and for Eulerian mean currents as well. Without mixing, however, steady eddy fluxes cannot drive Lagrangian mean circulation, according to non-acceleration theorems.

Introduction

Boundary regions are well known as sites of stronger flow than in the interior of bodies of water, whether lakes or oceans. Waters immediately adjacent to topographic features within lakes and oceans, tend likewise to be somewhat more active flow regions than farther in basin interiors. While boundary currents and such phenomena as Taylor caps have been explored extensively, both theoretically and observationally, oscillatory motions associated with boundary regions have received less attention in the literature of oceanography and limnology. Freely propagating waves in the open interior of basins must adjust to the presence of generally sloping lateral boundaries. Because baroclinic waves disperse anisotropically, those waves with ray slopes that are near bottom boundary slopes tend to reflect in a manner that induces nonlinear transports, thus be prone to breaking. The breaking process both redistributes wave energy and dissipates it in turbulent mixing. Turbulent mixing from reflection-produced breaking is confined near the boundary, but well outside the very thin boundary layer associated with wall turbulence. Mean circulation in the vertical plane normal to isobaths as well as along isobaths can be induced by such mixing. Together the magnitude and scales of this mixing determine the structure and strength of induced circulation. Waves which do not break but vary in amplitude may also induce circulation, but only in an Eulerian sense. These Eulerian mean currents are exactly compensated by the Stokes drift of the waves, so that there is no net transport in a Lagrangian sense without some mixing of momentum or density.

Physical Processes in Lakes and Oceans
Coastal and Estuarine Studies Volume 54, Pages 417-440

Mixing near sloping topography has received substantial attention in recent years through observation, theory, and laboratory experiment. Munk (1966) is generally credited with the opening salvo in an ongoing debate over the potential importance of boundary mixing in the ocean. The debate extends to lakes as well, where the effect of boundary mixing may be more important than in the ocean because of the smaller spatial scales over which turbulent mixing varies. Munk applied a one-dimensional advective-diffusive balance to the world ocean, the basis of which was an estimate of the renewal rate of deep water from winter convection in high latitude oceans (mostly in the Atlantic at polar extremes). The basin-average vertical stratification implied in this model a net vertical diffusivity of water properties on the order of 10^{-4} m^2 s^{-1}, often taken as a "canonical value" for the open deep ocean. Microstructure observations in the ocean interior rarely imply diffusivities as high as this value (Gregg, 1987, 1989). In contrast, these observations suggest values an order of magnitude or more weaker than the canonical value demanded by a one-dimensional advection-diffusion balance. Recent experiments with deliberate releases of dye in the open ocean point to diffusivities that are more or less uniform with depth of 10^{-5} m^2 s^{-1} (Ledwell et al., 1993). These are corroborated by collocated microstructure profiler estimates that are indistinguishable from the dye result (Toole et al., 1994). The order of magnitude gap between the overall ocean basin wide effective vertical mixing rate and *in situ* ocean interior values suggests that either the canonical value is a misleading figure or places other than the ocean interior provide much higher mixing than 10^{-5} m^2 s^{-1} in order that the canonical value be achieved.

Armi (1978) suggested that turbulent drag of benthic currents over topographic features provided the mixing that seemed lacking in the ocean interior. Garrett (1979) objected on energetic grounds, pointing out that benthic circulation was much too sluggish to provide the level of potential energy production necessary to obtain a basin averaged diffusivity that approached the canonical value. A process that promised to provide the necessary mixing rate is breaking of internal waves. Eriksen (1985) showed that if only a few percent of the incident flux of the open ocean internal wave spectrum (represented by the Garrett-Munk kinematic model, whose latest version is given by Munk (1981)) were lost to mixing at the ocean boundaries, the implied rate of potential energy production would be high enough to account for a basin-wide effective diffusivity at the canonical level. Internal wave spectra are observed to distort considerably near boundaries (Eriksen, 1982, 1995) and overturns and low gradient Richardson numbers are found to occur frequently. Gilbert and Garrett (1989) and Müller and Xu (1992) present estimates of mixing rates based on theories for internal wave reflection and scattering, respectively, at sloping boundaries, but these are based on an assumption that mixing is entirely at the expense of waves whose wavenumber exceeds some threshold. A more physically likely scenario is that waves of all wavenumbers superpose randomly to produce instability and subsequent mixing. Recent observational evidence for mixing is given by Eriksen (1998) and is summarized in the following section.

Phillips et al. (1986) showed in a laboratory experiment that mechanical mixing near a sloping boundary in a basin with nonuniform vertical stratification sets up a slow circulation in the vertical-onslope plane. The pycnocline is broadened slowly by less stratified fluid intruding into it from the boundary. This circulation is set up as the overall convergence within the mixing layer of upslope flow close to the boundary and downslope flow near its outer edge (Figure 1). Irrespective of the details of the mixing process, the induced circulation caused by boundary mixing can take a variety of patterns depending on

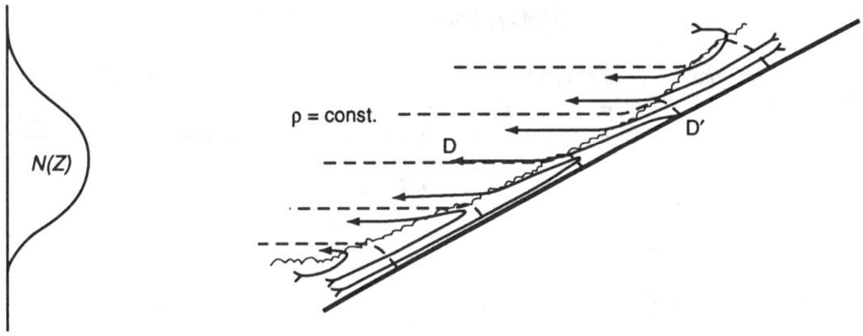

Figure 1. A schematic sketch of isopycnals (dashed) and mean streamlines (continuous curves) in the boundary layer region resulting from mixing a stratified fluid at a sloping boundary. The curve at left defines the ambient buoyancy frequency profile. Fluid having undergone mixing is injected into the pycnocline according the circulation indicated. Adapted from Philips et al. (1986).

the vertical structure of ambient stratification, the bottom slope shape, and the distribution of buoyancy flux in the mixing region (Garrett, 1991). McDougall's (1989) consideration of the overall advective-diffusive balance of the mixing region is particularly instructive. This balance is:

$$AwN^2 = \frac{\partial}{\partial z}(AKN^2) \quad (1)$$

where z is the upward coordinate, w is average vertical velocity through a boundary layer with horizontally projected area A, N is the buoyancy frequency outside the layer, and K is the effective vertical diffusivity within the layer (taking account of secondary flows in the boundary layer that redistribute fluid that has already been subject to mixing). This relation shows that upward (and by necessity upslope) flow is induced by spatial variations in the product of the projected area (a function of bottom slope), effective diffusivity, and stratification. The Phillips et al. (1986) experiments focused on variations in stratification to produce intrusions into the interior due to mixing, but variations in bottom slope or mixing intensity can have the same effect. Imberger and Ivey (1993) developed an analytic model for the strength of the induced interior circulation based on expanding in the aspect ratio of boundary layer thickness to the projection of pycnocline thickness on a sloping boundary. They find an effective basin vertical diffusion that varies as the ninth power of boundary layer thickness, the square of density gradient, roughly the cube of bottom slope, and inversely as the cube of boundary layer diffusivity in this limit. The extreme sensitivity to all four of these parameters, none of which is easy to estimate with great accuracy in any given physical system, suggests that boundary mixing may induce circulations that are either negligible or dominant in physical contexts that are not so very different.

Ivey and Nokes (1989) demonstrate in a laboratory experiment how effective internal wave breaking can be in mixing fluid in a boundary layer on a slope. The experiment used a single vertical mode wave at the critical frequency incident on a uniform slope to demonstrate a breaking process, described as the development of a surge up and down slope that quickly becomes nonlinear, then turbulent. Their experiment does reach much higher Reynolds numbers than earlier studies, but still somewhat smaller than expected in an oceanic context. Recent turbulence measurements on the flank of a seamount (coincident and concurrently with the moored data discussed below) indicate that mixing rates do

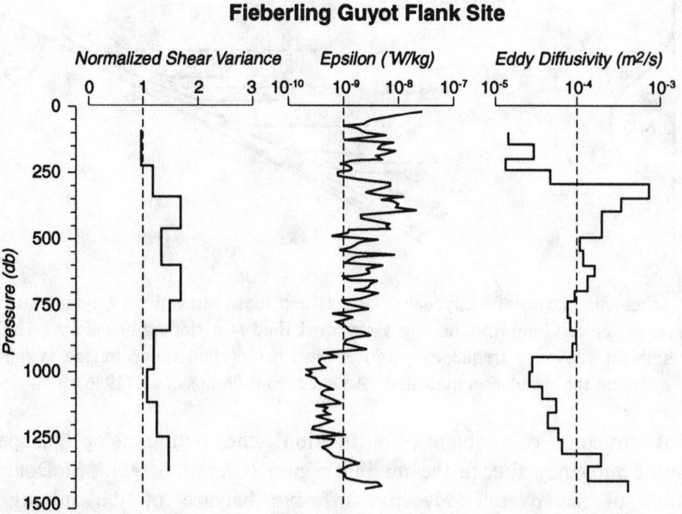

Figure 2. Vertical profiles above the sloping flanks of Fieberling Guyot of (left) the variance of fine-scale shear (3 to 128 m wavelength) normalized by the local N^2, (center) the turbulent kinetic energy dissipation rate ε averaged over 10 m vertically, and (right) vertical eddy diffusivity given by $\Gamma\varepsilon N^2$ where $\Gamma = 0.25$. This figure is from J. Toole, R. Schmitt, and K. Polzin as reported in Garrett et al. (1993). Turbulence is enhanced near the bottom (near 1500m) but also in the depth range 350 to 500m. The deep enhancement may be ascribed to internal wave breaking caused by reflection from the steep slope, while the shallower enhancement is associated with the presence of wavelike motions excited near the seamount summit.

indeed increase near a sloping bottom (Toole et al., 1997). Diffusivities within a few hundred m of the steep seamount flank exceed 10^{-4} m^2 s^{-1} (Figure 2). Evidence from moored measurements, summarized in the following section, demonstrates that the observed turbulence is induced by the internal wave reflection process.

Internal wave breaking is not the only wave process that can lead to enhanced mixing and induced mean circulation in the vicinity of sloping boundaries. Besides the increase in turbulence near the bottom evident in Figure 2, there is an even more prominent maximum in the depth range 350 to 500 m. This maximum is associated with wavelike motion that appears to emanate from the summit plain of Fieberling Guyot. The waves are dominated by diurnal tidal frequency oscillations. These are described variously as seamount-trapped waves, vortex-trapped effectively superinertial internal waves, and as vertically propagating evanescent waves. Ultimately all three of these descriptions are of a wave associated with topography and, in any case, the net result is mixing near a topographic feature.

The outline of this paper is to summarize the internal wave reflection process, then the diurnal (and, it turns out, broader band) motions trapped near a seamount, present the theory of motions trapped to a uniform slope, and to discuss the implications for mean flows. The observational evidence presented comes exclusively from detailed studies of flow at Fieberling Guyot, a large seamount in the eastern North Pacific Ocean. The processes at work there can be expected at any seamount. The same processes or their analogs can be expected in any stratified basin. Forcing scales of the waves in question determine resulting levels of turbulence.

Internal Wave Reflection and Breaking

Reflection dominates internal wave spectra over relatively simple topography in the ocean. This process is presumably at work over complicated topography as well, but is more difficult to recognize in the typically sparse field measurements that are affordable. The signature of reflection becomes less recognizable with distance from the bottom. One of the remarkable features of the open ocean internal wave spectrum is that it is so uniform from place to place and time to time. Another equally remarkable feature is that it can be so extremely distorted near topography, yet approach the open ocean quasi-universal spectrum as described by Garrett and Munk (see Munk, 1981) over very short distances in the ocean. The particular anisotropy of the internal wave dispersion relation forces waves that encounter boundaries of any slope whatever to change wavenumber. This alteration is due to the need to satisfy an inviscid boundary condition of no normal flow at the boundary. The change in wavenumber effects a change in group speed (shorter waves travel more slowly) and the transport of energy along phase surfaces (as opposed to normal to them) together cause energy density of linear waves to amplify proportionally to the square of the wavenumber amplification. The amplification is most at the internal wave critical frequency, where incident rays are inclined identically to the bottom slope. Here, the linearly predicted amplifications are infinite, while in a real fluid, and most especially in the real ocean, the amplification is held finite by nonlinearities which lead to turbulence which, m turn, implies dissipation.

Depending on incidence angle, internal waves can be reflected to lower as well as to higher wavenumber, hence to lower as well as to higher energy. Higher wavenumber waves dominate the sum of incident and reflected waves so that incident waves are masked by reflected (shorter) waves near the internal wave critical frequency. The range over which amplification is evident is relatively broad. For simple planar geometry, a doubling of energy is predictable more than an octave below and above the critical frequency. Reflection also changes the direction of waves since the alongslope (parallel to isobath) wavenumber component is invariant while the vertical and onslope wavenumber components amplify (for details see Eriksen, 1982). More energetic reflected waves are turned more parallel to the slope (closer to normal to isobaths) than the incident waves that excite them. Hence the internal wave field near slopes is highly anisotropic, in contrast to the open ocean interior where the quasiuniversal spectrum is horizontally isotropic. Near a slope, low frequency wave energy flux is focused toward deeper water while high frequency flux is toward shallower water. Given the robust constraints imposed by reflection, it would be surprising not to observe evidence of the phenomenon in natural bodies of water.

Indeed, it is difficult not to observe the manifestations of internal wave reflection near slopes. Eriksen (1982) presented evidence from a variety of latitudes and slopes to show that near-bottom enhancement of the internal wave field is a general feature of oceanic variability. These observations were found in a collection of available time series records of near-bottom currents in the open deep ocean. They motivated the need to describe the reflection process m greater detail. The observational program at Fieberling Guyot in 1990-1991 provided that opportunity. The desire was to set a three-dimensional array on as close to a smooth planar slope as possible well below the thermocline and deep enough that density fluctuations could be inferred from temperature measurements by virtue of a tight temperature-salinity correlation.

Spectra of current in the onslope direction (i.e. normal to local isobaths in the direction of shallower water) calculated from the records on a heavily instrumented mooring on the southwest flank of Fieberling Guyot indicate substantial departure from the Garrett-Munk model spectra that characterize deep open ocean spectra (Figure 3). The departure is strongest near the bottom and takes the form an enhancement at the local critical frequency $\sigma_c = 0.42$ cph. The enhancement around the critical frequency is evident even several hundred m above the bottom, while the spectrum from 95 m depth closely matches the Garrett-Munk prediction. A single complex empirical orthogonal function (CEOF) describes half or more of the variance in each frequency band in the bottom 300m, from an octave below to an octave above the critical frequency (Eriksen, 1998). When records from all depths on the mooring are decomposed into CEOFs, two frequency ranges stand out as being dominated by a single mode: a band from 24 through 16 h and a band from about 4 to 1.5 h (Figure 4). Discussion of the longer period band motions is deferred to the next section. The shorter period band is centered on the local internal wave critical frequency.

Both spectra and the complex eigenfunctions that dominate variance near the critical frequency indicate that linear theory accounts for much of the behavior of motions in the internal wave band. Linear features include the transition between prominent upward and offslope phase propagation at subcritical frequencies to downward and onslope propagation at supercritical frequencies. Waves are aligned across isobaths near the critical frequency, as expected by the linear theory of internal wave reflection off a sloping bottom. Linear theory is also consistent with the observed ratio of vertical to horizontal wavenumber for these eigenfunctions.

Linear theory fails to account for the finite enhancement of spectra at the critical frequency itself, for the decay of spectral enhancement with height off the bottom, and the vanishing of wavenumber at the critical frequency rather than the presence of only very fine scales. Statically unstable conditions are frequently found in the bottom few hundred m on the seamount flank. More than 11% of Richardson number Ri estimates over a 10 m separation are found to be negative while altogether more than 25% are less than 0.25. Shear and density gradient fluctuations near the critical frequency dominate variance in the internal wave band as well as overall variance. Internal wave reflection is responsible for wave breaking, hence loss from the wave field to dissipation and to production of potential energy (buoyancy mixing). The rate of potential energy production diminishes with distance from the bottom, suggesting a convergence of turbulent fluxes.

The potential energy production rate can be estimated from the distribution of density gradients, where the energy difference between neutrally stable and unstable conditions is assumed lost to mixing or, alternatively, where shear unstable conditions ($Ri<0.25$) are presumed to lead to mixing following the estimator of Kunze et al. (1990) (Figure 5). The magnitude of the potential energy production rates calculated suggest a diffusivity K = (potential energy production rate)/N^2 that is in the range $1.6–6.4 \times 10^{-4}$ m^2 s^{-1}. This range is similar to what is found from direct microstructure measurements at the same site (Figure 2). Taking as a typical value of the gradient in diffusivity of 10^{-4} m^2 s^{-1} over 100 m, the implied upwelling rate from (1) is about a quarter meter per day, assuming N=1 cph = constant and A constant. This rate implies a net onslope circulation on the steep flank of Fieberling Guyot of only about 0.5 m/d, which is immeasurably small using conventional current meters. The induced onslope flow conceivably could be much higher on a gentler slope for two reasons: 1) the reflection critical frequency for a gentle slope is closer to the inertial frequency, hence internal wave energy levels are presumably higher and 2) flow near the bottom must be parallel to the bottom slope so that onslope flow is stronger on a gentler slope for a given upwelling rate. In a shallow lake, for example, slopes may be on

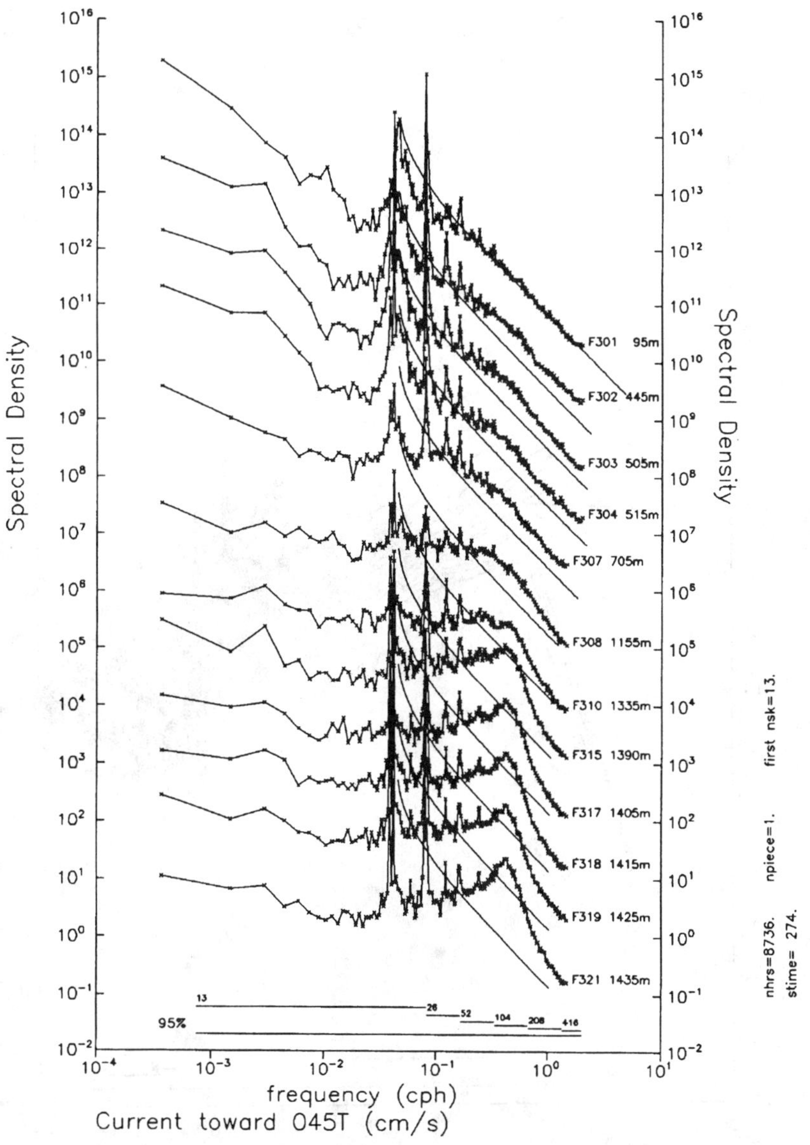

Figure 3. Spectra of the onslope (northeast) component of current on mooring F3 on Fieberling Guyot. Spectra are labeled by instrument number and depth and are arranged in order of depth from 95 m to 1435 m (where the anchor depth of the mooring was 1455 m). Scales are correct for the deepest spectral estimate (F321) and are successively offset by one decade for spectra at shallower depths. Smooth curves superimposed are the Garrett-Munk model estimates for the open deep ocean and serve as references for the observed spectra. Their endpoints are at the inertial frequency f and the local buoyancy frequency N. Spectra are enhanced about the internal wave critical frequency $\sigma_c = 0.42$ cph. Intervals of 95% confidence are based on frequency averaging of 13 to 416 independent spectral estimates, depending on frequency, as indicated.

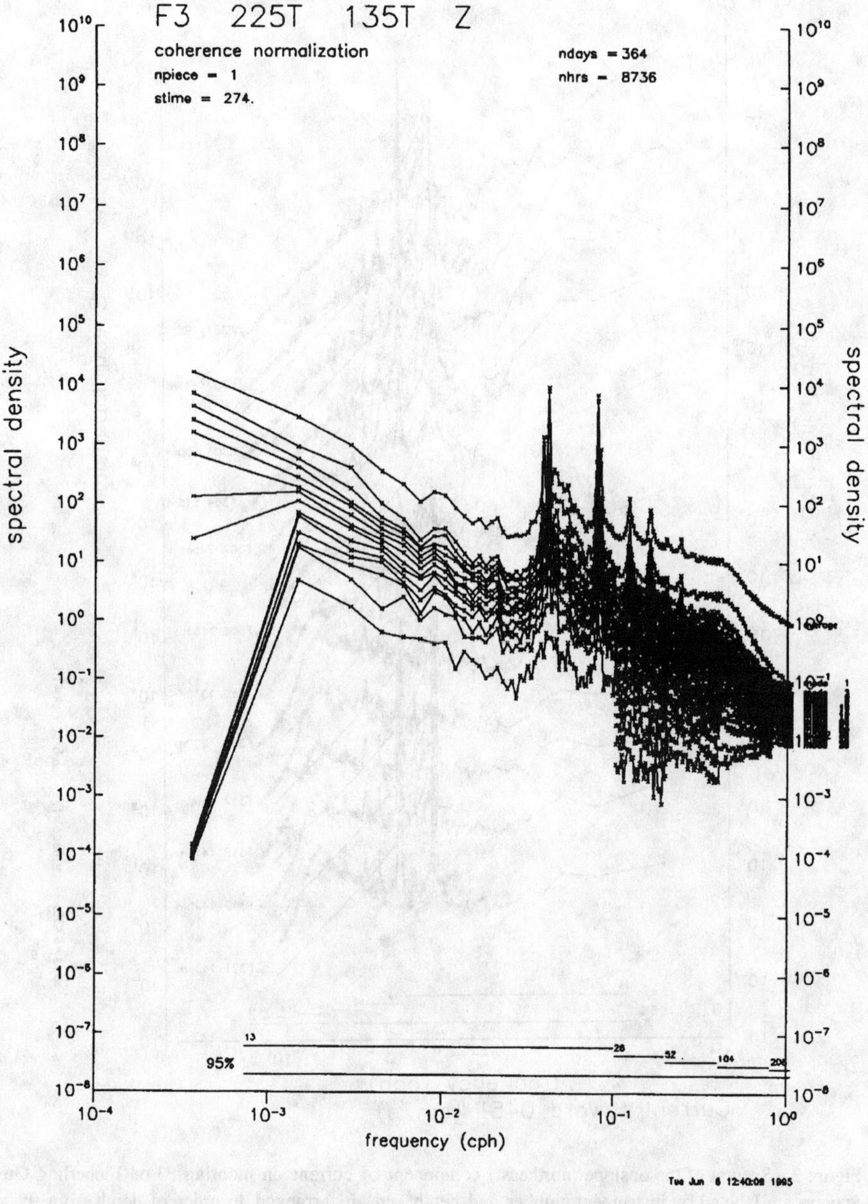

Figure 4. Spectra of complex empirical orthogonal functions calculated from the eigenvalues and eigenfunctions of the coherence matrix of current and temperature records on mooring F3 on Fieberling Guyot. The top curve gives the average spectrum and each curve below it depicts the energy density accounted for by a successively higher empirical mode. A single CEOF dominates structure over a band from the diurnal peak to a period of roughly 16 h and also over a band from about 4 to 1.5 h period, as is evident in the distinct separation of the second curve from the top from all other curves below it over these ranges.

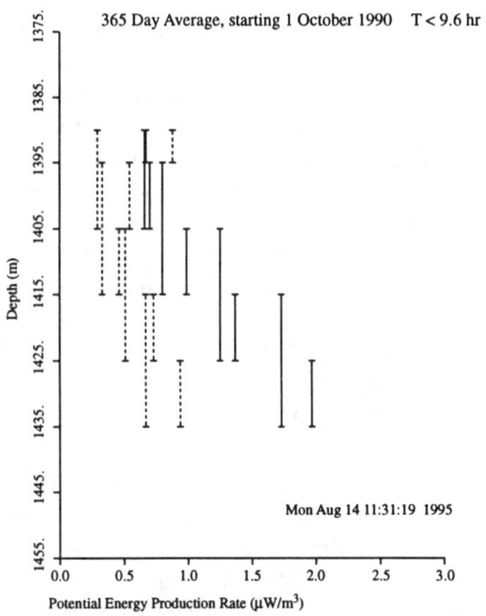

Figure 5. Potential energy production rate estimated from occurrence rates and amplitudes of unstable conditions based on sensors 5. 10, 15, and 20 m apart. Solid bars are based on statically unstable conditions alone and dashed bars are based on conditions of shear instability (calculated as in Kunze et al., 1990). The bottom depth for the mooring is 1455m, 20 m deeper than the deepest instrument used in any pair for calculation.

the order of a percent and stratification may change substantially over a few m near the pycnocline. Oceanic sites that can be expected to have higher mixing rates are those at low latitude with gentle bottom slope.

Wave motions can induce eddy fluxes that can drive mean currents as well, at least in the Eulerian frame of reference. The inviscid one dimensional analog of (1) is:

$$wN^2 = \frac{\partial}{\partial z} \langle u'b' \rangle \tag{2}$$

Where $\langle u'b' \rangle$ is the onslope buoyancy flux calculated as a covariance between fluctuations of horizontal current u' and buoyancy b'. Observed covariances between buoyancy and onslope flow on the flank of Fieberling Guyot are order 10^{-7} m^2 s^{-3} and vary over nearly the same scale laterally as do vertical buoyancy fluxes inferred from the distribution of statically unstable stratification. The two order of magnitude difference between vertical and horizontal buoyancy fluxes suggests that any mean flows detected due to inviscid wave processes will dominate those due to actual mixing. Without here presenting the full set of relevant dynamical balances, it suffices to state that without mixing or transients, any wave-induced Eulerian mean circulation is compensated by Stokes drift to produce no net Lagrangian flow (McIntyre, 1980).

Trapped Wave Observations

Coastal oceanographers have long recognized subinertial trapped waves as free modes of rotating stratified oceans with sloping lateral boundaries. These waves are nearly always taken as modes standing in the vertical-offshore plane. If the waves are trapped sufficiently tightly to the bottom, a boundary condition at the sea surface can be effectively ignored so that the waves may propagate not just parallel to isobaths but may cross them. Rhines (1970) illustrated such a solution, calling it an edge wave, and pointed out that it was a generalization of a Kelvin wave to a uniformly stratified ocean over a sloped planar bottom. A feature of this solution that has not been found for coastal trapped modes is that it may exist over a limited range of superinertial as well as subinertial frequencies. The observations at Fieberling Guyot reveal trapped wave modes that share two of the edge wave characteristics: a component of vertical propagation and existence at both subinertial and superinertial frequencies. The observational characteristics are summarized here.

A single empirical mode dominates flow structure at Fieberling Guyot in the diurnal and semidiurnal tidal frequency bands, but also over a broad range of periods from diurnal (subinertial) to about 16 h (superinertial). The dominant mode has similar, but not identical, structure throughout this range. CEOF spectra illustrate the dominance of a single empirical mode at the tidal peaks O_1, P_1-K_1 (the latter are indistinguishable in a one year record), M_2, and S_2 (Figure 4). They also illustrate the less pronounced, but still evident, dominance of a single mode from slightly subinertial ($0.94f$) to somewhat superinertial ($1.46f$) frequencies (where $f = 1/22.385$ cph is the inertial frequency). The inertial frequency divides this range dynamically, since free internal gravity waves are possible only at superinertial frequencies. Despite this division, coherent bottom-trapped Flow patterns of similar structure are found over a broad range of frequencies spanning it.

The existence of a roughly cylindrically symmetric first azimuthal mode diurnal oscillation trapped near the summit at Fieberling Guyot has recently been documented. Brink (1995) gives an interpretation in terms of radially and vertically standing free modes whereas Kunze and Toole (1995) noted the predominance of downward propagating phase in the clockwise component of current, consistent with the pilot mooring results (Eriksen, 1991). Similar flow structures appear over a broad range of frequencies, not simply at the diurnal tidal lines. Moreover, downward propagating phase is a consistent feature of these oscillations.

Above the seamount summit and near the summit rim, currents are nearly circularly polarized in the clockwise sense both at the diurnal tidal frequencies and across the continuum from diurnal to a frequency 50% above diurnal. The higher the frequency in this band, the more eccentric the current ellipse. By contrast, currents near the bottom on the flank of the seamount are nearly rectilinearly polarized with flow nearly parallel to the local isobaths. Current component and vertical displacement amplitudes tend to be highest at depths between those of the summit and the summit plain rim, even when scaled in a WKBJ sense by the local stratification, with the exception that amplitudes tend to rise sharply at the bottom. The total range of amplitudes is rather small, only a factor of two or so.

The CEOF decomposition describes the temporal relationship between measured quantities through their relative phases. Current component and displacement phases all tend to increase with depth (Figure 6) indicating downward phase propagation (not necessarily upward group speed). (Following the stretching convention of Kunze and Toole (1995) the reference buoyancy frequency is 3 cph and depths are stretched from the ocean surface; the seamount summit (444 m) is at 570 stretched m and the summit plain rim (700 m) is at 750 stretched m.) The rates of change of phase are different for the different flow components because their relative phases differ with position. For example, offslope (southwestward) current lags alongslope (southeastward) current by roughly 90° near the summit but is nearly in phase with it at depth, hence phase change rates with depth are different for the two components. The rate of phase increase with depth for the azimuthal component of flow gives a stretched vertical wavenumber of about 1 cycle per stretched vertical km downward at 24 h period. At 16 h period, the wavenumber magnitude is reduced to about 2/3 cycle per stretched vertical km.

The amplitudes and relative phase between offslope (equivalent to radial in cylindrical geometry) current and vertical displacement determine the contribution to eddy buoyancy flux from each frequency band. Kunze and Toole (1995) note that eddy buoyancy flux from diurnal period oscillations is in the offslope direction. The CEOF formalism allows eddy fluxes to be calculated from the coherent fluctuations of current and temperature across the moored array. These estimates are presented in the next to last section below.

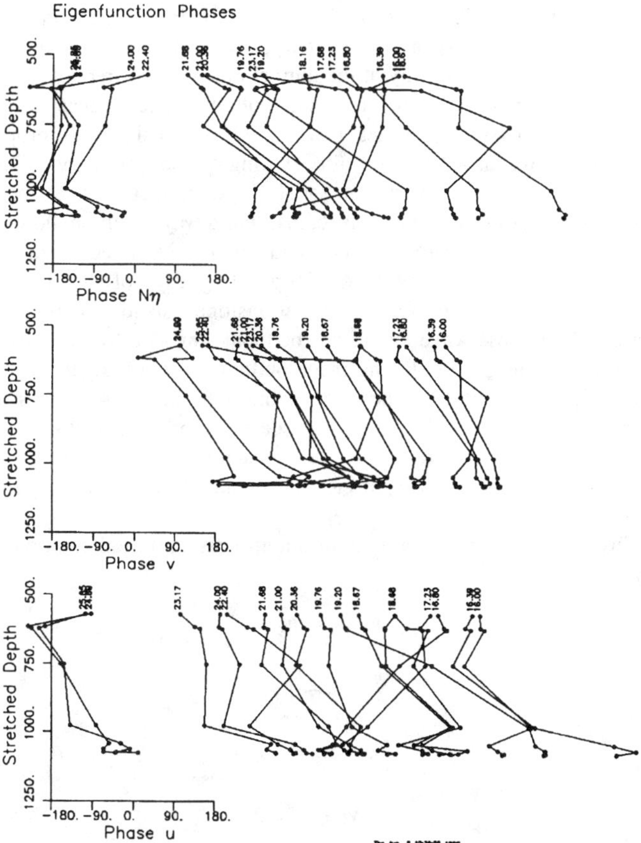

Figure 6. First CEOF mode current and vertical displacement eigenfunction phases on mooring F3 plotted against WKBJ stretched depth. The three panels give relative phases for the offslope component u (bottom panel), the alongslope component in the anticlockwise direction around the seamount v(middle panel) and upward displacement times buoyancy frequency $N\eta$ (top panel). Phases have been adjusted by integral cycles to minimize implied wavenumber magnitude. Independent frequency estimates, labelled by period in h, are successively offset by 45° in phase. The phase convention is that more positive phases lag.

Evanescent and Edge Waves

The fluctuations observed at Fieberling Guyot span a relatively broad frequency range that includes the inertial frequency. While Brink (1989) has found free resonant solutions trapped to a seamount, these are restricted to subinertial frequencies. Kunze and Toole (1995) have suggested that the diurnal frequency fluctuations at Fieberling are dynamically superinertial waves confined to the anticyclonic vortex near the seamount summit. Although vortex trapped waves support a radial buoyancy flux and vertical phase propagation, both observed features, the observed waves extend to the bottom on the seamount flanks well beyond where the vortex is found and exist at superinertial frequencies where they could not be trapped by a vortex with only anticyclonic vorticity. Radial-vertical seamount trapped modes support neither radial buoyancy flux nor vertical phase propagation, but they do extend to the bottom. Brink (1990) considered a forced problem, where the seamount response is a superposition of standing modes each excited off resonance, but the forcing was only at the diurnal (subinertial) frequency.

The purpose of this section is to point out that bottom-trapped waves can exist at both subinertial and superinertial frequencies that exhibit vertical phase propagation and support onslope heat flux, at least in infinite sloping plane geometry. Rhines (1970) found the special case of these waves that satisfies no normal flow at the bottom and propagates freely both along and across isobaths, but always with shallow water to the right. Since motion is everywhere parallel to the bottom and rectilinear, these waves cannot support an onslope buoyancy flux. Propagating rays of this type at the same frequency can be combined to form vertical-onslope standing modes. Whether propagating or standing, these wave modes can be though of as a Kelvin wave generalized to continuous stratification over a sloping plane bottom. The more general form of these waves does not satisfy the simple slip boundary condition, hence these must be forced by normal flow at the boundary. They are evanescent waves. They do support onslope buoyancy flux, appear to propagate along the boundary, and can exist over a range of superinertial and at all subinertial frequencies. They are the response to periodic normal flow forced at the boundary.

The linear inviscid equations of motion in a uniformly stratified fluid over an infinite planar bottom can be written as:

$$u_t - fv = -p_x \tag{3}$$

$$v_t + fu = -p_y \tag{4}$$

$$N^2 w + w_{tt} = -p_{zt} \tag{5}$$

$$u_x + v_y + w_z = 0 \tag{6}$$

where (u,v,w) specifies the onslope, alongslope, and upward current components of flow in a right-handed coordinate system, p is reduced pressure, and N and f are the buoyancy frequency and Coriolis parameter. Propagating plane wave solutions that are trapped to a planar bottom with slope s specified by z = sx exist when u,v,w and p all vary as $e^{i(kx + ly + mz - \sigma t)} e^{K(sx - z)}$ when (k,l,m) specify (real) wavenumber components and Ks and K are the (real) offslope and upward decay rates. The decay scale normal to the boundary is $(K\sqrt{1 + 5^2})^{-1}$ or cos α/K where s = tanα.) The system of equations (3-6) reduces to :

$$Q\left(p_{xx} + p_{yy}\right) = p_{zz} \tag{7}$$

where $Q = N^2 - \sigma^2/\sigma^2 - f^2$. Evanescent solutions to (7) are possible because two wavenumber components (onslope and vertical) are complex, while the third (alongslope) is real. Such solutions describe waves trapped to the boundary when the dispersion relations

$$\left(\frac{Ks}{\kappa}\right)^2 = -Qs^2\left[\cos^2\varphi + \frac{\sin^2\varphi}{\left(1 - Qs^2\right)}\right] \tag{8}$$

and

$$\frac{m}{k} = -Qs \tag{9}$$

are satisfied, where $\kappa = \sqrt{k^2 + l^2}$ is horizontal wavenumber magnitude and the angle $\varphi = \mathrm{Cos}^{-1}(k/\kappa) = \mathrm{Sin}^{-1}(l/\kappa) = \mathrm{Tan}^{-1}(l/k)$ specifies the real wavenumber direction in the horizontal plane ($\varphi = 0$ is an onslope wave, $\varphi = \pi/2$ is an alongslope wave with shallow water on the right). For a given pressure signal p, the current components are :

$$u = \frac{\kappa p}{\sigma^2 - f^2}\left(\sigma\cos\varphi + i\left(f\sin\varphi - \frac{\sigma Ks}{\kappa}\right)\right) \tag{10}$$

$$v = \frac{\kappa p}{\sigma^2 - f^2}\left(\sigma\sin\varphi - \frac{fKs}{\kappa} - if\cos\varphi\right) \tag{11}$$

$$w = \frac{\kappa p}{\sigma^2 - f^2}\left(\sigma s\cos\varphi - \frac{i\sigma K}{Q\kappa}\right). \tag{12}$$

Solutions are valid for all direction angles φ for subinertial frequencies $\sigma < f$ but for only the restricted range of angles for which K remains real and positive in the range $f < \sigma < \sigma_c$, that is, at superinertial frequencies less than the internal wave critical frequency σ_c where $\sigma_c = \sqrt{N^2 \sin^2 + f^2 \cos^2 \alpha}$. While these waves decay normal to the sloping bottom, phase lines are tilted upward and offslope at subinertial frequencies and upward and onslope at superinertial frequencies. They are level in the onslope-vertical plane at the inertial frequency. In the limit of vanishing frequency, phase lines attain an offslope-upward tilt of f^2/N^2s. In the limit of the critical frequency, phase lines are parallel to the sloping bottom in the onslope-vertical plane.

The special case of these waves for which flow is everywhere parallel to the boundary, i.e., w = su, requires that:

$$\frac{Ks}{\kappa} = \frac{Qs^2\frac{f}{\sigma}\sin\varphi}{Qs^2 - 1} \tag{13}$$

This restriction recovers the Rhines (1970) solution for edge waves. These are restricted to positive alongslope wavenumbers $l>0$ and to frequencies $\sigma < N\sin\alpha$, a more restrictive range than for evanescent waves.

The quantity Ks/κ appearing in (8) and (10)-(12) is the ratio of the horizontal length scale of the wave to its offslope decay scale. Large values of this ratio indicate that waves

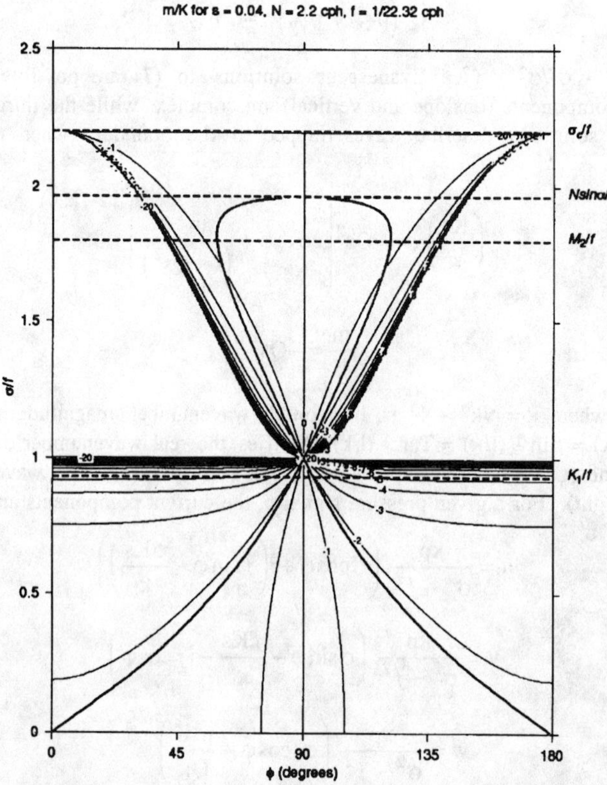

Figure 7. Ratio of vertical wavenumber to vertical decay scale for evanescent waves as a function of frequency σ and horizontal wavenumber vector orientation φ. Dispersion curves for edge waves are superimposed as heavy curves. Parameters are relevant to the summit region of Fieberling Guyot. Dashed lines mark the internal wave critical frequency (the maximum evanescent wave frequency), the maximum edge wave frequency, and the lunar semidiurnal and the lunisolar diurnal tidal frequencies. The ratio is not contoured where evanescent waves are not possible. Small ratios correspond to strong trapping. The orientations φ=0° and 90° correspond to pure onslope and alongslope (with shallow water to the right) propagation, respectively. Superinertial-onslope and subinertial-offslope waves propagate downward, and conversely.

decay in only a fraction of a horizontal wave scale, thus are tightly bottom trapped. Alternatively, the ratio of vertical trapping scale to vertical wave scale is given by

$$\left(\frac{m}{K}\right)^2 = -Qs^2\left(1+\frac{\tan^2\varphi}{1-Qs^2}\right)^{-1} . \qquad (14)$$

Contours of m/K calculated for parameters relevant to the summit region of Fieberling Guyot indicate that trapping is relatively strong (i.e. |m| < K) for near alongslope wave orientations at all frequencies and quite weak for frequencies just below the inertial frequency and for superinertial waves with orientations close to limit where free internal waves propagate exactly parallel to the bottom ($Qs^2 \cos^2 \varphi = 1$) (Figure 7). While evanescent waves of any direction are possible at subinertial frequencies, the range of directions is tightly confined to alongslope at slightly superinertial frequencies, but broadens to nearly any direction close to the internal wave reflection critical frequency. The azimuths excluded for superinertial evanescent waves correspond to those for which free internal waves are possible (the uncontoured region in Figure 7). Note that directions with an onslope component ($-\pi/2 < \phi < \pi/2$) correspond to upward propagation at subinertial frequencies and downward propagation at superinertial frequencies and conversely for directions with an offslope component ($\pi/2 < \varphi < 3\pi/2$) by (9) since Q changes sign (from $-\infty$ to $+\infty$) across the inertial frequency. Figure 7 is drawn only for the first two quadrants in direction because of these symmetries.

Evanescent waves have smaller vertical wavenumbers than free internal waves of the same frequency and horizontal wavenumber. If an internal wave of frequency σ has vertical wavenumber m_i, then

$$m_i^2 = m^2 + \left(Qs^2 - 1\right)K^2 = \frac{m^2}{Qs^2 \cos^2 \varphi} \tag{15}$$

where all terms are positive semi-definite. Evanescent waves exist for those wavenumber orientations for which incident free internal waves are precluded (those orientations corresponding to incident internal wave energy coming up through the sloping bottom).

Edge waves are restricted to the range of directions plotted (all alongslope with shallow water to the right for $f>0$). These dispersion curves are drawn on the same set of axes as heavy curves in Figure 7. For the slope, stratification, and rotation parameters relevant to the summit region of Fieberling Guyot (s=0.04, N=2.2 cph, f=1/ 22.32 cph), edge waves are moderately trapped over their complete frequency range.

Both diurnal and semidiurnal tidal frequencies at Fieberling Guyot fall within the range of possible evanescent and edge wave frequencies (Figure 7). Diurnal frequency motions are only slightly subinertial and, as such, will be weakly vertically trapped for all orientations except those nearly alongslope. Diurnal edge waves are aligned 3.6° onslope and offslope from the alongslope direction. At superinertial frequencies, the range of possible orientations broadens with increasing frequency. Semidiurnal edge waves attain nearly the most cross-isobath orientation possible over the possible range of superinertial frequencies. These and evanescent waves are strongly to at least moderately vertically trapped to the bottom at superinertial frequencies (Note $0 < |m|/K < 2$ for superinertial frequencies in this case).

Evanescent and edge wave amplitudes in pressure p can be expressed in terms of the energy density of the waves. The average energy per unit frequency per unit surface area normal to the slope E is

$$E = \frac{\rho_0 \kappa^2 \langle PP^* \rangle}{8K\left(\sigma^2 - f^2\right)^2} \left\{ \left(\sigma^2 + f^2\right)\left(1 + \frac{K^2 s^2}{\kappa^2}\right) - 4\sigma f \frac{Ks}{\kappa} \sin\varphi + \left(\sigma^2 + N^2\right)\left(s^2 \cos^2 \varphi + \frac{K^2}{Q^2 \kappa^2}\right) \right\} \tag{16}$$

where $\langle PP^* \rangle$ is the variance in reduced pressure per unit frequency at the bottom $z = sx$. The energy density E is a sum of potential and horizontal and vertical kinetic energies averaged over a wave period. Given the energy density E, the component amplitudes (u,v,w) appearing in (10-12) can be found by interpreting the amplitude of p as the standard deviation of pressure found in a specified frequency band from the spectrum $\langle PP* \rangle$ since $\langle pp* \rangle = \langle PP* \rangle e^{2K(sx - z)}$. Normalization by energy density E allows currents associated with waves of different frequency or wavenumber to be compared, as with the current ellipses discussed next. Note that for a given energy E, the variance in pressure (hence all flow variables) is higher for waves directed alongslope with shallow water to the right ($0<\varphi<\pi$) than for waves travelling with shallow water on the left due to the $\sin\varphi$ term in (16).

The current ellipse signature of evanescent waves varies considerably with wave orientation and frequency. The current ellipses in the horizontal and vertical-onslope planes for diurnal evanescent waves over a small range of wavenumber vector orientations ($79.2° < \varphi < 90°$) that includes the edge wave orientation are given in Figure 8. Horizontal ellipses and horizontal projections of wavenumber vectors are given in the top row of this figure for waves directed onslope. Vertical-onslope current ellipses (with a vertical exaggeration of 10) are given in the second row. The lower two rows display the corresponding ellipses for waves with offslope senses. The sense of rotation of the current vectors changes in both the horizontal and onslope-vertical planes depending on whether waves are directed more or less onslope or offslope than the edge wave. In general, current ellipses in the vertical-onslope plane intersect the bottom, demonstrating the need for motions to be forced normal to the bottom in order to excite evanescent waves. Edge waves have flow everywhere parallel to the bottom, so can exist as free waves. The ellipses of Figure 8 indicate how minor departures from no normal flow at the bottom can induce substantial horizontal current fluctuations.

Current ellipses for evanescent waves generally are not aligned with the wavenumber vector orientation. Waves travelling parallel to isobaths have current ellipses oriented normal to them, but in general horizontal ellipses and wavenumber vectors are not normal to one another. Edge waves not only have flow everywhere parallel to the bottom, but also have rectilinear flow (ellipses collapse to straight lines). Pairs of rays with horizontal wavenumber vectors symmetric about the alongslope direction (i.e. $\varphi = \pm \pi/2 \pm \theta$) can be summed to form standing modes in the vertical-onslope plane. In the case of standing modes, horizontal and vertical ellipses are oriented either parallel or normal to isobaths and the horizontal plane, respectively. Whereas individual rays can carry onslope or offslope momentum and buoyancy fluxes, standing modes cannot.

The horizontal buoyancy flux $F^b \equiv \langle u'b' \rangle$ of a propagating evanescent wave component is

$$F^b = -\frac{N^2 \langle pp* \rangle \sigma K \kappa \cos\varphi}{2\left(\sigma^2 - f^2\right)^2} \left\{ \frac{f\kappa s}{\sigma K} \sin\varphi - s^2 + \frac{1}{Q} \right\} \tag{17}$$

The expression in curly brackets vanishes for edge waves, by (13). If the mean buoyancy balance is given by $WN^2 = -\partial/\partial x \, F^b$ as in (2), then the mean Eulerian upwelling induced by the waves is $W = -KsF^bN^{-2}$ where U and W are Eulerian mean flow components. By continuity, $U = W/s$. This Eulerian flow parallel to the bottom and decreasing exponentially from it with a scale $\cos\alpha\,(2K)^{-1}$, is exactly offset by the Stokes drift components (U^S, W^S) which are equal and opposite to (U,W), in accordance with the predictions of nonacceleration theorems (McIntyre, 1980). So, while a mean Eulerian onslope or offslope flow can be induced by linear inviscid evanescent waves, there is no net Lagrangian circulation so induced. If mixing is introduced, the resulting nonzero Lagrangian mean circulation can be expected to have scales comparable to the domain in which the evanescent oscillations are found.

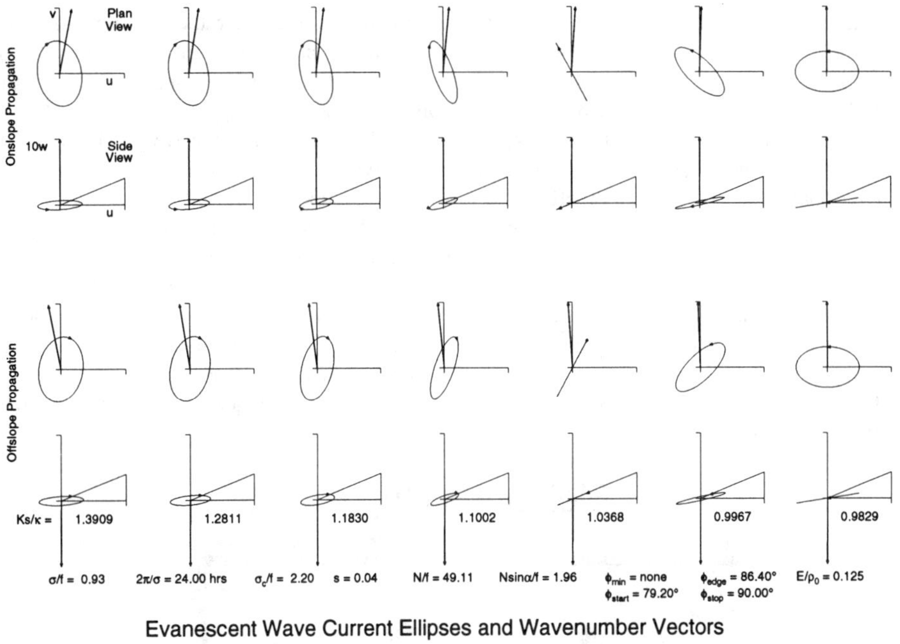

Figure 8. Current ellipses for evanescent waves of various orientation relative to isobaths at a fixed (diurnal) frequency for parameters relevant to the summit rim region of Fieberling Guyot. The top row shows current ellipses in the horizontal plane along with the horizontal projection of the wavenumber vector. The second row shows corresponding current ellipses and the bottom slope in the onslope-vertical plane, exaggerated tenfold vertically. From right to left in each row, waves are directed more onslope and upward in the top two rows. The bottom two rows are the corresponding ellipses for waves directed increasingly offslope and downward from right to left. Arrowheads in the ellipses indicate direction of current vector rotation with time. Arrows with tails at the origin of each set of axes indicate wavenumber vector directions (but not magnitudes). Waves are normalized to have equal energy.

Mean Flows over Sloping Boundaries

The one year mean flows near Fieberling Guyot indicate an anticyclonic swirl centered at a depth just below the summit peak depth (Brink, 1995, Kunze and Toole, 1995). This swirl tends to have an offslope component over the summit plain (Figure 9). At greater depth over the seamount flanks, where flows are much weaker, mean flows have an onslope component. Mean currents a few hundred m above the bottom on the flanks tend to be more parallel to isobath than flow immediately adjacent to the bottom, although the orientation of bottom contours is depth dependent due to departures of the topography from perfect radial symmetry and the mean flow is likewise not perfectly radially symmetric. The conventional expectation of Ekman dynamics is for flow to turn cyclonically with depth, so at first glance, the sense of offslope flow near the summit plain is not particularly surprising. Although the structure of rotating stratified boundary layers over sloping topography are not generally understood, two exceptions to the conventional expectations for flow over a flat bottom stand out in the observations. First, current vectors can be seen to have stronger flow magnitudes near the bottom than just above it in some cases (see the vectors drawn for sites P and C in Figure 9). Second, onslope flow in the cluster of moorings on the southwest flank of the seamount is to the right (of an onslope component) of (anticyclonic) flow above it.

The radial flows found at Fieberling Guyot may be due in part to wave processes. The subtidal frequency temporal dependence of the anticyclonic swirl on the spring-neap tidal cycle has been shown statistically (Eriksen, 1991, Brink, 1995, and Kunze and Toole, 1995). Alongslope currents near the bottom tend to be modulated with a fortnightly signal in phase with the diurnal tidal current amplitude, although they vary also with low frequency flow impinging on the seamount. That is, when diurnal fluctuations are strong, the anticyclone atop Fieberling is strong and conversely. This suggests that circulation near the seamount is, at least in part, due to rectification of tidal fluctuations. The onslope component may also be influenced by fluctuations, diurnal and otherwise. Codiga (1993) found subinertial oscillations to rectify into producing a low frequency anticyclonic swirl in a laboratory simulation.

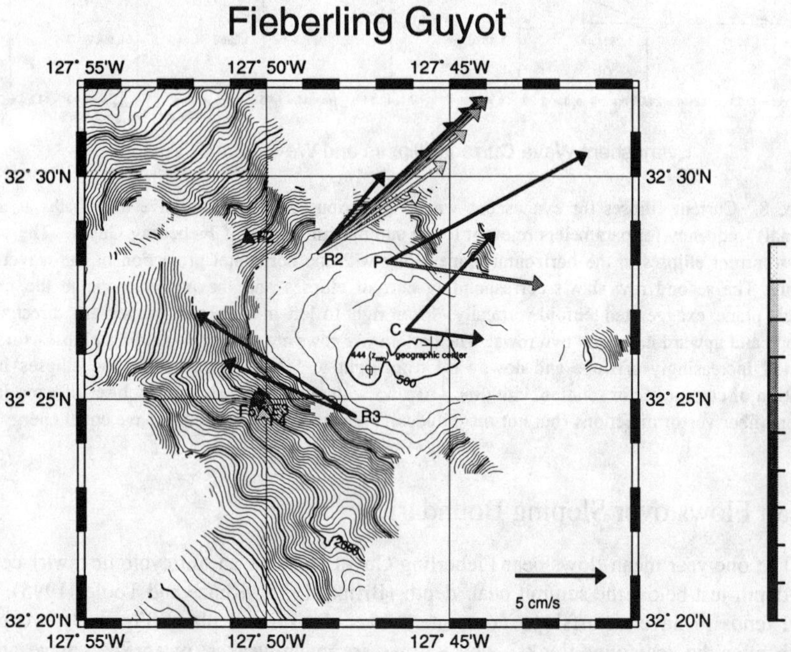

Figure 9. Bathymetry and one-year mean currents on the western summit region of Fieberling Guyot. The moorings designated C, R, and F were placed near the center, on the rim of the summit plain, and on the flanks of the seamount. The triangle of moorings F3, F4, and F5 formed an internal wave array. The mooring P was from a pilot study thirteen months before the rest of the array was set. Depth contours are from a partial Hydrosweep survey of the seamount on the mooring recovery cruise on *R/V Thompson*. Depth contours are drawn every 50 m with every tenth contour drawn with heavy curve. The summit, a rocky spur is at 444 m depth. The seamount rises from a surrounding region of 4500 m depth. A least-squares fit of the measured topography to a radially symmetric form gives the geographic center as marked. The depth of current vectors is indicated by shading.

Currents at site R2 on the northwest side of the seamount (Figure 9) show the mean anticyclonic circulation above the summit region as a flow most intense about 105 m above the bottom (535 m depth). These current estimates, each 20 m apart vertically, were collected with an acoustic Doppler current profiler. The current spiral turns leftward with depth at this site and current within the strongest part of the anticyclone appears to be parallel to the bottom contours. Flow closer to the bottom is offslope at roughly 1 cm s^{-1} while flow above the anticyclone maximum appears onslope at about the same rate. Mean currents at sites P, and R3, also on the summit plain rim, are similarly sheared, with an offslope component near the bottom of 1-5 cm s^{-1} as a departure from otherwise alongslope flow in the anticyclonic sense around the seamount. At the mooring nearest the geographic center, mooring C, flow is likewise sheared and offslope flow near the bottom is pronounced, as at P.

A description of currents on a finer horizontal scale is available from the cluster of moorings on the southwest flank of the seamount. While flow is weaker, it is directed slightly in the onslope direction, in contrast to the flow near the summit plain rim (Figure 10). Records from various locations within the internal wave array on the southwest flank have onslope flow components of up to about 0.2 cm s^{-1}. None of the currents measured within 160 m of the bottom over one year indicates an offslope component of flow. Moreover, the observed current vectors suggest an onslope component of flow even when compared to nearby isobaths of corresponding depth. Progressive vector diagrams indicate that onslope flow exists even when the alongslope current is in the opposite sense for a few days or more.

The eddy fluxes in the vicinity of the summit plain rim are dominated by tidal, especially diurnal, fluctuations. The gradients in offslope buoyancy flux above the summit plain and the flank imply mean Eulerian downwelling of about 50 m/day and upwelling of 5 m/day, respectively, in the two regions according to the balance in (2), neglecting vertical mixing. Given the slope on the summit plain of about 0.05, about 1.2 cm s^{-1} of offslope flow is implied by mass conservation in the radial vertical plane, rather close to the flow observed (darkest shaded arrows, Figure 9). The linearized version of horizontal momentum balances that accompany (2) imply that the vertical gradient of vertical eddy transport and radial gradient of radial transport of alongslope momentum to balance Coriolis acceleration. Estimates of these flux gradients are too uncertain to make a reliable estimate of radial flow. The buoyancy flux structure, however, appears of the correct sign and magnitude to explain the offslope flow found near the bottom at the rim.

The eddy flux structure near the bottom on the flank also suggests an Eulerian mean flow of the sign observed, but due to internal wave rather than tidal processes. Radial buoyancy flux is in the offslope sense due to internal wave fluctuations and diminishes with distance from the bottom, whether offslope or vertical. The magnitude of vertical eddy buoyancy flux by wave breaking and the scale over which it changes implies that it can be neglected as a source of induced upwelling in the balance (1). The convergent pattern of radial buoyancy flux implies an upwelling of roughly 13 m/day. This, together with the bottom slope within the internal wave array of 0.45 implies an onslope flow of about 0.03 cm s^{-1}. This estimate is at least a factor of 3 weaker than what is observed directly. The linear version of the azimuthal momentum balance on the flank is clearer than it is near the summit rim. The two flux gradient terms partially offset one another, but vertical gradients of vertical transport are somewhat larger than offslope gradients of offslope transport of alongslope momentum. The difference implies an onslope mean flow of about 0.3 cm s^{-1}, a factor of 3 or so larger than what is measured directly. The mass and azimuthal momentum balances predict onslope flows that bracket that is observed directly, while neither balance agrees well with the current meter observations.

Summary

The example of Fieberling Guyot demonstrates the importance of wave processes in determining mean circulation over topography. Internal gravity and evanescent waves are dynamically quite similar; they are neighbors across the topographic boundary that a sloping bottom represents. While free internal gravity waves can reflect at a sloping planar boundary, evanescent waves are the motions that can be excited in the directions that gravity waves are precluded. Evanescent waves can be thought of as internally-reflected waves trapped to the boundary from which they are forced. Wavelike motions at superinertial frequencies can take either form at a given site, depending on slope, stratification, and orientation. Edge waves are the special internally reflected waves whose motions are everywhere parallel to the boundary so that they may exist as free waves. Depending on how these waves are excited by externally imposed flows, they may become quite nonlinear and break, thus induce a mean circulation. At Fieberling Guyot, it appears that internal and evanescent waves are responsible for different features of the mean circulation.

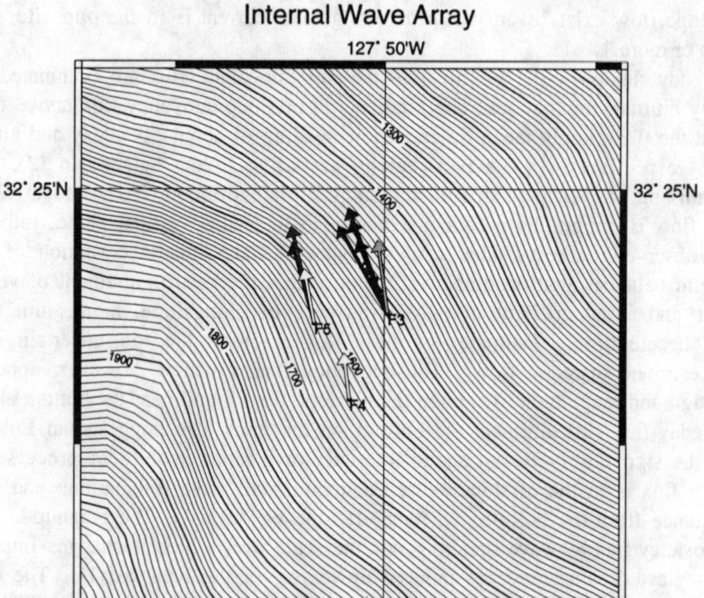

Figure 10. Mean flow over one year starting October 1, 1990, at 20 to 160 m above the bottom at the internal wave array sites. Currents 20 m from the bottom are drawn with a solid vector with currents farther off the bottom shaded more lightly to a height of 160 m, where vectors are unshaded. Instrument heights from the bottom are 20, 30, 40, 60, and 120 m off the bottom on F3, 160 m off the bottom on F4, and 40, 90, and 160 m off the bottom on F5. Depth contours at 10 m intervals are based on a Hydrosweep survey and the mooring position is based on acoustic and undithered GPS navigation.

In other topographic contexts, the relative roles of different wave processes may be quite different. For example internal wave reflection may be very important for lakes because of the small scales, while evanescent motions forced by weather may take a different role than do tidally forced motions in the ocean.

The flow found near Fieberling Guyot is dominated by relatively narrow-band signals, and in this it is probably not unusual for a seamount. It is in strong contrast to open deep ocean flows, where mesoscale eddy motions tend to dominate current variance. The three strong characteristics of currents at Fieberling are 1) the mean anticyclone near the summit, 2) the diurnal-inertial-semidiurnal band of structurally similar flows, and 3) the near- critical reflection internal wave band. That fluctuations in the summit anticyclone modulate with the spring-neap cycle suggests that wave processes are important to its existence. Estimates of mean circulation in the vertical-radial plane based on eddy flux gradients, albeit crude, are in rough agreement with the horizontal currents measured directly. These suggest a pattern of downwelling above the summit plain and upwelling on the seamount flanks. There is no strong evidence that these flows are other than Eulerian means induced by the presence of bottom intensified wave motions. The level of mixing found by Eriksen (1998) and the vertical scale over which it varies together are too small to produce mean flows even as big as the Eulerian flows observed.

Both internal wave band and diurnal motions effect measurable offslope buoyancy fluxes. Internal waves reflect off the steeply sloped flanks of the seamount in such a way as to generate rotary motions of current vectors in the vertical-onslope plane to generate offslope buoyancy flux near the bottom that decays with distance from it. Since these motions are strongest near the bottom, their decay implies a convergence of buoyancy flux, hence an induced (at least Eulerian) upwelling to compensate for it. The vertical gradients vertical buoyancy flux implied by the observed rate of density overturns are substantially weaker than observed horizontal gradients of offslope buoyancy flux.

Diurnal motions in the vertical onslope plane reach a maximum a hundred or more m above the seamount summit plain and at a finite radial distance from the seamount center, however roughly it can be defined. The offslope buoyancy flux these motions carry has gradients of both signs, implying downwelling at small radii (over the summit plain to about the rim) and upwelling outside this region, with the upwelling region more concentrated than that of downwelling. While there is considerable turbulent mixing reported above the summit plain, the magnitude of its gradients are weak compared to those of offslope buoyancy flux.

Neither the free seamount trapped wave whose resonant frequency is closest to the diurnal frequency (29 h period, according to Brink, 1995), nor the vortical trapped wave of Kunze and Toole (1995) completely explains the nature of the diurnal and up-to-slightly-superinertial motions observed at Fieberling Guyot. The principal shortcomings of the seamount trapped wave model of Brink (1990) are the failure to account for the observed phase propagation, offslope buoyancy flux, and frequency bandwidth (superinertial as well as subinertial) of the motions at Fieberling. The vortex-trapped wave model of Kunze and Toole (1995) accounts for downward phase propagation and offslope buoyancy flux, but cannot account for these features at superinertial frequencies nor at depths and radii well removed from the near-summit intensified anticyclone.

An examination of uniformly stratified rotating flow over a planar sloping boundary demonstrates that, at least locally, evanescent and edge waves are possible at both subinertial and superinertial frequencies. These waves are bottom trapped with scales that can be comparable to the degree of trapping observed at Fieberling. They can propagate vertically and horizontally with scales that are also comparable to those observed at Fieberling. The possibility of trapped wave motions at superinertial frequencies occurs because even though the governing equation (6) is hyperbolic for $\sigma > f$ ($Q>0$), two complex wavenumber components can be offset by a real third component, hence waves can be trapped in two dimensions, and propagate in the third. These waves are possible in addition to free internal gravity waves (which propagate in all three dimensions). The possible generalization of this solution to arbitrary stratification and bottom boundary shape is not clear. Nevertheless, evanescent waves can be expected to be supported where stratification and bottom slope are locally uniform.

The response of a seamount to externally imposed flow is equivalent to forcing flow normal to the bottom at the bottom. The prominent responses found at Fieberling are at the diurnal and semidiurnal tides. The forced response of a linear, inviscid system can be calculated by the projection of forcing onto the normal modes of the system. This approach was followed by Brink (1990) where the response away from resonance was calculated as the sum of phase-locked free modes. Seamount trapped waves form a complete discrete basis set and the projection of forcing at arbitrary frequencies defines the contribution of each mode to the total response. While individual modes are standing in the vertical-offslope plane, the phase-locked sum of such modes should lead to apparent vertical-radial phase propagation at arbitrary locations. Curiously, the Brink (1995) solutions do not indicate sufficiently robust phase differences to match observed phase changes. Brink's normal modes are subinertial only so that if superinertial trapped free modes are possible, they may contribute to forced response as well.

The ray solutions of evanescent waves at a particular frequency can be summed over a variety of wavenumber orientations to produce a response that not only exhibits phase propagation, but varies the shape of current ellipses with distance from the bottom. Such a response could be formed by forcing with an arbitrary waveform at a particular frequency.

Acknowledgments. I thank Robert Reid, John Dahlen, and their colleagues at the C. S. Draper Laboratory, the captains and crews of R/V Washington and R/V Thompson, and Neil Bogue for help in carrying out the moored measurements. The analysis benefited from discussions with Chris Garrett, Dan Codiga, Eric Kunze, and Ken Brink. This study was part of the U. S. Office of Naval Research Accelerated Research Initiative on Flow over Abrupt Topography and was supported under ONR grant numbers N00014-89-J-1621 and N00014-94-I-0081.

References

Armi, L., Some evidence for boundary mixing in the deep ocean. *J. Geophys. Res.*, *83*, 1971-1979, 1978.

Brink, K. H., The effect of stratification on seamount-trapped waves, *Deep-Sea Res.*, *36*, 825-844, 1989.

Brink, K. H., On the generation of seamount-trapped waves, *Deep-Sea Res.*, *37*, 1569-1582, 1990.

Brink, K. H., Tidal and lower frequency currents above Fieberling Guyot, *J. Geophys. Res.*, *100*, 10817-10832, 1995.

Codiga, D. L., Laboratory realizations of stratified seamount-trapped waves. *J. Phys. Oceanogr.*, *23*, 2053-2071.

Eriksen, C. C., Observations of internal wave reflection off sloping bottoms, *J. Geophys. Res.*, *87*, 525-538, 1993, 1982.

Eriksen, C. C., Implications of ocean bottom reflection for internal wave spectra and mixing. *J. Phys. Oceanogr.*, *15*, 1145-1156, 1985.

Eriksen, C. C., Observations of amplified flows atop a large seamount, *J. Geophys. Res.*, *96*, 15227-15236, 1991.

Eriksen, C. C., Internal wave reflection and mixing at Fieberling Guyot, *J. Geophys. Res., 103*, 2977-2994, 1998.

Garrett, C., Comment on "Some evidence for boundary mixing in the deep ocean" by Laurence Armi. *J. Geophys. Res., 84*, 5095, 1979.

Garrett, C., Marginal mixing theories. *Atmosphere-Ocean, 29*, 313-339, 1991.

Garrett, C., P. MacCready, and P. Rhines, Boundary mixing and arrested Ekman layers: Rotating stratified flow near a sloping boundary. *Annu. Rev. Fluid Mech., 25*, 291-323, 1993.

Gilbert, D., and C. Garrett, Implications for ocean mixing of internal wave scattering off irregular topography. *J. Phys. Oceanogr., 19*, 1716-1729, 1989.

Gregg, M. C., Diapycnal mixing in the thermocline: a review. *J. Geophys. Res., 92*, 5249-5286, 1987.

Gregg, M. C., Scaling turbulent dissipation in the thermocline. *J. Geophys. Res., 94*, 9686-9698, 1989.

Imberger, J., and G. N. Ivey, Boundary mixing in stratified reservoirs. *J. Fluid Mech., 248*, 477-491, 1993.

Ivey, G. N., and R. I. Nokes, Vertical mixing due to the breaking of critical internal waves on sloping boundaries. *J. Fluid Mech., 204*, 479-500, 1989.

Kunze, E., and J. M. Toole, Fine- and micro-structure observations of trapped diurnal oscillations atop Fieberling sea-mount, In: Proc. Eighth 'Aha Huliko'a Hawaiian Winter Workshop on Flow-Topography Interactions, Hawaii Institute of Geophysics, Honolulu, 1995.

Ledwell, J. R., A. J. Watson, and C. S. Law, Evidence for slow mixing across the pycnocline from an open-ocean tracer release experiment. *Nature, 364*, 701-703, 1993.

McDougall, T. J., Dianeutral advection. In: Proc. Fifth 'Aha Huliko'a Hawaiian Winter Workshop on Parameterization of Small-Scale Processes, Hawaii Institute of Geophysics, Honolulu, pp. 289-315, 1989.

McIntyre, M. E., An introduction to the generalized Lagrangian-mean description of wave, mean-flow interaction, *Pure and Applied Geophys., 118*, 152-176, 1980.

Müller, P., and N. Xu, Scattering of oceanic internal gravity waves off rough bottom topography. *J. Phys. Oceanogr.*, 22, 474-488, 1992.

Munk, W. H., Abyssal recipes. *Deep-Sea Res., 13*, 207-230, 1966.

Munk, W. H., Internal waves and small scale processes. In: *Evolution of Physical Oceanography, Scientific Surveys in Honor of Henry Stommel*, edited by B. A. Warren and C. Wunsch, MIT Press, Cambridge, Mass., pp. 264-291, 1981.

Phillips, O. M., J.-H. Shyu, and H. Salmun, An experiment on boundary mixing: mean circulation and transport rates. *J. Fluid Mech., 173*, 473-499, 1986.

Rhines, P., Edge-, bottom-, and Rossby waves in a rotating stratified fluid, *Geophys. Fluid Dyn., 1*, 273-302, 1970.

Toole, J. M., K. L. Polzin, and R. W. Schmitt, Estimates of diapycnal mixing in the abyssal ocean. *Science, 264*, 1120-1123, 1994.

Toole, J.M., R.W. Schmitt, K.L. Polzin and E. Kunze. Near-boundary mixing above the flanks of a midlatitude seamount. *J. Geophys. Res., 102*, 947-959, 1997.

[illegible]

[illegible] and [illegible], [illegible] flow over sloping boundary [illegible], 1991.

[illegible]

[illegible]

McIntyre, M. E., [illegible] Lagrangian [illegible] of waves on mean flow [illegible], J. Fluid Mech., [illegible]

[illegible], Currents, [illegible] and [illegible] waves in [illegible] sloping [illegible], [illegible], 1990.

31

Some Dynamical Effects of Internal Waves and the Sloping Sides of Lakes

S. A. Thorpe

Abstract

Processes associated with the presence of sloping lake boundaries are reviewed. Some are hypothetical, yet to be positively identified by field observations. Few are fully documented or quantified.

The shape of a lake determines the form of internal seiches and, consequently, the regions in which they are most likely to cause mixing. An example is given in which the maximum seiche amplitude and the lowest vertical density gradients (a trend towards static instability) occur in the shallowest water. Intrusions, resulting from boundary mixing, offer a mechanism by which fluid mixed at the lake boundaries is transported into the body of a lake, at least to a distance from the slope of order L_R, the internal Rossby radius of deformation, when this is less than the horizontal dimension of the lake. Intrusions occur as a consequence of instability of the turbulent boundary layer or as a consequence of mixing caused by internal waves. Stokes drift in internal waves in the thermocline may enhance the intrusive spread of mixed fluid. Internal waves encountering a sloping boundary lose some or most of their energy in breaking and creating turbulence at the boundary, interact with bed roughness to generate secondary waves which break in the velocity and density field of the primary wave, or may reflect and, on reflection, interact resonantly with themselves to create off-slope turbulence. A novel conclusion is that this resonant self-interaction can occur for waves with group velocity vector inclined at angle θ to the horizontal less than the slope angle, α, as well as at $\theta > \alpha$. In the former case resonance appears possible at large α, whereas in the latter the range is limited to moderate α. The internal wave field in a lake may be determined in part by the efficiency of reflection, which itself depends on the slope angle, α, and its distribution in depth and position around the lake boundary. Internal surges or non-linear wave fronts travelling along the sloping boundaries of a lake produce moving mixing regions which generate and radiate internal wave wakes. The stratified flows associated with such surges, with wind forcing or gravity currents, passing over rough topography on the lake slope, results in internal lee waves which, as a consequence of the presence of the slope, tend predominantly to carry their momentum into shallow water and perhaps enhance mixing in the thermocline close to its intersection with the lake bed. On a sloping boundary, the proximity of the lake bed to a shear layer in which Kelvin-Helmholtz instability is

Physical Processes in Lakes and Oceans
Coastal and Estuarine Studies Volume 54, Pages 441-460

developing may reduce growth rates, or even stabilise the flow. Recent laboratory experiments show that the turbulence generated by billows in the shear flow further from the bed spreads into the stable zone where the shear layer, e.g. the thermocline, meets the sloping bed. This particular region is one which would reward field observations, as it is likely to be here that the interactions between the stratification, the flow, and the sloping topography are most intense.

1. Introduction

Topography constrains water motion in lakes. The presence of the surrounding boundaries are obviously important in establishing the forms of the larger (standing or seiche mode) waves and the circulation patterns or modes of instability which are possible. Many lakes have horizontal dimensions which are not large in comparison with their baroclinic and barotropic radii of deformation; generally motion throughout is influenced by the presence and shape of their sides.

There are many important effects associated with lake basin topography which have received some (if often insufficient) attention. These include the promotion of wind-induced up- and down-welling at the sides and ends of lakes and the return flows under the influence of the Earth's rotation, the generation of fronts between mixed water in the shallows at a lake edge and deeper stratified water (including the near-shore, late winter or autumnal development of 'thermal bars'; for references see e.g. Farrow, 1995), the generation and propagation of internal surges or solitons, exchange-flow between lake basins and across sills or through narrow passages joining them, downslope flow of dense riverine waters, and flow around islands. Some complex effects still await detailed study in lakes, such as flow separation and the generation of eddies around promontories, and the modification of wind fields and local lake forcing resulting from the presence of surrounding orography, a consequence of which is that in large lakes the scale at which much of the energetic motion in a lake is forced is often much less than the characteristic dimension of the lake or the synoptic scale of atmospheric motions (see Bohle-Carbonell, 1991). Study of some of these effects could be aided by numerical simulations.

One effect of topography is the promotion of mixing. There are however few *in situ* observations which offer strong evidence either for the mixing or for its effects. Caldwell et al. (1978) observe step-like, 0.1 to 10 m scale, thermal microstructure to a depth of 50 m extending out 500 m from the steep (22 deg) sloping boundary of Lake Tahoe, California, possibly evidence of intrusive layers resulting from boundary mixing. They postulated that the source of mixing was breaking high-frequency internal waves at critical slopes or heating of the lake bed as a result of the transmission of solar radiation in the clear waters of the lake. Imberger and Hamlin (1982) show an echo sounder record containing a set of scattering layers adjacent to the side of the Wellington Reservoir in western Australia and ascribed the effect to 'sideways double diffusion' (see for example Thorpe et al., 1969; Turner, 1973) driven by substances being released into solution from the lake boundary. Imberger (1989) shows a dissipation profile in the Wellington Reservoir in which, although not extending closer than 2 m off the bottom, the elevated dissipation rates suggest the existence of a benthic layer. Imberger and Ivey (1991) provide stronger evidence of a bottom boundary layer about 5 m thick on a slope of 0.6 deg in the Wellington Reservoir at a depth of 27 m. The rate of dissipation of turbulent kinetic energy, ε, in the layer is about $5\times10^{-9}\,m^2\,s^{-3}$, greater than that measured in mid-water, and the layer is capped by an increase in temperature. Imberger and Ivey (1993) also show two examples of 'benthic boundary layer mixing' in shallow lakes producing bottom mixed layers. One has a layer 1 m thick at 6 m depth in the Lago di Lago (Venice) believed to be caused by internal wave breaking (see section 4 below). The other has a layer about 1.5 m thick at 12 m depth in Lake Argyle (W. Australia), where the bed slope is 0.6 deg and N is about 3×10^{-2} s^{-1}, caused by wind-driven bottom currents. Neither show signs of being capped by temperature steps and, in each, ε is about 2×10^{-7} m^{-2} s^{-3}. Imberger and Ivey's theoretical study of boundary mixing concludes that it is significant in driving changes in the interior density gradient of some lakes. The strongest evidence for the importance of boundary mixing in lakes comes from the observations of microstructure and SF6 diffusion by Wuest et al. (1994) in Lake Alpnach. They find that the benthic boundary layer is the most turbulent zone in the hypolimnion. Examples of measurements which extend into the boundary layer on the side of a lake are however few, and there appear to have been no intensive studies of the variability of, and in, such layers either in time or in space.

In contrast to lakes, the effects of lateral boundaries in the ocean have received considerable interest mainly because their possible importance in promoting diapycnal mixing. Much prominence has been given to the suggestion that such boundary mixing is an important contributor of diapycnal transport for the Global Ocean (Munk, 1966; Armi, 1978). This idea has stimulated observational and theoretical studies (Phillips, 1970; Wunsch, 1970; Weatherley and Martin, 1978; Wunsch and Hendry, 1972; Thorpe, 1987a; Thorpe et al., 1990; Trowbridge and Lentz, 1991; Garrett, 1990, 1991; MacCready and Rhines, 1991, 1993; Woods, 1991; White, 1994; van Haren et al., 1994; see also review by Garrett et al., 1993) as well as laboratory experiments (e.g. Phillips et al., 1986), to examine the flow and structure of the stratified sloping boundary layer. The oceanic environment most closely representing the hypolimnion of a lake is that of an ocean basin. SF6 tracer diffusion experiments have been made in such deep basins by Ledwell and Watson (1991), Ledwell and Bratkovich (1995) and Ledwell and Hickey (1995). The second and third of these report a dramatic increase in the rate of diapycnal dispersion once the tracer reaches the basin side walls, raising the average basin diffusion coefficient by a factor of 10. Particular attention has been drawn to the effects of internal waves reflected from slopes (Wunsch, 1971; Baines, 1971 a, b; Mied and Dugan, 1976; Eriksen, 1982, 1985, 1995; Thorpe, 1987b; Gilbert and Garrett, 1989) or steepening and breaking on the slopes (Ivey and Nokes, 1989; Thorpe et al., 1991; Thorpe, 1992a; Taylor, 1993). Internal wave generation, particularly by the flow of tidal currents across the shelf break, has also been well studied (Maxworthy, 1979; Hibiya, 1986, 1988; Lamb, 1994). The absence of shelf breaks and tides makes this mode of generation less relevant in lakes except when they contain sills or 'shelf' areas, but recent studies of internal wave generation by currents flowing along isobaths (see section 6) may have application in lakes.

Here I discuss some examples of processes which occur at or near the sloping lake boundaries. It is not intended to offer a comprehensive and detailed review - for this the reader should consult the references given - but some processes are selected which have not previously been given sufficient attention. An overview is offered in Figure 1 which will be explained in the course of the following sections. The processes affect the currents in the lake or result in local mixing which may change the vertical density structure of the interior water column of the lake by the formation of intrusions of mixed, or partially mixed, water. In the ocean, intrusive layers formed at the boundaries are restricted to spread to a distance of order of the internal Rossby radius of deformation, L_R, before

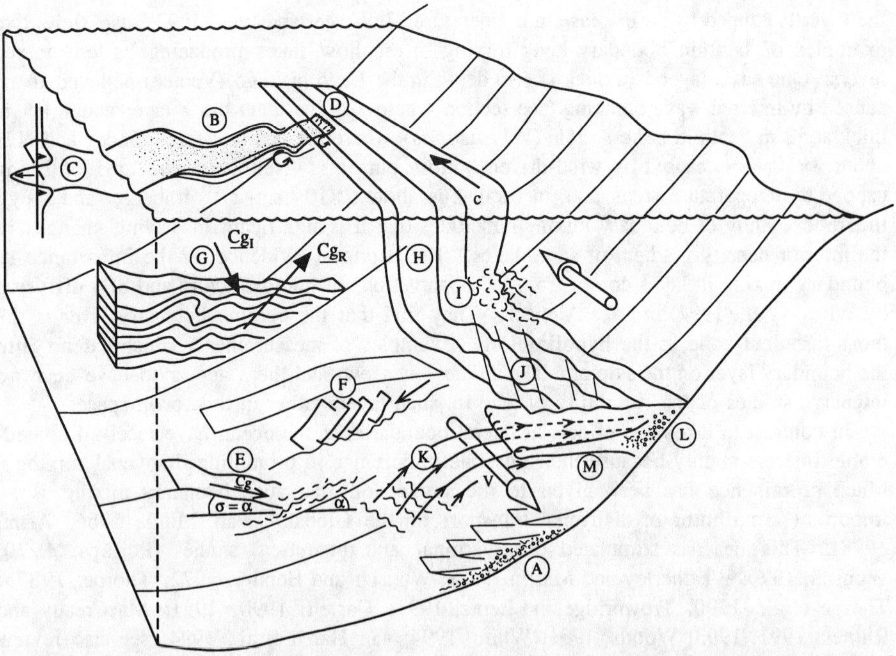

Figure 1. Sketch showing some processes of internal wave generation, mixing and intrusions on the side of a stratified lake. (A) Turbulent boundary layers may be unstable, naturally generating intrusions (section 3). (B) Intrusive layer resulting from internal wave breaking with off-slope flow augmented by (C) Stokes drift in the incident internal wave in the thermocline (see section 5). (D) The process of run-up in both 2-layer and continuously stratified fluids results in the formation of three-dimensional 'cleft and lobe' structures (see section 5). (E) Progressive internal waves reflecting at a critical slope generate boundary mixing (see section 4). (F) The movement caused by internal waves over rough topography may generate lee waves which are refracted and may break in the motion field of the primary wave (section 5). (G) Progressive internal waves reflecting at non-critical slopes may result in resonant interactions and overturn at a distance from the slope (section 5). (H) Non-linear edge or Kelvin waves, or standing internal waves in long basins propagate with surges or internal hydraulic jumps and cause rapid down-slope flow at lake boundaries, and (I) a mixing zone which propagates along the lake side with the wave front(section 6). From this emanates (J) an internal wave wake, which may locally enhance shear in the body of the lake (section 6). (K) Long internal waves and wind or river-driven circulation cause along-slope flows over the rough topography of the sloping boundary in the hypolimnion and internal lee waves which propagate preferentially towards shallower water (section 7). (L) Trapped beneath the epilimnion, these may break, creating (M) intrusive layers and locally transfering their momentum, or a drag, to the mean flow (section 7).

reaching a geostrophic balance in which further spread under gravity is limited; deep boundary nepheloid layers are found extending only to a distance of order of the Rossby radius (e.g. Thorpe and White, 1988). Unless spreading is augmented by some process other than gravitational collapse, such intrusions do not generally reach far into the ocean interior since basin widths are much greater than L.

In lakes however, L_R, may be of the same order or even greater than the width of a lake. For example, in Loch Ness where boundary slopes reach 45 deg, L_R is about 2 km whereas the lake width, L, is about 1.5 km and in Lake Geneva, L (about 14 km) does not greatly exceed L_R (about 3 km). Even in the Great Lakes a significant area is directly affected by the sides; in Lake Erie, L_R is about 5 km and L about 75 km. Whilst the precise extent of the horizontal spread of a mixed layer originating at a sloping boundary requires more research (the value of L_R is only locally defined on a slope), it appears that mixing at the sides will directly affect a large fraction of the total area of the World's lakes, and it is therefore of great importance to establish its causes and the associated fluxes.

Section 2 describes a simple example which illustrates how the shape of the lake bed can affect the form of internal seiches and hence determine the region of a lake in which mixing is most likely to occur as a consequence of their overturn and breaking. Our main concern, however, is with the processes which lead to fluxes of heat and momentum in a lake. The generation of intrusive layers by turbulence on a slope is discussed in section 3. Other processes described are directly associated with internal waves. The first is that of internal waves reflection from slopes (section 4). The slopes modify the internal wave field and the waves may generate turbulence by breaking (section 5). The second involves the generation of high frequency waves by the rapid lowering of a lake thermocline as low frequency non-linear waves or surges propagate along the side of the lake (section 6), and the third is internal wave generation by a mean current, for example that following a surge, flowing over rough topography (section 7). Unless dissipated, these boundary-generated internal waves will subsequently encounter other sloping boundaries and promote mixing there. Reference is made in section 8 to the effects of the presence of boundaries on the local stability of shear flows in lakes.

2. The Effect of Bed Shape on Internal Seiches

For simplicity, let us consider a stratified lake with uniform buoyancy frequency, N, and with rectangular plan form of length L and width W. A coordinate system with z upwards is selected; the lake boundaries are at $x = 0$ and L and $y = -W/2$ and $W/2$. The surface of the lake is at $z = H$ and the bottom is at $z = - H - d \sin ly$, where $d << H$. The shape of a more realistic lake bed might in principle be produced by a Fourier combination of solutions for this sinusoidal bedform type, but our objective here is to illustrate how this

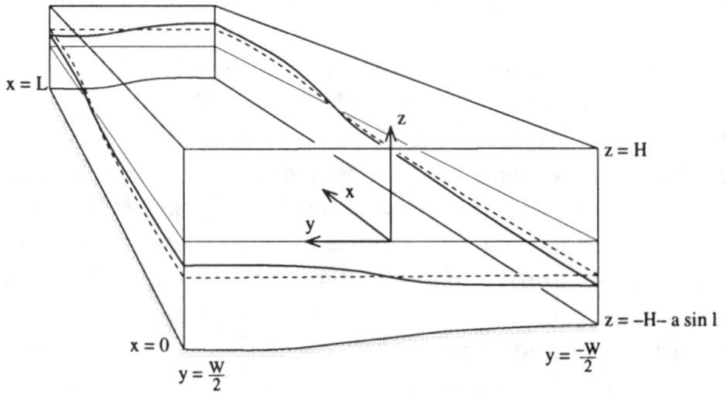

Figure 2. The coordinate axes x, y, z and the form of an isopycnal surface at mid-depth in the first longitudinal (x) and vertical (z) mode oscillation at maximum displacement (time, $t = 0$) in a rectangular lake, $0 < x < L$, $-W/2 < y < W/2$, of depth $2H + a \sin ly$ with $a << H$, $l = \pi/2W$. The buoyancy frequency N is constant. The isopycnal surface at mid-depth (full lines) is twisted relative to that in a lake of constant depth (dashed lines), resulting in larger vertical displacements in shallow water than in deep water when L/W is in the range $\sqrt{[(4n+1)^2-1]} < L/W < \sqrt{[(4n+3)^2-1]}$; $n = 0, 1, 2,...$

simple bed shape will affect the form of the longitudinal, x-directed, modes. Recent work by Thorpe and Holt (1995; see also section 7) has shown that an expansion technique in which the upper and lower boundary shapes are taken as a perturbation on a wave in a container with horizontal upper and lower boundaries provides a useful guide to the form of waves developing in a shear flow. I have adopted this technique to investigate the effect of a lake bottom on the dominant seiche, the primary internal standing wave in the x-z plane.

The solution is described in the Appendix and the leading order effect of the bed on the form of the wave is illustrated in Figure 2; the wave is twisted by the need to conform to the new bottom boundary conditions. Thorpe (1994) showed that first-mode standing waves strongly forced in a long rectangular container filled with fluid having a uniform density gradient 'break' after only 2 - 3 wave periods of forcing by developing z-shaped overturning isopycnals along an almost horizontal axis at mid-depth. The regions in which such breaking occurs are those in which the vertical density gradient becomes zero and then positive, density then increasing upwards so that the fluid is statically unstable (possibly - but not necessarily - also dynamically unstable). The regions in which such development will first occur are those in which the standing wave reduces the density gradient most. For the first mode internal seiche considered, these are found to be where the water depth is least, showing that, in this case at least, the zones in which mixing is most likely to occur through the development of static instability from the presence of the internal seiche are in the shallow water regions at the lake sides. The example shows what is perhaps obvious; in general the bed shape affects both the seiche amplitude distribution and the consequent perturbations to the density field. Although the study may strictly be regarded as providing an example of the effect of variable basin depth rather than of sloping sides, it suggests that the form of internal seiche modes in numerical models of real lakes should be examined to establish where they are most likely to induce mixing and thereby to promote, for example, the transfer of nutrients or pollutants between the epilimnion and the hypolimnion, and to direct observations accordingly. We shall see later (section 4) that more dramatic consequences - the absence of standing wave modes - may result if the lake sides are not vertical.

3. Intrusions Generated by Turbulence at the Sides of Lakes

Let us suppose for a moment that, through processes yet to be described, turbulence is generated on the sloping bottom of a stratified lake; the observations described in section 1 show that the boundary layers are indeed often found to be turbulent. Ivey and Corcos (1992) made a laboratory experiment which demonstrated the surprising fact that a vertical grid oscillating rapidly in a uniformly stratified fluid produces layers in the relatively quiescent outer region surrounding the near-grid turbulent zone. The layers spread out from the turbulent region with intrusive lenses of fluid moving outwards, being replaced by an intermediate flow of the outer region fluid towards the turbulent region. In this process, the mechanism of turbulence generation, the oscillating grid, is insignificant; it is the existence of turbulent motion which appears to be important. Indeed the experiment has been repeated in rather different geometries by Thorpe (1982) and by Park et al. (1994) but with similar results; a vertically periodic layer structure forms as a result of vertically uniform weak mixing in a uniformly stratified fluid.

Park et al. relate these observations to the stability theory of Phillips (1972) and Posmentier (1977), showing that layers appear to form under conditions in which, in accord with the theory, the flux Richardson number decreases with increasing gradient Richardson number (suitably-defined). In simple terms the instability results from a relatively large vertical flux of density (mass) in weak gradients and a lower flux where gradients are large. This leads to flux convergence, enhancing density below regions in which the gradient is large and so enhancing the gradient, reducing density in the weak gradients above high gradients, again enhancing the high gradient region. Weak turbulence, such as that found at the outer edge of a turbulent mixing region, leads to conditions favouring the instability and layering. Its spread into the outer region is presumably a consequence of the horizontal pressure gradients resulting from the adjusted density field.

Thorpe (1982) reports that layers are formed not only by a vertical oscillating grid but also when a grid lying on an inclined plane is oscillated rapidly up and down the slope to produce turbulence. This represents more closely the geometry of the situation in a lake. White (1990) finds that the turbulent layer on the slope produced by the grid grows in thickness to a near-steady value in about 2.4 to 3.2 buoyancy periods. Intrusions are visible after 10 buoyancy periods. The vertical separation between intrusions is approximately that of the thickness of the turbulent layer, measured in a direction normal to the slope. It therefore seems possible that intrusions form naturally at the sloping boundaries of lakes whenever a turbulent layer is produced, perhaps by the shear stress caused by currents induced by seiches or wind-induced circulation flowing over the rough lake bed (A, Figure 1). What is presently lacking is a definite measure of the size of the layers, although this may be related to the scale of eddies at the edge of the turbulent region (themselves determined by layer structure and bed roughness). It would also be useful to know the efficiency of the intrusions in transfering dissolved or particulate material from the edges to the central regions of a lake.

In reality, the intensity of turbulence produced on a lake bed may vary with position or depth as a consequence of different bed roughness or the processes leading to its generation. More intense mixing will tend to reduce the local density gradient and so produce horizontal imbalance in the pressure field which will promote local intrusions to develop.

Intrusions also occur in existing density interfaces (e.g. the thermocline) when there is mixing induced on sloping boundaries (e.g. see Phillips et al., 1986, and section 5). The slope will have an effect in modifying the amount of stretching, and hence the vorticity, of the column below the intruding layer; Hall (1990) has estimated the horizontal extent to which intrusions will spread on the interface between two layers. He considers intrusions resulting from the collapse of equal volumes of fluid produced at a density interface (e.g. at the thermocline) by boundary mixing at a sloping side wall. The extent of penetration of the fluid is essentially the same as that from a vertical wall if the slope angle α exceeds 5 deg, but becomes larger as α decreases, being 10% larger when $\alpha = 3$ deg. The region of influence of such boundary mixing will therefore depend on the slope of the side of a lake when this is less than 5 deg, but not otherwise.

4. Internal Waves Reflecting From Lake Slopes

Although often dominated by variability which appears to be associated with normal seiche modes, the frequency spectrum of temperature fluctuations in stratified lakes is generally broad-band, indicating the presence and activity of internal waves. The sloping sides of a lake will have a strong effect on the internal wave field. Consider for simplicity a lake, uniformly stratified with buoyancy frequency N and consisting of a channel of uniform triangular cross-section with sides sloping at angle α to the horizontal (Figure 3). The effects of the Earth's rotation will be ignored. Internal waves of frequency σ will propagate with group velocity inclined at angle θ to the horizontal, where $s_\theta = \sigma/N$. (Here and later $s_\theta = \sin\theta$ etc.). Eriksen (1982) has shown that if $\theta > \alpha$, internal waves will turn towards the upslope direction in reflecting from the sloping boundaries. If travelling towards deeper water their angle to the upslope direction will decrease. On reflecting at the water surface (or the foot of a mixed near-surface layer), the waves maintain their horizontal orientation but reverse their inclination to the horizontal. Figure 3a illustrates the trajectories of such waves. The full line shows the ray paths and the dashed lines the projection of these on the surface. Wave A begins in mid-water, moving over isobaths of increasing depth and towards the surface where it reflects at A_1. It meets the sloping bottom at A_2 and here turns towards the upslope direction, now moving over shallower and shallower isobaths. After a second surface reflection at A_3 it again meets the bottom at A_4 and again turns more directly upslope, so turning at each subsequent reflection from the bottom and propagating into shallower water. Wave B begins with a component of its group velocity directed upslope and continues towards shallower water, again turning more directly upslope at each subsequent reflection from the bottom. If wave A had crossed the deepest isobath of the lake before meeting the bottom at A_2, it would continue as does ray B following reflection B_1. Reflection from the vertical ends of the lake simply return the waves at the same angles to the vertical, reversing the normal component of group velocity. The overall effect is that all the waves with $\theta > \alpha$ eventually propagate into the shallow water regions at the edges of the lake where, with accumulating energy, they will break unless significant dissipation occurs during propagation. If $\theta < \alpha$, the waves turn towards the downslope direction on reflection, as illustrated in Figure 3b, and propagate towards the deepest isobath, again possibly breaking in its vicinity.

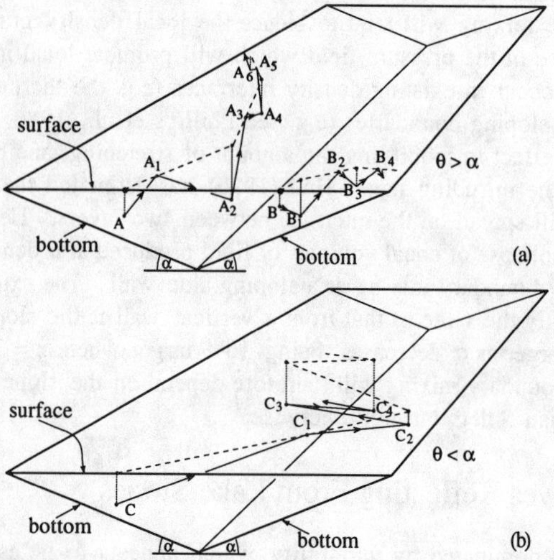

Figure 3. The propagation paths of internal waves in a lake composed of a channel of uniform triangular cross-section and sides sloping at angle α. The buoyancy frequency, N, is uniform. In (a) the wave frequency, σ, > N sinα, and in (b) σ < N sinα. The full lines show the ray paths of waves A, B, C and the dashed lines their projection onto the water surface.

The lake sides therefore act as a filter, with higher frequencies, those with $\sigma > Ns_\alpha$, accumulating in the shallows at the lake edges, and the low frequency waves with $\sigma < Ns_\alpha$ accumulating in the deepest water. Unless supplied by the generation of new waves, the central regions of the lake will become devoid of waves, or dominated by large-scale and slowly-damped seiches. Although lakes with rectangular section and vertical sides do have normal modes, this simple example suggests that there may be no normal modes of linear oscillation in some lakes having topography of particular shape, at least when the stratification is uniform. This appears to be true if a necessary condition for a standing wave to occur is that there exist ray paths which, perhaps after multiple reflections, are closed i.e. there exist waves originating from some location which, after propagating and reflecting from the sides, bottom or surface of the lake basin, arrive back at their original location moving in the same direction as they started out, and if the ray paths are reversible so that waves can also propagate in the opposite direction along a ray (see also Maas and Lam, 1995). In reality stratification is rarely uniform, but the effects described here will play some part in the modification of the internal wave field. In particular, the efficiency of reflection on slopes may depend on α, and so the variation and distribution of slope angles will contribute in determining the nature of the internal wave field (and through it, the mixing rates) in the body of a lake. A two-layer approximation to lake stratification may fail to include processes which, in a realistic density profile, may prevent the existence of pure standing waves even in 'regularly shaped' basins. We shall see later that when the bottom topography is irregular internal lee waves may be generated which drain momentum from the larger scale standing waves. It should also be noticed that the effects of the Earth's rotation, which permits the existence of Kelvin wave modes which may travel around the boundaries of a lake, may make it possible for periodic waves to occur, even when the non-rotating basin cannot support standing waves.

5. Internal Waves Breaking on Lake Slopes

One way by which boundary mixing occurs is through the processes mentioned above, by breaking internal waves. Taking for first consideration interfacial waves propagating along the thermocline, laboratory experiments show that turbulence is generated as progressive waves impinge on a sloping lake bed (Thorpe, 1966; Wallace and Wilkinson, 1988). The internal waves steepen as the slope is approached, producing a rotor with flow contrary to the wave propagation direction at the wave crest, unlike the 'forward overturn' of breaking surface waves (B, Figure 1). Close to the intersection of the interface and the slope the wave advances with a form like that of a gravity current (D, Figure 1). Kelvin-Helmholtz instability occurs in the shear flow at the wave crest as it does in a gravity current. Mixed fluid from the run-up of the proceeding wave flows over that of the following wave. This raises the mixed fluid from the bottom and, augmented by fluid from the mixing produced by successive waves, an intrusion develops. The rate of spread of an intrusion of thickness h along an interface between two deep layers of density $\rho_1 < \rho_2$ is about $2[g\Delta h]^{1/2}$, where $\Delta=(\rho_2-\rho_1)/(\rho_1+\rho_2)$ (e.g. Thorpe, 1973; Britter and Simpson, 1981).

It seems possible that intrusive spread at the interface may be enhanced by the Stokes drift in the incident wave (C, Figure 1); for progressive internal waves of the first mode and amplitude, A, on a density interface of density $\rho = \rho_0(1 - \Delta \tanh az)$ i.e. of thickness scale a^{-1}, the Stokes drift current at the interface is $A^2akc/2$ directed *contrary* to the direction in which the wave is propagating at speed $c = [g\Delta a/k(k+a)]^{1/2}$ and wave number k (Thorpe,

1968; Craik, 1982, discusses the Stokes drift speed of interfacial waves, referring to work done by Dore and co-workers, who find that the currents are much higher than those in a single layer, and considers its possible instability). When a>>k, the Stokes drift is comparable to the intrusion speed if $A^4ak/h \approx 16$, which is possible, for example if $Ak \approx 0.1$, $A/h \approx 2$ and $Aa \approx 9$. Although not yet observed, the processes of wave breaking and intrusive interlayering should occur in lakes whenever a thermocline is present.

Only two-dimensional interfacial waves - those having normal incidence on the slope - have so far been studied in the laboratory. Hogg (1971) pointed out that non-normal waves generate along-slope currents in a manner similar to the generation of along-shore currents in a coastal surf zone.

Returning to the case when the stratification is uniform, as it may be in the hypolimnion of a deep lake, there are several processes by which turbulence is generated as internal waves impinge and reflect from the sloping boundary. The best studied of these is reflection at critical slopes, when the group velocity vector of the incident wave is in a plane normal to the slope and is inclined at an angle θ to the horizontal equal to that of the slope, α (see E, Figure 1). Linear theory predicts that the wavenumber and amplitude of the reflected wave becomes increasingly large as θ tends to α, and fails to give useful predictions at $\theta = \alpha$. Laboratory experiments by Cacchione and Wunsch (1974) and Ivey and Nokes (1989) show that turbulence is generated at the slope when θ is at, or near, α. Eriksen (1982, 1985) has argued for and observed that the internal wave spectra in the ocean near sloping boundaries are modified in a frequency band surrounding that at which critical conditions are found, $\sigma_c = (N^2 s_\alpha^2 + f^2 c_\alpha^2)^{1/2}$, where $s_\alpha = \sin \alpha$ etc. Recent observations conducted by Dr U. Lemmin (LRH/EPFL, Lausanne, Switzerland) and the author in Lake Geneva at 20-30m depth on a 17° slope also show the prominence of waves at critical frequency in the bottom 2m.

Internal waves reflecting from non-critical slopes also promote the development of turbulence on the slope (Taylor, 1993) and up-slope propagating density fronts like those observed in interfacial waves (Thorpe, 1992a) with a cleft and lobe structure (D, Figure 1; Simpson, 1969). When the topography is rough and of wavelength scale smaller than that of the incident internal waves measured in a direction parallel to the slope, secondary internal lee waves are generated (F, Figure 1). Refracted by the fluctuating velocity and density field of the primary long reflecting wave, these may themselves break, contributing to turbulence (Thorpe, 1989).

Thorpe (1987b) considers the reflection of plane internal waves incident on a slope in the particular case $\theta > \alpha$. The buoyancy frequency, N, is constant. The incident and reflected waves interact. If, in a frame of reference with axes up and normal to the slope the incident wave is in a plane normal to the slope with wavenumber (k, n_I) and frequency σ, then the (first order) reflected wave has wavenumber (k, n_R) and frequency σ; both wavenumbers and frequencies satisfy a dispersion relation which, in the given frame of reference, is $\Sigma^2 = n^2 (Kc_\alpha - Ms_\alpha)^2/(K^2 + M^2)$, where Σ is the frequency of wavenumber (K,M). The incident and reflected waves each satisfy the equations of motion exactly, but together they produce unbalanced terms in the equation of motion which can only be satisfied by the addition of forced components. These may form resonant triads with the incident and reflected waves, or higher order components e.g. wavenumbers $[(m+n)k, (mn_I + nn_R)]$ and frequencies $(m+n)\sigma$, where m and n are integers. The phase-locking of the incident and reflected waves contributes to the strength of their interaction. The angles θ at which resonance is possible for given slope angle α are estimated up to third order.

Resonance was found only for moderate angles α, less than 8.4° for second order interactions and 11.8° for third order interactions. The effect of the Earth's rotation is to decrease these 'resonant slope angles'. Resonance at larger slope angles is however possible when the plane of the incident wave is not normal to the slope. A consequence of the resonant, or near resonant, interactions is that the wave field develops local regions of scale (k^{-1}, n_R^{-1}) in which isopycnals become very steep or may overturn. These occur at a distance from the slope depending on the incident wave and α, leading to static instability and probably dynamic instability (G, Figure 1).

From the earlier work, it appeared possible that resonant interaction might occur only when $\theta > \alpha$. This is however not so. In a non-rotating system with waves in the plane normal to the slope, resonance occurs when the forced m,n components satisfy the dispersion relation. It may be shown algebraically that the dispersion relation cannot be satisfied by higher order components $[(m+n)k, (mn_I + nn_R)]$ and frequencies $(m+n)\sigma$ with $m,n > 0$ for any $\theta > \alpha$. However when $n < 0$, $m > 1-n$ and $(m-n) > (m+n)^2$, there are interactions for $\theta < \alpha$, and these are possible up to $\alpha = \pi/2$. For small θ, the 'resonant slopes' are at $t_\alpha = (m+n)(1+m-n)t_\theta/[m-n+(m+n)^2]$, where $t_\alpha = \tan\alpha$ etc. The largest t_θ at which resonant interaction may occur is $[(m+n)^2-1]^{-}$ e.g. at $\theta = \pi/6$ when $m+n = 2$ is the largest such θ. Values of α tending to $\pi/2$ are found when $t_\theta = [(m-n)^2 - (m+n)^4]/[(m-n)^2\{(m+n)^2 -1\}]$. There may be other possible resonances between harmonics leading to the generation of steep mid-water isopycnals. The main conclusion however is that there are conditions for resonant interaction for waves with direction θ which are less than and greater than α, but that in each case there are limits to the range of α and θ in which these resonances may occur. Evidence in the ocean (e.g. Gilbert, 1990) suggests that even the second-order resonance, when it may occur, is not as strong as that when $\theta = \alpha$; it is a contributory and detectable, but not dominant, process in the modification of the internal wave field. It has yet to be established how important the resonant interactions are in promoting mixing, but this is one process by which mixing caused by the presence of the slope is created not just at the boundary itself, but some distance away.

The following sections describe ways in which internal waves of frequencies higher than those of the main seiche modes of a lake can be generated at the sloping boundaries.

6. Secondary Motions Induced by the Passage of Steep Non-linear Internal Waves Along a Sloping Boundary

In the summer and autumn, when mixing between the upper and lower layers is inhibited by strong thermal stratification, there are many observations in lakes of rapid changes in thermocline depth which, occurring in calm and warm weather, cannot be attributed to the direct effects of wind mixing or convection. Such changes are caused by the passing of internal surges or hydraulic jumps (Thorpe, 1971, 1977; Thorpe et al., 1972; Hunkins and Fliegel, 1973; Farmer, 1978; Jiao et al., 1993) or 'non-linear coastally trapped internal waves' (referred to here for simplicity as Kelvin waves although modified by the slope and shape of local topography; Poincaré waves are also common in large lakes). These result from the relaxation of the thermocline following periods of strong wind, or perhaps are generated by a partial resonance with atmospheric forcing (Thorpe, 1974). The waves are particularly notable in long narrow lakes in the late summer and early autumn when stratification is large and when the natural modes of standing oscillations in the closed basins take the form of undular bores.

It is known that in single-layer flows in channels with sloping side walls, there is significant run-up and turbulence at the sides as non-linear waves or jumps propagate down-channel (e.g. see Fenton, 1973, and Mathew and Akylas, 1990). Similar effects may be expected in waves on the thermoclines of lakes. There the tendency for internal hydraulic jumps to form as waves in which a shallow thermocline is lowered as the wave passes - rather than a surface being raised, as in surface hydraulic jumps - so that the epilimnion flows down (and not up) the sloping side of the lake as the jump passes, may modify this effect. Recent observations in Lake Geneva (Thorpe et al., 1996) have pointed to the interaction of non-linear internal Kelvin wave fronts with the sloping sides of lakes as moving sources of secondary waves and mixing. The rapid descent of the thermocline down the gentle slope on the side of the lake as the steep front of the long wave passes (H, Figure 1), rapidly increasing the thermocline depth and often reversing and increasing the magnitude of the currents, produces a mixing zone propagating with the wave front along the sloping side of the lake (I, Figure 1). This zone radiates short internal waves in its wake (J, Figure 1) in a manner similar to that of a solid body moving in a stratified fluid (Keller and Munk, 1970; Watson et al., 1992; Bonneton et al., 1993). Dispersion of the waves is affected by the shear flow associated with the Kelvin wave. They contribute to the short wavelength, high frequency, part of the internal wave spectrum, and perhaps enhance mixing in the centre of the lake by enhancing the level of shear. The sidewall mixing produced by the long wave may also result in a collapsing and intruding layer following the short internal waves.

7. Generation of Internal Lee Waves or Wakes by Flow Over Topography

Along-slope flows persist in lakes after a non-linear Kelvin wave has passed and often continue for many hours, depending on the period of the seiche or the changing wind field. We now consider the nature of the internal waves produced by such stratified flows, or those produced by gravity currents and by wind (often associated with up- or down-welling events), as they pass over the rough sidewall topography in a uniformly stratified hypolimnion of a lake. The rough topography may result, for example, from the slumping of sediment, the erosion of the lake bed by cold river water cascading down the side of the lake in winter, or from outcropping rocks and headlands.

Thorpe (1992b) has considered a steady flow of a uniformly stratified deep fluid at speed V along a slope tilted at angle α to the horizontal which is covered by roughness ripples set with crests aligned at an angle β to the line of greatest slope (Figure 4). It is supposed that the Coriolis parameter, f, is less than the buoyancy frequency N. The ripple wavenumber is l. Stationary lee waves, those having a component of their phase speed, σ/lc_β, which is equal and opposite to V, are generated only when the Froude number χ = Vl/N, satisfies the equation $1 > \chi c_\beta > [1 - (f/N)^2]s_\alpha{}^2c_\beta{}^2 + (f/N)^2$. Here σ is the intrinsic frequency of the waves and $c_\beta = \cos\beta$ etc. The physical significance of these limits are explained by Thorpe (1996).

The geometry of the internal waves is shown in Figure 5. The slope (stippled) is represented by a uniform plane ABCF inclined at an angle, a, to the horizontal plane AOEF. The (x,y,z) coordinate axes are shown in part a. A mean along-slope flow, V, is in the -y direction. The linear ripples (not shown here) are superimposed on the slope with crests aligned parallel to AC at an angle b to AB, the line of greatest slope. The angle b is

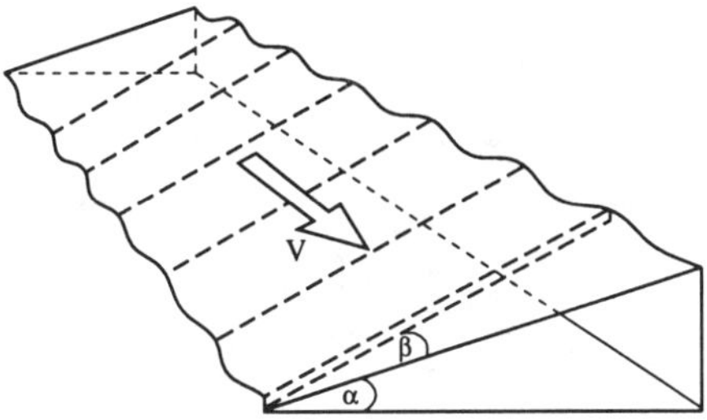

Figure 4. The analytical model geometry. The mean flow, V, is uniform along a slope which is at an angle, α, to the horizontal. The slope is covered with sinusoidal ripples with crests inclined at angle β to the line of steepest ascent up the slope.

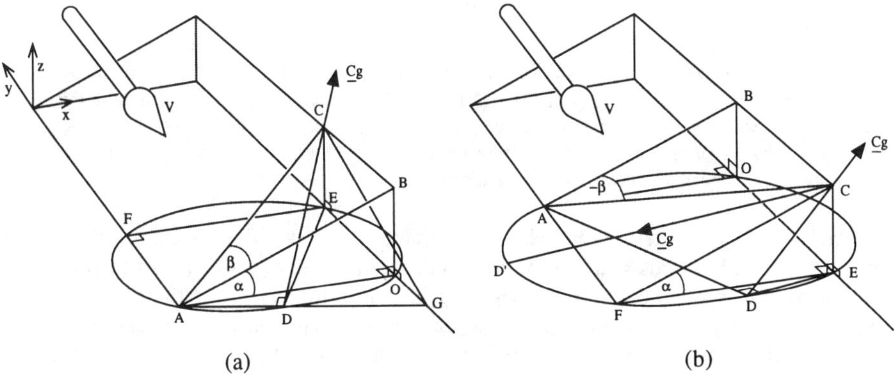

Figure 5. The geometry of the internal lee waves produced by along-slope flow. The slope is stippled. See text for an explanation.

positive in (a), negative in (b.) Internal waves formed as lee waves by flow over the topography form with constant phase surfaces which cut the plane in lines parallel to AC e.g. ACGD is a constant phase surface which meets the horizontal plane in the line AG. The group velocity of the internal waves, $\underline{c}_g$, viewed in a frame of reference moving with the mean flow, V, is directed in the constant phase surface normal to its intersection with the horizontal plane, i.e. along the line DC. DCE is therefore a vertical plane and angle ADE = $\pi/2$, so the locus of the point D is a circle of diameter AE passing through O and F. The inclination of $\underline{c}_g$ to the horizontal is θ, where $t_\theta = [(\sigma^2 - f^2)/(N^2 - \sigma^2)]^{1/2}$. The phase speed of the internal waves is directed normal to the relative group velocity in the vertical plane DCE, with alongslope speed $\sigma/lc_\beta = V$. In (a) (β>0) the phase speed has a positive component opposing the mean flow, offering possible stationary phase solutions, if D lies on the sector of the circle AO (when $\underline{c}_g$ has a positive component towards shallower water) or on sector OE (when $\underline{c}_g$ has a positive component towards deeper water). In (b) the phase speed has a positive component opposing the mean flow if D lies on sector FE (when $\underline{c}_g$ has a positive component towards shallower water) or on sector AF (shown as D$'$ with $\underline{c}_g$ in direction CD$'$ with a positive component towards deeper water).

Figure 6 shows the area, between curves a and b, in the χ - β plane in which stationary lee waves can exist when $s_\alpha = f/N = 0.1$ (i.e. for bed slopes of about 6 deg and $N \approx 10^{-3}$ s^{-1}

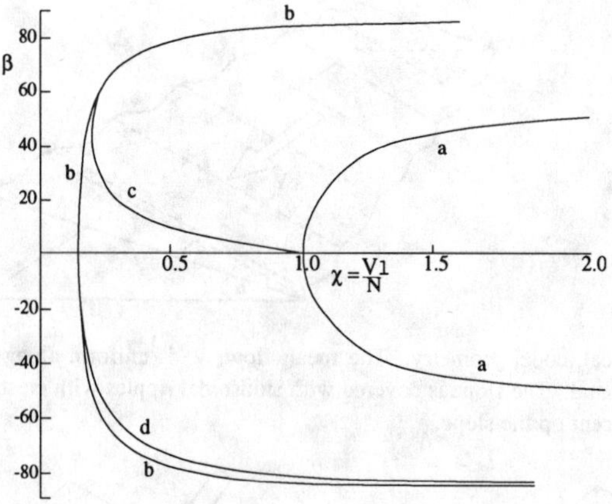

Figure 6. χ (or Vl/N) - β plane form wave generation on a slope when $s_\alpha = f/N = 0.1$. (a) is the curve $\chi = c_\alpha^{-1}$, (b) is $\chi c_\beta = [1 - (f/N)^2]s_\alpha^2 c_\beta^2 + (f/N)^2$, and the two curves bound the area within which stationary lee waves can be generated. Curve (c) is $\chi c_\beta = \{[s_\alpha^2 c_\beta^2 + (f/N)^2 s_\beta^2]/[s_\beta^2 + s_\alpha^2 c_\beta^2]\}$- and (d) is $\chi c_\beta = [s_\alpha^2 + (f/N)^2 c_\alpha^2]$-. Between these the group velocity of the stationary internal lee waves has a positive component directed towards shallower water.

when $f \approx 10^{-4}\ s^{-1}$. The area in which they have a positive component of their group velocity directed towards shallower water lies between curves a, c and d. When β is small, and in particular when $0 \geq \beta > -20$ deg, this includes virtually all stationary waves. In general the topography of the side of the lake may be represented by a superposition of Fourier components. Provided that these are dominated by those with small β, there will be a tendency for waves to travel towards shallow water (K, Figure 1).

Thorpe et al. (1995) have considered the effects of having finite depth and non-uniform stratification. The former limits the range of possible vertical modes and result in wave reflection at the upper surface; in subsequent bottom reflections the waves turn towards the upslope direction. Variation in N causes the generation condition to be satisfied only in certain depth zones and results in refraction of the waves. Waves maintain their intrinsic frequency, s, and are bent towards the horizontal. Those propagating towards shallower water may encounter the sloping lake sidewall and contribute to local mixing or be reflected. Oscillating flows cause periodic wave generation (Thorpe, 1996).

The effect of these slope-generated internal waves on lake dynamics depends on where the waves loose their energy by breaking, perhaps contributing to the local mixing of that region, and where they loose their momentum. They propagate against the mean flow V at the location of their generation, and so on breaking or otherwise losing their momentum (perhaps through interaction with the currents e.g. see Thorpe, 1978) they will impose a force in the direction contrary to V. This wave drag may produce counter currents and contribute to the separation of the flow from the upper levels of the slope. If, for example, the waves break in an area of the slope where they are focussed as sketched in Figure 7 (see also L, Figure 1) they would there reduce the density gradient, thus leading to an offslope moving intrusion (M, Figure 1), and reduce the flow so contributing to across-lake shear. The size and time-scales estimated by Thorpe (1996) in conditions appropriate to the oceanic shelf-break suggest that regions in which the topographic rms height is of order 10 m will produce a significant transfer of momentum in a period of a few days.

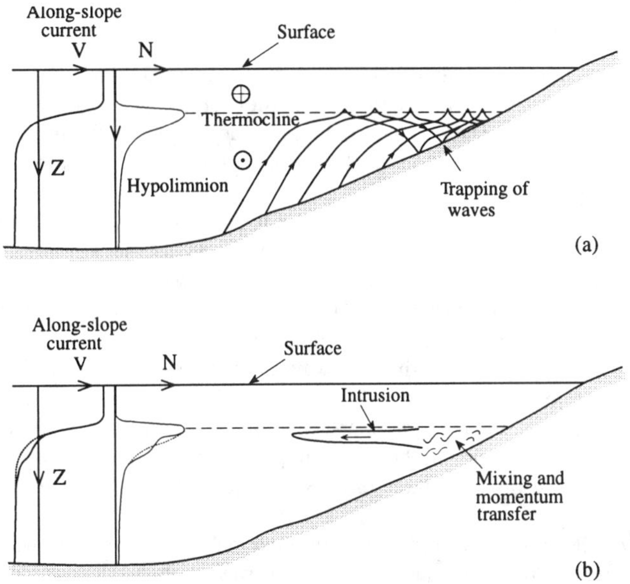

Figure 7. (a) shows ray-paths of internal waves produced by along-slope flow, V, in the hypolimnion in conditions favouring propagation towards shallower water (see Figure 5). Waves are trapped between the mixed epilimnion layer and the sloping lake boundary. (b) shows the consequences of mixing by the internal waves. It reduces the N profile and creates an intrusion in the thermocline. The along-slope current in the region of wave breaking is also modified as sketched.

8. Stability of Flow Along Sloping Topography

The presence of a neighbouring boundary may help to stabilise a stratified shear flow which, in the absence of boundaries, would be unstable. Miles (1967) showed that an inviscid flow with speed and density varying exponentially away from a plane horizontal boundary is stable to infinitesimal disturbances for all values of the flow Richardson number, Ri. Hazel (1972) considered the effects of symmetrically placed boundaries at $z = \pm H$ on either side of a stratified shear flow with speed and density varying as $\tanh(z/z_0)$; the flow is unstable for all Ri < 1/4 in deep fluids when $z_0/H \ll 1$, but the flow is stabilised, with a critical Ri < 1/4 when z_0/H is sufficiently large (> about 0.4). Similar results are found for a profile with error function distribution of speed and density with depth (Thorpe and Holt, 1995). This suggests that a uniform shear flow at a density interface which intersects a slope (e.g. a shear flow directed along isobaths, as in section 6, across a thermocline which intersects the sloping side of a lake) may be unstable in deep water far from the slope, but could be stable close to the boundary (disregarding of course the generation of instability and turbulence by the stress at the boundary itself). Laboratory experiments by Holt (1995, following those described in Thorpe and Holt) show that although the flow near such a sloping boundary may indeed be locally stable and will not develop growing waves and billows, it is strongly affected by the turbulence and spreading intrusions originating from the overturning Kelvin-Helmholtz billows in deeper water.

9. Conclusions

After a period of calm weather long enough for the internal wave field in the centre of a lake to decay to a level in which the probability of its generating local Richardson numbers which are < 1/4 is very small, the regions adjoining the sloping boundaries of the lake may be the only locations at which internal wave activity remains large and where diapycnal mixing persists.

Few measurements have been made of the temperature and current structure in the boundary layers on the sides of lakes. Such measurements are a priority if it is to be convincingly determined that boundary mixing is indeed important, and its effects adequately measured and parameterized. The level of the most energetic motions resulting from internal wave activity near a slope and therefore where activity is likely to be most intense, is where the amplitude of internal waves is greatest. This is usually in the thermocline. It is therefore suggested that where observations are most needed and might most profitably be made is at and near the intersection of the thermocline and the sloping sides of lakes. It would help establish the physical processes at work if measurements were to be made which can resolve the three-dimensional coherent structures to scales of 0.1 to 1 m.

In view of the possibility of generating internal waves by long-wave topography interaction (section 6) and because the internal wave field (and hence its influence on mixing in the body of a lake) may depend on the distribution and magnitude of boundary slopes (section 4), it would be profitable to undertake a study of the variation of the form of the internal wave spectrum, especially the high frequency end of the internal wave spectrum near N, as a function of depth and distance from the side of lakes, and to relate the variations in high frequency energy to the long wave activity.

Few numerical models of lakes have been constructed, even of those lakes where models would be useful in managing and controlling fisheries or preventing pollution. This is perhaps why rather little attention has been given to process studies in recent years; there is a need to devise or improve models suitable for lake management and for physical limnologists to describe and quantify processes such as those described above in ways which can be used in numerical models. I have attempted here to do little more than to identify some of the processes which might occur, and which still need quantification.

References

Armi, L., Some evidence for boundary mixing in the deep ocean. *J. Geophys. Res. 83*, 1971-1997, 1978.

Baines, P. G., The reflection of internal/inertial waves from bumpy surfaces. *J. Fluid Mech., 46*, 273-291, 1971a.

Baines, P. G., The reflection of internal/inertial waves from bumpy surfaces: Part 2. Split reflection and diffraction. *J. Fluid Mech., 49*, 113-131, 1971b.

Bohle-Carbonell, M., Wind and currents: response patterns of Lake Geneva. *Ann. Geophysicae, 9*, 82-90, 1991.

Bonneton, P., J. M. Chomaz, and E. J. Hopfinger, Internal waves produced by the turbulent wake of a sphere moving horizontally in a stratified fluid. *J. Fluid Mech., 254*, 23-40, 1993.

Britter, R. E. and J. E. Simpson, A note on the structure on the head of an intrusive gravity current. *J. Fluid Mech., 112*, 459-466, 1981.

Cacchione, D. and C. Wunsch, Experimental study of internal waves over a slope. *J. Fluid Mech., 66*, 223-240, 1974.

Caldwell, D. R., J. M. Brubaker and V. T. Neal, Thermal microstructure on a lake slope. *Limnol. Oceanogr., 23*, 372-374, 1978.

Craik, A. D. D., The drift velocity of water waves. *J. Fluid Mech.*, *116*, 187-205, 1982.
Eriksen, C. C., Observations of internal wave reflection off sloping bottoms. *J. Geophys. Res. 87*, 525-538, 1982.
Eriksen, C. C., Implications of ocean bottom reflection for internal wave spectra and mixing. *J. Phys. Oceanogr. 15*, 1145-1156, 1985.
Eriksen, C. C., Internal wave reflection and mixing at Fieberlong Guyot. Proceeding of the Hawaiian Winter Workshop, edited by Peter Muller, University of Hawaii, Honolulu, HI, (in press), 1995.
Farmer, D. M., Observations of long waves in a lake. *J. Phys. Oceanogr. 8*, 63-73, 1978.
Farrow, D. E., An asymptotic model for the hydrodynamics of the thermal bar. *J. Fluid Mech.*, *289*, 120-140, 1995.
Fenton, J. D., Cnoidal waves and bores in uniform channels of arbitrary cross-section. *J. Fluid Mech.*, *58*, 417-438, 1973.
Garrett, C. J. R., The role of secondary circulation in boundary mixing. *J. Geophys. Res.* 95, 3181-3188, 1990.
Garrett, C., Marginal mixing theories. *Atmosphere-Ocean. 219*, 313-339, 1991.
Garrett, C., P. MacCready and P. Rhines, Boundary mixing and arrested Ekman layers: rotating stratified flow near a sloping boundary. *Ann. Rev. Fluid Mech.*, *25*, 291-323, 1993.
Gilbert, D. Theory and observations of internal wave reflection off sloping topography. PhD thesis, 180 pp., Dalhousie University, Halifax, Nova Scotia, Canada, 1990.
Gilbert, D. and C. J. R. Garrett, Implications for ocean mixing of internal waves scattering off irregular topography. *J. Phys. Oceanogr. 19*, 1716-1729, 1989.
Hall, P., A model of the geostrophic adjustment of a three-layer fluid on a sloping boundary. *Ocean Modelling*, *89*, 7-10, 1990.
Hazel, P., Numerical studies of the stability of inviscid stratified shear flows. *J. Fluid Mech.*, *51*, 39-61, 1972.
Hibiya. T., Generation mechanism of internal waves by tidal flow over a sill. *J. Geophys. Res.*, *91*, 7697-7708, 1986.
Hibiya, T., The generation of internal waves by tidal flow over Stellwagen Bank. *J. Geophys. Res.*, *93*, 533-542, 1988.
Hogg, N. G., Longshore currents generated by obliquely incident internal waves. *Geophys. Fluid Dyn.*, *2*, 361-376, 1971.
Holt, J. T., The effect of boundaries on stratified shear flows. PhD dissertation, 300 pp., University of Southampton, UK, 1995.
Hunkins, K. and M. Fliegel, Internal undular surges in Seneca Lake: a natural occurrence of solitons. *J. Geophys. Res. 78*, 539-548, 1973.
Imberger, J., Vertical heat flux in the hypolimnion of a lake. In Proc. 10th AFMC, Melbourne, 1989, vol. *1*, 2.13-2.16, 1989.
Imberger, J. and P. Hamblin, Dynamics of lakes, reservoirs and cooling ponds. *Ann. Rev. Fluid Mech.*, *14*, 153-187, 1982.
Imberger, J. and G. N. Ivey, On the nature of turbulence in a stratified fluid. *J. Phys. Oceanogr.*, *21*, 659-680, 1991.
Imberger, J. and G. N. Ivey, Boundary mixing in stratified reservoirs. *J. Fluid Mech.*, *248*, 477-491, 1993.
Ivey, G. N. and G. M. Corcos, Boundary mixing in a stratified fluid. *J. Fluid Mech.*, *121*, 1-26, 1982.
Ivey, G. N. and R. I. Nokes, Vertical mixing due to the breaking of critical internal waves on sloping boundaries. *J. Fluid Mech.*, *204*, 479-500, 1989.
Jiao, C., M. Kumagai and K. Okuba, Solitary internal waves in Lake Biwa. *Bull. Disaster Prevention Res. Inst., Kyoto Univ.*, *43*, 61-72, 1993.
Keller, J. B. and W. H. Munk, Internal wave wakes of a body moving through a stratified fluid. *Phys. Fluids 13*, 1425-1431, 1970.
Lamb, K. G., Numerical experiments of internal wave generation by strong tidal flow across a finite amplitude bank edge. *J. Geophys. Res. 99*, 843-864, 1994.
Ledwell, J. R. and A. Bratkovich, A tracer study of mixing in the Santa Cruz basin. *J. Geophys. Res.*, *100*, 20,681-20,704, 1995.
Ledwell, J. R. and B. M. Hickey, Evidence of enhanced boundary mixing in the Santa Monica Basin. *J. Geophys. Res.*, *100*, 20,665-20,679, 1995.
Ledwell, J. R. and A. J. Watson, The Santa Monica Tracer Experiment; a study of diapycnal and isopycnal mixing. *J. Geophys. Res.*, *98*, 8698-8718, 1991.

Lighthill, J., *Waves in Fluids*. CUP, 504pp, 1978.
Maas, L. R. and F. P. A. Lam, Geometric focussing of internal waves. *J. Fluid Mech., 294*, 1-41, 1995.
MacCready, P. and P. B. Rhines, Buoyant inhibition of Ekman transport on a slope and its effects on stratified spin-up. *J. Fluid Mech. 223*, 631-661, 1991.
MacCready, P. and P. B. Rhines, Slippery boundary layers on a slope. *J. Phys. Oceanogr. 23*, 5-22, 1993.
Mathew, J. and T. R. Akylas, On three dimensional long water waves in a channel with sloping side walls. *J. Fluid Mech., 215*, 289-307, 1990.
Maxworthy, T., A note on the internal solitary waves produced by tidal flow over a three-dimensional ridge. *J. Geophys. Res. 84*, 338-346, 1979.
Mied, R. D. and J. P. Dugan, Internal wave reflection from a sinusoidally corrugated surface. *J. Fluid Mech., 78*, 763-784, 1976.
Miles, J. W., Internal waves in a continuously stratified atmosphere or ocean. *J. Fluid Mech., 28*, 305-317, 1967.
Munk, W. H., Abyssal recipes. *Deep-Sea Res., 13*, 207-230, 1966.
Park, Y. -G., J. A. Whitehead and A. Grenadeskiant, Turbulent mixing in stratified fluids: layer formation and energetics. *J. Fluid Mech., 279*, 279-313, 1994.
Phillips, O. M., On flows induced by diffusion in a stably stratified fluid. *Deep-Sea Res., 17*, 435-443, 1970.
Phillips, O. M., Turbulence in strongly stratified fluid - is it unstable? *Deep-Sea Res., 19*, 79-81, 1972.
Phillips, O. M., J. H. Shyu and H. Salmun, An experiment on boundary mixing; mean circulation and transport rates. *J. Fluid Mech., 173*, 473-499, 1986.
Posmentier, E. S., The generation of fine structure by vertical diffusion. *J. Phys. Oceanogr., 7*, 292-300, 1977.
Simpson, J. E., A comparison between laboratory and atmospheric density currents. *Quart. J. R. Met. Soc., 95*, 758-765, 1969.
Taylor, J. R., Turbulence and mixing in the boundary layer generated by shoaling internal waves. *Dyn. Oceans Atmospheres, 19*, 233-258, 1993.
Thorpe, S. A., Internal gravity waves. PhD dissertation, University of Cambridge, 1966.
Thorpe, S. A., On the shape of progressive internal waves. *Phil. Trans. R. Soc. Lond.* A, *263*, 563-614, 1968.
Thorpe, S. A., Asymetry of the internal seiche in Loch Ness. *Nature, 231*, 306-308, 1971.
Thorpe, S. A., Turbulence in stably stratified fluids; a review of laboratory experiments. *Boundary-Layer Meteorology, 5*, 95119, 1973.
Thorpe, S. A., Near-resonant forcing in a shallow two-layer fluid: a model for the internal surge in Loch Ness? *J. Fluid Mech. 63*, 509-527, 1974.
Thorpe, S. A., Turbulence and mixing in a Scottish loch. *Phil. Trans. R. Soc. Lond.* A, *286*, 125-181, 1977.
Thorpe, S. A., On internal gravity waves in an accelerating shear flow. *J. Fluid Mech., 88*, 625-639, 1978.
Thorpe, S. A., On the layers produced by rapidly oscillating a vertical grid in a uniformly stratified fluid. *J. Fluid Mech., 124*, 391-409, 1982.
Thorpe, S. A., Current and temperature variability on the continental slope. *Phil. Trans. R. Soc., Lond.*, A, *323*, 471-517, 1987a.
Thorpe, S. A., On the reflection of a train of finite amplitude internal gravity waves from a uniform slope. *J. Fluid Mech., 178*, 279-302, 1987b.
Thorpe, S. A., The distortion of short internal waves by a long wave, with application to ocean boundary mixing. *J. Fluid Mech. 208*, 395-415, 1989.
Thorpe, S. A., Thermal fronts caused by internal waves reflecting from a slope, *J. Phys. Oceanogr. 22*, 105-108, 1992a.
Thorpe, S. A., The generation of internal waves by flow over the rough topography of a continental slope. *Proc. R. Soc. Lond.* A, *439*, 115-130, 1992b.
Thorpe, S. A., Statically unstable layers produced by overturning internal gravity waves. *J. Fluid Mech., 260*, 333-350, 1994.
Thorpe, S. A., The cross-slope transport of momentum by internal gravity waves generated by along-slope currents over topography. *J. Phys. Oceanogr., 26*, 191-204, 1996.
Thorpe, S. A., M. Cure and M. White, The skewness of temperature derivatives in oceanic boundary layers. *J. Phys. Oceanogr. 21*, 428-433, 1991.
Thorpe, S. A., A. Hall and I. Crofts, The internal surge in Loch Ness. *Nature, 237*, 96-98, 1972.

Thorpe, S. A., P. Hall and M. White, The variability of mixing at the continental slope. *Phil. Trans. R. Soc. Lond.* A, *331*, 183-194, 1990.

Thorpe, S. A. and J. T. Holt, The effects of laterally sloping upper and lower boundaries on waves and instability inn stratified shear flows. *J. Fluid Mech.*, *286*, 49-65, 1995.

Thorpe, S. A., P. K. Hutt and R. Soulsby, The effect of horizontal gradients on thermohaline convection. *J. Fluid Mech.*, *38*, 375-400, 1969.

Thorpe, S. A., J. M. Keen, R. Jiang and U. Lemmin, High frequency internal waves in Lake Geneva. *Phil. Trans. R. Soc. Lond.* A, *354*, 237-257, 1995a.

Thorpe, S. A., D. Jiang and J. M. Keen, Shelf break momentum transport by internal waves generated by along-slope currents over topography. Proceeding of the Hawaiian Winter Workshop, edited by Peter Muller, University of Hawaii, Honolulu, HI, 111-118, 1995b.

Thorpe, S. A. and M. White, A deep intermediate nepheloid layer. *Deep-Sea Res.*, *35*, 1665-1671, 1988.

Trowbridge, J. H. and S. J. Lentz, Asymmetric behavior of an oceanic boundary layer above a sloping bottom. *J. Phys. Oceanogr.*, *21*, 1171-1185, 1991.

Turner, J. S., *Buoyancy Effects in Fluids*. Cambridge University Press, 367 pp, 1973.

van Haren, H., N. Oakey and C. Garrett, Measurements of internal wave band eddy fluxes above a sloping bottom. *J. Mar. Res.* *52*, 909-946, 1994.

Wallace, B. C. and D. L. Wilkinson, Run-up of internal waves on a gentle slope in a two-layer system. *J. Fluid Mech.*, *191*, 419-442, 1988.

Watson, G., R. D. Chapman and J. R. Apel, Measurement of internal wave wake of a ship in a highly stratified sea loch. *J. Geophys. Res.* *97*, 9689-9703, 1992.

Weatherley, G. L. and P. J. Martin, On the structure and dynamics of the oceanic bottom boundary layer. *J. Phys. Oceanogr.*, *8*, 557-570, 1978.

White, M., The temperature and current structure of the oceanic boundary layer. PhD dissertation, 184 pp, University of Southampton, 1990.

White, M., Tidal and sub-tidal variability in the sloping benthic boundary layer. *J. Geophys. Res.*, *99*, 7851-7864, 1994.

Woods, A. W., Boundary driven mixing. *J. Fluid Mech.*, *226*, 625-654, 1991.

Wuest, A., D. C. Van Senden, J. Imberger, G. Piepke and M. Gloor, Dyapycnal diffusivity mearured by microstructure and tracer techniques - a comparison. Paper presented at *the Proc. 4th International Symposium on Stratified Flows*, Grenoble, 29 June - 2 July, 1994.

Wunsch, C., On oceanic boundary mixing. *Deep-Sea Res.* *17*, 293-301, 1970.

Wunsch, C., Note on some Reynolds stress effects on internal waves on slopes. *Deep-Sea Res.* *18*, 583-591, 1971.

Wunsch, C. and R. Hendry, Array measurements of the bottom boundary layer and the internal wave field on the continental slope. *Geophys. Fluid Dyn.*, *4*, 101-145, 1972.

Appendix

I consider small internal standing waves in an inviscid, non-diffusive, stably stratified fluid with uniform buoyancy frequency, N. The fluid is confined in a basin of rectangular plan-form with vertical boundaries at $x = 0, L$ and $y = -W/2$, $W/2$ (W smaller than L), and fixed upper and lower boundaries at $z = H$ and $z = - H - d \sin ly$, where $lW = \pi$ and $d << H$. When $d = 0$, a first mode internal wave with motion confined in the x-z plane can be found (Thorpe, 1994) with $u = (am\sigma/k) \sin kx \sin mz \sin \sigma t$, $v = 0$, $w = a\sigma \cos kx \cos mz \sin \sigma t$, and $\rho = \rho_0 [1 - N^2/g(z + a \cos kx \cos mz \cos \sigma t -(a^2m/8)\sin 2mz (1 + \cos 2\sigma t))]$, correct to second order in the wave amplitude, where (u,v,w) is the velocity and ρ is the density. Terms a and ρ_0 are constant, g is the acceleration due to gravity, and $kL = \pi$, $m = \pi/2H$. The dispersion relation is $\sigma^2 = k^2N^2/(k^2+m^2)$.

The equations of motion, continuity and conservation of density, and the boundary conditions $w = 0$ at $z = H$, and $w = v\, d\zeta/dy$ at $z = - H + \zeta$ where $\zeta = - d \sin ly$, can be solved following Thorpe and Holt (1995). Transfering to a coordinate x' at the centre of the basin, $x = L/2$, so that $x' = x - L/2$, and supposing that the y-variation of the bottom can be accommodated by introducing terms which are of an order proportional to the product of a and d, an equation for ρ is found with additional terms of order ad:- $\rho = \rho_0 \{1 - N^2/g[z - a \sin kx' \cos \sigma t (\cos mz + (dm/\sin 2qH) \sin ly \sin q(z-H)) - (a^2m/8)\sin 2mz (1 + \cos 2\sigma t)]\}$, where $q = m(1 + l^2/k^2)^-$. The displacement, η, of isopycnal surfaces from their mean position, z_0, is given by $\eta = a \sin kx' \cos \sigma t [\cos mz_0 + \{dm/\sin 2qH\} \sin ly \sin q(z_0-H)]$, correct to order ad. The mid-depth ($z_0 = 0$) isopycnal displacements are > a in shallow ($y < 0$) water when $\cos qH < 0$ i.e. when $\sqrt{[(4n+1)^2-1]} < L/W < \sqrt{[(4n+3)^2-1]}$; n = 0, 1, 2,... The vertical structure of the order dm isopycnal displacements has n turning points, where n is the largest integer less than $\sqrt{(1 - L^2/W^2)}$, so that the site (i.e. $y>0$ or $y<0$) of enhanced isopycnal disploacements depend on the mean level, z_0, and on L/W. Resonances, when the (1, 0, 1) mode oscillations and the (1, 1, 2n+1), n = 1, 2, 3,... modes favoured by the bottom shape have the same frequency, are possible at $L/W = \sqrt{[(2n+1)^2 -1]}$. Here large transverse components of oscillation are likely.

Examination of the second order terms in the wave amplitude (following the analysis described in Thorpe, 1994) shows that the effect of the bed is to reduce the density gradient most in a region where $z > 0$, where the displacement, η, is negative (i.e. over a wave trough) and where $y < 0$. Differentiating the second order equation for ρ with respect to z and substuting $X = \sin kx'$, $Z = am \sin mz$, at $t = 0$ (when the density gradients are weakened most by the wave), it is found that $Z^2 - RXZ + P = 0$, where $P = (g/N^2\rho_0)d\rho/dz + 1 - am^2/2$ and $R = 1 - [dm/\sin 2qH] \sin ly$. This is the equation of hyperbolae with asymptotes $Z = 0$ and $Z = RX$. The smallest density gradients (i.e. the regions where the wave is most reducing the vertical density gradients and where, with sufficient increase in wave amplitude, static instability is most likely to occur) are found between the asymptotes. These gradients (and the volume of fluid containing them) increase as R increases and so in $\sqrt{[(4n+1)^2-1]} < L/W < \sqrt{[(4n+3)^2-1]}$, where y becomes large and negative. Laboratory experiment (Thorpe, 1994) confirm that the z-shaped overturns at large values of am occur tilted up from the x-axis so as to lie between the asymptotes. The region in which the z-form overturning layers are first formed will be where R is smallest i.e. where the fluid is most shallow.

At higher order (terms in ad^2) the bottom shape also affects the wave period.

32

Finescale Dynamics of Stratified Waters Near a Sloping Boundary of a Lake

U. Lemmin, R. Jiang and S. A. Thorpe

Abstract

Measurements are made in Lake Geneva to examine internal waves and flow characteristics in the stratified water overlying a 17° sloping bed in a water depth of about 30m. Even after a period of calm weather lasting three weeks, the water column is observed to be active, with internal oscillations of amplitude as great as 3 m. Internal waves are known to generate sheets and layers in the thermal structure of the water column, as described by Lazier (1973); those observed here have a predominantly downward phase propagation, indicating upward energy propagation. Such effects caused by enhanced internal wave activity through reflection near sloping boundaries may account for the increase in layering reported by Caldwell et al. (1978) as the side of a lake is approached, rather than the presence of intrusions. Internal waves with up- and down-slope currents reaching 10 cm s^{-1} and frequencies at, or close to, the 'critical' frequency on the slope, are observed in the bottom two metres of the water column; fluctuations with frequencies near critical are important in determining the nature of the slope boundary layer in lakes as well as in the ocean, where internal wave amplification has been observed by Eriksen (1982). Measurements of vertical velocity using an acoustic Doppler velocity profiler in this benthic layer, and ancillary observations, reveal a density current carrying sediment-laden warm water down the slope. The current is finely layered in its velocity structure, changes occurring over vertical scales of about 10 cm.

1. Introduction

Recent observations have pointed to the importance of the mixing which occurs in the stratified water column near the bottom and sides of lakes and small ocean basins. Wuest et al. (1994) studied diffusion in Lake Alpnach using tracers and microstructure measurements. They report that the bottom boundary is the most turbulent zone in the hypolimnion. Ledwell and Bratkovich (1995) find a dramatic increase in the rate of vertical diffusion of a tracer released into the mid-depth waters of the Santa Cruz Basin once it reaches the lateral boundaries. Thorpe (1998) reviews the processes which may produce mixing, and points to the lack of observations in the 'thermoslope', the region of contact of the highly stratified water in the thermocline with the sloping lateral boundaries of lakes or seas.

Physical Processes in Lakes and Oceans
Coastal and Estuarine Studies Volume 54, Pages 461-474

This is an account of observations of processes active in the thermoslope of a lake. It was anticipated that internal waves might be important; they may be amplified as they propagate into the wedge of shallowing water between the surface and the sloping bottom of the lake (Wunsch, 1968, 1969; Wallace and Wilkinson, 1988). They may cause boundary mixing when their frequency, σ, is near to $\sigma_c = N \sin \alpha$, where N is the local buoyancy frequency and α is the angle between the bottom slope and the horizontal (see for examples, Eriksen, 1982; Ivey and Nokes, 1989; Thorpe, 1992; Taylor, 1993). Supercritical incident waves with $\sigma > \sigma_c$ continue to propagate into the wedge whilst subcritical waves with $\sigma < \sigma_c$ are reflected back into deeper water. (The use here of super- and sub-critical is the opposite to that used by Cacchione and Wunsch, 1974, who link the terms to the bottom slope; they describe a slope with $\alpha > \sin^{-1} (\sigma/N)$ as supercritical). Measurements of currents and the thermal structure of the water column form the basis of the observations described below.

2. The Site and Instruments

The site chosen for observation is near Buchillon on the northern shore of Lake Geneva. The shoreline at the site is oriented roughly east-west. The mean bottom slope determined by repeated transects in the off-shore direction using an echosounder and GPS is 12.2° between 6 m and 60 m depth, with 18° between 10 m and 20 m, 17° at 20 m and 30 m, and 8° at 50 m (the possible error in bottom slopes is about 6%). The lake bottom in this area consists mainly of silt and sand. High resolution sonographs show the bed is smooth with no large boulders. Echo sounder transects in the east-west direction show depth contours are nearly parallel to shore.

Observations were made on three days in July 1995, all in conditions of calm weather. Diurnal winds with speeds between 1 and 3 m s^{-1} were the principal forcing agent; the measurements may be taken to characterise the processes in the boundary layer in calm and warm summer conditions.

Measurements are made with a CTD, profiling or suspended at constant depth near the bed, and with an Aanderaa thermistor chain sampling at 2 min intervals with eleven thermistors with a resolution of 0.1 K and with 2.5 m nominal separation. The lowest sensor is 0.1 m off the bottom. Other instruments, an acoustic Doppler velocity profiler (ADVP), Neil Brown (EG&G) current meters, and a small array of thermistors, added vital ancillary information.

3. Observations

a. Near-critical internal waves

Figure 1 shows three typical CTD profiles. Below the thermocline at 5 m a steppy structure is often observed. The mean profile between 10 m and the bottom at about 26 m gives $N = 0.015 \pm 0.002$ s^{-1}, and $T_c = 23.8 \pm 3.5$ min, and here the water column is stably stratified everywhere except in the lowest 2 m where some 0.1 - 0.2 m high temperature inversions are apparent.

Isotherms derived from an Aanderaa thermistor chain moored in 26.5 m depth on 17 July are shown in Figure 2. Regular oscillations are recorded in the bottom two sensors (up to 1 m from the bed) with a mean period of 25 ± 3 min. Their apparent amplitude is about 4 m and they have no direct link to fluctuations at higher levels. The critical period, T_c =

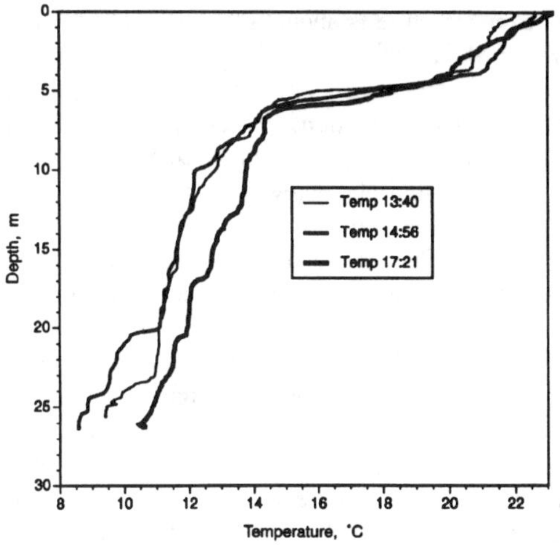

Figure 1. Temperature profiles on 12 July 1995. The times are indicated in the diagram.

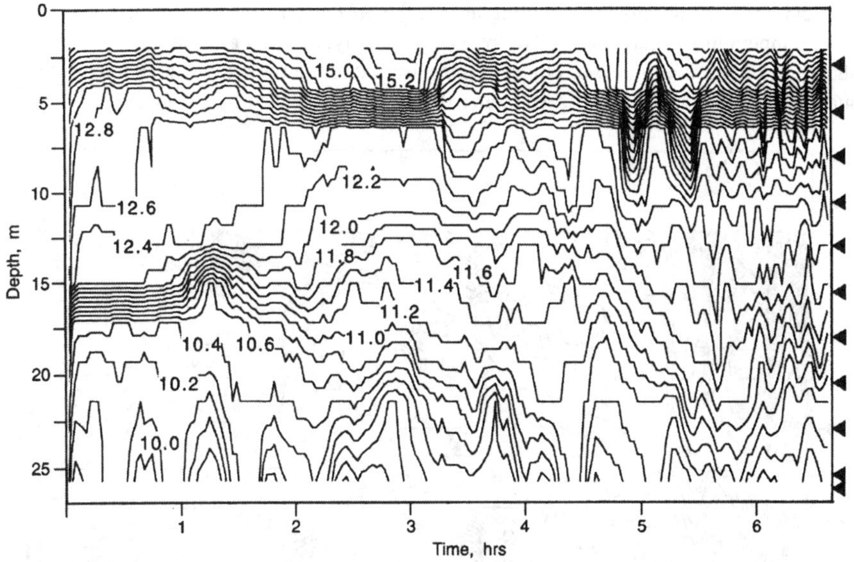

Figure 2. Isopycnals at 0.2°C spacing from thermistor chain data on 17 July 1995. The water depth is 26.5 m. The sensor depths are marked on the right and are 26.4, 25.5, 23.0, 20.5, 18.0, 15.5, 13.0, 10.5, 8.0, 5.5, and 3.0 m.

$2\pi/\sigma_c$, estimated between 8 and 25.5 m depth is 27.7 ± 2.0 min, close to the observed period. There is evidence of the vertical migration of regions of high and low density gradient, regions where the temperatures at neighbouring sensors are well separated or close together, respectively. Downward migration of the 'sheets and layers' in the temperature profiles (e.g. as shown in Figure 1) at a speed, c_z, of about 0.25 cm s^{-1}, is more clearly apparent in a series of CTD profiles made at 5 min intervals from 13:40 to 17:30 on 17

July. The periodicity of the structure is about 100 min giving a vertical wavelength of 15m (from $c_z = \sigma/m$, where σ is the frequency and m is the vertical wavenumber).

A 6 hr record from the thermistor chain moored in 28 m depth on 12 July is shown in Figure 3. The first 2.5 hrs show oscillations exceeding 2 K in the upper part of the chain (4.5 m to 14.5 m depth) and of isotherm displacement height exceeding 4 m. The oscillation period is about 40 min and shows coherence throughout the upper 8 sensors with a predominently first mode wave structure. Higher frequency waves are however apparent in the bottom 3.5 m. Spectral analysis of the record from a thermistor at 1.1 m off the bed reveals a spectral peak at 23 ± 2 min, close to the critical period and to the first harmonic of the overlying oscillations.

A Neil Brown current meter, 35 m from the thermistor chain, 0.8 m off the bottom and in 28 m depth, also captured these oscillations (Figure 4). The current fluctuations are predominantly up and down the slope (i.e. north-south) and are 90° out of phase with the temperature fluctuations; the temperature fluctuations are caused by water being carried up and down the slope with little along-slope motion. The largest horizontal north (or up-slope directed) current component is 10.6 cm s^{-1} and that to the south is 6.5 cm s^{-1}.

b. Bottom temperature pulses

A vertical array of thermistors sampling at 0.3, 0.6, 1.0 and 1.5 m off the bed was used to resolve temperature variations at 1 Hz frequency. One typical 'event' lasting about 10 min is shown in Figure 5. A pulse of colder water moves through the array; the temperatures fall sequentially through the sensors by some 0.6 K in 1 min. The fall at individual sensors

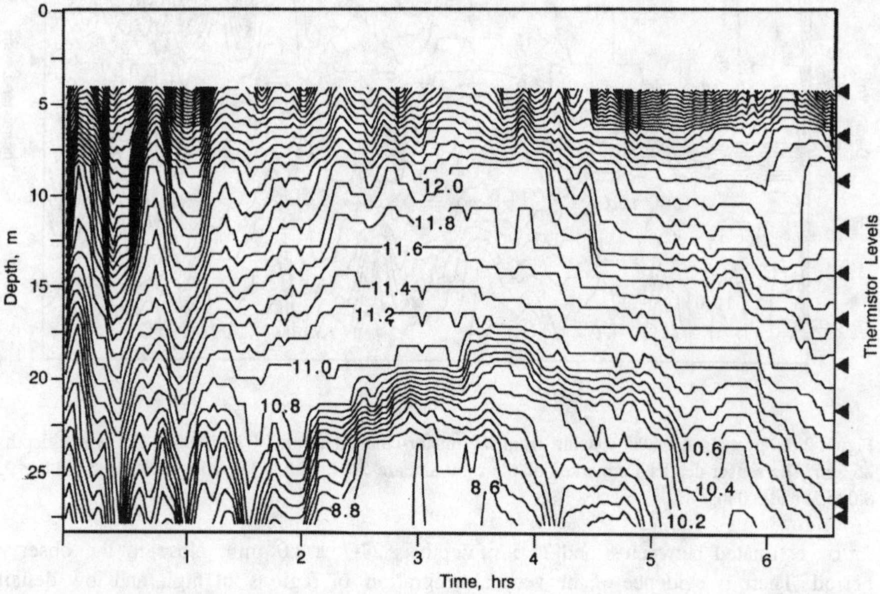

Figure 3. Isopycnals at 0.2°C spacing from the thermistor chain data on 12 July 1995 (see also Figure 1). The water depth is 28 m. The sensor depths (shown on the right) are 27.9, 27.0, 24.5, 22.0, 19.5, 17.0, 14.5, 12.0, 9.5, 7.0, and 4.5 m.

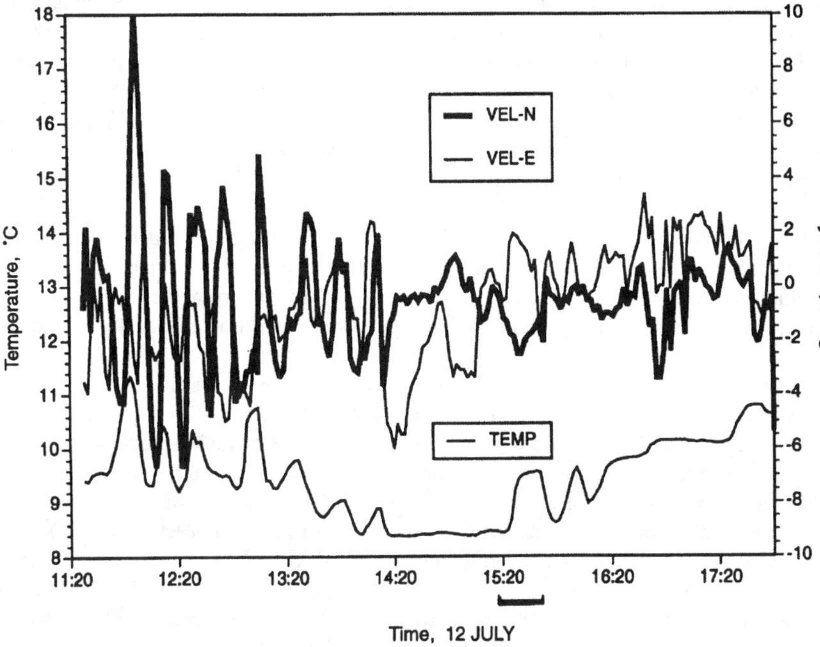

Figure 4. Time series of temperature and north and east (up-slope and along-slope, respectively) components of current measured by a Neil Brown (EG&G) current meter moored in 28 m water depth and 0.8 m off the bottom on 12 July 1995. Data are averaged over periods of 2 mins. The section includes the recovery of the instrument at 17:50. The horizontal line marks the period of data shown in Figure 6.

occurs more rapidly, at a rate of 0.3 K in 10 s. The fall at 0.3 m occurs 40 s before that at 1.5 m, giving an upward vertical component of velocity the cold water of 3 cm s^{-1}. If the isopycnals are horizontal this implies an up-slope encroachment speed of the cold water of 12 cm s^{-1} on a 17° slope. (It is more likely that the isopycnals are tilted, but the speed is within the range of those seen in Figure 4.) The motion is transient; the temperatures rise back to within 0.1 K of their original values in a further 8 min. The asymmetry of the temperature variation, the rapid fall in temperature and relatively slow rise, has the character of the interfacial waves studied in the lab by Wallace and Wilkinson (1988) and of reflecting internal waves (Thorpe, 1992). The large change at 15:20 has a smaller precursor which begins 20 s after 15:18. It extends only to 0.6 m off the bed but has a structure similar to the main event.

c. Near-bed vertical currents

A pulse-to-pulse coherent acoustic Doppler velocity profiler (ADVP) was mounted on gimbals on a 1 m square plate to point vertically upwards. The plate was placed on the sloping lake bed at 25.8 m depth. The ADVP (see Lhermitte and Lemmin, 1994) operates at a frequency of 1 MHz and with a pulse repetition frequency of 400 Hz. The transducer face was about 25 cm above the lake bottom. The first sampling gate is 15 cm above the transducer face. The 50 gates of the profiler, each with a height of 1.2 cm, therefore measure the vertical velocity between about 40 cm and 100 cm above the lake bed. Twenty-five minutes of data were recorded on the boat to which the transducer is linked via a coax-cable. During this period starting at 15:16, 12 July temperatures are rising (Figures 3 & 4), indicating a general motion of water down the slope and consistent with the currents to the SE (Figure 4). Although the record is short, the data are of such an interesting nature that a brief account must be given.

A contour plot of the time series is given Figure 6a. Data are averaged over 5 s intervals. Negative vertical velocities are directed upward, away from the slope. Throughout the recording period there is a layer between 40 cm and 48 cm from the bed in which the vertical velocity is continually upwards. The speed varies with a periodicity of about 2 to 2.5 min. The maximum speed exceeds 2.5 cm s^{-1}. Immediately above this layer of upward motion is another layer in which the velocity is predominantly positive (downward). This has a thickness of about 6 cm. Maxima occur in this layer, sometimes concurrently with those in the lower layer. Further from the lake bed, the velocity becomes small and uniform; this uppermost layer has little temporal structure. Figure 6b shows the relative backscattering intensity on a relative logarithmic scale. (Absolute values of acoustic scat tering cross-section cannot be derived from the present data). A region of higher scattering is seen in the first 5 min of the record extending to 45 cm range (85 cm off the bed), but this collapses to lower levels, although a band of high scattering persists in the lower layer below 8 cm range (40 to 48 cm off the bed). The vertical layering of the profile is also seen in mean profiles from sections of the recording shown in Figure 7.

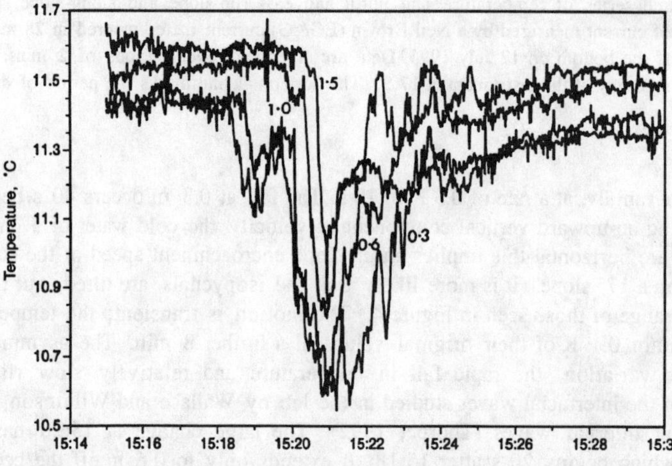

Figure 5. Temperatures recorded at 0.3, 0.6, 1.0 and 1.5 m off the lake bed for 15 min starting at 15:15, 17 July 1995. The very rapid fall in temperature and more gradual recovery is typical of the records made close to the bed. (The original data was 'spikey', a consequence of instability in the signal amplifiers. Spikes were 'single-valued' and well-spaced. Removal was achieved by identifying values greater than their right and left neighbours by more than 0.05K, a value found suitable by experience, and replacing them by the mean of the right and left neighbours. This process was repeated once. It did not affect the pattern of change illustrated in the figure).

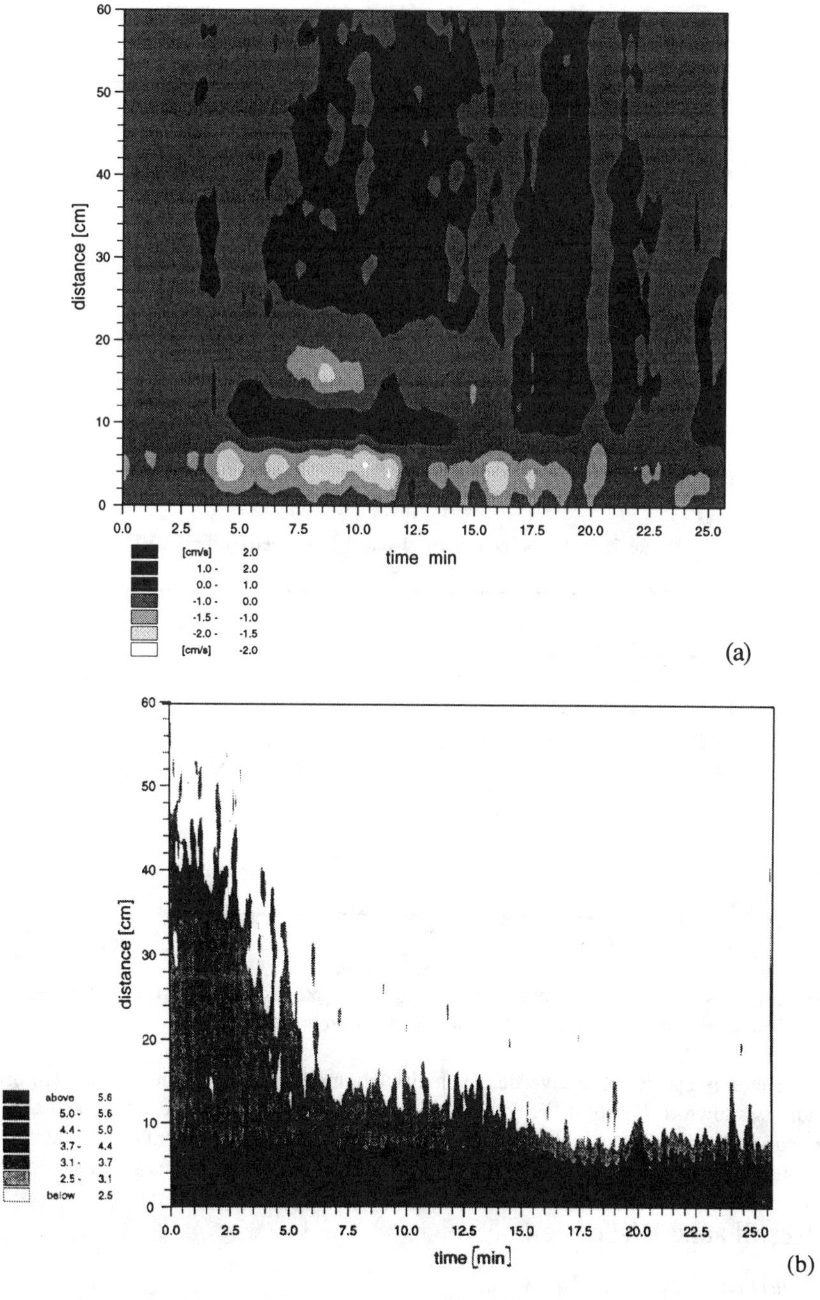

Figure 6. Contour plots of (a) the vertical velocity, and (b) the acoustic backscatter (log scale) from the ADVP. The horizontal scale is time measured from 15:16, 12 July 1995. The vertical scale is distance measured from 40 cm above the sloping lake bed at a depth of 25.8 m. The recording period is marked on Figure 4.

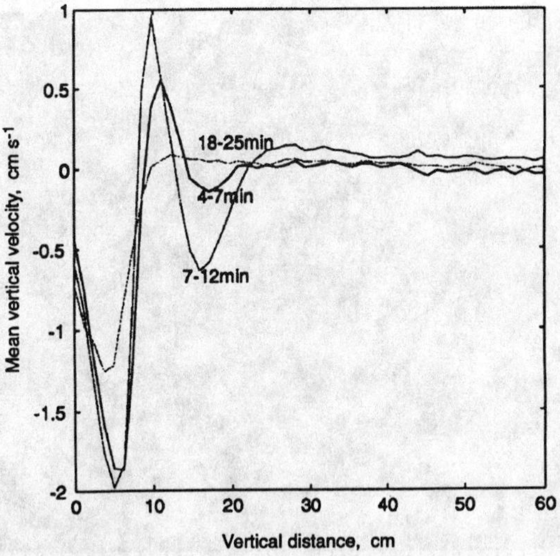

Figure 7. Profile of mean vertical velocity obtained by the ADVP.

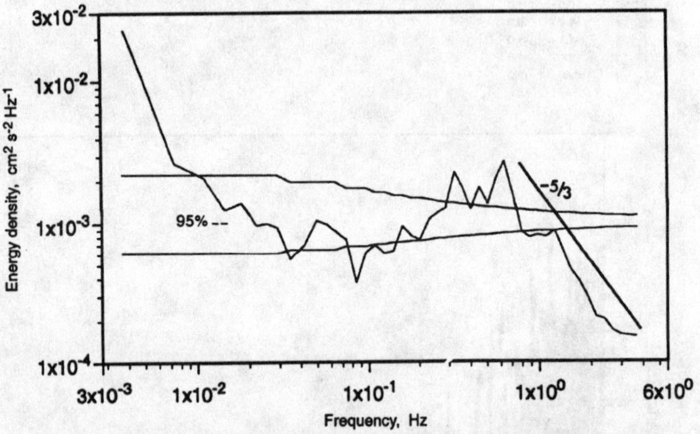

Figure 8. Energy density spectrum of the vertical velocity at gate seven of the ADVP calculated for the whole time series. Gate seven is located at a height of 8.5 cm in Figure 6.

The energy spectra of the vertical velocity component at a height of 48.5 cm off the bottom is shown in Figure 8. Peaks occur at periods between 1.8 and 3.5 s, and at 0.9 s. The spectrum follows a -5/3 law at higher frequencies, consistent with an inertial subrange. The implications of these, and other, observations are discussed in section 4d.

4. Results and Discussion

a. Internal Waves in Mid-water

The vertical temperature profiles in the wedge of water between the lake surface and the sloping bed are strongly modulated by alternating layers of high and low gradient which appear to migrate vertically downward through the water column (section 3a). Such vertical propagation has been observed before in lakes by Lazier (1973; see also Lazier and Sandstrom, 1978) and is known to mark the presence of vertically propagating internal waves which distort the thermal structure through which they travel. The low frequencies observed, $\sigma < \sigma_c$, implies that these waves are reaching the wedge from greater depths, travelling upwards before reflection from the thermocline. Vertical propagation of density structure is seen in lab experiments as waves reflect from the sloping boundary (e.g. see Thorpe and Haines, 1987, Figure 12, although there $\sigma > \sigma_c$). It is possible that the layers near the side of Lake Tahoe observed in individual vertical profiles by Caldwell et al. (1978) are caused by reflecting internal waves rather than by intrusions from the lateral boundary.

b. Critical Internal Waves

The observations shown in Figures 2, 3 and 4 are evidence for energetic waves at frequencies close to the critical in the lower 2 m of the water column. The magnitude of the horizontal current components in Figure 4 indicates up-slope (parallel to the bed) currents reaching 10.5 cm s^{-1}. The mean of about 6 cm s^{-1} with period 23 min indicates up-slope displacement amplitudes of 13.2 m, or vertical displacements of 3.9 m. The data displayed in Figure 3 suggest that the component amplified on reflection from the sloping boundary is a first harmonic of the first mode internal wave seen in the upper part of the water column. It appears that although the frequency of a low frequency incident wave (composed of primary components and subharmonics; Thorpe, 1968) is sub-critical, leading to off-shore reflection, its harmonic components may be close to 'critical' as found here. Cases must occur when the harmonics are super-critical so that, on reflection, they will continue to propagate towards shallow water. A significant effect is that all waves with frequencies near the critical, σ_c, *whatever* their direction of incidence on the slope (which could even be parallel to isobaths), reflect to a direction close to the line of greatest slope. The large current fluctuations resulting from near-critical reflected components are therefore almost directly up and down the line of greatest slope, here in the north-south direction, in accordance with the observations.

c. Evidence of Pore Water Heating

CTD profiles were made down to the very bed of the lake. The inversion in the bottom few centimetres of TEMP 1725 in Figure 1 was a commonly observed feature when the overlying temperature gradient is very stable. It is believed to be a consequence of the warmer water remaining in the loose and porous sediment as it is covered by colder water during the upslope phase of the wave-induced motion. As the CTD sensor reaches the bed, it penetrates the upper sediment, recording its slightly higher temperature. The warmer water in the sediment is of lower density than that of the overlying colder water, and a state of slow convective rise of water from the warm sediment may occur close to the sediment-water interface.

d. Turbidity Current

The continued vertical upward and downward motions recorded by the ADVP (Figure 7), the largest being upwards and averaging about 1.5 cm s^{-1} for 25 min suggest, at first sight, vertical displacement of some 22.5 m. This, not to mention the apparent convergence of the vertical flow, is plainly impossible; the mean vector velocity must parallel to the slope with components up and down the 17° slope, of which only the vertical components are sensed by the ADVP.

The presence of decaying oscillations of the flow with height above the lake bed, albeit for a short period of recording, is intriguing. Various mechanisms which might have contributed or led to the structure include intrusive flow from mixing at the boundary (see for example Phillips et al., 1986), the layers which form over a plate moved horizontally or oscillating in a stratified fluid (Martin and Long, 1968; Kildal, 1970), and the presence of internal waves with very high vertical wavenumber. Laboratory experiments on internal waves in uniformly stratified fluids 'breaking' on slopes have revealed two patterns of secondary flow, vortices with axes parallel to isobaths (Cacchione and Wunsch, 1974; Ivey and Nokes, 1989) and vortices with axes directed parallel to the line of greatest slope (Thorpe, 1987). Neither pattern appears to conform to the present observations.

The temperature is rising throughout the ADVP observation period, the arrival of warmer water implying a down-slope movement (if we neglect the possibility of along-slope advection of warmer water or solar heating) which is confirmed by the current meter record, Figure 4. Further evidence comes from

i) the CTD, 6 m east of the ACDP and 70 cm off the bottom at 25.8 m: a very sudden rise of 0.5 K occurred at 15:15 (Figure 9), just before the ACDP record started, and a more gradual rise continued afterwards; the light transmission falls sharply at 15:15 (Figure 9). The light transmission record is variable but it is notable that, during the period of the ADVP record, 'spikes' are present with a strong bias towards lower values.
ii) the Neil Brown current meter moored about 12 m away from the ADVP, and deeper, 0.8 m off the bottom at 28 m: it shows a rapid temperature increase of 0.3 K at 15:27 (Figure 4; the sensor response time is about 1 min) and a flow at about 3 cm s^{-1} down the slope to the southeast;
iii) the thermistor chain, also in 28 m depth, about 27 m away from the ACDP and 35 m west of the current meter (Figure 3): the lower sensors at 0.1 m and 1.1 m off the bottom show a gradual rise in temperature of about 0.7 to 1.0 K during the ACDP period, but not the abrupt rises seen in the other instruments.

If the ADVP is recording the vertical component of the upslope current, then an apparent vertical shear (dw/dz) which reaches about 2.5 cm s^{-1} over 5 cm, becomes a shear parallel to the slope of about 8.5 cm s^{-1} over 5 cm. The average is about 1 s^{-1}, and this persists for many minutes. For this large shear to be definitely linearly stable (i.e. gradient Richardson number, Ri > 1/4) a buoyancy frequency of at least 0.5 s^{-1} is required, or a temperature difference over the 5 cm shear layer of about 11 K. This is far greater than measured or possible. The flow must therefore be stabilised by suspended sediment. The concentration gradient required is 25×10^{-4} kg m^{-3} m^{-1}. Sediment flow would help account for the general downslope motion bringing the abrupt 'front' of warm water, first to the CTD and *later* to the deeper instruments, as well as providing the lower light transmission and evidence of occasional light-reducing particles or flocs (Figure 9) and the higher acoustic scattering in the ACDP record (Figure 6b). The first 5 min of this record appear to show the passing of the density current front. Scattering extends upwards almost twice the thickness of the following flow, a feature found in density currents (Simpson and Britter, 1979). The estimated gradient is only just enough to ensure marginal stability of the shear flow. The 2 min pulses seen in the ADCP speed record (Figure 6a) may be transient unstable billows advected by the mean flow indicating that the mean flow at 10 cm scales

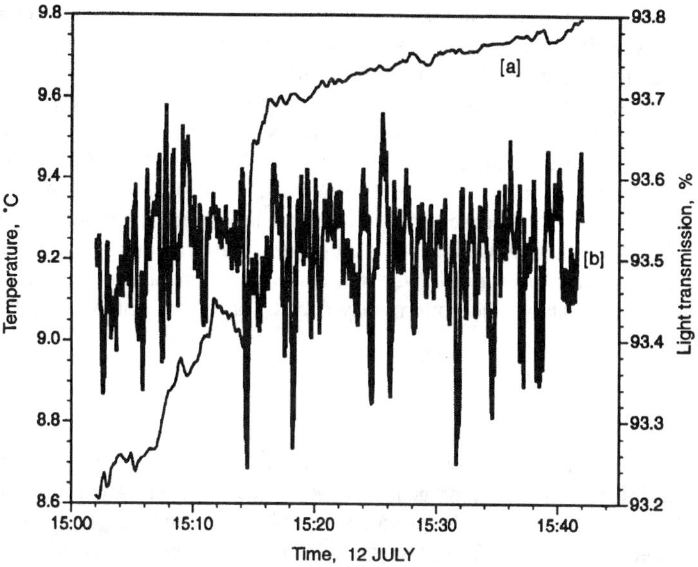

Figure 9. (a) the temperature and (b) the light transmission measured by a CTD suspended 70 ± 10 cm off the sloping lake bed from 15:02 to 15:42, 12 July 1995 in a depth of 26.3 m and about 6 m east of ADCP position. Light transmission is relative; the instrument is not calibrated.

is indeed marginal. The inertial subrange at high frequencies (Figure 8) suggests that at smaller scales the motion is turbulent.

The data captured in this short period are consistent with the movement downslope of a sediment-driven density or turbidity current, carrying with it fluid of greater-than-ambient temperature from shallower depths. The approximate speed of propagation of the temperature rise down the slope estimated from its passage from the CTD to the current meter is about 1.0 cm s^{-1} (2.2 m divided by sin 17° in 12 min). This appears small in comparison with the currents parallel to the slope inferred from the ADVP and those recorded by the current meter, but the two instruments were not directly in line down the slope, and the turbidity current front is unlikely to be two-dimensional even over the along-slope dimensions of 10 m separating the instruments. No abrupt temperature rise was seen at the thermistor chain, some 20 m to the eastsoutheast of the CTD.

Little quantitative information is however available for comparison about the motion of sediment-laden density currents on slopes underlying a stratified water column. McCave (1972) describes slow low density 'lutite' flows of sediments of concentrations of (1 to 10)$\times10^{-2}$ kg m^{-3} down slopes, and their tendency to run out along isopycnal surfaces as they loose negative buoyancy at increasing depth. McCave and Jones (1988) suggest that it may be possible to generate non-turbulent density currents when the sediment concentration is sufficiently high for the Richardson number of the current to be large. The concentrations inferred as being necessary to stablise the ADVP-measured flows are modest, at least by oceanographic standards. Studies of sediment-laden density currents travelling down uniform slopes in a uniform density fluid are described by Dade et al. (1994). Also of relevance to the present observations is the paper by Sparks et al. (1993) of a density current propagating over a horizontal plane in a stratified fluid. The fluid in the current has lower density than that of its surroundings, but the current is of greater net density because of its sediment load. As is shown in their Figure 7d, the deposition of sediment from the current results in the lifting-off and interleaving of fluid mixed in the shear flow behind the front of the density current, and of an eventual rise and intrusion of the front itself at a still higher level. On a slope, such release of the less dense, warmer fluid in the density current (inferred to be about 0.5 K higher than its surrounding from Figure 9) would lead to rising water and intrusions, and upslope and downslope flows above the density current, as observed here. Further laboratory experiments promised by Dade et al. (1994) but not yet made (Prof H.Huppert, private communication), are needed to clarify the physics and to establish whether, and in what conditions, flows of direction changing with height above the bottom may occur. The source of the turbidity current is not known.

5. Conclusions

Two separate and apparently distinct phenomena have been observed, internal waves with bottom intensification at or near critical frequencies, and a density current. The asymetrical 'events' illustrated in Figure 5 are reminiscent of surface gravity waves lapping at the water's edge on a beach. The presence of fluctuations at or near the critical frequency, documented here for the first time in a lake, demonstrates that there are processes present in the boundary layer which are driven by internal waves in the overlying water column with near critical frequencies. Under the calm conditions and for the relatively steep slopes investigated here, mixing from breaking of these waves did not appear to be very active, and is limited to a layer about 2 m thick adjacent to the lake bed. The bottom slope in the area of study, averaging 12.2°, is larger than the upper limit of 8.4° for second-order resonance between incident and reflected internal waves normal to a slope (i.e. lying in a plane perpendicular to the slope through a line of greatest slope) in a uniform density gradient, or of the 11.8° required for third-order resonance (Thorpe, 1987). Only waves approaching the 12.2° slope obliquely from a horizontal direction within an angle of 27° from the isobaths may be 'self-resonant' at second order (Thorpe, 1997). The results may therefore not be entirely typical in the more general conditions of small bottom slope. It may also be important that, if the spectral energy density of the internal waves is proportional to σ^{-2}, where σ is the wave frequency, the energy at the critical frequency ($\sigma_c = N \sin \alpha$) on an $\alpha = 17°$ slope will be about 1/9th of that on a 5° slope, and the corresponding incident wave amplitudes only 1/3rd. There is therefore reason to suppose that reflection at smaller slopes may have more violent consequences than observed here.

The chance observation of the sediment-driven current raises questions about the frequency of such phenomena. Following as it did a series of large amplitude waves near or at critical frequency in the lower layers of the water column, it is natural to suppose that the generation of the current may be a consequence of large currents produced by the internal waves. The effect of turbidity flows on the distribution of sediments, the generation of nepheloid layers and, since they transport heat downwards through the thermocline, their influence on the distribution of heat in lakes, and the exchange of water and heat with the underlying sediment, all deserve further study.

Acknowledgments. C. Perrinjaquet was in charge of the field work and much of the data processing and programming. We are grateful for his care and efficiency, and for funding for the project by the Swiss National Science Foundation, Grant 21-34 116.92.

References

Cachione, D. and C. Wunsch, Experimental study of internal waves over a slope. *J. Fluid Mech., 66*, 223-239, 1974.

Caldwell, D. R., J. M. Brubaker and V. T. Neal, Thermal microstructure on a lake slope. *Limnol. Oceanogr., 15*, 372-374, 1978.

Dade, W. B., J. R. Lister and H. E. Huppert, Fine-sediment deposition from gravity surges on uniform slopes. *J. Sedimentary Res. A 64*, 423-432, 1994.

Eriksen, C. C., Observations of internal wave reflection off sloping bottoms. *J. Geophys. Res. 87*, 525-538, 1982.

Ivey, G. N. and R. I. Nokes, Vertical mixing due to breaking of critical internal waves on sloping boundaries. *J. Fluid Mech., 204*, 479-500, 1989.

Kildal, A. On the motion generated by a vibrating plate. *Acta Mech., 9*, 78-84, 1970.

Lazier, J. R. N., Temporal changes in some freshwater temperature structures. *J. Phys. Oceanogr., 3*, 226-229, 1973.

Lazier, J. and H. Sandstrom, Migrating thermal structure in a freshwater thermocline. *J. Phys. Oceanogr., 8*, 1070-1079, 1978.

Ledwell, J. R. and A. Bratkovich, A tracer study of mixing in the Santa Cruz basin. *J. Geophys. Res., 100*, 20,681-20,704, 1995.

Lhermitte, R. and U. Lemmin, Open channel flow and turbulence measurement by high-resolution Doppler sonar. *J. Atm. Oceanic Tech., 11*, 1295-1308, 1994.

Martin, S. and R. R. Long, The slow motion of a flat plate in a viscous stratified fluid. *J. Fluid Mech., 31*, 669-688, 1968.

McCave, I. N., Patterns of fine sediment dispersion. In *Shelf sediment transport*, edited by D. J. A. Swift, D. B. Duane and O. H. Pilkey, 656pp, Dowden, Hutchinson & Ross, 1972.

McCave, I. N. and K. P. N. Jones, Deposition of ungraded muds from high-density non-turbulent turbidity currents. *Nature, 333*, 250-252, 1988.

Mortimer, C. H. Lake hydrodynamics, *Mitt. Intern. Verein. Limnol., 20*, 124-197, 1974.

Phillips, O. M., J. -H. Shyu and H. Salmun, An experiment on boundary mixing; mean circulation and transport rates. *J. Fluid Mech., 173*, 473-499, 1986.

Simpson, J. E. and R. E. Britter, The dynamics of the head of a density current advancing over a horizontal surface. *J. Fluid Mech., 94*, 477-495, 1979.

Sparks, R. S. J., R. T. Bonnecaze, H. E. Huppert, J. R. Lister, M. A. Hallworth, H. Mader and J. Phillips, Sediment-laden gravity currents with reversing buoyancy. *Earth & Planetary Sci. Letts., 114*, 243-257, 1993.

Taylor, J. R., Turbulence and mixing in the boundary layer generated by shoaling internal waves. *Dyn. Atmos. Oceans, 19*, 233-258, 1993.

Thorpe, S. A., On the shape of progressive internal waves. *Phil. Trans. R. Soc. A., 263*, 563-614, 1968.

Thorpe, S. A., On the reflection of a train of finite amplitude internal gravity waves from a uniform slope. *J. Fluid Mech., 178*, 279-302, 1987.

Thorpe, S. A., Thermal fronts caused by internal waves reflecting from a slope. *J. Phys. Oceanogr., 22*, 105-108, 1992.

Thorpe, S. A., On the interactions of internal waves reflecting from slopes. *J. Phys. Oceanogr., 27*, 2072-2078, 1997.

Thorpe, S. A. Some dynamical effects of internal waves and the sloping sides of lakes, in *Physical Processes in Lakes and Oceans,* edited by J. Imberger, pp 441-460, AGU, this volume, 1998.

Thorpe, S. A. and A. P. Haines, Appendix to Thorpe, S. A., On the reflection of a train of finite amplitude internal gravity waves from a uniform slope. *J. Fluid Mech., 178*, 279-302, 1987.

Wallace, B. C. and D. L. Wilkinson, Run-up of internal waves on a gentle slope in a two-layer system. *J. Fluid Mech., 191*, 419-442, 1988.

Wuest, A., D. C. Van Senden, J. Imberger, G. Piepke, and M. Gloor, Dyapycnal diffusivity meaured by microstructure and tracer techniques - a comparison. *Fourth International Symposium on Stratified Flows*, Grenoble, *Vol. 3*: B5 1-8, 1994.

Wunsch, C., On the propagation of internal waves up a slope. *Deep-Sea Res., 15*, 251-258, 1968.

Wunsch, C., Progressive internal waves on slopes. *J. Fluid Mech., 35*, 131-144, 1969.

33

Breaking of Super-Critically Incident Internal Waves at a Sloping Bed

I. P. D. De Silva, J. Imberger and G. N. Ivey

Abstract

A laboratory experiment was carried out to investigate the mixing on a sloping bed caused by a localised ray of internal waves breaking with super-critical frequencies. Previous work has shown that for near critical waves the turbulent boundary layer was confined to a thin region just above the bed. As a result, relatively small overturning turbulent lengthscales were observed. On the other hand, larger turbulent lengthscales and more mixing occurred at moderately super-critical waves, consistent with some earlier laboratory results. Calculation of the energy budget of the incident and reflected wave rays showed the vertical diffusivities as high as 10^{-4} m^2 s^{-1} observed in field studies could be realised by the internal wave breaking process at the boundaries.

Introduction

Field studies on natural lakes show significant vertical fluxes in the benthic boundary layer generated by the breaking of internal waves which can be responsible for much of the vertical transport of mass and momentum in the hypolimnion (Saggio and Imberger, 1995; Lemckert and Imberger, 1998). The lack of measurements in the abyssal seas and the mismatch in vertical diffusivities betweem those required for the abyssal budget and those frequently measured directly in the open ocean (e.g. Gregg, 1998), suggest that most of the mass flux may take place near the ocean boundaries. Armi (1978) first suggested that the turbulence was generated by mesoscale current drag against the sea-floor. This suggestion was shown to be unlikely by Garrett (1979 as the typical ocean stratifications and the conversion efficiency could not permit the high mixing rates suggested by Armi (1978) due to mean flows. Thus, internal wave breaking at the boundaries seems to be the most likely cause of the observed net vertical fluxes.

Armi (1978) suggested that the effective vertical diffusivity of a stratified basin may be written as $K = K_b A_r$, where K_b is the vertical diffusivity in the boundary layer, A_r is the ratio of horizontal area occupied by the boundary layer to the horizontal area of the basin. However, Garrett (1979) and Ivey (1987) pointed out that this required the bottom boundary layer to be stratified like the interior. For a basin of diameter L, with a boundary layer of thickness δ and slope angle β, the area ratio becomes $A_r \sim \delta/(L \sin \beta)$. Using

Physical Processes in Lakes and Oceans
Coastal and Estuarine Studies Volume 54, Pages 475-484

typical values for the ocean of $\beta \sim 4°$ (Bell, 1975), $\delta \sim 50$ m, $L \sim 4 \times 10^6$ m (Armi, 1978) we find that in the ocean $A_r \sim 1.8 \times 10^{-4}$; for lakes, $\beta \sim 10°$, $\delta \sim 2$ m, $L \sim 10 \times 10^3$ m, and thus $A_r \sim 1 \times 10^{-3}$. Thus, assuming comparable values of K_b in the ocean and in lakes, the boundary mixing process has a bigger impact in lakes than in oceans.

The no flux condition at the solid boundary requires the frequency of the reflected wave to be the same as the incident wave. The wave frequency ω of a small amplitude internal wave ray is related to the stable buoyancy frequency N by $\omega = N \sin \alpha$, where α is the inclination of the wave group velocity vector to the horizontal, and N is the background buoyancy frequency, defined by

$$N^2 = -\frac{g}{\rho_r}\frac{d\rho}{dz} \qquad (1)$$

Here ρ is the density, ρ_r is a reference density, g is the acceleration due to gravity and z is the vertical coordinate. Thus both the incident and reflected wave rays make equal angles to the vertical. Phillips (1977), Eriksen (1982, 1985) and Imberger (1994) have shown that an increase in energy density of the reflected ray results upon reflection. For the critical case where the group velocity of the reflected wave becomes parallel to the slope ($\alpha = \beta$), the reflected wave amplitude increases without bound, indicating a singularity in the linear theory. Eriksen (1982, 1998) illustrated this condition in the field, where an intensification of horizontal kinetic energy was observed. In what follows, for a given incident wave field of frequency ω, we define the parameter,

$$\gamma = \omega/\omega_c, \qquad (2)$$

where $\omega_c = N \sin \beta$ is the critical frequency.

At the boundary, a fraction of the energy of the incident wave is radiated away as a reflected ray and a part of the energy is lost to dissipation. The difference is converted into a buoyancy flux which accounts for the observed vertical transport of mass. The ratio of the amount of energy carried away by the reflected wave ray to that in the incident wave is defined as the reflection coefficient C_r. The exact dependence of C_r on the incident wave parameters such as the wave steepness and the geometry γ is not known. However, the numerical simulations of Javam et al. (1997a) showed that for a wide range of γ, C_r is approximately given by

$$C_r = -0.122\,\gamma^2 + 0.855\,\gamma - 0.613. \qquad (3)$$

Another parameter which is not accurately known is the mixing efficiency R_f, defined as the ratio of the increase in potential energy due to mixing divided by the incident energy flux. Ivey and Nokes (1989, 1990) demonstrated that R_f is strongly dependent on γ. Further, their experiments showed that $R_f \approx 0$ for $\gamma < 0.7$, $R_f \approx 0.2$ for $\gamma \approx 1$ (cf. Taylor 1993) and a maximum of R_f was observed around $\gamma \approx 1.2$. Numerical simulations of Slinn and Riley (1996) found that for $\gamma \approx 1$, $R_f \approx 0.35$ and $C_r \approx 0.10$.

The non-linear analysis of Thorpe (1987), showed that the interaction of incident and reflected waves can produce higher modes with frequencies larger than the ambient buoyancy frequency, thus generating evanescent modes. The same feature has also been observed for two intersecting wave rays in the numerical simulations of Javam et al. (1997b) and laboratory experiments of Teoh et al. (1997). These studies showed that wave reflection at a boundary possessed many similarities to the non-linear interaction of two intersecting wave beams.

Experiments

The experiments were conducted in a tank of dimensions 590 cm × 54 cm × 60 cm. At one end of the tank a 2 cm thick plexiglass sheet was used as the sloping bed. This sheet was pivoted about a horizontal axis 12 cm above the tank bottom, so that the slope angle could be varied. The internal wave rays were generated using a triangular shaped folding paddle, adapted from the original design of McEwan (1973). The paddle was made up of eight 5 cm long by 53 cm wide, hinged blades that spanned the width of the tank (for details see Teoh et al., 1997 and De Silva et al., 1997). The internal wave ray tube radiating from the paddle was approximately 1.5 wavelengths wide (Figure 1) while the wavelength was approximately 20 cm. The paddle was mechanically linked to an eccentric wheel, driven by a heavy-duty, adjustable D.C. motor. By varying the eccentricity of the wheel, the paddle amplitude could be varied. In this study, we report a series of runs with a paddle amplitude of 3.1 cm.

The tank was filled with a linearly stratified salt solution using a standard two-tank technique and the depth of the water column was approximately 48 cm. The temperature of the fluid column was measured using a fast response FP07 thermistor while the salinity was measured using two sensors: a fast response micro-scale conductivity probe for turbulence measurements and a suction probe for mean conductivity measurements. Vertical casts of temperature and conductivity were obtained by traversing the sensors through the fluid column at a set speed of 10 cm s^{-1} using a computer controlled linear translator. Both the direct and differentiated output from each sensor was recorded at a sampling frequency of 100 Hz through a 16-bit analog-digital converter.

The paddle oscillating frequency was chosen according to the stratification and α. The experiments were started by commencing the paddle oscillations from a rest state and the oscillations were maintained steadily throughout the duration of the experiment.

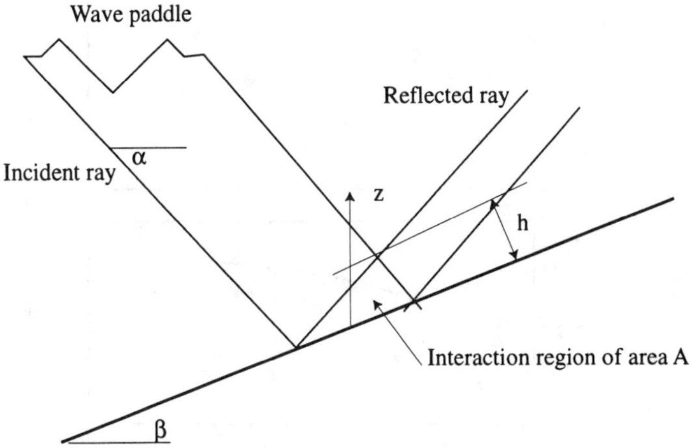

Figure 1. Schematic of internal wave reflection.

Results

For forward reflecting waves ($\gamma > 1$), the non-dimensional height normal to the bed h/λ, and the non-dimensional area A/λ^2, of the incident and reflected wave interaction region, can be shown to be

$$\frac{h}{\lambda} = \frac{\sin(\alpha - \beta)}{2\cos\alpha}, \tag{4a}$$

$$\frac{A}{\lambda^2} = \frac{\tan\alpha \sin(\alpha - \beta)}{4\sin(\alpha + \beta)}, \tag{4b}$$

by geometric linear ray theory, where A is the area of the interaction region in the vertical plane and λ is the horizontal wavelength of the incident wave field (Figure 1). Thus, for a given α the depth of the interaction region increases from zero at the critical condition to a maximum of $(\lambda \tan \alpha)/2$ when $\beta = 0$. Figure 2 shows how h/λ and A/λ^2 increase with increasing γ for a given incident wave field. In this example $\alpha = 56°$, the condition relevant to the runs described later in the paper. Both h and A initially increase rapidly as the incident wave becomes progressively super-critical, asymptoting to constant values for large γ. For example, from $\gamma = 1.5$ to $\gamma = 3$ the increase in h and A is approximately two-fold, whereas from $\gamma = 3$ to $\gamma = 6$ the increase is relatively small.

Flow visualisation images of Ivey et al. (1995) and De Silva et al. (1997) indicated that near critical conditions the motion was dominated by a thin, sheared, parallel flow just above the bed and superimposed on the background ambient wave field. This observation is consistent with the result in (4), that the interaction area in such situations becomes rather small. On the other hand, at super-critical frequencies the visualisation images

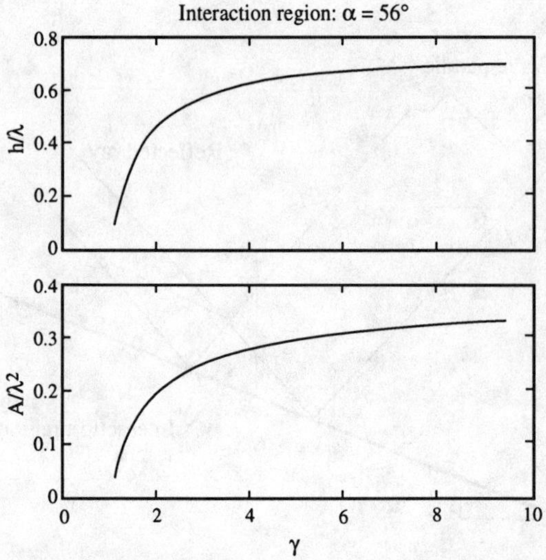

Figure 2. The variation of (a) the non-dimensionless extent of the interaction area (h/λ) and (b) the non-dimensionless area (A/λ^2) of the interaction region with γ.

showed that this strong shearing flow extended farther away from the bed. Work by Thorpe (1987) and Javam et al. (1997a) have shown that the boundary reflection process may be considered as the interaction of waves which comprise the incident, the reflected and the higher harmonics in the interaction region. According to Teoh et al. (1997), such interactions produce waves at frequencies higher than the ambient buoyancy frequency and these evanescent wave modes can lead to instabilities by absorbing energy from the primary waves through a non-linear, non-resonant interaction mechanism.

Turbulent Quantities

The turbulent quantities described here were obtained from the vertical casts of the temperature, density and their respective gradient profiles. The estimation of turbulent dissipation was central to the analysis, hence it is described first.

The estimation of turbulent kinetic energy dissipation rate ε by fitting the Batchelor spectra to the spectra of the measured temperature gradient signal using ε as a free parameter has been common in field studies (Dillon and Caldwell, 1980; Imberger and Boashash, 1986; Luketina, 1987) and in laboratory studies (Teoh, 1997; De Silva et al., 1997). The portion of the wavenumber spectra from the maximum down through the roll-off region was used in a curve-fitting procedure (e.g. Luketina, 1987). A typical measured temperature gradient spectra and the fitted spectra are shown in Figure 3 where the number of points used for the spectra was 128. With the sensor travelling at speeds of 10 cm s^{-1} and a sampling rate of 100 Hz, the Nyquist wavenumber cut-off was 500 cpm.

An overturning lengthscale characterising the turbulence was estimated using the vertical casts from the density profiles. This process involved two stages. Firstly, the recorded profile $\rho(z)$ with overturns was monotonised to obtain the statically stable density profile $\rho_0(z)$, which was associated with the minimum potential energy. The vertical distances l_d each fluid parcel has to be displaced to achieve this monotonised profile are known as the Thorpe displacements (Thorpe, 1977). Using the values of l_d, a single lengthscale l_c characterising the energy containing eddies in a single turbulent patch was calculated, as outlined in Imberger and Boashash (1986). This centralised lengthscale l_c,

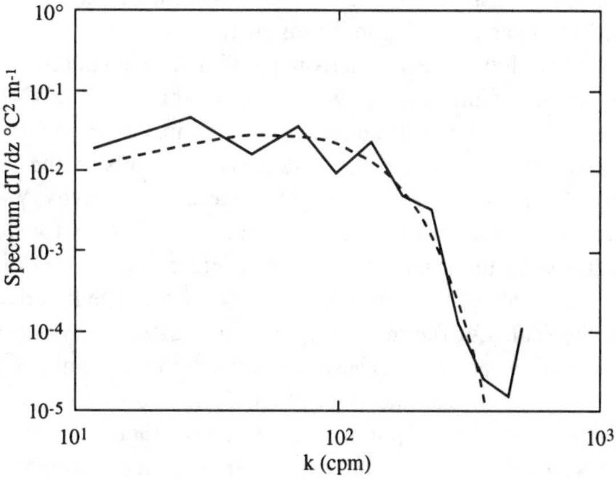

Figure 3. The wavenumber spectra of the temperature gradient signal and the fitted Batchelor spectra. The experimental conditions are, a = 3.1 cm, N = 0.615 rad/s, $\alpha = 32.5°$ and $\beta = 10.2°$.

is calculated by displacing the individual displacement scales l_d to the centre of the event by one-half of the l_d values themselves and taking the r.m.s of all values.

Based on the turbulent lengthscale l_c and dissipation rate ε, a turbulent velocity scale may be defined as $u = (\varepsilon l_c)^{1/3}$. The non-dimensional numbers associated with stratified turbulence may then conveniently be defined as (Ivey and Imberger, 1991; Imberger and Ivey, 1991; Imberger, 1994) as the turbulent Reynolds number, $Re_t = ul_c/\nu$, the strain Froude number, $Fr_\gamma = (\varepsilon/\nu N^2)^{1/2}$ and the turbulent Grashof number, $Gr_t = N^2 l_c^4/\nu^2$. These numbers can be successfully used to infer the mixing efficiency in stratified turbulence as shown by Imberger and Ivey (1991), for example. The above non dimensional numbers can also be interpreted as lengthscale ratios formed from l_c, the Ozmidov scale $l_O = (\varepsilon/N^3)^{1/2}$, where the buoyancy affects the motion, and the Kolmogorov scale $l_K = (\nu^3/\varepsilon)^{1/4}$, where the viscosity suppresses the motion, as $Fr_\gamma = (l_O/l_K)^{2/3}$ and $Re_t = (l_c/l_K)^{4/3}$. In general, a buoyancy flux occurs when there exists a range of overturning scales l_c between l_O and l_K. Gr_t describes the ratio of the buoyancy force which tries to bring a displaced fluid particle back to its neutral buoyancy level to the restraining forces due to viscous effects. Alternatively, it can also be related to the ratio of two timescales, the buoyancy timescale $t_b \sim N^{-1}$ and the viscous diffusion timescale $t_\nu \sim l_c^2/\nu$; thus $t_b/t_\nu \sim [l_c/(\nu/N)^{1/2}]^{-2} \sim Gr_t^{-1/2}$. When $t_b < t_\nu$, or $Gr_t > 1$, the buoyancy forces dominate over the viscous forces (Imberger, 1994).

Thorpe (1987) showed the importance of incident wave steepness as a parameter in defining the existence of statically unstable regions in the flow domain. His results indicated that for given β, the incident wave steepness at which a statically unstable region occurs increases as γ increases. The present incident wave parameters (the steepness, α, β) are, however, beyond the range covered in Thorpe (1987), and no direct comparison can be made. In this paper, we present a series of experiments keeping the incident wave steepness the same but varying slope angles to investigate the dependence of turbulent quantities on γ.

Results of De Silva et al. (1997) and Taylor (1993) indicated that there is a strong variability of turbulent properties within a wave cycle. This variability can be as high as two orders of magnitude (see Figure 12 of Taylor, 1993, for example). However, a useful overview may be obtained from the cycle averaged quantities, which are shown in Figure 4. As pointed out before, near critical conditions with $\gamma \approx 1$ the dominant feature of the interaction region was a highly sheared narrow parallel flow just above the bed as expected from (4). The thickness of this region was small so that relatively small displacement scales l_c were observed (practical difficulties associated in traversing the probe very close to the slope may have also contributed to the decrease of levels of the lengthscales in this case). The cause of the decrease in Fr_t for highly supercritical waves ($\gamma \approx 3$) is not clear, but it could have been due to the decrease in energy density available for the fluid to generate overturns, associated with the rapid increase in the interaction area (according to Figure 2b, the increase in area is about a factor three from $\gamma \approx 1.5$). The fact that $Gr_t < 1$ near $\gamma \approx 1$ clearly shows that the scales of the motion are strongly affected by the viscosity, whereas in the range $2 < \gamma < 3$, they are not. These non-dimensional numbers are cycle averaged, and one should not discount the fact that there is a strong variation in the turbulent quantities over a wave cycle. The point to note is that there is a certain portion of the wave cycle during which Fr_γ is high and it is this time period during which mixing likely takes place. More mixing seem to take place in the range $2 < \gamma < 3$ than it does for near critical waves.

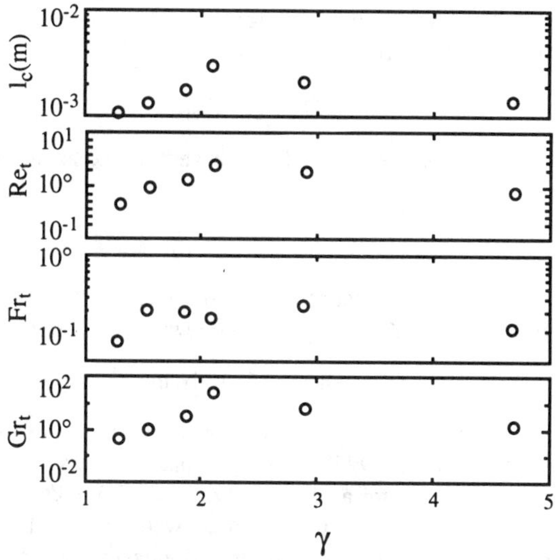

Figure 4. Variation of cycle averaged l_c, Re_t, Fr_t and Gr_t with the parameter γ. The first three panels are extracted from De Silva et al. (1997). The experimental conditions are a = 3.1 cm, N = 0.615 rad/s and $\alpha = 56°$.

Effective Vertical Diffusivity

It is of interest to oceanographers and limnologists to quantify the effective vertical exchange coefficients for mass and momentum for the turbulence generated by the internal wave breaking process. Relatively high diffusivities, for example of the order of 10^{-4} m^2 s^{-1} compared to the typical interior values of 10^{-5} m^2 s^{-1}, have been reported near the boundaries (Ledwell and Hickey, 1995; Lemckert and Imberger, 1998). Due to the horizontal pressure gradient between the locally mixed region at the boundary and the ambient stratification, boundary mixed fluid can intrude horizontally into the fluid interior along isopycnals. For the present configuration, De Silva et al. (1997) showed that the intrusion follows a viscous-buoyancy balance and spreads into the main fluid body with a length $l \propto t^{5/6}$ power law, as schematically shown in Figure 5. In natural water bodies, these intrusions carry nutrient rich benthic water into the interior enhancing the plankton growth.

The amount of vertical mixing generated by the breaking waves depends on the nature of the incident wave field, the strength of the stratification and the bottom geometry. Here we present a simple energy argument to estimate an effective turbulent vertical diffusivity for the wave breaking process. Consider an incident wave field of amplitude a and frequency ω, approaching a sloping bed; the wave energy per unit mass is of the order a^2N^2, so that the amount of wave energy available at the bed per unit mass per unit time is of the order $a^2N^2\omega$. For super-critical waves part of this energy is reflected in the form of a reflected wave at the bed. Denoting the ratio of the amount of energy taken away by the reflected wave to that incident at the bed by C_r, the amount of energy trapped in the interaction region would be $a^2N^2\omega(1 - C_r)$. Following Ivey and Imberger (1991), the mixing efficiency R_f is defined as the ratio of the buoyancy flux to the amount of energy available for mixing. Thus for the internal wave breaking in the interaction region, the buoyancy flux b can be written as,

$$b \sim a^2 N^2 \omega (1 - C_r) R_f. \tag{5}$$

Now it is possible to determine the local, effective vertical exchange coefficient for mass as $K_{eff} = b/N^2$, thus,

$$K_{eff} \sim a^2 \omega (1 - C_r) R_f. \tag{6}$$

The difficulty in using the above equation lies in the fact that the exact dependence of R_f and C_r are not known. Equation (6) for K_{eff} may be rewritten in the form,

$$K_{eff} \sim a^2 \gamma (1 - C_r) R_f N \sin\beta. \tag{7}$$

The only available information on the R_f dependence on γ is due to the work of Ivey and Nokes (1990) (see also Ivey et al, 1995). Taking a conservative value of $R_f = 0.2$, C_r from (3) and using the typical values of for a lake, a = 2 m, N = 10-3 rad/s, $\beta = 10°$ and $\gamma \approx 2$, we get $C_r \approx 0.6$, and $K_{eff} \approx$ 10-4 $m^2\ s^{-1}$, which agrees well with the high diffusivity levels observed near boundaries in some field experiments (Lemckert and Imberger, 1998; Wüest and Gloor, 1998).

Conclusions

We have described an experimental study of the breaking of a ray of internal waves on a sloping bed at varying bed angles. Previous work (De Silva et al., 1997) showed that near critical conditions the flow was confined to a thin region in the vicinity of the slope superimposed on the ambient wave motion. As a result, relatively small turbulent lengthscales, and hence small turbulent Reynolds numbers, were observed. The mechanism for off slope instabilities seems to be due to the interaction of incident and reflected waves (Thorpe, 1987; Teoh et al., 1997; Javam et al., 1997a,b); this was absent at critical conditions. Within the range of bed angles considered, the largest lengthscales occurred in the range $1.5 < \gamma < 2$. For highly supercritical waves ($\gamma > 3$), the overturning length-scales again decreased. Simple energy budgets in the interaction region reveal a possibility of realising enhanced vertical diffusivity for the wave breaking process.

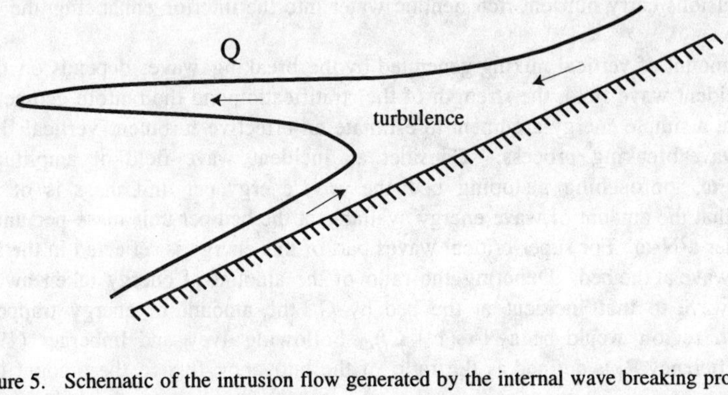

Figure 5. Schematic of the intrusion flow generated by the internal wave breaking process.

However, the comparison of field experiments to the present laboratory experiments should be done with caution, as there is a strong variability in the turbulence field within a wave period.

Acknowledgments. This work was supported by the Australian Research Council and the Centre for Environmental Fluid Dynamics program. Most of the work presented here was presented at the 1995 IUTAM symposium in Broome, Australia. The authors gratefully acknowledge constructive comments by the participants of the conference. One of us, IPDeS would like to thank Joe Fernando for his continued encouragement. This paper is Centre for Water Research reference ED-1004-PDS.

References

Armi, L. Some evidence for boundary mixing in the deep ocean. *J. Geophys. Res., 83*, 1971-1979, 1978.

De Silva, I. P. D., J. Imberger, and G. N. Ivey. Localised mixing due to a breaking internal wave ray at a sloping bed. *J. Fluid Mech.*, 350, 1-27, 1997.

Dillon, T. M. and D. R. Caldwell, The Batchelor spectrum and dissipation in the upper ocean. *J. Geophys. Res., 85*, 1910-1916, 1980.

Eriksen, C. C. Observations of internal wave reflection off sloping bottoms. *J. Geophys. Res., 87*, 525-538, 1982.

Eriksen, C. C. Implications of ocean bottom reflection for internal wave spectra and mixing. *J. Phys. Oceanogr., 15*, 1145-1156, 1985.

Eriksen, C. C. Waves, mixing, and transports over sloping boundaries, in *Physical Processes in Lakes and Oceans,* edited by J. Imberger, pp 402-424 this volume, 1998.

Garrett, C. Comments on "Some evidence for boundary mixing in the deep ocean." by L. Armi. *J. Geophys. Res., 84*, 5095, 1979.

Gregg, M. C. Estimation and geography of diapycnal mixing in the stratified ocean, in *Physical Processes in Lakes and Oceans,* edited by J. Imberger, pp 294-326, this volume, 1998.

Imberger, J. Transport processes in lakes: A review, in *Limnology Now: A Paradigm of Planetary Problems*, edited by R. Margalef, pp 99-193, Elsevier, 1994.

Imberger, J. and B. Boashash, Application of the Wigner-Ville distribution to temperature gradient microstructure: A new technique to study small-scale turbulence. *J. Phys. Oceanogr., 16*, 1997-2012, 1986.

Imberger, J. and G. N. Ivey, On the nature of turbulence in a stratified fluid. Part II: Application to lakes. *J. Phys. Oceanogr., 21*, 659-680, 1991.

Ivey, G. N., The role of boundary mixing in the deep ocean. *J. Geophys. Res., 92*, 11,873-11,878, 1987.

Ivey, G. N. and R. I. Nokes, Vertical mixing due to the breaking of critical internal waves on sloping boundaries. *J. Fluid Mech., 204*, 479-500, 1989.

Ivey, G. N. and R. I. Nokes, Mixing driven by the breaking of internal waves against sloping boundaries. *Proc. Intl. Conf. on Physical Modelling of Transport and Dispersion*, MIT, Boston, 11A3-11A8, 1990.

Ivey, G. N. and J. Imberger, On the nature of turbulence in a stratified fluid. Part I: The energetics of mixing. *J. Phys. Oceanogr., 21*, 650-658, 1991.

Ivey, G. N., I. P. D. De Silva, and J. Imberger, Internal waves, bottom slopes and boundary mixing. *Aha Hulikoa winter workshop*. Honolulu, Hawaii, 199-205, 1995.

Javam, A., J. Imberger, and S. Armfield, Numerical study of internal wave overturning on a sloping boundary. Revised and resubmitted to *J. Fluid Mech.*, 1997a.

Javam, A., J. Imberger, and S. Armfield, Numerical study of internal wave-wave interactions. Revised and resubmitted to *J. Fluid Mech.*, 1997b.

Ledwell, J. R. and B. M. Hickey, Evidence for enhanced boundary mixing in the Santa Monica basin. *J. Geophys. Res., 100*, 665-679, 1995.

Lemckert, C. and J. Imberger, Turbulent boundary layer mixing events in fresh water lakes, in *Physical Processes in Lakes and Oceans,* edited by J. Imberger, pp 503-516, AGU, this volume, 1998.

Luketina, D. A. Frontogenesis of freshwater overflow. PhD dissertation. University of Western Australia, 1987.

McEwan, A. D. Interactions between internal gravity waves and their traumatic effect on a continuous stratification. *Boundary-Layer Met., 5*, 159-175, 1973.

Phillips, O. M., Dynamics of the upper ocean. Cambridge University Press., 1977.

Saggio, A. and J. Imberger, Internal wave climatology in lake Biwa. In Preparation, 1995.

Slinn, D. N. and J. J. Riley, Turbulent mixing in the oceanic boundary layer due to internal wave reflection from sloping terrain. *Dyn. Atmos. Oceans, 23*, 51-62, 1996.

Taylor, J. R., Turbulence and mixing in the boundary layer generated by shoaling internal waves. *Dyn. Atmos. Oceans, 19*, 233-258, 1993.

Teoh, S. G., G. N. Ivey, and J. Imberger, Laboratory study of the interactions between two internal waves. *J. Fluid Mech., 336*: 91-122, 1997.

Thorpe, S. A., Turbulence and mixing in a Scottish loch. *Phil. Trans. Roy. Soc. London, A, 286*, 125-181, 1977.

Thorpe, S. A., On the reflection of a train of finite-amplitude internal waves from a uniform slope. *J. Fluid Mech., 178*, 279-302, 1987.

Wüest, A. and M. Gloor, Bottom boundary mixing in lakes: The role of the near-sediment density stratification, in *Physical Processes in Lakes and Oceans,* edited by J. Imberger, pp 485-502, AGU, this volume, 1998.

34

Bottom Boundary Mixing: The Role of Near-Sediment Density Stratification

A. Wüest and M. Gloor

Abstract

The turbulent dynamics and stratification of bottom boundary layers, as well as the net diapycnal buoyancy flux in the deep water, have been observed to vary strongly among lakes. The most relevant parameters governing the different regimes are the bottom current stress and the rate of release of dissolved solids from the sediment. The ratio of boundary-induced mixing to the density flux associated with the flux of ions from the sediment determines whether the bottom boundary layer is extremely stably stratified or well-mixed. The aim of this contribution is (1) to demonstrate these two boundary phenomena, (2) to give a physical criterion for assessing the two mixing regimes, (3) to present a potential model to quantify the boundary-induced buoyancy flux and the basin-wide diapycnal diffusivity, and (4) to test the model with data from two representative lakes with significantly different deep-water mixing characteristics.

The Stratification of the Bottom Boundary in Lakes

Limnologists often observe a distinctly different vertical stratification of water properties within the region directly above the sediment of lakes. Even though it is evident that this fact has implications for the ecology of the entire aquatic system, little attention has been paid so far to the detailed physical mechanisms within lacustrine bottom boundaries. Therefore, compared to the ocean, where the role of the bottom boundaries for basin-wide diapycnal mixing and transport has frequently been recognized and investigated (Wunsch, 1970; Armi, 1978; Garrett, 1979; Phillips et al., 1986; Ivey, 1987; Thorpe, 1988; Garrett, 1990; Toole et al., 1994), the number of studies in lakes is rather small (Imberger and Patterson, 1990; Imboden and Wüest, 1995).

In the first part of the paper we present data sets of currents and stratification from the bottom boundary of two lakes which are subjected to distinctly different external forcing and hence show drastically different deep-water mixing characteristics. The bottom boundary layer is considered here as the zone above the sediment which is predominantly affected by the stress due to bottom friction. The first example is from Alpnachersee (Figure 1), which is representative for lakes in which the bottom boundary layer is mostly

Physical Processes in Lakes and Oceans
Coastal and Estuarine Studies Volume 54, Pages 485-502

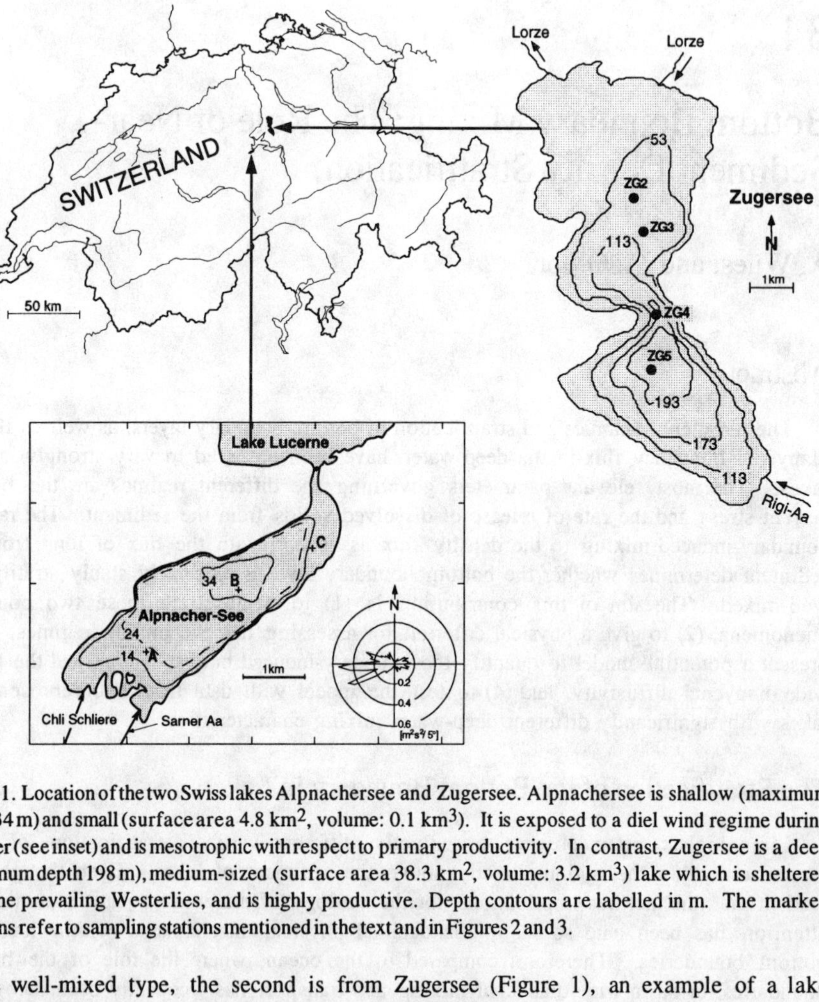

Figure 1. Location of the two Swiss lakes Alpnachersee and Zugersee. Alpnachersee is shallow (maximum depth 34 m) and small (surface area 4.8 km^2, volume: 0.1 km^3). It is exposed to a diel wind regime during summer (see inset) and is mesotrophic with respect to primary productivity. In contrast, Zugersee is a deep (maximum depth 198 m), medium-sized (surface area 38.3 km^2, volume: 3.2 km^3) lake which is sheltered from the prevailing Westerlies, and is highly productive. Depth contours are labelled in m. The marked positions refer to sampling stations mentioned in the text and in Figures 2 and 3.

of the well-mixed type, the second is from Zugersee (Figure 1), an example of a lake showing very strong near-sediment density stratification.

In the second part, we discuss the physical characteristics of the two boundary layers with extremely different stratification. Finally, we evaluate the relevance of bottom boundary stratification for the basin-wide diapycnal fluxes by presenting and applying a boundary mixing model. As the results of the model are consistent with basin-wide tracer diffusivity observations, we conclude that the model provides a reasonable description of the diapycnal fluxes in stratified enclosed water bodies with simple shapes.

(a) The "Alpnachersee case": Internal Seiches as the Source of Bottom Boundary Mixing

Alpnachersee is a small and relatively shallow (max. depth = 34 m) side-basin of Lake Lucerne (see morphometry in Figure 1). In summer, the mountain and valley breezes along the nearby mountains result in a predominantly diel wind that blows regularly parallel to the major axis of the lake (Figure 1; for details see Münnich et al., 1992). Under such conditions, internal seiching and corresponding deep-water currents are excited in this lake (Gloor et al., 1994), as illustrated in Figure 2 for a two-month period during the summer of 1994.

In June and July 1994, the bottom boundary current oscillated with an amplitude of about 3-5 cms^{-1} and sporadically reached maximum speeds of up to 6-7 cms^{-1}. Power spectral analysis (Press et al., 1986) of the current and isotherm time series reveal that the two dominant periods (of about 8 and 24 hours: Gloor et al., 1994), correspond to the first vertical first horizontal and second vertical first horizontal seiche modes, schematically shown in Figure 3 and described mathematically by Münnich et al. (1992). For both modes, bottom currents along the major axis of the lake correlate well with the displacements of the hypolimnetic isotherms, as shown in detail by Gloor et al. (1994).

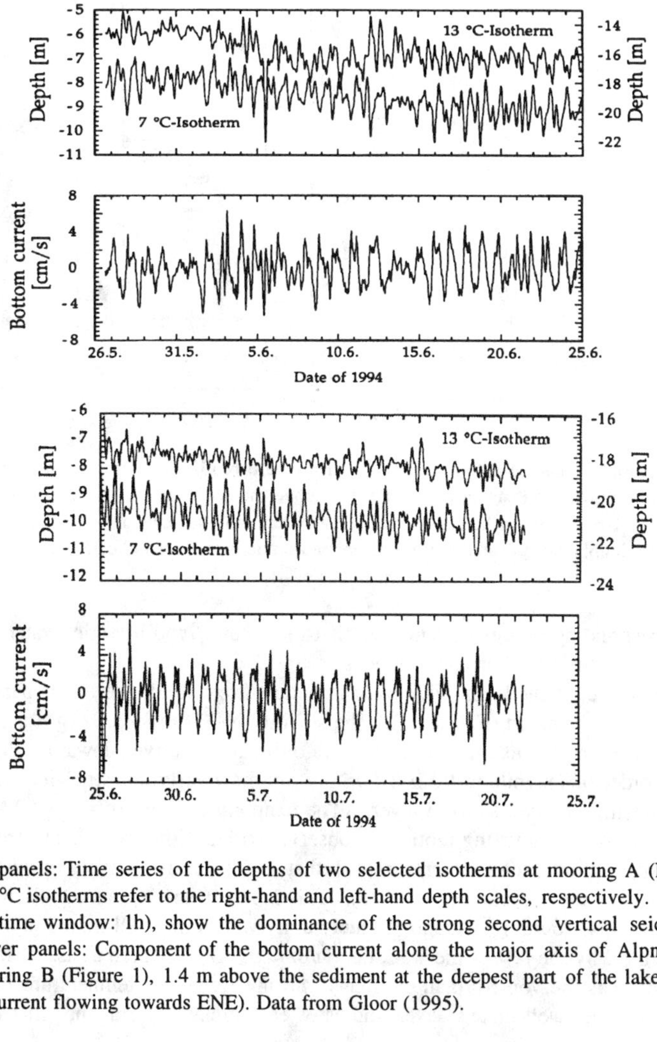

Figure 2. Upper panels: Time series of the depths of two selected isotherms at mooring A (Figure 1). The 7 °C and 13 °C isotherms refer to the right-hand and left-hand depth scales, respectively. The data, slightly filtered (time window: 1h), show the dominance of the strong second vertical seiche mode (Figure 3). Lower panels: Component of the bottom current along the major axis of Alpnachersee, measured at mooring B (Figure 1), 1.4 m above the sediment at the deepest part of the lake (positive values indicate current flowing towards ENE). Data from Gloor (1995).

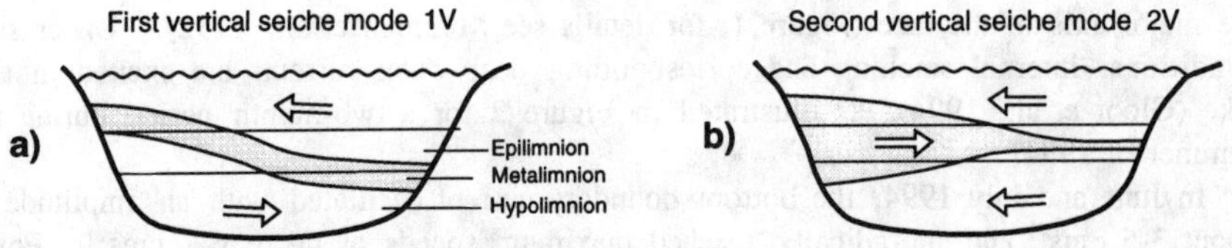

Figure 3. Schematic illustration of the vertical structure of the displacement of the first vertical (two-layer), and the second vertical (three-layer) seiche modes revealed in the data of Figure 2. The arrows indicate flow directions after maximum displacement of the layers.

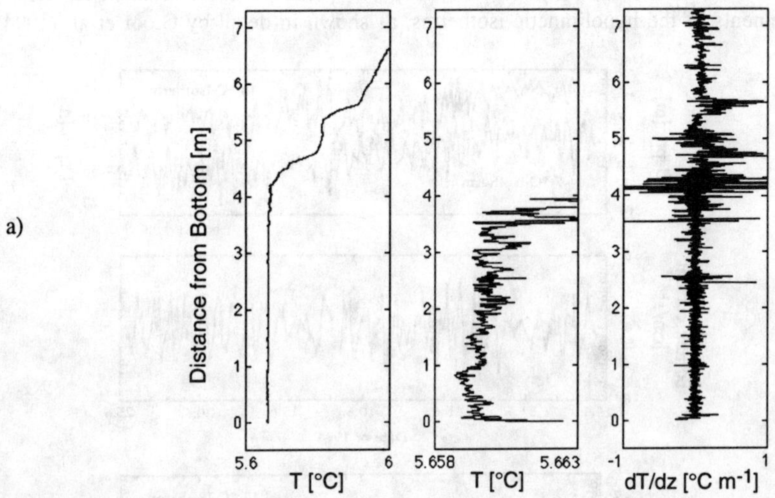

Figure 4a. An example of a temperature microstructure profile measured at mooring B (Figure 1) on June 2, 1994, down to a distance of about 8 cm above the sediment. In the well-mixed bottom layer of thickness $\delta_{mix} \approx 4.2$ m (left-hand panel) the temperature is homogeneous to within about 1 mK (middle panel). The double-sided gradient of the temperature microstructure indicates active mixing (right-hand panel).

The corresponding phases demonstrate that the whole hypolimnetic water body is excited by seiches of the first horizontal modes.

As an effect of the basin-wide hypolimnetic current field, the lower part of the bottom boundary is well-mixed (Figure 4a). The stability $N^2 = -g\rho^{-1}\,\partial\rho/\partial z$ (g = acceleration due to gravity; ρ = water density; z = vertical co-ordinate, positive upwards) dropped by more than an order of magnitude from $2\cdot10^{-5}$ s^{-2} in the overlying stratified water column to $7\cdot10^{-7}$ s^{-2} within the well-mixed layer. (The temporal average of $\approx 1.5\cdot10^{-6}$ s^{-2} is twice as large as this due to seiching motion as observed from within a Eulerian frame). Temperature microstructure measurements reveal that the layer is actively mixed by turbulence generated near the sediment (Figure 4a). This is not surprising, as the Richardson gradient number $Ri = N^2\,(\partial u/\partial z)^{-2}$ is approximately unity within the well-mixed layer if the shear is approximated by the law of the wall, i.e. $\partial u/\partial z = u_*(kz)^{-1}$ (where u_* is the friction velocity and k = 0.41 is the von Kàrmàn constant). Strong signals of temperature microstructure at the top of the well-mixed layer indicate entrainment from the thermally stratified overlying water column. In the thermocline, double-sided signals, characteristic of active mixing, are only occasionally observed (as in Figure 4a).

For 300 temperature microstructure profiles taken along the main axis of the lake, the thickness δ_{mix} of the well-mixed layer was determined semi-automatically (and checked visually) by a gradient criterion for an abrupt change in the temperature profile. The thickness was observed to vary with the phase and intensity of the seiching motion and the depth of the lake bottom at the site. The frequency distribution of the occurrence of the mixed layer thicknesses (Figure 4b) shows that the arithmetic mean of the thickness of the well-mixed layer $<\delta_{mix}> = \int \delta'_{mix} \cdot P(\delta'_{mix})\, d\delta'_{mix}$ was about 2.6 m in the deepest part and 0.6 m at mid-depth in Alpnachersee (where $P(\delta'_{mix})\, d\delta'_{mix}$ is the frequency of occurrence of the thickness δ'_{mix} to $\delta'_{mix} + d\delta'_{mix}$ of the well-mixed layer). Due to the seiching motion, averaging δ_{mix} at maximum depth underestimates the true thickness, which is maximally about 4 to 5 m (Figure 4a,b).

(b) The "Zugersee Case": Stratification by Dissolved Solids Emanating from the Sediment

A completely different density structure has been found in Zugersee, a medium-sized lake with a maximum depth of 198 m (see morphometry in Figure 1), characterized by high primary productivity and ineffective vertical mixing at large depths (Wehrli et al., 1994). Due to the correspondingly high gross sedimentation and mineralization rates of algal matter, significant quantities of dissolved solids (mainly HCO_3^- and Ca^{2+}) are released into the deep hypolimnion. Combined with the low wind exposure (the lake is sheltered from

b)

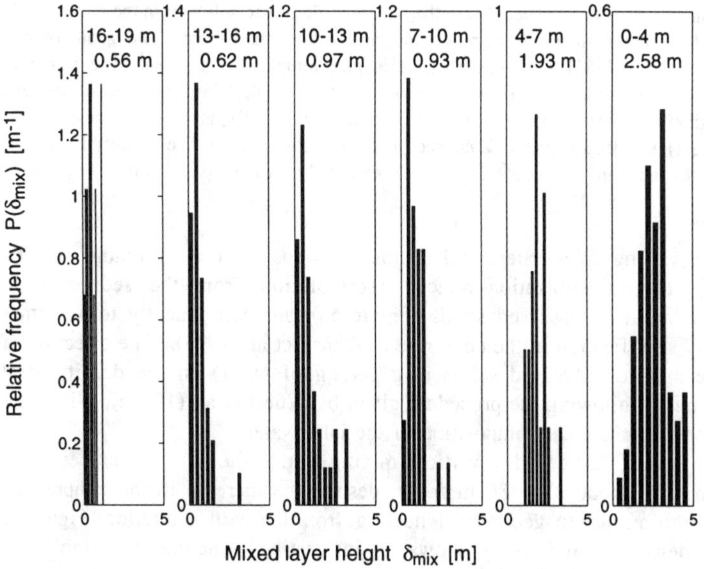

Figure 4b. Relative frequency of occurrence of the thickness δ_{mix} of well-mixed bottom layers such as that depicted in (a). The upper numbers indicate the height above maximum depth (i.e. 16-19 m) and the lower number represents the average mixed layer thickness found in that depth range (i.e. 0.56 m). Data from Gloor (1995).

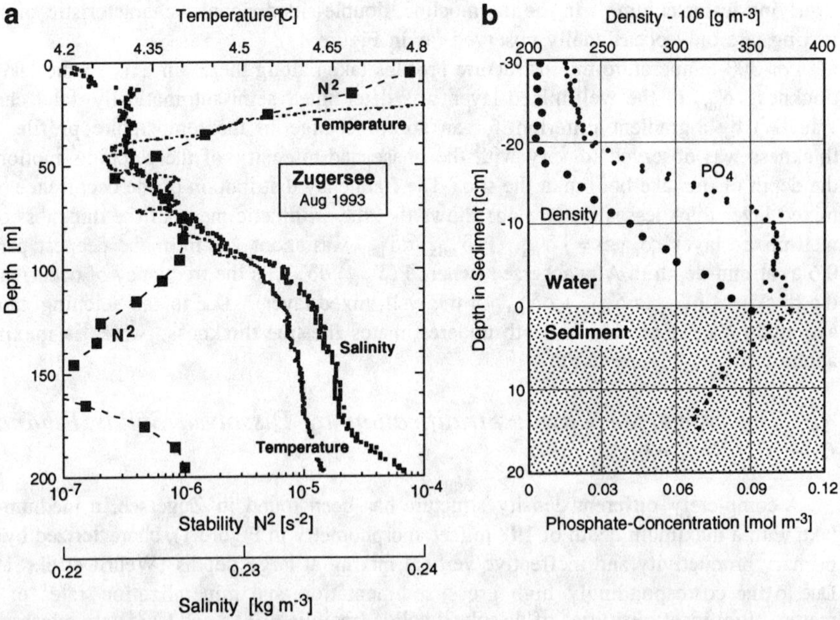

Figure 5. (a) Vertical profiles of temperature, salinity and stability N^2 (Position ZG5; Figure 1), measured in August 1993. The deep water of Zugersee, below about 80 m depth, is permanently biogenically density-stratified due to the release of dissolved solids from the sediment. (Structures above 80 m depth are remnants from mixing during the previous winter). The geothermal heat flux ($\approx$ 0.1 Wm^{-2}; Finckh, 1981) is the cause of the inverse temperature gradient. The stabilizing effect of the dissolved solids is however several times greater than the destabilizing effect of the geothermal heat flux (Wehrli et al., 1995). Figure 5. (b) Profiles of density and PO_4 at the sediment interface at position ZG5 (Figure 1), the deepest part of Zugersee (note the vertical scale). The stability N^2 within the first 20 cm above the sediment is $\approx 8 \cdot 10^{-3}\ s^{-2}$, which is extremely high compared to the open-water stability.

the prevailing Westerlies, and southerly winds are only sporadic: see Imboden et al., 1988), the mineralization-induced redissolution from the sediments has led to an accumulation of dissolved solids (Figure 5a), and consequently to a permanent biogenic density stratification in the deep water (Wehrli et al., 1995). The effect of salinity (defined as the mass of dissolved solids in g per kg of water) on the density stratification was determined following the procedure given by Wüest et al. (1996b), using molal volumes of the specific chemical composition of the lake water.

In the last few decades, vertical mixing during the cold winter period has not reached further than about 50 to 80 m depth, despite the increase in the temperature in the lower hypolimnion due to geothermal heating from the earth's interior (Figure 5a). The deep-water density stratification shows an interesting structure: the stability N^2 reaches a minimum of $\approx 10^{-7}\ s^{-2}$ at 160 m depth, but increases again to $\approx 10^{-6}\ s^{-2}$ towards the maximum depth (198 m). A microscopic view of the sediment-water interface reveals an even stronger gradient just above the sediment surface. Within an extremely stratified 20 cm thick near-sediment layer (Figure 5b), the stability N^2 jumps by several orders of magnitude to $\approx 8\cdot10^{-3}$ s^{-2}. The density profile in Figure 5b was determined based on HCO_3^- and Ca^{2+} concentrations measured by the peeper technique (Brandl and Hanselmann, 1991) with a 1.5 cm vertical resolution.

Such strong gradients indicate that turbulent mixing above the sediment is being drastically reduced to nearly molecular level. This can be concluded by two arguments: (1) The time required to accumulate the measured ion concentration of HCO_3^- and Ca^{2+} within the 20 cm thick stratified near-sediment layer is comparable to the seasonal time scale. Given a rate of release of dissolved solids from the sediment of $F_{sed} \approx (2 \text{ to } 3)\cdot10^{-6}$ $gm^{-2}s^{-1}$ (Wehrli et al., 1995) and the amount of ions (HCO_3^- and Ca^{2+}) contained within the 20 cm thick bottom layer shown in Figure 5b, a time scale of 2 months is obtained. (2) The molecular flux associated with the gradient of that layer is compatible (within a factor $\lesssim$ 2) to F_{sed}. This observation is also consistent with the phosphorus flux from the sediment and its accumulation shown in Figure 5b.

Having observed such strong stratification above the sediment, we were curious about the magnitude of the bottom current velocities in the deep hypolimnion. Measurements conducted at the deepest location during the period of low overall stratification in winter 1992-1993 revealed oscillating currents with typical amplitudes smaller than 2 cms^{-1} (Figure 6a,b). Unfortunately, during summer (the sampling time of the profile in Figure 5b), data collection at the deepest position failed. However, measurements near the mean lake depth during summer indicate that the current at the deepest location must have been even smaller (Figure 6c).

The Effect of the Release of Dissolved Solids on Mixing

As shown in the previous section, the near-sediment density gradients in Alpnachersee and Zugersee are completely different. In order to evaluate these differences we make use of the turbulent kinetic energy (TKE) balance. In non-stratified water, the production of TKE, driven by bottom friction, can be characterized by the scaling law (Dewey and Crawford, 1988)

$$\varepsilon = \frac{u_*^3}{k\ z} = \frac{C_{1m}^{3/2}}{k\ z}\, u_{1m}^3, \qquad [Wkg^{-1}] \qquad (1)$$

which quantifies dissipation ε as a function of the bottom velocity u_{1m} (1 m above bottom) and as a function of the distance z from the sediment ($u_* = C_{1m}^{1/2}\, u_{1m}$, and C_{1m} is the bottom drag coefficient). In turn, "negative" TKE is introduced by the buoyancy flux (local rate of change of potential energy) due to the flux of dissolved solids F_{sed} from the sediment. For $F_{sed} \approx 2.5\cdot10^{-6}$ gm^{-2} s^{-1}, the related buoyancy flux $J_b^{sed} = (g\beta/\rho)\cdot F_{sed} \approx 2\cdot10^{-11}$ Wkg^{-1} (β is the coefficient of haline contraction, with $\beta \approx 0.79$ ‰$^{-1}$ for calcareous water: Chen and Millero, 1986; Wüest et al., 1996b). The Monin-Obukhov length scale

$$L_{MO} = \frac{u_*^3}{J_b^{sed}}, \qquad [m] \qquad (2)$$

the distance above the bottom where the production of turbulence due to the bottom current u_{1m} and its consumption due to the buoyancy flux J_b^{sed} are equal, provides a measure for the suppressing effect of the flux of dissolved solids (Figure 7). The comparison for Alpnachersee and Zugersee yields:

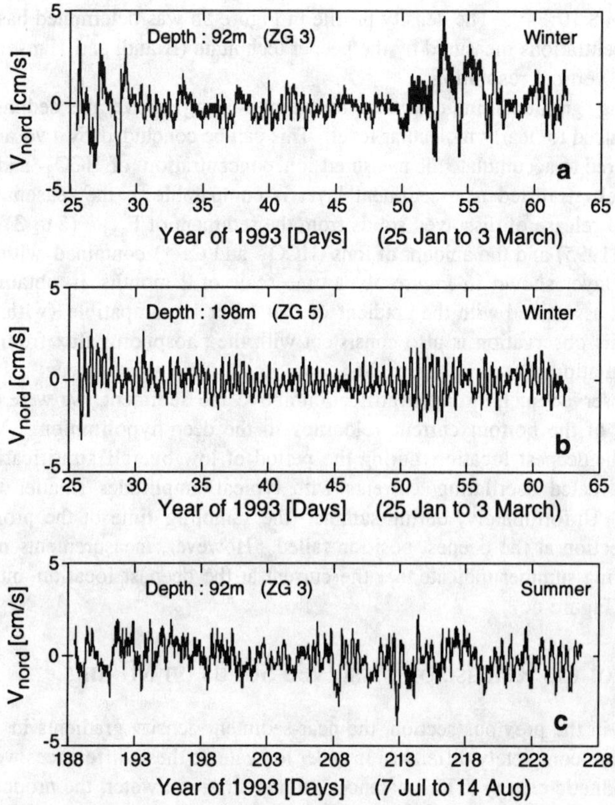

Figure 6. The component of the bottom current along the major axis of Zugersee during the winter considered in this paper (upper two panels) and the following summer period (lower panel). Measurements in the deepest part were carried out in the southern basin (position ZG5; Figure 1), whereas measurements near the mean depth were performed in the northern basin (position ZG3; Figure 1). The current speed at the deepest location was slightly lower than that measured at the mean depth.

a) The "Alpnachersee case": For a typical bottom current velocity $u_{1m} \approx 2.6$ cms^{-1} (as determined from Figure 2, see details below), (2) yields $L_{MO} > 100$ m (Figure 7), a length which is several times larger than the depth of the hypolimnion. Consequently, biogenic stratification cannot build up in such a system, since bottom-induced mixing overcomes the stratifying effect of the dissolved solids.

b) The "Zugersee case": During the winter period under consideration, $u_{1m} \approx 1.7$ cms^{-1} (as determined from Figure 6a, see details below), and (2) yields $L_{MO} \approx 35$ m (Figure 7). The fact that L_{MO} is indeed much smaller than the thickness of the permanently stratified deep-water body, which typically extends from about 80 m down to the lake bottom (Figure 5a), is consistent with the long-term stability of the water mass. It means that deep bottom currents are not energetic enough to erode the permanent biogenic stratification over a long period of time. Indeed, an extreme storm event in 1982 was able to homogenize the deep water body – beginning at maximum depth – only minimally (Imboden et al., 1988).

The Monin-Obukhov argument explains the difference in large-scale vertical structure in the deep water between the two lakes, but not the difference in near-sediment density gradient. In order to understand why the internal seiche-induced bottom currents in Alpnachersee are apparently capable of redistributing the released ions, whereas in Zugersee the dissolved solids become accumulated above the sediment surface, we use another energy criterion. Laboratory experiments with density-stratified water have shown that the dissipation of turbulent kinetic energy has to fulfil the condition

$$\varepsilon > (15 \text{ to } 25) \cdot \nu N^2 \approx 20 \nu N^2 \qquad [\mathrm{Wkg^{-1}}] \qquad (3)$$

(Stillinger et al., 1983; Rohr et al., 1987) to produce "active" turbulence; i.e. irreversible diapycnal mixing with a non-vanishing buoyancy flux J_b (= $K_d N^2 > 0$). Equation (3) expresses the fact that for irreversible diapycnal mixing to occur (diffusivity $K_d > 0$), the turbulent eddies have to overcome the suppressing effects of viscosity ν and stability N^2.

In Figure 7, the energy criterion of (3) has been applied to the near-sediment stability N^2 of the two situations discussed above (Figures 4a and 5b); i.e. to $N^2 \approx 7 \cdot 10^{-7} s^{-2}$ (Alpnachersee) and $N^2 \approx 8 \cdot 10^{-3} s^{-2}$ (Zugersee). The corresponding turbulence levels $20\nu N^2$ (Alpnachersee: $\approx 2 \cdot 10^{-11}$ $\mathrm{Wkg^{-1}}$; Zugersee: $\approx 2 \cdot 10^{-7}$ $\mathrm{Wkg^{-1}}$) are compared to the turbulent dissipation of a steady, logarithmic boundary layer, as described by (1). It turns out that in the case of Alpnachersee, even for small currents of 1 $\mathrm{cms^{-1}}$, $\varepsilon = u_*^3/(kz)$ overcomes the critical activity level $20\nu N^2$ at a distance from the sediment as great as the maximum thickness $\delta_{mix} \approx 4$ m (Figure 7). However, in the case of Zugersee, a current of 3 $\mathrm{cms^{-1}}$ (which we regard as a long-term maximum in the deepest part of this lake) would not be

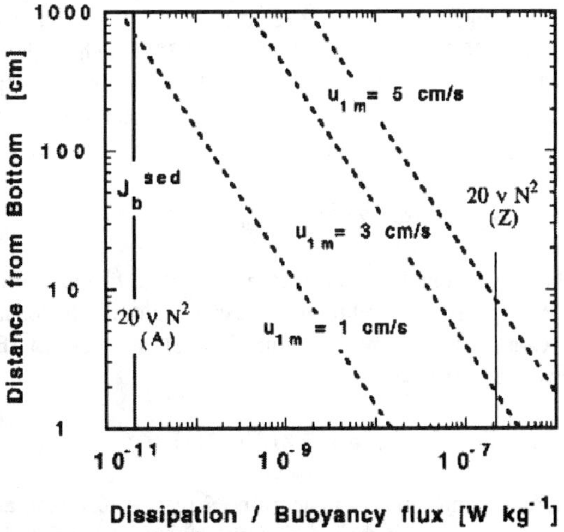

Figure 7. Rates of dissipation of turbulent kinetic energy calculated by similarity scaling (drag coefficient $C_{1m} = 1.6 \cdot 10^{-3}$) as a function of distance from the sediment, and as a function of the bottom current velocity u_{1m} (dashed lines), are compared to: (1) $J_b^{sed} \approx 2 \cdot 10^{-11}$ $\mathrm{Wkg^{-1}}$, the buoyancy flux due to the release of dissolved solids (black line, left), and (2) $20\nu N^2$ (thin lines), the criterion for active turbulence in stratified water, applied to the near-sediment stratification of Alpnachersee (A) and Zugersee (Z) (20 cm thick suppression layer) (note: J_b^{sed} coincides with $20\nu N^2$ in Alpnachersee). Monin-Obukhov lengths L_{MO}, the intersections of the lines of J_b^{sed} and ε, lay beyond the upper boundary of the figure (see text).

able to mix more than a layer of 2 cm above the sediment (Figure 7). As this scale is equal to the thickness δ_v of the viscous sublayer of a non-stratified fluid, given by $\delta_v = 10\nu/u_*$ (Hinze, 1975), we conclude that turbulence is suppressed in the 20 cm thick stratified near-sediment layer of Figure 5b. Consequently, the exchange within this layer is not turbulent, but drops progressively to the molecular level and thereby sustains the accumulation of dissolved solids. In the following, we will call this zone the "suppression" layer, the thickness of which will be denoted by δ_{sup}.

As an effect of suppressed turbulence and high density stratification above the sediment, the vertical exchange in the deepest part of Zugersee is expected to decrease. Indeed, analysis of the temperature profiles in Zugersee for a six-month winter period in 1992-93 (Müller, 1993) reveals a decrease in the diapycnal diffusivity K_d towards the sediment at the deepest part of the lake. In contrast, in Alpnachersee K_d is increasing in that zone. This difference will be discussed in the following two sections.

The Effectiveness of Bottom Boundary Mixing

In this section we evaluate the effect of the near-sediment stratification on the basin-wide (total bulk) diapycnal diffusivity in the deep-water body. The idea of expressing the basin-wide diapycnal diffusivity as a function of the mixing rate at the sediment boundary is not new, since Munk (1966) has already suggested that much of the ocean mixing might take place at the sloping boundary. Based on the observation of signatures of well-mixed layers advected away from sloping ocean bottom boundaries, Armi (1978, 1979) suggested estimating the basin-wide oceanic vertical diffusivity K_d by distinguishing between an interior diffusivity, K_d^I, and a boundary layer diffusivity, K_d^B. He proposed scaling the near-sediment diffusivity K_d^B, determined by the Rn-222 method (Broecker et al., 1968; Sarmiento et al., 1976; Chung and Kim, 1980), by the aspect ratio $\partial A(z)/\partial z \cdot \delta_{mix}(z) \cdot A^{-1}(z)$, which expresses the ratio of the ocean sediment surface of the mixed bottom layer $A_{sed}(z) = \partial A(z)/\partial z \cdot \delta_{mix}(z)$ at depth z (Figure 8) to the ocean cross-sectional area A(z):

$$K_d(z) = K_d^I(z) + K_d^B(z) \cdot \{\partial A(z)/\partial z \cdot \delta_{mix}(z) \cdot A^{-1}(z)\} \qquad [m^2\, s^{-1}] \qquad (4)$$

Garrett (1979, 1990) questioned the correctness of this formulation, since mixing of already well-mixed water near the sediment is less efficient. Instead, he proposed first determining the area-averaged buoyancy flux $J_b^{tot}(z)$ at depth z (Figure 8) and then calculating the diapycnal diffusivity

$$K_d(z) = J_b^{tot}(z) \cdot N^{-2}(z) \qquad [m^2\, s^{-1}] \qquad (5)$$

by using the background (interior) stratification $N^2(z)$. A different approach was made, following Phillips' (1970) solution, to express the diapycnal diffusivity K_d as a function of the thickness δ_{mix} of the mixed bottom layer and the diffusivity K_d^B within that layer. Garrett (1990) showed that the total buoyancy flux $J_b^{tot}(z)$ is reduced by an effectiveness factor (defined below), which takes into account the fact that the stratification is not maintained all the way to the sediment. As this effectiveness factor is strongly dependent on K_d^B, which is difficult to determine experimentally, alternative models have to be considered. Imberger and Ivey (1993) extended Phillips' (1970) solution, valid for a sloping boundary, to a horizontal sediment bottom, and found a dependence of $K_d \sim \delta_{mix}^9$.

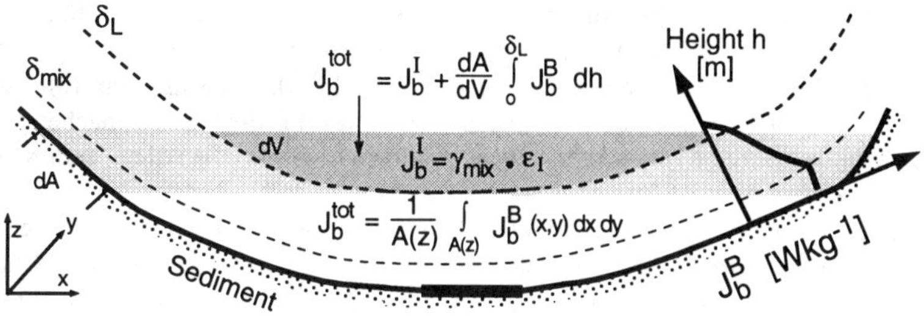

Figure 8. Schematic illustration of the boundary-layer model describing the total buoyancy flux J_b^{tot} as a function of the buoyancy fluxes J_b^I and J_b^B from the interior and from the bottom boundary, respectively. δ_L and δ_{mix} denote the thicknesses of the logarithmic and the well-mixed layer, respectively. h is the local height above the sediment. $\partial A/\partial V$ describes the sediment surface per unit volume of lake water as a function of depth.

This relation is of limited applicability, as the uncertainties in δ_{mix} present in spatially and temporally complex natural systems (Figure 4b), are amplified dramatically (e.g. a shift of 30% in δ_{mix} would lead to a change of K_d of an order of magnitude).

Faced with this unsatisfactory situation, we evaluate another boundary mixing model that is less sensitive to K_d^B and δ_{mix}. We follow Garrett's approach for the estimation of the total buoyancy flux $J_b^{tot}(z)$ at depth z by horizontally integrating the contributions from the interior $J_b^I(z)$ and the boundary $J_b^B(x,y,z)$, respectively, along the cross-sectional

$$J_b^{tot}(z) = J_b^I(z) + \frac{1}{A(z)} \int_{A(z)} J_b^B(x, y, z)\, dx\, dv \quad [\mathrm{Wkg^{-1}}] \tag{6}$$

area A(z), as depicted in Figure 8. Equation (6) is a simple balance for the potential energy and so far not specific to the model. The formulation of $J_b^B(x,y,z)$ is, however, dependent on the model and has to be specified. The basin-wide diapycnal diffusivity K_d is finally determined using (5).

To determine the buoyancy flux $J_b^B(x,y,z)$ in the boundary layer, we make the following model assumptions:

(a) The current in the bottom boundary follows a logarithmic profile up to a height δ_L, and the rate of dissipation is given by the scaling law $\varepsilon = u_*^3(kz)^{-1} = C_{1m}^{3/2}\, u_{1m}^3\, (kz)^{-1}$. Water beyond δ_L is considered as interior water, with a dissipation rate ε_I defined by interior processes.

(b) The buoyancy flux J_b^B in the stratified part of the water column (interior or boundary layer) is a constant fraction γ_{mix} of the dissipation rate ε; i.e. $J_b^B = \gamma_{mix} \cdot \varepsilon$.

(c) Within the extremely stratified suppression layer of thickness δ_{sup}, $J_b^B = 0$, ($\gamma_{mix} = 0$).

(d) The heat flux between sediment and water is neglected. Consequently, the buoyancy flux J_b^B decreases linearly to zero within a well-mixed layer when approaching the sediment (as does the heat flux).

Assumption (a) is based on several experimental observations (e.g. Dewey and Crawford, 1988; Weatherly and Martin, 1978; Weatherly and Wimbush, 1980), and assumption (c) is motivated by profiles of the type shown in Figure 5b. The mixing efficiency γ_{mix}, required for assumption (b), is surely not a constant. However, since no practical parameterization is available, we use the constant $\gamma_{mix} = 0.12$ (Peters and Gregg,

1988), which lies well within the range of other estimates (Osborn, 1980; Ivey and Imberger, 1991). In addition this choice is justified by the finding that γ_{mix} does not depend on the level of turbulence (Peters and Gregg, 1988). Assumption (d) has to be tested separately for each individual water basin (it is well fulfilled for Alpnachersee as the heat flux within the relatively shallow water column is much larger than sediment-water heat exchange).

The weak point among assumptions (a) to (d) is the poorly defined choice of the thickness of the logarithmic layer, which, based on experiment (Wüest et al., 1996a), has been chosen as $\delta_L = 10$ m. However, changes in δ_L by a factor of 2 modify the dissipation, and therefore the buoyancy flux $J_b^{tot}(z)$ (6), integrated over the logarithmic layer, by only 10%, as can easily be calculated from (7). Given all the uncertainties in the available data and the inhomogeneities of such natural basins, we consider this deficiency to be acceptable.

Based on the above assumptions (a) to (d), on the constants γ_{mix}, δ_L and $C_{1m} = 1.6 \cdot 10^{-3}$ (smooth bottom, Elliott, 1984), and on a specified topography, we are now able to determine the horizontally-averaged buoyancy flux $J_b^{tot}(z)$ from a given horizontal velocity field $u_{1m}(x,y,z)$ as a function of depth z using (6). In order to integrate (6), it is helpful to distinguish between (1) different distances from the maximum lake depth and (2) three different types of stratification.

Close to the maximum depth of the lake, the total buoyancy flux of (6) is made up entirely of contributions from within the logarithmic layer (Figure 8, lower layer). It is therefore not possible to simplify (6), and the integration has to be carried out by an appropriate procedure over the entire cross-sectional area A(z).

If, on the other hand, the vertical distance between z and the maximum depth is larger than the thickness δ_L of the logarithmic layer, (6) is simplified considerably. The horizontal area within the bottom boundary can be approximated by the sediment surface within the layer being considered (i.e. $dA \cdot A^{-1} = \partial A/\partial V \cdot dz$). In addition, the horizontal integration can be replaced by vertical integration over the thickness δ_L, as shown schematically in Figure 8. Consequently, the total buoyancy flux is then calculated by

$$J_b^{tot}(z) = J_b^I(z) + \frac{\partial A}{\partial V}(z) \int_0^{\delta_L} J_b^B(h,z)\, dh \qquad [\mathrm{Wkg^{-1}}] \qquad (7)$$

where h is the vertical co-ordinate originating at the local lake bottom (Figure 8).

For a bottom boundary layer stratified all the way to the sediment (i.e. to the viscous sublayer $\delta_v = 10\nu/u_*$), integration of (7) (neglecting small terms proportional to the sine of the bottom slope) yields

$$J_b^{tot}(z) = \gamma_{mix} \cdot \left[\varepsilon_I(z) + \frac{\partial A}{\partial V}(z) \frac{u_*^3}{k} \ln\left(\frac{\delta_L}{\delta_v}\right) \right] \qquad [\mathrm{Wkg^{-1}}] \qquad (7a)$$

where $J_b^B(h)$ has been replaced by $\gamma_{mix} \cdot \varepsilon(h)$ and $\varepsilon(h)$ by $u_*^3\,(kh)^{-1}$ (the dependence of u_*^3 and δ_L on depth z is not indicated). If the bottom boundary layer consists of a stratified and a well-mixed layer of thicknesses $\delta_L - \delta_{mix}$ and δ_{mix}, respectively, as observed in Alpnachersee (Figure 4a), then integration of (7) leads to

$$J_b^{tot}(z) = \gamma_{mix} \cdot \left[\varepsilon_I(z) + \frac{\partial A}{\partial V}(z) \left\{ \int_0^{\delta_{mix}} J_b^B(h,z)\, dh + \int_{\delta_{mix}}^{\delta_L} J_b^B(h,z)\, dh \right\} \right]$$

$$= \gamma_{mix} \cdot \left[\varepsilon_I(z) + \frac{\partial A}{\partial V}(z) \left\{ \frac{\delta_{mix}}{2} J_b^B(\delta_{mix}) + \frac{u_*^3}{k} \ln\left(\frac{\delta_L}{\delta_{mix}}\right) \right\} \right]$$

$$= \gamma_{mix} \cdot \left[\varepsilon_I(z) + \frac{\partial A}{\partial V}(z) \frac{u_*^3}{k} \left\{ \frac{1}{2} + \ln\left(\frac{\delta_L}{\delta_{mix}}\right) \right\} \right] \qquad [\mathrm{Wkg^{-1}}] \quad (7b)$$

In cases such as Zugersee, where the bottom boundary layer is made up of a stratified layer and a suppression layer (Figure 5b), integration of (7) yields

$$J_b^{tot}(z) = \gamma_{mix} \left[\varepsilon_I(z) + \frac{\partial A}{\partial V}(z) \frac{u_*^3}{k} \ln\left(\frac{\delta_L}{\delta_{sup}}\right) \right] \qquad [\mathrm{Wkg^{-1}}] \quad (7c)$$

In order to make equation (7a,b,c) easier to interpret, we show in Figure 9a the vertical structure of the buoyancy fluxes $J_b{}^B(h)$ for the three cases discussed. Whereas $J_b{}^B(h)$ increases toward the sediment in the stratified case, $J_b{}^B(h)$ decreases linearly within the well-mixed layer and drops to zero within the suppression layer.

The effect of both the mixed and the suppression layers is to lead to a reduction in the effectiveness of boundary mixing with increasing thicknesses δ_{mix} and δ_{sup}, respectively, of the layers. We define effectiveness identically to Garrett (1990) as the ratio $J_b{}^{tot\text{-}mix}$, the total buoyancy flux including a well-mixed layer, to $J_b{}^{tot\text{-}strat}$, the total buoyancy flux that would occur if there were stratification all the way down to the sediment. Figure 9b shows quantitatively the effect of the well-mixed layer thickness δ_{mix} to the effectiveness of boundary mixing. For very thin well-mixed layers ($\delta_{mix} \rightarrow \delta_V$), the effectiveness is 1 by definition. It decreases to about 7% in the other extreme case, when the whole logarithmic layer is mixed ($\delta_{mix} \rightarrow \delta_L$).

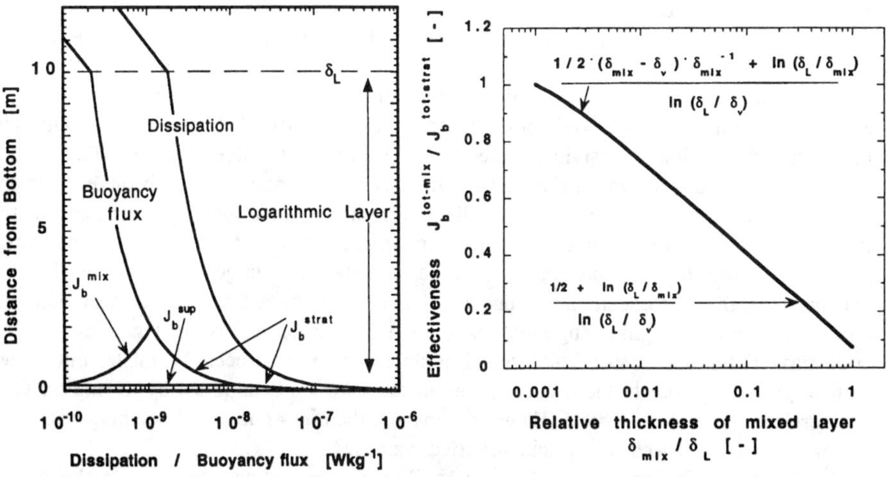

Figure 9. (a) Buoyancy flux as a function of height above the sediment within the logarithmic layer for three cases, *viz.* "stratified", "well-mixed" and "suppression" layers. The break-off to interior values at δ_L is drawn arbitrarily. Figure 9.(b) Effectiveness (definition in text) as a function of the relative thickness δ_{mix}/δ_L of the mixed layer (here we have set $\delta_L = 10$ m and $\delta_V = 1$ cm, but these values are not critical).

Application of the Boundary Mixing Model

In this section, the model predictions (Equations 5 and 6) are compared to observations from Alpnachersee and Zugersee. Since these two lakes have relatively simple basin shapes (Figure 1), but are very different in terms of stratification, external forcing and mixing, we consider them as ideal candidates to test the model. As detailed information on the horizontal flow field u_{1m} (x,y,z) is missing, we simplify the problem by using the two near-bottom current records $u_{1m}(t)$ shown in Figures 2 and 6. We consider this simplification to be appropriate in the deep hypolimnion of morphologically simple basins, as long as bottom currents are dominated by first horizontal internal seiche modes (Münnich, 1993). In addition, we assume for both lakes that internal mixing can be neglected in the deep hypolimnion and therefore set $\varepsilon_I = 0$. This approximation is appropriate, as it has been shown for Alpnachersee by tracer experiments (Goudsmit et al., 1996) and by microstructure measurements (Wüest et al., 1996a) that mixing in the interior is an order of magnitude weaker than the horizontal basin average. This is even more pronounced in the very deep part, where $\partial A/\partial V$ is large.

a) *Alpnachersee*: As input to the model we use the current data shown in Figure 2, which were collected 1.4 m above the sediment at the deepest part of the lake (Figure 1) during a period of 56 days in June-July 1994. From this record, a mean value for u_{1m}^3 of $(2.6 \text{ cm s}^{-1})^3$ was determined, which is used in (1), (6) and (7). The lower part of the bottom boundary in Alpnachersee is a well-mixed layer, with a thickness δ_{mix} of 4 to 5 m near maximum depth (Figure 4a). Upslope, the thickness is continuously decreasing (Figure 4b).

The total buoyancy flux $J_b^{tot}(z)$ is calculated using (6) and (7b), integrating over the entire cross-sectional area A(z) in vertical steps of 20 cm, and using the lake-specific topographic function $\partial A/\partial V(z)$ and the thickness $\delta_{mix}(z)$. The diapycnal diffusivity $K_d(z)$ is finally determined by applying (5) and using a representative water column stability $N^2(z)$ for the period during June-July 1994.

Within the deepest 10 m of the lake, the diapycnal diffusivity in Alpnachersee (Figure 10a) shows two main features. The diffusivity is (1) low due to the overall strong stratification N^2 and (2) increasing towards the sediment, since the stability N^2 is extremely low in the well-mixed bottom layer. The comparison between model results and long-term basin-wide diffusivities determined by the heat-budget method (Powell and Jassby, 1974) is also shown in Figure 10a. It turns out that both (1) the absolute value and (2) the vertical shape of $K_d(z)$ are in excellent agreement without having to tune any of the model parameters. Nevertheless, the agreement in the vertical structure of $K_d(z)$ is partly accidental, since the heat-budget method has a relatively large error within 4 m of maximum depth. The uncertainty is due to (1) the unknown sediment-water heat flux (the lake is not meromictic like Zugersee) and (2) the poorly defined vertical gradients in the "well-mixed" bottom layer. This second deficiency may be especially misleading when vertical gradients close to the deepest point in the lake are determined by averaging CTD or thermistor profiles in vertical (Eulerian) bins and thereby - due to the seiching motion - averaging over both well-mixed and stratified water layers.

b) Zugersee: In applying the model to Zugersee we follow the same procedure as above. As input we use the current data shown in Figure 6, which were measured at 1.8 m above the sediment at the deepest point in the lake in February 1993 (during 37 days). Averaging the third power of the bottom current u_{1m} leads to a value of $(u_{1m})^3 = (1.7 \text{ cms}^{-1})^3$, which is again introduced into (1), (6) and (7). Since the hypolimnetic water column is stratified all the way down to the sediment, as shown in Figure 5a, the stratified bottom boundary case is applicable. Consequently, the total buoyancy flux is determined by (7a) again using the topographic function $\partial A/\partial V(z)$. Finally, the diapycnal diffusivity $K_d(z)$ was calculated using (5) with the stability $N^2(z)$ determined from CTD profiles taken in March 1993.

The result for $K_d(z)$ is shown in Figure 10b, where it is also compared to the values determined by the heat-budget method for a six-month period during winter 1992-1993. Again, it turns out that both (1) the absolute value and (2) the approximate vertical shape of $K_d(z)$ are in good agreement without tuning any of the model parameters. Near maximum depth, the model apparently slightly overestimates the diffusivity. Besides the experimental uncertainties, this disagreement is most probably due to the suppression layer. Given the observed thickness of the suppression layer of about 20 cm (Figure 5b), application of (7c) would indeed reduce the model diffusivity by the factor $\ln(\delta_L/\delta_v)$ / $\ln(\delta_L/\delta_{sup}) \approx 1.8$ (arrow in Figure 10b). Since the lateral structure of the suppression layer is not known, this correction is not defined well enough to be included in Figure 10b. Even though the good agreement may to some extent be accidental, the decrease in the diapycnal diffusivity towards the sediment at maximum depth is consistent with the model.

Although one should be cautious not to overinterpret the agreement between model and data (there are numerous possible sources of deviations: e.g. varying drag coefficients, irregular topography, etc.), we can conclude that the model is excellently consistent with the absolute values as well as with the vertical profile of diapycnal diffusivity, both of

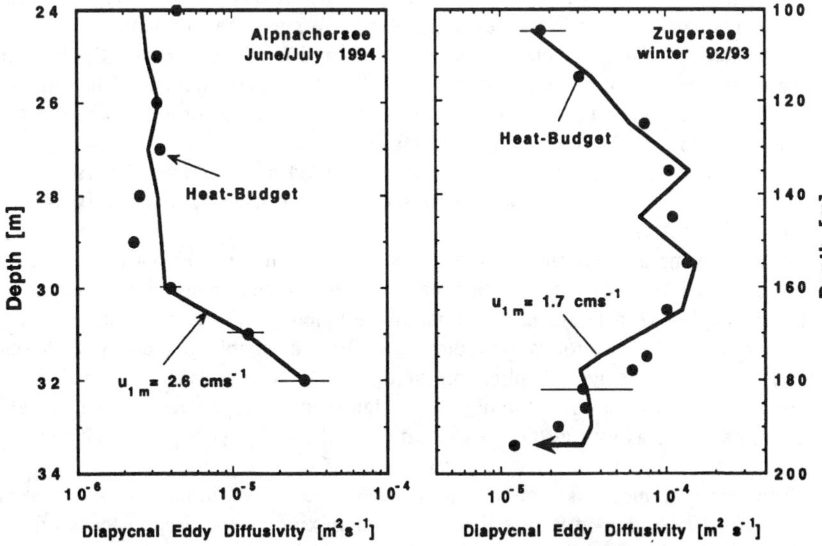

Figure 10. Comparison of diapycnal diffusivity K_d in Alpnachersee (a) and Zugersee (b), determined by the heat budget method (black dots), with the boundary mixing model (line, Equations 5 and 6; see text). The heat budget was applied over periods of 6 months in Zugersee (winter 92/93) and 2 months in Alpnachersee (June/July 94) (note the different vertical scales). The value of 1.1 cm^2 s^{-1} obtained for K_d in the weakly stratified part of the hypolimnion of Zugersee ($\approx$ 120 to 160 m depth) is supported by the long-term oxygen balance of the deep water (Wehrli et al., 1994). The effect of the suppression of turbulence (within a 20-cm inactive bottom layer) on the model result is shown by the arrow (factor 1.8). Data from Gloor (1995) and Müller (1993) for Alpnachersee and Zugersee, respectively.

which are caused by boundary mixing alone. Interestingly, near maximum depth the K_d's of the two lakes have about the same value, even though the mechanisms are very different: weak but efficient mixing in one case (Zugersee), strong and inefficient in the other case (Alpnachersee).

Conclusions

Based on experimental observations, theoretical considerations and modelling, an analysis of diapycnal (vertical) mixing was conducted in the deep water of two lakes exhibiting significantly different characteristics of density stratification and external forcing. The following conclusions can be drawn:

(1) The degree of stratification of the deep-water column is not only the result of external physical forcing at the surface, but also of the rate of release of dissolved solids from the sediment. The latter is determined mainly by long-term photosynthetic productivity in the epilimnion.

(2) For lacustrine systems with low kinetic energy input over periods of weeks, the accumulation of dissolved solids above the sediment-water interface leads to the formation of a highly stratified layer. Within this layer turbulence is almost completely suppressed and diffusion drops to molecular levels. Consequently, the basin-wide diapycnal diffusivity decreases with increasing thickness of the suppression layer.

(3) Aquatic systems with high kinetic energy input - relative to the buoyancy flux due to the release of dissolved solids - reveal well-mixed layers near the sediment.

(4) The most relevant parameters governing the two different regimes are the bottom current speed and the rate of release of dissolved solids from the sediment. The ratio of boundary-induced mixing to the flux of ions from the sediment determines whether the near-sediment layer is highly stratified or well-mixed.

(5) The presence of such well-mixed or highly stratified near-sediment layers reduces the effectiveness of boundary mixing and diminishes the basin-wide buoyancy flux as well as the diapycnal diffusivity.

(6) With a simple boundary mixing model, assuming (a) smooth flow in a logarithmic boundary layer of constant thickness, where the rate of turbulent dissipation follows the scaling law (Eq. 1), (b) constant mixing efficiency γ_{mix} in the stratified water column, and (c) no heat flux through the sediment, we have been able to quantify the basin-wide rate of diapycnal mixing. Application of the model to two lakes in which both density stratification and external forcing differ significantly reproduced astonishingly well the absolute values as well as the vertical structure of the diapycnal diffusivity in the deep hypolimnion.

(7) The presented model is consistent with tracer observations in medium-sized lakes (Goudsmit et al., 1996), demonstrating that diapycnal mixing in the deep water is driven entirely by boundary mixing. Application of the model to other systems in which horizontal current structure is better defined will hopefully show the model to have a wide range of applicability.

Acknowledgments. We are indebted to many members of the Environmental Physics Department at EAWAG, too numerous to mention, for making the collection of field data possible; our special thanks, however, go to M. Schurter for his unfailing presence on the lakes during the entire sampling program. Figure 5b was kindly provided by B. Wehrli. Part of the drawing and typing work was done by H. Bolliger and V. Graf, respectively.

The constructive criticism of two anonymous reviewers helped to improve the manuscript. D. M. Livingstone "styled" the English. This study was supported by Swiss National Science Foundation grants 20-32700.91 and 20-36364.92.

References

Armi, L., Some evidence for boundary mixing in the deep ocean, *J. Geophys. Res.*, *83*, 1971-1979, 1978.

Armi, L., Effects of variation in eddy diffusivity on property distributions in the ocean, *J. Mar. Res.*, 37, 515-530, 1979.

Brandl, H., and K. Hanselmann, Evaluation and application of dialysis porewater samplers for microbiological studies at sediment-water interfaces, *Aquat. Sci.*, *53*, 55-73, 1991.

Broecker, W. S., J. Cromwell, and Y. H. Li, Rates of vertical eddy diffusion near the ocean floor based on measurements of the distribution of excess Rn-222, *Earth Planet. Sc. Lett.*, 5, 101-105, 1968.

Chen, C. T., and F. J. Millero, Precise thermodynamic properties for natural waters covering only the limnological range, *Limnol. Oceanogr.*, 31, 657-662, 1986.

Chung, Y., and K. Kim, Excess Rn-222 and the benthic boundary layer in the western and southern indian ocean, *Earth Planet. Sc. Lett.*, 49, 351-359, 1980.

Dewey, R. K., and W. R. Crawford, Bottom stress estimates from vertical dissipation rate profiles on the continental shelf, *J. Phys. Oceanogr.*, *18*, 1167-1177, 1988.

Elliott, A. J., Measurements of the turbulence in an abyssal boundary layer, *J. Phys. Oceanogr.*, *14*, 1779-1786, 1984.

Finckh, P., Heat-flow measurements in 17 perialpine lakes: Summary, *Geological Soc. Am. Bull*, Part I, 92, 108-111, 1981.

Garrett C., Mixing in the ocean interior, *Dyn. Atmos. Oceans*, *3*, 239-265, 1979.

Garrett, C., The role of secondary circulation in boundary mixing, *J. Geophys. Res.*, *95*, 3181-3188, 1990.

Garrett, C., Marginal mixing theories, *Atmosphere Ocean*, *29*, 313-339, 1991.

Gloor, M., A. Wüest, and M. Münnich, Benthic boundary mixing and resuspension induced by internal seiches, *Hydrobiologia*, *284*, 59-68, 1994.

Gloor M., Methode der Temperaturmikrostruktur und deren Anwendung auf die Bodengrenzschicht in geschichteten Wasserkörpern, *Dissertation ETH Zürich*, Nr. 11'336, p. 159, 1995.

Goudsmit, G. H., F. Peeters, M. Gloor and A. Wüest, Boundary versus internal diapycnal mixing in stratified waters, 1996, *J. Geophys. Res. 102*, 27903-27914, 1997.

Hinze, J. O., *Turbulence*, McGraw-Hill, New York, 618 pp., 1975.

Imberger, J., and G. N. Ivey, Boundary mixing in stratified reservoirs, *J. Fluid Mech.*, *248*, 477-491, 1993.

Imberger, J., and J. C. Patterson, Physical limnology, in *Advances in applied mechanics*, edited by J. W. Hutchinson and T. Y. Wu, Academic Press, Cambridge, 303-475, 1990.

Imboden, D. M., B. Stotz and A. Wüest, Hypolimnic mixing in a deep alpine lake and the role of a storm event, *Verh. Internat. Verein, Limnol.*, *23*, 67-73, 1988.

Imboden, D. M., and A. Wüest, Mixing mechanisms in lakes, p. 83-138, In *Lakes: Chemistry, Geology, Physics*, edited by A. Lerman, D. M. Imboden and J. Gat, Springer, New York, 1995.

Ivey, G. N., The role of boundary mixing in the deep ocean, *J. Geophys. Res.*, *92*, 11'873-11'878, 1987.

Ivey, G. N., and J. Imberger, On the nature of turbulence in a stratified fluid. Part I: The energetics of mixing, *J. Phys. Oceanogr.*, *21*, 650-659, 1991.

Müller, B., Sauerstoffentwicklung im Zugersee, *Diplomarbeit EAWAG/ETH*, 1993.

Münnich, M., A. Wüest, and D. M. Imboden, Observations of the second vertical mode of the internal seiche in an alpine lake, *Limnol. Oceanogr.*, 37, 1705-1719, 1992.

Münnich, M., On the influence of bottom topography on the vertical structure of internal seiches, *Dissertation EAWAG/ETH Zürich*, Nr. 10434, 1993.

Munk, W. H., Abyssal recipes, *Deep-Sea Res.*, *13*, 707-730, 1966.

Osborn, T. R., Estimates of the local rate of vertical diffusion from dissipation measurements, *J. Phys. Oceanogr. 10*, 83-89, 1980.

Peters, H., and M. C. Gregg, Some dynamical and statistical properties of equatorial turbulence, In *Small-scale Turbulence and Mixing in the Ocean, Proc. 19th Intnl Liège Colloquium on Ocean Hydrodynamics*, edited by J. C. J. Nihoul and B. M. Jamart, Elsevier, Amsterdam, 185-200, 1988.

Phillips, O. M., J. H. Shyu, and H. Salmun, Experiment on boundary mixing: mean circulation and transport rates, *J. Fluid Mech., 17*, 473-499, 1986.

Phillips, O. M., On flows induced by diffusion in a stably stratified fluid, *Deep-Sea Res., 17*, 435-443, 1970.

Powell, T., and A. Jassby, The estimation of vertical eddy diffusivities below the thermocline in lakes, *Water Resour. Res., 10*, 191-198, 1974.

Press, W. H., B. P. Flannery, S. A. Teukolsky, and W. T. Vetterling, Numerical recipes, Cambridge University Press, Cambridge, 1986.

Rohr, J. J., K. N. Helland, E. C. Itsweire, and C. W. Van Atta, Turbulence in a stably stratified shear flow: A progress report, In *Turbulent Shear Flows 5,* Springer-Verlag, Berlin, 1987.

Sarmiento, J. L., H. W. Freely, W. S. Moore, A. E. Bainbridge, and W. S. Broecker, The relationship between vertical eddy diffusion and buoyancy gradient in the deep sea, *Earth Planet. Sci. Lett., 32*, 357-370, 1976.

Stillinger, D. C., K. N. Helland, and C. W. Van Atta, Experiments on the transition of homogeneous turbulence to internal waves in a stratified fluid, *J. Fluid Mech., 131*, 91-122, 1983.

Thorpe, S. A., The dynamics of the boundary layers of the deep ocean, *Sci. Prog., Oxford, 72*, 189-206, 1988.

Toole, J. M., K. L. Polzin, and R. W. Schmitt, Estimates of diapycnal mixing in the abyssal ocean, *Science, 264*, 1120-1123, 1994.

Weatherly, G. L., and P. J. Martin, On the structure and dynamics of the oceanic bottom boundary layer, *J. Phys. Oceanogr., 8*, 557-570, 1978.

Weatherly, G. L., and W. Wimbush, Near-bottom speed and temperature observations on the Blake-Bahama outer ridge, *J. Geophys. Res., 85*, 3971-3981, 1980.

Wehrli, B., A. Wüest, and D. M. Imboden, Grundlagen für die Sanierung des Zugersees: Untersuchungen des Stoffhaushaltes von Tiefenwasser und Sediment, EAWAG, Dübendorf, Auftrag 37-4840, 1994.

Wehrli, B., A. Wüest, and D. M. Imboden, Sind biogen meromiktische Seen intern sanierbar? Fallbeispiel Zugersee, in *Limnologie Aktuell,* edited by D. Jaeger and R. Koschel, *8*, 29-37, 1995.

Wüest, A., D. C. van Senden, J. Imberger, G. Piepke, and M. Gloor, Comparison of diapycnal diffusivity measured by tracer and microstructure techniques, *Dyn. Atmos. Oceans 24*, 27-39, 1996a

Wüest, A., G. Piepke and J. D. Halfman, Combined effects of dissolved solids and temperature on the density stratification of Lake Malawi (East Africa), In *The Limnology, Climatology and Paleoclimatology of the East African Lakes*, edited by Johnson, Gordon and Breach Scientific Publishers, NY, p. 183-202, 1996b.

Wunsch, C., On oceanic boundary mixing, *Deep-Sea Res., 17*, 293-301, 1970.

35

Turbulent Benthic Boundary Layer Mixing Events in Fresh Water Lakes

C. Lemckert and J. Imberger

Abstract

The nature of the turbulent benthic boundary later (TBBL) within thermally stratified lakes was investigated using data collected from two field studies. A profiling instrument was used to record the turbulent properties of the layers, while thermistor chains were used to monitor the internal wave field. The results obtained indicated that in these studies the TBBLs were generated by internal waves as they progressed over the sloping beds of the lakes. The mixing efficiency of the turbulent events associated with the TBBL was found to be dependent on the turbulent Froude number. The findings indicate that the TBBL's were very important vertical turbulent transport mechanism.

Introduction

Evidence is mounting to suggest that active turbulent benthic boundary layers (TBBLs) are important mechanisms for the vertical transport of mass and heat within stratified lakes and the ocean. These layers, which are defined as being those regions adjacent to the bed where turbulent activity keeps the vertical density structure nearly constant, have been observed in both lakes (e.g., Gloor et al., 1994; Wuest et al., 1994) and the ocean (e.g., Eriksen, 1992; Ledwell and Hickey, 1995). For TBBL's to remain active there must be a source of turbulent kinetic above that level which will be dissipated through viscous shear. There are a number of possible sources of energy that are capable of driving the mixing process (Imberger, 1994, 1998). These include mean flows over hydraulically rough beds (such as bottom intrusions and turbidity currents), and the breaking of internal waves on or adjacent to the bed. For the latter case, two possible mechanisms exist. One is the result of bottom/wave interactions, whereby internal wave energy is increased and instabilities develop as a result of incident/reflective wave interactions (Phillips, 1977; Thorpe, 1987). The other results from the breaking of shoaling internal waves on the boundary (Thorpe et al., 1990; Gloor, 1994). The mechanism that dominates will depend on the ambient and forcing conditions. In both instances active mixing is induced, which in turn generates a net vertical mass flux near the boundary.

The vertical mass flux associated with TBBLs is extremely important for the biology, heat budget, sediment and chemical nature of lakes. This is because, TBBLs act as cycling

Physical Processes in Lakes and Oceans
Coastal and Estuarine Studies Volume 54, Pages 503-516

mechanisms, whereby water is moved initially vertically at the boundary and then injected horizontally through the ambient (De Silva et al., 1997). For example, Frechette et al. (1989) postulated that TBBLs acts as a path of enhanced movement which increases the exchange between food rich sea surface water and the food-impoverished water located near the sea bed. Consequently, the more active the TBBL is, the greater the time averaged food concentration will be for benthic organisms.

Dewey et al. (1987) used an ascending profiler equipped with shear sensors to partially document the dissipation rate above and within an oceanic TBBL. They found significantly higher dissipation rates within the TBBL, which is also in agreement with results presented in Garrett et al. (1993). However, both studies did not report on the mixing characteristics of the turbulent events contained within TBBLs. More recently, Ivey and Nokes (1989), using laboratory experiments, examined the breaking of internal waves on steep beaches and found that the mixing efficiencies (i.e., the conversion of kinetic energy into increasing the potential energy of the water column) could be as high as 20%. In this investigation we will examine the case where the TBBL is generated by the movement of internal waves over a sloping bottom boundary within two large stratified freshwater lakes. Specifically we will examine the nature of turbulent events in TBBLs located within thermally stratified lakes, with particular emphasis on the efficiency of the induced mixing process.

Parameterisation of Turbulence

Turbulent events may be characterised by examining their respective turbulent Froude, Fr_t and Reynolds, Re_t, numbers (e.g., Gibson, 1980; Ivey and Imberger, 1991). The turbulent Froude number is defined as:

$$Fr_t = \frac{u'}{NL_c} \sim \left(\frac{L_R}{L_c}\right)^{2/3} \sim \frac{u'}{(g' L_c)^{1/2}} \sim \left(\frac{\varepsilon^2}{g'^3 L_c}\right)^{1/6} \sim \left(\frac{\varepsilon^2}{g' u'}\right)^{1/2} \tag{1}$$

where $u' = (\varepsilon L_c)^{1/3}$ (Ivey and Imberger, 1991) is the rms turbulent velocity scale of the event, ε is the rate of dissipation of turbulent kinetic energy, L_c is a statistical average measure of the vertical length scale of the energy bearing eddies within the event (Imberger and Boashash, 1986), g is the acceleration due to gravity, $L_R = (\varepsilon/N^3)^{1/2}$ is the Ozmidov length scale, $N^2 = (g/\rho)(d\rho/dz)$, ρ is the fluid density and Z is elevation. The modified gravitational acceleration, g', is obtained by firstly monotonising the recorded density profile, and then subtracting the monotonised profile from the original. The difference in density at each point represents a measure of the out of balance mass, ρ', which when multiplied by g and divided by the mean density yields g' at each point. The length scale, L_c, is obtained by firstly determining the distance a fluid element has been displaced from its level of neutral buoyancy in a monotonised profile (this distance is referred to as the Thorpe displacement scale, L_d). Following this, the displacement estimates, L_d, are moved vertically by one half of their value, thereby placing them at the centre of the overturn event. The absolute values of all the estimates at a particular depth are then summed and averaged at their new position; giving a measure of the eddy displacement, L_c, which is now positioned at the event centre.

The turbulent Reynolds number, Re_t, is defined as:

$$Re_t = \frac{u' L_c}{\nu} \sim \left(\frac{L_c}{L_K}\right)^{4/3} \quad (2)$$

where ν is the viscosity, $L_K = (\nu^3/\varepsilon)^{1/4}$ is the Kolmogorov length scale representing the smallest overturn scale in the turbulent field.

While Fr_t and Re_t are commonly used length scale ratios, a third possible ratio exists. This is between the buoyancy and viscous forces, or L_R and L_K, which is written as the small scale Froude Number:

$$Fr_\gamma = \left(\frac{\varepsilon}{\nu N^2}\right)^{1/2} \sim \left(\frac{L_R}{L_K}\right)^{2/3} \quad (3)$$

The three dimensionless groupings Fr_t, Re_t and Fr_γ are related by $Fr_\gamma = Fr_t\, Re_t^{1/2}$, and are commonly used to assess the state of the recorded turbulent events at the time of measurement. For example, when Fr_t is large inertia dominates over the buoyancy restoring force; while for small Re_t viscosity dampens inertia.

The efficiency at which kinetic energy is transferred to increasing the potential energy of the water column is referred to as the mixing efficiency, or the Flux Richardson number (Ri_f), which is defined by (Ivey and Imberger, 1991):

$$Ri_f = \frac{b}{m} = \frac{b}{b+\varepsilon} = \frac{1}{1+(\varepsilon/b)} \quad (4)$$

This represents the ratio of the flux of the vertical buoyancy, $b = -(g/\rho)\,\overline{(w'\rho')}$, to the rate of change of available mechanical energy, m, where w' is the vertical turbulent velocity component and the overbar denotes ensemble averaged values. Using laboratory and numerical data Ivey et al. (1998) have shown that Ri_f can be readily related to the turbulent Froude and Reynolds numbers. Generally, the peak efficiency (of 25%) occurs when Fr_t = 1, with the efficiency decreasing the more Fr_t varied from 1. This result required that all the data were analysed over statistically stationary segments.

Recent findings by Etemad-Shahidi and Imberger (1996) have shown that, in the thermocline regions of lakes, the average mixing efficiency is approximately zero throughout the Fr_t range. Consequently, the vertical exchange coefficient is approximately zero. This result suggests that mixing processes in the thermocline, away from solid boundaries, are an inefficient energy transfer mechanism. The question remains as to what is the efficiency of energy transfer like in a lake benthic boundary layer generated by shoaling internal waves. This is the question addressed in this investigation.

Experimental Methodology

Study Site and Instrumentation

The experimental work of this investigation was primarily conducted in Lake Kinneret (Sea of Galilee), Israel, from the 17th to the 21st July, 1994, inclusive (Days of the year 198 to 202). At the time of the study the elliptical shaped lake (see Figure 1) had a maximum depth of 50 m and a corresponding surface area of 165 km^2. As this study was performed during the summer months the water column was dominated by a strong seasonal thermocline, with a generally well mixed surface layer. In addition to the Kinneret experiment data presented later were also used from a study performed on Lake Biwa, Japan

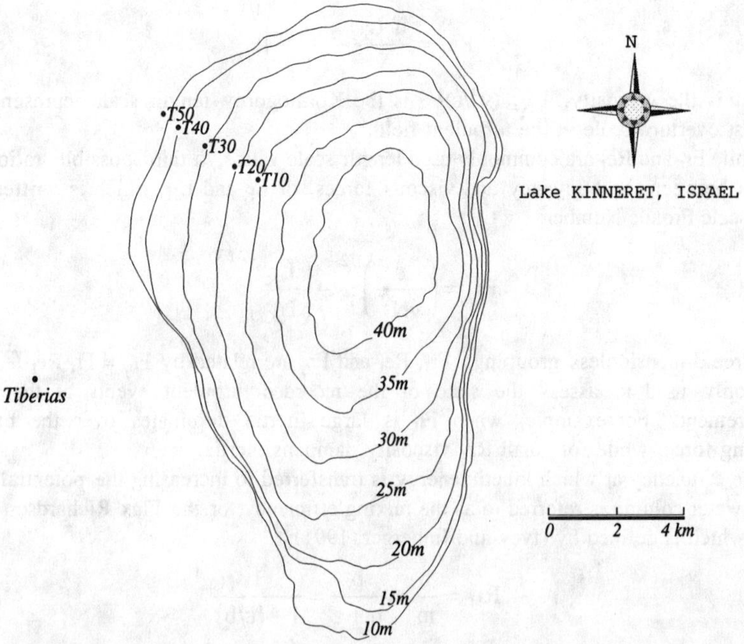

Figure 1. Bathymetric map of Lake Kinneret showing the thermistor chain station.

(see Saggio and Imberger, 1998). Like Lake Kinneret, Lake Biwa also experienced large scale internal waves that resulted in the generation of TBBLs.

Continual monitoring of the thermal structure of Lake Kinneret was achieved using a thermistor chain positioned at site T10 (see Figure 1), where the bedslope was 0.2° with a normal direction of 113° downslope. Thermistors, each with resolutions of 0.01°C, were set at depths of 1, 3, 5, 6.5, 8, 9, 10, 11, 12, 13, 14, 15, 16, 17, 18, 19, 20, 21, 22 and 22.5 m, and sampled every minute. This site was chosen as it corresponded closely to where the seasonal thermocline made contact with the lake bed; thus giving a the optimal location for internal wave shoaling.

A Portable Flux Profiler (see Figure 2), which was developed by the Centre for Water Research, The University of Western Australia, was used to simultaneous collect profiles of temperature, conductivity and water velocities, all on a microscale. In brief, this battery-powered profiling instrument comprised of a two-component Laser Doppler velocimeter (LDV) system, compass and x-y tilt sensors, positioned around a smaller microstructure probe equipped with two combined four-electrode conductivity/temperature sensors and a pressure sensor. The resolutions for the temperature, conductivity and pressure sensors were 6×10^{-4} °C, 2×10^{-4} Sm^{-1} 2×10^{-3} m, respectively. The LDV, which has a velocity resolution of 1.2×10^{-3} ms^{-1}, measured the vertical and one horizontal component of the water velocity, relative to the PFP, 3 mm above a temperature/ conductivity sensor pair. The compass and tilt sensors had resolutions of 3° and 0.1°, respectively. In this study the PFP was deployed at the surface and set to descend at a rate of 0.1 ms^{-1}, while all data channels were sampled at 100 Hz. From this instrument we were able to assess the state of turbulence within the TBBL and the associated mixing efficiency level.

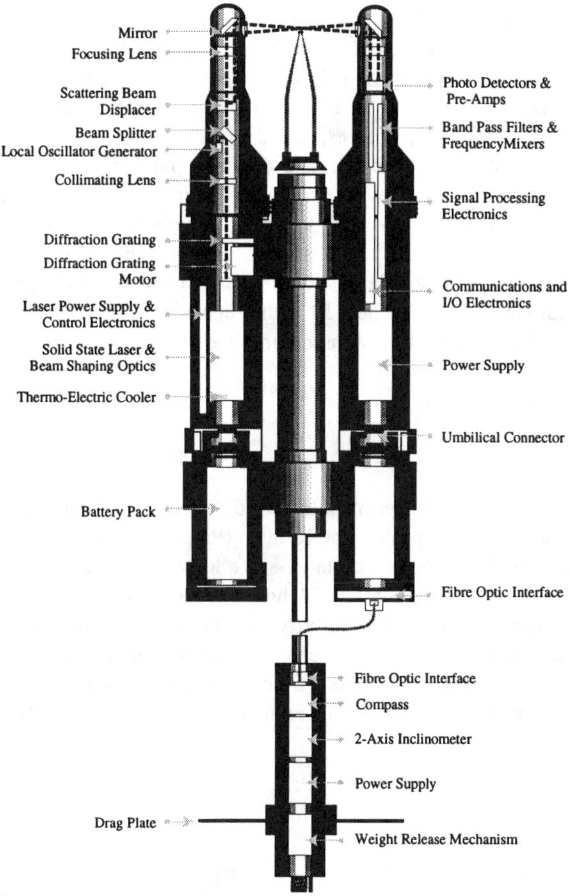

Figure 2. Schematic of the portable flux profiler

Data Processing

During the collection of profile data a number of independent events were often sampled. So as to ensure statistical significance when analysing the turbulent behaviour of particular events data were only analysed within stationary segments. These segments were chosen by applying the stationary test proposed by Imberger and Ivey (1991) to the temperature gradient signals following the necessary preconditioning required to compensate for sensor response. Dissipation estimates within a stationary segment were derived using a Batchelor curve fitting technique. This was chosen as it represents a readily measurable quantity by microscale profiling instrument, and because Oakey (1982) and Lemckert and Imberger (1995) have shown that dissipation estimates made by this technique are consistent with those derived from the variance of the horizontal velocity gradient signal. Lemckert and Imberger (1995) have shown that data collected the PFP represents a snapshot of the turbulence field, and that the frozen flow hypothesis (which is inherent in the analysis of the data) was justified.

Profiles of the vertical turbulent velocity fluctuations, w', were derived by firstly low pass filtering the raw vertical velocity profile data with a filter based on the size of the turbulent events the instrument passed through (see Etemad-Shahidi and Imberger, 1996). This filtered signal was then subtracted from the raw signal to leave the fluctuating components w'. The derived quantities, ρ and w' were then multiplied together at each sample point in a profile to give an estimate of the vertical density fluxes, on a point by point basis, and then averaged over a stationary segment for flux Richardson number determination. Turbulent Froude and Reynolds numbers were also determined using stationary segment averaged results. Note that data within the stationary segment was used only if it was free of any large noise spikes induced by underwater objects and satisfied the condition $L_S/L_C > 2$ where L_S is the segment length. This latter condition was applied to ensure that we actually captured complete events, and not a portion thereof.

Experimental Observations

Large Scale Thermal Features

During the Lake Kinneret experiment the wind pattern was dominated by a diurnal cycle with strong afternoon breezes peaking at 12 ms^{-1} (Figure 3). Figure 3 also shows that the seasonal thermocline also experienced large-scale long-period internal wave activity with an oscillatory period of approximately 24 hours, which corresponds to the wind forcing period. The smaller internal wave activity observed in the isotherm structure, as the thermocline moved past the thermistor chain, indicated localised mixing events in the regions of strong temperature gradients. Since temperature was the dominant stratifying agent, this also represents the region of the strongest density gradients.

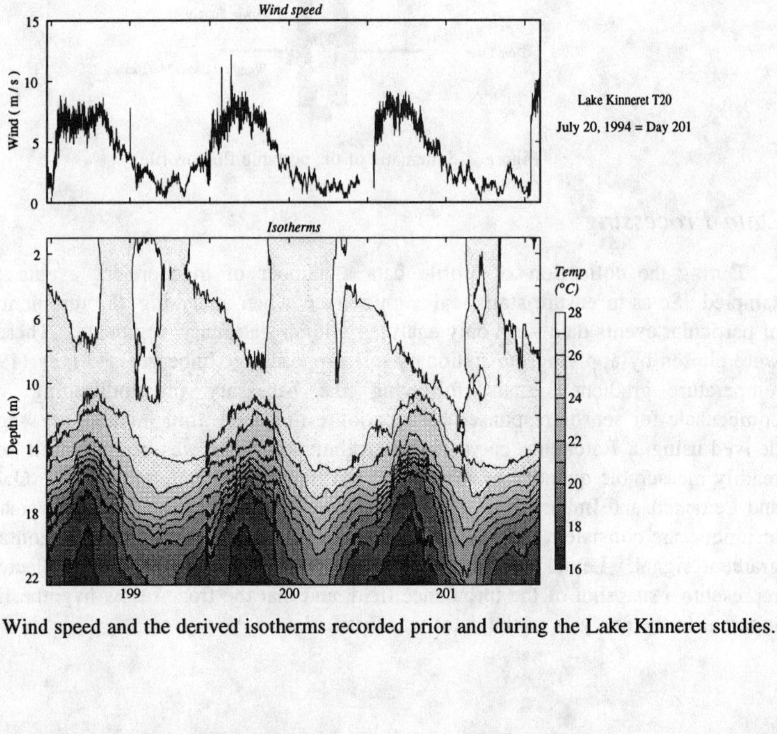

Figure 3. Wind speed and the derived isotherms recorded prior and during the Lake Kinneret studies.

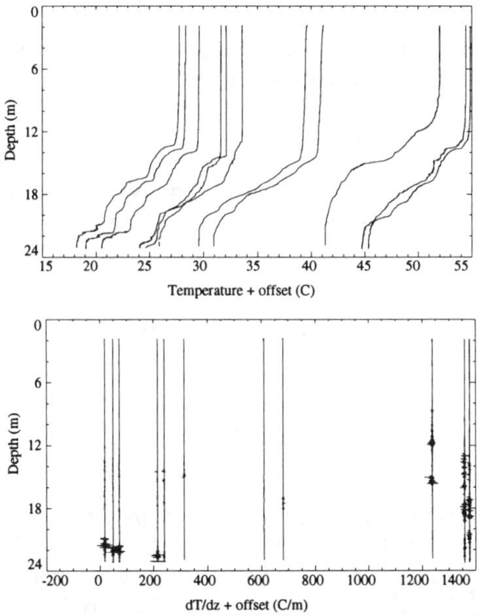

Figure 4. Stagger plots of temperature and the corresponding temperature gradient profiles collected at site T20 in Lake Kinneret on day 201. The temperature profiles have been offset by x*0.02, where x is the time in minutes from 0000 hrs on day 201, while the temperature gradient signals were offset by x only. The time of data collection were, from left to right, 0018, 0049, 0112, 0335, 0400, 0514, 1009, 1121, 2038, 2238 and 2257 hrs, respectively. The dashed line indicates the upper level of the turbulent benthic boundary layer as defined by the temperature signal.

While the experiment extended over a period of 3 days data presented here will primarily come from day of the year 201. On this day a series of descending PFP casts were performed over a 24 hours period near site T20 (see Figure 1) in a water depth of 23 m. Figure 4 shows a stagger plot of the derived temperature and corresponding temperature gradient profiles collected during this time. From this figure it is evident that, by our previous definition of a TBBL, an active layer existed near the lake bed and that its character changed over time. As time progressed from the commencement of the profiling operation the TBBL was found to decrease in thickness until at 0400 hours (corresponding to the 5th profile in Figure 4) it was no longer evident. Following this there was a period where there was no evidence of a TBBL. However, after 2038 hrs (corresponding to profile 8) a new TBBL rapidly developed with its thickness increasing with time. Note that the gradient levels depend critically upon the background temperature stratification, and thus in the TBBL we expect lower signals as the water is nearly homogeneous, However, at the boundary between the TBBL and the ambient fluid high gradient signals are expected as the TBBL mixes in warmer water from above. Therefore, the gradient signals are good indicators of the mixing boundaries.

By comparing the profile data of Figure 4 with isotherm data in Figure 2 it can be seen that the TBBLs developed as the metalimnitic water moved over the lake bed. Recent analysis of the dynamics of Lake Kinneret by Saggio and Imberger (1998) reveal that the lake experiences significant wind forced Kelvin wave motions. This results in the development of the TBBLs. Throughout the existence of the TBBLs significant turbulent activity and high temperature gradient signals were evident, especially at the TBBL/ambient interface, where the well mixed TBBL water interacted with the warmer ambient water across a noticeable density gradient.

Mean Flows Within the TBBL

By using data from two microstructure PFP profiles, collected close together both temporally and spatially, it can be shown that for slowly varying flow fields the mean flow field velocity, U, and direction, α, at any given depth are given by:

$$U = \frac{u_1}{\cos(\theta_1 - \alpha)} = \frac{u_2}{\cos(\theta_2 - \alpha)} \tag{5}$$

and

$$\alpha = \arctan\left(\frac{u_2 \cos\theta_1 - u_1 \cos\theta_2}{u_1 \sin\theta_2 - u_2 \sin\theta_1}\right) \tag{6}$$

where u is the component of water velocity measured by the LDV, θ is the corresponding probe compass orientation and the subscripts 1 and 2 denote the first and second profiles respectively. As can be seen from the above equations close probe orientations will result in significant errors. Consequently, data can only be considered when the difference in probe orientations were greater than 10°.

An example of a resultant velocity profile derived from the above technique is presented in Figure 5. From this it is evident that the velocity expectably increases with elevation until a mid TBBL depth is reached, whereupon it decrease once more. In conjunction with this is a rotation of the flow in the clockwise direction with increasing elevation, in a manner like an Ekman spiral. This flow pattern indicates the shoaling internal wave in progressing with a sweeping type motion around the lake.

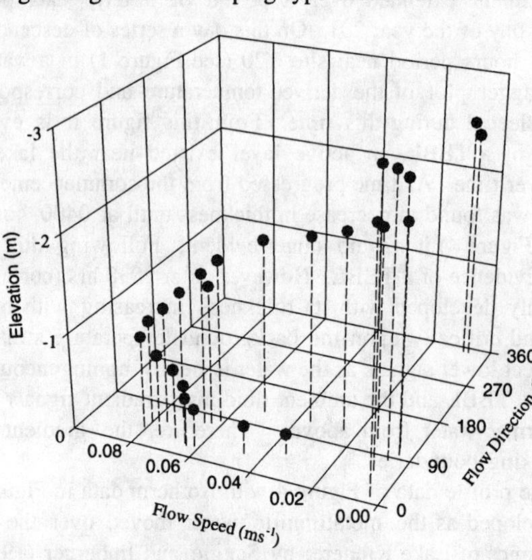

Figure 5. Plot of the horizontal velocity within the TBBL presented as a function for depth and direction as determined from the profiles collected at 0049 and 0053 hrs.

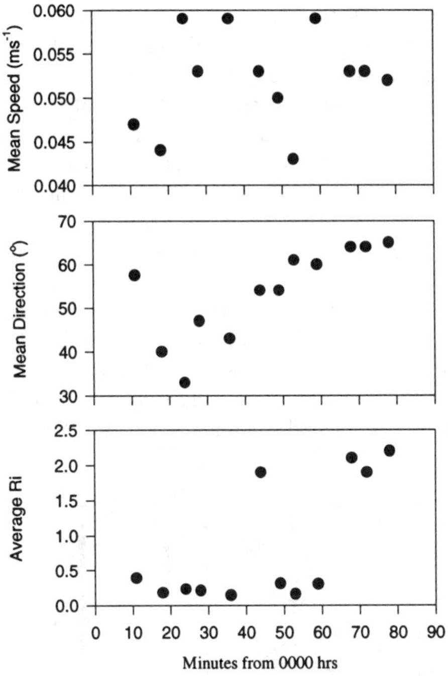

Figure 6. Plots of the mean flow direction, mean speed and segment averaged gradient Richardson number of the TBBL for the first 90 minutes on day 201. The water depth was 23 m.

To build up a picture of the flow behaviour of the TBBL the depth averaged flow and direction were derived for the first 90 minutes on day 201. The results, presented in Figure 6 clearly show the mean flow velocity remains nearly constant at 0.05 ms^{-1}, while the direction is slowly increasing with time. By translating this information onto the bathymetric map it reveals that the TBBL flow is primarily tangential to the bottom contours; indicating that the thermocline is oscillating as a Kelvin wave.

Turbulent Features

A typical PFP microstructure profile is displayed in Figure 7. This figure reveals a well defined constant temperature and high flow region adjacent to the bed which marks the extent of the TBBL. Associated with this are large temperature gradient signals and significant horizontal and vertical turbulent velocities. These results indicate high levels of active turbulent activity within the TBBL. Above the TBBL a strong non-turbulent jet like flow was present that was most likely generated by a seiching type process (e.g., see Gloor et al., 1994). This figure also shows that both up-gradient and down-gradient mass transport occur in the boundary layer.

Realisations of up-gradient transport are likely to be the result of restratification following an overturn event, since the area was not subject to natural convection activity. These realisations also strongly depend on the position and time at which the micro-structure sensors pierce through turbulent overturn events. For example, if a profile was recorded directly through the centre of an isolated overturn event that underwent solid body rotation, then w' = 0 and hence w' ρ' = 0, even though ρ' may be at a maximum. Similarly, if the eddy was pierced at an edge after a quarter revolution, then w' and ρ' would both be at high values and their multiple would be an overestimate of the average. Therefore, when assessing buoyancy flux data collected by profiling instruments, a number of casts will be required to derive an estimate under similar forcing conditions. From the experiments performed data from 7 Lake Biwa profiles and 25 Lake Kinneret profiles were found to contain usable data.

Presented in Figure 8 are the segment averaged properties of measured turbulent events associated with the TBBL, as recorded by the PFP and which satisfied the quality conditions. This figure shows that the majority of the events were very active with inertia dominating over both viscosity and buoyancy. While it has been found in laboratory investigations that turbulent mixing appears to cease when $Re_t < 15$ (e.g., see Ivey and Imberger, 1991), the data collected in these field studies revealed that active events, with well defined Batchelor spectra, existed when $Re_t < 15$. For this reason they were retained for further processing.

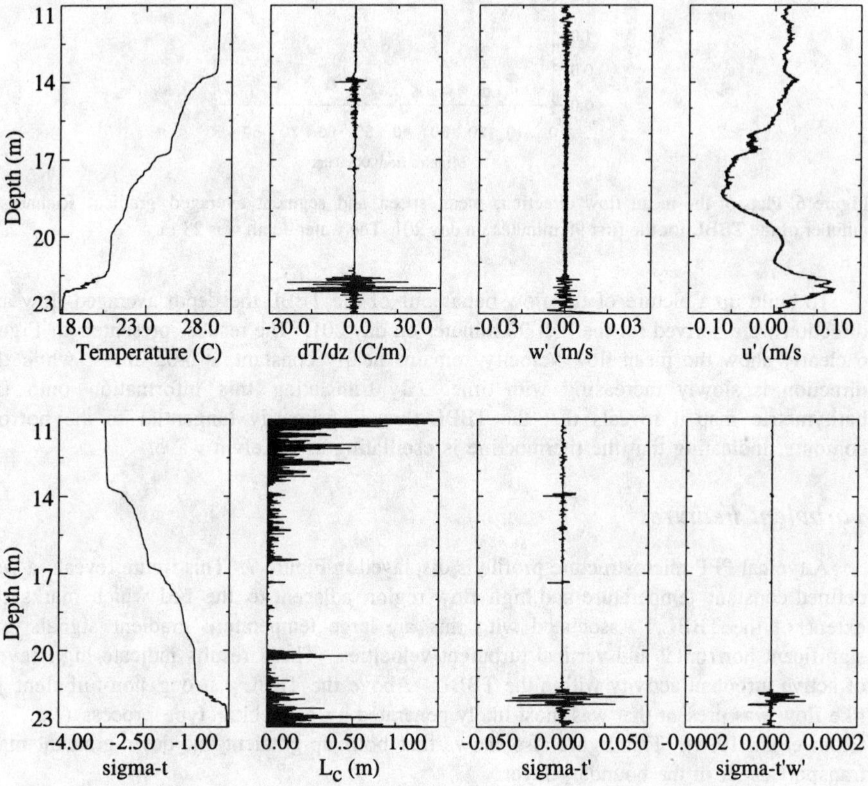

Figure 7. Example of a microstructure profile recorded with the PFP as it penetrated and active benthic boundary layer in Lake Kinneret at 0053 hrs on day 201, 1994. The data presented are Temperature, gradient temperature, w', u (as recorded by the PFP), sigma-t, L_C, sigma-t' (ρ') and sigma-t'w' (ρ' w').

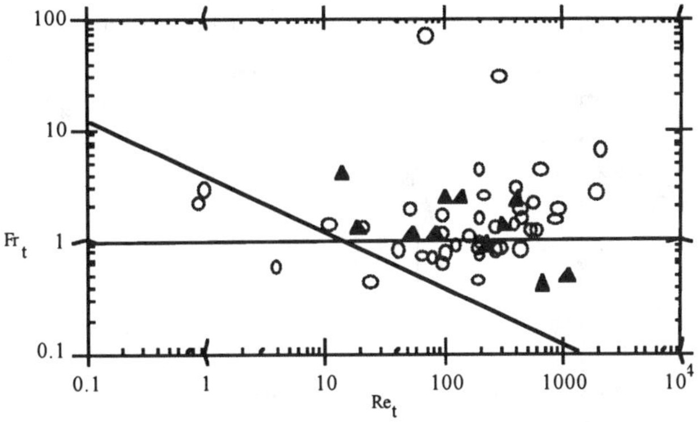

Figure 8. Activity diagram (plot of Fr_t vs Re_t) showing the nature of the turbulent events recorded within turbulent benthic boundary layers generated by shoaling internal waves. The horizontal line represents the expected transition from inertia dominated to buoyancy dominated turbulence, while the sloping line indicates the proposed transition to internal wave motions. The data are from O Lake Kinneret and from ▲ Lake Biwa, 1993, where an international experiment was performed to look at the overall lake dynamics.

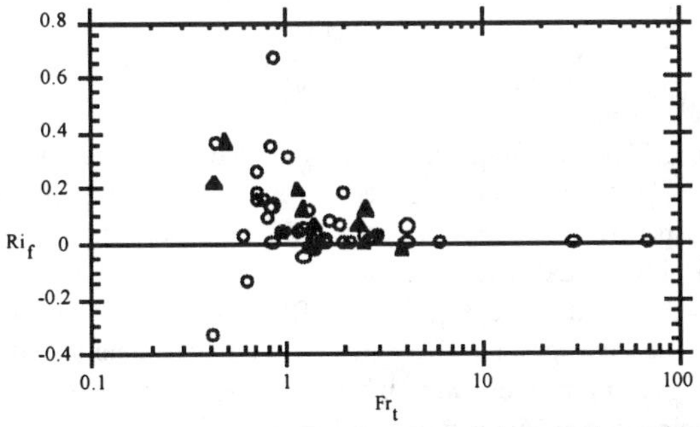

Figure 9. Plot of the mixing efficiencies, Ri_f, as a function of the turbulent Froude number, Fr_t. Symbols are defined in Figure 6.

By using [4], and the required segment averaged values, estimates of the mixing efficiency, Ri_f, were obtained. The individual results, presented in Figure 9, show a large scatter in the data, with larger estimates congregating around $Fr_t = 1$. Note that this corresponds closely to that found by Ivey and Imberger (1991) and Lemckert et al. (1995), but varies significantly from that of Etemad-Shahidi and Imberger (1996). To ascertain whether any discernible trends existed between Ri_f and Fr_t in the TBBL Ri_f data estimates were binned together over small integer ranges of Fr_t (each bin containing at least 5 points) and an average derived. The result is presented in Figure 10, with the associated caption showing the binning ranges used. While there was significant scatter in the raw data this binning process reveals that the maximum efficiency occurs close to $Fr_t =1$. In addition to this, Figure 10 also shows Ri_f estimates smoothed with a 10 point box car running average filter. The resultant curve is very similar to that predicted by Ivey et al. (1998) for environments with different Prandtl numbers.

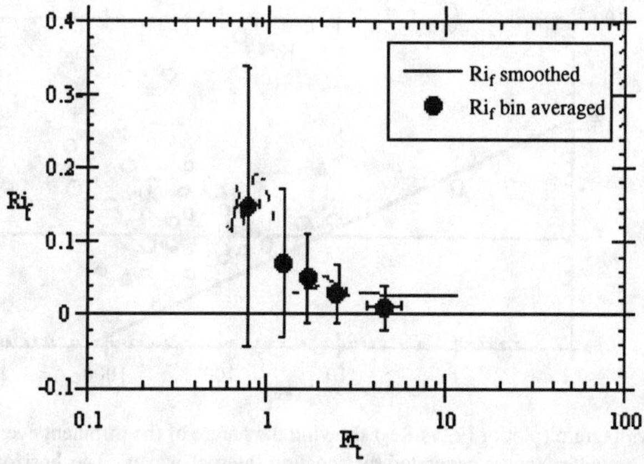

Figure 10. Plot of Ri_f as a function of Fr_t for smoothed (10 point box smoothing filter) and bin averaged data with bin ranges of 0.5-1.0, 1.5-2.0, 2.0-3.-, 3.0-10.0). The error bars indicate correspond to 1 standard deviation about the mean value.

A measure of the stability of the TBBL flow can be obtained by determining the gradient Richardson number. From profile U and α data, northerly and easterly velocities were derived. The gradient Richardson number, Ri, can then be determined from:

$$Ri = \frac{M^2}{u_z^2 + v_z^2} \tag{7}$$

where M is the Brunt Vaisalla frequency as determined from the monotonized density profile, with u_z and v_z representing the vertical velocity shears in the northerly and easterly direction respectively. The analysis was then performed on PFP data files, collected on the morning of day 201, with the derived Ri values averaged over the same segments as were used to derive the Fr_t, Re_t and Ri_f estimates. The results, presented in Figure 11, show that the layer gradient Richardson number is generally close to the critical level of 0.25, where shearing motions are able to generate and sustain turbulent events in stratified fluids at their most efficient mixing level.

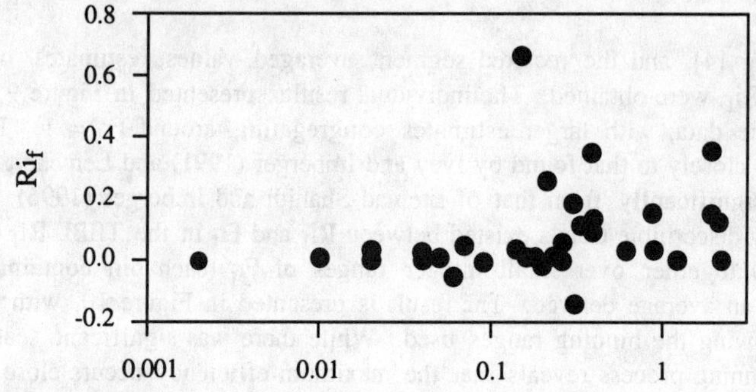

Figure 11. Plots of segment averaged flux Richardson numbers against the segment averaged gradient Richardson number.

Conclusions

From the studies performed to date, it is evident that turbulent benthic boundary layers exist within lakes, and that they can be generated by internal waves moving across the bed. Microstructure data showed that TBBLs are active mechanisms by which mass can be transported vertically within stratified lakes. The efficiency of energy transfer from turbulent kinetic energy to the raising of the potential energy is a function of the turbulent Froude number, with the efficiency reaching a peak value of 20% at a Froude number of unity. These results show that the TBBL is an important and efficient mixing mechanism within stratified lakes.

Acknowledgments. The authors would like to thanks Amir Etemad-Shahidi, Cathy Thomson and Angelo Saggio for all their assistance in preliminary data processing. This project was partially supported by funds from the Australian Research Council (ARC) large grant A89531551. This study could not have been carried out without the assistance and facilities of the Y. Alon Kinneret Limnological Laboratory, Israel Oceanographic & Limnological Research Ltd., and the active support by many of its staff members.

Reference

Dewey, R. K., W. R. Crawford, A. E. Gargett, and N. S. Oakey, A microstructure instrument for profiling oceanic turbulence in coastal bottom boundary layers, *J. Atmos. Oceanic Technol., 4*, 288-297, 1987.

De Silva, P., J. Imberger, and G. N. Ivey, Localised mixing due to a breaking internal wave ray at a sloping bed, *J. Fluid Mech. 350*, 1-27, 1997.

Erikson, C. C., Observation of internal wave reflection off sloping bottoms, *J. Geophys. Res., 87*, 525-538, 1982.

Erikson, C. C., Implications of ocean bottom reflection of internal wave spectra and mixing, *J. Geophys. Res., 87*, 1145-1156, 1985.

Etemad-Shahidi, A. and J. Imberger, Anatomy of turbulent in thermally stratified bodies, Report ED-952, Centre For Water Research, University of Western Australia, 1996.

Fréchette, M., C. A. Butman, and Geyer, W. R. The importance of boundary-layer flows in supplying phytoplankton to the benthic suspension feeder, M*ytilus edulis L. Limnol. Oceanogr., 34*, 19-36, 1989.

Garrett, C., P. MacCready, and P. Rhines, Boundary mixing and arrested Ekman layers: rotating stratified flow near a sloping boundary, *Ann. Rev. Fluid Mech., 25*, 291-323, 1993.

Gibson C. H. Fossil temperature, salinity and vorticity in the ocean, in *Marine Turbulence*, edited by J. C. T. Nihoul, Elsvier, 1980.

Gloor, M., Wüest, A., and M. Munnich, Benthic boundary mixing and resuspension induced by internal seiches, *Hydrobiologia*, 284, 59-68, 1994.

Gregg, M. C., H. E. Seim, and D. B. Percival, Statistics of shear and turbulent dissipation profiles in random internal wave fields, *J. Phys. Oceanogr., 2*, 1777-1799, 1993.

Imberger, J. and B. Boashash, Application of the Wigner-Ville distribution to temperature gradient microstructure: A new technique to study small-scale turbulence, *J. Phys. Oceanogr., 16*, 1997-2012, 1986.

Imberger, J. Transport processes in lakes: A review, in *Limnology Now: A Paradigm of Planetary Problems*, edited by R. Margalef, Elsevier Sciences, B. V., 1994.

Imberger, J. Flux path in a stratified lake: A Review, Byuoancy fluxes in a stratified fluid, in *Physical Processes in Lakes and Oceans*, edited by J. Imberger, pp 1-18, AGU, this volume, 1998.

Imberger, J. and G. N. Ivey, On the nature of turbulence in a stratified fluid. Part II: Application to lakes, *J. Phys. Oceanogr., 21*, 659-680, 1991.

Ivey, G. N. and J. Imberger, On the nature of turbulence in a stratified fluid. Part I: The energetics of mixing, *J. Phys. Oceanogr., 21*, 650-658, 1991.

Ivey, G. N., J. Imberger, and J. R. Koseff, Byuoancy fluxes in a stratified fluid, in *Physical Processes in Lakes and Oceans*, edited by J. Imberger, pp 377-388, AGU, this volume, 1998.

Ivey, G. N. and R. I. Nokes, Vertical mixing due to the breaking of internal waves on sloping boundaries, *J. Fluid Mech., 204*, 479-500, 1989.

Ledwell, J. R. and B. M. Hickey, Evidence of enhanced boundary mixing in the Santa Monica basin, *J. Geophys. Res., 100*, 20,665-20679, 1995.

Lemckert, C. J., A. Etemad-Shahidi, and J. Imberger, Mixing efficiencies in thermally stratified fluids, Proc. 12th Australasian Fluid Mechanics Conference, University of Sydney, Australia, December, 1995, 359-362, 1995.

Lemckert, C. J. and J. Imberger, Turbulence within inertia-buoyancy balanced axisymmetric intrusions, *J. Geophy. Res. (Oceans), 100*, 22649-22666, 1995.

Oakey, N. S. Determination of the rate of dissipation of turbulent kinetic energy from simultaneous temperature and velocity shear microstructure measurements, *J. Physical Oceanogr., 12*, 256-271, 1982.

Phillips, O. M. *The Dynamics of the Upper Ocean*, Cambridge University Press, 1977.

Saggio, A. and J. Imberger, Internal wave weather in a stratified lake, *Limnol and Oceanogr.*, in press, 1998.

Taylor, J. R. Turbulence and mixing in the boundary layer generated by shoaling internal waves, *Dynamics of Oceans and Atmospheres, 19*, 233-258, 1993.

Thorpe, S. A. Current and temperature variability on the continental slope, *Phil. Trans. R. Soc. Lon. A, 323*, 471-517, 1987.

Thorpe, S. A., P. Hall, and M. White, The variability of mixing at the continental slope, *Phil. Trans., R. Soc. Lon. A, 331*, 183-194, 1990.

Wüest, A., D. C. van Senden, J. Imberger, G. Piepke, and M. Gloor, Diapycnal diffusivity measured by microstructure and tracer techniques - A comparison. Preprints of Fourth International Stratified on Flows Symposium, Grenoble, France, June-July 1994. Vol. 3: B5 1-8, 1994.

36

Using Measurements of Variable Chlorophyll-*a* Fluorescence to Investigate the Influence of Water Movement on the Photochemistry of Phytoplankton

R. L. Oliver and J. Whittington

Abstract

The concepts of active fluorometry are introduced and their application to the measurement of variable chlorophyll-a fluorescence in phytoplankton is described. The technique provides information on the major processes involved in photochemistry including light capture and electron transport which in combination give an estimate of the rate of photosynthesis. The fluctuations in these variables in response to changing irradiance and nutrient conditions are illustrated using field and laboratory data on phytoplankton from four major taxa, and the results used to identify responses with time scales similar to those of mixing processes. Fluorometric estimates of photosynthetic rate are compared with oxygen electrode measurements showing excellent agreement and confirming the validity of fluorometric measurements. Effective absorption cross-sections were surprisingly stable under the measuring conditions with those of cyanobacteria generally being smaller than other algae. The freshwater species had significantly smaller cross-sections than those reported for marine phytoplankton. Mechanisms of fluorescence quenching responded to changing light intensities at time scales either in the micro- to milli-second range or hours to days, the longer time scales having potential for tracking phytoplankton movement. In particular the maximum quantum yield of fluorescence differed between species and responded to changing light conditions with time scales similar to those of mixing processes. The impact of mixing was demonstrated by monitoring changes in the quantum yield of natural phytoplankton populations over a range of depths under stratified and mixed conditions. The results clearly demonstrate that quantum yield measurements can provide a useful tracer of the vertical mixing of phytoplankton.

Introduction

Water movement is a predominant feature of the aquatic environment and plays a central role in establishing the fluctuating patterns in energy and nutrients encountered by the phytoplankton community. By modifying the availability of resources, and also through its role in cell suspension, water mixing plays a crucial part in setting the environmental

Physical Processes in Lakes and Oceans
Coastal and Estuarine Studies Volume 54, Pages 517-534
Published in 1998 by the American Geophysical Union

conditions to which the phytoplankton must respond if they are to grow successfully (Reynolds, 1998). The different capabilities of organisms comprising the phytoplankton lead to particular species being more successful under certain environmental conditions, resulting in changes of community composition over both seasonal and annual cycles. The intricacies of the interplay between water mixing, the supply of algal growth requirements, and changes in the biomass and composition of the phytoplankton community, are still poorly understood. This dearth of information is largely due to a mismatch in the scales at which physical and biological processes are currently measured. Investigations of mixing processes are generally underpinned by detailed temperature measurements that can be made rapidly and precisely, providing extensive detail at a range of scales. In comparison, measurements of biological processes are generally difficult and laborious so that data sets are limited in scale both temporally and spatially. Data interpretation is then vexed by problems of scaling from the measurement period to both smaller and larger scales, especially as the underlying mechanisms controlling the biological processes are not usually well understood. An example of particular importance in this context is the measurement of phytoplankton primary production rates.

Detailed measurements of phytoplankton photosynthesis have been restricted by the intensive labour and equipment requirements of the available techniques, and their need for extensive incubation periods. The enclosure of samples in bottles, and their exposure to fixed light intensities for extended periods of time, has been the source of considerable controversy regarding the degree to which the measurements accurately reflect in-situ rates of photosynthesis occurring in fluctuating light regimes as a result of water movement. Furthermore, although both the oxygen and 14C-bicarbonate techniques can be used to measure rates of photosynthesis under suitable conditions, they do not directly describe the mechanisms that underlie the photosynthetic process. Instead, the photosynthetic characteristics of the cell are derived by fitting empirical equations that describe the photosynthetic response to light intensity and from which "empirical" estimates of cellular attributes are deduced. This is a considerable weakness, as the ability of different organisms to adjust their photosynthetic system to changing environmental conditions influences their competitive success, and as a consequence influences the community composition (Reynolds, 1998). As adjustments in photophysiology may well occur in response to the altered environmental conditions caused by water movement, these characteristics should be useful in tracking the motion of phytoplankton and providing a link between water movement and cell function. Recent advances in techniques for measuring the variable chlorophyll-a fluorescence of phytoplankton, have provided means of circumventing some of the difficulties associated with current methods for measuring photosynthesis.

In the sixty-six years since Kautsky and Hirsch (1931) first reported that the fluorescence of chlorophyll-a varied in response to changes in the rate of photosynthesis, substantial progress has been made in understanding the cause of the variable fluorescence yield and in describing its connection with photochemistry (Govindjee, 1995). In the following sections we will briefly review the major principles of active fluorometry; describe techniques for using the variable fluorescence yield of chlorophyll-a to probe the condition of the photosystem; demonstrate that fluorescence techniques can be used to determine rates of photosynthesis; and illustrate the use of these measurements for investigating the response of phytoplankton to changing environmental conditions, especially changes caused by water mixing.

Active Fluorometry

The photosynthetic unit is divided into two major sections, Photosystem II (PS II) and Photosystem I (PS I), that are connected by an electron transport chain (Figure 1). Each of the photosystems is associated with an antenna of pigment molecules that captures the incident light and funnels the energy to reaction centres at the core of each photosystem. This energy is used to transfer an electron from the reaction centre to an acceptor molecule. In the case of PSII the primary acceptor is a quinone denoted as Q_A which is positioned at the start of the electron transport chain. Following the reduction of Q_A the newly oxidised reaction centre is again reduced by electrons derived from water, resulting in the evolution of oxygen. Meanwhile the electron at Q_A moves down the electron transport chain to PS I where, following a second excitation by the absorption of a photon of light, the electron is used in further reactions to reduce NADP. The rate of electron transfer down the transport chain is a function of the concentration and turnover time of the intermediates, and the rate of processing at PSI. The size and oxidation-reduction status of the plastoquinone (PQ) pool appears to be of particular importance in regulating the onward passage of electrons. Because the transfer of electrons from the PQ pool to PSI is a rate limiting step the PQ pool acts as a capacitor and its size is critical to the continued reduction of Q_A under fluctuating high light conditions (Kolber and Falkowski, 1993). Despite the presence of two reaction centres with associated chlorophyll molecules, the fluorescence signal that is obtained from intact cells at room temperature is almost solely from the chlorophyll associated with PS II.

The relationship between photosynthesis and variable chlorophyll-a fluorescence can be described using a simple model (Figure 2). In this model, energy captured by chlorophyll-a molecules can be dissipated through two major pathways, either it can be used in photosynthesis or it can be re-emitted by chlorophyll-a molecules in the form of light i.e. via fluorescence. Other competing pathways for energy dissipation do occur which result in thermal loss, but for this simple model they are initially considered to be constant.

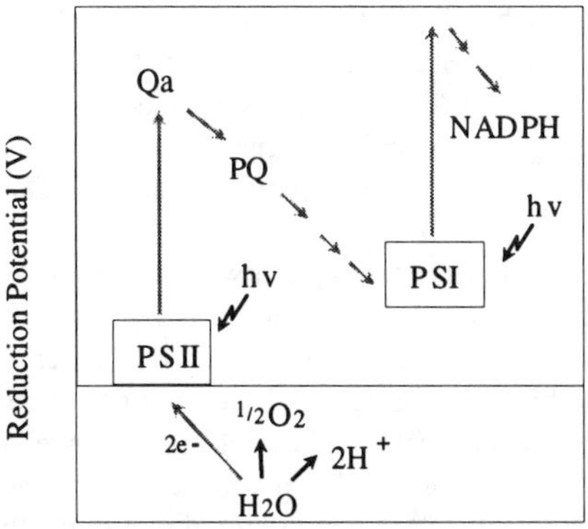

Figure 1. The light reactions of oxygenic photosynthesis showing Photosystem II (PSII) and Photosystem I (PSI) connected by an electron transport chain that commences with Q_A , the primary electron acceptor of PS II, and terminates at PSI. An important intermediate in the electron transport chain is plastoquinone (PQ) which plays a major role in regulating the flow of electrons to PSI. The absorption of a quanta of light (hv) by the antenna of each photosystem provides the energy for electron excitation. The reaction centre in PSII is oxidised by the transfer of an electron to Q_A, and is then reduced by electrons derived from water resulting in the evolution of oxygen.

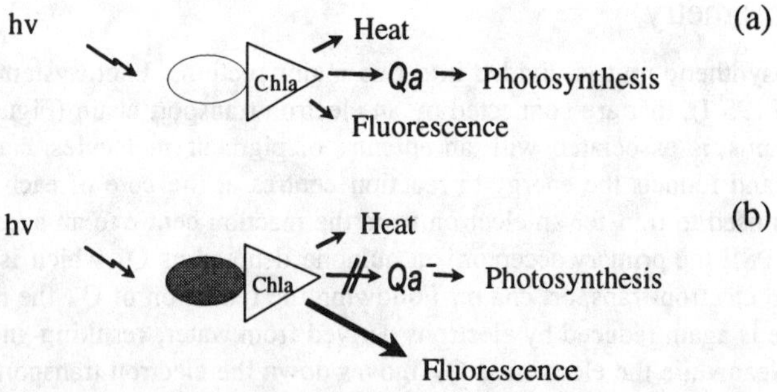

Figure 2. A simple conceptual model of the relationship between photosynthesis and the variable yield of chlorophyll-a fluorescence. The three competing dissipative pathways for light energy absorbed by the antenna molecules of PSII are heat, photosynthesis and fluorescence. In this simple model heat loss is constant and the intensity of the fluorescence output is a function of the reduction status of Q_A. If Q_A is oxidised then it can accept the energy from the absorbed photon of light (the trap is open) and energy is passed to photosynthesis with a minimal loss through chlorophyll-a fluorescence. If Q_A is reduced (the trap is closed) then energy cannot be passed on for use in photosynthesis and there is a maximal loss through chlorophyll-a fluorescence.

During the first step of photosynthesis, energy captured in the antenna of PS II drives the transfer of an electron from the reaction centre of the photosynthetic unit to Q_A (Figure 2a). If the antenna of PS II captures a photon of light while Q_A is still reduced by an electron from a previous photon hit, then the energy of the second photon cannot be transferred to Q_A (the trap is shut) and instead it is dissipated giving a maximal fluorescence signal (Figure 2b). Conversely, if an electron is passed from Q_A into the electron transport chain prior to the antenna capturing the next photon (the trap is open), then the captured energy is used in the reduction of Q_A and the fluorescence signal is minimal (Figure 2a). As each cell contains many photosynthetic units, the intensity of the fluorescence signal will depend on the relative proportion of open and closed traps, and as a result the chlorophyll-a fluorescence signal is variable. The presence of open traps effectively quenches the fluorescence signal and the analysis of these relationships is frequently referred to as fluorescence quenching analysis (Duysens and Sweers, 1963; Schreiber et al., 1986).

In the technique of "active fluorometry", a weak flash of light is used to probe the status of the photosystem which can be conditioned by a preceding bright flash of light (Figure 3). The intensity of the weak flash is such that a very small proportion of the photosynthetic units are hit by photons, and the majority of those hit intercept only a single photon. In dark adapted cells, where electrons have drained from the electron transport chain and all traps are open, the photochemical quenching of fluorescence is maximal. In this case the probe-flash stimulates a minimum fluorescence signal (f_o) as most energy goes to the reduction of Q_A (Figure 3). The proportion of closed traps can be increased by pre-conditioning the dark-adapted cells with a flash of light before measuring the fluorescence response to the weak probe light. By increasing the intensity of the pre-conditioning flash (Figure 3) the number of closed traps is increased until the fluorescence signal reaches a maximum (f_M) when all traps are closed and photochemical quenching is minimal. The relationship between trap closure and the intensity of the conditioning flash is used to assess the physiological status of the photosystem and to identify the impact of environmental conditions on primary production. These measurements can be made while cells are incubated under a background light intensity (an actinic light), so that information is obtained on the capacity of the cells to utilise the incident irradiance.

Only recently have fluorometers that are capable of making these measurements on natural phytoplankton populations become available commercially. Equipment suitable for use with higher plants has been available for a decade or more but has not been sensitive enough for use with natural phytoplankton populations, and even required algal cultures to be highly concentrated prior to use. Fluorometers specifically designed for phytoplankton studies have been constructed "in-house" by several research groups, particularly the group at Brookhaven National Research Institute, New York. This team has built benchtop and submersible versions of both Dual Flash Fluorometers (Kolber et al., 1990) and Fast Repetition Rate Fluorometers (Kolber and Falkowski, 1992), but unfortunately these are not available commercially.

In the last year the PAM Fluorometer (Pulse Amplitude Modulated fluorometer, WALZ Pty. Ltd.), a benchtop instrument that was originally designed for work with higher plants, had its range increased by the addition of a high-sensitivity cuvette holder (Schreiber, 1994, Schreiber et al., 1995) enabling it to be used on samples with low to moderate concentrations of phytoplankton cells (> 7 μg chlorophyll-a L^{-1}). This instrument can measure changes in variable chlorophyll-a fluorescence over a wide range of conditions, and in the presence of continuous background illumination. It is flexible enough to be configured to make many of the necessary fluorescence measurements, although some measurements are awkward as the equipment is not arranged to adjust automatically the intensity of the xenon conditioning flash. Measurements from our laboratory were made using this equipment.

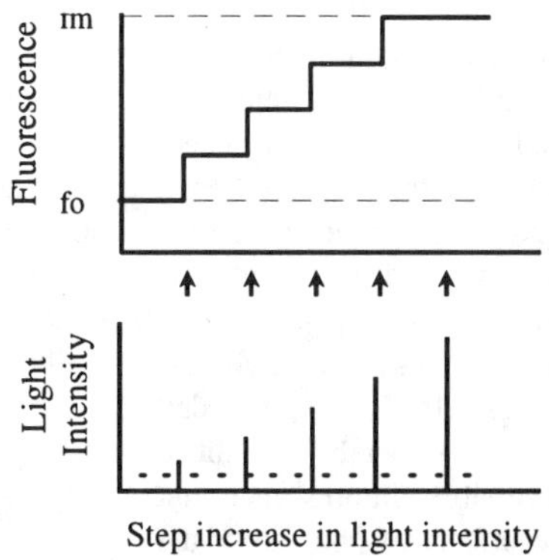

Figure 3. A diagrammatic representation of the application of active fluorometry to measure variable chlorophyll-a fluorescence. The upper panel depicts the fluorescence readings measured in response to the light conditions depicted in the lower panel. The measured fluorescence output is in response to the low intensity probe light (-) after pre-conditioning the photosystem with a flash of light (I). In fully dark adapted cells the fluorescence signal elicited by the probe light is minimal (f_O). The fluorescence signal from the probe light increases as the intensity of the pre-conditioning flash increases until all photosynthetic units are closed and the fluorescence signal reaches a maximum (f_M).

Characteristics of Photosynthesis Measured Using Active Fluorometry

Active fluorometry can be used to measure several major components of the photosynthetic process. Amongst these are three important characteristics; photochemical quenching (q_P) which is a measure of the relative number of RC II sites that are "open" and indicates the capacity of cells to carry out photosynthesis under the experimental conditions; the effective absorption cross-section (σ) of photosystem II which is a measure of the size of the antennae that the photosystem uses to capture light; and the maximum change in quantum yield ($\Delta\Phi_{MAX}$) which is a measure of the number of functional reaction centres. Individually each of these measurements provides information on the effectiveness of major steps in the light reactions of photosynthesis, and they can be used to assess the impact of changing environmental conditions on these processes. Together they provide a means of estimating the gross rate of photosynthesis.

Photochemical and non-photochemical quenching (q_P and q_N)

Photochemical quenching is a measure of the proportion of open reaction centres at an instant in time, i.e. the proportion of reaction centres able to process a photon and carry out photosynthesis. It is calculated as:

$$q_P = (f'_M - f_t) / (f'_M - f_O)$$

Here f_O is the fluorescence signal stimulated by the probe light immediately following dark adaptation of the sample (Figure 4), when all active reaction centres are open so that photochemical quenching is at a maximum; f'_M is the maximum fluorescence signal stimulated by the probe light and is measured in conjunction with a conditioning light flash of sufficient intensity to saturate (close) all of the open reaction centres so that photochemical quenching is at a minimum; f_t is the level of variable fluorescence stimulated by the probe when the sample is under a continuous background irradiance, the proportion of closed reaction centres depending on the intensity of the actinic irradiance (Figure 4).

The fluorescence signal can also be quenched as a result of energy loss through competing dissipative pathways that release the energy as heat. This leakage of energy away from fluorescence and photosynthesis is termed non-photochemical quenching (or sometimes "energy quenching") and is attributed to a number of processes. Under most conditions the major component of non-photochemical quenching is dependent on the formation of a pH gradient across the thylakoid membranes of the photosynthetic apparatus (Ruban and Horton, 1995). In the dark or under low light this gradient is dissipated relatively quickly. Under high light conditions other processes involving slowly reversible or irreversible fluorescence quenching can increase in importance. These are attributed to the effects of photo-oxidative damage to PSII and to a protective energy dissipation in the antennae (Long et al., 1994; Ruban and Horton, 1995).

The effect of non-photochemical quenching can be observed in a reduction of the level of f'_M (Figure 4). The amount of non-photochemical quenching is estimated as:

$$q_N = (f_M - f'_M) / (f_M - f_O)$$

where f_M is the maximum level of fluorescence observed after a period of dark adaptation sufficient to remove all non-photochemical quenching.

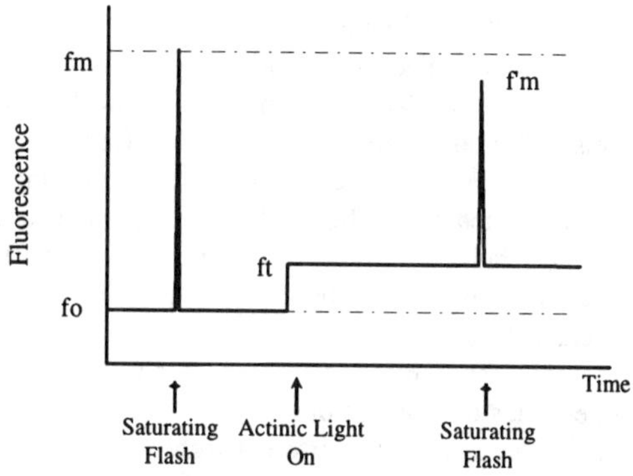

Figure 4. Definition of the measurements used in the analysis of variable chlorophyll-a fluorescence. In completely dark adapted cells the minimum fluorescence elicited by the probe light is f_O, while the maximum signal following a flash of light of sufficient intensity to close all PSII reaction centres without inducing non-photochemical quenching is f_M. The steady-state fluorescence signal measured from the probe light in the presence of a continuous background actinic light is f_t and is determined by the proportion of PSII reaction centres closed by the intensity of the actinic light. Momentarily overlaying the actinic light with a saturating flash yields a fluorescence signal f'_M which may be lower than f_M due to reduction of the fluorescence signal by non-photochemical quenching.

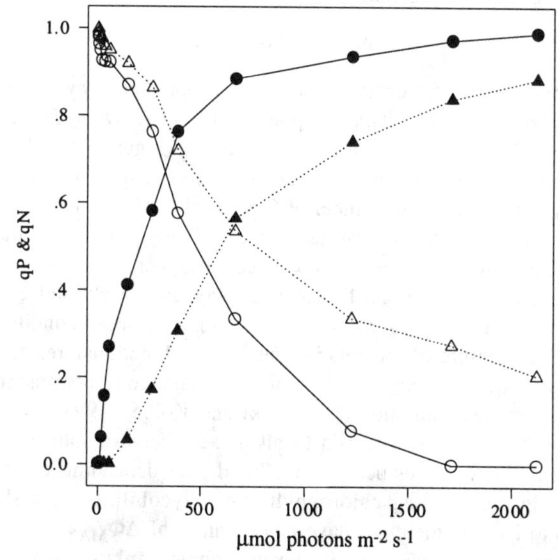

Figure 5. Changes in photochemical quenching (open symbols) and non-photochemical quenching (closed symbols) with increasing light intensity for cultures of the green alga *Selenastrum capricornutum* grown at either 16 μmol photons m^{-2} s^{-1} (solid lines) or 210 μmol photons m^{-2} s^{-1} (broken lines) incident irradiance.

Cells acclimatise to different light conditions by changing components of their photosystem so that both the onset, and the degree, of photochemical and non-photochemical quenching are altered. Experiments with the green alga *Selenastrum capricornutum* showed that cells acclimatised to a light intensity of 210 µmols photons m^{-2} s^{-1} were better able to process electrons through the electron transport chain than cells acclimatised to a light intensity of 16 µmols m^{-2} s^{-1} (Figure 5). The high-light grown cells had higher levels of photochemical quenching (more reaction centres open) particularly at higher light intensities, indicating a greater capacity for photosynthesis. In contrast, the cells grown under low light showed a rapid decline in photochemical quenching with increasing irradiance, indicating a reduced capacity for photosynthesis that was especially evident at higher irradiances. Rapid changes in q_P in response to fluctuations in light intensity occur in time ranges of microseconds to milliseconds, while adaptation of the photosystem, so that the level of q_P is altered at a given light intensity, requires changes on time scales of hours to days.

Adjustment of *Selenastrum* to the two different light intensities was also reflected by changes in non-photochemical quenching (Figure 5). In the low light grown cultures non-photochemical quenching commenced at 10 µmols photons m^{-2} s^{-1} whereas in the high-light adapted cells non-photochemical quenching was not apparent until ca. 90 µmols photons m^{-2} s^{-1}. At all light intensities non-photochemical quenching was less in the high light adapted cells.

Maximum quantum yield of fluorescence ($\Delta\Phi_{MAX}$)

The maximum change in quantum yield of fluorescence is calculated as :

$$\Delta\Phi_{MAX} = (f'_M - f_O) / f'_M$$

Experimentally this is measured on short-term dark adapted cells by measuring f_O with the probe light immediately after dark adaptation and f'_M with the probe light after conditioning the cells with a saturating light flash (Figure 4). The length of dark adaptation depends upon the previous light exposure and ranges from seconds to minutes. $\Delta\Phi_{MAX}$ provides a measure of the number of functional PSII reaction centres. It has been suggested from measurements on marine species of phytoplankton including diatoms, chlorophytes and haptophytes, that $\Delta\Phi_{MAX}$ has an upper value of ca. 0.65 under ideal growth conditions (Kolber, Zehr and Falkowski, 1988; Falkowski and Kolber, 1993). By comparing the value of $\Delta\Phi_{MAX}$ measured under experimental conditions to the upper maximum of 0.65 a measure of the relative number of functional reaction centres can be obtained. This ratio then accounts for physiological damage to the reaction centres due to photoinhibition or nutrient limitation (Falkowski and Kolber, 1993). However, the upper value of 0.65 for $\Delta\Phi_{MAX}$ does not extend to all algae. For example, Buchel and Wilhelm (1993) report that $\Delta\Phi_{MAX}$ varies between 0.37 and 0.83 depending upon the algal group. They argued that algae containing chlorophyll-c or phycobilins often show very high f_O values in relation to f_M, and therefore have lower values of $\Delta\Phi_{MAX}$.

We have observed in cultures of freshwater phytoplankton that species from the different major divisions do have quite different upper limits to their $\Delta\Phi_{MAX}$ values. The cyanobacteria had lower values with *Anabaena circinalis* producing an upper value for $\Delta\Phi_{MAX}$ of 0.6, and *Microcystis aeruginosa* a maximum of 0.54. The diatom *Melosira granulata* gave a value of 0.69, somewhat higher than that observed in marine diatoms and similar to the green algae *Selenastrum capricornutum* and *Oocystis* sp which had upper values for $\Delta\Phi_{MAX}$ of 0.7. In contrast the green alga *Volvox* sp. had an upper value of 0.81. These differences suggest major variations in the strategies used by these organisms to adjust the number and function of their reaction centres in response to changing light conditions.

Changes in $\Delta\Phi_{MAX}$ can occur over a range of time scales depending on the irradiance and nutrient conditions in the environment. Under high light conditions a rapid onset of non-photochemical quenching, including photoinhibitory responses in the reaction centres, causes an almost instantaneous decrease in $\Delta\Phi_{MAX}$. Under reduced light intensities $\Delta\Phi_{MAX}$ declines more slowly as non-photochemical quenching increases over time, while at low light intensities, where non-photochemical quenching does not occur, $\Delta\Phi_{MAX}$ is unaffected. The extent of these responses depends on the intensity of the irradiance, and the duration of exposure, and is evident in the kinetics of the systems recovery when returned to lower light intensities. In Figures 6 and 7 are shown the changes in $\Delta\Phi_{MAX}$ prior to, during and after exposure to a high light intensity. These responses were measured in samples dominated by the green alga *Oocystis* sp. collected from Chaffey Reservoir, Australia (31° 21'S, 151° 8'E). Maximum quantum yield measured under the prevailing experimental irradiances (Figure 6) showed a rapid decline with the onset of high light, reducing from 0.65 to 0.15 in < 2 minutes. On return to low light (20 μmol photons m^{-2} s^{-1}) the $\Delta\Phi_{MAX}$ recovered in several stages. Initially there was an almost instantaneous recovery of 75% of the original $\Delta\Phi_{MAX}$ to 0.49, followed by a much slower recovery over the following 40 minutes to a value of 0.62, which was about 95% of the original value. The immediate recovery in $\Delta\Phi_{MAX}$ is due to the relaxation of photochemical quenching and the faster components of non-photochemical quenching, while the continued recovery is due to slower components of non-photochemical quenching.

In a second experiment a sample was given a longer dose of more intense light, but in this case $\Delta\Phi_{MAX}$ was not measured under the prevailing light conditions, but was measured after a 12 minute period of dark adaptation (Figure 7). This removed from the

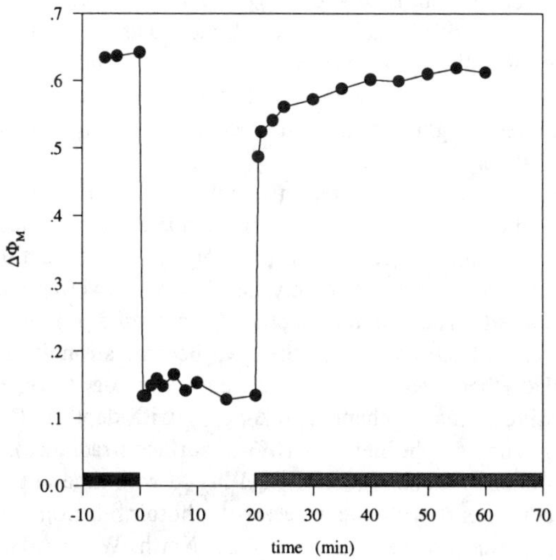

Figure 6. Changes with time in the maximum change in quantum yield of chlorophyll-a fluorescence measured in the dark, under an incident irradiance of 1100 μmol photons m^{-2} s^{-1}, and under an incident irradiance of 20 μmol photons m^{-2} s^{-1}. The three periods are indicated by time bars at the bottom of the graph. The sample was from Chaffey Reservoir and was dominated by the green alga *Oocystis* sp.

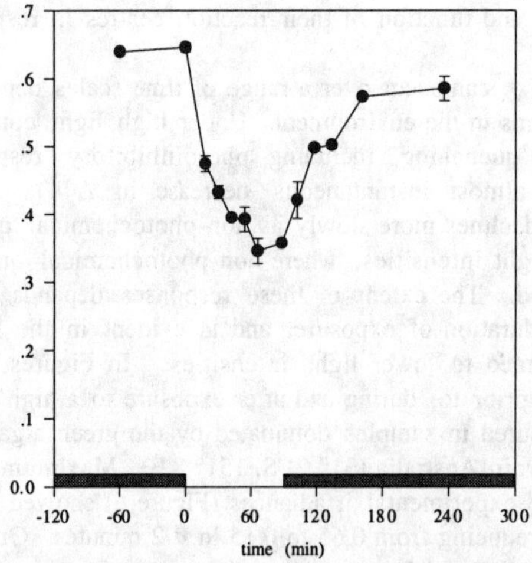

Figure 7. Changes with time in the slowly recovering components of the maximum change quantum yield of fluorescence measured after 12 minutes dark adaptation following incubation at 20 μmol photons m^{-2} s^{-1}, 1800 μmol photons m^{-2} s^{-1}, and on return to 20 μmol photons m^{-2} s^{-1}. The three periods are indicated by time bars at the bottom of the graph. The sample was from Chaffey Reservoir and was dominated by the green alga *Oocystis* sp.

measurements those components that rapidly recovered so that changes in the slower components, including photo-oxidative damage to PSII and energy dissipation in the antennae, could be followed. During the period of bright illumination the effects of the slow components increased with time and $\Delta\Phi_{MAX}$ reduced from 0.65 to 0.35 in 60 minutes (Figure 7). Following transition back to a low light intensity (20 μmol photons m^{-2} s^{-1}) the minimum value of $\Delta\Phi_{MAX}$ (0.35) was smaller than that observed in the previous experiment (0.49) where the intensity and duration of the light treatment was less (Figure 6). After the more intense light treatment the recovery of $\Delta\Phi_{MAX}$ was slower taking 2.5h to reach ca. 95% of its original value.

The size of $\Delta\Phi_{MAX}$ was also measured in samples collected from a number of depths in Chaffey Reservoir and compared with changes in temperature stratification (Figures 8a and b). The time between sample collection and its analysis by active fluorometry was of the order of 15 to 30 minutes, so that it was only the slowly recovering component of $\Delta\Phi_{MAX}$ that was being monitored. The euphotic depth (1% transmission) was ca. 3.7m.

During the first half of the day, when the lake became strongly stratified and vertical mixing was curtailed, there was a maximum depression of $\Delta\Phi_{MAX}$ in the near surface sample and progressively smaller changes in $\Delta\Phi_{MAX}$ with depth until ca. 1.8m, the mid-point of the euphotic zone (light intensity 10% of surface irradiance). Below this $\Delta\Phi_{MAX}$ did not change significantly. The reduction of $\Delta\Phi_{MAX}$ with time in the upper layers of the euphotic zone is consistent with the effects of photoinhibition and is similar to the phenomena described for phytoplankton in the North West Atlantic (Falkowski and Kolber, 1990) and from the eastern equatorial Pacific (Greene et al., 1994).

Later in the day the isotherms indicate that two brief surface mixing events occurred, one at 15:30 and the other at 17:00, both penetrating to a depth of ca. 1m (Figures 8a and b). The first mixing event corresponded to an increase in the $\Delta\Phi_{MAX}$ of the near surface sample to a value equivalent to that observed in the sample from 0.55m depth. The reformation of stratification after 17:00 corresponds with a slight reduction in the $\Delta\Phi_{MAX}$ of the surface sample measured at 17:50, even though $\Delta\Phi_{MAX}$ in the deeper layers was

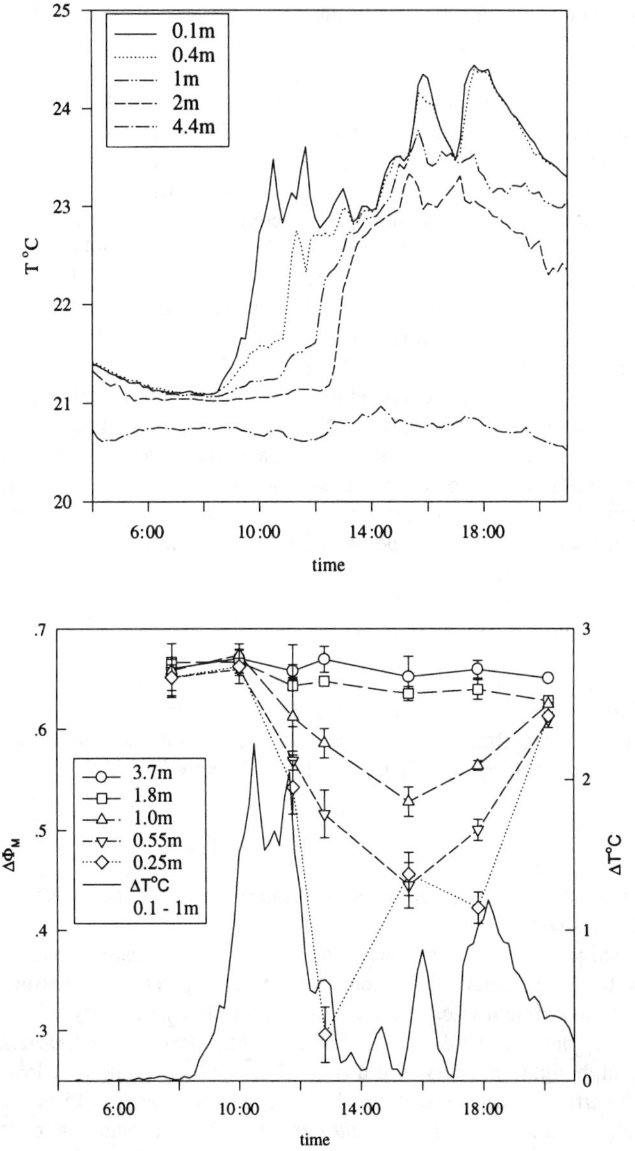

Figure 8. Changes with time in (a) the thermal stratification and (b) the temperature change between 0.1m and 1m depths along with the maximum change in quantum yield of chlorophyll-a fluorescence at five depths between 0.25m and 3.7m as shown on the key. Measurements were made at Chaffey Reservoir when the algal community was dominated by the green alga *Oocystis* sp.

increasing as a result of the afternoon decline in incident irradiance. The details of the length and time scales of the mixing processes for this period have yet to be assessed, however even from this initial data it is evident that changes in the $\Delta\Phi_{MAX}$ will be useful for appraising the light environment experienced by phytoplankton and from this estimating phytoplankton mixing trajectories for comparison with mixing patterns derived from thermal changes.

In general, the recovery times of $\Delta\Phi_{MAX}$ appear to range from seconds to minutes when energy quenching is largely due to effects associated with the formation of a trans-thylakoid proton gradient and state transitions. Recovery time is extended to tens of minutes or hours when exposure to light is sufficient to cause protein breakdown in the reaction centres, or following relatively large exposures when protective pathways such as the xanthophyll cycle are substantially triggered. In general, recovery of $\Delta\Phi_{MAX}$ from non-photochemical quenching is faster in low light than in complete darkness.

Changes in $\Delta\Phi_{MAX}$ can provide a useful tool for assessing the light history of phytoplankton, however care is required as it can be affected also by nutrient availability (Kolber et al., 1988). In addition the ability of nutrient deficient phytoplankton to recover from non-photochemical quenching is impaired relative to nutrient sufficient algae (Greene et al., 1994). In continuous cultures of the cyanobacterium *Microcystis aeruginosa* grown under phosphorus-limited conditions, the maximum change in $\Delta\Phi_{MAX}$ decreased in response to phosphorus limitation as a function of relative growth rate. This response is similar to that observed in several species of marine phytoplankton under nitrogen limited conditions (Kolber et al., 1988). The time required for nutrient induced alterations in $\Delta\Phi_{MAX}$ appears to be on the order of days, although definitive data is not available to confirm this.

Effective absorption cross-section (σ)

The effective absorption cross-section is determined from the rate of increase in variable chlorophyll-a fluorescence with an increase in intensity of the conditioning light flash (Figure 3). The flash intensity saturation curve follows the cumulative single-hit Poisson distribution (Ley and Mauzerall, 1982; Falkowski et al., 1986; Falkowski and Kolber, 1993):

$$(f_T - f_O) / (f_M - f_O) = 1 - e^{-\sigma I}$$

where I is the flash intensity in quanta Å^{-2} flash^{-1}, and σ is the effective (or functional) absorption cross-section in Å^2 quanta^{-1}.

The effective absorption cross-section of the light harvesting antenna can change in response to the light climate encountered by the cells, generally increasing in size as cells acclimatise to lower light intensities, and decreasing in size as they adjust to higher light intensities. Ley and Mauzerall (1982) showed that the green algae *Chlorella vulgaris* when grown at a high light intensity had a σ of 44 Å^2 quanta^{-1} while at a low light intensity increased the effective target size per PSII trap to 115 Å^2 quanta^{-1}. In our measurements on the green alga *Selanastrum capricornutum* σ was 101 Å^2 quanta^{-1} in cells grown at 211 μmol photons m^{-2} s^{-1} and increased to 137 Å^2 quanta^{-1} in cells grown at 16 μmol photons m^{-2} s^{-1}.

Measurements on freshwater phytoplankton suggests that the average cross-section size, and its range of adjustment, varies between different species. The common bloom forming cyanobacteria *Anabaena circinalis* and *Microcystis aeruginosa* generally have smaller σ than commonly observed eukaryotic phytoplankton such as the diatom *Melosira granulata* or green algae such as *Selenastrum capricornutum*, *Oocystis* sp., or *Ankistrodesmus* sp. *Volvox* sp. is an exception to this pattern having a small cross-section comparable in size to that of the cyanobacteria (Table 1).

If the size range of the effective absorption cross-section is indicative of the light conditions to which a cell is best suited, then these results indicate that the bloom forming cyanobacteria are better adapted to higher light intensities than the eukaryotic algae. The exception in Table 1 is the green alga *Volvox* sp., which is motile, and under stratified conditions can migrate towards the surface. Perhaps the smaller cross-sections of the cyanobacteria and *Volvox* are associated with their comparable ability to move into the well illuminated surface layers, and this could provide one explanation for their demise under mixed conditions where the average irradiance is reduced.

Effective absorption cross-sections measured in marine phytoplankton species range in size from 200 to > 900 $Å^2$ $quanta^{-1}$ (Kolber and Falkowski, 1992; 1993; Greene et al., 1991) and appear in general to be considerably larger than for freshwater species (Table 1). The reason for this difference is not apparent. It might be due to more frequent nutrient limitation in oceanic waters, or perhaps to limitation by specific nutrients that especially impact on the photosynthetic unit, such as iron and nitrogen (Greene et al., 1991; Kolber et al., 1988). However the differences might also indicate that marine species are generally adapted to utilising a poorer mean light climate that freshwater species.

Physiological changes, such as State 1 - State 2 transitions, that result in alterations to the effective absorption cross-section through a redistribution of absorbed excitation energy can occur within minutes (Mullinaux et al., 1986). In contrast structural changes such as pigment synthesis require time scales of hours or days to occur.

As with $\Delta\Phi_{MAX}$, changes in effective absorption cross-section also respond to nutrient limitation (Kolber et al., 1988). It has been observed in marine phytoplankton that iron or nitrogen limitation can lead to an increase in effective absorption cross-section (Kolber et al., 1988; Falkowski, 1992) despite a reduction in cellular pigment concentration. This has been attributed to a greater relative loss of reaction centres than antenna pigments and a pooling of antennae between reaction centres.

TABLE 1. Absorption cross-section measurements of cells grown under a range of light conditions in laboratory cultures and sampled from the phytoplankton community in Chaffey Reservoir and Green Pond.

Organism	Origin	Absorption cross-section ($Å^2$ $quanta^{-1}$)
Anabaena circinalis	Culture	29 to 54
Anabaena circinalis	Chaffey R.	32
Microcystis aeruginosa	Culture	31 to 63
Microcystis aeruginosa	Chaffey R.	35 to 37
Melosira granulata	Culture	86 to 178
Selenastrum capricornutum	Culture	101 to 137
Oocystis sp.	Chaffey R.	75 to 85
Ceratium sp.	Chaffey R.	59 to 60
Ankistrodesmus sp.	Green Pond	74
Volvox sp.	Chaffey R.	24

Photosynthesis

Assessing the condition of particular photosynthetic processes through the use of chlorophyll fluorescence measurements provides insight to the adaptability of the cells and their responses to changing environmental conditions. However a critical test of these fluorescence measurements is that they can be combined together to determine the rate of photosynthesis. An equation to do this was proposed by Falkowski and Kolber (1993):

$$P = I \sigma q_P (\Delta\Phi_{MAX}/0.65) \Phi_e \eta_{PSII}$$

where P is the rate of photosynthesis (mol O_2 mol chla^{-1} h^{-1}), I is the irradiance intensity (mols photons m^{-2} h^{-1}), σ the effective absorption cross-section (m^2 mol quanta^{-1}), $\Delta\Phi_{MAX}/0.65$ the fraction of functional reaction centres, Φ_e the mols of oxygen evolved per photon processed by the reaction centres, and η_{PSII} the number of functional PSII reaction centres per mol of chlorophyll-a or the number of photons processed per mol of chlorophyll-a. To date this equation has not been extensively tested despite its importance in demonstrating the validity of the fluorescence measurements and its usefulness in assessing primary production rates. We examined the relationship under defined laboratory conditions using cultures of *Anabaena* sp., *Microcystis* sp. and *Melosira* sp. Rates of photosynthesis estimated from the fluorescence measurements, and expressed in terms of oxygen evolution, were successfully matched with rates measured directly using an oxygen electrode (Figure 9).

Fluorescence estimates of the rates of phytoplankton photosynthesis were also made on field samples, to provide better insight to some of the causes in variation of phytoplankton primary production. A common observation when measuring primary production over 24 hour periods, is a depression in the rate of photosynthesis during the afternoon. Although suggestions have been made about the likely reason for this decrease in photosynthesis a clear demonstration of its cause has not been provided. Measurements were made on depth integrated samples taken from a pond that contained a mixture of green algae dominated by *Ankistrodesmus* sp. The pond remained isothermal throughout a warm

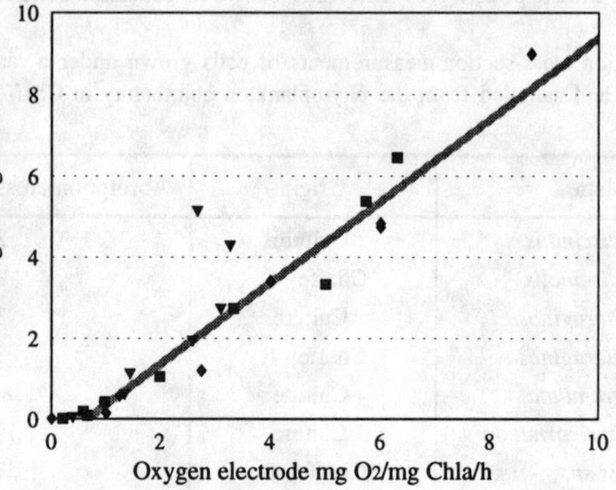

Figure 9. A scatter plot showing the correlation between rates of photosynthesis measured using an oxygen electrode and a PAM fluorometer for the algal species Microsystis sp. (■), Anabaena sp. (◆), and Melosira sp. (▼).

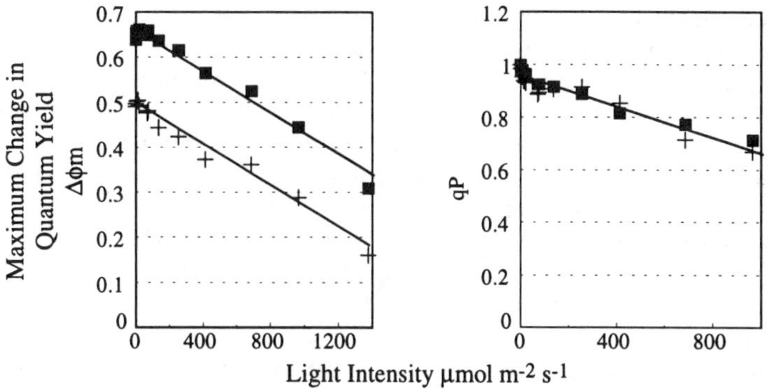

Figure 10. The effect of light intensity on the maximum change in quantum yield and photochemical quenching in samples taken at 08:20h (squares) and at 16:05h (crosses) from Green Pond and dominated by the green alga *Ankistrodesmus* sp.

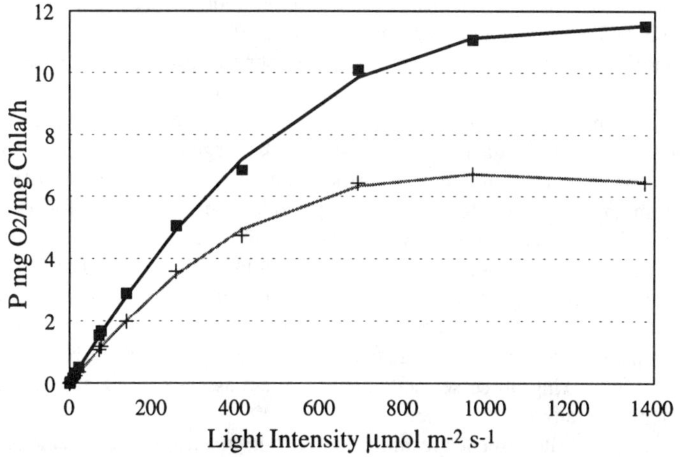

Figure 11. Photosynthesis-irradiance relationships for samples taken from Green Pond at 08:20h (dark line) and 16:05h (light line). The phytoplankton community was dominated by the green alga *Ankistrodesmus* sp. and rates of photosynthesis were estimated using fluorescence measurements.

day suggesting a reasonable degree of mixing, and sunlight penetrated to the bottom. Fluorescence measurements indicated that the depression in photosynthetic rate observed during the afternoon was due to a reduction in $\Delta\Phi_{MAX}$ (Figure 10). There was no change in the effective absorption cross-section (data not shown) or photochemical quenching (Figure 10). It appears that these organisms respond to high light intensities by reducing the number of functional reaction centres over time. As the effective cross-section remained constant it would appear that the reduction in $\Delta\Phi_{MAX}$ was not due to a change in the rate of thermal dissipation in the antennae, but that the antennae-reaction centre complex was disabled. The effect of this change on photosynthesis-irradiance curves was to reduce the initial slope and the maximum photosynthetic rate in the same proportion (Figure 11).

The quantitative influence of photoinhibition on the primary production of phytoplankton communities is complicated by assessing the rate of vertical mixing *in situ* (Long et al., 1994, Platt et al., 1980). The diurnal reduction in the photosynthetic rate of the phytoplankton in Green Pond clearly shows that photoinhibition can considerably reduce carbon fixation rates and we are now attempting to increase the spatial and temporal scales of this work by incorporating measurements of mixing processes. The likely significance of these physiological adjustments to algal growth are well described by Reynolds (1998). The techniques of active fluorometry provide the means to measure these changes both in natural populations and in algal cultures and to assess responses to changing environmental conditions. This information is critical to the development of algal growth models capable of describing the competition between species and accounting for seasonal changes in species dominance, as well as longer term shifts in community structure to less desirable species.

Conclusions

The technique of active fluorometry provides a powerful tool for investigating the photobiology of phytoplankton using the variable fluorescence signal of chlorophyll-a. It is a rapid and non-invasive approach enabling the physiological condition of the phytoplankton to be assessed with a minimum amount of prior manipulation, in both the field and the laboratory. It provides measurements on a number of specific photochemical processes so that changes in their characteristics can be related to changing environmental conditions, and to the growth of the organisms. As the photochemical attributes respond to changing environmental conditions with characteristic time scales ranging from microseconds to days, they provide useful markers for connecting the physiology of the organisms to physical and chemical conditions. In particular, light induced changes in the maximum quantum yield of fluorescence, the degree of photochemical quenching, and the size of the photosystem II antenna, provide cell markers that can be used to trace cell movement through the light field. This information can be compared with implied changes in the light climate derived from physical measurements of mixing processes and enable a direct assessment of the influence of water motion on phytoplankton movement.

The value of the fluorescence technique is further enhanced by the demonstration that it can be used to measure rates of phytoplankton photosynthesis, providing a connection between photosynthesis-irradiance relationships and major characteristics of the photochemical system. These fluorescence techniques enable rates of photosynthesis to be measured more quickly and with less likelihood of experimental artefacts than in methods that are commonly used for measuring photosynthesis in the field. This improved capacity will enable measurements of photosynthesis to be made over more appropriate time and space scales than could be achieved in the past, improving our knowledge of production rates in aquatic ecosystems.

A major impediment to the progress of this research is the dearth of suitable fluorometers for measuring the variable chlorophyll-a fluorescence of phytoplankton samples. Although we have used a benchtop PAM fluorometer to appraise many of the major fluorescence characteristics of phytoplankton, and have successfully applied it to estimate rates of photosynthesis, it requires some improvements to extend its usefulness and to increase its efficiency. However at this point in time we are not aware of any other commercially available equipment that is suitable for making these measurements. The problem is further exacerbated by a complete lack of submersible instruments capable of making the appropriate measurements for fluorescence quenching analysis *in situ*. The only submersible fluorometers available for this work have been built by research groups and are not commercially available. Until suitable equipment becomes widely available these techniques will continue to be restricted in their application.

Acknowledgments. This research was supported by grants to RLO from the Land and Water Resources Research and Development Corporation (MDR8), the Natural Resources Management Strategy of the Murray-Darling Basin Commission (M302), the Commonwealth Scientific and Industrial Research Organisation and the Co-operative Research Centre for Freshwater Ecology. We are grateful to M. Wood, M. Fink, G. Linde and D. Green for assistance and valuable discussion.

References

Buchel, C. and C. Wilhelm, In vivo analysis of slow chlorophyll fluorescence induction kinetics in algae: progress, problems and perspectives. *Photochemistry and Photobiology*. *58*, 137-148, 1993.

Duysens, L. N. and H. E. Sweers, Mechanisms of two photochemical reactions in algae as studied by means of fluorescence. In *Studies on Microalgae and Photosynthetic Bacteria,* edited by S. Miyachi 353-372, 1963.

Falkowski, P. G., Molecular ecology of phytoplankton photosynthesis, in *Primary productivity and biogeochemical cycles in the sea*, edited by P. G. Falkowski and A. D. Woodhead, 47-67, 1992.

Falkowski, P. G. and Z. Kolber, Phytoplankton photosynthesis in the Atlantic Ocean as measured from a submersible pump and probe fluorometer *in situ*, In *Current Research in Photosynthesis*, V. 4, 923-926, 1990, Kluwer.

Falkowski, P. G. and Z. Kolber, Estimation of phytoplankton photosynthesis by active fluorescence. ICES mar. Ser. Symp., *197*, 92-103, 1993.

Falkowski, P., K. Wyman, A. C. Ley and D. Mauzerall, Relationship of steady-state photosynthesis to fluorescence in eucaryotic algae, *Biochimica et Biophysica Acta*, *849*, 183-192, 1986.

Govindjee. Sixty-three years since Kautsky: Chlorophyll-a fluorescence, *Aus. J. Plant Physiol.*, *22*, 131-160, 1995.

Greene, R. M., R. J. Geider and P. G. Falkowski., Effect of iron limitation on photosynthesis in a marine diatom, *Limnol. Oceanogr. 36,* 1772-1782, 1991.

Greene, R. M., Z. S. Kolber, D. G. Swift, N. W. Tindale, and P. G. Falkowski, Physiological limitation of phytoplankton photosynthesis in the eastern equatorial Pacific determined from variability in the quantum yield of fluorescence, *Limnol. Oceanogr. 39*:1061-1074, 1994.

Kautsky, H. and A. Hirsch, Neue Versuche zur Kohlensaureassimilation, *Naturwissenschaften*, *19*, 964, 1931.

Kolber, Z. and P. G. Falkowski, Fast repetition rate (FRR) fluorometer for making in situ measurements of primary productivity. In *Mastering the Oceans Through Technology*, Proc. Oceans '92, IEEE, p 637-641, 1992.

Kolber, Z. and P. G. Falkowski, Use of active fluorescence to estimate phytoplankton photosynthesis *in situ*, *Limnol. Oceanogr.*, *38*, 1646-1665, 1993.

Kolber, Z., K. Wyman and P. G. Falkowski, Natural variability in photosynthetic energy conversion efficiency: A field study in the Gulf of Maine, *Limnol. Oceanogr.*, *35*, 72-79, 1990

Kolber, Z., J. Zehr, and P. Falkowski, Effects of growth irradiance and nitrogen limitation on photosynthetic energy conversion in photosystem II, *Plant Physiol.* 88, 923-929, 1988.

Ley, A. C. and D. C. Mauzerall, Absolute absorption cross-sections for photosystem II and the minimum quantum requirement for photosynthesis in Chlorella vulgaris, *Biochimica et Biophysica Acta.*, *680*, 95-106, 1988.

Long, S. P., S. Humphries, and P. G. Falkowski, Photoinhibition of photosynthesis in nature, *Annu. Rev. Plant Physiol. Plant Mol. Biol.*, *45*, 633-662, 1994.

Platt, T., C. L. Gallegos and W. G. Harrison., Photoinhibition of photosynthesis in natural assemblages of marine phytoplankton, *J. Mar. Res.*, *38*, 687-701, 1980.

Reynolds, C. S., Plants in motion: physical-biological interaction in the plankton, in *Physical Processes in*

Lakes and Oceans, edited by J. Imberger, pp 535-560, AGU, this volume, 1998.

Ruban, A. V. and P. Horton, Regulation of non-photochemical quenching of chlorophyll fluorescence in plants. *Aust. J. Plant Physiol., 22*, 221-230, 1995.

Schreiber, U., New emitter-detector-cuvette assembly for measuring modulated chlorophyll fluorescence of highly diluted suspensions in conjunction with the standard PAM fluorometer, *Zeitschrift fur Naturforschung, 49c*, 646-656, 1994.

Schreiber, U., H. Hormann, C. Neubauer and C. Klughammer, Assessment of Photosystem II photochemical quantum yield by chlorophyll fluorescence quenching analysis, *Aust. J. Plant Physiol., 22*, 209-220, 1995.

Schreiber, U., U. Schliwa and W. Bilger, Continuous recording of photochemical and non-photochemical chlorophyll fluorescence quenching with a new type of modulation fluorometer, *Photosynth. Res., 10*, 51-62, 1986.

37

Plants in Motion: Physical - Biological Interaction in the Plankton

C. S. Reynolds

Abstract

The special constraints posed on the evolutionary and functional adaptations of plant life in the open water of the pelagic are rehearsed. Microscopic size exploits the mechanical support and entrainment properties of water but avoids the stresses imposed by even the smallest turbulent scales. Obtaining carbon, nutrients and, especially, light in mixed and often rarefied environments is reviewed. The availabilities of each set upper limits to the carrying capacities of given systems, although the maximum rates of uptake comfortably saturate the requirements for growth. Unless one or other of these requirements is effectively exhausted, the plant processes its resources to an even rate of self-replication, which will often be the fastest that can be supported at that temperature and for that photoperiod. This principle is fitted into contemporary growth models and assists in the estimation of production rates from biomass determinations.

Introduction

Agriculturalists are interested in the soil that supports their crops. Terrestrial plant ecologists are interested also in the way that latitude and aspect affect the microclimates in which plants grow. Among plankton ecologists, the concern focussed upon the importance of water chemistry and upon the competition for nutrients has outweighed the attention afforded to the physical quality of the environment. This is perhaps a pity. Because the immediate environment of planktonic plants is, at once, in three-dimensional, fluid motion, it might be deduced that it is the first and most important environmental feature to which the plants must adapt. The first objective of this paper is to emphasise how the biological form and function of planktonic plants are embedded into the fluidity of the physical environment. Attention is given to the criteria for entrainment of plankton within the mixed layer and to the reciprocal argument that mixing processes most influences size in planktonic organisms. Secondly, it seeks to recount the ways in which planktonic plants deal with the scales of variability they experience, as a direct consequence of their entrainment within open-water environments. These mechanisms permit planktonic plants not merely to survive in dynamic environments but also to increase their populations; indeed, they are shown to be at the heart of the natural regulation of what is generally referred to as "primary production".

Physical Processes in Lakes and Oceans
Coastal and Estuarine Studies Volume 54, Pages 535-560
Published in 1998 by the American Geophysical Union

It is, of course, fully acknowledged that quantitative relationships between production and such physical properties of the aquatic environment as its temperature and underwater light-attenuation have been studied extensively in the past (see, for instance, Harris, 1978). Recent renewal of interest in these relationships perhaps owes most to advances in satellite-based remote-sensing and imagery of phytoplankton and the possibility of its use in estimating remotely the primary production of the sea (Kirk, 1994). This progress coincides with rapid advances in the understanding of the contributory processes, at both the smallest (the molecular biology of the reactivity of the photosynthetic apparatus: e.g., Falkowski, 1992) and largest scales (biological oceanography and limnology: e.g. Mann and Lazier, 1991). Thus, the third objective of the paper is to show how the *regulation* of primary production is related to the *scales* of physical variability. Besides providing a context to the presentations in this segment of the IUTAM Symposium, I hope that this paper may reinforce the profitable linkages which have already been forged among physical limnology and oceanography, on the one hand, and plankton biology, on the other.

Plant Life of Pelagic Habitats

It is still rather unconventional to refer to micro-organisms as plants or even vegetation so my first task is to clarify usage. Away from peripheral wetlands (dominated by vascular macrophytes) and marginal shallows (where both macrophytes and macrophytic algae may be abundant), the primary producers of open water are exclusively microphytic and planktonic. Depending upon the authority followed, there are between seven and ten recognized phyla from two great kingdoms represented in the freshwater phytoplankton. There are probably in the order of 4 000 or so described species. Despite being universally microscopic, they cover a four-orders of magnitude range of size, from picoplanktonic cells of 0.2 μm in diameter to mesoplanktonic colonies of 2 mm or more. They share with land plants the universal property of containing chlorophyll *a*, plus a range of accessory pigments. When incorporated into light-harvesting organelles, these intercept light energy and use it to catalyse the most powerful of all biological reduction reactions, namely that which splits water molecules to supply the reductant required to build carbon dioxide molecules into high-energy carbohydrates. They can accumulate condensation products, like starch and oils, and they synthesise amino-acids and the proteins which build the cells of the next generations. Also as on the land, oxygen gas is liberated in consequence of primary production: around 25% of the atmospheric oxygen is generally reckoned to be derived from the photosynthesis of planktonic micro-organisms of the oceans, lakes and rivers. This emphasises the scale of planktonic plant production. In addition to supplying oxidant for animal respiration, primary producers also supply organic carbon to animals as food and to bacterial processes as substrate. In this respect, too, the phytoplankton mirrors the functions of terrestrial vegetation and is just as much the base of pelagic food webs as the comparable spectrum of trees and shrubs, herbs and grasses, spermatophytes, pteridophytes and bryophytes constitute that of the land.

From biochemical, physiological and ecological standpoints, then, there is no clear distinction between terrestrial and pelagic vegetation, save that of scale – the spatial scales of organism size, the physical scales of the habitat variability and the temporal scales in which living processes can be accommodated. The evolutionary significance of these differences has largely been ignored, which is perhaps why the popular understanding of "vegetation" does not intuitively embrace the plant life of pelagic habitats. In the following sections, I seek to establish that the nature of the functional differences is directly related to the nature of the physical environments to which pelagic plant life is necessarily adapted. With this base, it is much easier to understand how the primary productivity of the seven-tenths of the planet's surface is both organised and regulated.

The Constraints of Pelagic Habitats

There is a widespread notion (innocently promulgated in older publications) that lakes and seas provide an ideal habitat for living organisms. The high specific heat of water, its solvent properties and its transparency are correctly cited as factors beneficial to planktonic life-forms. The density of water (ρ_w ~1000 kg m^{-3}, vs air, <1.3) offers mechanical (Archimedean) support not available to land plants. Water is also non-compressible: whereas the atmosphere complies with the Ideal Gas Laws and its temperature changes with pressure, such changes are negligible in liquid water. Water is some fifty times more viscous than air; water is less easily set in motion but, correspondingly, its inertia delays its coming to rest. Indeed, it is the fluidity of water, though low in comparison with the air, which most conditions the evolutionary adaptation and morphological selection of planktonic plants, in ways very different from the factors selecting among land plants. Moreover, the scales and variability of motion impinge heavily upon the levels of primary production that can be achieved by those plants which are able to inhabit the pelagic. The nature of water movements is treated elsewhere in this volume; the scales most relevant to the lives of phytoplankton have been reviewed before (Reynolds, 1994). It is the adaptations of the plants themselves, to survive and even exploit an immediate environment in substantially perpetual motion, which are the focus of attention here.

In a strictly teleological speculation, were one given the task of designing the ideal terrestrial primary-producer, it would surely include the provision of a photosynthetic surface for maximum light interception, which was simultaneously thin (to facilitate gas and vapour exchanges), elevated (to be higher than those of neighbours), yet, dissected (to allow the free passage of wind), and arranged to maximise light interception. It would also have to have an appropriate means of mechanical support and irrigation of the photosynthetic surface. The exercise results in quite a good approximation to a tree. The analogous challenge to design an aquatic primary producer might also result in a tree-like structure, devoid of weight-bearing tissue, but nevertheless flexible and toughened to resist tearing and folding by large turbulent eddies. The architecture of kelps and fucoids embodies just these traits. However, the length of anchoring tissue required in deep water soon becomes prohibitive. Despite some exaggerated claims, the authenticated maximum length attained by the Pacific giant kelp (*Macrocystis pyrifera*) is scarcely more than 60 m (McWhirter, 1994). Supposing unrooted, disconnected structures make teleological sense, why aren't pelagic plants like carpets or footballs drifting around under the surface? What is it about open water that, since the habit has arisen among several plant phyla, plainly favours microscopic size?

The Planktonic Habit

Several factors contribute to the solution to this conundrum. One is to do with the density difference between cells and water, which, due to the variability in the latter, can never be continuously and precisely matched by the plant, even supposing that to be desirable. Though predominantly of water, plant cells also comprise heavy materials like proteins, carbohydrates and condensates and, sometimes, other skeletal materials (calcite, opaline silica). Among fresh waters, at least, the plants are likely to be more dense (1020 - 1100 kg m^{-3}) than water and, on balance, to settle through it. Stokes' Law has it that large objects sink out faster than smaller objects of the same density, subject to certain conditions about shape and drag factors. This naturally affords an advantage to small cells over large ones, which would sink out of the thin, insolated (photic) layer in a correspondingly shorter time period. However, the real advantage of a slow-sinking rate lies in the ease of entrainment within the motion of the water. The defined surface boundaries and non-compressibility of water impart a certain conservation of its mass and, as is now well-recognised, kinetic energy applied to lakes and seas is necessarily absorbed within their bodies, or else is spilled out (Smith, 1979). Motion, induced by gravity in a channel or by wind at the water surface, is absorbed through the (Kolmogorov) spectrum of decaying turbulent eddies, from the basin-circulation scale to the point where it is overwhelmed at the level of molecules. The measurement of turbulent intensity $[(u^*)^2]$ is not the concern of the present paper and we shall simply accept its approximation as the quotient of the applied horizontal force (τ, in kg m s^{-2}) and the density of the water (ρ_w, in kg m^{-3}). Then,

$$\tau \rho_w^{-1} = (u^*)^2 \quad (1)$$

The units are $m^2 s^{-2}$; the square root, u^*, has the dimensions of velocity and it is known as the shear velocity or friction velocity. If the imparted velocity much exceeds the settling velocity of the plant cell (w_s), then it will tend to entrain it within the motion. This can never be complete but is substantial if u^* is an order of magnitude or two greater than w_s (Reynolds, 1984). If the condition is not satisfied, the organism will be disentrained, and its own sinking (or floating, or swimming) movements will prevail. Following Humphries and Imberger (1982), I have since approximated the extent of entrainment, ψ, as the quotient w_s / 15 u^*, which must solve to less than 1 for all organisms dependent upon turbulence for continued suspension (Reynolds, 1994). It may be deduced that, subject to the shear velocities generated by a wind of (say) 8 m s^{-1} ($u^* \sim 1 \times 10^{-2}$ m s^{-1}), suspended cells must have settling velocities $< 0.67 \times 10^{-3}$ m s^{-1}. Given that the density difference between the cell (ρ') and the water (ρ_w) is ~ 100 kg m^{-3}, we may calculate the largest possible nominal size to satisfy the entrainment condition. Since the Stokes Equation has it that:

$$w_s = g\, d^2 (\rho' - \rho_w)(18\,\eta)^{-1} \quad (2)$$

where g is gravitational acceleration (9.81 m s^{-2}), d is the diameter of the cell and η is the absolute viscosity of the water (~10^{-3} kg m^{-1} s^{-1}). Rearranging, we may solve

$$d = [\,(18 \times 10^{-3} \times 0.67 \times 10^{-3})\,(9.81 \times 100)^{-1}]^{1/2}$$
$$\sim 1.11 \times 10^{-4} \text{ m.}$$

This simple calculation helps us to explain the fact that non-motile plankton is, ideally, of the order of 100 µm in diameter or less.

A second factor relates to the problem of maintaining the photosynthetic surface. Unless it is mechanically strengthened, turbulent eddies will quickly crumple a flat plate into an untidy ball. However, this will not occur in a viscous fluid. In water, we have to look towards the end point of the eddy spectrum, where viscous forces finally overwhelm the inertia, to find the appropriate conditions. Modern theory concerning how the input energy is dissipated has been used to calculate this end point (e.g. Spigel and Imberger, 1987). In the example proposed, the energy introduced into the largest eddies generated by a constant wind is fed progressively through smaller and smaller eddies into deeper and deeper water. The resultant vertical gradient of horizontal velocities (du/dz, in m s^{-1} m^{-1}) dissipates the energy at the same rate; the dimensions (ℓ_e, in m) and friction velocities (u*, in m s^{-1}) of the larger eddies are correlated, such that:

$$u^* \sim \ell_e (du/dz) \tag{3}$$

The rate of dissipation (E) is equivalent to the product of the turbulent intensity and the velocity gradients:

$$E = (u^*)^2 (du/dz) \tag{4}$$

By substitution of (3) in (4),

$$E = (u^*)^3 \ \ell_e^{-1} \quad \text{in } m^2 s^{-3} \tag{5}$$

The smallest eddy size (ℓ_m) is calculated from the relation of the viscosity to the dissipation rate:

$$\ell_m = [\eta \rho_w^{-1})^3 E^{-1}]^{1/4} \tag{6}$$

These equations have been used to determine that eddies generated in the deep, unstratified Bodensee (Lake of Constance) by winds of 5 - 20 m s^{-1} penetrate to depths of 45 - 180 m and dissipate at rates of between 1.4×10^{-8} to 2.2×10^{-7} m^2 s^{-3}, down to eddies varying in size between 2.9 and 1.5 mm, respectively (Reynolds, 1994). In stratified lakes, where the density gradient acts as a barrier to the penetration of weak eddies, and in shallow lakes where the water column is unable to accommodate the downward eddy propagation, the energy fluxes associated with similar wind conditions must be dissipated within the limited water depth. The energy of a 20-m s^{-1} wind is dissipated twenty times faster in Lough Neagh, Northern Ireland (mean depth <9m) than in the Bodensee (E = 4.3×10^{-6} m^2 s^{-3}) and the smallest eddy sizes are barely 0.7 mm. In the most rapidly dissipative tidal estuaries, E may approach 5.5×10^{-4} m^2 s^{-3} with eddies as small as 0.2 mm across. In order to take advantage of the viscosity of mobile water masses, planktonic algae need to be smaller than the eddies likely to be generated within their lifetimes. Bearing in mind that the cells of most planktonic algae fall below 200 μm and most colonies below 2 mm (and, then, generally in only very calm or stratified water layers), the hypothesis that eddy size places an upper size constraint on planktonic plants is hardly disproved.

The Productive Capacities of Pelagic Habitats

However speculative it may considered to assert that microscopic size has been positively selected by the nature of the medium, it remains a hard fact that planktonic plants are of a size which makes them readily entrainable. Simply, the smaller is w_s, then the lesser is the value of u* necessary to satisfy the entrainment criterion ($\psi < 1$). If the fastest buoyant velocity of a colonial cyanobacterium is in the order of 3 mm s^{-1}, then the minimum entraining friction velocity (45 mm s^{-1}) is that driven by a wind of 36 m s^{-1}. For larger diatoms with sinking rates of generally < 100 μm s^{-1} the maximum threshold value of u* requires a wind of 1.2 m s^{-1}. For most other non-motile phytoplankton (w_s <10 µm s^{-1}), entrainment is always almost complete. Most planktonic plants are effectively "embedded" in any motion (gravitational, convectional, wind-driven or inertial) to which the water is subject. Phytoplankton will be randomised through any gently-mixed water layer, being broadly dispersed everywhere within the bounds of the motion. Individual algae are thus transported in eddies propagating throughout the surface mixed layer and at approximately the same rates. Following Denman and Gargett (1983), if it requires some 25 minutes for wind-generated eddies to propagate through the full extent of a mixed layer and it requires (generally less) a time proportional to mix a layer restricted by a physical boundary (the bottom or a stable density gradient), then it takes the same time for entrained planktonic algae to become randomised through the layer and, probabilistically, for any to be carried through its full depth.

The task is to determine what responses this embedding evokes from the algae and, in turn, what effects they have in regulating planktonic production. Planktonic primary production is taken here to refer to the photoautotrophic assembly of cells and populations and, by implication, the rates at which it can be achieved. This is a wider understanding than the usage which is synonymous with photosynthetic production, though the latter is vital to the consideration and is acknowledged to set a limiting capacity on the rate of cell growth. It is not the only such limit, however, and it is important to assert what they are.

Photosynthetic carbon fixation is a fundamental process of all ecosystems. The familiar chemical equation (7) summarises the outcome of the fundamental reactions. Light energy is used to bring about the reduction of carbon dioxide to carbohydrate:

$$6CO_2 + 6H_2O \overset{\text{light}}{\rightarrow} C_6H_{12}O_6 + 6O_2 \tag{7}$$

Photoautotrophy embraces the conversion of carbohydrate into amino-acids and the assembly of amino-acids into specific proteins and proteins into new cytoplasm and organelles; eventually, daughter cells are produced. In addition to carbon, hydrogen and oxygen, another score of elements are required, usually in some stoichiometric ratio to carbon, and they have to be obtained from the water as *nutrients*. Some are readily available in relative abundance (such as sulphur and chlorides), while others (such as copper and molybdenum) are required in such relatively tiny amounts that availability is rarely problematic. Yet others (notably, nitrogen and phosphorus), are required in relatively large proportions but are not always available. If the healthy, living protoplasm of cells is represented by a chemical formula representing these elements in their typical stoichiometric (or "Redfield") ratio to carbon, viz., $C_{106}N_{16}P$....., then equation (8) provides a rough summary of the reactions underpinning algal growth:

$$106\ CO_2 + 16\ [NO_3]^- + [PO4]^{3-} = C_{106}N_{16}P \tag{8}$$

Supplies of one or other may well fail to satisfy the potential demand of plankton growth, in which case the deficient element places a capacity limit on the size of plankton crop that can be supported and, eventually, a rate-limitation on the time taken to attain it. Nitrogen and phosphorus are not the only limiting nutrients, however. For diatoms, the availability of dissolved, reactive silicon in the water may also fall to levels inadequate to sustain the formation of the siliceous exoskeletons of the next generation (Lund, 1950), while strong interest currently surrounds the role of iron availability in the regulation of planktonic primary production in the great oceans (Martin and Fitzwater, 1988).

In water, as on land (Grime, 1979), there are four essential requirements for autotrophic plant growth: a supply of water, a supply of carbon, a supply of each of the 20-or-so other elements used in the assembly of plant cells, and an adequate supply of light energy to drive their synthesis into new plant tissue. Although the apparent sufficiency of water is not guaranteed (few freshwater species have the osmoregulatory capacity to resist water loss in even mildly saline solution), it is assumed here never to be a limiting resource. Attention is confined to the transparent and solvent properties of water and how temporal variability impinges upon the availability of light, carbon and nutrients. These are well-studied topics and require only the briefest of rehearsals here.

Light

The nature of light has been extensively studied by optical physicists and its role in aquatic photosynthesis has been explored thoroughly by Kirk (1994). "Light" refers to the visible wavelengths ($\lambda \sim 400\text{-}700$ nm) of the electromagnetic radiation emanating from the sun, which band accounts for almost half (46 - 48%) the solar energy reaching Earth. Light beams comprise a continuous flow of indivisible quanta (or, in the visible spectrum, *photons*), travelling at a constant velocity, c, of $\sim 3 \times 10^8$ m s^{-1} (differences in wavelength are reciprocal to frequency: $v = c/\lambda$). The intensity of light (I) equates to the number of photons, arriving at a surface normal to the beam per unit time, or the photon-flux density (PFD). It is a highly variable quantity, which, at any given location, is dependent upon the solar elevation and its variability through the day and through the year. At the distance of the earth from the Sun (~150 million km), the theoretical maximum photon flux is related to the *solar constant* (being the energy flux reaching a notional surface placed above the atmosphere and perpendicular to the sun's rays: approximately 1350 W m^{-2}),which delivers about 1.75×10^{21} photons m^{-2} s^{-1}; division by the Avogadro number (6.023×10^{23}) renders the quantity in mol photons, i.e. $\sim 2.9 \times 10^{-3}$ mol photons m^{-2} s^{-1}. During their passage through the atmosphere, absorption of photons and scattering by dust and by water vapour reduce the total flux by scarcely less than 20%: the brightest sunlight that might be experienced at the tropical zenith during very dry and cloudless conditions is about 2.3 mmol photons m^{-2} s^{-1}. Over the 12 h between sunrise and sunset of one day, the daily dose is equivalent to about 75 mol photons m^{-2} d^{-1}. The maximum intensity at the summer solstice at a latitude of 50° N (1.35×10^{-3} mol photons m^{-2} s^{-1}) is part-compensated by a 16-h day (~58 mol photons m^{-2} d^{-1}). Elevation, latitude, albedo and, especially, cloud cover ensure that daytime light intensities experienced at ground level are often much lower than this (0.2 - 1.0×10^{-3} mol photons m^{-2} s^{-1} is a much more familiar range).

A portion (generally < 5%) of the light penetrating the atmosphere and striking a smooth water surface (now identified as I_0) is reflected back to the sky. However, this portion increases steeply at low angles of incidence (< 30°); close to sunrise or sunset, wind-driven waves actually increase the passage of light across the water surface (Larkum and Barrett, 1983). Once it has penetrated immediately beyond the water surface, the remaining flux (I_0') is subject to rapid wavelength-selective absorption and back-scattering by particles in suspension. Even in some of the clearest alpine lakes and seas (Laguna Negra, Chile: see Cabrera and Montecino, 1984; the eastern Mediterranean Sea: Berman et al., 1985) light of all wavelengths, red light especially, is absorbed much as it is in pure water. The vertical attenuation of light is exponential with depth, conforming to the Beer-Lambert formulation for a uniform liquid, so that the light (I_z) penetrating to a given depth, z m beneath the surface is calculated from:

$$I\,z = I_0\,'.\ \exp(-z\varepsilon_\lambda) \tag{9}$$

where ε_λ is the coefficient of light extinction at a specified wavelength or within a spectral block (units: m^{-1}): absorbance in pure water is least at ~440 nm (< 0.015 m^{-1}); the average across the visible spectrum is ~0.165 m^{-1}. An idealised depth profile of attenuation is included in Figure 1a. Note that when the water is being actively mixed, the light level falling on a given entrained cell fluctuates with every adjustment of its depth in the light gradient. A hypothetical case is shown in Figure 1b. If light is attenuated rapidly with depth and extinguished within the full mixed depth, then there is some portion of the randomised population that is in effective darkness and all of it can be expected to experience a night/day transition within every mixing period. Reynolds (1987) introduced the term I* to represent the mean irradiance in a mixed column, being the logarithmic mean of the intensities obtaining at the surface and at the base of the mixed layer, (I_m):

$$\ln I^* = (\ln I_0' + \ln I_m)\,/\,2 \tag{10}$$

ε_λ has several components; that part due to the water and any residual colour from dissolved leachate from soils or organic plant remains is ε_w; that due to non-living suspensoids (e.g. clay) is ε_p. Both can severely restrict light penetration in organically- or humic lakes or in ones routinely charged with fine clay particles; Kirk (1994) cites numerous examples. Frequently the greatest absorption and scatter is from algal chlorophyll, the principal functions of which include the harvesting of light energy and

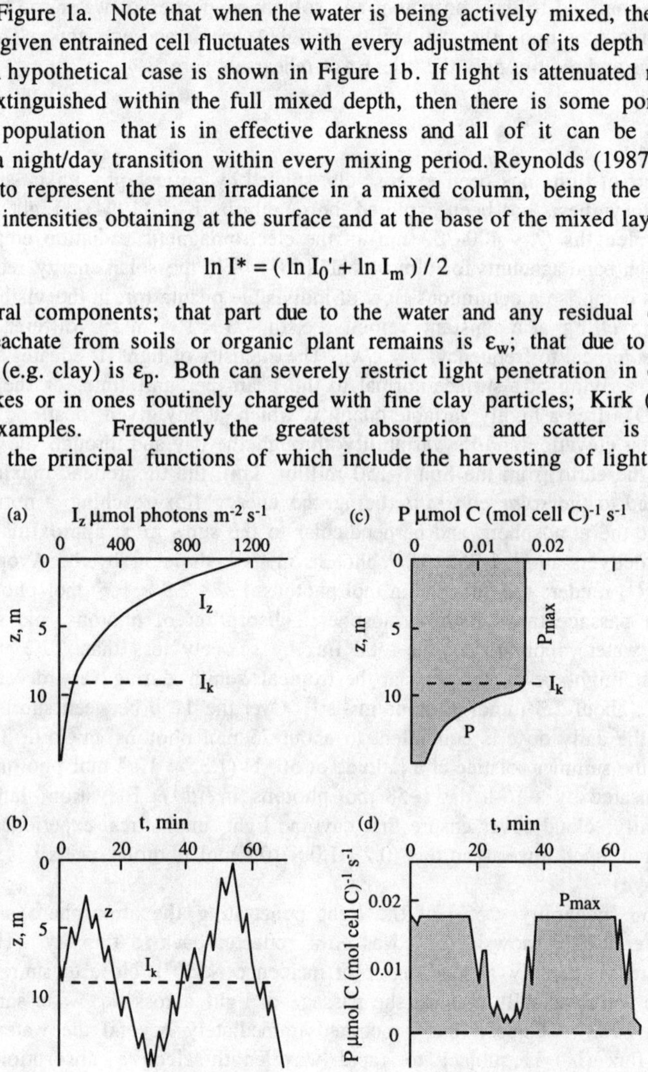

Figure 1. (a) Typical vertical profile of light through water column, with the limit of light saturation, I_k, inserted; (b) "random walk" plot of entrained cell through one hour in water column (a) mixed to 15 m; (c) instantaneous photosynthetic rate sustainable by light availability in light-gradient (a), and (d) instantaneous photosynthetic rate maintained by cell while mixed through light gradient (a) on track (b).

to effect the photosynthetic reduction of carbon dioxide. Algal cells contain a typical average mass of chlorophyll equivalent to about 1% of the ash-free dry mass. From Reynolds' (1993) calculations of the areas subtended by different algae [2 - 26 m^2 (mol cell $C)^{-1}$] and supposing carbon to account for 50% of the ash-free dry mass, then in life, 1 mg of algal will cover a maximum area, a, of anything up to the same area as 100 mg dry matter, i.e. between 0.01 and 0.1 m^2 (in practice, chlorophyll does not occupy the full cross-section; most derivations fall in the range 0.004 and 0.021 m^2). Absorption is scaled to concentration, so the attenuation due to algae, ε_A, is the product of the concentration, N, and a.

The full evaluation of ε_λ is:

$$\varepsilon_\lambda = \varepsilon_A + \varepsilon_p + Na \tag{11}$$

We may estimate the impact of the plankton on light absorption and the impact of mixing on the plankton concentration. As an example, we might suppose that the planktonic population, N, filled the remaining capacity for light absorption in the top z meters, such that as (I_z) approached zero, $(-z\varepsilon_\lambda) \sim -\ln(I_0{}')$, and then Nz was equivalent to $\{\ln I_0{}' - (\varepsilon_w + \varepsilon_p)\}/a$. Then, assuming $a = 0.01 m^2$ (mg chlorophyll$)^{-1}$, we might deduce a self-shade capacity for an otherwise clear water ($\varepsilon_w + \varepsilon_p \sim 0.2\ m^{-1}$) of z = 1 m depth and receiving an average daily insolation of 1.0×10^{-3} mol photons $m^{-2}\ s^{-1}$ to be 670 mg chlorophyll m^{-3}.

The impact of dispersive mixing on the light-limitation of the algal carrying capacity is two-fold. One is that greater entrainment may suspend more particulate matter (ε_p is increased). The other is that the greater is the depth of mixing, then the greater is the additive impact of ($\varepsilon_w + \varepsilon_p$) on the carrying capacity and the more dilute is N: when z = 10 m, N falls to 49 mg chlorophyll m^{-3}; when z = 30m, it drops to 3 mg chlorophyll m^{-3}.

Carbon

Planktonic photoautotrophs mostly obtain their carbon through the uptake of carbon dioxide dissolved in the adjacent water. The high solubility of carbon dioxide compensates for the low partial pressure exerted by atmospheric concentration of the gas (3×10^{-4} atmosphere): in the relevant temperature range (0 - 25°C), the air-equilibrated concentration in pure water varies between 1.0 and 0.5 mg $CO_2\ l^{-1}$. As is well-known, however, aqueous concentrations are further regulated by equilibria involving the bicarbonate alkalinity and the pH of the water: according to the nomogram of Mackereth et al. (1978), the circumneutral equilibrium concentrations of carbon dioxide across a spectrum of alkalinities of from 0.4 to 4.0 meq l^{-1} may be an order of magnitude greater (3 - 35 mg $CO_2\ l^{-1}$). Depending upon the relative magnitudes of the relevant fluxes, these levels may be raised still further by respiration and the decomposition of organic carbon, or be depleted by photosynthetic uptake. Based on the supposition that the ash-free biomass is 50% carbon, consumption of all the CO_2 might yield a nominal phytoplankton mass of up to 1.6 - 18 mg dry mass l^{-1} or 16 - 180 mg chlorophyll m^{-3}.

Net removal of free carbon dioxide results in rising pH, compensatory dissociation of bicarbonate above pH 8.8; ultimately, however, the response to photosynthetic withdrawal and equilibrial imbalance is that more carbon dioxide invades across the air-water interface. Thus, the limitation upon the biomass capacity is really determined by the rate of air-water exchange. The direction and magnitude of the flux (F) conforms to the relationship:

$$F = G \gamma \Delta p_{CO2} \qquad (12)$$

where γ is the solubility coefficient (in mol m^{-3} atmosphere^{-1}), Δp_{CO2} is the difference in partial pressure of carbon dioxide between water and air, and G is the gas exchange coefficient, or linear migration rate (m s^{-1}). In fact, the magnitude of the potential exchanges are difficult to establish but Frankignoulle (1988), Upstill-Goddard et al. (1990) and Watson et al. (1991) are among those who have attempted to make accurate field measurements of gas-transfer rates. In productive waters, seasonal variability in p_{CO2} (up to 9×10^{-4} atmosphere) is, indeed, a particularly important driver (the fastest consumption stimulates the most rapid invasion!) but the transfer velocity is accelerated as a function of wind-speed and surface roughness, from $\sim 10^{-5}$ m s^{-1}, at wind velocities beneath the critical value of 3.5 m s^{-1}, to an order of magnitude greater at 15 m s^{-1}. At high values of Δp_{CO2}, the corresponding invasion fluxes are likely to be within the order of 3 - 30 $\times 10^{-8}$ mol m^{-2} s^{-1}, or between 31 and 310 mg C m^{-2} d^{-1}. Using the same conversion factor (C:chlorophyll = 50), this is theoretically sufficient to sustain a productive increment of only 0.6 to 6 mg chlorophyll m^{-2} d^{-1}.

Large crops of algae, especially in soft-water lakes, are known to be sensitive to rate-limitation by carbon-dioxide depletion or to the sharp pH rise that may accompany it (Talling 1973, 1976). Limitation of biomass increment by the supply of carbon is probably not a rare event. On the other hand, these phases are generally brief. Maberly (1996) has shown how even small, quite eutrophic but soft-water lakes experience net annual CO_2 loss to the atmosphere if the inflow streams carry predictable concentrations of the gas.

Nutrients

In many lakes and in the open ocean, the non-availability of one of the nutrients required to elaborate new cells will intervene in the rate of specific biomass increase. As with light and with carbon, the amount of nutrient available, or likely to be available, will determine a maximum capacity of the water to support a crop of primary producers. The one with the lowest capacity, relative to probable demand, is supposed to determine the maximum crop that, eventually, may be supported. Until that point is approached, however, the growth rate is not likely to be regulated by the concentrations of nutrients obtaining.

There is a vast literature on the nutrient requirements of algae and on the concentrations said to be "limiting". It is not appropriate to do more than summarise those paradigms relevant to the development of this essay.

So far as lakes are concerned, the relative deficiency of phosphorus among temperate waters is often critical to sustained phytoplankton growth and this central role in governing the magnitude of plankton crops, at least among oligotrophic and mesotrophic lakes, is acknowledged in the well-known "Vollenweider model" (see Vollenweider and Kerekes, 1980). The relationship shows, within a defined confidence interval, that the average chlorophyll content of each of a number of lakes is a function of the phosphorus availability. The relationship is not continuous, however, and the averaging conceals a great deal of fluctuation in the biomass supported in any one of the lakes (Reynolds, 1992). Given the level of stoichiometry anticipated between C and P in healthy populations (equation 8; 41:1 by weight), the ratio of the cell-specific mass of chlorophyll to cell phosphorus is tolerably close to 1. Experience upholds a slope of less than 1 (Reynolds, 1992), which predicts that a concentration of biologically-available phosphorus equivalent to 10 mg P m^{-3} has the potential to support a maximum development of phytoplankton equivalent to 24 mg chlorophyll m^{-3} At 100 mg P m^{-3}, the potential is 94 mg chlorophyll m^{-3}. At 1000 mg P m^{-3}, it is about 360 mg chlorophyll m^{-3}.

Following an analogous argument for the numerous continental lakes at lower latitudes, wherein the availability of assimilable combined nitrogen (nitrate, ammonia, urea) is sparse, the biological stoichiometry between C and N (equation 8; 6:1 by weight) suggests a chlorophyll yield to biologically available nitrogen of ~1/7, every 100 mg N m^{-3} potentially supporting another 14 mg chlorophyll m^{-3}. Most seas are probably more dilute than lakes with respect to phosphorus and, especially, nitrogen, yet investigations of their concentrations in cells and in sea water (~2.5 mg P m^{-3}, 100 mg N m^{-3} but chlorophyll <1 mg m^{-3}), rarely suggest their quantities to be exhaustively exploited (Martin and Fitzwater, 1988). Their own experiments showed that the addition of iron in solution to a concentration of up to 10 nmolar Fe (650 μg Fe m^{-3}) was sufficient to increase the biomass ten fold and take up almost all the nitrogen. Some experiments which were carried out in my laboratory (with K. Button, G.H.M. Jaworski and E. Tipping) had attempted to determine a half-saturation constant for iron in a cultured freshwater chlorophyte. In fact, no compelling regression was detected, although it was shown that concentrations of 10^{-8} molar (i.e. 650 μg Fe m^{-3}) regularly saturated growth rates but 10^{-11} (0.65 μg Fe m^{-3}) was always too little to revive cultures in ostensibly iron-free media (the problem of impurities is significant at these concentrations).

The impact of mixing on nutrient availability and uptake is not necessarily a direct one but it does have the effect of homogenising the concentrations of each nutrient through the mixed layer itself. Entrained plankton cells effectively share similar nutrient environments. This is in contrast with stratifying columns in which significant density gradients form. These have the immediate effect of distinguishing the environments for growth through the column, with differing levels of activity modifying the nutrient-resource concentrations to differing extents. As a result, significant gradients of nutrient availability develop and these are enhanced by settling of disentrained algae and the faecal pellets of any planktonic herbivores. As Margalef (1958) realised, there is thus a potentially strong segregation of the water column, with a nutrient-depleted upper part perhaps overlying a dark but enriched lower part. Such physical structures may be exploited by migrating algae or specialised plate-formers which themselves stratify at depth. If, at any time during the development of segregation, renewed mixing disrupts the structure and re-randomises its resources into a single whole again, either at once or soon after, biological production benefits from the restoration of limiting nutrients into the well illuminated part of the water column. Mixing events have long been regarded as stimuli to renewed planktonic productivity in the plankton (Kemp and Mitsch, 1979) and to initiate new successional sequences.

Interacting Limiting Factors

The last example demonstrates that capacity limitations on biomass production are not immutable and there is often opportunity for alternation. Many anthropogenically enriched lakes and rivers have available-phosphorus concentrations in the order of 1 g P m^{-3} but the chlorophyll concentration does not come close to achieving this supportive potential. If the water is mixed over 2m, there will not be sufficient light to sustain the potential areal concentration of 720 mg chlorophyll m^{-2}: the light is already the limiting factor. If an oligotrophic lake is frequently or continuously mixed through 30m, no matter that its content of dissolved phosphorus is only 1 mg P m^{-3}, this is more than sufficient to support the light-limited biomass. Its poor production owes to the physical environment, not to the lack of nutrient. Unless the water is very turbid anyway, when the mixed layer of the same lake shrinks to <10m, the productive constraint posed by light energy is overcome and nutrient limitation will surely soon be imposed. Deep cold lakes and high-latitude seas may suffer light limitation all year-round.

In Figure 2, the axes of light (energy) limitation and nutrient (resource) limitation are used to show that many systems alternate in being severely deficient in one or the other. An important deduction to be made is that the pelagic, far from being an ideal medium for plant life, it is often a highly rarefied and unpromising medium.

The further down or the further to the right among the contingencies represented in Figure 2 are the co-ordinates of the field conditions represented, then the more selective is the environment and the more specialised are the survival adaptations of the successful species likely to be. To the right, a good light harvesting ability becomes paramount and species subtending a large area per unit mass (a, see above) would appear to be advantaged (but see later). Downwards, it is the ability to take up (by virtue of a high affinity for) the limiting nutrient or to have storage capacity for it when the opportunity to do so arises, or to conserve the biomass in which it is incorporated. The argument is repeated for each nutrient but the requirements of planktonic plants are not identical. Interacting gradients can be extremely important to species selection. The obligate skeletal requirement of diatoms for silicon is selectively disadvantageous when aqueous concentrations are depleted but nitrogen and phosphorus are present in excess. Cyanobacteria of the heterocyst-forming class Nostocales have the potential advantage over other members of the freshwater plankton in being able to fix atmospheric nitrogen dissolved in the water, and so raise their potential biomass to the limits of (e.g.) the phosphorus available.

This behaviour should not be confused with the recognition that when nitrogen and phosphorus (or any other pair of interacting resources) are at chronically low levels, or are reduced to them, differing competitive abilities to gather them distinguish among the successful species (Rhee, 1982; Tilman et al., 1982). Perception of nutrient limitation of plankton populations is dominated by Michaelis-Menten kinetics, as modified by Droop (1974). Because at steady state, the rate of uptake matches the rate of consumption and

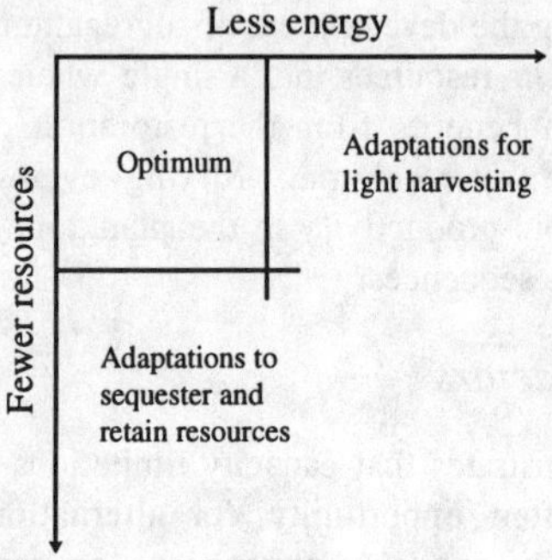

Figure 2. Contingencies of planktonic habitats defined by axes of deficient energy and deficient nutrient resource. Primary adaptations are required to survive either towards the upper right or to the lower left but the lower right is untenable (cf. Grime, 1979).

because the maximal rate of uptake of nutrient by cells is found to be an asymptotic function of concentration, it is reasonable to suppose that growth rate is a continuous function of the element present in least concentration relative to its minimal demand. The models describe well events at low concentrations, where they are consistent with resource-competition theory. To the top left-hand side of Figure 2, however, is represented simultaneous resource- and energy-saturation, as might be encountered in clear waters of a shallow pond or of a calm river, where environmental adequacy approaches that provided for stock cultures in the laboratory. Measurements of growth rate turn out not to be the fastest predicted by the nutrient uptake or interception efficiencies but a reproducible maximum for that species under those conditions. In short, give the cell everything it needs and there are still finite, temperature-dependent rates of assimilation, assembly and cell replication. The maximum carbon fixation rates and nutrient uptake rates have not only to be capable of supplying the maximum sustainable growth; they are in considerable excess (Reynolds, 1990). This is an important principle to emphasise, for it means that (i), in resource- and light-replete environments, neither nutrients nor light need be limiting the rate of growth and the cells present are NOT directly competing for the resources. Moreover (ii), growth rate does not become limited by these components until after the nutrient uptake rate or the carbon fixation rate falls below those which will sustain the maximum rate of cell replication. Further, until the rate of cell growth falls to the maximum that can be sustained by the resource in weakest supply, it is not reasonable to claim (iii) that growth rate is limited by light or by any nutrient, or to infer that (iv) the growth rate in situ is less than the maximum that could be achieved in culture, at the same temperature and under the corresponding photoperiod.

For the production physiologist, as well as the plankton ecologist, it is important that these thresholds of limitation be defined. That they exist is apparently well-known. Droop's (1974) formulation is based upon the premise that there is a minimum cell quota, for each of the required nutrients, below which it must not be allowed to fall, least of all through cell replication. Microbial cells do not grow at full rate until nutrient is exhausted but, rather, enter a state of limitation during which they adapt their nutrition, through a complex series of physiological close-downs, translocation of residual stocks of the element to key structures and some morphological re-organisation (including spore-formation), which will allow the cell to survive starvation and retain viability for the duration (Kjelleberg et al., 1993). This behaviour is observable in algae and is likely to be a general adaptation to life in environments where nutrient depletion is likely to occur eventually and periodically (Mann, 1995). The corresponding principle is demonstrated among phytoplankton dealing with low light levels. Once harvestable light levels (I_z or I^*) fall substantially and continuously below that required to maintain the maximum growth ($r < P$), the cell will usually attempt to raise its photosynthetic efficiency, α, that is, the slope of the curve of photosynthetic rate (P) on light (I), as included in Figure 1c.

It would be extremely useful to nominate the concentrations of nutrient and the levels of I^* at which growth thresholds occur. In the wider definition of primary productivity adopted above, they are, indeed, the critical ranges within which any direct relation between the concentration of organism and the flux of primary products becomes uncoupled. Unfortunately, there are not many data upon which to make a general proposition but it is valuable to explore here, from current knowledge, where these ranges might fall. Attention is given first to light, because the time scales of sufficiency/deficiency are generally much shorter than for carbon or other nutrients.

Regulating Production

Maximum Growth Rates

The data on maximum growth rates are both more accessible and in better mutual agreement than those on critical limits. The thorough surveys of (then) published algal growth rates by Hoogenhout and Amesz (1965), together with later information collected for tabulation in Reynolds (1984), were used to derive statistically good relationships between the maximum replication rates of common species of freshwater phytoplankton at 20°C and the surface/volume ratios (sv^{-1})of their sizes, and for the sensitivity of specific replication rates to altered temperature in terms of the same property; Reynolds (1989) showed:

$$r_{20} = 1.142\ (sv^{-1})^{0.325} \qquad (13)$$

and

$$\log r_\theta = \log r_{20} + \beta\ [1000/(273 + 20) - 1000/(273 + \theta)] \qquad (14)$$

where r_{20} is the regression-predicted replication rate at 20°C and r_θ is the predicted rate of growth at another temperature. The units of equation (13) are d^{-1} but the result could be divisible by 86 400 to derive the more useful base, s^{-1}. The equations predict that small cells grow faster than large ones, unless their shapes are distorted far from the spherical (like rods and fusiforms), and much more so at low water temperatures. Equation (13) predicts that the maximum growth of a 4-μm diameter *Chlorella* (s = 50 μm^2, v = 33 μm^3, $sv^{-1} = 1.5\ \mu m^{-1}$) at 20°C is 1.31 d^{-1}, or $\sim 15 \times 10^{-6}$ per second of light period. To precisely double the biomass (i.e., $r_{20}\, t = 0.693$) within 24 h, the *Chlorella* cells would require to be exposed to growth-saturating light for a total of 12.7 h. In conventional units, this is to increase mass by a factor of 1.000015 s^{-1}, or to be assembling carbon at a rate of 15×10^{-6} mol carbon (mol cell carbon)$^{-1}$ s^{-1}.

The problem is now to determine how the cell is able to deliver nutrients and photosynthetically-fixed carbon at the rates required to maintain the rate of growth in environments where the driving requirements are pulsed at relatively high frequencies.

Molecular Basis of Nutrient Assimilation

Investigations of nutrient requirement and the dependence of the rate of uptake upon the external requirement have long been dominated by the Monod-type model, based on Michaelis-Menten enzyme kinetics (Dugdale, 1967) and described in terms of the quantities V_{max}, for the maximum uptake rate of starved cells, S for the external concentration, and K_v, for the concentration just sustaining half the maximum uptake rate. Then, the rate of uptake, V_S, is given by:

$$V_S = V_{max}\ S / (K_v + S) \qquad (15)$$

The Droop (1974) model has more general acceptance since it contains a term recognising the internal stores and a minimum cell quota. Whereas this adjustment is more practical for measuring uptake, the argument has to be advanced that it will not predict maximum growth rate under all eventualities. If a fast-growing alga like *Chlorella* increases at a rate of 15×10^{-6} mol C (mol cell C)$^{-1}$ s^{-1}, and it incorporates phosphorus in the Redfield ratio, then it requires to consume 0.14×10^{-6} mol P (mol cell C)$^{-1}$ s^{-1}, which it is able to take up at least this rate from external concentrations as low as 6×10^{-8} molar, without change to the internal store. It seems erroneous to consider that the growth rate (at least at 20°C) of *Chlorella* could be considered to be "phosphorus-limited" while external phosphorus concentrations remain above 6×10^{-8} molar, i.e. 2 μg P l^{-1} (Reynolds, 1990).

This argument extends to most other algae for which phosphorus uptake data are available, and there is an analogous argument for the uptake of nitrogen. They all have a capacity to take up nutrient far faster than they can ever use it; this possibly reflects a common circumstance that nutrients are often encountered in pulses or small patches. This is not to say that they cannot run out or compete interspecifically for quantities in chronically short supply. When nutrients are abundant, it is likely to be those with the absolutely the fastest growth rates which have the best opportunity to dominate. Presumably, species with a very high V_{max} relative to r derive some advantage from pulsed nutrient supplies ("storage adapted": Sommer, 1984); similarly, species ("affinity"-) adapted to take up nutrient from very low concentrations (having a low K_v relative to growth-rate saturation) seem better equipped to environments with chronic nutrient deficiency. Self-regulation and biomass-conservation provide one route for survival of such habitats, traits which favour motility and generally larger size ranges among planktonic organisms. On the other hand, if nutrients had not been abundant at the time of growth, such organisms would never have become numerous in the first place. In environments without even phases of relative nutrient abundance, it is the very smallest photoautotrophs of the picoplankton which seem to maintain dominance.

Molecular Basis of Photosynthetic Production

Present understanding of the photosynthetic process is that it comprises an arrangement of both light and dark reactions. According to the accepted "z-scheme" of Hill and Bendall (1960), two linked photochemical systems (PS I and II) accommodate the key light reactions. In the first of these (actually PSII), electrons are stripped from a donor substance (usually water, which is photochemically oxidised to oxygen). In the second, light energy is used to transfer electrons to carbon dioxide, through the reduction of nicotinamide adenine dinuceotide phosphate (NADP to $NADPH_2$) and the generation of adenosine triphosphate (ATP) from phosphate ions and adenosine diphosphate (ADP). The photosystems are linked by a regulatory electron transport system in which the charge is separated and through which electrons are passed first to phaeophytin (Pheo), then to a primary quinone acceptor (Q_A), and then to another quinone, Q_B; once two electrons have been accepted, Q_B dissociates to enter a pool of plastoquinone (PQ). Reduced PQH_2 is eventually oxidised by the cytochrome, b_6/f, and the electrons are passed to PSI (see e.g. Walker, 1992; Kolber and Falkowski, 1993). The pool of PQH_2 acts as a capacitor, like a surge tank, collecting any sudden burst of electron flow, holding them back from a damaging rush into PSI and, as noted below, also feeding-back a resistance to the acceptance of further electrons.

In the dark reactions, the reduced NADP and the ATP are used to reduce the carbon dioxide. The essential steps, each catalysed by the appropriate enzyme, include the carboxylation of ribulose-1,5 biphosphate (RuBP) to 3-phosphoglyceric acid (3-PGA). This is mediated by the enzyme ribulose biphosphate carboxylase (RUBISCO). This so-called "C_3 photosynthesis" is typical of the planktonic algae but contrasts with the "C4 photosynthesis" of some terrestrial vascular plants. The 3-PGA reacts with ATP and $NADPH_2$ to yield 3-phosphoglyceralde (3-P-GAP), which now incorporates the high energy phosphate bond of the ATP synthesised in PSI. The 3-P-GAP is further metabolised to triose, hexose and polysaccharides (such as starch or glycogen) in which form, if photosynthesis proceeds faster than growth and respiration can consume its products, it may be stored in the cell.

The end products of normal photosynthesis, the sugars upon which the proteins and lipids and the cell structures of living cells are founded and which will yield the energy to sustain all the relevant metabolic pathways, are thus quantitatively related to the flux of photons harvested by the cell's photosynthetic apparatus in the first place. Photons are intercepted by specialised membranous organelles, the thylakoids. These are distributed either through the body of the cell, as among the Cyanobacteria, or within the discrete, membrane-bound chloroplasts of eukaryotic algae. Closely associated with the thylakoid surface are the granules corresponding to the two photosystems. Those associated with PSII comprise a complex of 200 - 300 chlorophyll molecules (mostly chlorophyll *a*, up to 30% may be of chlorophyll *b*), certain other pigments (carotenes and xanthophylls) and associated proteins. The entire complex has a combined molecular mass of about 400 kDa (Dau, 1994). Each PSII unit has a light-harvesting core (referred to as an LHC). When exposed to light, the energy of the photons raises electrons in special ("P680") chlorophyll molecules in the reaction centres from their ground-state to excited-state orbitals. Associated phaeophytin acceptors transfer the electrons singly towards PSI, through the series of reactions involving the quinones Q_A and Q_B, and the plastoquinone (PQ) pool. At quiescence, the reaction centre is said to be "open": P680 is in its reduced state, phaeophytin, Q_A, and Q_B are all oxidised. Excitation of chlorophyll P680 initiates its oxidation and, thus, the serial reduction of the phaeophytin and the quinones as electrons are transferred through to the plastoquinone pool. At the same time, the otherwise uncomplemented positive charge created by the P680 excitation is satisfied by an electron from water (that is, it is re-reduced, via a separate donor system, based upon a tyrosine-protein and usually signified as Z). While all this is taking place, the reaction centre is unable to accept further electrons. It remains "closed" until re-oxidation of Q_A clears the pathway for the next electron. In this way, the capacity of the electron transport is self-regulated.

Meanwhile, simultaneous excitation of the analogous chlorophyll-protein complex of PSI (known as P700) and its acceptor (usually denoted as A) permits the passage of electrons beyond the plastoquinone pool. Raising the energy level of P700 to the point where its electrons can be passed to A requires the equivalent flux of photons. A further cascade of transfers and regenerative oxidations passes the electrons to ferredoxin and thence to the reduction of NADP and the dark reduction of carbon dioxide to carbohydrate $[(CH_2O)_n]$.

A more helpful empirical statement of the overall photosynthetic reaction than equation (7) might then be:

$$2\,H_2O + CO_2 \overset{\text{photons}}{\rightarrow} CH_2O + H_2O + O_2 \qquad (16)$$

This reaction requires the transport of four electrons ($4\ H^+ + 4e^- \rightarrow 4H$) through each photosystem. It follows, theoretically, that 8 photons are required for the fixation of each carbon atom. This stoichiometry is indisputable according to the Z-scheme (e.g. Walker, 1992). The reciprocal, the quantum yield (ϕ), thus has a sound theoretical maximum yield of 0.125 mol carbon (mol photon)$^{-1}$. Most attempts to measure ϕ, in a variety of photosynthetic organisms, have come up with values below this theoretical maximum:

most determinations for algae fall between 0.07 and 0.09 mol carbon (mol photon)$^{-1}$ (Bannister & Weidemann, 1984; Walker, 1992).

Light-dependent Photosynthetic Rates

The speed of the overall reaction is relevant to the number of such reactions required per cell per unit time to sustain the cell growth. Kolber and Falkowski (1993) approximated the aggregate of reactions linking excitation (occupying less than 100 fs or 10^{-13} s: Knox, 1977) to re-oxidation of Q_A^- (0.6 ms). The principal rate-limiting step is the onward passage of electrons from the plastoquinone pool, which falls between 2 and 15 ms, depending upon water temperature. A single linked pathway might process up to 66 reactions per second at near-freezing temperatures and up to 500 at around 30°C (note, close to one doubling in the rate for every 10°C), if the next electron was always to arrive at the reaction centre the instant that it re-opened. At this point the apparatus would be filled to capacity and the rate of photosynthesis would be on the point of light saturation by the photon flux density.

The flux of light just able to saturate the photosynthetic capacity is referred to as I_k. According to its molecular basis, it may be defined as:

$$I_k = (\sigma\, t_p)^{-1} \text{ photons m}^{-2}\text{ s}^{-1} \qquad (17)$$

where t_p is the limiting reaction time (in s photon^{-1}) and σ is the area of the photon-absorbing antenna. The area, σ, may be approximated from the mass of chlorophyll associated with a single antenna (200 - 300 molecules, each 894 Da or, from the Avogadro number, 148.4×10^{-23} g), times the area of light field subtended by 1 g chlorophyll ($1000 \times \alpha$), to be scarcely more than 4.5×10^{-18} m^2. The photon flux density required to activate the antenna $1/t_p$ times per second is therefore around 15×10^{18} m^{-2} s^{-1}, at the lower temperature, about 111×10^{18} m^{-2} s^{-1} at the upper and, hence, about 56×10^{18} m^{-2} s^{-1} at 20°C. These values convert to orthodox measures of light availability of about 25, 185 and 92 µmol photon m^{-2} s^{-1} respectively.

If the quantity of chlorophyll anticipated to cover 1 m^2 is 1/a = 100 mg chlorophyll *a* and it intercepted 92 µmol photon m^{-2} s^{-1}, then the maximum light-saturated chlorophyll-specific carbon fixation rate is the product (ϕ I_k) = 115 µmol C (g chlorophyll)$^{-1}$ s^{-1}, if ϕ = 0.125 mol C (mol photon)$^{-1}$ or 74 µmol C (g chlorophyll)$^{-1}$ s^{-1}, if ϕ = 0.08 mol C (mol photon)$^{-1}$. Putting the ratio of cell carbon to chlorophyll at 50, these chlorophyll-specific rates are respectively equivalent to 2.3 and 1.5 µmol C (g cell C)$^{-1}$ s^{-1}, and between 27.6 and 17.8 µmol C (mol cell C)$^{-1}$ s^{-1}.

Allowing for a basal metabolism of even 10%, these calculations show the carbon fixation rate to be capable of maintaining the growth-saturating requirement for reduced carbon. The deduction requires to be tempered by the recognition that, at higher average intensities, chlorophyll-*a* absorbance is not uniform across the spectrum, that the maximum absorbance is in the red wavebands and that, in terms of the total irradiance penetrating the surface, a higher I'_0 value is required to provide a flux of photons in the appropriate waveband in order to saturate the photosynthetic reaction. The 4-µm *Chlorella* cell, equivalent (say) to 7.4 pg C (Reynolds, 1990) and containing ~0.149 pg chlorophyll *a* in a cross section of [$\pi(d/2)^2$ =] 12.6 µm^2, has an interceptive capacity over ~20.4 m^2 (mol cell C)$^{-1}$ and a maximum absorptive capacity over 0.085 m^2 (mg chlorophyll *a*)$^{-1}$. In order to saturate a growth rate of 15 µmol C (mol cell C)$^{-1}$ s^{-1} it requires a minimum light absorbance 120 µmol photons (mol cell C)$^{-1}$ s^{-1} and, realistically, nearer to 180 µmol photons (mol cell C)$^{-1}$ s^{-1}. These values are theoretically sustainable at intensities of appropriate wavelength of 6 - 9 µmol photons m^{-2} s^{-1}. Converting in respect of the proportional absorption of visible light by chlorophyll (f = 0.137; Reynolds 1990), experimentally validated I_k values for growth are derived (44 - 66 µmol photons m^{-2} s^{-1})

Reynolds (1994) used similar logic to derive theoretical growth-rate saturating irradiances for other species of phytoplankton. For the much larger spherical colonies of *Microcystis* [a = 2.2 m^2 (mol cell C)$^{-1}$], less efficient interference with the light field raises I_k to ~ 143 µmol photons m^{-2} s^{-1}. For filaments of *Planktothrix agardhii* [11-3 m^2 (mol cell C)$^{-1}$], I_k ~ 126 µmol photons m^{-2} s^{-1}; for stellate 8-celled colonies of the diatom *Asterionella* [26.4 m^2 (mol cell C)$^{-1}$], I_k ~ 46 µmol photons m^{-2} s^{-1}. The light dependence of growth in each case is the slope of r on subsaturating irradiances, $I_z < I_k$, α_r, which approximates to r_{max} / I_k. On the basis of the above calculations, α_r is equivalent to (15 × 10^{-6} mol C per mol cell C per second divided by 66 × 10^{-6} mol photons per m^2 per second =) ~ 0.23 mol C (mol cell C)$^{-1}$ (mol photon)$^{-1}$ m^2 in the case of *Chlorella*, (14.4 / 46 =) ~ 0.31 for *Asterionella*, (10.7 / 126 =) ~ 0.085 for *Planktothrix* and (4.2/143 =) 0.03 mol C (mol cell C)$^{-1}$ (mol photon)$^{-1}$ m^2 for *Microcystis*.

Light Saturation of Photosynthesis

By definition, cells transported to depths receiving photon flux densities > I_k cannot increase their growth rate. If photosynthesis alone is considered, there is ultimately a point at which photosynthetic rate is saturated. Between these points is a range of irradiances wherein the cell may accumulate fixed carbon and produce excess oxygen, especially if growth rate is constrained by a lack of nutrients. The photosynthetic apparatus would be liable to be stressed by either of these eventualities, were it not for the fact that planktonic plants have a number of biochemical, physiological, behavioural and adaptive mechanisms which operate at a variety of time scales to protect them from the risk of damage (summarised in Figure 3). These manifestly bring about an adjustment of the photosynthetic capacity to the needs of the cells. Biomass-specific yield reductions are well-known and have been generically referred in the past as photoinhibition but, in the majority of instances, they are the result of internal adjustment and not merely passive effects of exposure to severe light. The topic has developed rapidly but there has been a good number of excellent reviews (see especially Demmig-Adams and Adams, 1992; Long et al., 1994).

From receiving a subsaturating photon flux, the upward moving cell will experience two almost concurrent responses. The greater bombardment of the light harvesting complexes by photons means that some now arrive to reaction centres which are still closed, while the accelerated accumulation of PQH_2 in the plastoquinone pool limits the re-reduction of Q_A and, hence, reactivation of the centres in the LHCs. The energy absorbed from unused photons continuing to arrive at P680 is re-radiated as fluorescence. This is readily measurable and the spectral signal of emitted fluorescence has been used as an index of plankton biomass (an analogue of an analogue: Lorenzen, 1966) and as a sensor of pigments associated with major components, at least to phylum level (Hilton et al., 1989). More recently, the measurement of light-stimulated changes in *in vivo* fluorescence from cells exposed to a flash of weak light in the dark (F_0, when all centres are open) and then to a saturating flash (F_m, corresponding to total closure) has been used to calculate photosynthetic rates (Kolber and Falkowski, 1993). The variable chlorophyll fluorescence ($F_v = F_m - F_0$) is equivalent to the increase in the quantum yield of fluorescence

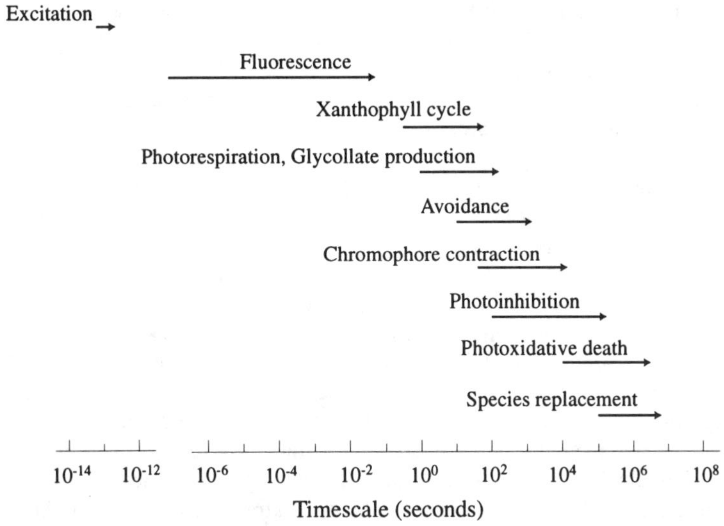

Figure 3. Time scales for contending with high irradiance (PFD) in algae. Note logarithmic scale.

and is used to calculate the maximum rate of photosynthetic electron transport. The reciprocal of this is that the quantum yield of photosynthesis is reduced under high light intensities.

Planktonic cells are generally too small for plastid relocation to have the significance it does in the cells of higher plants (Long et al., 1994) but, over periods of minutes to hours, contraction of the chromophores and the cross-sectional areas presented by chlorophyll is commonly observed in planktonic diatoms subjected to prolonged doses of high light (Neale, 1987). It has also been observed that sinking rates of these cells is sharply enhanced (Reynolds and Wiseman, 1982; Neale et al., 1991), supposedly as a light-avoidance mechanism. Many flagellates migrate quickly out of high light, while the function of buoyancy-regulatory mechanism of planktonic cyanobacteria normally achieves an analogous movement away from the surface, which is more effective in larger colonies (Reynolds et al., 1987).

In cells exposed to frequent or continuing high light intensities over a generation time or more, over-excitation of the PSII LHCs is avoided by changes in the xanthophylls. These oxygenated carotenoids are subject to a series of light-dependent reactions, which, among the Chlorophytes as among other green plants, results in the accumulation of zeaxanthin under excess light conditions and its reconversion to violoxanthin on the return of normal light conditions. Among the Chrysophyta (sensu lato, to include the diatoms) and the Dinoflagellates, the analagous reaction involves the conversion of diadinoxanthin to diatoxanthin, when light is excessive, and the epoxidation back to diadinoxanthin in darkness. Significantly, perhaps, the reaction is said to be about ten times faster than the analogous reaction in higher plants (Long et al., 1994, quoting the work of M. Olaizola). The principal function of the xanthophyll cycle is to protect PSII from excessive PFD by siphoning off its energy as heat. Many details of the cycle and its fine-tuning are considered by Demmig-Adams and Adams (1992); it is sufficient here to see its contribution to the adaptations of phytoplankton to high-light and how these contribute to sustained productivity. Ibelings et al. (1994) have recently demonstrated just this sort of acclamation of planktonic species, especially in *Microcystis*, where the sustained presence of zeaxanthin contributes to an on-going ability to dissipate excess excitation energy as heat. The mechanism protects cells from overexposure to light during surface-bloom formation, as originally proposed by Paerl et al. (1983).

For nutrient-limited cells at even modest light levels, wherein photosynthate is scarcely consumed in growth, the potential difficulties of accumulating fixed carbon and of photooxidative damage by accumulating free oxidant in the cell are met principally by the production of antioxidants, such as ascorbate and glutathione. Reduction of oxygen (to peroxide, then water) can occur in PSI (Mehler Reaction). Beyond PSI, ribulose 1,5-biphosphate is oxidised instead of being carboxylated, leading to the formation of phosphoglycollic acid. In green algae, as in higher plants, the latter may be further oxidised to produce carbon dioxide at an accelerated rate (photorespiration). Release of glycolate and other photosynthetic intermediates into the water is one of the "healthy" (cf. Sharp, 1977) ways in which cells of other algal groups can vent unusable dissolved organic carbon into the medium, at reported rates equivalent to up to 92% of the contemporaneous carbon uptake rates (see Geider and Osborne, 1992, for review and full discussion).

Nevertheless, prolonged exposure of phytoplankton cells to high light intensities over periods of days to weeks usually results in pigment loss, photooxidation, loss of enzyme activity and, ultimately, death. If photoinhibition is considered to be synonymous with damage to the photosystems, then such events are clear instances of severe photoinhibition. It is no less clear that many responses to environmental variability which have resulted in photosynthetic rate reductions, previously referred to photoinbition, should be seen as homeostatic protective mechanisms to counter the effects of exposure to the full range of environmental extremes consequent upon turbulent embedding and as necessary adaptations to planktonic survival (see discussion in Long et al., 1994).

Photoadaptation to Low Light Intensities

As stated above, prolonged exposure to high light is atypical of pelagic habitats. It may be a feature of clear, shallow waters or in the near-surface layer of a well stratified lake with an extremely stable density gradients. The persistence of a cyanobacterial scum or hyperscum places cells in conditions of lethal surface irradiances but it is clear that the cells beneath the scum surface enjoy a very different physico-chemical environment which may be dark and microaerophilous, if not anaerobic (see especially Zohary and Pais-Madeira, 1990). The vertical gradient of light through the scum is a greatly exaggerated version of that through the entire water column of a deep but still lake. The analogy serves as a pointer to the fact that low light intensities can obtain in unmixed water deep in the light gradient and they may characterise the water column mixed to depths beyond the limits of maximal photosynthesis. The latter case may be regarded as being much the nearer of the two to the typical conditions experienced by phytoplankton.

Judged against the thrust of the previous two sections, it is immediately apparent that it is precisely the mobility provided by entrainment through the full depth of the upper mixed layer which avoids the prolonged exposure to high light discussed. The timescale of mixing also makes it unlikely that the population as a whole is over-exposed, nor much of it for periods averaging much longer than the probabilistic mixing time. Having a photosynthetic response which allows the carbon fixation to follow the PFD-determined rate when $I < I_k$ and a series of short-term metabolic mechanisms to control production in the periods when $I > I_k$ is likely typical. That plankton should respond as they do to being put into bottles for several hours at fixed points in the light gradient, with near-surface exposures turning in low aggregates of fixed carbon and deeper ones returning light-limited values, is now recognised to manifest the mechanisms of optimising photosynthesis in the pelagic zone (Harris, 1983; Long et al., 1994).

Mixing, however, does not guarantee the right balance of saturating and subsaturating irradiances over a time integral, nor even sufficient to balance out the replication rate at the temperature- and daylight- determined maximum. Hour-to-hour changes in surface PFD, in mixing depth and in relative light penetration exert their own impacts, singly or in concert, upon the underwater photosynthetic environment. There are strong seasonal trends, however, with major alternations in temperate waters among deep-mixed, unstratified cold water in winter, stratifying clear water in spring, warm, stratified but often turbid water in summer and cooling, increasingly deep-mixed and relatively turbid water in autumn. Annual fluctuations in water temperature may be less severe in low-latitude lakes but weather-influenced changes in density gradients and wind-mixing impart cycles which are just as distinctive. These changes are accompanied by differences in overall productivity, in most cases, coupled with alternations among quite different dominant plankton associations (Reynolds, 1984).

It is interesting that the groups of species well-represented in predominantly deep-mixed water columns are those having what are arguably the most appropriate preadaptations - those both optimally entrained and making efficient antennae (having a high a value): many of the attenuated diatoms, including *Asterionella* and the filamentous *Aulacoseira*, fulfil these criteria, as does the filamentous cyanobacterium, *Planktothrix*. These genera also provide some of the best examples of the obvious physiological opportunity of increasing the number of light-harvesting complexes in the cell, that is, to increase the amount of chlorophyll *a* in the cell, absolutely and in relation to other chlorophylls and certainly to cartenoid pigments (the chlorophyll:carotenoid pigment ratio is one of the longest standing indices to the condition of phytoplankton: Margalef, 1958). I have found (Reynolds, 1984) almost two-fold variations in the chlorophyll content of *Asterionella* during one population cycle (1.3 - 2.3 pg $cell^{-1}$), of about 1.5 in *Cryptomonas* cf. *ovata* and a remarkable nine-fold between deep-stratified and free-mixed phases in a population of *Planktothrix mougeotii* (unpublished observations). Merely increasing chlorophyll does not increase the antennal effect (a is a fairly inflexible species-specific quantity), unless the increment is better dispersed through the cell. Potentially, the same chlorophyll-specific photosynthetic rate stands to support a doubled cell-specific photosynthetic rate if the chlorophyll content per cell is doubled. The great value is less the faster cell-specific maximum *per se* so much as the raised steepness of the slope α_r, for this permits faster growth at all the same subsaturating light intensities.

A further mechanism is to increase the spectral range of light absorption by diversifying the of accessory pigments. While chlorophyll *a* is common to all oxygenic photautotrophs, many also contain other chlorophylls, carotenoids and phycobiliproteins (and which, together, may make up the larger proportion of the light-absorbing capacity). Their peak photon-absorbance is at alternative (shorter) wavelengths to chlorophyll *a* but their presence in close association with the LHCs facilitates the transfer of excitation energy to chlorophyll *a*. The corollary of widening the spectral bands of absorption is a change of algal colour, which is explicitly acknowledged in the term, "chromatic adaptation" (Tandeau de Marsac, 1977). Some of the best examples involve cyanobacteria of the erstwhile genus, *Oscillatoria*, including the performance of the additional chlorophyll and 3-fold increase in c-phycocyanin pigment in low-light grown *Planktothrix agardhii* (Post et al., 1985), observations on deep stratified populations of *P. agardhii* (Utkilen et al., 1985) and changes befalling a non-buoyant population of *Tychonema bourrelleyi* as it slowly sank through the water column of a deep lake (Ganf et al., 1991). The behaviour reaches its highest development in populations of *Planktothrix rubescens* (Meffert, 1971) and *Lyngbya* (Reynolds et al., 1983), all of which achieve a colour of claret or plum after several months of remaining deep in the light gradient of stably stratified lakes. In those cases where measurement has been made, the chlorophyll-specific photosynthetic rate has been raised, the cell-specific photosynthetic efficiency sharply increased (for a proportionate decrease in the I_k of growth) and generally to have taken steps to maintain substantial cell growth at low light intensities (raised α_r).

Low-light adapted plankton cells, whether brought to the surface from deep-stratified layers or from deep, wind-mixed surface layers to be enclosed in glass bottles, are analogous to shade leaves (Demmig-Adams and Adams, 1992) and they quickly undergo analogous responses to shade leaves placed in the light. Clearly, enhanced electron transport occurs at once, probably briefly achieving the maximum chlorophyll-specific photosynthetic rates that naturally-circulating cells achieve near the surface. Very soon, however, and at lower absolute light thresholds than would be applicable in high-light grown cells of the same species, photoprotective reactions are triggered which decrease photon absorption, dissipate fluorescence and heat and remove active oxygen in the cell. In extremes, these responses of the photosynthetic apparatus can occur within minutes (as was the case in the low-light adapted *Anabaena* studied by Robarts and Howard-Williams, 1989), though in significantly longer periods than they could have been exposed to surface light intensities during the mixing to which they had previously adapted. The accelerated sinking-rate response of low-light adapted diatoms "stranded" near the surface by abrupt weakening of the vertical circulation (Reynolds and Wiseman, 1982: Reynolds, 1983) accompanies reduced light absorption and photosynthetic "close-down" and should be seen as defensive. Nevertheless, such cells are still most likely to become genuinely photoinhibited and to suffer damage to the photosynthetic apparatus, photo-oxidative bleaching, lysis and death. That such events are not commonly observed may be considered to illustrate how flexible and how efficient are the mechanisms that the cell has, not only to counter the extremes to which it may be exposed but also to average out the extreme environmental heterogeneity into a smoothed, even growth rate response. Such adaptations are fundamental to the plankton of well-mixed and/or oligophotic environments, in which moderate-sized, attenuated and generally high-s/v species of diatom and filamentous cyanobacteria.

Conclusions

The foregoing sections have examined, once again, how the plant-life of open water is adapted to survive and, indeed, exploit a fluid environment. In particular, it has examined some of the consequences to resource and energy gathering in environments which are typically dilute and crepuscular. Moreover, the nature of water movement, whether generated by wind, by tide or by gravity introduces a high frequency of fluctuation in the energy but tends to suppress resource variability. Reduced motion may restore circadian cycles of energy but lead to the spatial partitioning of resource gradients. Among these possibilities is an array of organisms, each attempting to perform at its optimum, armed with a battery of molecular and physiological response mechanisms to assist it to deal with almost continuous environmental variability. If the common pigment, chlorophyll *a,* is taken as base, or if the common processes of carbon fixation or oxygen generation or the uptake of phosphorus or of nitrogen or even of iron are taken as the measure of biological productivity, then certain generalities emerge about the rates of production sustainable in different pieces of water. If the question is why is species x the one to do it, or how much more successfully x does it than y or why z does not do it at all, then a little more needs to be known about the various further adaptations of x, y and z.

The secondary question may be considered academic but it influences the result sought by the first for an important reason. If the problem is to determine how productive is a patch of planktonic plants, one may need to know the instantaneous biomass as a reference. If it could be simultaneously assumed that cell replication rate would be the fastest that the species present could maintain at that temperature, in that daylength and subject to that daily light integral, then the net primary production level (*sensu* new cell carbon added per unit carbon present per unit time) is predictable as a function of the capacities of cells present (Reynolds, 1989). The assumption of capacity-filling is not applicable to severely carbon- or nutrient-limited material. Later modelling approaches (Hilton et al., 1992) have addressed these quantities but for the moment it is sufficient to observe that under these conditions, the fluorescence signal will be high and the net productivity will be deduced to be low.

The foregoing sections have also pointed to the morphological properties of planktonic algae which predispose them to rapid growth (high s/v), to those which protect acquired resources (large size, motility), to those which assist light harvesting at low average PFD (attenuation) and to those which facilitate nutrient uptake from low concentrations (high affinity). There has been a good number of studies made to quantify these performance criteria and from a collection of these results, Reynolds (1993) was able to make comparisons of the idealised growth of about a dozen species. The data set is processed one stage further to produce Figure 4a, which, against the gradients used in Figure 2, shows boxes defining the range of light integral and phosphorus concentration at which maximum replication rate could still be maintained. In the other three, some contours are inserted in respect of species.

The plots go some way to revealing that certain species which contend with energy limitation may never face severe nutrient depletion. Others are better adapted to operate under low-phosphorus conditions or to survive ones of forced deficiency, but where the light income is not problematic. Where resources and energy are abundant, a fast growth rate is the principal selective asset.

The complete Figure 4 gives the distributions of illustrative pelagic plant species. In the context of the introductory analogy, the diagram may have only about the predictive value of a corresponding plot distinguishing the ranges of a cactus, a cattail and a cushion moss on axes of water availability and sun tolerance. Were it not enough merely to establish the concept, however, there are intuitive deductions about the productivites of the three types of higher plant named, which also transfer well to the more rarefied poles of the pelagic ranges.

Acknowledgment. Circumstances prevented me from attending the meeting at Broome but I am extremely grateful to Rod Oliver, who not only presented my paper for me at rather short notice but very carefully checked all my mathematics.

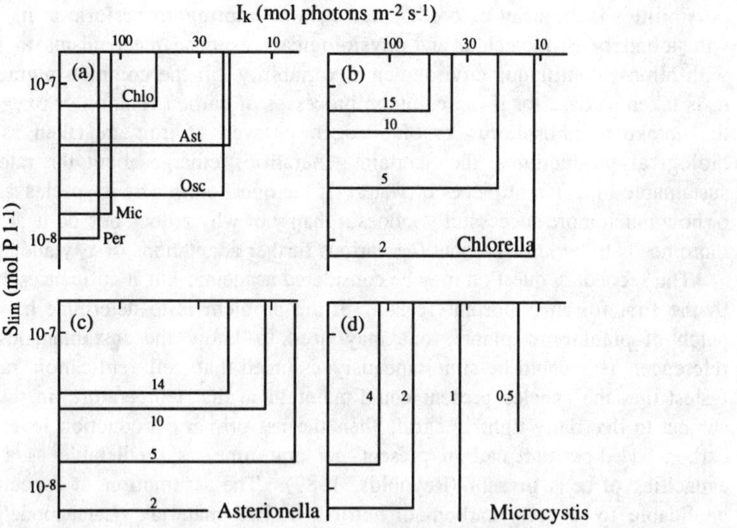

Figure 4. (a) Limits of PFD and of phosphorus concentration to maximum growth rate at 20°C of selected algae for which data are available (dataset compiled by Reynolds, 1993: Chlo = *Chlorella*, Ast = *Asterionella formosa*, Osc = *Limnothrix agardhii*, Mic = *Microcystis aeruginosa*, Per = *Peridinium willei*); (b) as (a) but showing growth-rate contours for light-limited and phosphorus-limited growth in *Chlorella*; contours scaled x 10^{-6} s^{-1}; (c) ditto for *Asterionella*; (d) ditto for *Microcystis*.

References

Bannister, T. T. and A. D. Weidemann, The maximum quantum yield of phytoplankton photosynthesis *in situ*. *J. Plankton Res. 6*, 275-295, 1984.

Berman, T., P. D. Walline, A. Schneller, J. Rothenberg and D. W. Townsend, Secchi-disk depth record: a claim for the eastern Mediterranean. *Limnol. Oceanogr. 30*, 447-448, 1985.

Cabrera, S. and V. Montecino, The meaning of the euphotic chlorophyll *a* measurement. *Verh. int. Verein. theor. angew. Limnol.* 22, 1328-1331, 1984.

Dau, H., Molecular mechanisms and quantitative models of variable photosystem II fluorescence. *Photochem. Photobiol* . *60*, 1-23, 1994.

Demmig-Adams, B. and W. W. Adams, Photoprotection and other responses of plants to high light stress. *Ann. Rev. plant Physiol. plant mol. Biol. 43*, 599-626, 1992.

Denman, K. and A. E. Gargett, Time and space scales of vertical mixing and advection of phytoplankton in the upper ocean. *Limnol. Oceanogr. 28*, 801-815, 1983.

Droop, M. R., The nutrient status of algal cells in continuous culture. *J. mar. biol. Ass.* 54, 825-855, 1974.

Dugdale, R. C., Nutrient limitation in the sea: dynamics, identification and significance. *Limnol. Oceanogr. 12*, 685-695, 1967.

Falkowski, P. G., Molecular ecology of phytoplankton photosynthesis. In *Primary Productivity and Biogeochemical Cycles in the Sea*, pp. 47-67, Plenum Press, New York, 1992.

Frankignoulle, M., Field-measurements of air-sea CO_2 exchange. *Limnol. Oceanogr. 33*, 313-322, 1988.

Ganf, G. G., S. I. Heaney and J. Corry, Light absorption and pigment content in natural populations and cultures of a non-gas-vacuolate cyanobacterium, *Oscillatoria bourrellyi* (= *Tychonema bourrellyi*). *J. Plankton Res. 13*, 1101-1121, 1991.

Geider, R. J. and B. A. Osborne, *Algal Photosynthesis: The Measurement of Algal Gas Exchange*. Chapman and Hall, New York, 1992.

Grime, J. P. *Plant Strategies and Vegetation Processes*. Wiley-Interscience, Chichester, 1979.

Harris, G. P., Photosynthesis, productivity and growth: the physiological ecology of phytoplankton. *Ergebn. Limnol. 10*, 1-163, 1978.

Harris, G. P., Mixed-layer physics and phytoplankton populations: studies in equilibrium and non-equilibrium ecology. In *Progress in Phycological Research*, v. 2, 1 -52, Elsevier, Amsterdam, 1983.
Hill, R. and F. Bendall, Function of the two cytochrome components in chloroplasts: a working hypothesis. *Nature 186*, 136-137, 1960.
Hilton, J., A. E. Irish and C. S. Reynolds, Active reservoir management: a model solution. In *Eutrophication: Research and Application to Water Supply*, pp. 185-196, Freshwater Biological Association, Ambleside, 1992.
Hilton, J., E. Rigg and G. H. M. Jaworski, Algal identification using in vivo fluorescence spectra. *J. Plankton Res. 11*, 65-74, 1989.
Hoogenhout, H. and J. Amesz, Growth rates of photosynthetic microorganisms in laboratory cultures. *Arch. Mikrobiol. 50*, 10-25, 1965.
Humphries, S. E., and J. Imberger, The influence of the internal structure and dynamics of Burrinjuck Reservoir on phyoplankton blooms. Centre for Water Resarch Reference ED 023 JI, 1982.
Ibelings, B. W., B. M. A. Kroon and L. Mur, Acclimation of photosystem II in a cyanobacterium and a eukaryotic green alga to high and fluctuating photosynthetic ohoton flux densities, simulating light regimes induced by mixing in lakes. *New Phytol. 128*, 407-424, 1994.
Kemp, W. M. and W. J. Mitsch, Turbulence and phytoplankton diversity : ageneral model of the paradox of the plankton. *Ecol. Modell. 7*, 201-222, 1979.
Kirk, J. T. O., *Light and Algal Photosynthesis* (2nd edn). Cambridge University Press, Cambridge, 1994.
Kjelleberg, S., N. Albertson, K. Flärdh, L. Holmquist, Å. Jouper-Jaan, R. Marouga, J. Östling, B. Svenblad and D. Weichart, How do non-differentiating bacteria adapt to starvation? *Antonie van Leeuw. 63*, 333-341, 1993.
Knox, R. S., Photosynthetic efficiency and exciton trapping. In *Primary Processes of Photosynthesis*, Elsevier, Amsterdam, 55-97, 1977.
Kolber, Z. and P. G. Falkowski, Use of active fluorescence to estimate phytoplankton photosynthesis in situ. *Limnol. Oceanogr. 38*, 1646-1665, 1994.
Kuhlbrandt, W. and D. N. Wang, Three-dimensional structure of plant light-harvesting complex determined by electron crystallography. *Nature, 351*, 130-131, 1991.
Larkum, A. W. D. and J. Barret, Light harvesting processes in algae. In Advances *in Botanical Research*, Vol. *10*, 1-219, Academic Press, London, 1983.
Long, S. P., S. Humphries and P. G. Falkowski, Photoinhibition of photosynthesis in nature. *Ann. Rev. plant Physiol. plant mol. Biol. 45*, 633-662, 1994.
Lorenzen, C. J., A method for the continuous measurement for *in-vivo* chlorophyll concentration. *Deep-Sea Res. 13*, 223-227, 1966.
Lund, J. W. G., Studies on Asterionella formosa Hass. II. Nutrient depletion and the spring maximum. *J. Ecol. 38*, 1-35, 1950.
Maberly, S. C., Diel, episodic and seasonal changes in pH and concentrations of inorganic carbon in a productive lake. *Freshwat. Biol. 35*, 579-598, 1996.
Mackereth, F. J. H., J. Heron and J. F. Talling, Some methods of water analysis for limnologists. *Sci. Publs Freshwat. Biol. Ass. 36*, 1 -120, 1978.
Mann, K. H. and J. R. N. Lazier, *Dynamics of Marine Ecosystems*. Blackwell Scientific Publications, Oxford, 1991.
Mann, N. H., How do cells express nutrient limitation at the molecular level? In *Molecular Ecology of Aquatic Microbes*. Springer-Verlag, Berlin, 171-190, 1995.
Margalef, R., Temporal succession and spatial heterogeneity in phytoplankton. In *Perspectives in Marine Biology*, University of California Press, Berkeley, 323-349, 1958.
Margalef, R., Life-forms of phytoplankton as survival alternatives in an unstable environment. *Oceanol. Acta 1*, 493-509, 1978.
Martin, J. H. and S. E. Fitzwater, Iron deficiency limits phytoplankton growth in the north-east Pacific subarctic. *Nature 331*, 341-343, 1988.
McWhirter, N., *The Guinness Book of Records*. Guinness Superlatives, London, 1994.
Meffert, M. -E., Cultivation and growth of two planktonic *Oscillatoria* species. *Mitt. int. Ver. theor. angew. Limnol. 19*, 189-205, 1971.
Neale, P. J., Algal photoinhibition and photosynthesis in the aquatic environment. In *Photoinhibition*, Elsevier, Amsterdam, 39-65, 1987.
Neale, P. J., S. I. Heaney and G. H. M. Jaworski, Responses to high irradiance contribute to the decline of the spring diatom maximum. *Limnol. Oceanogr. 36*, 761-768, 1991.
Paerl, H. W., J. Tucker and P. T. Bland, Carotenoid enhancement and its role in maintaining blue-green

algal (Microcystis aeruginosa) surface blooms. *Limnol. Oceanogr. 28*, 847-857, 1983.

Post, A. F., R. de Wit and L. R. Mur, Interactions between temperature and light intensity on growth and photosynthesis of the cyanobacterium, *Oscillatoria agardhii*. *J. Plankton Res.*, *7*, 487-495, 1985.

Reynolds, C. S., A physiological interpretation of the dynamic responses of populations of a planktonic diatom to physical variability of the environment. *New Phytol. 95*, 41-53, 1983.

Reynolds, C. S., *The Ecology of Freshwater Phytoplankton*. Cambridge Univ. Press, Cambridge, 1984.

Reynolds, C. S., The response of phytoplankton communities to changing lake environments. *Schweiz. Z. Hydrol. 49*, 220-236, 1987.

Reynolds, C. S., Physical determinants of phytoplankton succession. In *Plankton Ecology*, Brock-Springer, Madison, 9-56, 1989.

Reynolds, C. S., Temporal scales of variability in pelagic environments and the response of phytoplankton. *Freshwat. Biol.*, *23*, 25-55, 1990.

Reynolds, C. S., Eutrophication and the management of lakes: what Vollenweider couldn't tell us. In *Eutrophication: Research and Application to Water Supply*. Freshwater Biological Association, Ambleside, 4-29, 1992.

Reynolds, C. S., Swings and roundabouts: engineering the environment of algal growth. *In Urban Waterside Regeneration*. Ellis Horwood, Chichester, 330-349, 1993.

Reynolds, C. S., The role of fluid motion in the ecology of phytoplankton in lakes and rivers. In *Aquatic Ecology; Scale Pattern, Process*, Blackwell Scientific Publications, Oxford, 141-187, 1994.

Reynolds, C. S., R. L. Oliver and Walsby, A. E., Cyanobacterial dominance: the role of buoyancy regulation in dynamic lake environments. *N. Zealand J. mar. freshwat. Res. 21*, 379-390, 1987.

Reynolds, C. S., J. G. Tundisi and K. Hino, Observations on a metalimnetic *Lyngbya* population in a stably stratified tropical lake (Lagoa Carioca, Eastern Brasil). *Arch. Hydrobiol. 97*, 7-17, 1983.

Reynolds, C. S. and S. W. Wiseman, Sinking loses of phytoplankton in closed limnetic systems. *J. Plankton Res. 4*, 489-522, 1982.

Rhee, G. -Y., Effects of environmental factors on phytoplankton growth. *In Advances in Microbial Ecology*, Vol. 6. Plenum Press, New York, 33-74, 1982.

Robarts, R. D. and Howard-Williams, C., Diel changes in fluorescence capacity, photosynthesis and macromolecular synthesis by Anabaena in response to natural variations in solar irradiance. *Ergebn. Limnol. 32*, 35-48, 1989.

Sharp, J. H., Excretion of organic matter by marine phytoplankton: do healthy cells do it? *Limnol. Oceanogr. 22*, 381-399, 1977.

Sommer, U., The paradox of the plankton: fluctuations of phosphorus availability maintain diversity of phytoplankton in flow-through cultures. *Limnol. Oceanogr. 29*, 633-636, 1984.

Smith, I. R., Hydraulic conditions in isothermal lakes. *Freshwat. Biol.*, *9*, 119-145, 1979.

Spigel, R. H. and J. Imberger, Mixing processes relevant to phytoplankton dynamics in lakes. *N. Zealand J. mar. freshwat. Res.*, *21*, 361-377, 1987.

Talling, J. F., The application of some electro-chemical methods to the measurement of photosynthesis and repiration in fresh waters. *Freshwat. Biol. 3*, 335-362, 1973.

Talling, J. F., The depletion of carbon dioxide from lake water by phytoplankton. *J. Ecol. 64*, 79-121, 1976.

Tandeau de Marsac, N., Occurrence and nature of chromatic adaptation in cyanobacteria. *J. Bacteriol. 130*, 82-91, 1977.

Tilman, D., S. S. Kilham and P. Kilham, Phytoplankton communiy ecology: the role of limiting nutrients. *Ann. Rev. Ecol. System. 13*, 349-372, 1982.

Uppstill-Goddard, R. C., A. J. Watson, P. S. Liss and M. I. Liddicoat, Gas transfer velocities in lakes measured with SF_6. *Tellus 42B*, 364-377, 1990.

Utkilen, H. -C., P. M. Skulberg and A. E. Walsby, Buoyancy regulation and chromatic adaptation in planktonic *Oscillatoria species*: alternative strategies for optimising ligt absorption in stratified lakes. *Arch. Hydrobiol.*, *104*, 407-417, 1985.

Vollenweider, R. A. and J. Kerekes, The loading concept as basis for controlling eutrophication philosophy and preliminary results of the OECD programme on eutrophication. *Progr. Wat. Technol. 12*, 5-38, 1980.

Walker, D., Excited leaves. *New Phytol. 121*, 325-345, 1992.

Watson, A. J., R. C. Uppstill-Goddard and P. S. Liss, Air-sea gas exchange in rough and stormy seas measured by a dual-tracer technique. *Nature 349*, 145-147, 1991.

Zohary, T. and A. M. Pais-Madeira, Structural, physical and chemical characteristics *of Microcystis aeruginosa* hyperscums from a hypertrophic lake. *Freshwat. Biol.*, *23*, 339-352, 1990.

38

Turbulent Mixing and Resource Supply to Phytoplankton

S. MacIntyre

Abstract

The temporal and spatial variability of turbulence was determined by profiles of temperature-gradient microstructure or from surface energy budgets in three shallow and two deep lake basins. Rates of dissipation of turbulent kinetic energy ε, length scales of turbulent eddies, the coefficient of eddy diffusivity, and irradiance measurements were used to infer the light climate of phytoplankton. Time scales of mixing τ_{mix} are contrasted with time scales for photoadaptation τ_a. On calm days in a stratified basin, phytoplankton circulated within eddies that were at most 10% of the depth of the euphotic zone and could have experienced fluctuations that were up to 20% of surface irradiance. τ_{mix} was rapid enough that phytoplankton in the surface layer would not have had time to develop differences in photoadaption if $\tau_a \geq 15$ minutes. On windy days, overturns extended throughout and even beyond the depth of the euphotic zone. When overturns at the surface were larger than 1 meter, differences in the extent of photoadaptation would develop within overturns if $\tau_a = 15$ minutes. For $\tau_a = 6$ hours, photoadaptation was unlikely within weakly stratified surface overturns.

The turbulence data also were used to assess physical mechanisms important for fluxes of biologically important materials. Convective motions due to heat loss at night were more important than wind mixing in the day for vertical transport of solutes in a shallow, equatorial lake. Either intrusions or the inefficiency of mixing where turbulent Froude numbers are less than one led to persistent stratification during the day of heat, particles and solutes in a eutrophic lake despite moderate wind forcing. Wind forcing in a seasonally stratified lake led to upwelling and mixing in the thermocline. Because concentrations of PO_4-P increased below the depth of mixing, algae remained phosphorus limited. The muted biological response to high winds was in contrast to the five-fold increase in algal biomass in a shallow, attached basin in which phosphorus limitation was briefly alleviated due to sediment resuspension. Boundary mixing in a meromictic lake led to values of ε and nutrient fluxes that were two to four orders of magnitude higher at an onshore site than they were at an offshore site at similar depths.

Physical Processes in Lakes and Oceans
Coastal and Estuarine Studies Volume 54, Pages 561-590

Introduction

Ecological and physiological processes are affected by turbulent flow fields on both large and small scales. On large scales, tens of centimeters to tens of meters, turbulence affects populations as well as individuals. Phytoplankton, other particles, and solutes are dispersed (Yamazaki and Kamykowski, 1991; Lewis et al., 1984b, 1986), with effects on nutrient supply, light availability to phytoplankton (MacIntyre, 1993), sediment resuspension, and grazing by zooplankton and larvae. Turbulence on this scale can redistribute marine snow leading to localized accumulations with important implications for oceanic community structure and chemical cycling (MacIntyre et al., 1995a). Similarly, eddies generated in the benthic boundary layer affect the number of larvae which settle and metamorphose to adults (Pawlik and Butman, 1993; Eckman, 1983). This paper will emphasize turbulence on the larger scales as it impacts on light and nutrient supply to phytoplankton.

On the small scales, less than the Kolmogoroff microscale (L_k) and up to ten times L_k, the turbulent flow field affects algae, marine snow, and patches of solutes directly. For instance, the vorticity of a turbulent flow field may distort the boundary layer around a phytoplankton cell, increasing the concentration gradient of dissolved solutes, and thereby affecting nutrient uptake or release of wastes (Lazier and Mann, 1989; Pasciak and Gavis, 1974, 1975). The strains created may be harmful to the individual cell (Thomas and Gibson, 1990; Gibson and Thomas, 1995) or may act to disaggregate particle flocs (Alldredge et al., 1990). Contact rates between particles embedded in individual eddies may be increased due to the relative motions of the eddies. The resulting increased rate of aggregation of particles such as marine snow has implications for the rates at which different pollutants are absorbed or desorbed from the particle surface (Alldredge and Gotschalk, 1988). Similarly, contact rates between zooplankton and their prey may be increased (Rothschild and Osborn, 1988, Yamazaki et al., 1991, Denman and Gargett, 1995). Alternatively, turbulent eddies could distort the wake or chemical trail left by prey and make capture more difficult (Andrews, 1983; J. R. Strickler, pers. comm.) or dilute waterborne settlement cues and reduce recruitment of benthic larvae (Turner et al., 1994). The eddies at the air-water interface, many of which are small, move dissolved gases from the interior of the fluid to the boundary and determine the rates of gas flux across the air-water interface (MacIntyre et al., 1995b).

Temperature-gradient and shear microstructure profilers have great potential for addressing problems related to dispersion of particles and nutrients as well as for defining the strains and shears acting on the small scale. In particular, they indicate where the water column is turbulent and provide turbulent velocity and length scales from which mixing rates can be calculated. Lewis et al. (1984, 1986) first used microstructure profilers to address the problem of light and nutrient supply to phytoplankton in the ocean. Since Lewis et al.'s pioneering measurements, microstructure profilers have only been applied in a few biological studies (Carr et al., 1992; Carr et al., 1995; Crawford and Dewey, 1989; Haury et al., 1992; MacIntyre, 1993).

Microstructure profilers have allowed development of a deeper understanding of the dynamics of the upper mixed layer, that portion of the water column above the seasonal thermocline, as well as within the thermocline. These insights are essential for proper design and interpretation of biological and chemical studies. Imberger (1985) showed that, despite its name, the upper mixed layer is not always mixing, and both his and Shay and Gregg's (1986) measurements indicated that only small temperature differences (e.g.

0.02°C) separate regions with different mixing intensities. The insights gained from these measurements led to new terminology, applicable to lakes and oceans, to describe the upper mixed layer (Imberger and Patterson, 1990; Imberger, 1994). Its uppermost portion, the surface layer, also called a diurnal surface layer or surface mixing layer, is that part of the water column directly effected by wind and surface heat fluxes. The extent of mixing depends on the balance between wind forcing and heat flux. If the mixing is insufficient to penetrate to the seasonal thermocline, a diurnal thermocline will form below the surface layer, with a subsurface layer extending to the seasonal thermocline. The extent of these layers varies seasonally as well as diurnally. Because these regions are generally part of the euphotic zone, an understanding of their dynamics is critical for describing the circulation of phytoplankton, nutrient fluxes, and ultimately predicting phytoplankton growth.

In the following, I discuss results from studies in which turbulence from the surface layer and into the seasonal thermocline or to the sediment-water interface was measured to predict light and nutrient supply to phytoplankton. Data include time series of profiles of temperature-gradient microstructure in two seasonally stratified lakes and two diurnally mixing ones. Reference is made to a time series of surface energy fluxes calculated from a fifth basin. Results are used to identify the percentage of the euphotic zone that was mixing during different meteorological conditions in two morphologically distinct lake basins with similar attenuation coefficients. In addition, the likelihood of differences in photoadaption occurring is inferred based on thermal structure and from time scales of photoadaption relative to time scales of mixing. Examples from different lakes are used to illustrate the importance of mixing due to heat loss in warm water lakes, the difficulty of mixing stratified water bodies while heat flux is into the lake, the importance of sediment resuspension to nutrient flux, differing responses of two morphologically distinct basins to similar meteorological forcing, and localized nutrient fluxes due to boundary mixing.

Description of the Lakes and Methodology

The four lakes studied (Table 1) included Lake Biwa, Japan (35°11'N, 135°58'E), which was the site of the multi-disciplinary Lake Biwa Transport Experiment (BITEX) designed to assess the effect of both horizontal and vertical transport on biogeochemical processes in large lakes. It took place from late August through mid-September 1993, a time when strong wind forcing due to typhoons was expected. With its seasonally stratified North Basin and its shallow South Basin which was essentially a surface layer, it was an excellent site to contrast the affects of similar meteorological forcing on both photoadaptation

TABLE 1. Characteristics of the four lakes. Attenuation coefficients are those at the time of the studies reported here.

	Surface area (km^2)	Maximum depth (m)	Mean depth (m)	Attenuation coefficient (m^{-1})
Lake Biwa, Japan				
North Basin	616	104	44	0.3 — 0.45
South Basin	58	8	3.5	0.4 — 1.5
North Lake, W. Australia	0.29	2.6		2.1
Lago Calado, Brazil	2—8	1—12		1.49 (January 1983) 1.35 (February 1983)
Mono Lake, California	160	45	17	0.42

and nutrient fluxes. North Lake, Western Australia, is also a shallow lake with the whole water column constituting a surface layer. North Lake is located 14 km south of Perth, W.A., and 7 km from the Indian Ocean in the Cockburn chain of wetlands. With its typically high attenuation coefficient, stronger thermal stratification developed during the day than in the South Basin of Lake Biwa, with greater resistance to wind mixing. Lake Calado, Brazil (3°15'S, 60°34'W), is a seasonally inundated lake on the Amazon floodplain. At low water, it diurnally stratifies and mixes with the entire water column constituting a surface layer (MacIntyre and Melack, 1984). As the water level rises, the lake develops a seasonal thermocline. The measurements reported here were taken just after the lake made the shift to seasonal stratification. Mono Lake, California, USA (38°N, 119°W) is a chemically stratified, hypersaline closed basin lake in the Great Basin (Jellison and Melack, 1993). In the past decade and a half, the lake has shifted between several years of monomixis to several years of meromixis depending on the magnitude of the freshwater inputs. Another meromictic period began during 1995, the year of our study.

Temperature-gradient microstructure profiling was performed to provide a realistic description of the mixing activity within all the lakes except Lake Calado. The temperature-gradient microstructure profilers, modified from Carter and Imberger (1985), provided temperature and temperature-gradient measurements from which were computed overturning scales (Thorpe, 1977; Imberger and Boashash, 1986) and rates of energy dissipation (Imberger and Boashash, 1986; Dillon and Caldwell, 1980; Imberger and Ivey, 1991). Both the maximum overturning scale, which provides an estimate of the length scale of the maximum vertical extent of a turbulent eddy, and the root mean square (rms) displacement scale, which provides an estimate of the scale of the energy containing eddies for mixing calculations, were computed. Energy dissipation rates were computed in segments in which the temperature gradient signal was stationary and by implication the turbulence was statistically stationary. Within the segments, the power spectrum of the temperature gradient signal was matched to the theoretical Batchelor spectrum to validate that the flow was turbulent. The coefficient of eddy diffusivity was calculated as $K_z = (R_f/1-R_f)\,\varepsilon N^{-2}$ (Osborn, 1980) with the flux Richardson number R_f computed as in Ivey and Imberger (1991) and Ivey et al. (1998). To calculate the buoyancy frequency, $N = (-g/\rho\ \partial\rho/\partial z)^{1/2}$ where g is gravity and ρ is density, the density profile was first reordered so as to be monotonic with depth. The turbulent Froude $Fr_t = u/\ell N$ and Reynolds $Re_t = u\ell/\upsilon$ numbers were calculated as in Ivey and Imberger (1991), where u and ℓ are turbulent velocity and length scales respectively and υ is kinematic viscosity. In this paper, ℓ refers to the maximum overturning scale.

The euphotic zone is defined as the depth to which 1% of incident surface irradiance penetrates. It is calculated from the relation $I_z = I_o\ e^{-k_s z}$ where I_z and I_o are irradiance at a depth z and at the surface, respectively, and k_s is the diffuse attenuation coefficient.

Circulation of Phytoplankton

Whether phytoplankton tend to stay at one location in a water body or are frequently circulated throughout the euphotic zone has implications for ecosystem productivity. For instance, mixing affects the light climate of phytoplankton and resulting photosynthesis (Marra, 1978a,b; Patterson, 1991). Reynolds (1998) discusses the adaptive strategies of phytoplankton when exposed to both high and low irradiances. When mixing rates are low and patches of phytoplankton persist, growth of zooplankton and larvae is favored.

There are several critical questions to answer when relating turbulence to phytoplankton circulation:

Where is the water column mixing? Mixing near the surface will reduce photoinhibition that often accompanies the formation of diurnal thermoclines (Vincent et al., 1984). Similarly, it may reduce the deleterious effects of ultraviolet-B radiation in regions where ozone levels are reduced (Vincent and Roy, 1993). Mixing near the pycnocline or near the benthic boundary layer may replenish nutrient supply to phytoplankton.

How long does the mixing last? How frequent are the mixing events? Even brief episodes of mixing can alter the position of phytoplankton in the water column in ways that can reduce patchiness and lessen overall grazing success. Persistent mixing will circulate phytoplankton continuously through a light gradient; short mixing events may only alter the position of a subset of the population and have minimal effect on irradiance experienced by the population.

How large are the overturns? While conceptual models of the upper mixed layer scale the overturn size with mixed layer size, it is more likely that an actively mixing layer is comprised of many smaller overturns (MacIntyre, 1993; Moum, 1996). If the overturns are contiguous in space, phytoplankton may circulate through the entire region. If the mixing layers in the upper mixed layer are discontinuous in space (Imberger, 1985; MacIntyre, 1996), the irradiance experienced by phytoplankton will be different near the surface and base of the mixed layer. Many overturns in stratified regions are on scales of tens of centimeters and, depending on the attenuation coefficient, the variability in irradiance experienced by the phytoplankton may be high or low.

What are the turbulent velocity scales? Velocity and length scales together determine the rate that patches are dispersed and the rapidity with which phytoplankton experience changes in irradiance. Time scales in a diurnal surface layer were on the order of 5 minutes for circulation through a 1.5 m overturn (MacIntyre, 1993). If a phytoplankton cell had circulated through the overturn, it would have gone from 85% to 3% of the light incident upon the lake's surface in a few minutes. Most of the overturns were on the order of tens of centimeters, and time scales were on the order of tens of seconds to minutes. Recent measurements of quantum yield of fluorescence in a similarly stratified lake in eastern Australia predicted similar time scales of mixing (J.W. Whittington, R.O. Oliver, and B.S. Sherman, pers. comm.)

Fundamental to the problem of understanding how turbulence affects exposure of phytoplankton to irradiance are good measurements of the attenuation coefficient of light. Similarly, the biological aspects of the problem cannot be ignored. For each water body, we must know which of the potentially dominant phytoplankton species follow the flow passively and which ones may have sinking or rising speeds greater than turbulent velocity scales. Diatoms with their siliceous frustules have sufficient mass that they may have sinking speeds in excess of turbulent velocity scales, and buoyant or flagellated species may propel themselves at speeds also in excess of turbulent velocity scales. The sinking and rising speeds of the organisms must be included when assessing their overall circulation.

The difficulty in addressing this problem was observed with cyanobacteria in a hypereutrophic lake in East Africa. Abundance peaks of buoyant cyanobacteria occurred even when turbulent velocity scales, computed as 0.001 times the mean wind speed, exceeded their sinking and rising speeds by one to two orders of magnitude (MacIntyre and Melack, 1995). One approach for determining whether phytoplankton cells will follow the flow involves computing the ratio of the time scale for mixing τ_{mix} to the time scale for sinking τ_s. $\tau_{mix} / \tau_s = (\ell^2/2K_z) / (\ell/v)$, K_z is the coefficient of eddy diffusivity, and v is the sinking or rising speed of the phytoplankton cells. When $\tau_{mix} / \tau_s << 1$, cells will follow the flow. When the ratio >> 1, the cells' motility will determine their position. For values of the ratio near 1, assessing circulation would require including both the cells' speed and the turbulent velocity scales. Analysis using microstructure data from North Lake, Australia, and typical ranges of rising and sinking speeds for the cyanobacteria which dominated the phytoplankton in this lake, illustrated that whether cells followed the flow or not depended on the turbulent Froude number (Fr_T) (MacIntyre, 1993). The turbulent Froude number (Ivey and Imberger, 1991) is a ratio of the inertial forces causing mixing to the buoyancy forces resisting mixing and can be thought of as an inverse Richardson number. For highly mobile cells with rising speeds of 100 to 250 m day^{-1}, their motility was more important than the turbulence in determining their position for $Fr_T < 1$ where buoyancy was suppressing turbulence, but both contributed at higher Fr_T where the water column was actively mixing. In contrast, for cells which sank at speeds of 10 m d^{-1}, both their sinking and the turbulence affected their position at all Fr_T.

Mixing and Photoadaptation in a Shallow, Diurnally Stratifying Basin (South Basin, Lake Biwa)

Surface forcing was a major determinant of the size of overturning regions in the South Basin of Lake Biwa (Figures 1 and 2). A typhoon occurred on 4 September 1993 with winds in the North Basin reaching 22 m s^{-1}. When sampling commenced in the South Basin after winds had dropped but were still higher than 4 m s^{-1}, overturns ranged from 5% to 100% of the depth of the basin, with most overturns similar to the lake's depth. However, on the other days, when wind speeds were less than this, overturning scales were sensitive to irradiance. Maximum overturns ranged between 30% and 100% of the lake's depth at night, and in the morning and mid to late afternoon on days when maximum solar irradiance was less than 800 W m^{-2}. Otherwise, maximum overturns were less than 25% of the lake's depth. Overturns tended to be less than 20% of the euphotic zone depth on calm days near midday, and only extended beyond the euphotic zone on the day of the typhoon (MacIntyre, 1996).

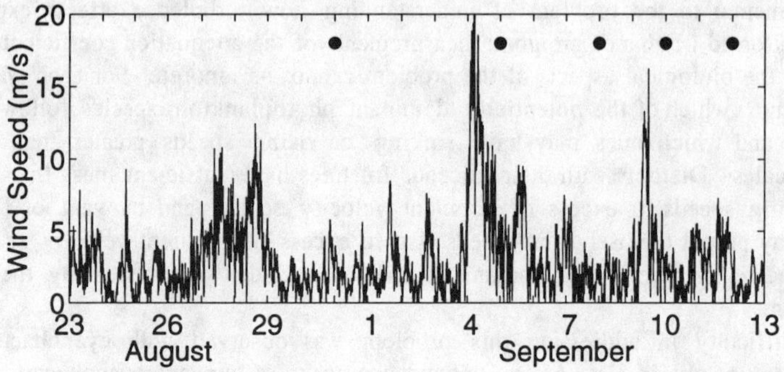

Figure 1. Windspeed obtained from 23 August 1993 (day 235) to 13 September 1993 (day 256) in Lake Biwa at Stn K (see Figure 4). Sampling days in the North Basin are marked with a •. Data were taken at 10 minute intervals. The peak windspeed was 22 m s^{-1}. Meteorological data were provided by the Lake Biwa Research Institute.

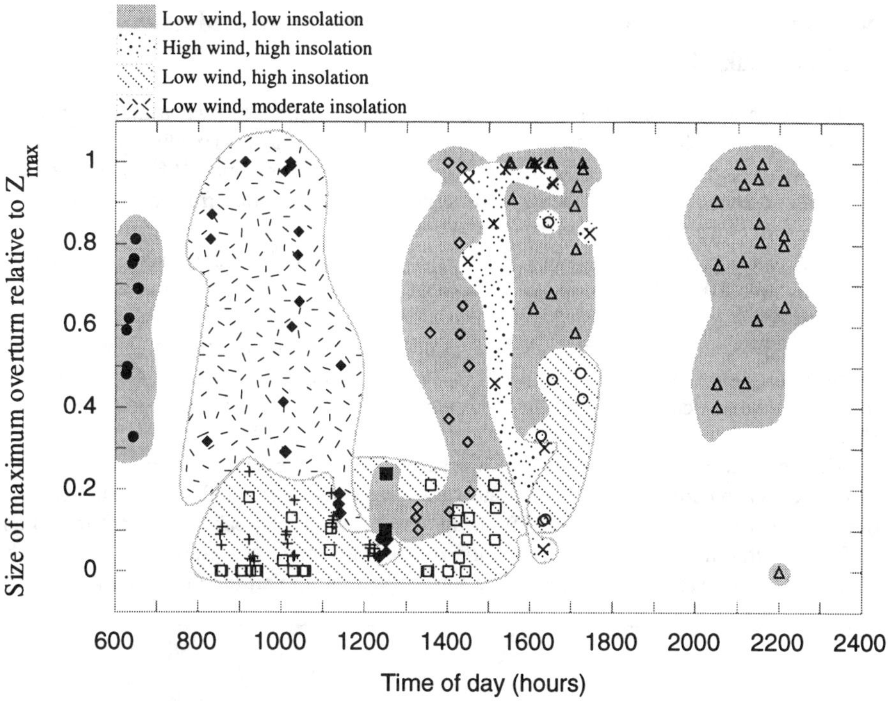

Figure 2. Size of the maximum displacement in each profile divided by the local depth in the South Basin, Lake Biwa, plotted against time of day. Lake depth was determined by the cessation of sinking of the microstructure profiler; on upcasts, 0.5 m was added to the depth. Depth of the water column at the sampling stations ranged from 3 to 5 m. Low wind, hourly average < 3 m s^{-1}; high wind, hourly average > 4 m s^{-1}; high insolation, maximum insolation > 0.8 kW m^{-2}; moderate insolation, 0.2 kW m^{-2} < maximum insolation < 0.8 kW m^{-2}; low insolation, maximum insolation < 0.2 kW m^{-2}. 1 Sept. (o); 2 Sept. (□);3 Sept. (◊);4 Sept. (x); 5 Sept. (+); 6 Sept. (Δ); 7 Sept. (•);8 Sept. (■); and 10 Sept. (◆). From MacIntyre, 1996.

These microstructure results indicate that mixing driven by wind and/or heat loss would be likely to circulate phytoplankton through an appreciable portion of the euphotic zone on diurnal time scales in the South Basin. These frequent movements would be likely to preclude development of differences in photoadaptation that develop over time scales of hours. In fact, Frenette et al. (1996b) observed comparable values for the maximum specific photosynthetic rate (P_{max}) and for the photosynthetic efficiency α at the surface and 0.8 m above the bottom for both picoplankton and total phytoplankton sampled in the morning throughout the BITEX period. P_{max} changes on a time scales of hours; α changes on time scales of tens of minutes (Cullen and Lewis, 1988). Only for periods of a few hours around mid-day was overturning reduced. In consequence, photoadaptive processes occurring on time scales less than an hour may have differentiated the populations at the surface and bottom of the basin at this time. For instance, to reduce the damaging effects of high solar radiation for phytoplankton trapped in a shallow surface mixing layer, rapid changes could have occurred in pigments in the xanthophyll cycle as discussed by Reynolds (1998).

Mixing in the Euphotic Zone of a Seasonally Stratified Lake (North Basin of Lake Biwa)

Overturns were small relative to the depth of the euphotic zone when the North Basin was heating (days 243, 253, and 255) (Figure 3). On these days, maximum displacements averaged 2% of the euphotic zone depth with maximal overturns only 10% of the depth of the euphotic zone. Given that the euphotic zone depth (Zeu) varied from about 9 to 13 m, the largest overturns were approximately 1 m in size. In contrast, on a day of high winds after a typhoon as well as a morning after two overcast days, the average size of the largest overturns was 10% of the euphotic zone depth and a number of overturns equalled or exceeded Zeu.

In the following, I contrast thermal structure, depths of overturning, and rates of energy dissipation for 31 August, 1993, a typical calm day, and for 5 September, the moderately windy day following the typhoon. Locations of sampling are indicated in Figure 4.

On the calm day, the largest overturns occurred within the surface layer. Its evolution from mid-day to early evening is illustrated in Figure 5. Around midday, thermal stratification extended to the lake's surface. With the combination of weak winds (< 3 m s^{-1}) and surface cooling in the afternoon, the surface layer developed and grew to 2 m depth by 1708 h. Mixing was first observed in the surface layer in the profile taken at 1445 h. The

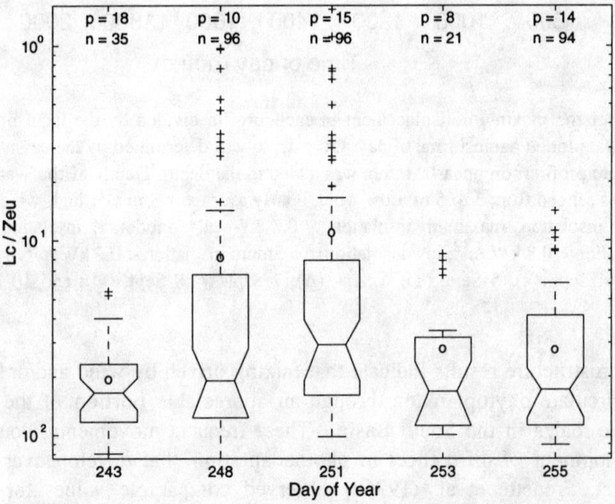

Figure 3. Box and whiskers plot of maximum overturning scale (Lc, i.e. maximum displacement scale) within each profile divided by the depth of the euphotic zone. Data are from the North Basin, Lake Biwa. Overturning regions were defined within depth intervals where Lc was everywhere ≥ 0.1 m; maximum values of Lc for each depth interval are used in the analysis. Number of profiles, p; number of overturning regions with maximal Lc ≥ 0.1 m in length, n. Within box and whiskers plots, the box indicates the size of overturns from the 75th to the 25th quartile with the median at the notch. Whiskers are 1.5 times the interquartile range, and +'s indicate outliers. Mean size of Lc is indicated by o. Sampling occurred at the following times: 31 August (day 243), 1059-1804 h; 5 Sept. (day 248), 1544-1725 h; 8 Sept. (day 251), 0956-1227 h; 10 Sept. (day 253), 1601-1739 h; 12 Sept. (day 255), 1514-1656 h. From MacIntyre, 1996.

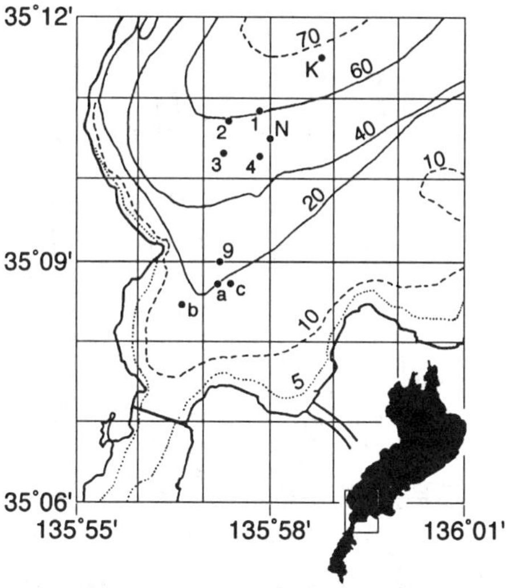

Figure 4. Map indicating sampling stations in North Basin, Lake Biwa on 31 August and 5 September 1993. Profiles on 31 August were taken at locations 1, 2, 3, and 4 near Station N. Profiles on 5 September were taken at Station 9 (1544, 1549, and 1553 h), station a (1618 h), station b (1646, 1650, and 1655 h), and station c (1716, 1720, and 1725 h). K marks the location of the meteorological station.

extent of the water column that was mixing at 1459 h is illustrated in Figure 6. Within the surface mixing layer, the maximum overturning length scale was 0.3 m and ε was modest, 1×10^{-7} m^2s^{-3}. The subsurface layer was weakly stratified, with only a few small overturns and a few regions with moderate to low levels of turbulence. Based on the size of the depth intervals in which the power spectra of the temperature gradient spectra matched a Batchelor spectra and $\varepsilon > 10^{-10}$ m^2 s^{-3}, only 15% of the euphotic zone was turbulent. Turbulence was also intermittent in the thermocline. The pattern was similar in the other profiles taken from 1445 h until 1804 h, with mixing occurring through much of the surface layer but intermittently below.

In contrast, mixing extended throughout much of the euphotic zone on 5 September, the windy day after the typhoon. The temporal variations in thermal structure, overturning scales and ε occurring at one location and the spatial variations in these parameters at four stations separated by at most 1.4 km (Figure 4) are presented in Figures 7, 8, and 9. The close proximity of stations with markedly different thermal structure suggests that other processes besides surface heating and cooling affected the depths of the surface and subsurface layer and diurnal thermocline. As a result, depth of mixing of the phytoplankton would have been quite different at each station.

At station 9, the upper mixed layer, which extended beyond the 13 m deep euphotic zone to a depth of 14 to 15 m, was weakly stratified (Figure 7). The maximum sizes of overturns in the three profiles were 3.2 m, 1.9 m, and 2.9 m, respectively. While the largest overturns tended to be at the surface with considerably smaller ones below, the 4 large ones at 1549 h were similar in size. All were constrained by small temperature steps. In contrast to the calm day, most of the euphotic zone and upper mixed layer was turbulent. Values of ε tended to be modest and, with a few exceptions, were on the order of 10^{-8} to 10^{-7} m^2s^{-3}.

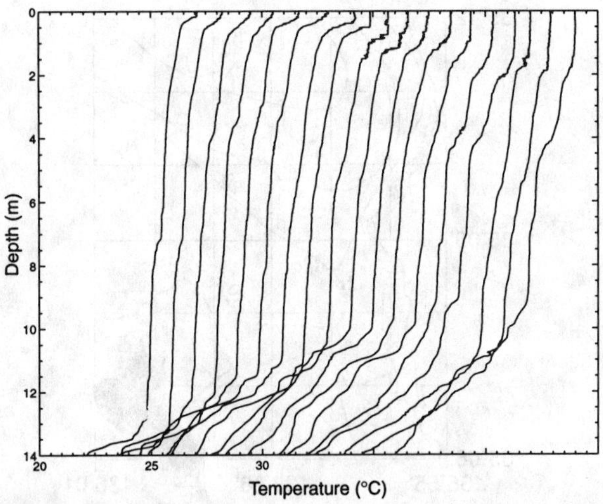

Figure 5. Temporal evolution of the surface layer and diurnal thermocline on 31 August 1993 in the North Basin, Lake Biwa. Only the upper 14 m are presented. Profiles were taken at 1059 (1), 1110 (1), 1141 (2), 1159 (2), 1230(3), 1253 (3), 1445 (4), 1459 (4), 1519 (1), 1540 (1), 1550 (1), 1606 (2), 1618 (2), 1646 (3), 1708 (3), 1726 (4), 1734 (4) and 1804 (2) h, with the number in parenthesis matching the location in Figure 4. The diurnal thermocline extended to the surface until 1253 h; it descended to a depth of 2 m by 1708 h. The depth of the seasonal thermocline depended upon internal wave activity, but was below 10 m depth. Considerable thermal structure is evident in the subsurface layer.

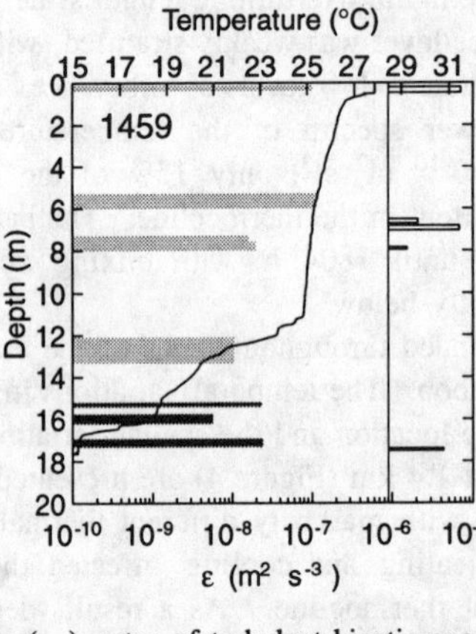

Figure 6. Profiles of temperature (—), rates of turbulent kinetic energy dissipation (filled bars), and overturning scale (boxes to right) at 1459 h on 31 August 1993 near Station N, North Basin of Lake Biwa. The vertical extent of an overturning region is indicated by the length of the overturn on the depth axis; the rms overturn scale is proportional to the width. The largest rms overturn at this time is 0.15 m. Filled bars demarcate segments in which the turbulence was statistically stationary; lighter grey bars are in the euphotic zone.

At each of the other three stations, the thermal structure in the water column was quite different. Six tenths of a kilometer away (Station a), temperatures decreased by nearly a degree across the diurnal thermocline at 7 m depth (Figure 8). The 6.7 m deep surface layer

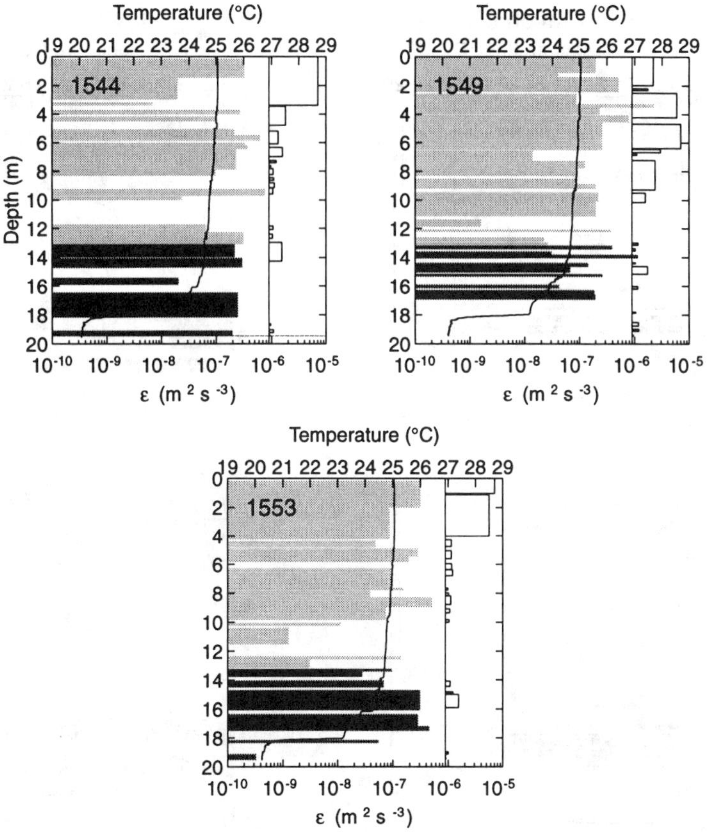

Figure 7. As in Figure 6, but for Station 9 on 5 Sept. 1993. Rms overturning scale is proportional to the width of the box and varies with each panel; maximum rms overturning scale at 1544 h was 1.6m, at 1549 h was 0.8 m, at 1553 h was 1.9 m. Data from 1549 h redrawn from Robarts et al. (1998).

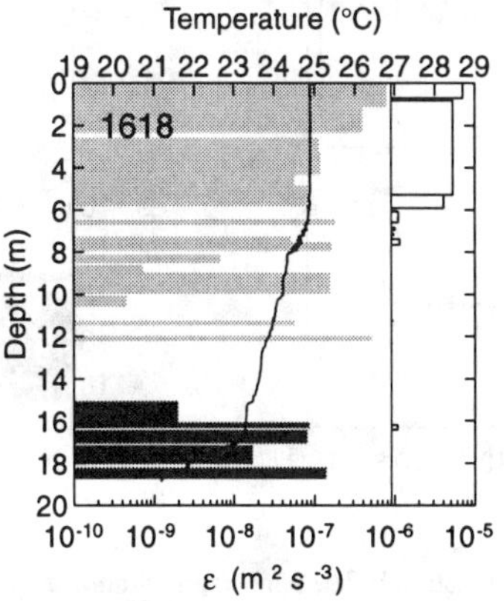

Figure 8. As in Figure 6 but for Station a at 1618 h on 5 Sept. 1993. Maximum rms overturn scale was 1.4 m.

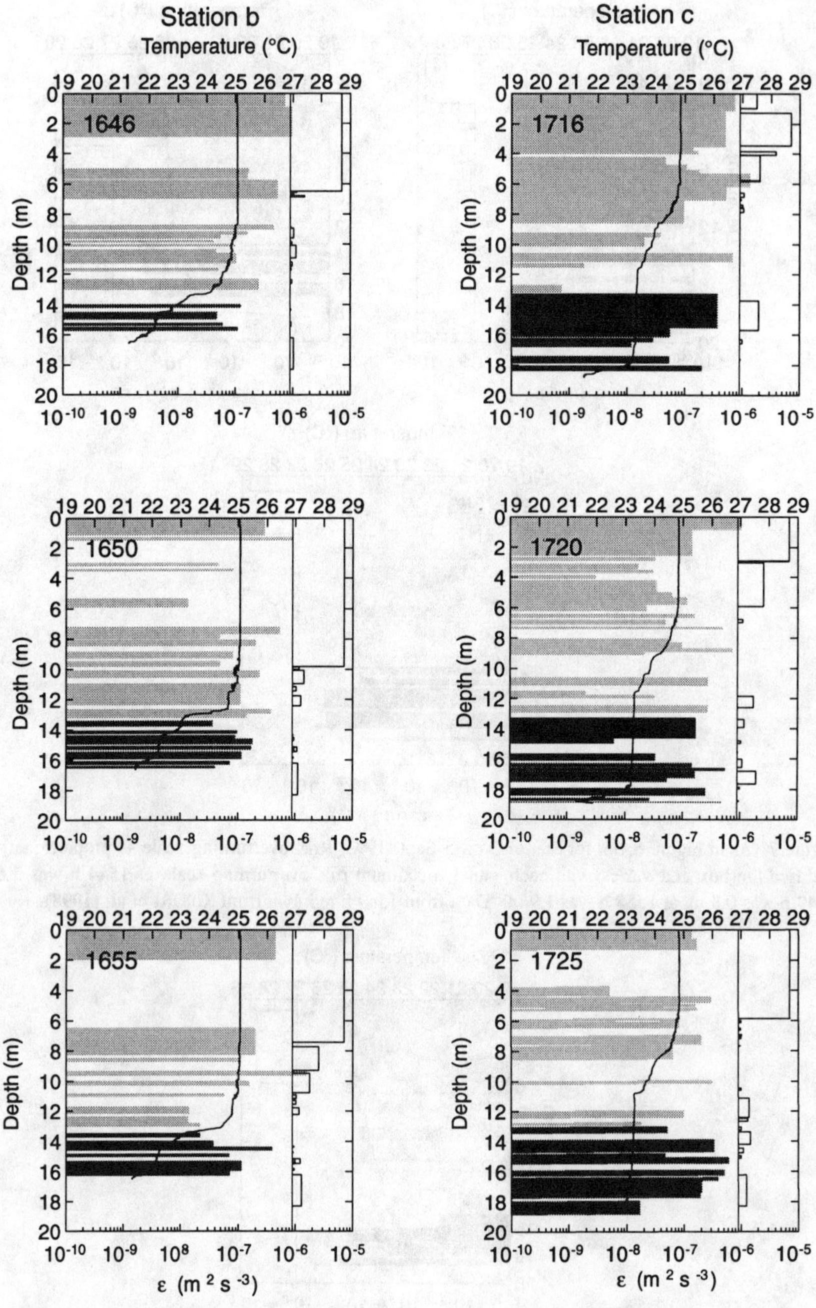

Figure 9. As in Figure 6, but for Station b (left panels) and Station c (right panels) on 5 September 1993. Maximum rms overturn scales: 1646 h, 3.7 m; 1650 h, 2.4 m; 1655 h, 1.5 m; 1716 h, 1.9 m; 1720 h, 1.4m; 1725 h, 1.7 m.

was turbulent. The largest values of ε occurred at the surface, 10^{-6} m^2s^{-3}, and decreased by an order of magnitude in the bottom half of the layer. Overturning occurred within the diurnal thermocline with ε equal to 10^{-7} m^2s^{-3} in overturns whose vertical dimension ranged from 0.1 m to 0.7 m. Turbulence was intermittent in the euphotic zone below the diurnal thermocline.

At Station c, the depth of the surface layer was similar to that at Station a, but the diurnal thermocline was nearly 6 m thick with near isothermal water extending below it to the seasonal thermocline at 18 m (Figure 9, right panels). In two of the three profiles, the surface layer was subdivided into two or more overturns; only in the last one did overturning extend throughout the layer. The water column was turbulent throughout much of the surface layer and in at least the upper half of the diurnal thermocline. The water at the base of the euphotic zone and below was cooler than at stations 9 and a, suggesting some upwelling.

Energy dissipation rates were not measurable in the middle of the surface mixing layer at 1725 h (Figure 9). This problem, typical when the water column is nearly isothermal and mixing is partially driven by heat loss at the surface, also occurred in the three profiles at Station b (Figure 9, left panel). Temperature gradients were small and the temperature-gradient microstructure profiler had insufficient sensitivity to measure ε. However, when ε has been measured with a shear profiler in convectively mixing surface layers in the ocean, values of ε were highest near the surface due to the additional component of breaking surface waves (Anis and Moum, 1994). Below that uppermost layer, ε either was uniform throughout the rest of the convectively mixing layer or decreased slightly as a function of depth. Due to infrequent wave breaking at the time measurements were made in Lake Biwa and the similarity of ε values at the top and bottom of the surface layers, it is reasonable to assume that ε tended to values between 10^{-7} m^2s^{-3} and 10^{-6} m^2s^{-3} throughout the surface layers at 1725, 1646, 1650, and 1655 h.

The three profiles taken from 1646 to 1655 h were obtained at Station b, 1.4 km away from Station 9 in water about 17 m deep (Figure 9). In each, the surface layer extended 9 m with at least one large overturn throughout much of the layer. Surface temperatures were as warm as those at the first station sampled. As mentioned above, ε was on the order of 10^{-7} to 10^{-6} m^2s^{-3}. Below the surface mixing layer, temperatures decreased in a steppy fashion with one of the largest temperature steps at the base of the euphotic zone. Smaller overturns occurred throughout the stratified base of the euphotic zone with ε ranging from 10^{-8} to 10^{-6} m^2s^{-3}. Temperatures at 16 m were colder than at the other stations, again suggesting upwelling.

The data from these four stations show that the magnitude of overturning in the upper part of the euphotic zone is variable at one location at one time, but that even larger variations can occur over relatively short distances. Cooling occurring over the 1.75 hour sampling period may have led to some deepening of the surface layer. However, the stratification across the diurnal thermocline varied between sites and could have limited or enhanced the rate of entrainment. These differences in the stratification and thickness of the diurnal thermocline and subsurface layer are likely due to basin scale movements, such as upwelling. Subsequent horizontal movements due to gravitational adjustments may have contributed to the small temperature steps at Station 9 which limited the scale of vertical overturning.

Differences in irradiance experienced by phytoplankton circulating the full extent of an overturn on the calm and windy day are illustrated in Figure 10. Phytoplankton circulating throughout overturns on the windy day experienced fluctuations as large as 80%

of the surface irradiance I_o, whereas the maximum fluctuations on a calm day were about 20% of I_o. In both cases, the largest fluctuations were due to large overturns at the surface. On both days, many of the surface fluctuations were small. On the calm day, overturns originating below 4 m depth, where irradiance was 20% of that at the surface, led to fluctuations of at most 1% of surface irradiance. Similarly, on the windy day, eddies originating below 6 m caused fluctuations of only a few percent of surface irradiance. Many of the surface overturns were deep enough to entrain some of the phytoplankton previously exposed to low irradiance. Variations in irradiance experienced by phytoplankton within surface overturns on 7 September, when mixing was driven primarily by surface cooling, were similar to those on the windy day.

The importance of these changes in irradiance depends on whether the irradiances are in the light limited portion of a photosynthesis versus irradiance curve (P vs I) curve, the light saturated portion, or the photoinhibited portion. Laboratory experiments indicate that subtle shifts in irradiance where phytoplankton are light limited may induce substantial changes in growth and in processes affecting photosynthetic rates. For instance, in cultures grown at constant light, growth rates of the marine diatom *Thalassiosira pseudonana* were 2.5 times higher when grown at irradiances of 100 µmol m^{-2} s^{-1} than at 20 µmol m^{-2} s^{-1} (Cullen and Lewis, 1988). Given that maximal surface irradiances are on the order of 2000 µmol m^{-2} s^{-1}, these two irradiances represent 5% and 1% of surface irradiance near midday. Growth rate only increased by an additional factor of 1.6 for irradiance of 2200 µmol m^{-2} s^{-1}. The two key parameters in modeling photosynthesis, P_{max}, the maximum photosynthetic rate, and α, the initial slope of the P vs I curve, are also sensitive to small shifts in irradiance. P_{max} was a factor of 1.4 higher

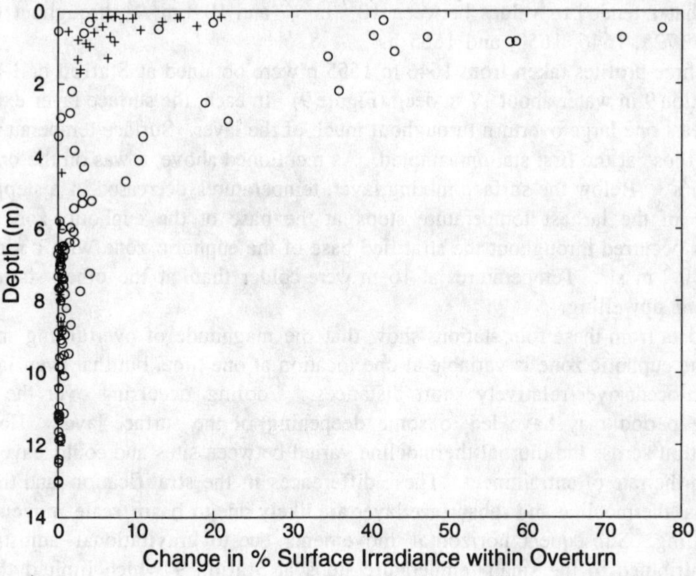

Figure 10. Change in percent of surface irradiance experienced by phytoplankton cells circulating through the full extent of an overturn plotted against depth of the top of the overturn. Data are from the North Basin, Lake Biwa; + for the 18 profiles obtained 31 August 1993 in the North Basin of Lake Biwa, o for the 15 profiles obtained 5 Sept 1993.

when grown in cultures with irradiances of 100 as opposed to 20 μmol m^{-2} s^{-1}, and doubled as irradiance increased from 100 to 2200 μmol m^{-2} s^{-1}. Circulation to depths where irradiances are less than those that induce photoinhibition can also increase photosynthetic rates.

Phytoplankton at both the top and bottom of the mixed layer were light limited up to an irradiance of 250 μmol m^{-2} s^{-1} (J.J. Frenette, pers. comm.). At midday on sunny days, such irradiances would have occurred below 5 m depth, where changes in irradiance in overturns originating at those depths were at most 5%. Overturning at the surface would have increased productivity by reducing exposure to damaging light levels. At other times of day and on cloudy days, algae would have been light limited higher in the water column where the likelihood of being entrained in large overturns was higher.

When exposed to a given irradiance for a long enough period, phytoplankton will adapt to that irradiance in order to maximize photosynthetic rates or to reduce damage from high irradiances. That surface overturns did not extend beyond the depth of the surface layer (Figures 5 and 6) indicates that phytoplankton cells at the top and bottom of the euphotic zone in the North Basin would have adapted to high and low irradiances when wind speeds were less than 4 m s^{-1}. Differences in total maximum photosynthetic uptake rate (P_{max}-C), maximum specific uptake rate (V_{max}-C), and photosynthetic efficiency (α-C) clearly show the phytoplankton at the top and base of the euphotic zone were differentiated physiologically prior to the typhoon (Frenette et al., 1996a).

While not entirely unequivocal, physiological data indicates the picoplankton and > 2 μm sized phytoplankton continued to be differentiated at the surface and base of the mixed layer on 5 September despite the wind forcing on September 4 and 5 (Frenette et al., 1996a). The limitation of overturning scales by the thermal structure in the upper mixed layer on 5 September may explain this surprising result (Figures 7-9).

Calculation of the time scale of photoadaptation τ_a relative to the time scale for mixing, $\tau_{mix} = \ell^2/K_z$, can also be used to determine whether phytoplankton will photoadapt as a function of depth in a mixing layer (Lewis et al., 1984a, b). If τ_{mix} exceeds τ_a, phytoplankton will differentiate in the water column. The time scales for photoadaption vary from seconds to days (Ferris and Christian, 1991; Prezelin et al., 1991). Photoadaptive parameters such as *in vivo* fluorescence and α, which are modified via redistribution of energy in the photosynthetic apparatus, change on time scales of tens of minutes; changes in P_{max} and chlorophyll concentration occur on time scales of hours (Cullen and Lewis, 1988).

Realistic application of Lewis et al.'s model to the data from Lake Biwa required two adaptations. One, I let ℓ be the length of overturning originating at the surface as opposed to the depth of the upper mixed layer. The mixing data from Lake Biwa show that on calm days, only the surface layer is mixing, whereas on windy days, when the entire upper mixed layer is mixing, overturning is constrained by temperature steps in the water column. Overturning deeper in the water column was not considered as Lewis et al.'s (1984a) analysis is only applicable when mixing proceeds from the surface downwards.

Computing the time scale of mixing using K_z is not appropriate under conditions of convective cooling when the mixing layer is nearly isothermal. In addition, equations for calculating R_f in Ivey et al. (1998) have not been tested for $Re_t >> 100$. For the large overturns in Lake Biwa, Re_t was often in excess of 1000. Imboden and Wuest (1995) propose the relation $t = 5/2\ N^2\ell^2/\ J_b{}^o$ where $J_b{}^o$ is surface heat flux. This relation takes into consideration the change in potential energy as the mixing penetrates into stratified waters and the inefficiency of mixing. Following Imberger (1985), $\varepsilon = 0.45\ J_b{}^o$. The same time scale is obtained from ℓ^2/K_z when $K_z = 0.2\varepsilon N^{-2}$. This latter equation assumes maximum mixing efficiency. Imboden and Wuest's approach was used when $N < 0.003\ s^{-1}$. The time scale for convective overturning, $(\ell^2/\ J_b^o)^{1/3}$, which is based on dimensional analysis (Anis and Moum, 1984), gives time scales of mixing up to an order of magnitude faster.

Figure 11 shows the ratio of the time scale for mixing τ_{mix} to τ_a for τ_a = 15 min and 6 h. Following Lewis et al. (1984a, b), the ratio is plotted against the ratio of the mixing length scale to the optical length scale, $k_s\ell$ where k_s is the attenuation coefficient. If ε varied within overturns, the logarithmic average of K_z within an overturn was used.

On August 31, 1993, K_z ranged from 10^{-6} to $10^{-4}\ m^2\ s^{-3}$ in the surface layer and overturns were up to 0.7 m. Due to the mixing, phytoplankton in the overturns would not have differentiated vertically even for processes with time scales of 15 minutes. Because mixing was primarily within the surface layer photoadaptive properties would have differentiated with depth in the euphotic zone.

On 5 September, 1993, overturns ranged from 1 to nearly 10 m depth. K_z ranged from 10^{-4} to $10^{-3}\ m^2\ s^{-1}$ when $N > 0.003\ s^{-1}$. Only one large surface overturn occurred at this stratification; τ_{mix} was 12 h. In contrast, τ_{mix} ranged from 20 minutes to 5 h in less stratified waters. For τ_a = 15 minutes, differentiation with depth of the physiological state of phytoplankton began when overturns were 0.5 m deep. For τ_a = 6 h and $N < 0.003\ s^{-1}$, the crossover point between well mixed phytoplankton cells and ones mixing slowly enough for photoadaptation to occur with depth occurred at 10 m. Patterns were similar using the time scale for convective overturning (not shown).

This model predicts that on September 5, phytoplankton within surface overturns would have been well mixed and would not show differences in photoadaption with depth. However, given that many overturns did not extend the full extent of the upper mixed layer, it is not surprising that Frenette et al.'s (1996a) data do not indicate the phytoplankton were fully homogenized on 5 September.

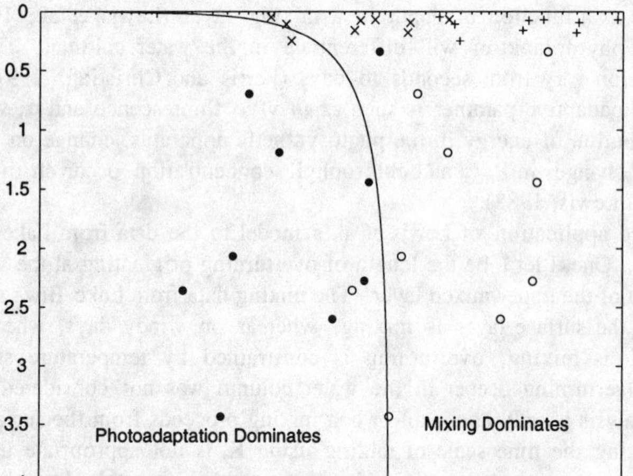

Figure 11. Ratio of time scale for photoadaptation to time scale for mixing plotted against $k_s\ \ell_{max}$ where k_s is the attenuation coefficient of irradiance and ℓ_{max} is the maximum displacement scale. Ratio for 31 August for τ_a = 15 minutes (x), 6 h (+). Ratio for 5 September for τ_a equals 15 minutes (•), 6 h (o).

Lewis et al. (1984a) developed their model as a way to predict mixing rates when vertical profiles indicating photoadaptive state were available for several processes with different rate constants. Because phytoplankton exhibit diurnal periodicities in photoadaption as well as changes due to the extent of mixing, interpretation of mixing rates based on physiology alone is difficult. Microstructure profiling enables a direct assessment of mixing rates. In addition, microstructure profiling indicates where the water column is mixing, and time series measurements indicate how long it has been mixing. This information, coupled with the perspective gained from Lewis et al.'s model and measured values of P_{max}, α, and other photoadaptive measures during periods of calm and of mixing, enable more exact modeling of photosynthesis and ultimately primary productivity.

Vertical Mixing and Nutrient Supply

Nocturnal Cooling and Solute Fluxes

Data from Lake Calado, a seasonally flooded lake in the Amazon basin, illustrate the importance of mixing due to heat loss at night (MacIntyre and Melack, 1995). Data from a six week study in January and February, 1983, illustrate the water column dynamics during the time when this lake was thermally stratified. The mixed layer depth varied diurnally, with depths ranging from 1 to 6 m at night. It tended to shallow during the morning, with mixed layer depths ranging from only 0.5 to 1 m by early afternoon. Wind speeds ranged up to 6 m s^{-1}, with maximal wind speeds occurring during daylight hours. The ratio of the turbulent kinetic energy flux due solely to buoyancy flux to that due to both wind and buoyancy flux was calculated as in Imberger (1985). Between 70 and 100% of the flux of turbulent kinetic energy into the lake between sunset and sunrise was due to buoyancy flux, i.e. surface cooling. While there were times during the heating period that all the energy input was due to wind, surface cooling did frequently contribute, with a maximal contribution of 75%. Given the shallow mixed layer depths in the daytime, most of the turbulence would have worked against the pronounced stratification in the upper 2 m. It is at night and early morning, when most mixing is driven by heat loss at the surface, that entrainment of nutrient-rich waters from the thermocline or hypolimnion is likely.

Measurements of dissolved methane, obtained after the lake had been stratified for several months, indicated that methane concentrations increased with depth in the water column, with larger increases below 3 m (Crill et al., 1988). During a diurnal study of gas flux, fluxes were highest at sunrise when the density gradient in the upper 2 m was smallest and lowest at midday when the density gradient in the upper 2 m was maximal. Results of these two studies indicate penetrative convection due to heat loss is critical for gas flux in stratified, warm-water lakes. Nutrient fluxes are also likely to vary over diurnal time scales and may be maximal during times of convective mixing.

Buoyancy-affected Turbulence, Advection, and Nutrient Fluxes

While moderate to strong afternoon winds would be expected to mix a shallow lake, data from North Lake, Western Australia, illustrate persistent thermal structure when turbulence is damped by stratification and when differential heating occurs (MacIntyre, 1993). North Lake is small with a surface area of 29 ha and a maximum depth of 2.6 m. Wind patterns are largely predictable in summer, with calm mornings and commencement of a sea breeze during late morning or early afternoon. Temperature-gradient microstructure sampling was undertaken in December, 1987. Development of stratification in the water column during the morning and the subsequent changes when wind speeds increased from 3 to 6 m s^{-1} around 1100 h the first day are illustrated in Figure 12. The surface cooled in conjunction with the increase in wind speeds. The subsequent abrupt increases in temperature at 82, 121, 160 and 199 cm depth are indicative of the turbulent front's penetration to those depths. That thermal stratification persisted until 1510 h despite strong wind forcing indicates the difficulty in mixing a highly stratified lake while it is continuing to gain heat. While ammonium and chlorophyll *a* were stratified during the morning heating period and mixed somewhat thereafter; concentration differences persisted even after 4 h of wind mixing and were strongest below 1 m (Figure 13).

Advection of a water mass with different thermal structure led to the abrupt change in thermal stratification after 1510 h and may have contributed to the persistent differences in ammonium and chlorophyll *a* concentrations. Since surface waters in the water mass were even warmer than those previously recorded at the meteorological station despite

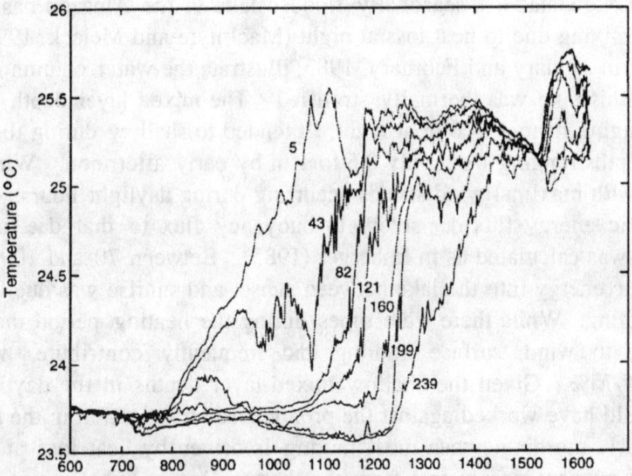

Figure 12. Water temperatures (°C) in North Lake on 16 December 1987 with thermistors located at depths in cm shown on the figure. From MacIntyre (1993).

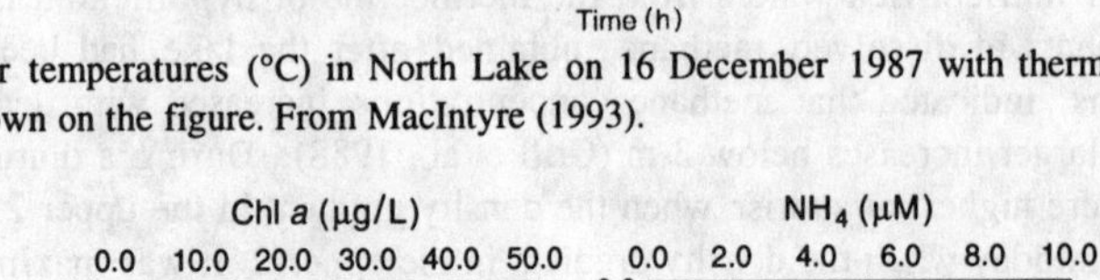

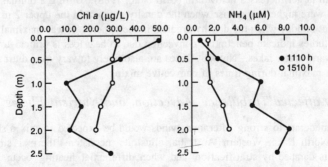

Figure 13. Profiles of chlorophyll *a* (µg L^{-1}) and ammonium concentration (µM) in North Lake on 16 December 1987. From MacIntyre and Melack (1995).

decreasing solar radiation, differential heating (Imberger and Parker, 1985) presumably occurred with the water mass likely originating near-shore. In shallow, stratified water bodies, complete mixing requires not only mixing within the pelagic zone but also of the warmer or cooler layers of water formed in the littoral that propagate into the pelagic zone. Pronounced diurnal thermal stratification which resists wind mixing during the day has been noted in other water bodies with high attenuation coefficients (MacIntyre and Melack, 1982, 1984). Holloway (1980) predicts the time scale to mix a stratified lake for a given wind speed and attenuation coefficient.

The gradients in chlorophyll and ammonia may also have persisted due to the inefficiency of mixing in highly stratified flows. The time scales of mixing discussed below provide evidence for this alternate explanation.

Accurate assessment of the coefficient of eddy diffusivity K_z is essential for ecosystem studies. Nutrient fluxes can be calculated using the expression $F = K_z\, \partial C/\partial z$ where C is concentration. When microstructure measurements are available, K_z can be estimated using $Kz = (Rf/(1\text{-}Rf))\, \varepsilon N^{-2}$ (Osborn, 1980). The flux Richardson number Rf indicates the efficiency of mixing. Ivey and Imberger (1991) show that when the turbulent Froude number Fr_t, drops below 1, the flux Richardson number decreases. Imberger and Ivey (1991) describe turbulence when $Fr_t < 1$ as buoyancy-affected.

Recent direct measurements of buoyancy flux indicate that R_f can be predicted from Fr_t and Re_t in boundary layers but not in the main thermocline away from the boundary (Lemckert and Imberger, 1998; Imberger, 1998; E. Shahidi and J. Imberger, personal communication). Hence Ivey and Imberger's approach should be valid in shallow water bodies such as North Lake which constitute surface boundary layers.

Following Imberger and Ivey (1991), the data from North Lake were pooled to create a Froude-Reynolds number diagram (MacIntyre, 1993). While the data from the layers nearest the surface indicated that the water column was actively mixing ($Fr_t > 1$), data from the subsurface layers indicated they were affected by buoyancy ($Fr_t < 1$). As Fr_t decreased below 1, values of K_z, calculated following Osborn (1980) with the flux Richardson number obtained as in Ivey and Imberger (1991), also decreased from their maximum near $Fr_t = 1$. K_z varied from 10^{-7} to 10^{-4} $m^2\ s^{-1}$ where buoyancy affected the turbulence (MacIntyre, 1993). Using an upper bound for K_z of 10^{-4} $m^2\ s^{-1}$, the time for mixing, $\tau_{mix} = \ell^2/2K_z$, a 2 m water column was on the order of 5.5 h. In contrast, if the turbulence were not buoyancy-affected, the highest value of K_z would have been an order of magnitude larger and the mixing time would have been 0.6 h. That thermal stratification persisted in this lake for 4 h, the time scale taking into account the decrease in R_f, strongly suggests that the turbulence was affected by buoyancy. Similarly, the chemical stratification may also have persisted for this reason. In fact, the ammonia distributions as well as the chlorophyll *a* concentrations were nearly uniform only in the upper meter where $Fr_t > 1$. There, cyanobacteria would have become homogenized as the turbulent velocity scales were equal to or greater than their sinking and rising speeds. In this lake, as in others with pronounced stratification, mixing may be inefficient and more accurate assessments of nutrient and particle fluxes are likely to be obtained by taking into account Fr_t and Re_t when computing K_z.

Differential Response of Large and Small Lakes to Similar Wind Forcing

The Lake Biwa Transport Experiment (BITEX) provided a means to explore the importance of different pathways of nutrient flux into the euphotic zone in large, deep lakes as well as shallow, diurnally mixing basins. This interdisciplinary experiment took place in August and September 1993, a time when typhoons were expected to occur. Short, intense mixing events such as would be generated by a typhoon are likely to cause changes in phytoplankton and bacteria on similar time scales. Physiological changes are expected.

A time series of assays of ^{32}P uptake rate, alkaline phosphatase activity, and sestonic C:N:P ratios were undertaken to assess the physiological state of the phyto- and bacteria-plankton (Robarts et al, 1998). Values of these measures before and after a typhoon with winds to 20 m s^{-1} are summarized below in the context of the hydrodynamical, chemical, and biological measurements made by ourselves and other BITEX investigators.

The greater depth and thermal stratification of the North Basin relative to the South led to differences in the flux of nutrients due to mixing during the typhoon. Prior to the typhoon, phytoplankton were phosphorus (PO_4-P) limited in both the North and the South Basin. While they remained so immediately after the typhoon in the North Basin, they briefly were phosphorus sufficient in the South Basin. The response of the North Basin to the strong wind was a dramatic tilting of the thermocline and mixing within it (Hayami et al., 1996), but only entrainment and upwelling of waters from the upper two meters of the thermocline. Concentrations of dissolved phosphate increased below 16 m (Hashitani et al., 1996), well below the depths of entrainment and upwelling. Microstructure data from the day after the typhoon (Figures 7, 8, and 9) indicated moderate to high values of ε and small overturning scales in the thermocline. Similar results were obtained in a microstructure profile taken late afternoon the day of the typhoon (J. Imberger, unpublished data). On 5 September, K_z values were high in the thermocline, ranging from 10^{-6} to 10^{-3} m^2 s^{-1}, but mixing was intermittent (not shown). Again, because PO_4-P concentrations began below 16 m, little PO_4-P would be expected to mix directly into the mixed layer. Any PO_4-P mixed upwards would have been rapidly taken up by phytoplankton within the thermocline, some of whom may have been mixed into the euphotic zone. In consequence, vertical mixing through the thermocline was only sufficient to cause a reduction in the degree of phosphorus limitation. Because total available nitrogen (TAN) concentrations did increase beginning at the top of the thermocline, TAN increased at the base of the mixed layer after the typhoon. Five days after the typhoon, the biomass of phytoplankton > 2μm doubled and biomass of picoplankton in the surface waters increased 2.5 times (Frenette et al., 1996a).

In contrast to the North Basin, PO_4-P concentrations increased in the South Basin after the typhoon. Using measured wind speeds and a model of sediment resuspension, we calculated that wave induced shear stresses in the South Basin were sufficient to entrain sediments and associated pore waters. While the tilting of the thermocline in the North Basin led to a surge of waters into the South Basin (Hayami et al., 1996), the upwelled water had low PO_4-P content. Some nutrients were entrained from the sediments, and the high shears associated with the surge led to intense mixing between waters from the two basins, (MacIntyre, 1996), but the surge only penetrated 20% of the length of the South Basin. The wave induced sediment resuspension and entrainment of pore waters, predicted to occur within 60% of the basin, was a more likely cause of the increased loading of dissolved phosphate in the South Basin. Advection of phosphate from a nutrient rich embayment was another pathway of nutrient supply to the main body of the South Basin. In response to these inputs, ^{32}P uptake rates increased, alkaline phosphatase activity decreased, and seston ratios indicated PO_4-P sufficiency on the day following the typhoon. There was a shift from bacterial dominance to phytoplankton dominance. Algal biomass increased five-fold in the South Basin over the course of the next few days.

Boundary Mixing and Nutrient Flux

Temperature-gradient microstructure profiling with concurrent nutrient profiling was undertaken at both a near-shore and a mid-lake site on Mono Lake, California, to determine whether boundary mixing occurred and the magnitude of the resulting nutrient fluxes (MacIntyre et al., 1998). The following is a synopsis of that work. As for the North Basin of Lake Biwa, the location of the nutricline relative to the depths of mixing determines the consequences of mixing for nutrient supply to the euphotic zone. At the time of sampling, the density difference across the pycnocline, located between 9 and 16 m depth, was 9 kg m^{-3} and the buoyancy frequency was 81 cph. Because Mono Lake is chemically stratified, complicated thermal structures can develop. For instance, at the time of sampling, a warm tongue of water was located near 10 m depth. Ammonia and chlorophyll were strongly stratified, with pronounced increases in chlorophyll concentrations below 12 m and in ammonia concentrations below 15 m.

Meteorological data, temperature-gradient microstructure profiles, conductivity-temperature-depth (CTD) transects, and nutrient profiles were collected from October 11—14, 1995. Thermistor chains were also deployed. Due to the passage of a front, high wind speeds occurred on 11 and 12 October. Mean wind speeds were in excess of 10 m s^{-1} from late morning until early evening on 11 October, with an increase to 14 m s^{-1} just after midnight (Figure 14). Gusts reached 15 m s^{-1}. Wind speeds were also high on the 12th from early afternoon until just before midnight. Winds averaged 8 m s^{-1} with gusts to 11 m s^{-1}. In contrast, winds were less than 5 m s^{-1} on 13 October with the exception of a brief period with gusts to 8 m s^{-1} at sunset. All days were sunny. Tilting of the thermocline was associated with the onset of winds on the morning and late evening of 11 October; Wedderburn and Lake Numbers dropped to values of 5 and 2, respectively. A pronounced internal wave field developed with the largest amplitude temperature variations between 10 and 17 m.

Representative profiles of energy dissipation rate (ε) and temperature are presented for each site (Figures 15, 16). On 11 October, intense mixing occurred near the surface at both sites and in association with the tongue of warm water near 10 m depth. Values of ε at the surface reached 10^{-5} m^2 s^{-3}, values among the highest measured at the air-water interface in lakes and oceanic waters. ε decreased below 10 m with little mixing occurring below 15 m.

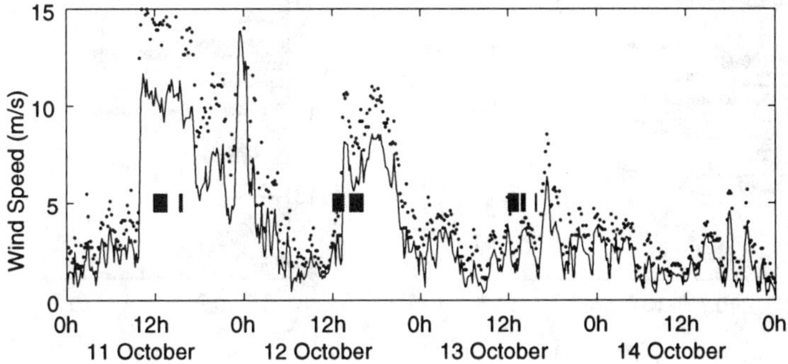

Figure 14. Wind speed at Mono Lake from midnight on 10 October to midnight on 14 October 1995. Solid line is 10 minute average, • indicate the highest gust in the 10 minute period. Microstructure profiles were obtained during sampling periods indicated by ▪. Data were collected onshore during the first sampling period on the 11th, the second sampling period on the 12th, and the first and third sampling periods on the 13th .

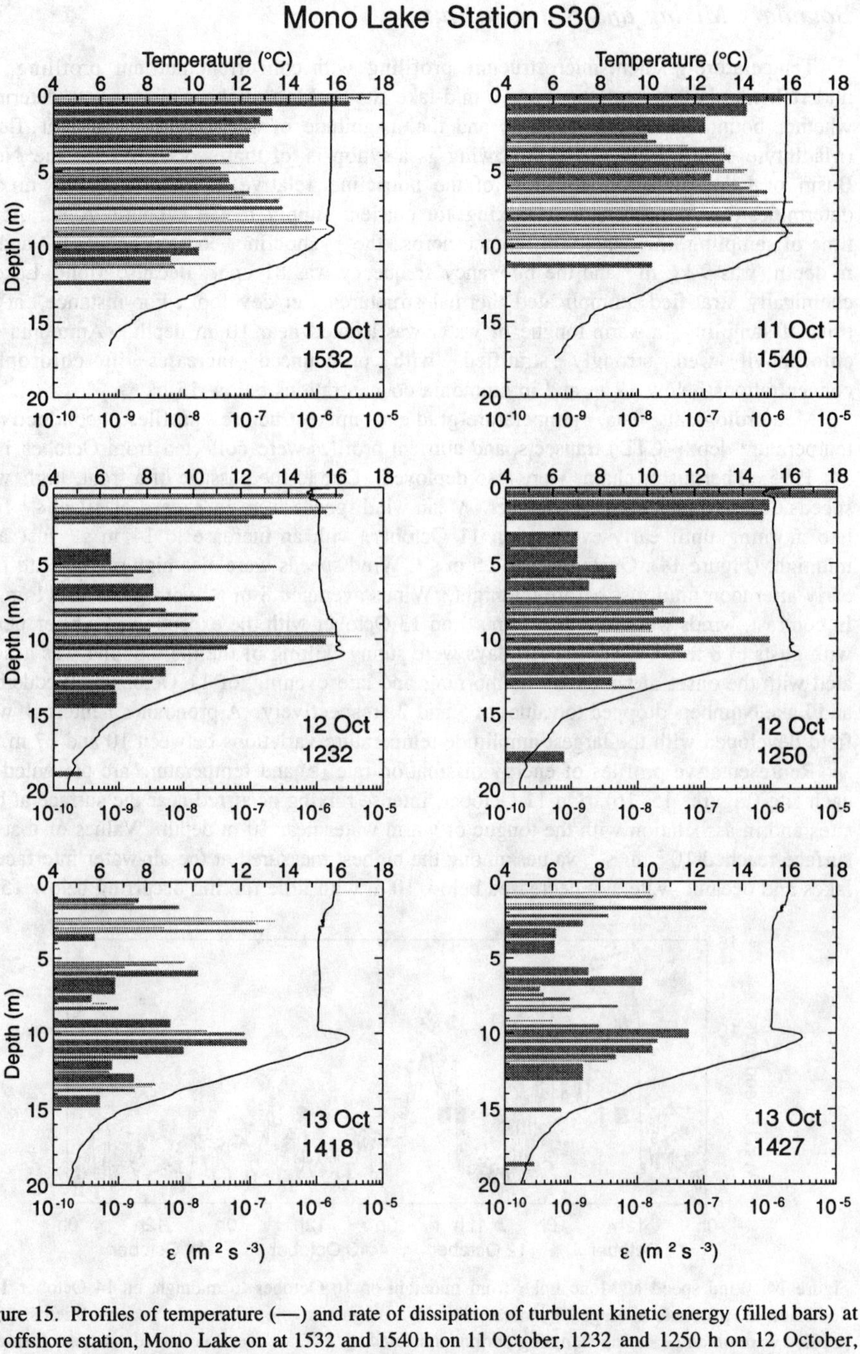

Figure 15. Profiles of temperature (—) and rate of dissipation of turbulent kinetic energy (filled bars) at the offshore station, Mono Lake on at 1532 and 1540 h on 11 October, 1232 and 1250 h on 12 October, and 1418 and 1427 h on 13 October. Data are only plotted for segments whose spectra fit a Batchelor spectra. Only the upper 20 m are illustrated. Profiles were downcasts on 13 October.

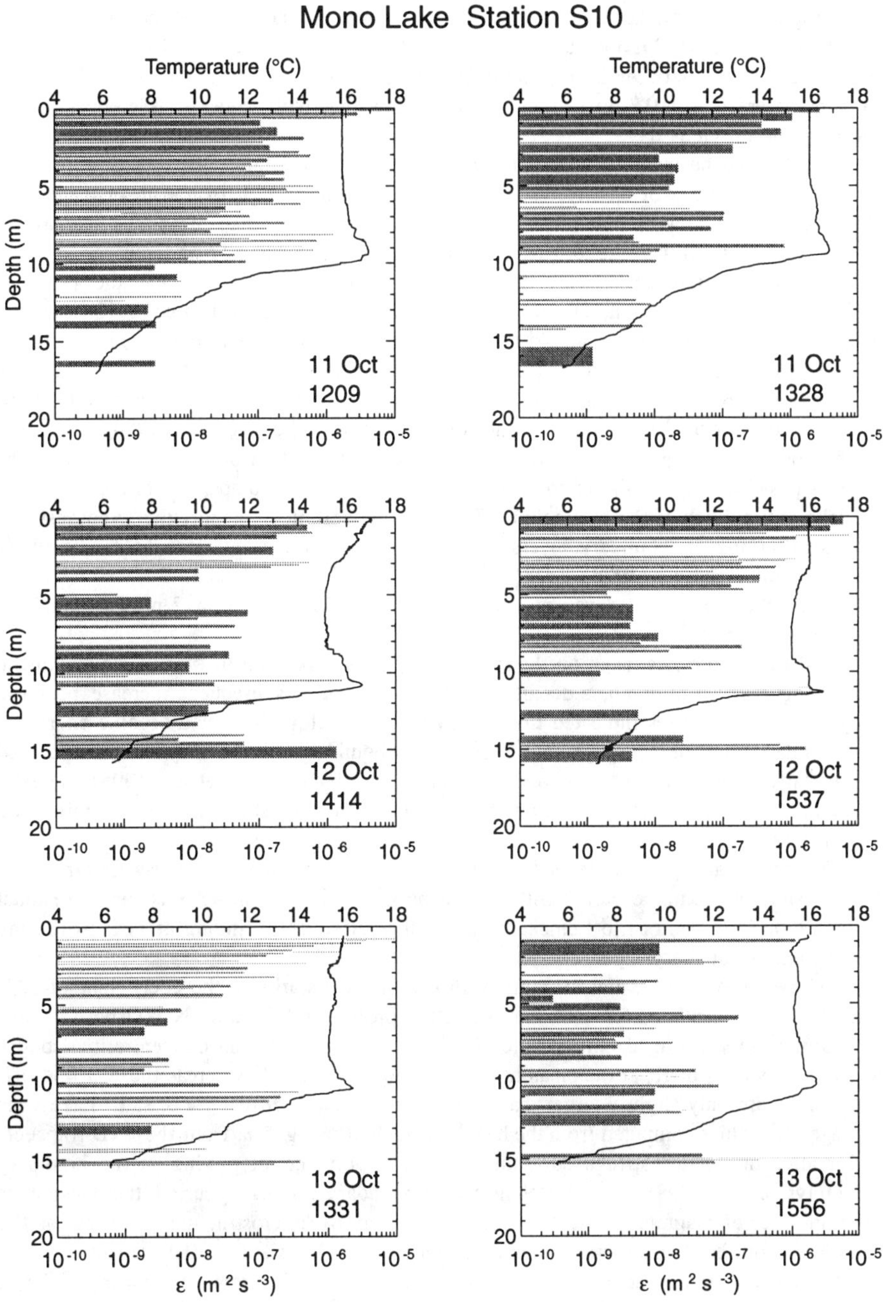

Figure 16. As for Figure 15 for data from the onshore station. Profiles shown are from 1209 and 1328 h on 11 October, 1414 and 1537 h on 12 October, and 1331 and 1556 h on 13 October.

Sampling at the offshore station occurred just prior to and after the increases in wind speed on 12 October. Mixing at the surface was as intense as on the previous day after the winds picked up. Between 3 and 9 m depth, ε ranged from 10^{-9} to 10^{-7} $m^2 s^{-3}$. ε was maximal near 10 m (10^{-6} $m^2 s^{-3}$) and tapered below that with values of ε comparable to those on the previous day.

Sampling at the inshore station on 12 October occurred while wind speeds were high. In consequence, rates of energy dissipation were extremely high in the upper meter, and the water column between 3 and 9 m depth was more energetic than offshore. Increased ε was again associated with the tongue of warm water between 8 and 12 m depth.

The major difference between the inshore and offshore stations was the presence inshore of a mixing region, up to 3 m thick near 15 m depth, with ε up to 10^{-6} $m^2 s^{-3}$. These profiles were upcasts and did not sample the bottom meter of the water column. Because the thermocline intersected the sediment-water interface in this region, the enhanced values of ε may have resulted from the shoaling of internal waves (Taylor, 1993), from shear due to thermocline displacements (Gloor et al., 1994), or from interactions of a Kelvin wave with the local topography (Maxworthy et al., 1998). The latter is possible because the internal Rossby radius is 10 km, comparable to the length of the lake at the depth of the pycnocline. These results definitely support the hypothesis of enhanced mixing in boundary layers near the sediment-water interface as discussed by Lemckert and Imberger (1998).

On October 13, elevated values of ε again occurred in the bottom 2 m of the inshore site with ε up to 10^{-5} $m^2 s^{-3}$.

Whilst most profiles taken on the two days show a maximum in ε associated with the warm tongue at 10 m depth, a decline in ε to 13 m, and a subsequent increase near the boundary, a few profiles indicated that the pycnocline below 10 m had become energized with similar values of ε throughout. Energy from boundary mixing may have propagated horizontally or at an angle to the vertical as suggested by De Silva et al. (1998). Greater variability in thermal structure below 10 m inshore than offshore further supports this hypothesis.

The temperature profiles tended to remain stratified near the sediment-water interface when boundary mixing occurred with an exception at 1331 h on 13 October. Residual stratification is in contrast to Lemckert's and Imberger's (1998) finding of near isothermy in the mixing boundary layer.

To determine the potential for nutrient flux at the two stations, the coefficient of eddy diffusivity was calculated for each microstructure profile as $K_z = 0.2\ \varepsilon N^{-2}$ (Osborn, 1980) and subsequently averaged. From 15 to 16 m depth, where ammonia concentrations begin to increase, values of K_z at the onshore site averaged 10^{-6} $m^2 s^{-1}$ whereas those at the offshore site were only slightly larger than values for molecular diffusivity of salt. Lake wide averaged K_z values computed from the heat flux method using data from the CTD transects and moored thermistor strings substantiated the importance of mixing at the boundary (MacIntyre et al., 1998). At the depths where boundary mixing occurred, these averages were an order of magnitude less than those obtained by microstructure profiling at the inshore station, but two orders of magnitude larger than those at the offshore site.

Ammonia fluxes computed from the two stations using the relation, $F = K_z\ (\partial C/\partial z)$, were two to three orders of magnitude higher at the inshore station. Phytoplankton immediately above the nutricline would rapidly consume the ammonia that was mixed upwards, indicating that boundary mixing at the base of the pycnocline in Mono Lake was likely to lead to new production within the deep chlorophyll layer.

Final Remarks

Data on overturning scales and their relation to physical forcing are one of the major steps for further development of physical-biological models (Denman and Gargett, 1995). As the data from Lake Biwa show, in both shallow and deep basins, phytoplankton are only circulated in a small percentage of the euphotic zone during periods of calm and high insolation. During moderate winds, surface overturns are deeper; in Lake Biwa they ranged from 1 to 10 m in vertical extent in a 13 m deep euphotic zone. The depth of overturns at the surface can vary horizontally; in Lake Biwa they increased from ca. 4 m to 10 m in a distance of 1.4 km. Some of the differences may have been caused by cooling, some by subsurface processes which influenced the stratification. Fluctuations in irradiance were up to 80% of surface irradiance. Near midday, the overturns could circulate cells away from damaging irradiances and reduce photoinhibition; at other times overturns would have circulated phytoplankton within the light limited portion of the P vs. I curve. In both cases, growth rates would have changed. Phytoplankton within overturns in less stratified parts of the water column ($N < 0.003\ s^{-1}$) were circulated at rates more rapid than changes in P_{max}. However, they would have shown differences in α or *in vivo* fluorescence with depth. In more stratified regions, they were circulated more slowly and would have developed differences in P_{max} as well.

When basic information about the frequency of mixing and nutrient supply rates are available, predictions about the dominant algal species can be made as well. For instance, Reynolds (1994) predicts that picoplankton will dominate in environments which mix on time scales from hundreds to thousands of seconds but in which nutrient supply is low, but that larger cells will become more prevalent as either more light or more nutrients become available. Before the typhoon in Lake Biwa when the shallow South Basin was nutrient limited, picoplankton were dominant. Diatoms dominated after the period of brief nutrient supply. In this case the frequent overturning at night could help maintain these species in suspension.

Through straining on small scales, turbulence affects the rate of flocculation of particles, contact between zooplankton predators and their prey, nutrient uptake rates, and growth rates. The strain rate γ equals $(\upsilon/\varepsilon)^{1/2}$ where υ is kinematic viscosity. On calm days, straining only occurs intermittently in the water column (Figure 6); on windy days it occurs throughout much of the upper mixed layer and thermocline (Figure 7—9). When overturning scales are large, transport and mixing will occur concurrently with these processes. When overturns are on centimeter scales or less, the ecological implications will be different. Contact rates between zooplankton and their algal prey will be increased without dispersion of any patches of prey and rates of flocculation of particles will be increased without dispersion. Phytoplankton could take up nutrients without being transported away from locations with elevated concentrations.

The mixing regimes in the four lakes described here range from polymixis to meromixis. However, the water columns in the polymictic lakes, North Lake, Western Australia, and South Basin of Lake Biwa, are surface layers, so the descriptions of mixing within them apply to the uppermost part of the seasonally stratified lakes as well. North Lake and the South Basin undergo diurnal stratification and mixing. The near isothermy in early morning due to mixing associated with heat loss at night and the subsequent build up of stratification in the day is illustrated for North Lake in Figure 12. The associated change in overturning scales throughout a diurnal cycle for such lakes is given in Figure 2. That at least four hours with wind speeds in excess of 6 m s^{-1} were required for the water column of North Lake to reach near-isothermy illustrates the difficulty of mixing surface layers with high attenuation coefficients during the day. Nocturnal mixing in these shallow lakes is generally considered a mechanism whereby nutrients released from the sediments may be mixed through the water column. For instance, PO_4-P may be released when the overlying waters become anoxic, as in hypereutrophic Lake Nakuru, Kenya (Melack and MacIntyre, 1992). This simple picture, however, can be complicated when lateral intrusions with colder water prevent mixing to the bottom at night, as was observed in the one diurnal study in the South Basin (MacIntyre, 1996). When the chemistry at the sediment-water interface is such that nutrients are not released on a regular basis, or if benthic populations of phytoplankton take up the nutrients, more intense mixing events, such as the strong winds at Lake Biwa that resuspended sediments, may be needed for nutrient supply. Depending on the oxygen dynamics, PO_4-P release may be a two step process. For instance, in Lake Okeechobee, increases in dissolved phosphorus did not occur on the day when sediments were resuspended, but on a subsequent day when oxygen concentrations decreased and desorption was enhanced (Sheng et al., 1998). Lake Calado, which had developed a seasonal thermocline by the time of sampling, also illustrates the difficulty of mixing the surface layer in the day and attests to the greater possibility of nutrient supply when convective motions deepen the surface layer at night.

As lakes deepen and seasonally stratify, the nutricline may be found farther from the surface layer. Mechanisms leading to mixing in the pycnocline become important for nutrient supply. While nutrient supply is usually calculated using a coefficient of eddy diffusivity calculated over time scales of weeks, the data from the North Basin of Lake Biwa and Mono Lake indicate that nutrient release will be intermittent and spatially variable. Rates of nutrient release will depend on the frequency of periods with high winds that initiate upwelling, boundary mixing, or mixing within the thermocline or of nocturnal cooling events that penetrate into the nutricline. One approach for predicting the frequency of nutrient release is to compute the Lake Number (Imberger and Patterson, 1990), a ratio of the stability of a lake to the wind forcing. Both Romero et al. (1998) and Imberger (1998) show that the magnitude of K_z depends on the Lake Number L_N. Imberger and Patterson's (1990) analysis indicates upwelling will occur at the base of the pycnocline for $L_N < 1$, and MacIntyre et al. (1998) show boundary mixing at the base of the pycnocline for $L_N = 2$. Time series of L_N allow prediction of the frequency of events that may lead to upwelling or boundary mixing in the metalimnion and, depending on the depth of the nutricline relative to the mixing, potential supply of nutrients to the euphotic zone. Depending on the location of the nutrients that are limiting, these short intense events may or may not lead to rapid physiological changes in the plankton, shifts in community structure, and growth.

Acknowledgments. Financial support was provided by National Science Foundation grants DEB93-17986 to SM and DEB81-11398 to J. M. Melack and T. R. Fisher, the Centre for Environmental Fluid Dynamics, University of Western Australia, and the Agency of Industrial Science and Technology, Japan. The Centre for Environmental Fluid Dynamics provided equipment and software for the research on North Lake, Western Australia, and Lake Biwa, Japan. Logistic support was provided by the Lake Biwa Research Institute; the Instituto National de Pesquisas da Amazonia, Brazil; and the Sierra Nevada Aquatic Research Laboratory, University of California Natural Reserve System.

I am grateful for the help of John Melack in Brazil, Australia, and Japan, Darla Heil on Mono Lake, Hiroshi Hashitani, Shuji Nakamura, and Mark Trevorrow in Japan, and Terry Smith, Roger Fletcher, and Roger Head in Australia. I also thank Kevin Flynn and

Centre for Environmental Fluid Dynamics, University of Western Australia, and the Agency of Industrial Science and Technology, Japan. The Centre for Environmental Fluid Dynamics provided equipment and software for the research on North Lake, Western Australia, and Lake Biwa, Japan. Logistic support was provided by the Lake Biwa Research Institute; the Instituto National de Pesquisas da Amazonia, Brazil; and the Sierra Nevada Aquatic Research Laboratory, University of California Natural Reserve System.

I am grateful for the help of John Melack in Brazil, Australia, and Japan, Darla Heil on Mono Lake, Hiroshi Hashitani, Shuji Nakamura, and Mark Trevorrow in Japan, and Terry Smith, Roger Fletcher, and Roger Head in Australia. I also thank Kevin Flynn and Manuela Lorenzi-Kayser for assistance with data processing, programming and graphics, and Gabrielle Johnson for assistance with data processing. The paper benefited from the comments of David Hamilton, Kevin Flynn, John Melack, and two anonymous reviewers. I am especially grateful to Michael Head both for his redesign of the microstructure profilers used at Mono Lake and for his unstinting technical support and to Jörg Imberger for his invaluable guidance in Japan and Australia.

References

Alldredge, A. L., and C. C. Gotschalk, In situ settling behavior of marine snow. *Limnol. Oceanogr., 33*, 339-351, 1988.

Alldredge, A. L., T. C. Granata, C.C. Gotschalk, and T. D. Dickey, The physical strength of marine snow and its implications for particle disaggregation in the ocean. *Limnol. Oceanogr., 35*, 1415-1428, 1990.

Andrews, J. C., Deformation of the active space in the low Reynolds number feeding current of calanoid copepods. *Can. J. Fish. Aquat. Sci., 40*, 1293-1302, 1983.

Anis, A., and J. N. Moum, Prescriptions for heat flux and entrainment rates in the upper ocean during convection. *J. Phys. Oceanogr., 24*, 2142-2155, 1994.

Carr, M. -E., N. S. Oakey, B. Jones, and M. R. Lewis, Hydrographic patterns and vertical mixing in the equatorial Pacific along 150°W. *J. Geophys. Res., 97*: 611-626, 1992.

Carr, M. -E., M. R. Lewis, and D. Kelley, A physical estimate of new production in the equatorial Pacific along 150°W. *Limnol. Oceanogr. 40*: 138-147, 1995.

Carter, G. D., and J. Imberger, Vertically rising microstructure profiler. *J. Atmos. Oceanic Technol.,* 3, 462-471, 1986.

Crawford, W. R., and R. K. Dewey, Turbulence and mixing: Sources of nutrients on the Vancouver Island continental shelf. *Atmos. -Ocean, 27*: 428-442, 1989.

Crill, P. M., K. B. Bartlett, J. O Wilson, D. I. Sebacher, R. C. Harriss, J. M. Melack, S. MacIntyre, L. Lesack, and L. Smith-Morrill, Tropospheric methane from an Amazonian floodplain lake. *J. Geophys. Res., 93*, 1564-1570, 1988.

Cullen, J. J., and M. R. Lewis, The kinetics of algal photoadaptation in the context of vertical mixing. *J. Plankton Res., 10,* 1039-1063, 1988.

Denman, K. L., and A. E. Gargett, Biological-physical interactions in the upper ocean: the role of vertical and small-scale transport processes. *Ann. Rev. Fluid Mech., 27*, 225-255, 1995.

DeSilva, P., J. Imberger, and G. N. Ivey, Internal wave energized benthic boundary layer transport, in *Physical Processes in Lakes and Oceans*, edited by J. Imberger, Coastal and Estuarine Studies, pp. 475-484, this volume, 1988

Dillon, T. M., and D. R. Caldwell, The Batchelor spectrum and dissipation in the upper ocean. *J. Geophys. Res., 85*, 1910-1916, 1980.

Eckman, J. E, Hydrodynamic processes affecting benthic recuitment. *Limnol. Oceanogr.*, 28, 241-257, 1983.

Ferris, J. M., and R. Christian, Aquatic primary production in relation to microalgal responses to changing light: a review. *Aquat. Sci., 53,* 187-217, 1991.

Frenette, J. -J., W. F. Vincent, and L. Legendre, Size-dependent changes in phytoplankton C and N uptake in the dynamic mixed layer of Lake Biwa. *Freshwater Biol., 36*, 1996a,

Frenette, J. -J., W. F. Vincent, L. Legendre, and T. Nagata, Size-dependent phytoplankton responses to atmospheric forcing in Lake Biwa. *J. Plankton Res., 18,* 371-391, 1996b.

Gibson, C. H., and W. H. Thomas, Effects of turbulence intermittency on growth inhibition of a red tide dinoflagellate, *Gonyaulax polyedra* Stein. *J. Geophys. Res., 100,* 24,841-24,846, 1995.

Gloor, M., A. Wuest, and M. Munnich, Benthic boundary mixing and resuspension induced by internal seiches. *Hydrobiologia 284,* 59-68, 1994.

Hashitani, H., Y. Seike, A. Hirayama, and M. Kumagai, Temporal and spatial distribution of nutrients and Chl *a* during BITEX '93, in *Baseline data overviews of BITEX '93,* edited by M. Kumagai and S. Nakano, pp. 115-128, Lake Biwa Research Institute, Otsu, Japan, 1996.

Haury, L. R., H. Yamazaki, and C. L. Fey, Simultaneous measurements of small scale physical dynamics and zooplankton distributions. *J. Plankton Res., 14, 513-530, 1992.*

Hayami, Y., T. Fujiwara, and M. Kumagai, Internal surge in Lake Biwa induced by the strong winds of a typhoon. *Jpn. J. Limnol. 57,* 425-444, 1996.

Holloway, P. E., A criterion for thermal stratification in a wind-mixed system. *J. Phys. Oceanogr., 10,* 861-869, 1980.

Imberger, J., The diurnal mixed layer. *Limnol. Oceanogr., 30,* 737-770, 1985.

Imberger, J., Transport processes in lakes: A review, in *Limnology Now: A Paradigm of Planetary Problems, edited by* R. Margalef, pp. 99-193, Elsevier Science, 1994.

Imberger, J., Flux paths in a stratified lake: A review, in *Physical Processes in Lakes and Oceans,* edited by J. Imberger, Coastal and Estuarine Studies, pp. 1-18, AGU, this volume, 1998.

Imberger, J., and B. Boashash, Application of the Wigner-Ville distribution to temperature gradient microstructure: A new technique to study small-scale variations. *J. Phys. Oceanogr., 16,* 1997-2012, 1986.

Imberger, J. and G. Ivey, On the nature of turbulence in a stratified fluid. Part 2: Application to lakes. *J. Phys. Oceanogr., 21,* 659-680, 1991.

Imberger, J., and G. J. Parker, Mixed layer dynamics in a lake exposed to a spatially variable wind field. *Limnol. Oceanogr., 30,* 473-488, 1985.

Imberger, J., and J. C. Patterson, Physical limnology, in *Advances in Applied Mechanics,* edited by J. W. Hutchinson and T. Y. Wu, Cambridge Academic Press, 27, 303-475, 1990.

Imboden, D. M., and A. Wuest, Mixing mechanisms in lakes, in *Physics and Chemistry of Lakes,* edited by A. Lerman, D. M. Imboden, and J. R. Gat,. pp. 83-138, Springer-Verlag., 1995.

Ivey, G. N., and J. Imberger, On the nature of turbulence in a stratified fluid. Part 1: The efficiency of mixing. *J. Phys. Oceanogr., 21,* 650-658. 1991.

Ivey, G. N., J. Imberger, and J. R. Koseff, Buoyancy fluxes in a stratified flow, in *Physical Processes in Lakes and Oceans,* edited by J. Imberger, Coastal and Estuarine Studies, pp 377-400, AGU, this volume, 1998.

Jellison, R., and J. M. Melack, Meromixis in hypersaline Mono Lake, California. 1. Stratification and vertical mixing during the onset, persistence, and breakdown of meromixis. *Limnol. Oceanogr., 38,* 1008-1019, 1993.

Lemckert, C. and J. Imberger, Turbulent benthic boundary layers in fresh water lakes, in *Physical Processes in Lakes and Oceans,* edited by J. Imberger, Coastal and Estuarine Studies, pp 503-516, AGU, this volume, 1998.

Lewis, M. R., J. J. Cullen, and T. Platt, Relationships between vertical mixing and photoadaptation of phytoplankton: similarity criteria. *Mar. Ecol. Prog. Ser., 15,* 141-149, 1984a.

Lewis, M. R., E. P. W. Horne, J. J. Cullen, N. S. Oakey, and T. Platt, Turbulent motions may control phytoplankton photosynthesis in the upper ocean. *Nature,* 311, 49-50, 1984b.

Lewis, M. R., W. G. Harrison, N. S. Oakey, D. L. Hebert, and T. Platt, Vertical nitrate flues in the oligotrophic ocean. *Science, 234,* 870-873, 1986.

Lazier, J. R. N, and K. H. Mann, Turbulence and the diffusive layers around small organisms. *Deep-Sea Res., 36,* 1721-1733, 1989.

MacIntyre, S., Vertical mixing in a shallow, eutrophic lake: Possible consequences for the light climate of phytoplankton. *Limnol. Oceanogr., 38,* 798-817, 1993.

MacIntyre, S., Turbulent eddies and their implications for phytoplankton within the euphotic zone of Lake Biwa, Japan. *Jpn. J. Limnol., 57,* 395-410, 1996.

MacIntyre, S., A. L. Alldredge, and C. C. Gotschalk, Accumulation of marine snow at density discontinuities in the water column. *Limnol. Oceanogr., 40,* 449-468, 1995a.

MacIntyre, S., K. M. Flynn, R. Jellison, and J. R. Romero, Boundary mixing and nutrient flux in Mono Lake, CA. Submitted to *Limnol. Oceanogr.,* 1998.

MacIntyre, S., and J. M. Melack, Meromixis in an equatorial African soda lake. *Limnol. Oceanogr.*, *27*, 595-609, 1982.

MacIntyre, S., and J. M. Melack, Vertical mixing in Amazon floodplain lakes. *Verh. Internat. Verein. Limnol.*, *22*, 1283-1287, 1984.

MacIntyre, S., and J. M. Melack, Vertical and horizontal transport in lakes: linking littoral, benthic, and pelagic habitats. *J. N. Am. Benthol. Soc.*, *14*, 599-615, 1995.

MacIntyre, S., R. Wanninkhof, and J. P. Chanton, Trace gas exchange across the air-water interface in freshwater and coastal marine environments, in *Biogenic Trace Gases: Measuring Emissions from Soil and Water,* edited P. A. Matson and R. C. Harriss, by Chapter 3, Blackwell, 1995b

Marra, J., Effect of short-term variations in light intensity on photosynthesis of a marine phytoplankter: a laboratory simulation study. *Mar. Biol.*, *46*, 191-202, 1978a.

Marra, J., Phytoplankton photosynthetic response to vertical movement in a mixed layer. *Mar. Biol.*, *46*, 203-208, 1978b.

Maxworthy, T., J. Imberger, and A. Saggio, A laboratory demonstration of a mechanism for the production of secondary internal gravity-waves in a stratified fluid, in *Physical Processes in Lakes and Oceans*, edited by J. Imberger, Coastal and Estuarine Studies, pp 261-270, AGU, this volume, 1998.

Melack, J. M., and S. MacIntyre, Phosphorus concentrations, supply and limitation in tropical African lakes and rivers in *Phosphorus Cycles in Terrestrial and Aquatic Ecosystems,* edited by H. Tiessen, pp 1-18. Regional Workshop 4: Africa. SCOPE and UNEP, Nairobi 1991, 1992.

Moum, J. N., Energy containing scales of turbulence in the ocean thermocline. *J. Geophys. Res,. 101*, 14095-14109, 1996.

Osborn, T. R., Estimates of the rate of vertical diffusion from dissipation measurements. *J. Phys. Oceanogr.*, *10*,: 83-89, 1980.

Patterson, J. C., Modelling the effects of motion on primary production in the mixed layer of lakes. *Aquat. Sci. 53,* 218-238, 1991.

Pasciak, W. J., and J. Gavis, Transport limitation of nutrient uptake in phytoplankton. *Limnol. Oceanogr.*, *19*, 881-888, 1974.

Pasciak, W. J., and J. Gavis, Transport limited nutrient uptake rates in *Dytilum brightwellii. Limnol. Oceanogr.*, *20*, 604-617, 1975.

Pawlik, J. R., and C. A. Butman, Settlement of a marine tube worm as a function of current velocity: Interacting effects of hydrodynamics and behavior. *Limnol. Oceanogr,. 38*, 1730-1740, 1993.

Prezelin, B. B., M. M. Tilzer, O. Schofield, and C. Haese, The control of the production process of phytoplankton by the physical structure of the aquatic environment with special reference to its optical properties. *Aquat. Sci. 53,* 136-186, 1991.

Reynolds, C. S., The role of fluid motion in the dynamics of phytoplankton in lakes and rivers, in *Aquatic Ecology: Scale, Pattern, and Process,* edited by P. S. Giller, A. G. Hildrew, and D. G. Raffaelli, pp. 141-187, Blackwell Scientific, Oxford, 1994.

Reynolds, C. S., Plants in motion: physical - biological interaction in the plankton, in *Physical Processes in Lakes and Oceans*, edited by J. Imberger, Coastal and Estuarine Studies, pp. 535-560, AGU, this volume, 1998.

Robarts, R. D., M. Waiser, O. Hadas, T. Zohary, and S. MacIntyre, Contrasting relaxation of phosphorus limitation due to typhoon-induced mixing in two morphologically distinct basins of Lake Biwa, Japan. *Limnol. Oceanogr.* (in press), 1998.

Romero, J. R., R. Jellison, and J. M. Melack, Stratification, vertical mixing, and upward ammonium flux in hypersaline Mono Lake, California. *Arch. Hydrobiol.* (in press), 1998.

Rothschild, B. J., and T. R. Osborn, Small-scale turbulence and plankton contact rates. *J. Plankton Res.*, *10*, 465-474, 1988.

Shay, T. J., and M. C. Gregg, Convectively driven mixing in the upper ocean. *J. Phys. Oceanogr.*, *16*, 1777-1798, 1986.

Sheng, Y. P., X. Chen, and S. Schofield, Hydrodynamic vs. non-hydrodynamic influences on phosphorus dynamics during episodic events, in *Physical Processes in Lakes and Oceans*, edited by J. Imberger, Coastal and Estuarine Studies, pp. 613-622, AGU, this volume, 1998.

Taylor, J. R., Turbulence and mixing in the boundary layer generated by shoaling internal waves. *Dyn, Atmos. and Oceans, 19*, 233-258, 1993.

Thomas, W. H., and C. H. Gibson, Quantified small-scale turbulence inhibits a red tide dinoflagellate, *Gonyaulax polyedra* Stein. *Deep Sea Res. Part A*, *37*, 1583-1593, 1990.

Thorpe, S. A., Turbulence and mixing in a Scottish loch. *Phil. Trans. Royal Soc., London. Series A,. 286*,

125-181, 1977.

Turner, E. J., R. K. Zimmer-Faust, M. A. Palmer, M. Luckenbach, and N. D. Pentcheff, Settlement of oyster (*Crassostrea virginica*) larvae: Effects of water flow and a water-soluble chemical cue. *Limnol. Oceanogr,. 39*, 1579-1593, 1994.

Vincent, W. F., P. J. Neale, P. J. Richerson, Photoinhibition: algal responses to bright light during diel stratification and mixing in a tropical alpine lake. *J. Phycol. 20,* 201-211, 1984.

Vincent, W. F., and S. Roy, Solar ultraviolet-B radiation and aquatic primary production: damage, protection, and recovery. *Environ. Rev. 1,* 1-12, 1993.

Yamazaki, H., T. Osborn, and K. D. Squires, Direct numerical simulation of planktonic contact in turbulent flow. *J. Plank. Res. 13,* 629-643, 1991.

Yamazaki, H., and D. Kamykowski, The vertical trajectories of motile phytoplankton in a wind-mixed water column. *Deep-Sea Res. 38*, 219-241, 1991.

39

The Influence of Biogeochemical Processes on the Physics of Lakes

D. M. Imboden

Abstract

This review article deals with the question of how biogeochemical processes influence the physics of lakes. An important phenomenon is the steady transport of chemicals by settling particles to the deep waters of lakes and oceans accompanied by the production of negative buoyancy (gravitation-induced stability production). Meromixis develops if the vertical mixing resulting from the input of kinetic energy is not sufficient to overcome this stability. This is often the case in eutrophic lakes, exemplified here by Lake Baldegg (Switzerland), as a result of intense nutrient cycling. A second class of phenomenon is related to river inflow. As shown here for Lake Baikal and Lake Lucerne, river inflow may be important for deep-water renewal, especially during peak discharge, when suspended particles increase the bulk density of the river water. In addition, if the drainage basins of inflowing rivers have different geochemical characteristics, significant density gradients may be established between basins, possibly driving inter-basin exchange over sills or through narrows and sea straits (e.g. Black Sea, Mediterranean, North Atlantic). Situations also exist in which organisms influence the physics of their environment directly, some examples of this being the input of turbulent kinetic energy by swimming fish, the intensification of vertical density stratification resulting from light absorption by large concentrations of phytoplankton, and the vertical convection induced by sinking bacteria at the oxic/anoxic interface in a meromictic lake.

Biogeochemical Cycles

It is well known that in aquatic ecosystems physics plays a crucial role in controlling the biogeochemical cycles. The traditional view has evolved that processes in lakes and oceans are ordered hierarchically. According to this view, mixing and transport processes control chemical and biological processes such as redox reactions (Davison, 1985), photosynthesis, respiration and mineralization (e.g., Tilzer, 1990; Imboden, 1990). It is mainly by mixing that reaction partners with different spatial distributions of their sources are brought together, and biomass is exposed to light or moved into the dark where photosynthesis ceases. Lake pollution and eutrophication in particular have triggered an interest in mixing mechanisms in lakes. These practical problems have given rise to the

Physical Processes in Lakes and Oceans
Coastal and Estuarine Studies Volume 54, Pages 591-612

development of various lake models in which mixing processes always play an important part (Imboden and Gächter, 1979; Karagounis et al., 1993; Ulrich et al., 1995).

In contrast, the influence of biogeochemical cycles on the physics of lakes has seldom attracted the attention of limnologists, in spite of a number of situations that demonstrate the importance of this influence. The aim of this review is to describe these situations with a few selected examples and thus to sharpen the eye of the future researcher for the sometimes subtle, but important influence of biogeochemical processes on mixing and transport patterns in lakes.

Mineral Cycle

Lakes are influenced by biogeochemical cycles on very different time and space scales. On a global scale, minerals are formed in the earth's mantle and crust, transported to the surface by volcanism or tectonic movements of the crust, exposed to the eroding power of water and air, washed into lakes and oceans by rivers and deposited in the sediments, from where they may either be uplifted to form sedimentary surface rocks or undergo new metamorphic changes in the interior of the earth (Figure 1, processes A-E). These processes form the so-called *mineral cycle* which controls most elements, especially the heavier ones. Its typical time scales encompass millions to billions of years, but its effect on lakes is usually much faster. With a few exceptions (such as Lakes Biwa, Baikal, Tahoe, Titicaca, and others) the mineral cycle limits the age that a lake can reach to a few ten thousands of years (Meybeck, 1995). All lacustrine basins are eventually filled by sediments, unless tectonic processes cause a continuous enlargement and deepening of the lake basin that may compensate for the deposition of sedimentary material. Thus, on a long time perspective the effect of the mineral cycle on the physics of lakes is the most spectacular one since it continuously reduces the water volume until the lake ceases to exist. Although the aspect of the limited life-time of lakes is not further discussed here, examples will be presented which show that the chemical composition of the inflows – another trace of the mineral cycle – may be important for the physics of some lakes.

The Local Aquatic Nutrient Cycle

Embedded in the slow mineral cycle, much faster and more local cycles determine the flow of important elements in aquatic systems (see processes 1-7, Figure 1). Phosphorus, an essential nutrient for algal growth, is taken up by phytoplankton in the trophic zone of lakes and released by mineralization. Often, a single phosphate ion undergoes several uptake/release cycles before leaving the trophic zone in settling biomass (Imboden and Gächter, 1979). Other short-circuit cycles occur via the mineralization of biomass in the deep waters followed by upward mixing back into the trophogenic zone, as well as via input to the sediment, followed by mineralization, release from the sediments into the water column and upward mixing. Within the framework of the lake system these cycles are not completely closed: the external input of phosphorus compensates for that lost to the lake via the outflow and to the "permanent" sediments. This makes the *local aquatic nutrient cycle* a part of the larger mineral cycle.

Furthermore, certain elements that are essential for biological processes (e.g. C, O, N, S) undergo biogenic phase transitions between dissolved, particulate, and gaseous phases (process 8, Figure 1). These transitions are usually linked to redox reactions. A prominent example is that of carbon: its most highly oxidized form (CO_2), the major nutrient source

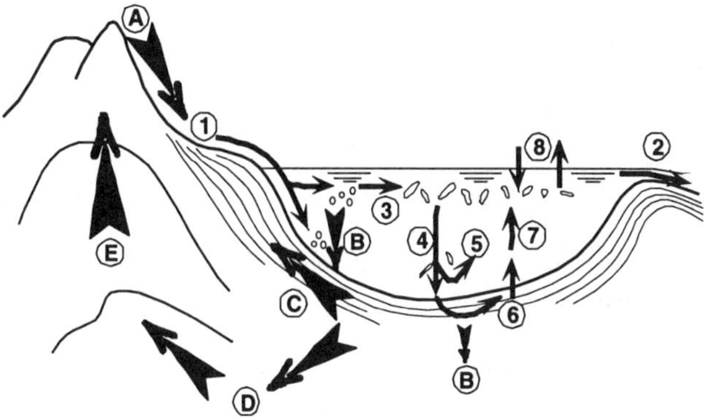

Figure 1. The flux of chemical elements through the geosphere occurs on very different space and time scales. Long-term mineral cycles include chemical weathering and erosion from the land surface (process A), transport to lacustrine and ocean sediments (B), followed either by uplift and new erosion of sedimentary rocks (C) or by metamorphism in the interior of the earth (D) and formation of new igneous or metamorphic rock (E). The local aquatic nutrient cycle is linked to the mineral cycle by external fluxes (1,2). It includes the uptake of nutrients (3), the settling of biomass (4), respiration/mineralization (5), re-dissolution from the sediments (6) and vertical transport back into the trophic zone (7). Some elements (C, O, N, S) are linked to the atmosphere by gas exchange (volatile cycle, 8).

for photosynthesis, is transported between different terrestrial and aquatic ecosystems mainly through the atmosphere. Incorporation of carbon into biomass via photosynthesis reduces its redox state, while mineralization can either lead back to carbon dioxide or to methane, the most reduced form of carbon. Thus, biology to a great extent determines the form in which carbon is present and how it moves through the geosphere.

The local nutrient cycles occur on time scales which may be extremely important for the physics of lakes. As shown in the next section, this occurs via the influence of water chemistry on the density of water. The chemical properties of inlets and their influence on vertical exchange are discussed in the following section. This includes discussion of the role of suspended particles, which form part of the biogeochemical cycles. Next, the interaction between adjacent lake basins influenced by different geochemical fluxes from their drainage basins is analysed. The final section is devoted to the special phenomenon of a biologically controlled boundary layer.

Water Chemistry and Density Stratification

Density of Water

The equation of state of water, i.e. the density of water as a function of its temperature, pressure, and chemical composition, is non-linear (Chen and Millero, 1986). Here we are mainly interested in the influence of solutes on water density. Due to the physicochemical interaction between different dissolved ions or molecules the equation is non-linear in the chemical concentrations C_i (Wüest et al., 1996). Yet for small concentrations, the water density ρ can be written as

$$\rho(T,C_i) = \rho_o(T)\cdot[1 + \sum_i \beta_i C_i] \qquad (1)$$

where $\rho_o(T)$ is the temperature-dependent density of pure water as given, e.g., by Chen and Millero (1986),

$$\beta_i = \frac{1}{\rho}\left(\frac{\partial\rho}{\partial C_i}\right)_{T,C_{j\neq i}} \tag{2}$$

is the specific expansion coefficient of water due to solute i, and the summation in Eq. (1) is extended over all solutes.

Since the relative abundance of the major ions in ocean water is very constant, Eqs. (1) and (2) are usually replaced by

$$\rho(T,S) = \rho_o(T)\,[1 + \beta_S S]\ ,\ \ \beta_S = \left(\frac{1}{\rho}\frac{\partial\rho}{\partial S}\right)_T \tag{3}$$

where S is salinity and β_S is the coefficient of haline contraction. There is no corresponding general expression for lakes. Often the electrical conductivity at 20°C, κ_{20}, is used as a substitute for S (Imboden and Wüest, 1995):

$$\rho(T,\kappa_{20}) = \rho_o(T)\,[1 + \beta_\kappa \kappa_{20}]\ ,\ \ \beta_\kappa = \frac{1}{\rho}\left(\frac{\partial\rho}{\partial\kappa_{20}}\right)_T \tag{4}$$

Note that Eq. (4) fails to recognize the influence of non-ionic species. For instance, silica may be essential for the size of the chemically induced density gradient as shown by McManus et al. (1992) for Crater Lake, Oregon, by Wüest et al. (1996) for Lake Malawi and by Hohmann et al. (1996) for Lake Baikal.

The influence of suspended particles (with concentration P) on ρ can be expressed correspondingly:

$$\beta_P = \frac{1}{\rho}\left(\frac{\partial\rho}{\partial P}\right)_{T,C_i} \tag{5}$$

Specific expansion coefficients for various components are summarized in Table 1.

TABLE 1. Influence of dissolved or suspended substances on the density of water (from Imboden and Wüest, 1995)

Substance		β (g/kg)$^{-1}$	
$Ca(HCO_3)_2$		$0.807\cdot10^{-3}$	a
$Mg(HCO_3)_2$		$0.861\cdot10^{-3}$	a
$Na(HCO_3)$		$0.727\cdot10^{-3}$	a
$K(HCO_3)$		$0.669\cdot10^{-3}$	a
$Fe(HCO_3)_2$		$0.838\cdot10^{-3}$	a
$NH_4(HCO_3)$		$0.462\cdot10^{-3}$	a
CO_2		$0.273\cdot10^{-3}$	a
CH_4		$-1.250\cdot10^{-3}$	a
air		$-0.090\cdot10^{-3}$	a
suspended particles	with $\rho_S = 2.65$ g·cm^{-3}	$0.623\cdot10^{-3}$	b
	with $\rho_S = 1.1$ g·cm^{-3}	$0.091\cdot10^{-3}$	b
electrical conductivity (β_κ) due to $Ca(HCO_3)_2$		$0.705\cdot10^{-6}$	(μS/cm)$^{-1}$

a) β_i-values for T = 25°C as defined in Eqs. 2, 3 and 6

b) β_P as defined by Eq. 5. ρ_S is dry density of suspended particles

A vertical water column is locally stable if a fluid parcel displaced isentropically from its initial position by an infinitesimal vertical distance dz experiences a restoring force. This is equivalent to the condition that the density change due to an infinitesimal isentropic transport must exceed the density change in the water column over the same distance dz:

$$\left(\frac{\partial \rho}{\partial z}\right)_{isen} - \frac{d\rho}{dz} > 0 \qquad (6)$$

where the subscript *isen* stands for "isentropic" (McDougall, 1984), i.e. for a process without exchange of heat and mass (salt). Note that the oceanographic convention is used to describe the coordinate system with the vertical (z-)axis pointing upwards. Stability is commonly quantified by the square of the Brunt-Väisälä frequency, N^2, which can be written as follows (e.g., Millard et al., 1990):

$$N^2 = \frac{g}{\rho}\left[\left(\frac{\partial \rho}{\partial z}\right)_{isen} - \frac{d\rho}{dz}\right] = g\left[\alpha\left(\frac{dT}{dz}+\Gamma\right) - \sum_i \beta_i \frac{dC_i}{dz}\right] = g\left[\alpha\frac{d\theta}{dz} - \sum_i \beta_i \frac{dC_i}{dz}\right] > 0 \qquad (7)$$

where Γ is the adiabatic lapse rate, i.e. the rate of temperature change due to adiabatic compression, and θ the potential temperature (see Gill, 1982). Note that if the water temperature is close to the temperature of maximum density, α is small. In such lakes the term that describes the influence of chemistry on stability, $\Sigma_i \beta_i dC_i/dz$, may dominate the size and sign of N^2. Therefore we expect the influence of the biogeochemical cycles on the physics of lakes to be especially pronounced in cold lakes.

Gravitationally-Induced Stability Production and Meromixis

The most apparent imprint of geochemical cycles on lakes is the steady transport of chemicals by sinking particles to the sediments, where the compounds partially redissolve and migrate back into the water column (Figure 1, processes 1, 4 and 5). This results in a negative salinity gradient above the sediments. (Remember that the vertical (z-)axis is pointing upwards.) Expressed in physical terms the formation, settling, and redissolution of particles results in a net downward transport of solute mass, i.e. in an increase of water density at the bottom and a decrease at the surface. Hence negative buoyancy is produced (Imboden and Wüest, 1995) and the vertical stability of the water column increases. Since gravitation acting on particles with a density greater than that of water is the driving force for the net mass flux, we call this process *gravitationally-induced stability production.*

The gravitationally-induced downward mass flux is partially compensated for by the vertical turbulent flux of dissolved substances against the concentration gradient, i.e. from the lake bottom to the surface. Yet, during the warm season when the vertical temperature gradient keeps the stratification strong, the diffusive flux is usually not very large, and chemical gradients remain. Vertical mixing is intensified by the annual lake circulation during the cool season, especially in temperate lakes. Thus, the annual turnover of the water column keeps the vertical chemical gradients from becoming too large. However, there are lakes in which this is not the case. Examples include lakes in which, because of natural circumstances, permanent salinity gradients have developed which inhibit complete turnover events. Such lakes are called *meromictic*. Natural meromictic lakes are either deep, terminal lakes (lakes with no outlet), and thus salty, or they have a large flux of chemicals from their sediments. Lake Tanganyika is a good example of a naturally meromictic lake. Its mixed surface layer, 50 to 100 m deep, is separated from the deep water by a chemocline (Craig, 1975). In the past, Lake Tanganyika has gone through periods when it did not have an outlet. During these periods enough salt accumulated in the lake and in the sediments to keep the salinity in the deep waters slightly above that of the surface waters, in spite of the diffusive loss of salt to the surface and out through its present outlet. Measurements of tritium in the lake document the fact that mixing across the chemocline is small (Craig, 1975).

Some lakes alternate between states of permanent stratification and regular annual mixing. The Dead Sea had been a meromictic lake for several decades before it turned over completely in February 1979 (Steinhorn, 1985). Prior to this event, the ongoing evaporative loss of water from the surface, combined with a reduced freshwater input, had caused a steady decrease in the vertical salinity gradient. As a consequence, the salt-induced vertical density gradients gradually disappeared until the first of a new series of mixing events took place in 1979. Since then, the lake has gone through intermittent periods of mero-mixis and regular overturn (Anati and Stiller, 1991). The Dead Sea demonstrates in an impressive manner the importance of geochemistry, coupled with hydrology, for the mixing regime.

In some cases the impact of human activities has brought about changes that have led to anthropogenic meromixis by the intensification of processes 4 to 6 shown in Figure 1. The acceleration of nutrient cycling in eutrophic lakes has caused the gravitationally-induced stability production to grow beyond the capacity of some lakes to erode it away by mixing (Culver, 1977). For instance, as shown by the sedimentary record of Lake Baldegg (Switzerland), a small (surface area 6 km^2) but rather deep lake (maximum depth 67 m), about 100 years ago increasing lake productivity turned the formerly holomictic lake into a meromictic one (Niessen and Sturm, 1987). In this particular case, artificial mixing in combination with a significant reduction in the phosphorus input to the lake were chosen as artificial measures to overcome anthropogenic meromixis (Figure 2).

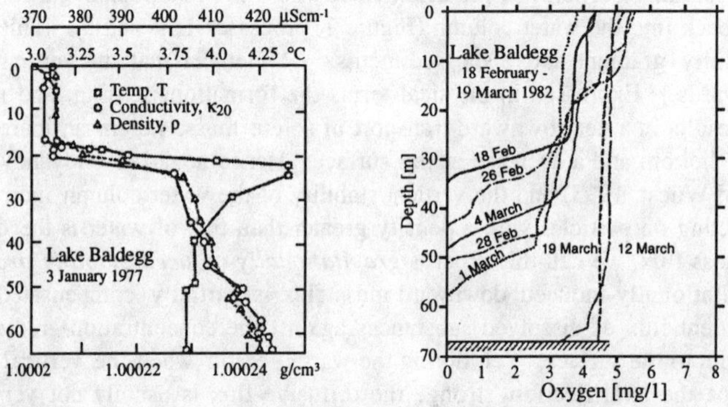

Figure 2. Left: The vertical profile of water density ρ in Lake Baldegg (Switzerland), calculated using Eq. (4) from measured water temperature T and electrical conductivity κ_{20}, show that the water column is stabilized by large solute concentrations near the bottom of the lake. Though water temperature had dropped below 4°C on January 3, 1977, vertical mixing could only penetrate to a depth of 18 m. The eutrophication-induced chemocline kept the hypolimnion anoxic during most of the year until the installations for artificial mixing by air bubbles came into operation on February 25, 1982. Right: Vertical profiles of dissolved oxygen concentration depict the movement of the mixing front to the bottom within about 3 weeks after the onset of artificial mixing. Note that on March 19, 1982 the oxygen concentration was nearly constant with depth, but saturation was only about 35%. At this time, the limiting step for re-aeration of the lake was the rate of air/water exchange at the lake surface. From Imboden (1985).

Geothermal Heat Flux and Double Diffusion

The increase of salinity (or electrical conductivity) with depth is sometimes accompanied by a temperature increase near the lake bottom that is caused by the geothermal heat flux. This situation may lead to *double diffusion*, as was, for instance, nicely demonstrated by Newman (1976) for Lake Kivu.

Using T and κ_{20} as the only parameters determining stability, Eq. (7) can be rewritten as

$$N^2 = g\,[\alpha \frac{\partial \theta}{\partial z} - \beta_\kappa \frac{\partial \kappa_{20}}{\partial z}] \tag{8}$$

Double diffusion in the so-called *diffusive regime* (where both terms in Eq. 8 are negative) can occur if the stability ratio

$$R_d = \frac{\beta_\kappa \frac{\partial \kappa_{20}}{\partial z}}{\alpha \frac{\partial \theta}{\partial z}} \gtrsim 1 \tag{9}$$

i.e., if the absolute value of the density gradient produced by the stabilizing component (salt) is only slightly larger than the destabilizing influence of the temperature gradient.

Wüest (1994) has analysed the temperature and salinity gradients in the hypolimnion of Lake Lugano (Switzerland/Italy) for the possibility of double diffusion (Figure 3). The value $R_d \sim 4.5$ in the deep water makes the presence of active double diffusion unlikely, but it may well be that this value reflects a limit which is controlled by sporadic double diffusion events that keep the value of R_d from becoming too small. Since the employment of high-resolution temperature and conductivity sensors has become more common for lake research, temperature and salinity profiles like those in Lake Lugano have been observed in other deep temperate lakes, e.g. in Lake Zug (Switzerland) (Imboden et al., 1988) and in Lake Baikal (Siberia) (Hohmann et al., 1996).

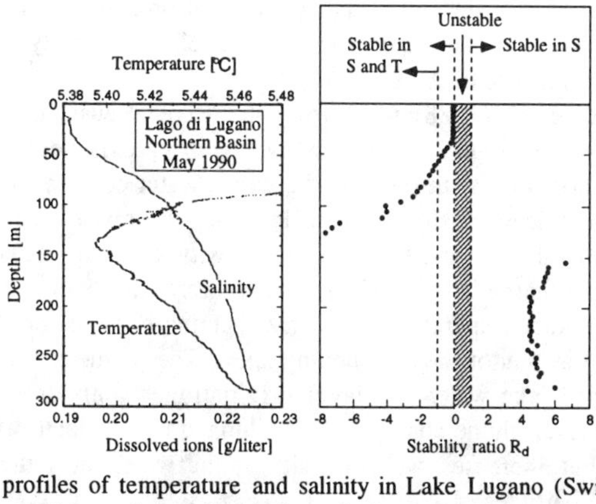

Figure 3. Vertical profiles of temperature and salinity in Lake Lugano (Switzerland/Italy) reflect the influence of geothermal heat flux and the flux of chemicals from the sediments into the quasi-permanently stratified deep hypolimnion of the lake. Below 150 m depth, salinity has a stabilizing and temperature a destabilizing influence on the water column, but the influence of salinity is at least 4 times stronger than that of temperature, as indicated by the stability ratio R_d (*right graph*) defined in Eq. 9. In principle, this situation could lead to double diffusion (diffusive regime), but R_d is too large for double diffusion to be active during the time of measurement. From Wüest (1994).

The Role of Inflows as the Principle Geochemical Conveyor

Inflowing rivers are an important source of solutes and suspended particles to lakes (Figure 1, process 1). The concentration of dissolved and suspended material, together with water temperature, determine the depth at which the inflowing water is spread horizontally within the lake. Often, for a given river/lake pair this depth varies seasonally according to the relative temperature and salinity variations occurring in both the river and the lake. This seasonal pattern may be interrupted by events of very large discharge (floods) that are often accompanied by large concentrations of suspended particles and by deep penetration of the turbid river water into the lake (see below). This phenomenon is illustrated by two examples.

Inflows into Alpine Lakes: The Example of Lake Lucerne (Switzerland)

Urnersee and Gersauersee, two adjacent basins of Lake Lucerne (Vierwaldstättersee, see Figure 11), are separated by a sill at 85 m depth. They have a similar topography (maximum depth 195 and 213 m, respectively), but different concentrations of dissolved oxygen near the bottom (van Senden et al., 1990). As shown by Wüest et al. (1988), one major reason for this is the influence of the River Reuss on Urnersee and the absence of an inflow of equal importance into Gersauersee.

The River Reuss drains an area dominated by crystalline rocks. Its electrical conductivity κ_{20} lies mostly between 80 and 90 $\mu S \cdot cm^{-1}$. In contrast, other less important inflows to Lake Lucerne originate from the calcareous part of the Alps. Their electrical conductivities are typically 250 $\mu S \cdot cm^{-1}$ or larger, while those in the lake are between 200 and 210 $\mu S \cdot cm^{-3}$. According to Table 1 the relative density difference between water from the River Reuss and the lake (assuming equal temperatures) corresponds to about $\Delta\rho/\rho = 85 \times 10^{-6}$. This would roughly correspond to the density difference between water at 4°C and 7.5°C. Water in Alpnachersee (see Figure 11), a basin of Lake Lucerne that receives water with large concentrations of calcium carbonate, has a conductivity of 310 $\mu S \cdot cm^{-1}$. This corresponds to a relative density difference between Alpnachersee and the main lake of $\Delta\rho/\rho = 70 \times 10^{-6}$.

As shown by Wüest et al. (1988) water from the River Reuss, in spite of its small conductivity, frequently penetrates to layers below 110 m in Urnersee. In April, the river water usually spreads more or less throughout the whole water column. In May, intrusions of river water, which are clearly visible by their low κ_{20}, are observed mainly between 50 m depth and the thermocline. During the summer, density currents occasionally penetrate to greater depth when discharge is large (Figure 4). Currents at 65 m reach values of up to 0.4 $m \cdot s^{-1}$. At this time, the temperature of the river plume exceeds the temperature of the pure river water as well as that of the surrounding lake. The plume represents a mixture of river water and entrained lake water (the latter originating mainly from the warm layers near the surface). The river plume could not have plunged to this depth without the help of another component that increases water density. The water must therefore have been loaded with suspended particles.

We can estimate a lower limit to the particle concentration using the salinity-induced relative density difference between river and lake water of Dr/r ~ 85 x 10-6 (see above). Since entrainment of surface lake water causes rising temperatures in the plume (Figure 4), the temperature at the lake surface must have been greater than in the river water. Thus initially, temperature helped to push the plume downwards. By assuming a dry particle

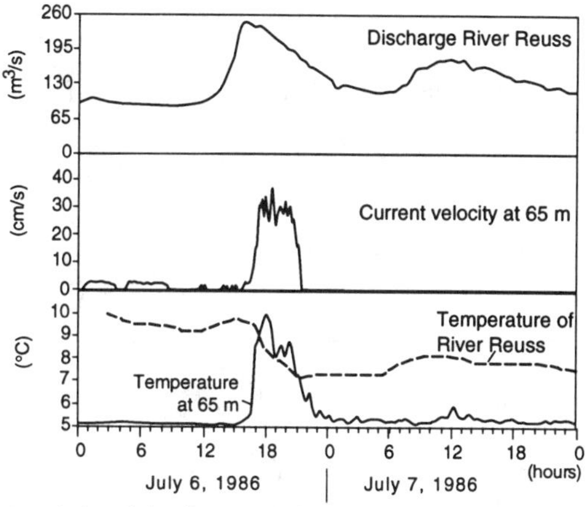

Figure 4. Record of a river-induced density current at the mouth of the River Reuss, Lake Lucerne (see Figure 11). The discharge peak leads to an immediate response of the current meter anchored off the river mouth at 65 m depth. Due to entrainment of water from the warm lake surface into the plunging river plume, water temperature recorded at the depth of the current meter temporarily exceeds the temperature of the river water. From Wüest et al. (1988).

density of 2.65 g·cm^{-3} (Table 1), an initial concentration of suspended solids in the River Reuss of not more than 140 mg·L^{-1} would have been needed to overcome the chemically-induced density difference. Peak values measured in the river Reuss exceed 1 g·L^{-1}. In contrast, at 65 m the difference between the temperature in the plume (up to 10°C) and the surrounding lake water (5°C) yields a negative relative buoyancy of 265 x 10^{-6}, which, by neglecting remaining salinity differences, still requires a suspended solid concentration of 430 mg·L^{-1} for its compensation. We can only speculate on how this can happen. Perhaps local resuspension of sediment particles in the delta triggers a flood-induced turbidity current. It could also be that at this depth the plume had attained positive buoyancy, but that the inertia of the flow kept it from rising to higher layers immediately .

River water on its way to the deep hypolimnion is diluted by at least a factor of 10, i.e. it entrains at least ten times its own volume from the surface layers and consequently gives rise to a significant flux of water to the hypolimnion (Wüest et al., 1988). Therefore, although the total duration of large discharge events was less than 60 hours in May and June 1986, penetrating river water combined with surface entrainment was able to replace about 20% of the water in Urnersee below 110 m depth.

The Role of Rivers in the Formation of Deep Water in Lake Baikal

An even more spectacular case of river-induced density flow can be observed in Lake Baikal (Siberia). Lake Baikal is the largest freshwater lake on earth with respect to both volume (V=23,000 km^3) and depth (max. 1,636 m). As Weiss et al. (1991) have concluded from vertical profiles of CFC-12, the water age in the lake, i.e. the time since the water at a given depth was last exposed to the atmosphere, nowhere exceeds 15 years. This value has been revised downwards (slightly) based on helium/tritium isotope measurements (unpublished results of the Dept of Environmental Physics, EAWAG/ETH, Switzerland). The rather large relative molecular oxygen saturation (75% or more throughout the lake) is consistent with the significant vertical exchange and the low primary productivity of the lake.

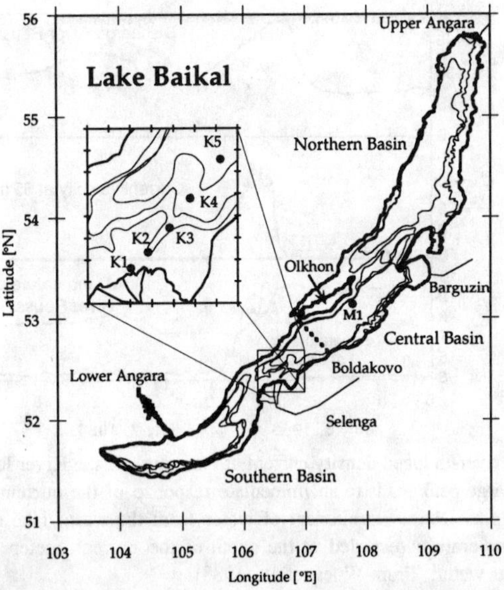

Figure 5. Map of Lake Baikal (Siberia) with depth contour lines at 400, 700, and 1000 m. The lake is divided by underwater sills into a Southern Basin, a Central Basin with the deepest point of the lake of 1636 m (black dot marked M1), and a Northern Basin. K1 to K5 represent stations along Kukui Canyon north of the Selenga delta, and the black dots in the Central Basin are CTD stations across the lake from Boldakovo to Olkhon. Irkutsk is located at the northern end of the dammed section of the Lower Angara, the only outlet of Lake Baikal. From Hohmann et al. (1996).

Although the detailed mechanisms responsible for the renewal of deep water at a rate of about 10% per year have not yet been fully explored, the Selenga River, the lake's major inflow, must play an important role. As shown in Figure 5, the large delta of the Selenga River is located at the sill (depth 290 m) which separates the Southern Basin (max. depth 1,423 m) from the Middle Basin (max. depth 1,636 m). An underwater canyon cut into the delta (Kukui Canyon) can be observed from the surface to about 1,000 m depth. Relative to the local topography the canyon is about 100 m deep.

Russian scientists have speculated that in spring the Selenga River must plunge into the deep part of the Middle basin. Current velocities in the canyon at 218 m depth, 3 m above ground, registered by a moored current meter at station K1 (see Figure 5) reached peak values of 0.10 $m \cdot s^{-1}$. The largest velocity averaged over 3 hours is 3.4 $cm \cdot s^{-1}$ (unpublished data, Dept. of Environmental Physics, EAWAG/ETH, Switzerland). Periods with large current velocities coincide with those of low water temperature. At 800 m the current signal was no longer detectable. In addition, CTD profiles taken in May 1995 along the Kukui Canyon, just after the ice had disappeared from the Middle Basin, show a water layer of 100 to 200 thickness above the bottom that is characterized by low temperature, large turbidity and large salinity (Figure 6). The concentration of dissolved oxygen in this layer is greater than in the water further from the bottom. These data indicate that denser water from the river mixed with water from the lake surface is flowing down the canyon (Hohmann et al., 1996).

The flow along Kukui Canyon may explain the sudden changes in the composition of the bottom water of the Middle Basin that are usually observed in early summer.

Sometimes in spring, cold water with higher salinity penetrates to the bottom layer of the basin (Figure 7). It seems that Selenga water (typical $\kappa_{20} = 150\ \mu S \cdot cm^{-1}$), perhaps aided by the effect of significant concentrations of suspended solids, is diluted by the cold water from the lake surface (0 to 2°C, $\kappa_{20} = 100\ \mu S \cdot cm^{-1}$) while flowing along the canyon. As a result, the river plume loses most of its extra salinity, but keeps enough negative buoyancy to find its way to the bottom of the lake.

Roughly one month later, increasing water temperatures in both the Selenga River and at the surface of the lake keep the temperature of the entraining river plume above the temperature of the deep water in the Middle Basin (3.2°C). The plume no longer reaches the bottom of the basin but spreads northwards at some intermediate depth.

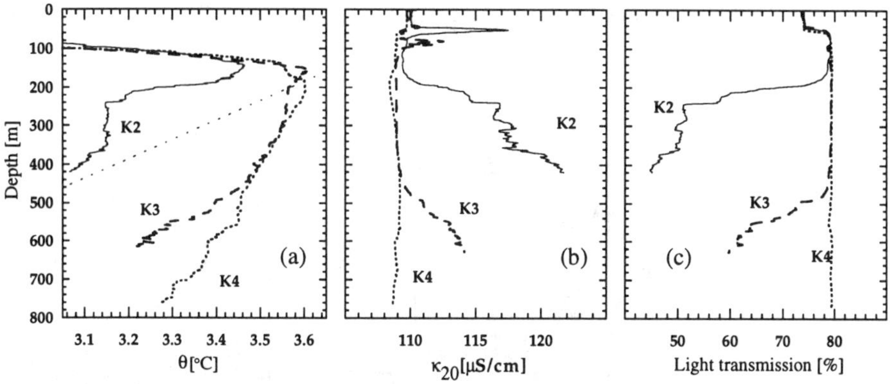

Figure 6. Vertical profiles of (a) potential temperature θ, (b) electrical conductivity κ_{20}, and light transmission measured at stations K2 to K4 (see Figure 5) in Kukui Canyon (Lake Baikal) on May 15, 1995. The bottom of the canyon drops from 420 m (K2, about 12 km from river mouth) to 630 m (K3, 19m km from mouth) and 760 m (K4, 30 km from mouth). The water flowing along the canyon is warmer, more saline, and more turbid than lake water. From Hohmann et al. (1996).

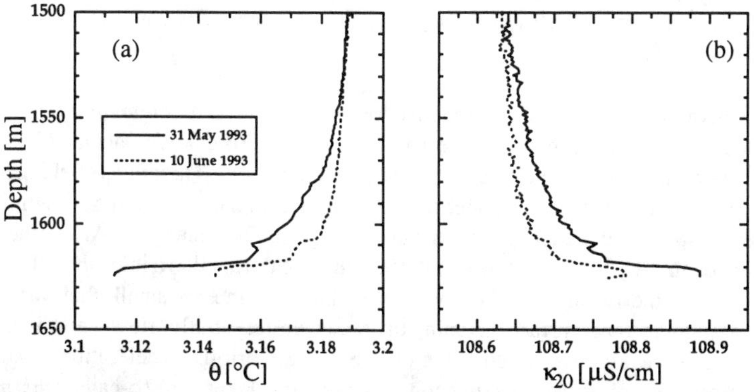

Figure 7. Vertical profiles of (a) potential temperature θ, (b) electrical conductivity κ_{20} measured at station M1 in Lake Baikal in spring 1993 (see Figure 5). Prior to May 31, a distinct water mass with low temperature and large salinity penetrated to the bottom of the Middle Basin. Based on measurements in the Kukui Canyon this water must be a mixture of Selenga water and water entrained at the then still rather cold lake surface. Ten days later, part of the bottom water signal had disappeared, probably due to vertical turbulent diffusion. Data from Hohmann et al. (1996).

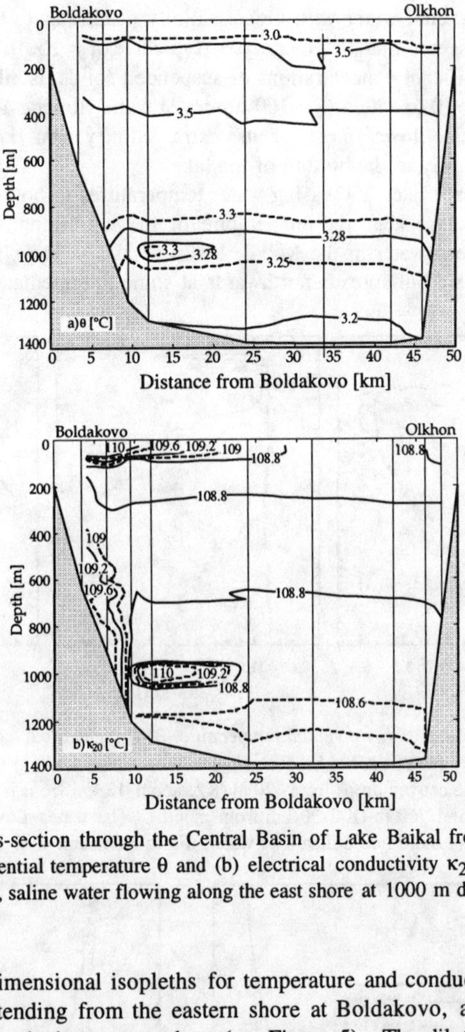

Figure 8. Vertical cross-section through the Central Basin of Lake Baikal from Boldakovo (east) to Olkhon (west) of (a) potential temperature θ and (b) electrical conductivity κ_{20} measured on May 23, 1993. The lens of warm, saline water flowing along the east shore at 1000 m depth originates from the Selenga River.

Figure 8 gives two-dimensional isopleths for temperature and conductivity in the Central Basin in a section extending from the eastern shore at Boldakovo, about 70 km north of the Selenga delta, towards the western shore (see Figure 5). The diluted Selenga water can be identified as a lens of both higher temperature and higher conductivity (salinity).

Later in spring, the Selenga water undergoes a further change. At a time when the traces left behind by the plume at about 1000 m depth are still visible (though probably no longer active), the density of the river water has become so small that the water now spreads along the lake surface. Again, the water preferentially flows northwards, either because of the prevailing wind direction or due to the Coriolis acceleration. At this time, another prominent physical phenomenon of large, cold lakes, the so-called *thermal bar*, is developing. In principle, lakes which freeze during the winter undergo two periods each year when surface temperatures are above 4°C in some areas and below in others. Consequently, interfaces must exist where water temperatures are exactly at 4°C, the temperature of maximum density. This defines the thermal bar. Water meeting at the interface sinks to larger depths thus separating areas outside and inside the thermal bar.

Shimaraev et al. (1993) have postulated that the formation of deep water in Lake Baikal is linked to the spring thermal bar. As confirmed by numerous temperature measurements as well as by infrared satellite data (unpublished data from the ATSR/ESA sensor provided by David Llewellyn-Jones, Leicester University, UK.), it is indeed true that at the end of May and beginning of June a zone develops with temperatures greater than 4°C along the eastern shore of Lake Baikal while surface temperatures in the middle of the lake are still around 2°C. This so-called spring thermal bar is also visible by eye since it separates more turbid water near the shore from highly transparent and cold off-shore water.

Figure 9 shows a two-dimensional transect of temperature and salinity taken perpendicular to the eastern shore of Lake Baikal at Boldakovo (see Figure 5). In June 1993, the surface temperature in the middle of the lake was still at 2°C while near the shore it had already reached 7°C. At about 2 km off shore, the temperature is 4°C and equal to the temperature of maximum density $T_{\rho max}$. There, water from both sides of the thermal bar sinks downwards, although it is stopped within the top 200 to 300 m, since $T_{\rho max}$ decreases with pressure at a rate of 0.2°C per 100 m depth increment.

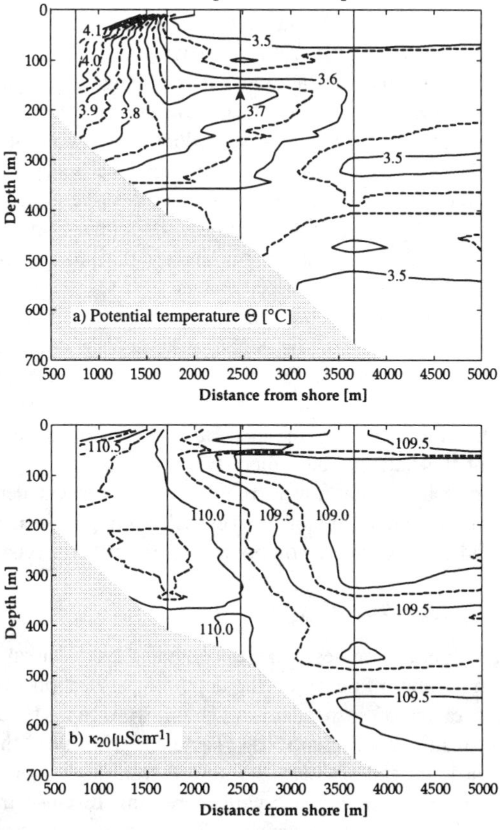

Figure 9. An enlargement of the eastern part of the cross-section shown in Figure 8 depicting the situation in Lake Baikal on June 5, 1993. As can be seen from the potential temperature θ, a thermal bar has developed about 1500 m from the shore. (Note: The thermal bar is located where the 4°C isotherm reaches the surface.) Electrical conductivity and temperature show that the vertical circulation induced by the thermal bar does not reach beyond a depth of about 300 m.

TABLE 2. Composition of water trapped by a thermal bar near the eastern shore of the Central Basin of Lake Baikal. C_M, C_S, C_{TB}: concentration in open water of Central Basin, in Selenga River and on the shore side of the thermal bar at Boldakovo. Unpublished data by Laura Sigg (EAWAG), taken in June 1993.

Parameter		C_M	C_S	C_{TB}	η_{mix} a)
Alkalinity	($mmol \cdot L^{-1}$)	1.065	1.435	1.205	0.38
Sr	($\mu mol \cdot L^{-1}$)	1.5	1.975	1.6	0.21
$SO_4^{=}$	($mmol \cdot L^{-1}$)	0.056	0.083	0.065	0.35
Cl^-	($mmol \cdot L^{-1}$)	0.0145	0.0375	0.026	0.50
					$\overline{\eta}_{mix} = 0.36 \pm 0.12$

a) $\eta_{mix} = \frac{C_M - C_{TB}}{C_M - C_S}$: relative volume fraction of Selenga water in surface water trapped by thermal bar

Chemical analysis of water samples taken (1) from the centre of the lake, (2) between the shore and the thermal bar, and (3) from the Selenga River (Table 2) shows that about 36% of the water trapped behind the thermal bar is Selenga water flowing from the river mouth northwards along the shore. Apparently, development of the thermal bar – at least at the eastern shore of the Middle Basin – is predominantly controlled by Selenga water that, at this time of the year, is warmer than the open lake water. Since the thermal bar impedes horizontal mixing between the open lake and the river water, the latter keeps its identity (including its density) over long distances. In contrast, in other large lakes, e.g. in the Laurentian Great lakes, the thermal bar is mainly the product of differential heating (or cooling) in near-shore (shallow) and deep waters.

Much remains to be unravelled in this story. However, as a preliminary result it seems very likely that the geochemistry of the Selenga River plays an important role in the formation of deep water. As long as these mechanisms are not fully understood, it is impossible to predict the effects of the anthropogenic changes that have occurred during the last 30 years in the drainage basin of Lake Baikal. For instance, it may turn out that the increasing load of the Selenga and other rivers with anthropogenic compounds may affect the ecosystem of Lake Baikal more strongly via the vertical density structure of the water column (followed by alteration of the vertical mixing pattern) than by the direct effects of nutrients and anthropogenic compounds on the aquatic ecosystem.

The Role of Suspended Particles

As demonstrated by both examples, Lake Lucerne and Lake Baikal, suspended solids in rivers may have an important influence on the depth of penetration of riverine waters into stratified lakes. The effect of suspended solids on water density can be calculated by defining specific expansion coefficients, β_P (Table 1). However, there is an important difference between particles and solutes: the former can settle out of the plume as the intensity of turbulence begins to decrease with increasing distance from the river mouth, leaving behind water with a lower bulk density. This process may proceed stepwise. The largest particles are lost from the plume first, followed by particles of smaller size. Sedimentologists use size-dependent settling of particles (visible in the sediment structure) to distinguish between river-induced sedimentation, the effect of turbidity currents, and normal particle settling (Håkanson and Jansson, 1983).

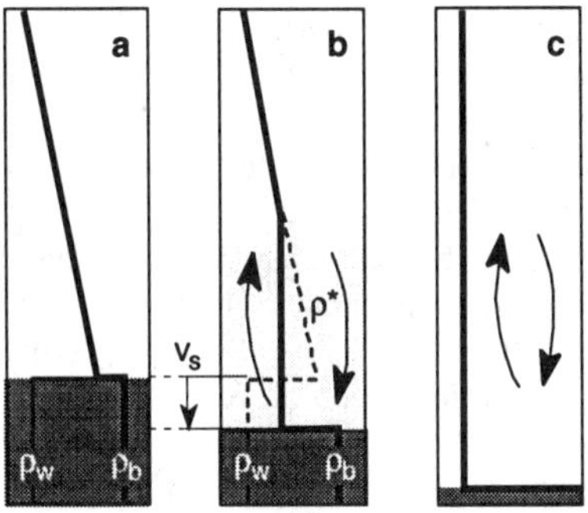

Figure 10. Schematic picture for the coupling between sedimentation and convection as observed in a laboratory tank. (a) Water with stable density stratification (density ρ_w) overlies a suspension composed of fresh water (density ρ_w smaller than ρ_w above) in which glass beads are suspended. The bulk density of the suspension is ρ_b. (b) When the glass beads settle out of the water they leave behind water of lower density. This causes vigorous convection in an overlying layer. (c) Eventually, the suspended particles completely disappear from the water column leaving a homogeneous layer of water. From Kerr (1991).

Sturm and Matter (1978) have observed that the River Aare, the major inflow to Brienzersee (Switzerland), sometimes separates into two plumes, one flowing along the bottom of the lake, the other spreading out horizontally at some intermediate depth of the lake. The former, which contains the larger particles, is stopped sooner than the latter, which retains the smaller particles and is able to deposit its load across the whole lake.

A similar, quasi-continuous process has been postulated by Kerr (1991) to cause erosion of a stable density gradient by sedimentation-driven convection. In his laboratory experiments, Kerr produced a stratified layer of pure water (density increasing from 0.999 to 1.020 $g \cdot cm^{-3}$) underlain by a suspension composed of dyed fresh water (density 0.999 $g \cdot cm^{-3}$) in which spherical glass beads (density 2.47 $g \cdot cm^{-3}$, mean diameter 23.5 μm) were suspended, resulting in a bulk density of 1.034 $g \cdot cm^{-3}$ for the suspension (Figure 10a). At the top of the suspension, sedimentation of the glass beads left behind water with positive buoyancy, which rose in tendrils from below and drove vigorous convection in an overlying layer (Figure 10b). The ongoing process continuously decreased the bulk density of the well-mixed convecting region until all the sediment had settled, leaving a homogenous region overlain by the remaining gradient (Figure 10c).

Such situations are met with in lakes whenever warm riverine water containing suspended solids flows down to the lake bottom (see Figure 4). Similar situations may also occur in the ocean. For instance, Quadfasel et al. (1990) observed that relatively light deep water flows over the sill in the northern part of the 5000 m deep Sulu Sea (eastern Indonesia). The authors conclude that the overflow events must result from turbidity currents combined with classical plume convection. Kerr (1991) interprets the 2000 m of homogeneous deep water overlying a stratified water body as observed by Quadfasel et al. as being consistent with his model derived from the tank experiments.

The Geochemical Interaction Between Adjacent Lake Basins

The influence of the Selenga River does not explain all phenomena responsible for deep-water renewal in Lake Baikal. As shown by Peeters et al. (1996), the different geological characteristics of the drainage basins of the three major basins, combined with gradients of the net heat flux along the lake axis, lead to different vertical temperature/salinity structures in these basins. At the sills, new water is formed by the mixing of water from adjacent basins. Owing to the non-linearity of the equation of state, this water may sink along so-called neutral tracks (Peeters et al., 1996) on either side of the sill into the deep layers of the basins. The corresponding density gradients that drive the exchange flow are extremely small, and so must be the induced flow velocities also. Yet, since the contact area at the inter-basin sills is large, the total water exchange may nonetheless be significant for the formation of deep water.

At present, postulation of the existence of an inter-basin exchange flow is based on an elaborated theoretical concept as well as on indirect evidence from tracer measurements. Relevant direct measurements (e.g., by current meters located at the sills) are yet to be made. However, even if this were to be done some day, it would still be difficult to separate the (weak) signal of the density-induced flow from other current signals (e.g., those from wind- or seiche-induced currents). To demonstrate the role of drainage-basin geology on inter-basin water exchange, a more straightforward example will therefore be employed in this review.

Inter-basin Exchange in Lake Lucerne

Figure 11 shows the map of Lake Lucerne (Switzerland) and its major inflows. These drain two geologically very different parts of the Alps. The River Reuss, entering the lake at its southern end, originates from the crystalline part of the central Alps; its conductivity is typically between 80 and 90 $\mu S \cdot cm^{-1}$. In contrast, the Rivers Muota, Engelberger Aa, and Sarner Aa contain larger concentrations of calcium carbonate; their conductivity lies between 250 and 350 $\mu S \cdot cm^{-1}$.

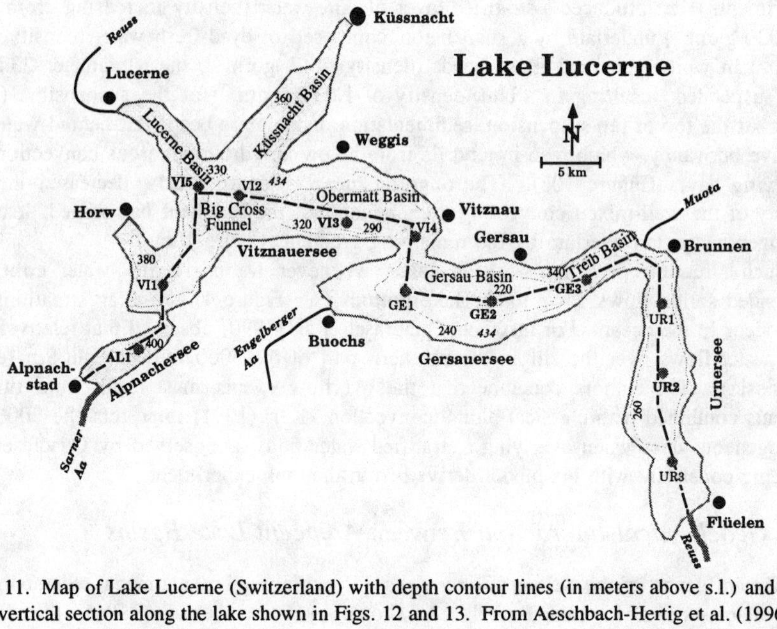

Figure 11. Map of Lake Lucerne (Switzerland) with depth contour lines (in meters above s.l.) and trace of the vertical section along the lake shown in Figs. 12 and 13. From Aeschbach-Hertig et al. (1996).

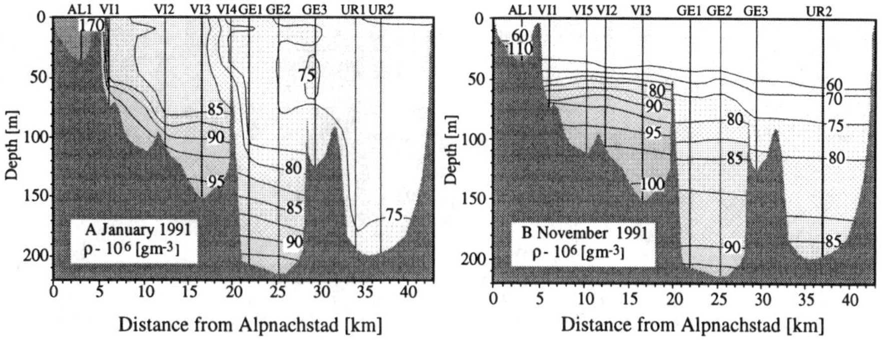

Figure 12. Section along Lake Lucerne (see Figure 11) of water density calculated from water temperature and electrical conductivity in January and November 1991. Isopycnals are labeled with the density minus 10^6 g·m^{-3}. From Aeschbach-Hertig et al. (1996).

Figure 12 shows two-dimensional density distributions along a track from Alpnachersee to Urnersee for January and November 1991, respectively. The graph shows clearly the separation of the deep waters by the sills. Early in the year (Figure 12a), the upper layers of the lake are stratified. The heavy water from the Sarner Aa causes a distinct horizontal density gradient at the sill to Vitznauersee and from there to Gersauersee. The structure of the isopycnals suggest that water must flow across the sills from west to east and penetrate into the deep layers of the adjacent basins.

Urnersee contains the water with the lowest density, not only because it receives chemically light water, but also because its exposure to strong winds results in intense vertical mixing, and thus in relatively warm surface waters during the winter (Wüest et al., 1988). The horizontal density gradient between Gersauersee and Urnersee is therefore a result of the combined influence of geochemistry and wind shear. In November (Figure 12b), at the end of the stratification period, horizontal density currents have completely eroded the horizontal density gradients above the sills, while the deep waters keep their different densities.

The exchange fluxes that can be inferred from the spatio-temporal evolution of the density field are nicely confirmed by the distribution of water ages as calculated from helium-3 and tritium (Aeschbach-Hertig, 1994; Aeschbach-Hertig et al., 1996). As shown in Figure 13, the oldest water is formed at the bottom of Gersauersee, while the deep waters of Urnersee are steadily ventilated by both the density-driven flow over the sill from Gersauersee and the penetrating waters from the River Reuss (see above). The largest density currents occur between Alpnachersee and the rest of Lake Lucerne. The overflow of water from Alpnachersee can be traced in Vitznauersee by its large electrical conductivity (Figure 14). In winter, the water from Alpnachersee flows along the bottom of the lake, while during the stratification period it appears as a distinct layer within or below the thermocline.

Inter-basin Exchange in Lakes and Oceans

The situation observed in Lake Lucerne resembles the exchange of water through sea straits. Such flows are often the combined result of two different cycles, the geochemical cycle, transporting salts, and the hydrological cycle, driving precipitation, river flow and

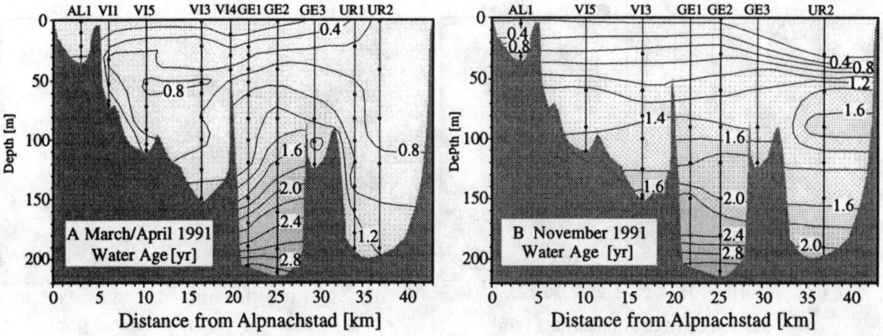

Figure 13. Section along Lake Lucerne (see Figure 11) of water age calculated from tritium and helium isotopes. Due to the density-driven exchange between the basins, the oldest water is found at the bottom of the intermediate Gersauersee basin, where ageing and the overflow of younger water from the adjacent Vitznauersee kept the age constant during summer 1991. From Aeschbach-Hertig et al. (1996).

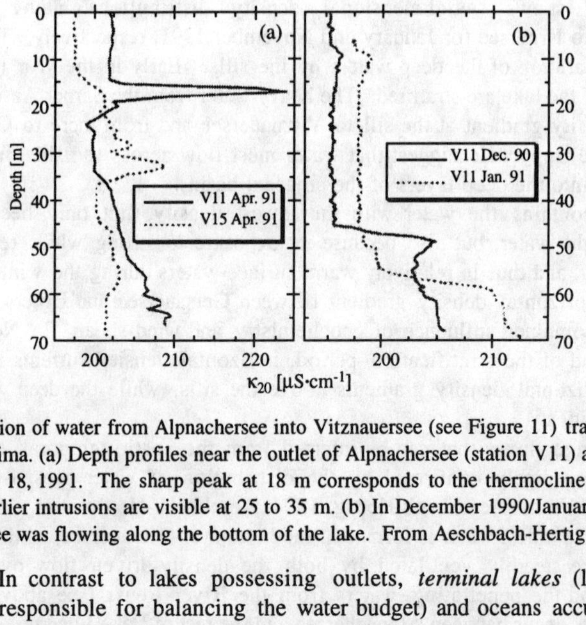

Figure 14. Intrusion of water from Alpnachersee into Vitznauersee (see Figure 11) traced by electrical conductivity maxima. (a) Depth profiles near the outlet of Alpnachersee (station V11) and at station V15 recorded on April 18, 1991. The sharp peak at 18 m corresponds to the thermocline depth. At station V15, traces of earlier intrusions are visible at 25 to 35 m. (b) In December 1990/January 1991, the water from Alpnachersee was flowing along the bottom of the lake. From Aeschbach-Hertig et al. (1996).

evaporation. In contrast to lakes possessing outlets, *terminal lakes* (lakes in which evaporation is responsible for balancing the water budget) and oceans accumulate solutes to such an extent that the influence of water temperature on density eventually becomes negligible in comparison to the influence of salt-induced density gradients. Wherever adjacent basins have different ratios of freshwater input to evaporation, significant salinity differences can develop. Where such basins are connected by straits or narrows, density-driven exchange fluxes occur. Depending on (1) the depth of the strait, (2) the salinity difference, and (3) the net flow of water, a two-layer exchange regime may develop in which the lighter water flows in the upper layer against the inter-basin salinity gradient, while the denser water flows in the opposite direction underneath.

The most striking examples of this are found between the open ocean and semi-enclosed seas, e.g., between the Black Sea and the Mediterranean, or between the

Mediterranean and the Atlantic Ocean (Tziperman, 1987). For instance, the outflow of Mediterranean water through the Strait of Gibraltar can be observed as a warm, salty tongue in large parts of the North Atlantic at a depth of about 1.5 km. On a smaller scale, estuaries and fjords receiving fresh water from inlets at one end while being connected to the open ocean by narrows (often also by sills) at the other, give rise to a wide variety of density currents. The exchange of water is often controlled by the so-called *critical overflow condition* at the smallest cross-section of the strait (Farmer and Armi, 1986). A review of exchange flows in lakes is given by Hamblin (1998).

Biologically-Controlled Physical Environments

Organisms That Create Their Own Mixing Regime

To conclude this article, let us briefly address the question of how organisms can directly determine mixing processes in aquatic systems. Organisms can be important physical agents if either their contribution to the energy flux is very large, or if the natural energy flux is extremely small. Examples of the former are well documented in the literature. For instance, Farmer et al. (1987) have evaluated the contribution of large fish populations to turbulence in the water column. Furthermore, the strengthening of the density stratification due to strong light absorption by large concentrations of phytoplankton in the surface layer of lakes has been observed, e.g., by Dickey and Simpson (1983).

Cases where the latter situation may occur are met at interfaces where turbulence is suppressed and only molecular processes remain. The physics of the sediment-water boundary is discussed in greater detail by Wüest and Gloor (1998). Here it shall only be pointed out that, due to the large gradients of redox potential and nutrient concentrations, biological processes at the sediment surface are often very intense. Furthermore, spatial gradients of turbulent kinetic energy at the sediment-water interface (Stolzenbach et al., 1992) and density discontinuities within the water column (MacIntyre et al., 1995) favour aggregation of particles. As shown by Wolanski et al. (1989), mixing across the so-called *lutocline* (the interface between clear water on top and a fluid-mud mixture underneath) is complex. Micro-organisms which are responsible for the characteristics of the particulate matter at the lutocline must have a significant influence on these processes.

A unique situation was observed by Wüest (1994) in meromictic Lago Cadagno, a small lake (0.27 km^2, max. depth 21 m) located on the southern slope of the Swiss Alps at an altitude of 1,923 m. In the stratified water column of this lake a biologically-controlled sharp interface is observed at a depth of 13.5 m. Subsurface springs containing large concentrations of sulphate and other solutes resulting from the weathering of gypsum in the drainage basin make the hypolimnetic water dense whereas the surface water is controlled by the input of relatively fresh water. The two water bodies are separated by a rather homogeneous interface which is sometimes nearly one meter thick (Figure 15). In this layer turbidity is high and the water has a reddish colour. Water temperature and salinity are remarkably constant, in spite of the large gradients existing just above and below the layer. The interface is densely populated by phototrophic sulfur bacteria which profit from the spatial overlapping of sulfide and molecular oxygen.

The question arises of how the interface can be kept homogeneous in spite of the eroding action of diffusion at its boundaries. Wüest (1994) estimated that diffusion should destroy the homogeneity of the layer within less than one day. Double diffusion is also excluded as a mixing agent since both temperature and salt act to stabilize the water

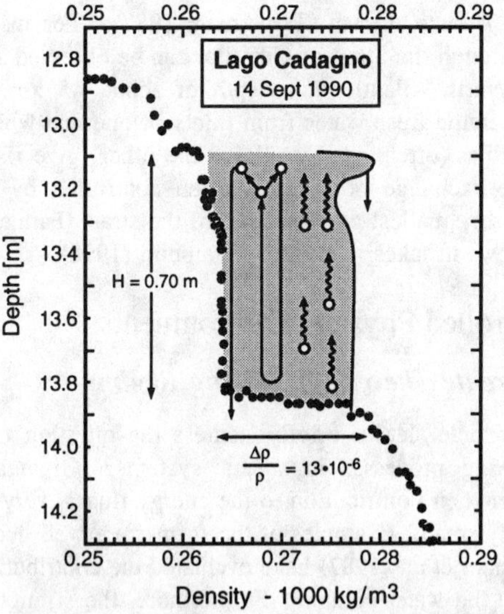

Figure 15. Convective mixing induced by sinking bacteria produces a layer of fairly homogeneous temperature at the interface between the less dense oxic surface water and the denser anoxic bottom water of a small Alpine lake. From Wüest (1994).

column. The same is true for the production of turbulence by the active movement of the bacteria, since the spatial scale of the induced water motion is far too small. The only remaining mechanism is convection. If the concentration of bacteria at the upper boundary of the interface becomes high enough, the resulting excess density induces buoyancy-driven vertical motion. The sinking water parcels are stopped at the distinct density gradient at the lower boundary of the layer, thus keeping this interface very sharp. However, growth of the bacteria at the upper boundary alone cannot produce the necessary buoyancy to keep the layer homogeneous. The bacteria must actively swim to the upper boundary from where they sink again, thus producing the necessary mixing energy.

From an ecological point of view, the extension of the chemical interface over several decimeters is certainly an advantage for the microbial population. It enlarges the water volume in which the bacteria find the right chemical ingredients for the redox reaction supporting their energy needs. Thus, one may ask the question whether the phenomenon described is just an odd side-effect of nature or whether in some cases biology is capable of actively shaping its physical environment according to its needs. The *Gaia Hypothesis* (Lovelock and Margulis, 1973; Lovelock, 1995) is centred around the idea that the latter possibility is correct - at least sometimes. For somebody who is merely interested in "hard" physical or chemical facts, such ideas must remain purely speculative. Yet, ideas such as the Gaia Hypothesis do teach us how to look at facts from different angles by combining knowledge from different disciplines.

Acknowledgments. This article greatly profited from investigations on deep-water formation in Lake Baikal, a project of the Baikal International Center of Ecological Research (BICER), conducted by the Environmental Physics group at EAWAG/ETH in close cooperation with the Limnological Institute at Irkutsk. The principal Russian investigators are M. Grachev, M. N. Shimaraev and N. Granin, and the main Swiss participants are R. Kipfer, D. M. Imboden, R. Hohmann, F. Peeters, G. Piepke, M. Schurter, M. Gloor, A. Wüest, and L. Sigg. The Swiss financial contribution to BICER was provided by the Swiss Federal Office of Science and Education (BBW), the Swiss Federal Institute of Technology (ETH), and the Swiss Federal Institute of Environmental Science and Technology (EAWAG). Furthermore, two unknown reviewers helped to improve an earlier version of this article. Last but not least I thank David Livingstone for his skilful editorial support.

References

Aeschbach-Hertig, W., Helium und Tritium als Tracer für physikalische Prozesse in Seen, Diss. ETH Zürich, Nr. 10714, 1994.

Aeschbach-Hertig, W., R. Kipfer, M. Hofer, and D. M. Imboden, Density-driven exchange between the basins of Lake Lucerne (Switzerland) traced using the ^3H-^3He method, *Limnol. Oceanogr., 41*, 707-721, 1996.

Anati, D. A., and M. Stiller, The post-1979 thermohaline structure of the Dead Sea and the role of double-diffusive mixing, *Limnol. Oceanogr., 36*, 342-354, 1991.

Chen, C. -T. A, and F. J. Millero, Precise thermodynamic properties for natural waters covering only the limnological range, *Limnol. Oceanogr., 31(3)*, 657-662, 1986.

Craig, H., Lake Tanganyika geochemical and hydrographic study, 1973 expedition, Scripps Inst. Oceanography Reference Series, La Jolla, CA, USA, 75-51, 1975.

Culver, D. A., Biogenic meromixis and stability in a soft-water lake, *Limnol. Oceanogr., 22*, 667-686, 1977.

Davison, W., Conceptual models for transport at a redox boundary, in *Chemical Processes in Lakes*, edited by W. Stumm, Wiley, New York, pp 31-53, 1985.

Dickey, T. D., and J. J. Simpson, The influence of optical water type on the diurnal response of the upper ocean, *Tellus, 35B*, 142-154, 1983.

Farmer, D. M., and L. Armi, Maximal two-layer exchange over a sill and through a combination of a sill and contraction with barotropic flow, *J. Fluid Mech., 164*, 53-76, 1986.

Farmer, D. M., G. B. Crawford, and T. R. Osborn, Temperature and velocity microstructure caused by swimming fish, *Limnol. Oceanogr., 32*, 978-983, 1987.

Gill, A. E., *Atmosphere-Ocean Dynamics*, Academic Press, 662 pp, 1982.

Hakanson, L., and M. Jansson, *Principles of Lake Sedimentology*, Springer, Berlin, 316 pp, 1983.

Hamblin, P. F., Exchange flows in lakes, in *Physical Processes in Lakes and Oceans*, edited by J. Imberger, Coastal and Estuarine Studies, pp 187-198, AGU, this volume, 1998.

Hohmann, R., R. Kipfer, F. Peeters, G. Piepke, M. Schurter, and D. M. Imboden, Processes of deep water renewal in Lake Baikal, *Limnol. and Oceanogr.*, to be published, 1996.

Imboden, D. M., and R. Gächter, The impact of physical processes on the trophic state of a lake, in *Biological Aspects of Freshwater Pollution*, edited by O. Ravera, Pergamon Press, pp 93-110, 1979.

Imboden, D. M., Restoration of a Swiss lake by internal measures: can models explain reality, in *Lake Pollution and Recovery*, European Water Pollution Control Association, 91-102, 1985.

Imboden, D. M., B. Stotz, and A. Wüest, Hypolmnic mixing in a deep alpine lake and the role of a storm event, *Verh. Internat. Verein. Limnol., 23*, 67-73, 1988.

Imboden, D. M., Mixing and transport in lakes, in *Large Lakes. Ecological Structure and Function*, edited by M. M. Tilzer and C. Serruya, Springer, Berlin, pp 47-80, 1990.

Imboden, D. M., and A. Wüest, Mixing mechanisms in lakes, in *Physics and Chemistry of Lakes*, edited by A. Lerman, D. M. Imboden, and J. R. Gat, Springer, Heidelberg, pp 83-138, 1995.

Karagounis, I., J. Trösch, and F. Zamboni, A coupled physical-biochemical lake model for forecasting water quality, *Aquatic Sci., 55*, 87-102, 1993.

Kerr, R. C., Erosion of a stable density gradient by sedimentation-driven convection, *Nature, 353*, 423-425, 1991.

Lovelock, J. E., and L. Margulis, Atmospheric homeostasis by and for the biosphere: the gaia hypothesis, *Tellus 26*, 2-10, 1973.

Lovelock, J. E., *The ages of GAIA*, Oxford University Press, 2nd edition, 255 pp, 1995.

MacIntyre, S., A. L. Alldredge, and C. C. Gotschalk, Accumulation of marine snow at density discontinuities in the water column, *Limnol. Oceanogr., 40*, 449-468, 1995.

McDougall, T. J., The relative roles of diapycnal and isopycnal mixing on subsurface water mass conversion, *J. Phys. Oceanogr., 14*, 1577-1589, 1984.

McManus, J., R. W. Collier, C. -T. A. Chen und J. Dymond (1992): Physical properties of Crater Lake, Oregon: A method for the determination of a conductivity- and temperature-dependent expression for salinity. *Limnol. Oceanogr. 37*, 41-53.

Meybeck, M., Global distribution of lakes, in *Physics and Chemistry of Lakes*, edited by A. Lerman, D. M. Imboden, and J. R. Gat, Springer, Heidelberg, pp 1-35, 1995.

Millard, R. C., Owens, W. B. and Fofonoff, N. P., On the calculation of the Brunt-Väisälä frequency, *Deep-Sea Res.*, 37, 167-181, 1990.

Newman, F. C., Temperature steps in Lake Kivu: A bottom heated saline lake, *J. Phys. Oceanogr., 6*, 157-163, 1976.

Niessen, F., and M. Sturm, Die Sedimente des Baldeggersees (Schweiz) – Ablagerungsraum und Eutrophierungsentwicklung während der letzten 100 Jahre, *Arch. Hydrobiol. 108*, 365-383, 1987.

Peeters, F., G. Piepke, R. Kipfer, R. Hohmann, and D. M. Imboden, Description of stability and neutrally buoyant transport in freshwater lakes, *Limnol. Oceanogr.*, 41, 1711-1724, 1996.

Quadfasel, D., H. Kudrass, and A. Frische. Deep-water renewal by turbidity currents in the Sulu Sea, *Nature, 348*, 320-322, 1990.

Shimaraev, M. N., N. G. Granin, and A. A. Zhdanov, Deep ventilation of Lake Baikal waters due to spring thermal bars, *Limnol. Oceanogr., 38*, 1068-1072, 1993.

Steinhorn, I., The disappearance of the long-term meromictic stratification of the Dead Sea, *Limnol. Oceanogr., 30*, 451-462, 1985.

Stolzenbach, K. D., K. A. Newman, and C. S. Wong, Aggregation of fine particles at the sediment-water interface, *J. Geophys. Res., 97*, C11, 17889-17898, 1992.

Sturm, M, and A. Matter, Turbidites and varves in Lake Brienz (Switzerland): deposition of clastic detritus by density currents, *Spec. Publs int. Ass. Sediment*, 2, 147-168, 1978.

Tilzer, M. M., Environmental and physiological control of phytoplankton productivity in large lakes, in *Large Lakes. Ecological Structure and Function*, edited by M. M. Tilzer and C. Serruya, Springer, Berlin, pp 339-367, 1990.

Tziperman, E., The Mediterranean outflow as an example of a deep buoyancy-driven flow, *J. Geophys. Res., 92*, C13, 14510-14520, 1987.

Ulrich, M. M., D. M. Imboden, and R. P. Schwarzenbach, MASAS - A user-friendly simulation tool for modeling the fate of anthropogenic substances in lakes, *Envir. Software, 10*, 177-198, 1995.

Van Senden, D. C., R. Portielje, A. Borer, H. Ambühl, and D. M. Imboden, Vertical exchange due to horizontal density gradients in lakes; the case of Lake Lucerne, *Aquatic Sci., 52*, 381-398, 1990.

Weiss, R. F., E. C. Carmack, and V. M. Koropalov, Deep-water renewal and biological production in Lake Baikal, *Nature 349*, 665-669, 1991.

Wolanski, E., T. Asaeda, and J. Imberger, Mixing across a lutocline, *Limnol. Oceanogr., 34*, 931-938, 1989.

Wüest, A., D. M. Imboden, and M. Schurter, Origin and size of hypolimnic mixing in Urnersee, the southern basin of Vierwaldstättersee (Lake Lucerne), *Schweiz. Z. Hydrol., 50*, 40-70, 1988.

Wüest, A., Interaktionen in Seen: Die Biologie als Quelle dominanter physikalischer Kräfte, *Limnologica, 24*, 93-104, 1994.

Wüest, A., G. Piepke, and J. D. Halfman, Combined effects of dissolved solids and temperature on the density stratification of Lake Malawi (East Africa), in *The Limnology, Climatology and Paleoclimatology of the East African Lakes*, edited by T. Johnson, pp 183-202, 1996.

Wüest, A., and M. Gloor, Bottom boundary mixing in lakes: The role of the near-sediment density stratification, in *Physical Processes in Lakes and Oceans*, edited by J. Imberger, Coastal and Estuarine Studies, pp 485-502, AGU, this volume, 1998.

40

Hydrodynamic vs. Non-Hydrodynamic Influences on Phosphorus Dynamics During Episodic Events

Y. P. Sheng, X. Chen and S. Schofield

Abstract

Hydrodynamics and sediment dynamics in Lake Okeechobee have been studied by Sheng et al. (1989) and Sheng (1993). Sheng (1993) analyzed field data of suspended sediment and phosphorus concentrations collected in 1988 and 1989 and suggested the possible role of sediment resuspension on phosphorus resuspension. Sheng (1993) estimated the rate of phosphorus resuspension from the rate of sediment resuspension and measured phosphorus concentration, and found the resuspension flux of phosphorus to be much greater than that of sediments. This paper presents results of a more recent field and modeling study aimed to quantify the roles of hydrodynamic vs. non-hydrodynamic factors on phosphorus and phytoplankton dynamics during an episodic event on February 2–4, 1993 in Lake Okeechobee. Continuous data of hydrodynamics, suspended sediment concentration, DO, pH, and phosphorus and chlorophyll-*a* concentrations were collected from 1–3 vertical levels at a fixed platform in the lake. Results indicated that during the first day of the event, strong wave-induced turbulence caused significant resuspension of bottom sediments, which led to elevated bottom concentrations of phosphorus and chlorophyll-*a*. During the second day, although the lake was relatively calm with little or no resuspension of sediments, there were significant temporal fluctuations in phosphorus concentrations. Using the data and a comprehensive one-dimensional modeling system of hydrodynamics, sediment transport, and water quality dynamics, we found that the observed temporal fluctuations in phosphorus were significantly influenced by temporal variations in DO and pH. Peaks in phosphorus concentrations on the second day were caused by the delayed desorption of SRP (Soluble Reactive Phosphorus) from suspended sediments due to lowered DO and higher pH. Fluctuations in phosphorus did not coincide with fluctuations in TSS (Total Suspended Solids), thus confirming that desorption and adsorption are not in equilibrium during episodic events.

Introduction

Lake Okeechobee (Figure 1) is a large shallow lake in south Florida, U.S.A. A series of studies have been conducted to quantify the nutrient loading into the lake from various sources: internal loading from bottom sediments, external loading from rivers and canals, flux from vegetation zone to open water, and atmospheric loading. These studies included

Physical Processes in Lakes and Oceans
Coastal and Estuarine Studies Volume 54, Pages 613-622

the wind-driven circulation (Sheng and Lee, 1991; Sheng, 1993), wind waves (Ahn and Sheng, 1990), suspended sediment transport (Sheng, 1989; Sheng et al., 1989) and water quality dynamics (Sheng, 1993). Sheng (1993) showed that phosphorus concentration and suspended sediment concentration in Lake Okeechobee are significantly correlated, and resuspension of sediments can lead to increased phosphorus concentration in the water column. A field and modeling study was thus designed to capture the dynamics of suspended sediments and phosphorus during episodic events. The primary purpose of this study was to determine quantitatively the influence of hydrodynamic vs. non-hydrodynamic factors on phosphorus dynamics. Another purpose was to examine the validity of the "equilibrium partitioning" concept which assumes that desorption and adsorption are in equilibrium.

A platform as shown in Figure 2 was used for the Lake Okeechobee study at a site near Station A in Figure 1 where the depth is about 350 cm. Continuous sampling of wind and hydrodynamic data (pressure, currents, suspended sediment concentration, and temperature) started in November 1992. On February 2, 1993, the monitoring of an episodic event commenced with simultaneous water quality data collection, in addition to the hydrodynamic and meteorological data. The instruments used for the field study consist of one real time data unit, three Seabird water temperature probes, three Marsh McBirney bi-directional current meters, three Downing turbidity measurement (OBS) probes, one Transmetrics pressure transducer, one R.M. Young wind anemometer, plus one Tattletale data logger. The real-time data unit was programmed to collect data from each instrument for a 10 minute period, every 15 minutes, at a one second bursting rate. The average for the 10 minute run was then calculated and stored for future retrieval. The instruments were mounted on three separate arms: Arm 1 (lower arm) at 30 cm above the bottom, Arm 2 (middle arm) at 127 cm above the bottom, and Arm 3 (upper arm) at 295 cm from the bottom. The pressure transducer was located at 10 cm above Arm 3. For sampling during the 3-day episodic event, a multi-parameter monitoring instrument (Hydrolab Datasonde-

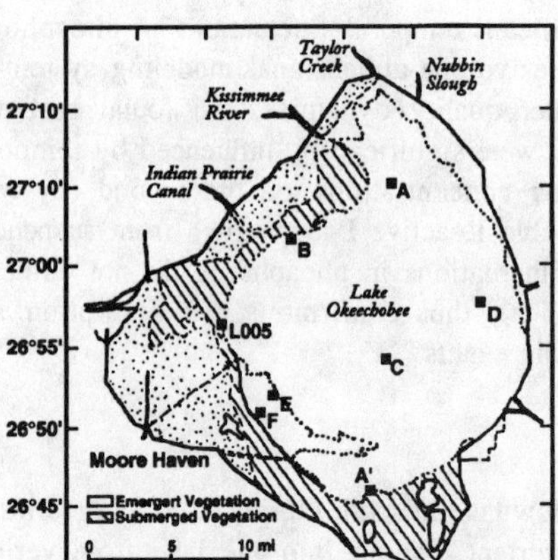

Figure 1. Map of Lake Okeechobee, Florida.

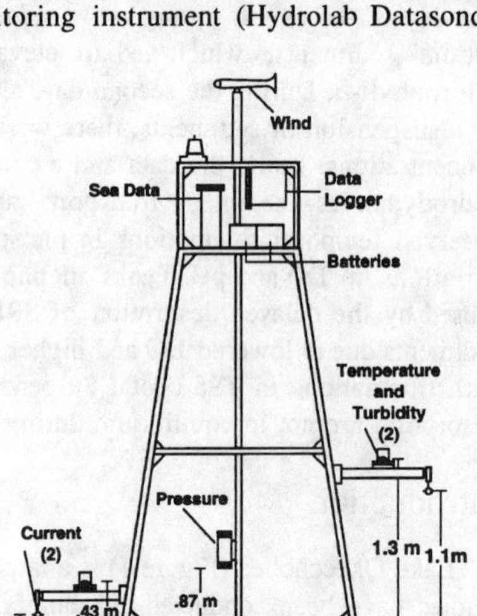

Figure 2. Instrument platform used for the Lake Okeechobee study. Instruments and their locations differ from the figure and are explained in the text.

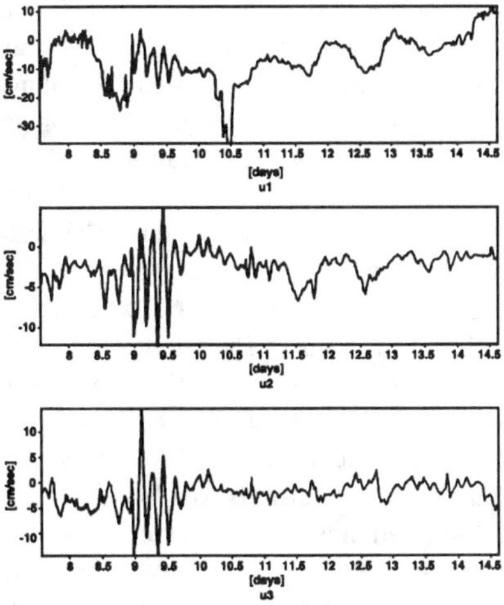

Figure 3. Measured north-south horizontal velocity at the Lake Okeechobee platform during January 7–14, 1993. Top panel: lower arm; middle panel: middle arm; lower panel: top arm.

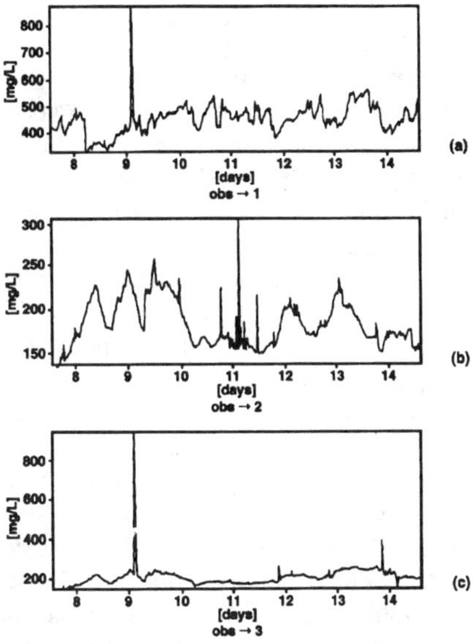

Figure 4. Measured suspended sediment concentration at the lower arm (top panel), middle arm (middle panel), and top arm (lower panel) at the Lake Okeechobee platform during January 7–14, 1993

3) was used to record the water quality parameters (DO, pH, etc.). A Sigma water sampler was also deployed to sample water at three levels (the same height as the instrument platform arms) every hour for the first 24 hour period, and every two hours thereafter for the remaining 40 hours.

Fifteen-Minute Averaged Data During January 7–14, 1993

As an example, some fifteen-minute averaged data during January 7–14, 1993 are presented here. Wind speed and direction during January 7–14, 1993 were generally between 0–7 m s^{-1} from the south during January 7–8, increased to 11 m s^{-1} from the north on January 9, and then decreased to 0–7 m s^{-1} from the southeast on January 10. The energy spectrum of the pressure signal showed noticeable energy peaks around 0.2–0.25 hr^{-1} which correspond to seiche motion in Lake Okeechobee with periods between 4–5 hrs. The predominant energy peaks occurs at 0.09 hr^{-1} corresponding to an 11.1-hour wave. Horizontal water currents were calculated and referenced to magnetic north. Figure 3 shows the north-south and east-west currents during January 7–14, 1993. As shown in Figure 4, the suspended sediment concentration data measured by the OBS probes at three vertical levels have similar general trends but with noticeable vertical stratification. During this period, there was insignificant thermal stratification at the platform. The lack of correlation between the suspended sediment concentration data and the mean current data indicates that seiche oscillation is not the cause for sediment resuspension. Cross-correlations between the wind speed and the three OBS probes showed that the three OBS readings lagged the wind by 0.25 hrs, 1 hr, and 1.5 hrs, from the bottom to the top levels.

Episodic Event Data During February 2–4, 1993

Fifteen-minute averaged data as well as 2-Hz bursting data were collected during the episodic event. The bursting data were collected during the first 24 hours. The real-time data unit recorded 2-Hz time series of the pressure and currents during the time of high wave activity, on the hour and half-hour. The 15-minute averaged data were recorded 15 minutes after the hour and 15 minutes before the hour. The real-time data unit program was changed and 2-Hz data logging restarted by 13:25 on February 2. The Hydrolab Datasonde-3 was initialized and started collecting samples at 13:00 with a 15-minute collection interval. Winds at the site during set-up were NE 4–5 m s^{-1} with seas less than 30 cm. The water sampling from the Sigma samplers was started on February 2, 1993 at 16:30 with a sampling interval of 1 hour. At the start of sampling the winds were 7–10 NE with seas at 30–100 cm. The first set of water samples were collected at 23:00. The wind at this time had begun to decrease to 5–7 m s^{-1}, with seas decreasing to 30–60 cm. The next set of samples were collected at 7:30 a.m. on February 3, 1993. The winds shifted to the east at 5 m s^{-1}, with seas of 30–60 cm. On February 3 at 16:30 the next set of samples were collected. Winds and seas at this time were similar to the conditions on February 2 at 23:00. At 6:00 a.m. on February 4, the next set of water samples were collected. The wind was from the east at 5–10 knots with seas of 30 cm. After 9:00 on February 4 the winds were calm with seas less than 30 cm. At 17:30, the last set of water samples were collected. The water sampling was stopped on February 5 at 8:30 a.m. Wind speed and direction during February 2–4, 1993 are shown in Figure 5. Figure 6 shows the pressure time series and energy spectra at 18:00 on February 2. The energy spectrum shows narrow bands centered around 0.348 Hz corresponding to a wave period of 2.86 sec. The significant wave height (SWH) increased from 21.2 cm at 14:00 to 37 cm at 17:00 and then decreased slightly to 34. 6 cm at 18:00.

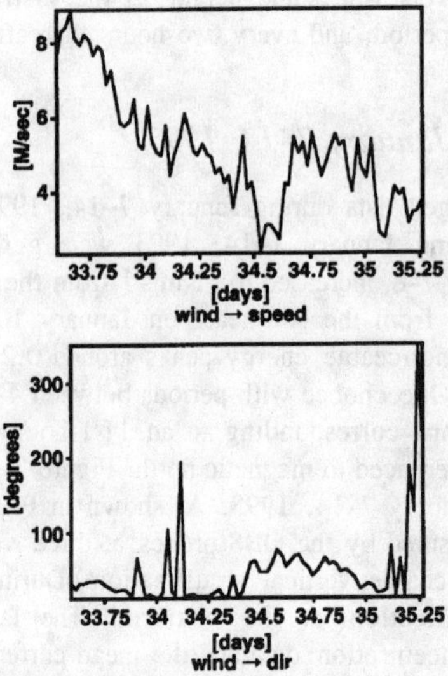

Figure 5. Measured wind speed (top panel) and wind direction (lower panel) at the platform in Lake Okeechobee during February 2–4, 1993.

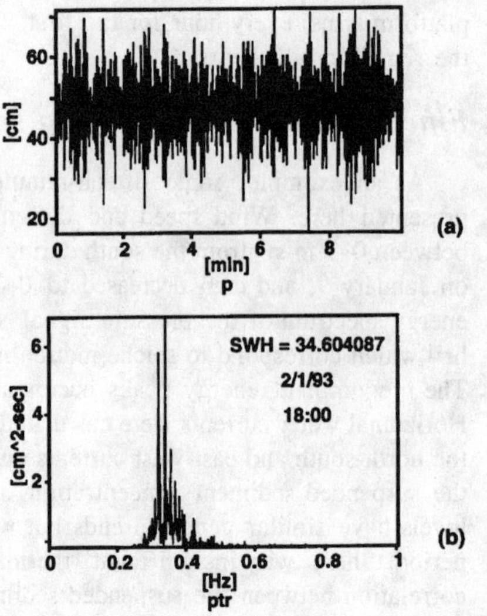

Figure 6. Bursting pressure time series (top panel) and energy spectrum (lower panel) at the platform during 18:00 on February 2, 1993.

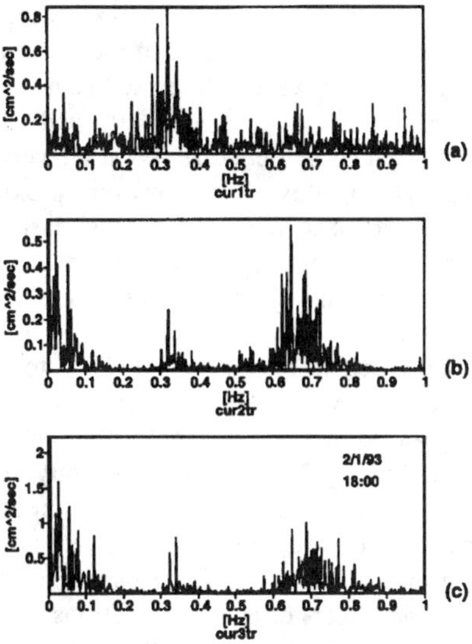

Figure 7. Current energy spectra at lower arm (top panel), middle arm (middle panel), and top arm (lower panel) at the platform during 18:00 on February 2, 1993..

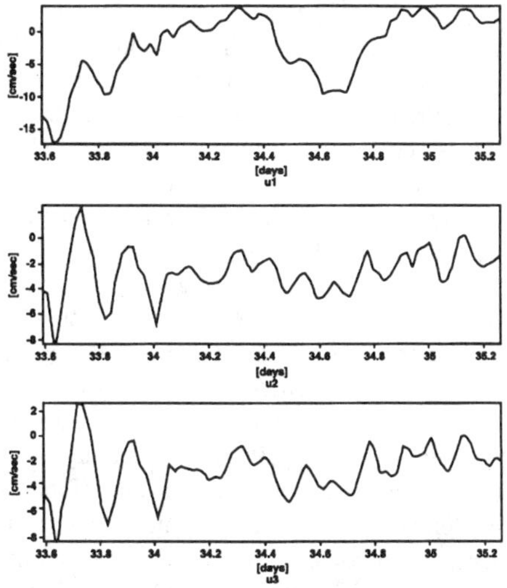

Figure 8. Measured north-south current at the lower arm (top panel), middle arm (middle panel), and top arm (lower panel) at the platform during February 2–4, 1993.

Current meter spectra at 18:00 are shown in Figure 7. The current spectra at the middle and top arms are similar and show narrow bands centered around 0.38 Hz corresponding to a wave period of 2.86 sec, similar to the pressure spectrum. The spectra at middle and top arms also show significant energy around 0.05 Hz (20 sec) and 0.83 Hz (1.2 sec). The bottom current spectra at 14:00 show energy content at all frequencies. Near the peak of the event at 18:00, the bottom current spectra show a dominant frequency at 0.38–0.4 Hz. At the peak of the event, energy associated with the 2.86 sec wave was the strongest at the bottom arm and the weakest at the top arm, suggesting the generation of bottom turbulence by orbital motion associated with the 2.86 sec wave. The north-south currents during February 2–4 are shown in Figure 8. Suspended sediment concentrations during this time are shown in Figure 9. Again, there is a lack of correlation between the mean currents and the suspended sediment concentration, confirming that seiche oscillation is not the cause for sediment resuspension. Figures 10 and 11 show measured concentrations of total phosphorus (TP) and soluble reactive phosphorus (SRP), respectively, during the episodic event. In both figures, the top panel show data measured at the top arm, the middle panel show data measured at the middle arm, and the bottom panel show data measured at the bottom arm. The TP, TDP and SRP concentrations are well correlated, as was found in the 1988 and 1989 data (Sheng, 1993).

Factors Influencing the Observed Phosphorus Dynamics

As shown in Figure 9, at the beginning of the 3–day episodic event (between 14:00 and 19:00 on February 2 which is Julian day 33), there were significant increases in suspended sediment concentration at all three vertical levels. Figures 10 and 11 show that the bottom phosphorus concentrations peaked at around 21:00. The chlorophyll-*a* concentration, however, peaked around midnight. The bottom phosphorus concentration peaked before those at the middle and top levels. These suggest that wave-induced sediment resuspension caused resuspension of phosphorus and chlorophyll-*a*. Between 00:00 and 19:00 on February 3 (Julian day 34), suspended sediment concentration decreased at all vertical levels, due to reduced wind and increased settling. During this time, however, peaks in phosphorus concentration occurred at the top level first and the bottom level last, due apparently to non-hydrodynamic processes, e.g., those related to DO, pH, and ionic concentration. The measured DO and pH at the bottom arm indicated significant fluctuations in DO and pH during the episodic event. Between Julian day 33.5 to 34.5, DO initially jumped from 6.5 to 7.5, then steadily decreased from around 7.5–7.75 to 6.5. During the first half day (33.5 to 34), pH went from 8.3 down to 8.2 at 4 p.m., went back to 8.3 at 5 p.m then down to 8.25 at 6 p.m., then peaked to 8.45 around 10 p.m., until it dropped to 8.35 at midnight. It stayed around 8.4 during day 34 and dropped slightly to 8.3 on day 35. During 19:00 on February 3 to early February 4, another resuspension event took place to cause resuspension of phosphorus and chlorophyll-*a*. Our experience in modeling phosphorus dynamics suggests that desorption of inorganic phosphorus from sediment particles is a major process controlling the phosphorus concentrations. Since temperature was over 17 deg Celsius, DO and pH are major factors controlling the release of phosphorus from sediment particles. Numerous studies (Pomeroy et al., 1965; Ishikawa and Nishimura, 1989) showed that increased DO could suppress the release of phosphate from sediment particles. In order to determine the effect of DO on the observed dynamics of suspended sediments and phosphorus, we can examine the correlation between SRP release and DO in terms of the following correlation coefficients among phosphorus

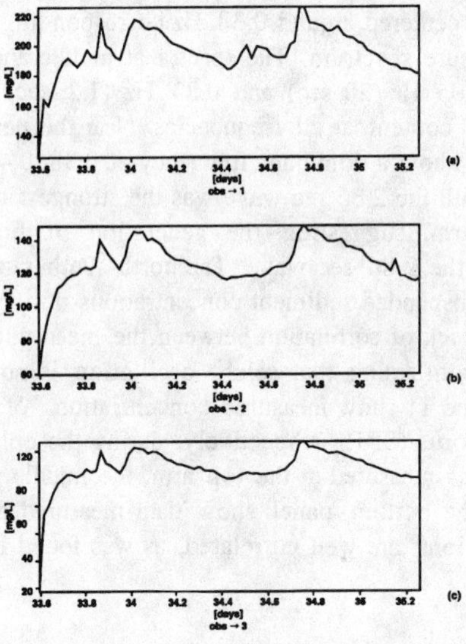

Figure 9. Measured suspended sediment concentration at the lower arm (top panel), middle arm (middle panel), and top arm (lower panel) during February 2–4, 1993.

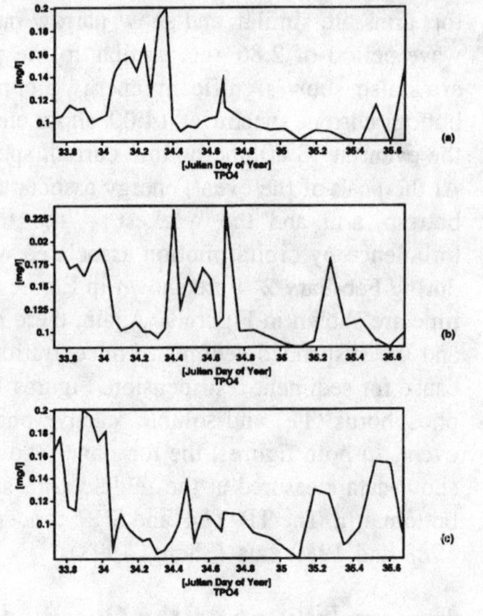

Figure 10. Measured total phosphorus concentration at the top arm (top panel), middle arm, and bottom arm (lower panel) during February 2–4, 1993.

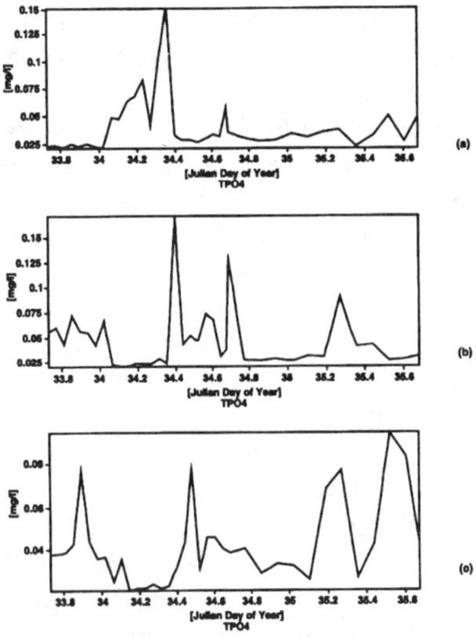

Figure 11. Measured soluble reactive phosphorus concentration at the top arm (top panel), middle arm, and bottom arm (lower panel) during February 2-4, 1993.

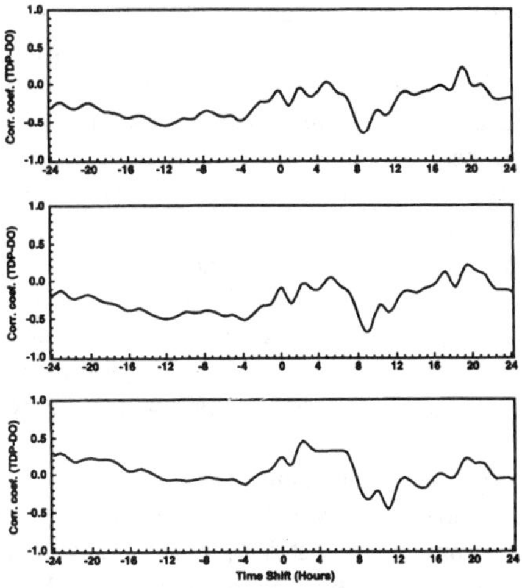

Figure 12. Correlation coefficients between DO and SRP , DO and TDP, and DO and TP at the Lake Okeechobee station during February 2-4, 1993.

concentration and DO concentration. Figure 13 shows the correlation coefficients between DO and SRP, DO and TDP, and DO and defined as:

$$\rho_{DO-P} = \frac{Cov[P(t),\ DO(t-h)]}{\sqrt{Var[P(t)]\ Var[DO(t-h)]}} \tag{1}$$

where P stands for SRP, TDP, or TP and $\rho_{DO\text{-}P}$ stands for $\rho_{DO\text{-}SRP}$ (top panel), $\rho_{DO\text{-}TDP}$ (middle panel), and $\rho_{DO\text{-}TP}$ (bottom panel) in Figure 14, respectively. Cov(x, y) is the covariance between x and y, respectively. As shown in Figure 12, at about 8 hours, $\rho_{DO\text{-}SRP}$ and $\rho_{DO\text{-}TDP}$ are about -0.7. This suggests that the decrease of DO precedes the increase of SRP and TDP by about 8 hours. During an episodic event, if the desorption of SRP is indeed due to the decrease in DO, then the desorption process would take about eight hours to complete. Olila and Reddy (1993) showed that as pH exceeds 7.5, the partition coefficient for phosphorus is increased as a nonlinear function of the deviation of pH above 7.5. Although there are insufficient data of DO, pH, and ferrous iron over the entire water column, we conducted numerical experiments on the observed phosphorus dynamics in Lake Okeechobee to confirm the effects of DO and pH on the desorption process. The numerical model is a comprehensive one-dimensional modeling system (Sheng, 1994; Chen and Sheng, 1994) which includes a hydrodynamics model, a sediment transport model, and a phosphorus dynamics model over the entire water column and sediment column. The one-dimensional model is valid here because the effect of horizontal advection is much less important than that of vertical turbulent mixing. The effect of so-called patchiness will be discussed in more detail in a future paper.

The one-dimensional modeling system was used to simulate the hydrodynamics, sediment transport processes, and phosphorus dynamics for the February 2–4 episodic event in Lake Okeechobee (Sheng, et al., 1993). SMB (Sverdrup-Munk-Bretschneider) wave model was used to calculate the significant wave heights and periods at various locations in the lake by using realistic wind and bathymetry data (Ahn and Sheng, 1990). In the 1-D model simulation, at every time step, results of the SMB model are first used to calculate the bottom shear stresses due to waves or wave-current interactions, before the sediment dynamics are calculated. In calculating the bottom shear stress due to waves or wave-current interactions, a very small time step (1/60 to 1/20 of the wave period) was used over a simulation period of at least four wave cycles. Also calculated during this process are the wave-cycle-averaged eddy viscosity/diffusivity, which represents the combined turbulence mixing due to waves and currents in the wave boundary layer. These wave-cycle-averaged eddy viscosity and diffusivity are used in calculating currents, sediment concentration and phosphorus dynamics at the new time step. Each of the one dimensional equations of motion for the horizontal velocities include a tendency term, a pressure gradient term, and a turbulent stress term. The pressure gradient includes a slowly varying part, which are related to the seiche and wind-driven circulation in the lake, and a fluctuating part, which varies within a wave cycle. The turbulent eddy viscosity in the equations is calculated by either the equilibrium second-order closure model or the TKE (Turbulent Kinetic Energy) closure model (Sheng and Villaret, 1989). Since the model computes the currents and turbulence over the entire water column using a time step generally 1/60 to 1/20 of the wave period, nonlinear current-wave interaction is thus directly computed without resorting to ad-hoc parameterization (Sheng et al., 1994). The

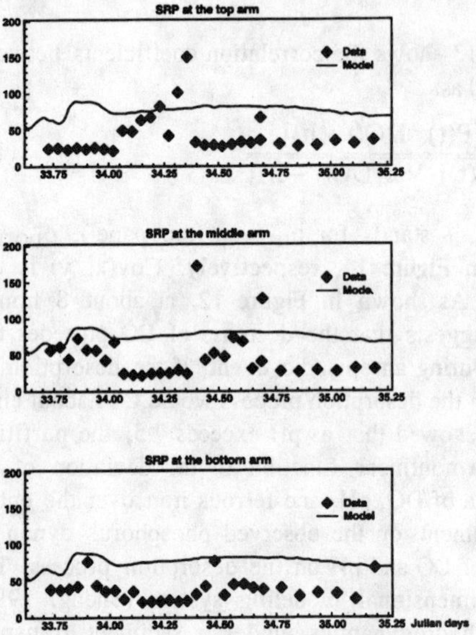

Figure 13. Measured and simulated SRP at the top arm (top panel), middle arm (middle panel), and bottom arm (bottom panel) at the Lake Okeechobee station during February 1993. Effects of DO and pH on desorption of SRP are not considered.

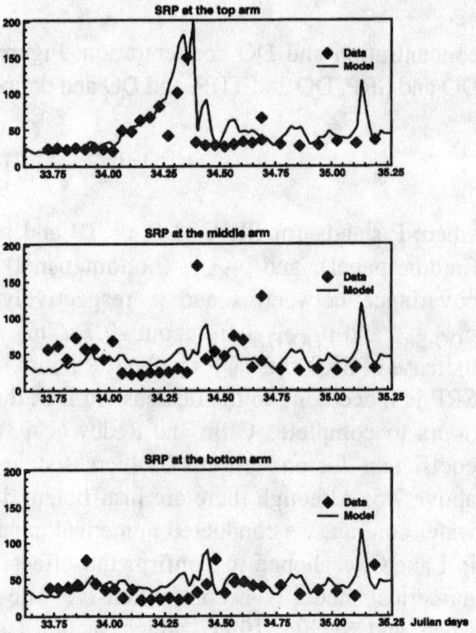

Figure 14. Measured and simulated SRP at the top arm (top panel), middle level (middle panel), and bottom level (bottom panel) at the Lake Okeechobee station during February 1993. Effects of DO and pH on SRP release are considered.

sediment model includes the processes of settling, vertical mixing, deposition, resuspension, and flocculation. Erosion/resuspension rate is determined to be a function of the excess bottom shear stress (Sheng et al., 1989). Deposition is modeled as a fluid dynamic process (Sheng 1986). Flocculation is parameterized terms of a settling velocity which varies with suspended sediment concentration and turbulence level (Chen and Sheng 1994).

The simulation of currents and suspended sediment concentrations for this storm event produced results which are almost identical to the data. The simulated sediment concentrations at Arms 2 and 3 have the same trend and comparable magnitude as measured data. The model parameters used in the sediment model are: the critical shear stress for erosion τ_{cr} = 1 dyn/cm^2 and the erosion rate constant $E_0 = 2.5 \times 10^{-7}$g/cm^2 sec. For the simulation of phosphorus dynamics, we solved the dynamic equations of soluble reactive phosphorus (SRP), dissolved organic phosphorus (DOP), algal particulate phosphorus (Alg-P), zooplankton particulate phosphorus (Zoo-P), particulate organic phosphorus (POP), and particulate inorganic phosphorus (PIP). We considered the entire water column and the entire sediment column for the water quality model. Thus, the benthic fluxes (due to diffusion and resuspension) at the sediment-water interface becomes a model output instead of having to be prescribed if the water column and sediment column were de-coupled from each other. Turbulent mixing is included in the equations for the dissolved species, while settling and turbulent mixing are included in the equations for the particulate species. Desorption and adsorption are modeled as kinetic (rather than equilibrium) processes. To simulate the February 1993 event, initially we did not consider the effects of pH and DO on SRP desorption. The results agreed very poorly with the data (Figure 13). We then considered the effect of pH on SRP desorption by an empirical equation, which is obtained by fitting the laboratory data of Olila and Reddy (1993) for Lake Okeechobee sediments:

$$P_{ip}(pH) = P_{ip}(10.5)\frac{(pH-7.5)^3}{64} \qquad (2)$$

where P_{ip} (10.5) is the partition coefficient at pH of 10.5. This improved the results somewhat but the effect of pH in the range of 8.2 to 8.5 is not very significant. After we added the following effect of DO on SRP desorption:

$$P_{ip}(DO) = \left(\frac{(DO_m)}{DO_m - DO}\right)^n P_{ip}(0) \qquad (3)$$

where Do_m is a model parameter having the same unit as DO and P_{ip} (0) is the partition coefficient at anaerobic condition (DO = 0), the model results improved significantly as shown in Figure 14.

Conclusions

Sediment resuspension and desorption kinetics of inorganic phosphorus play important roles in phosphorus dynamics in Lake Okeechobee. Current data and suspended sediment concentration data confirmed the previous finding (Sheng, 1993) that seiche oscillation, which significantly influences the currents, is not a major cause for sediment resuspension. During the first few hours of the February 1993 event, the high SRP values at the bottom and the middle arms were obviously due to wave-induced resuspension. On the second day with relatively calm weather conditions, desorption of SRP from sediments due to reduced DO and increased pH apparently caused significant peaks and fluctuations in phosphorus concentrations. Thus, phosphorus dynamics can be significantly affected by hydrodynamic as well as non-hydrodynamic factors, although the relative importance of these factors can vary with time during the episodic event. The one-dimensional model simulation of the episodic event strongly suggest the significant influence of DO and pH on phosphorus dynamics during the quiescent period. "Equilibrium partitioning", which assumes that desorption and adsorption are instantaneous equilibrium, is not valid during the episodic event.

References

Ahn, K. and Y. P. Sheng, Wind-wave hindcasting and estimation of bottom shear stress in Lake Okeechobee, Technical Report, Univ. of Florida, Gainesville, 1990.

Chen, X. and Y. P. Sheng, Effects of Hydrodynamics and Sediment Dynamics on Phosphorus Dynamics in Shallow Lakes and Estuaries, Technical Report, Univ. of Florida, Gainesville, 1994.

Ishikawa, M. and H. Nishimura, Mathematical model of phosphorus release rate from sediments considering the effect of dissolved oxygen in overlying water, *Water Res., 23*, 351-359, 1989.

Olila, O. G. and K. R. Reddy, Influence of pH on phosphorus sorption in Lake Okeechobee, Unpublished manuscript, Soil and Water Science Department, University of Florida, Gainesville, 1993.

Pomeroy, L. R., E. E. Smith, and C. M. Grant, The exchange of phosphorus between estuarine water and sediments, *Limnol. Oceanogr., 10*, 167-172, 1965.

Sheng, Y. P., Predicting the dispersion and fate of contaminated marine sediments, In *Contaminated Marine Sediments Assessment and Remediation*, Marine Board, National Research Council, National Academy Press, pp. 166-177, 1989.

Sheng, Y. P., Hydrodynamics, Sediment Transport and Their Effects on Phosphorus Dynamics in Lake Okeechobee, In *Coastal and Estuarine Studies 42 Nearshore and Estuarine Cohesive Sediment Transport*, edited by A. J. Mehta, pp. 558-571, AGU, Washington, DC, 1993.

Sheng, Y. P., Modeling hydrodynamics and water quality dynamics in shallow waters, Keynote Paper *International Symposium on Ecology and Engineering*, Taman Negara, Malaysia, November, 1994.

Sheng, Y. P., and C. Villaret, Modeling the effect of suspended sediment stratification on bottom exchange processes, *Geophys. Res. 95*, 14429- 1444, 1989.

Sheng, Y. P., V. Cook, S. Peene, D. Eliason, P. F. Wang, and S. Schofield, A Field and Modeling Study of Fine Sediment Transport in Shallow Waters, In *Estuarine and Coastal Modeling*, edited by M. L. Spalding, pp. 113-122, ASCE, Newport, RI, 1989.

Sheng, Y. P., X. - J. Chen, S. Schofield, and E. Yassuda, Hydrodynamics and sediment and phosphorus dynamics in lake Okeechobee during an episodic event, Technical Report, Coastal & Oceanographic Engineering Department, University of Florida, Gainesville, 1993.

Sheng, Y. P., X. Chen and E. A. Yassuda, Wave-induced sediment resuspension in shallow waters,- Proc. *International Conference on Coastal Engineering*, Kobe, Japan, ASCE, October 1994.

41

Coupling of Hydrobiology and Hydrodynamics: Lakes and Reservoirs

M. Straškraba

Abstract

Recently accumulated empirical evidence suggests significant differences in the behavior of limnological variables between lakes and reservoirs in the temperate region. The phosphorus retention capacity strongly depends on the theoretical water retention time (RT) but is significantly less in lakes than in reservoirs, particularly when their RT is shorter. The cause seems to be the differences in hydrodynamics with RT between lakes and reservoirs. Studies with the model DYRESM indicate that reservoir limnology is very sensitive to theoretical retention time up to the values of about one year, with the steepest sensitivity range below some 200 days. The hydrodynamic differences are due mainly to the position of the outflow in the two water body types: surface outflow from the lakes and deep outflow from the reservoirs.

A positive feedback exists between phytoplankton development and reservoir hydrodynamics. An increasing amount of phytoplankton results in shallower thermoclines, warmer surface layer and other corresponding differences in hydrodynamics. A simple model of phytoplankton growth is used to elucidate these relations quantitatively. Study with DYRESM indicates that the description of the light attenuation by a simple extinction coefficient does not reproduce well the empirical observations on the dependence between transparency and mixing depth. Mixing depth decreases as extinction increases.

The density dependence of specific photosynthetic rates in phytoplankton algae and the density dependence of respiration in bacteria cause an order of magnitude drop of activities between dense and diluted populations. Possible micro-hydrodynamic processes causing these differences are speculated.

Hydrodynamic effects on zooplankton are listed to provoke discussion and further closer collaboration among physical and biological limnologists.

Introduction

Physical limnology is a neglected field: this is documented is the survey by Bourget and Fortin (1995) who analyzed the subjects covered in the 253 papers published in Limnology and Oceanography in 1980, 1985 and 1990. Only about 8% of the papers

Physical Processes in Lakes and Oceans
Coastal and Estuarine Studies Volume 54, Pages 623-644

covered physical limnology and physical oceanography as contrasted to 30% dealing with chemical topics, 24% with zooplankton and 22% with phytoplankton. The authors also point out that limnology and oceanography are still dominantly descriptive (64% of papers), only 26% being classified by the authors as belonging to the experimental hypothesis based category of papers.

The low percentage of papers on physical limnology does not mean, that physics is unimportant for limnology, as shown in the historical development of limnology: an understanding of stratification, flows, mixing, temperature and light conditions played an important role in limnological progress. This has been particularly evident in standing waters where physical studies have been linked with chemical and biological investigations. More recently progress seems to be connected with the scale and resolution problem: whereas in early times relatively unsophisticated measurements and theory were useful for the macroscale processes, we now run into the period when much more detailed views, microscale processes, highly elaborate instrumentation and theories are necessary for significant progress at the physical and biological level. This makes it difficult for individuals to cope simultaneously with several fields, and biologists are more inclined to relate their processes with chemistry than with physics.

Also, there seems to be a bias in the physical limnology as applied to biological problems: a great deal of attention is devoted to the effect of physical variables and processes on phytoplankton (reviews, e.g., in Spigel and Imberger, 1987 from the hydrophysical point of view and Reynolds, 1994 from the hydrobiological viewpoint) but much less to zooplankton. Highly neglected are the feedbacks between biological and physical processes. Also, there is usually no distinction made between the two basic types of larger water bodies - lakes and reservoirs. Moreover, I see the use of physics in limnology to be broader than just hydrodynamics. Thermodynamics seems to give explanation not only to the limitations and performances of many biological phenomena at the organism level, but also at the ecosystem level. It also points to the forces structuring the biological world and offers methods useful in biological limnology. It is particularly the notion of optimality stemming from physics which is responsible for considerable recent progress in elucidating many hydrobiological phenomena.

I see the goal of this paper mainly to provoke physical limnologists to deal more intensively with biologically relevant physical processes and biologically oriented limnologists to deal more deeply with physical processes determining biological phenomena. I would like to touch mainly these less treated subjects: reservoirs, feedback effects between biology and physics, and zooplankton. As a biologist (oriented towards reservoir studies) I am not only biased but also ignorant of physics to do more than point to some problems.

Limnological Differences Between Lakes and Reservoirs

There are many features distinguishing reservoirs from lakes (Wetzel, 1990, Straškraba et al., 1993). In spite of this, usually not much distinction is made between the two and during more global analyses the data from both types of water bodies are pooled. The basic processes be they of physical, chemical or biological nature are identical, indeed. However, they are also basically identical for small waterbodies like ponds, periodic pools and phytothelms (small puddles in certain kinds of plants with volume of up to few tens of cm^3) but these are nevertheless distinguished as different water body types. A peculiar combination and action of the general forces characterizes the specific behavior of each of these water body types. The question therefore is if reservoirs behave limnologically differently enough from lakes to warrant specific treatment.

Difference Between Lakes and Reservoirs in Respect to their Phosphorus Retention Capacity

Some recent results of comparative lake and reservoir investigations which point definitely to different behavior of lakes and reservoirs are given in Figure 1. Observed data on the retention of total phosphorus by deep (i.e., stratified) temperate lakes and by deep reservoirs are plotted against the theoretical retention time of water. The phosphorus retention capacity (P-retention RP(%)) is calculated as the percentage of the estimated phosphorus output by the outflow from the water body as related to the load of the lake/

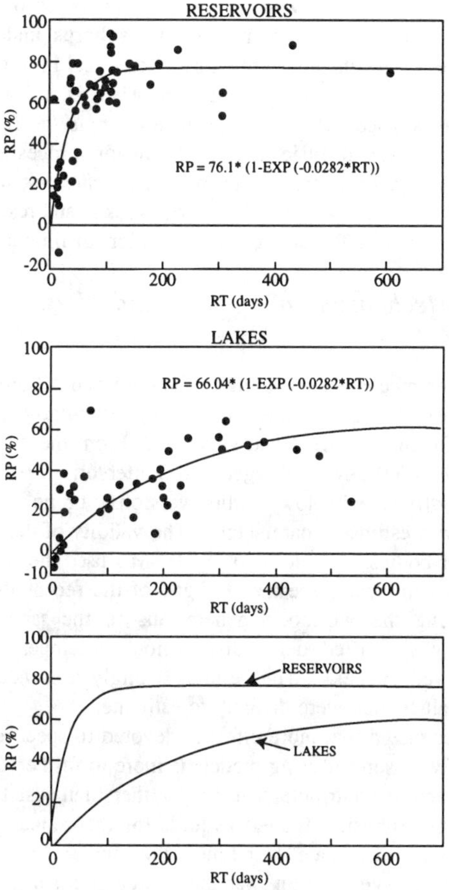

Figure 1. Phosphorus retention capacity of reservoirs (upper panel), lakes (middle panel) and their comparison (lower panel). The curves were fitted to the data by using the nonlinear regression procedure in QuatroPro (modified from Straškraba et al., 1995).

reservoir by the inflows. Each point represents an annual or seasonal average value. The accuracy of these estimates is difficult to judge in detail, as several sources of bias are included: inadequate hydrological budget, long sampling intervals, biased methods of time integration and others. The theoretical retention time of water (RT, expressed in days) is calculated simply as the ratio of the water body volume to the annual or seasonal average daily inflow or outflow. Precipitation and evaporation are assumed to be negligible and more or less balanced in these geographical latitudes. A number of reservoir data stems from the Czech Republic but additional data from the literature covering water bodies from Germany, Switzerland, Italy, Spain and USA were included (Straškraba et al., 1995). The results show that the P-retention of both lakes and reservoirs strongly depends on RT. Nearly no P-retention or even sometimes release of P is seen at low RT, the capacity rising rapidly with RT until the maximum P-retention of some 70 - 90% is reached. Reservoirs have on the average much higher P-retention than lakes, particularly for low RT. Phosphorus is retained in water bodies mainly by direct sedimentation of phosphorus bearing particles contained in the inflowing water and by sedimentation of particles formed within the water body, dominantly by phytoplankton. Phosphorus undergoes complex cycling in the water body, affected by chemical transformations and biological uptake and release processes by phytoplankton, zooplankton and fish. Also, chemical exchange between the sediment and water takes place. However, sedimentation is the final process responsible for phosphorus retention, and diffusion and resuspension processes are important for the exchange at the sediment water interface. Therefore, we can look for major limnophysical differences between lakes and reservoirs and between lakes and reservoirs with different RT as the possible explanation of the above differentiation in limnochemistry.

Processes and Mechanism in Lakes and Reservoirs with Different Retention Times

For the study of the mechanisms leading to the observed differences between lakes and reservoirs dynamic mathematical models are helpful, particularly when they are well based theoretically and sufficiently verified and validated. From the available one-dimensional hydrodynamic models DYRESM (Imberger and Patterson, 1981; 1990) seems almost totally theoretically driven, with low (earlier versions) or no (recent version) need for empirically or otherwise estimated parameters. The validity of the model has been verified on a number of water bodies. Previous to DYRESM I used the model by Markofsky and Harleman (1973) and can clearly see the strength of the recent development from direct comparison of applying the two model generations to the same data sets. The earlier model resulted in very simplified depth distribution of temperatures, not matching the complexity of observed shapes. The results strongly depended on the selection of different parameter values, that were difficult to estimate.

It is generally recognized that more effort is devoted to specific model application than to model analysis. My reason for using models is more to further the understanding of the processes going on and to contribute to theory rather than just to predict some specific situations. I use existing physical models as tools for theoretical studies of the dynamics of aquatic ecosystems to investigate their functioning under different conditions.

One such theoretical investigation was the systematic study of the effect of RT on reservoir temperatures, stratification, flow and mixing phenomena (Hocking and Straškraba, 1994; Straškraba, 1994; and Straškraba and Hocking, in prep.a). The investigation was directed to two major questions: a) to verify how far the model DYRESM reproduces the observations in reservoirs with retention times close to the critical values for the development of stratification, and, b) if it does this successfully what are the probable mechanisms underlying the observed changes. We took hydrometeorological data for 3 representative reservoirs, two in temperate conditions of the Czech Republic (Slapy and Rímov reservoirs) and one in the subtropical conditions of Western Australia (Canning Reservoir). For Slapy Reservoir we used input data for years characterized by major differences in hydrometeorology. A series of simulations was performed by systematically changing RT, without varying the meteorology or the basic morphometry. Two ways of changing the theoretical retention time were investigated: by keeping flow rates fixed and varying volumes, and by keeping the volumes fixed and varying the flow rates. The first method corresponds approximately to the situation when the flow rates of a reservoir are varying from year to year but its water levels remain constant and the second approximates the same flow rates but with changing water levels in different years. In both cases it was assumed that the seasonal pattern of the flow rates was not changed. The existence of correlations between flow rates and other meteorological variables (air temperatures, precipitation and others) as observed, e.g., by Straškraba and Javornicky (1973) was neglected. The differences between the results obtained by the two methods seemed negligible and the following results are based on changing the reservoir volumes. Individual reservoirs will have both different hydrometeorological conditions and morphometry and their RT will vary between years only within certain limits.

The effect of the depth of the reservoir outflow was investigated by simulating for each situation the annual changes that occurred if either a surface outlet (called below a "lake-like" situation) or a deep outlet (corresponding to the actual location of the outlet) operated during the whole season. Also, the effect of changed inflow temperatures was studied by modelling two situations: one when the reservoir was fed by a river with no dams on it and a second when it was fed by colder waters from the hypolimnion of an upstream larger reservoir.

In Figure 2 are schematically represented, for a reservoir with the natural inflow unaffected by an upperlying reservoir and with a near-bottom outlet (Rimov Reservoir), the average summer values of some of the main characteristics and processes going on when the retention time of a reservoir changes. The figure shows that major changes in the depth and thickness of the intrusion take place. This produces large differences in temperatures at all levels, including the surface temperature. The temperature of the outflow to the river below the reservoir is changed considerably. On a more continuous basis the temperature changes at different depths for a reservoir obtaining cold water from the hypolimnion of an upperlying reservoir and releasing water from deep strata (Slapy Reservoir) are shown in Figure 3. As seen, for short retention times the deep temperatures grow during summer quite high. For the retention times of 400 days (not shown) and more (see the situation for 800 days) the bottom temperature remains nearly constant all over the year.

The average summer differences of several main hydrodynamic characteristics of the reservoir with deep and surface outlet are shown in Figure 4. Some of the values for the bottom outlet situation are the same as in Figure 2. The temperature at the intrusion level (T_{int}) is by one half to one degree higher in the bottom outlet case and so is the surface temperature, too. Marked temperature differences are seen for low retention times, particularly for temperatures at 15m. The difference in Z_{mix} is not consistent and the difference in the volume of the intrusion (V_{int}) is not very pronounced.

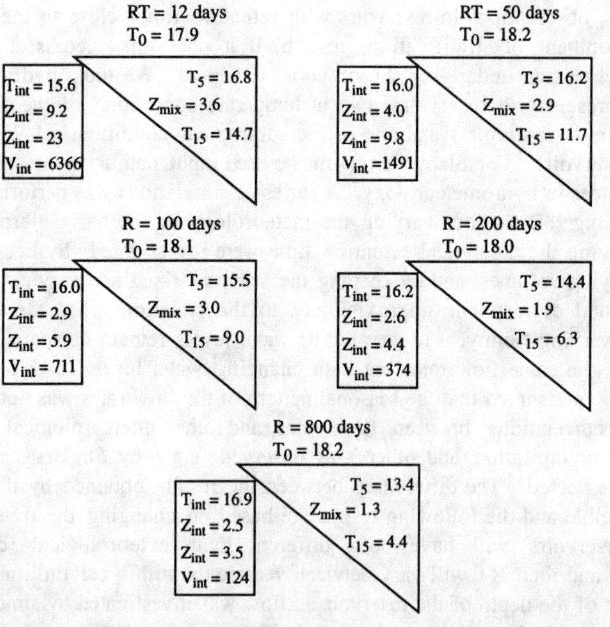

Figure 2. Average summer situation for a bottom outflow reservoir for different values of the theoretical retention time of water, RT. T_0, T_5 and T_{15} are the surface, 5 m and 15 m temperatures, respectively, Z_{mix} - mixing depth and V_{int} the volume of the intrusion. The two values of Z_{int} represent the position of the upper and lower boundary of the intrusion. Simulation results with DYRESM.

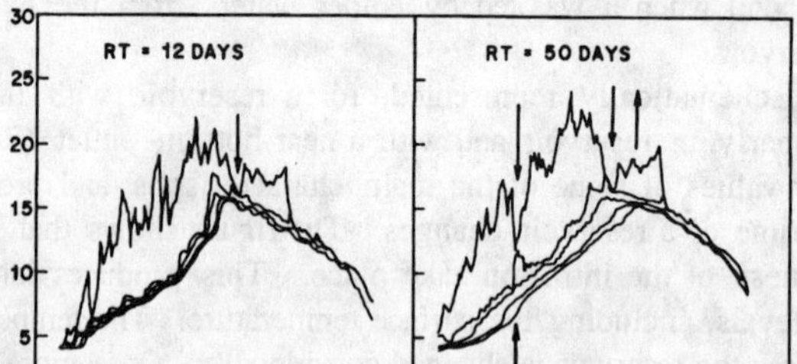

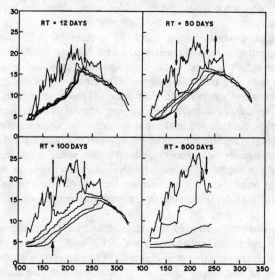

Figure 3. Annual changes of temperature at the depths 0, 5, 10, 20 and 30 metres in a reservoir with bottom outlet, obtaining cold inflow water from the hypolimnion of an upperlying reservoir. The arrows indicate weather caused unusual mixing events resulting in abrupt temperature changes at 0 and 5 metres. On the x-axis Julian days. Simulation results with DYRESM.

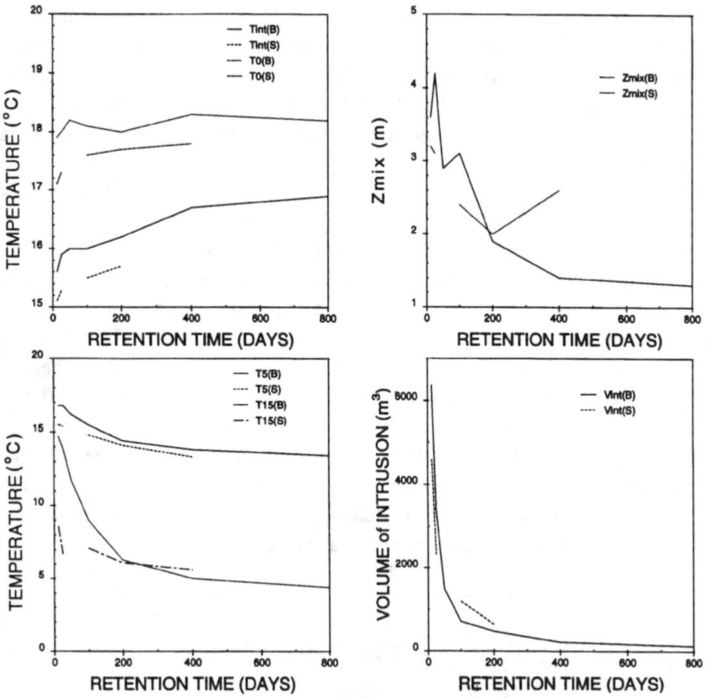

Figure 4. Comparison of the average summer values of several hydrodynamic characteristics in the same reservoir simulated with a bottom outlet (labelled by (B)) and a surface outlet (labelled by (S). Explanation of symbols as in Figure 3. Simulation results with DYRESM.

The question is, how far can the model results be trusted. We used two methods to assess the validity of the results. The ordinary method consisted of comparing field measurements for one locality with the corresponding simulations. We obtained a successful matching of the observed annual pattern not only for one year, but also for the same reservoir over several years of very contrasting hydrology, and therefore highly differing retention time. Moreover, we also used a method I called comparison of trends (Straškraba and Gnauck, 1985). The method uses much more than one or two years of observations from one or two water bodies. This model validation method consists of comparing the responses of a number of natural systems to some input variable with the responses of the model on the same variable. Observations from a number of localities are generalized into a trend function. Results of a number of simulation runs with the model are evaluated in the same way as the natural observations to see if the responses match. This is a much stronger validation method than one based on individual localities, as it is directed to the general rather than to the specific characteristics of responses.

The results of observations of a simple measure of the reservoir stratification against the theoretical water retention time were plotted for a number of reservoirs in the Czech Republic and Germany (Figure 5). The stratification measure used was the temperature difference between the surface and the depth of 30 m during the period of maximum surface temperatures. The reason for selecting a fixed depth and not the bottom for the difference was twofold: firstly to obtain a better comparability because of the depth dependence of hypolimnic temperatures, and secondly to exclude the increased temperatures at the bottom of some reservoirs due to the effect of submerged weirs (weirs used during reservoir construction for retaining and/or diverting the river). The results show that the reaction of reservoirs to changes in the retention time has an asymptotic character. The changes are steep at retention times below about 100 days and level off asymptotically as retention time increases above 200 to 300 days.

An evaluation using the simulation results demonstrates the similarity of model responses with reality. This correspondence increases the credibility that the model is able to help us understand what is going on in nature. In addition, the more rapid response calculated for the "lake-like" situations (surface outlet) indicates that the location of the outlet in a reservoir plays a major role and that the surface outflow in lakes and deep outflow in reservoirs might be one of the dominant differentiating factor between the two water body types. Undoubtedly lakes have other features differentiating them from a reservoir with surface outlet: different morphometry, usually no unidirectional flow from the inflow to the outlet, often several rivers flowing in. I have no direct evidence how these additional characteristics will modify the relation observed for surface outflow reservoirs. I have a feeling that the modification by these other characteristics will not be great. It is also shown in Figure 5 that this type of reaction is valid for subtropical reservoirs, where the difference between the simulations with surface and deep outlet is quite pronounced.

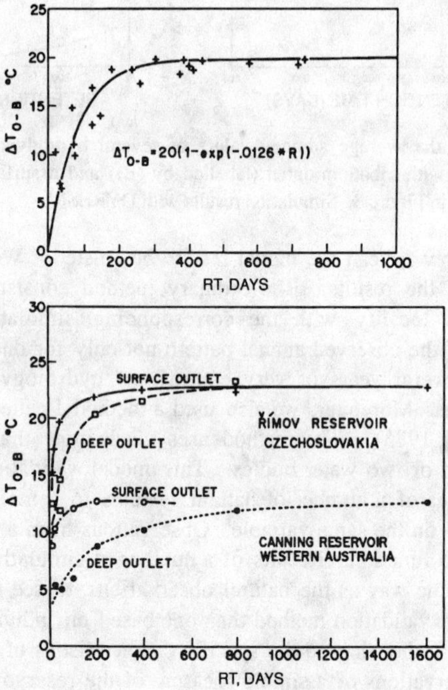

Figure 5. The simple measure of summer maximum stratification: temperature difference between surface and a depth of 30 m at the period of maximum surface temperatures plotted against the theoretical retention time of water. Upper panel - observations in reservoirs of the Czech Republic and Germany. Lower panel - result of simulation with DYRESM for situations corresponding to Rimov Reservoir, Czech Republic and to Canning Reservoir, Western Australia. Results for bottom and surface outflows.

Conclusions about the Effect of RT on Lake and Reservoir Limnology

The main conclusions of the investigation into the reactions of reservoirs to changing retention times can be summarized as follows:

- In all instances, RT has a marked effect on the temperature structure of the reservoir. This includes the "lake-like" situation, i.e., when the simulated reservoir has a surface outflow (Figures 2-5).
- In the surface outlet case the surface temperatures are affected significantly, but the hypolimnetic ones undergo only minor changes. Surface temperatures and temperatures at the intrusion level change with RT almost exactly as in the case of the deep outlet, in spite of the thermal structure being rather different (Figure 4).
- In the deep outlet cases the deep water temperatures are always significantly affected (Figures 2-3).
- In the case of natural river inflow, surface temperatures are nearly independent of RT, while the hypolimnetic ones are greatly affected by RT (Figure 2).
- The effect of RT on the surface-bottom temperature difference is strong in all observed instances up to RT = 200, and is sometimes seen up to RT = 400, but levels off for RT > 400. However, typical temperate "lake" situation with 4°C water in the hypolimnion is reached only at RT = 800 days (Figure 3).
- In all instances the temperature of the water intruding into the reservoir is changed significantly by entrainment of the reservoir water. This means that simplistic estimates of intrusion depth obtained from river and reservoir temperatures may be greatly in error. The intrusion depth decreases with increasing RT and the thickness of the intrusion decreases markedly.
- Temperature stratification in a reservoir seems to be very sensitive to the inflow temperature. This indicates major differences in the same geographical region due to differences in elevation.
- The seasonal average values of the Lake Number (Imberger and Patterson, 1990) change non-linearly and non-monotonically with retention time, the differences between shortest and longest retention time being about eightfold. Moreover, the values depend on the outflow location.
- The seasonal average values of the Inflow Lake Number rise non-linearly with RT, with almost two orders of magnitude difference between RT = 12 and RT = 800 days. Simultaneously, the Outflow Froude Number decreases exponentially with RT.
- It is impossible to predict these changes without taking the dynamics of the participating processes into account.

Coupling Between Reservoir Chemical, Biological and Physical Limnology

We now return to the dependence of the phosphorus retention capacity of lakes and reservoirs on the retention time and the difference between lakes and reservoirs in this respect. Comparing Figures 1 and 4, we see that a similar asymptotic shape describes the dependency of surface-bottom temperature difference and of phosphorus retention capacity on the theoretical retention time. Therefore, the changes in hydrodynamics with retention time evidently underlie the changes in phosphorus retention. Note that the degree of the reaction of the temperature stratification and phosphorus retention among the two water body types is reversed: more steep reaction of temperature stratification on the retention time in "lakes" than in reservoirs, while for phosphorus retention reservoirs react on retention time more steeply. We assume here, with some hesitation, that the results of the surface outlet simulations will not be principally changed when the other lake characteristics are included in the model.

The following hydrodynamic processes are important for phosphorus retention: the intrusion depth, the thickness of the intrusion, velocity of the inflow current, all important for sedimentation of phosphorus containing particles brought by the inflows. Changes in the temperature of the productive layers and the mixing depth are decisive for phytoplankton formation and sedimentation as the dominant phosphorus elimination mechanism. The following model results indicate an explanation for the generally increasing phosphorus retention with increasing retention time:

- With increasing RT the intrusion depth decreases and so does its thickness. Flow velocities decrease. The consequence for phosphorus-rich particles brought by the inflow is that with increasing retention time their sedimentation will be more intensive in water bodies with longer retention time.
- The summer surface temperature increases with RT, with positive effect on the phytoplankton development and therefore on phosphorus retention.
- With increasing RT the mixing depth decreases. The positive effect on phytoplankton development follows from more light being available to phytoplankton in shallow thermoclines.

As to the difference between reservoirs and lakes, the following results from the modelling exercise are relevant:

- In the "lake-like" situation the decrease of the intrusion depth and thickness with RT is less steep. Therefore, the sedimentation of P-rich particles will be less intensive than in reservoirs.
- In the hydrodynamic simulations the mixing depth is not different for the two situations. However, as we will see below, due to the positive feedback between phytoplankton development and mixing depths, Z_{mix} will be higher with less phytoplankton being developed as a consequence of deeper mixing. This will again have the consequence of higher sedimentation in reservoirs. The positive feedback was not covered by the simulations.

Moreover, one feature of reservoirs important for phophorus retention is not covered by the one dimensional simulation. This is the strong longitudinal differentiation of flow rates in the valley reservoir, with considerable consequences for phytoplankton development and phosphorus sedimentation. Both are highly increased near the inflowing stretch of the reservoir.

Feedback Between the Phytoplankton Development and Water Body Hydrodynamics

Until now we have demonstrated the importance of hydrodynamics for chemical and biological processes in reservoirs, in particular the effect of changed retention time. Is there not a feedback effect of biological - chemical processes on hydrodynamics? This part deals with this question.

We devoted another theoretical study with DYRESM to the effect of changes in the light attenuation in the water body on its hydrodynamics. Our results with the earlier hydrodynamic model by Markofsky and Harleman (1973) have shown an extreme sensitivity of the depth distribution of temperature to changes in extinction coefficient. The depth of the thermocline, Z_{mix}, considerably increased with the decreasing extinction coefficient. Mixing depth was doubled when the extinction coefficient ϵ was changed from 0.8 to 0.2 (see Straškraba and Gnauck, 1985). In fact this was already observed by the author of the model (Harleman, 1982). Simultaneously, the surface temperature remained nearly unchanged. Such an extreme sensitivity of the mixing depth and no sensitivity of the surface temperatures to changes in the extinction coefficient did not seem to agree with natural observations. The same question was studied with DYRESM. The vertical attenuation coefficient for visible light (= extinction coefficient ϵ), was changed systematically for a series of model runs. The coefficient was held constant during each one-season run. The annual changes of the hydrometeorological conditions for all simulations were set according to an observed situation in the modelled reservoir, Slapy Reservoir in the Czech Republic. Preliminary analysis of this study and the analysis of consequences for the coupling between phytoplankton development and water body hydrodynamics have been published in Straškraba (1994) and the final results are in preparation (Straškraba and Hocking, in prep.b).

Figure 6 demonstrates the calculated differences in the temperature stratification during the same day of the year, corresponding approximately to the period of maximum surface temperatures. The curves were obtained by keeping the indicated values of ϵ constant over the year. In comparison with the changes calculated by the earlier model the effect of ϵ is less pronounced. The location of the thermocline increases with the increase of ϵ, while the increase of surface temperature does not seem to be consistent. Also, the profiles are no longer smooth. The latter difference (smooth profiles obtained by the earlier model and more irregularities predicted by DYRESM) are not just the consequence of more adequate process description in DYRESM but also the consequences of the different water bodies modelled. In the first case a reservoir with long retention time was the prototype, while in the second case a short retention time reservoir with temperature profiles more disturbed by the high flows entering the reservoir, relative to the reservoir volume, was modelled. Nevertheless, when estimating

$$SD = 1.8/\epsilon \tag{1}$$

SLAPY 1963 DAY 220

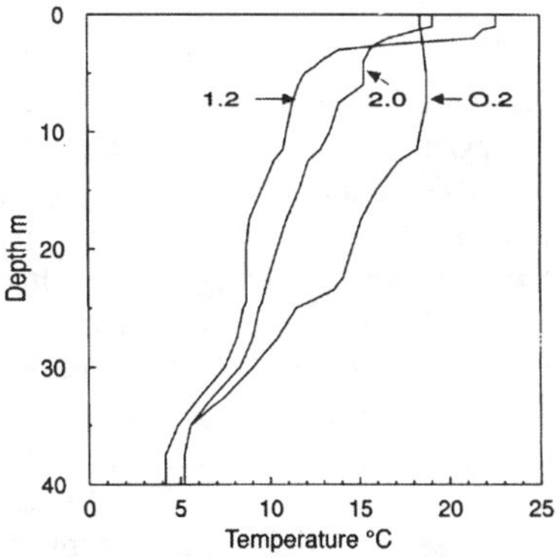

Figure 6. Depth profiles of temperature during the period of maximum surface temperatures calculated by DYRESM for different values of the vertical beam attenuation coefficient for white light ϵ.

it is possible to derive Z_{mix} from these profiles with the resulting rough approximation of the relation between the Secchi Disc reading, SD, and Z_{mix}:

$$Z_{mix} = 4.5 + 1.53 * SD. \qquad (2)$$

Indications of the dependence of water body temperatures on different kinds of light absorbing matter appeared in literature several times. Idso and Foster (1974) observed an increase of shallow pond temperatures when the pond contained dense phytoplankton blooms. Schiebe et al. (1975) compared temperature conditions in two comparable small, shallow reservoirs, one with very high suspended sediment concentrations. The annual average Secchi Disc reading was extremely low in both reservoirs, 35 and 10 cm, respectively. Temperatures were lower in the reservoir with the lower transparency, which the authors attributed to higher surface reflection. Based on this, Stefan et al. (1982) attempted to incorporate the relationship between turbidity, solar radiation reflection and solar radiation attenuation in the water into the model RESQUAL. They noted that reflectivity and attenuation affect water temperatures and the degree of the temperature stratification but did not comment on the changes in mixing depth. Their model of light penetration was explained in detail in Stefan et al. (1983).

Extensive empirical observations comparing the dependence of the mixing depth (Z_{mix}) on the Secchi Disc depth (SD, an indirect measure of the extinction) were made on lakes of the Canadian Shield by Mazunder et al. (1990). The two values, SD and Z_{mix}, were plotted for a number of lakes of different size. The positive dependence of the mixing depth on the water body size due to the nonlinear increase of wind fetch is well known (see, e.g., Straškraba, 1980). For this reason the authors also plotted the relationship between Z_{mix} and wind fetch. Unfortunately they did not use it for eliminating the effect of wind fetch, so that the results are biased by the changes in lake size and there is no tabulation of data enabling such calculation to be performed. Earlier, Wofsy (1983) also plotted Z_{mix} against SD for a few lakes but mainly for eutrophic oceanic waters and estuaries. Such large waters with correspondingly larger fetches cannot be directly compared with smaller lakes (or reservoirs) of the size range studied by Mazunder et al. (1990). Therefore, for the purposes of comparison with the simulated effect of $\in$ on mixing depths we use the latter data, approximated by the linear dependence

$$Z_{mix} = 0.7 + 1.82 * SD \qquad (3)$$

Another indication of this kind of dependence is evident from the changes of mixing depth observed in the Lohi Lake in Canada (Yan, 1983). The lake was acidified, later neutralized and then reacidified again. The observed changes in mixing depth were clearly related with the Secchi Disc changes attributed to precipitation of dissolved, colored organic matter or to a reduction in color of this matter during changes of pH.

Several causes can lead to decreasing Secchi Disc readings: water color caused by decomposing organic matter, usually of yellow-brown character, turbidity due to mineral particles, and phytoplankton. Size and shape of the particles as well as their optical properties play a role. Assuming that in lakes studied by Mazunder et al., the main cause of the changes in SD is phytoplankton, we have the following coupling between physical and biological processes:

- Mixing depth decreases when Secchi Disc readings decrease, i.e., with the increasing amount of phytoplankton the extinction of water increases and light penetration decreases due to the selfshading effect.

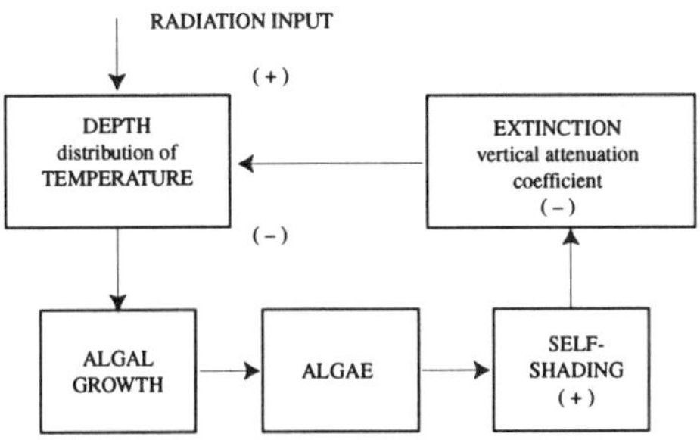

Figure 7. Positive feedback between phytoplankton growth, mixing depth and temperature, caused by the negative feedback of phytoplankton biomass on radiation attenuation.

- The decrease of light penetration means more light is absorbed in the upper water strata.
- As a consequence Z_{mix} decreases and temperature of the uppermost water strata increases.
- The decrease of Z_{mix} and increase of temperature positively affect phytoplankton growth.
- The increase of phytoplankton growth leads to more phytoplankton with the consequent decrease of Secchi Disc depth, decrease of Z_{mix} and increase of temperature.

Figure 7 indicates this kind of feedback effects: positive feedback between mixing depth and phytoplankton growth, and negative feedback between phytoplankton growth and light, temperature and mixing depth. The theoretical biological model of phytoplankton growth by Wofsy (1983) is able to elucidate these relationships both qualitatively and quantitatively.

The model represents the steady state solution for phytoplankton biomass in dependence on mixing depth and light extinction:

$$P = 1/(\propto_P * \propto_D) [B/(r_0 * Z_{mix}) - (\epsilon_w + \epsilon_s)] \quad (4)$$

where
P = phytoplankton biomass
D = detritus
B = daily integral of photosynthesis
$\propto_P, \propto_D$ = specific light extinction coefficients of P and of D
r_0 = sum of the ratios of maximal instantaneous rate of photosynthesis used for respiration and grazed
Z_{mix} = seasonal thermocline
ϵ_w, ϵ_s = light extinction coefficient of water and of nonliving material, respectively.

Consequences of the Feedback of Phytoplankton Concentration on Mixing Depth for Hydrodynamic Modelling

The above positive feedback between physical variables and growth of phytoplankton leading to substantial changes in reservoir/lake hydrodynamics is not covered in most hydrodynamic models. In a dynamic way this is possible only while coupling the physical and biological processes into one model.

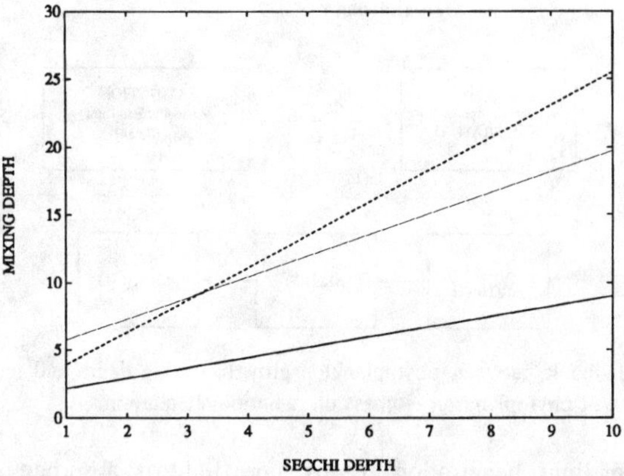

Figure 8. Comparison of the dependence of the mixing depth during maximum summer temperatures, Z_{mix}, and Secchi Disc depth as approximate from the results obtained by the hydrodynamic model of Markofsky and Harleman (1972) dotted line), by the model DYRESM (thin line) and from observations on Canadian lakes by Mazunder et al. (1990) (thick line).

However, there is one more consequence of this investigation for hydrodynamic modelling. As indicated in Figure 8, there is a major discrepancy between the mixing depths calculated by DYRESM and that represented by the existing series of observations. The calculated mixing depths are up to twice as high as the observed ones. Most other hydrodynamic models use the same simple description of radiation attenuation and it is to be expected that they have the same inadequacy. Our observation is not the first one of this kind. Simpson and Dickey (1981) noted for the ocean the same kind of inconsistency when the downward irradiance is modelled by means of a constant extinction coefficient. They obtained much improved model results when approximating the penetration of irradiance by means of two extinctions, one for the more rapidly absorbed longwave (red) wavelengths and the other for less rapidly absorbed wavelengths of the blue-green part of the spectrum. Zaneveld et al. (1981) found a pronounced effect of the optical water type (and therefore $\in$) on the heating rate of a constant depth mixed layer.

Therefore, the differential absorption of different wave-lengths and the effect of refractory dissolved organic matter, mineral particles and phytoplankton on radiation absorption and reflection has to be incorporated into the hydrodynamic models to adequately cover the heat absorption processes in the surface layers of the lakes/ reservoirs. Methods for incorporating the differences in spectral characteristics of the three different groups of causes of light extinction has to be investigated.

An Example of Microlevel Coupling of Hydrodynamics and Biology - Density Dependence of Microorganisms

The above sections stressed the need for macrolevel coupling between the fields of hydrodynamics and biology. In the following we will demonstrate one of the many examples illustrating the same requirement to connect microlevel biological and physical phenomena. In this case we do not have a model describing the underlying phenomena: we can only speculate about the processes involved.

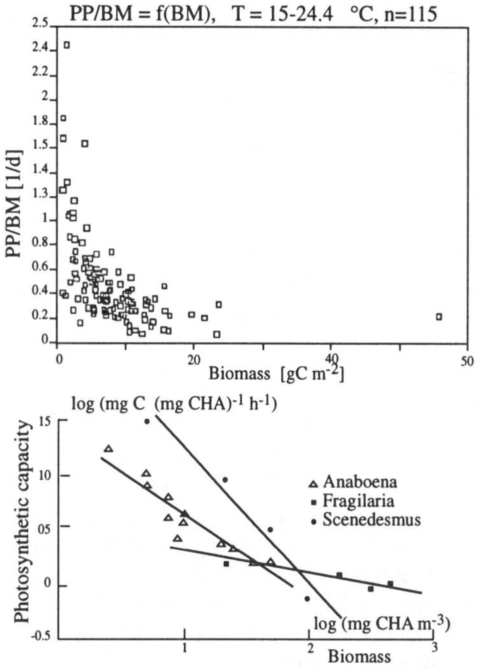

Figure 9. Density dependence of the specific phytosynthetic capacity of phytoplankton. Upper panel - natural observations from lake Constance based on data kindly supplied by Dr. U. Gaedke (Constanz, Germany). Lower panel - experimental observations with several species of algae taken from Slapy Reservoir and from cultures.

The underlying dependencies are shown in Figures 9 and 10. The first demonstrates for lake and reservoir phytoplankton the dependence between the specific photosynthetic activity as a measure of the organisms metabolism, and phytoplankton density. The specific photosynthetic capacity is integrated over the euphotic zone in field measurements, expressed in mg $C.h^{-1}$. (mg $CHA)^{-1}$, where CHA is chlorophyll-a as a convenient measure of the phytoplankton biomass. The upper part of Figure 9 covers natural observations on mixed phytoplankton populations from a lake. The observations are far from being unique. I have asked several colleagues measuring phytoplankton photosynthesis in lakes to investigate this kind of relationship with their data and I obtained several sets of observations, all showing the same relationship. In the lower part of Figure 9 experimental observations with different algae obtained from Slapy Reservoir (*Anabaena circinalis* and *Fragilaria crotonensis*) or from cultures (*Scenedesmus quadricauda*) are given. During the experiments done in a thermoluminostat at near-optimum temperatures and saturated light intensities the algal culture was diluted to several biomass concentrations and the photosynthetic oxygen evolution measured after one half to two hour exposures. Algae were in a rich cultivation medium so that limitation of the culture by average concentration of nutrients was not possible. The medium was also enriched by CO_2 by bubbling with air. The presence of non-limiting concentrations of CO_2 at the end of the experiment was checked by measuring pH. Both the natural observations and the concentration/dilution experiments show an extremely strong decrease of the specific photosynthetic activity at fairly low concentrations of algae. Different species show a different density-dependent reaction. The much higher spreading of the natural observations is attributable to the presence of different species, but also to different temperatures and other conditions in the water bodies observed. Also, there is no guarantee that in some cases algae were not limited by nutrients or light.

Figure 10 shows the same trend for the activity of another group of microscopic organisms, bacteria. The important life process in this case is specific respiration or specific uptake of organics as measures of the metabolic activity of bacteria. In the upper part results of the experimental determination of the amino acid uptake of bacteria taken on different occasions from Rimov Reservoir, Czech Republic (Straskrabová and Simek, 1984) are given, in the lower part natural observations by Santrucková and Straškraba (1991) for soil bacteria. The character of the relationship is fairly similar to that for phytoplankton. Again, other observations on both soil and aquatic bacteria not discussed here in detail show the same relationship.

One problem is often discussed in connection with specific rates of any kind, specifically in connection with the "self-thinning" process in higher plants: the statistical problem. The possible bias of the observed relationships as the result of statistical treatment is discussed in the paper by Santrucková and Straškraba (1991). The paper shows that the relationships given above, both for algae and for bacteria, are real.

One working hypothesis for the explanation of this kind of relationship is in the micro-hydrodynamic conditions around the phytoplankton cells in media with different densities of cells. Pasciak and Gavis (1974) developed a model concerning the diffusion envelope of the cells shown in Table 1. In the zone close to the cell there is a strongly decreased concentration of nutrients due to consumption by the algae. Pasciak and Gavis's solution is devoted to CO_2 and covers the processes of replacement of CO_2 from bicarbonates. For phosphorus as the limiting resource the only replacement is from excretion by algae, other microorganisms and also by higher aquatic organisms. In the model, there is no interference between the individual cells, and the cells are stationary.

TABLE 1. Summary of the model for diffusion into algal cells by Pasciak and Gavis (1974).

$$Q_R = 4 * \pi * R^2 * JR \quad [\mu mol.hr^{-1}] \qquad (5)$$

$$J_R = D/R\ [R\ \sqrt{(k'/D)} + 1]\ [C_e - C_R] \quad [\mu mol.cm^{-2}.hr^{-1}] \qquad (6)$$

where	D	=	diffusivity	$[cm^{-2}.hr^{-1}]$
	R	=	radius of the cell	[cm]
	k'	=	constant of kinetic transformation of C	
	C_e	=	external concentration	[μmol]
	C_R	=	internal concentration	[μmol]

$$C_R^{*2} + [1/F + 1 - C_e^*]\ C_R^* = 0 \qquad (7)$$

with $$F = (14.4 * \pi * R * D * K)/\ V_{max} \qquad (8)$$

where	K	=	half saturation constant	[μmol]
	V_{max}	=	maximum uptake rate	$[\mu mol.hr^{-1}.cell^{-1}]$

The steady state concentrations are labeled by a *

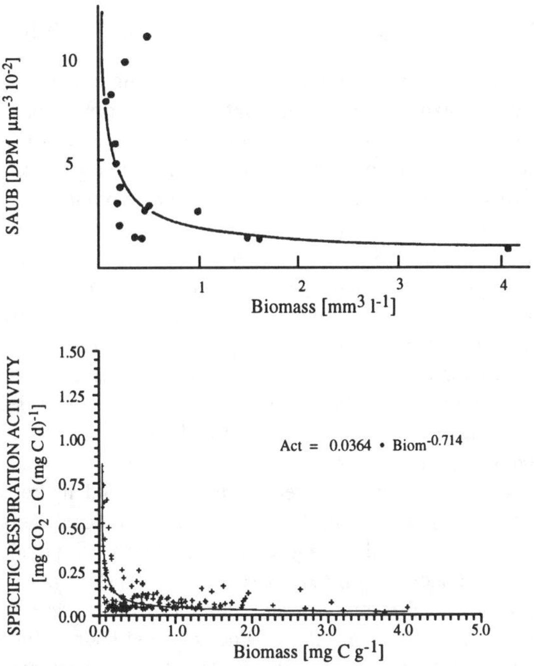

Figure 10. Density dependence of life processes of soil and aquatic bacteria. Upper panel - experimental measurements of the specific uptake rates of labeled organic substrates by aquatic bacteria by Straskabová and Simek (1984). Lower panel - natural observations on the specific respiration rate of soil bacteria by Santrucková and Straškraba (1991).

In comparison to CO_2, the replacement processes for phosphorus are spatially discontinuous, located just around the organisms. Moreover, most of the excreted phosphorus is in the organic form, therefore available to algae mostly only after decomposition to mineral form. Extracellular enzymes produced by the organisms help to break down such compounds.

Also it is known that particles can interfere with one another and are affected by their own movements and by water movements (e.g., O'Melia, 1985). The question arises as to how far the changes in water movements and both passive and active cell movements and cell interactions at different cell densities affect the nutrient uptake and liberation.

The competing hypotheses concern:

- In the case of algae, the possibility of light limitation by selfshading. With the generally accepted specific attenuation coefficients per algal biomass, selfshading produces measurable effects at much higher algal concentrations than those observed in the above observations.
- The inhibition of activities by life products of the organisms (O_2 in the case of algae and CO_2 in the case of bacteria). Processes of diffusion of these products from the cells can again play a role.
- Changes in algal and bacterial metabolism at low and high organism concentrations in the medium. If this is the case, the question of the mechanism triggering such changes still remains to be solved.

Relationship Between Hydrodynamics and Zooplankton

Phytoplankton is now becoming familiar to physical limnologists because of the well known effects of hydrodynamics on phytoplankton functioning. In this last part of the paper I would like to point towards the effects of hydrodynamic processes on zooplankton, which has received much less attention from the hydrodynamic point of view.

The effect of physical conditions in a water body on zooplankton is less straightforward than for phytoplankton. Mostly there is an accompanying effect of the physically determined phytoplankton behavior and distribution of zooplankton. As an example, zooplankton life processes are affected by temperature, and simultaneously by food. In the case of phytoplankton-eating zooplankton the amount of food present under different temperature conditions will also be determined by temperature directly. It is clear that the same is valid for phytoplankton: its biomass at any one time in the water body depends not just directly on temperature, but also on other conditions determining nutrient distribution and the intensity of grazing by the zooplankton.

Zooplankton heterogeneity in reservoirs. The horizontal and vertical distribution of zooplankton in a valley reservoir is related to biological and hydrodynamic processes. Two scales can be distinguished among the hydrodynamic forces driving the distribution:

- Macroscale, represented by the longitudinal gradients in physical conditions in reservoirs. These are connected with the decreasing flow rates, intrusion of the inflow water, mixing of the river and reservoir water along the valley reservoir. The gradients have consequences for phytoplankton that provide the food source for zooplankton.
- Microscale, realized in the form of spatial and temporal changes produced by a number of physical processes and associated local changes in physical, chemical and phytoplankton conditions. Physical conditions do not determine the microscale variability alone. An interplay between their effect on phytoplankton and on zooplankton and the interactions between phytoplankton, zooplankton and fish is occurring. We have to study simultaneously the physics and biology to understand how these variables together determine zooplankton distribution.

Zooplankton patchiness is caused by the Langmuir circulation and seiches. There is simultaneously a direct effect of flow pattern within the Langmuir cells on zooplankton distribution and an indirect effect through accumulation of phytoplankton (Ledbetter, 1979; Kamykowski, 1978). We have to know more about velocities in Langmuir cells and about internal movements within seiches under different conditions. We also need to know more about the swimming and accumulation abilities of the organisms to be able to understand how the direct and indirect effects are combined.

Zooplankton migration is triggered by light. Organisms migrate upwards during the night and downwards during the day. Certain organisms, however, have reversed migration which is recognized as a mechanism for avoiding the competition with some other organisms. It is known that physical mixing can prevent migration (Gliwicz, 1986; Gabriel and Thomas, 1988). To understand at what mixing intensities and flow rates migration is stopped we need to develop the ability to predict the hydrodynamic forces during migration. We also have to understand that the main mechanism producing migration is the presence of predators, mainly fish. Therefore, physical factors affecting the capturing mechanisms of fish and other predators, their visual capabilities, rates of attack etc. have to be studied.

Moon trap. Gliwicz (1986a) has made observations which present an interesting example of the interplay of physical and biological factors. He found in the Cabora Bassa

Reservoir in Africa an exponential increase of zooplankton density during periods of low moon shine, followed by an exponential decrease during the period of high moon shine. Zooplankton migrating during the day to deeper strata to avoid predation by fish appeared to be cropped by sardines during the night while migrating to feed near the surface. The cause was the visibility of zooplankton for sardines during high moon shine.

Cyclomorphosis of Daphnia. The formation of helmets in a number of species of *Daphnia* was recognized to be produced by the interplay of temperature and turbulence. It was found, however, that the populations with helmets are reproductively disadvantaged, having lower egg production than non-helmeted populations (Hrbácek, 1959). The question is, what is the hydrodynamic (or other) advantage of helmets. In biology, we have evidence that such energetically expensive and apparent appendages will not be formed without reason.

Zooplankton filtratory feeding. Two schools try to explain the function of "filtratory apparatuses" of zooplankton, in particular of *Daphnia* (Ganf and Shiel, 1985; Gerritsen et al., 1988). At the low Reynolds numbers prevailing in the food chambers of cladocerans, one school supports the classical meaning of filtration and considers it possible under these conditions. The other argues that filtration is impossible and other mechanisms of particle collections by the filter-like apparatuses is used. The true mechanism has not yet been definitely recognized. A positive relation between grazing rates and size of the food organisms was recognized recently. The relationship can be explained only by filtration.

Zooplankton filter adaptation to food density. Recent investigations have shown that the "filters" increase the density of setae when food is scarce (Korinek et al., 1986). This implies that the apparatus is capable of collecting food more effectively when increasing the "filter" density. Physical explanation of this improved functioning will be necessary should the apparatus not function as a filter.

Tasting of food by zooplankters. The capability to select "good" food organisms was recognized for several groups of marine and freshwater zooplankton (DeMott, 1989). Experiments show that the recognition is based on taste, which is detected in the water at a certain distance. It is expected that the detection depends on microflows which the organisms produce around them. The question about the relation to naturally occurring micromovements in water arises.

Excretion of phosphorus by swimming zooplankters. Lehman and Scavia (1982) recognized that individual *Daphnia* creates miniature patches of dissolved nutrients and constructed a model of how algae exploit the nutrient patches. It is not clear how phosphorus utilization depends on the plume behind *Daphnia* and how algae could track this plume.

Feedback Effect of Zooplankton on Hydrodynamics

We have demonstrated above that there is a strong feedback effect of phytoplankton development on temperature stratification and relevant hydrodynamic processes. The question appears, does some kind of feedback effect exist between zooplankton and hydrodynamics? The answer is yes. The effect is not direct but is mediated by the phytoplankton in the way described above. The "bottom-up" effect represents the causal chain; physical factors → chemical variables → phytoplankton → zooplankton → fish. Simultaneously, the "top down control" hypothesis which is now generally recognized states that there is a strong influence of the trophically higher organisms, in particular fish, on the development of zooplankton and of zooplankton on phytoplankton. These effects are due to grazing on the respective organisms. The influence is not only quantitative, by grazing a certain amount of zooplankton or phytoplankton, respectively but also selective with grazing leading to suppression of certain species and support of others. In this way the effect is quite strong. Fish grazing can result in the substitution of zooplankton species effectively grazing on phytoplankton for less effective species so as a consequence of successful zooplankton grazing small phytoplankton species are substituted with much larger ones. Simultaneously, the biomass of phytoplankton is decreased and water quality improved. From the hydrodynamic point of view both the decrease of phytoplankton and substitution for larger species are important. The light extinction produced by the same biomass consisting of larger species is significantly lower.

Summary

Valley reservoirs behave limnologically differently from lakes. This is shown by the differences between the phosphorus retention capacity of both water body types. The main cause seems to be differences in hydrodynamics. One major factor differentiating lakes and reservoirs is the different location of the outflows: surface ones in lakes and deep ones in valley reservoirs. Theoretical retention time of water plays an important role in both lakes and valley reservoirs. It produces differences of stratification, mixing and flows within the water body. The one-dimensional hydrodynamic model DYRESM validated both for a number of water body situations all over the world and for its ability to reproduce the trends of reservoir behavior under the influence of changing retention times, helps to elucidate the processes underlying such trends.

The higher phosphorus retention of valley reservoirs compared to lakes, particularly at low theoretical water retention time, is explained by the shallower intrusion depth and decreased thickness of the intrusion layer. There is a correspondingly higher sedimentation of phosphorus-rich particles brought in by the inflow. Together with these changes, higher temperatures cause higher phytoplankton growth and result in increased sedimentation as the main process of phosphorus retention. The different behavior of lakes and valley reservoirs needs to be recognised in comparative hydrophysical and limnological studies.

A systematic investigation of the effect of the vertical beam attenuation coefficient for white light, $\in$, suggests that the mixing depth does seem to be overestimated by the simple representation of exponential radiation attenuation in DYRESM and most other hydrodynamic models. There is a positive feedback between phytoplankton biomass and hydrodynamic conditions. Increased phytoplankton produces increased $\in$ with the consequent decrease of Z_{mix}. The shallower thermocline results in improved average light conditions for the phytoplankton populations and as a consequence increased growth. The representation of radiation attenuation in the models needs improvement. Also, the positive feedback between phytoplankton, mixing depth and temperature needs to be recognised in hydrodynamic models.

The density dependence of the activity of microorganisms that results in a steep drop of the specific activities of algae and bacteria with increasing biomass is a strong feature producing order-of-magnitude differences in activities between reduced and dense populations. It is a highly neglected phenomena in limnology. One hypothesis to explain this feature is based on changed micro-hydrodynamic conditions around the cells when their density is low and high. Nutrient micro-limitation or inhibition by life products are suggested to be responsible for the observed result. Collaboration between biological and physical limnologists is needed to deal with this topic.

The list of the possible effects of hydrodynamic conditions on zooplankton shows that both macro- and micro-level processes of flows, mixing, and light conditions determine the distribution, behavior, migration, feeding and predation by and of zooplankton, as well as the morphology of some zooplankton species.

Acknowledgments. The collaboration with the Water Research Center, University of Western Australia enabled the use of the model DYRESM. I am indebted to Graeme Hocking (Murdoch University, Perth) for collaboration with the application of the hydrodynamic model. The preparation of this paper was supported by the project No. 204/94/1672 of the Grant Agency of the Czech Republic. The participation in the symposium on Physical Limnology was partly supported by the Center of Water Research, UWA.

References

Bourget, E., and M. J. Fortin, A commentary on current approaches in the aquatic sciences, *Hydrobiologia, 300/301*, 1-16, 1995.

DeMott, W. R., Optimal foraging theory as a predictor of chemically mediated food selection by suspension-feeding copepods. *Limnol. Oceanogr., 34(1)*, 140-154, 1989.

Gabriel, W., and B. Thomas, Vertical migration of zooplankton as an evolutionarily stable strategy. *Amer. Natur., 132*, 199-216, 1988.

Ganf, G. G., and R. J. Shiel, Feeding behaviour and limb morphology of two Cladocerans with small intersetular distances. *Aust. J. Mar. Freshw. Res., 36*, 69-86, 1985.

Gerritsen, J., K. G. Porter, and J. R. Strickler, Not by sieving alone: observations of suspension feeding in *Daphnia. Bull. Mar. Sci., 43(3)*, 366-376, 1988.

Gliwicz, Z. M., Predation and the evolution of vertical migration in zooplankton. *Nature, 320*, 746-748, 1986.

Gliwicz, Z. M., A lunar cycle in zooplankton. *Ecology, 67(4)*, 883-897, 1986a.

Harleman, D. R. F., Hydrothermal analysis of lakes and reservoirs. *J. Hydraulic Engng Div. ASCE, 108*, 302-325, 1982.

Hocking, G., and M. Straškraba, An analysis of the effect of an upstream reservoir using a mathematical model of reservoir hydrodynamics. *Water Sci. & Technol., 30(2)*, 91-98, 1994.

Hrbácek, J., Circulation of water as a main factor influencing the development of helmets in *Daphnia* cucculata Sars. *Hydrobiologia, 13*, 117-185, 1959.

Idso, S. B., and J. M. Foster, Light and temperature relations in a small desert pond as influenced by phytoplanktonic density variations. *Water Resour. Res., 10, 129-132*, 1974.

Imberger, J., and J. Patterson, A Dynamic Reservoir Simulation Model - DYRESM5, in *Transport Models for Inland and Coastal Waters*, edited by H. B. Fisher, pp. 310-361, Academic Press, New York, 1981.

Imberger, J., and J. Patterson, Physical Limnology, in *Advances in Applied Mechanics*, edited by T. Wu, pp. 303-475, Academic Press, Boston, 1990.

Kamykowski, D., Organism patchiness in lakes resulting from the interaction between the internal seiche and planctonic diurnal vertical migration. *Ecol. Modelling, 4*, 197-210, 1978.

Korinek, V., B. Krepelová, and J. Machácek, Filtering structures of Cladocera and their ecological significance II. Relation between the concentration of the seston and the size of filtering combs in some species of the genera *Daphnia* and *Ceriodaphnia*. *Acta soc. zool. bohem., 50*, 244-258, 1986.

Ledbetter, M., Langmuir circulation and plankton patchiness. *Ecol. Modelling, 7*, 283-310, 1979.

Lehman, J. T., and D. Scavia, Microscale nutrient patches produced by zooplankton. *Proc. Natl. Acad. Sci. USA, 79*, 5001-5005, 1982.

Markofsky, M., and D. R. F. Harleman, Prediction of water quality in stratified reservoirs. *J. Hydraulic Engng Div. ASCE, 99*, 729-745, 1973.

Mazunder, A., W. D. Taylor, D. J. McQueen, and D. R. S. Lean, Effect of fish and plankton on lake temperature and mixing depth. *Science (Wash.), 247*, 312-325, 1990.

O'Melia, Ch. R., The influence of coagulation and sedimentation on the fate of particles, associated pollutants, and nutrients in lakes. In *Chemical Processes in Lakes*, edited by W. Stumm, pp. 207-224, Wiley, New York, 1985.

Pasciak, W., and J. Gavis, Transport limitation of nutrient uptake in phytoplankton. *Limnol. Oceanogr., 19*, 881-888, 1974.

Reynolds, C. S., The role of fluid motion in the dynamics of phytoplankton in lakes and rivers, in *Ecology of Aquatic Organisms: Scale, Pattern, Process*, edited by Giller P. S., Hildrew A. G., and D. Raffaelli, pp. 141-187, Blackwell Sci. Publ., Oxford, 1994.

Santrucková, H., and M. Straškraba, On the relationship between specific respiration activity and microbial biomass in soils. *Soil. Biol. Biochem., 23*, 525-532, 1991.

Schiebe, F. R., J. C. Ritchie, and J. R. McHenry, Influence of suspended sediments on the temperature of surface waters of reservoirs. *Verh. internat. Verein. Limnol., 19*, 133-136, 1975.

Simpson, J. J., and T. D. Dickey, The relationship between downward irradiance and upper ocean structure. *J. Phys. Oceanogr., 11*, 309-323, 1981.

Spigel, R. H., and J. Imberger, Mixing processes relevant to phytoplankton dynamics in lakes. *New Zealand J. Mar. Freshw. Res., 21*, 361-377, 1987.

Stefan, H. G., S. Dhamotharan, and F. R. Schiebe, Temperature/sediment model for a shallow lake. *J. Environ. Engng Div. ASCE, 108*, 750-765, 1982.

Stefan, H. G., J. J. Cardoni, F. R. Schiebe, and C. M. Cooper, Model of light penetration in a turbid lake. *Water Resour. Res., 19*, 109-120, 1983.

Straškraba M., The effect of physical variables on freshwater production: analyses based on models, in *The Functioning of Freshwater Ecosystems*, edited by LeCren E. D., and R. Lowe-McConnell, pp. 13-84, Cambridge Univ. Press., Cambridge, 1980

Straškraba M., Vltava Cascade as teaching grounds for reservoir limnology. Water Sci. & Technol., 30(10), 289-297, 1994.

Straškraba M., I. Dostálková, J. Hejzlar, and V. Vyhnálek, The effect of reservoirs on phosphorus concentration. *Intern. Revue. ges. Hydrobiol., 80*, 403-413, 1995.

Straškraba M., and A. H. Gnauck, *Freshwater Ecosystems. Modelling and Simulation*, 309 pp., Elsevier, Amsterdam, 1985.

Straškraba M., and G. Hocking, The effect of theoretical retention time on temperature stratification and hydrodynamics of deep river valley reservoirs, in preparation (a).

Straškraba M., and G. Hocking, The effect of light penetration on hydrodynamics of river valley reservoirs, in preparation (b).

Straškraba M., and P. Javornicky, Limnology of two re-regulation reservoirs in Czechoslovakia, in *Hydrobiological Studies*, Vol. 2, edited by J. Hrbácek, and M. Straškraba, pp. 249-316, Academia, Praha, 1973.

Straškraba M., Tundisi J. G., and A. Duncan (Eds), *Comparative Reservoir Limnology and Water Quality Management*, 291 pp., Kluwer Acad. Publ., Dordrecht, 1993.

Straskrabová, V., and K. Simek, Total and individual cell uptake of organic substances as a measure of activity of bacterioplankton. *Arch. f. Hydrobiol. Beihefte, Ergebnisse der Limnologie, 19*, 1-6, 1984.

Wetzel R. G., Reservoir ecosystems: conclusions and speculations, *in Reservoir Limnology: Ecological Perspectives*, edited by K. W. Thornton, B. L. Kimmel, and F. E. Payne, pp. 227-238, Wiley, New York, 1990

Wofsy S. C., A simple model to predict extinction coefficients and phytoplankton biomass in eutrophic waters. *Limnol. Oceanogr., 28*, 1144-1155, 1983.

Yan, N. D., Effect of changes in pH on transparency and thermal regimes of Lohi Lake, near Sunbury, Ontario. *Can. J. Fish. Aquat. Sci., 40*, 621-626, 1983.

Zaneveld, J. R. V., J. C. Kitchen, and H. Pak, The influence of optical water type on the heating rate of a constant depth mixed layer. *J. Geophys. Res., 86*, 6426-6428, 1981.

42

3D Modeling of Water Quality Transport Processes with Time and Space Varying Diffusivity Tensors

C. Dejak, R. Pastres, I. Polenghi, C. Solidoro and G. Pecenik

Abstract

A 3D seasonalized transport-water quality model is presented, capable of reaching realistic steady state or regime conditions through a pure turbulent diffusion mechanism. The model is applied to water basins with a complex morphology and negligible residual currents, after estimation of a 3D eddy diffusivity. In a second approximation, the constant diffusivity substituted by a space and time varying tensor of eddy diffusivities computed by means of an accurate statistical analysis of velocity fields given by advective models. A Lagrangian method for the estimation of the Fickian diffusivity tensor is described, based on an analogy with molecular transport kinetics. Established results of statistical mechanics of simple systems are used for linking the eddy diffusivities to the statistical properties of the circulation in real water bodies. Is has been attempted of computing also the shear stress, a contribution to diffusivity, using von Kàrmàn's formulation, but in the present case it is impossible to obtain reliable values, due to large fluctuations of the velocities computed by hydrodynamic models. Also described is the purposely elaborated explicit finite difference numerical method employed for integrating the diffusion equation, with variable coefficients, in presence of a continuous source and of open boundaries. The method is proved to be stable and conservative and therefore permits an extension of the 3D model to include the description of chemical and biological processes.

Introduction

Modeling the behaviour of a complex water environment can be accomplished only by combining in a unique theoretical framework the analysis of the most relevant physical, chemical and biological phenomena that occur in water bodies. The maintenance of life or, in general, of steady-state conditions in an ecosystem is made possible by the continuous inflow of valuable radiative and chemical free energy and outflow of degraded materials and infrared radiation with a higher entropic content. These time-irreversible inputs and outputs, require suitable transport mechanisms in order to describe appropriately the consequent fluxes, as hydrodynamic models follow the evolution of mean velocities and do not take into consideration their random components. A physical process, such as mean advection, is essentially time-reversible and does not cause entropy variation of the

Physical Processes in Lakes and Oceans
Coastal and Estuarine Studies Volume 54, Pages 645-662

system; this process, however, is able to reproduce details of the water body internal movements, not considered with time irreversible transport processes as given by a diffusive approach. Nevertheless, both entropy production and suitably detailed processes are needed in the modeling of water ecosystems. By taking into account such information, classical and irreversible thermodynamics appear as the proper link between physics and biology, as they offer the possibility of investigating time-reversible as well as irreversible phenomena (Dejak et al., 1992). Previous experiences in the field of statistical mechanics of strong electrolytes, (Dejak, 1960, Dejak et al., 1971, Dejak et al., 1972) and of Fourier analysis applied to structured and condensed phases (Dejak et al., 1970) provided a sound basis for developing a thermodynamic approach to this multidisciplinary environmental studies. The role of highly detailed eddy diffusion is here emphasized as it represents the only transport mechanism which causes dilution: the related entropy increase takes the system to a steady state condition. This is particularly simple if water basins are considered, with a hydrodynamics driven by tidal agitation and with residual currents that may be neglected. These conditions are very common in real world and have been shown to exist in the Venice lagoon to which a trophic-diffusion 3D model is applied (Dejak and Pecenik, 1987). As a matter of facts Fourier analysis of temperature and concentration fields of the most relevant chemicals in the lagoon of Venice, showed that this system is characterized by a regime steady-state (Dejak et al., 1975a). Moreover it was demonstrated by Dejak et al. (1987d), that, in presence of continuous sources and constant eddy diffusivity, a stationary spatial distribution of a conservative pollutant can be reached only by integrating the diffusion equation in a three-dimensional domain, while in one-dimensional, truly two-dimensional (Dejak et al., 1975b) or semi-three-dimensional ones (Dejak et al., 1981) the solution diverges. These considerations stem from the results yielded by a previously developed combined eutrophication diffusion 3D model of the central part of the Venice lagoon. The numerical integration of the 3D eddy-diffusion equation by means of explicit schemes, becomes computationally cumbersome and implicit methods must be employed at least in the vertical direction. Among them, it was chosen the fully implicit Laasonen scheme, which assures not only the stability, the convergence and the consistency, but also the conservativity of the solution, even for irregular bathymetries with continuous sources (Dejak et al., 1987d). Open boundary conditions were set according with a delocalized Gaussian extrapolation, based on the values of the concentration at three grid points adjacent to the boundary (Dejak et. al. 1987b): this method assures the continuity of the solution and of more realistic outward fluxes in the presence of a continuos source. After a first best fitting of the kinetic parameters that affect the rate of the chemical and biochemical reactions (Bertonati et al., 1987), the 3D model was used for short term simulation, which gave realistic results (Dejak et al., 1987a, and 1990).

Previous studies on the daily and seasonal trend of the two main forcing functions, light and temperature, (Mariotti and Dejak 1982), provided the basis for computing the variation of the water temperature during the whole year (Dejak et al., 1987c) and reproducing the vertical stratification (Dejak et al., 1992) that occurs in the deepest channels. The subroutines required by the new purposes of the research were tested in smaller 3D systems (Dejak et al., 1989), and then implemented in the model. The program was completely restructured for taking full advantage of the new generation of vector and parallel supercomputers (Pastres et al., 1995a). The extension of the model, implemented for a whole year with hourly time steps, has prompted for a recalibration of all the chemical-physical and trophic parameters against time series of environmental structural data. In order to estimate the most likely seasonal evolution, the meteorological and biological factors that cause large statistical fluctuations had to be removed from the data. This seasonal scenario was found by choosing the Fourier fitting which minimises a purposely defined estimation of the negentropy between the fitting and the multiannual data set (Dejak et al., 1993). The final calibration of the most important parameters, determined by a previous sensitivity analysis (Pecenik at al. 1990, Pastres et al., 1995b), on these time series, is one of the next goals of the research. Further improvements will concern also the modeling of the inputs due to minor or diffused sources and of the competition between microalgae and macroalgae for nutrients (Pecenik et al., 1991, Solidoro et al., 1995, 1997). To this regard, the expansion of macroalgae colonies has been modelled on the basis of remote sensing data. Even though the seasonal evolution, modelled with constant diffusivities, was consistent with the experimental data and was used for practical purposes (CRTSF, 1984 and 1989), still the model could not reproduce correctly minor features of the spatial pattern that emerge from a detailed analysis of the experimental data. In fact, the dilution of a solute takes place much more rapidly in the deepest channel than in the stagnant shallow areas and a better agreement with the reality could be achieved only by linking the diffusivities to the local circulation patterns.

Vertical Patterns

The ecosystem of the lagoon of Venice can only be described, in a water quality modeling approach, with a 3D model because it was demonstrated (Dejak et al., 1987b) that in three dimensions it is possible for the system to achieve a steady state, since, only in this case, as time tends to infinity, the persistent pollutant concentration tends to a finite value as it is shown in (1b):

$$C = \frac{\dot{Q}}{4K\pi r} \cdot \mathrm{erfc}\left(\frac{r}{\sqrt{4Kt}}\right) = \frac{\dot{Q}}{4\sqrt{K\pi}}\left(\frac{1}{r\sqrt{K\pi}} - \frac{1}{\pi K}\frac{1}{\sqrt{t}} + \frac{r^2}{12\pi K^2}\frac{1}{\sqrt{t^3}} - \frac{r^4}{160\pi K^3}\frac{1}{\sqrt{t^5}}\cdots\right)$$

$$\lim_{t \to +\infty} C = \frac{\dot{Q}}{4K\pi r} \qquad (1)$$

where r is the distance from the source point, K is the diffusivity and Q is the incoming flow at the source. Besides, only a 3D diffusion model is capable of describing appropriately the chemical and biological reactions strongly influenced by light and temperature that, inturn, depend on the depth and therefore also on stratification and other important vertical phenomena.

Modeling of vertical structure has always represented a rather difficult task. From experimental data (Rossi and Stoicovic, 1974), it was noticed the opposite sign of velocity vectors in adjacent layers, due to tidal oscillation delay. More precisely it was often seen that, when in the upper cell of a water column, the tidal velocity vector was ebbing, in the underlying cell it was still flooding, or viceversa. In order to describe the vertical structure, a lot of experimental campaigns were done during all seasons in deep channels of the lagoon (Dejak et al., 1991). From these experimental data, the presence of a stable pycnocline is evident. Since in the lagoon of Venice salinity, below the pycnocline, is always greater than above it, the sign of the temperature gradient does not seem to be relevant for stability conditions, as they are already determined by the salinity gradient. This is explained because stability implies that liquid parcels, that fluctuate vertically around the equilibrium position, spontaneously return to the equilibrium position. For these microscopic parcels, heat and salinity are diffused molecularly, and the exchange rate of heat is higher than the exchange rate of salt, so that temperature riequilibrium for a water parcel is faster reached than that of salinity. Both in summer time, when superficial water is hotter than deeper water, and in winter, when there is the opposite situation, stratification is therefore stable; consequently, no salt finger phenomenon is present in the lagoon, while it was verified to characterize a different area of the north of Italy i.e. the Po river mouth. There, the presence of a thermal power plant causes a stratification in three main layers due to the discharge of hot water on the surface, to the tidal income of a salted wedge from the sea and to the fresh water of the river. The first layer is hot and slightly salted, the middle one is colder and not salted, and in the deepest one a salt wedge is present. Between the first and second layer there is an instable double diffusion (finger regime), while between the second and the third there is a double diffusion stable regime, because, as, in winter in the lagoon, fresh water is colder than the salted water entering from the sea.

Molecular diffusivity is here applied because it characterizes the molecular phenomena that determine stability and also because it assumes different values according to the diffused property it is referred to: either salinity or heat. Molecular diffusivity can be described in statistical thermodynamics with an ultrasimplified theory of the perfect gas (Hirshfelder et al., 1966). In this context molecular diffusivity is defined as the rate between the flux and the gradient of an intensive property and so it is demonstrated to be proportional to velocity, averaged in a time between two molecular collisions, and to the mean free path. The mean free path requires a longer time to be averaged because it is inversely proportional only to the gas density and to the molecules cross section and therefore independent from fast varying forcing functions. These two entities (parameters) differ according to the diffusing property (salinity, temperature and so on) they are related to.

Macroscopically, it is possible to define turbulent diffusivity by approximating a non linear time averaged term, using an expression analogous to the one used in molecular diffusion theory. In this case one has to refer to eddy-diffusivity and phenomenological laws: Fick law for mass transport and Fourier law for heat transport. Considering that, in this case, averaged velocity does not depend on the diffusing property but on the medium, and that the mean free path depends on the width of the vortex, eddy-diffusivity is the same for salinity and heat (also for kinematic viscosity in the moments transport). Diffusivity must be described through a tensor as there are phenomena, depending on the motion of the water body, that detour the direction of the flux from the gradient direction. This tensor will be time and space varying.

The so called Fickian diffusion is not the only mechanism occurring, especially along the vertical direction where stratification causes a strong reduction of it. Examining the spreading of a pollutant patch, the enlargement of its area increases proportionally to the elapsed time, while experimental results show a faster increasing due to other contributions to dispersion. A very important one seems to be the shear stress (Okubo, 1971) that causes an even quicker diffusion than that shown by experience. However, the time scale is larger than that of phenomenological theory. The effect of a spreading patch due to shear stress is calculated and verified for a continuous enlargement of tens of hours (during the very first hours it seems to be closer to linearity) but it is very difficult to extend this effect to an oscillating regime steady state with a few hours tidal period. On the contrary Fickian diffusivity implies a shorter time scale, and it is also more correctly applicable to such a periodic regime: it requests, however, much more modulated eddy diffusivities for the tidal phenomenon.

Whereas molecular diffusivity is easy to calculate, turbulent diffusion was generally described by researchers resorting to dimensional analysis. One of these attempts was done by Prandtl, (1925), for the Fickian diffusion, but does not seem to be suitable for irregular bathymetries. For the calculation of vertical component of the diffusivity, given by the shear stress, the only formulation that can be applied to irregular bathymetries, is the one due to von Kàrmàn (1930)

$$K = k^2 \left| (\partial v_x / \partial z)^3 / \left(\partial^2 v_x / \partial z^2\right)^2 \right| \tag{2}$$

where v_x is the velocity vector in the direction perpendicular to z axis and k is the von Kàrmàn' s adimensional constant ($k = 0.36 \div 0.40$). As already mentioned above, the derivatives can assume high values, especially when velocities change direction above and below the pycnocline. This formulation was applied with available horizontal velocity fields, but it did not give acceptable results. Along the lagoon channels reasonable results were expected, but the discretization of the whole area is not enough detailed, if compared with the width of the channels, to give results which are meaningful when inserted in (2): the space step does not permit a very different localisation of the first and second velocity derivatives, the second having high values in the centre of the channel and the first along its borders. On the contrary, only in the larger very few bights of channels, results are more realistic. For this purpose a bathymetry with a space step of 50 m is presently elaborated for the application of the best hydrodynamic 3D models comprehensive buoyancy terms, for which salinity sources are inferred. Moreover a mathematical simplification of von Kàrmàn formulation, without the need of second derivatives was attempted with the same dimensional analysis, using eddy diffusivities instead of velocities. Before applying it, space varying eddy diffusivities were to be calculated, which is the aim of the present work.

Neither the vertical diffusion component given by the shear stress nor buoyancy could be taken into account with a confident theoretical treatment, so that vertical diffusion can be separated by the diffusion in the other two directions. Diffusion in the horizontal direction is governed by tidal phenomena, simulated using velocity fields in the different horizontal layers, while diffusion in the vertical direction is not yet completely theoretically explained and could be evaluated from experimental measurements (Dejak et al., 1991). To obtain these vertical diffusivities theoretical researches from old measurements was carried out (Dejak et al., unpublished) as well as new experimental campaigns (Imberger J., private communications) are presently being attempted. So far, space varying averaged vertical diffusivities could only be estimated by statistical examination of hundreds of temperature vertical profiles (Dejak and Pecenik, 1987, CRTS 1984 and 1989) interpolating temperature gradients with a generalized Gamma function, under constant vertical flux conditions (Dejak et al., 1991): the minimum value of eddy diffusivity is found at about 4 m deep, well simulating the real pycnocline.

In the actual framework, as already said, the empirical vertical diffusion must be separated by those modelled in the other two directions. The Lagrangian and Eulerian diffusivity tensor can be expressed as a reducible matrix, transforming the gradient vector into the flux one:

$$\begin{pmatrix}\Phi_u\\ \Phi_v\\ \Phi_w\end{pmatrix}=\begin{pmatrix}K_{uu} & 0 & 0\\ 0 & K_{vv} & 0\\ 0 & 0 & K_{ww}\end{pmatrix}\bullet\begin{pmatrix}\Gamma_u\\ \Gamma_v\\ \Gamma_w\end{pmatrix} \text{ and } \begin{pmatrix}\Phi_x\\ \Phi_y\\ \Phi_z\end{pmatrix}=\begin{pmatrix}K_{xx} & K_{xy} & 0\\ K_{yx} & K_{yy} & 0\\ 0 & 0 & K_{zz}\end{pmatrix}\bullet\begin{pmatrix}\Gamma_x\\ \Gamma_y\\ \Gamma_z\end{pmatrix} \quad (3)$$

in which, Φ_u,Φ_v,Φ_w are Lagrangian and Φ_x,Φ_y,Φ_z Eulerian fluxes respectively, whereas Γ_u, Γ_v, Γ_w and Γ_x, Γ_y, Γ_z are the gradients of the concentration along the Lagrangian direction of the velocities and along the Cartesian axis. $K_{xx}(x,y,z,t)$, $K_{yy}(x,y,z,t)$, $K_{zz}(z,t)$, $K_{xy}(x,y,z,t) = K_{yx}(x,y,z,t)$ represent the eddy diffusivities, with the conditions: $K_{xx} > 0$, $K_{yy} > 0$, $K_{xx}K_{yy} - K_{xy}{}^2 > 0$ that are verified through algebraic calculations, while in the Lagrangian reference system, whose orientation vary with the position within the grid, the eddy diffusivity tensor are diagonals, with components K_{uu}, K_{vv} and K_{ww} (Csanady, 1980)

Theoretical Considerations on the Concept of Diffusivity

As mentioned above, even though the molecular diffusivity is a steady-state property and not an equilibrium one, it can be computed on the basis of well-established statistical thermodynamic approaches that can be applied to ultrasimplified systems. Theoretical applications to complex systems and experimental observations are also possible and they both confirm that, the molecular diffusivity is proportional to the mean velocity and the mean free path. This is due to the extension of the statistical thermodynamic to local equilibria and to the decoupling of the two mean values, which are obtained by averaging over two not overlapping time periods.

As turbulence can not be described deterministically, the eddy diffusivity can be computed only by means of statistical approaches which asks for semiempirical closures that are all based on a dimensional analysis, such as the "Mischungsweg" of L. Prandtl (1925) and subsequent improvements by G.I. Taylor (1932), the extensions to any bathymetric situation due to Von Kàrmàn (1930), and the more recent suggestions of L.H. Kantha et al. (1977), of Smagorinsky et al. (1965) and of R.T. Cheng (1967). All the approaches listed above have are Eulerian, in the sense that all the partial derivatives of the velocities are computed at a fix point, even though, in some cases, tensorial properties are introduced.

At present, standard computer programs can produce velocity fields on an extremely detailed finite-difference mesh: these results can be used for approaching the problem form a Lagrangian point of view. In this frame, it is possible to associate a significative ensemble of realizations of trajectories at any point of the grid and at any time. The results obtained by the statistical analysis of the velocities and of the displacements along the trajectories are more rigorous and can be used for evaluating the importance of the residual currents and the range of applicability of Fick law. Further, they are a sound basis for computing the components of a Lagrangian tensor of eddy diffusivity, in analogy with the computation of the molecular diffusivity outlined above. In this way, it is possible to apply the results of advective models periodically forced by the tide to the problem of the transport of chemicals, heat and microorganisms, instead of using the results directly and introducing diffusivities that, not being based on semiempirical hypothesis, vary according with the different approaches proposed in the literature.

As a simple example, let us take a 1D system with fixed length, characterized by a constant velocity or diffusivity, in which an extensive variable is emitted with different modalities at a source point. With an instantaneous input, the evolution is described by the time-reversible translation of a δ function in the advective case, while in the diffusion case, a Gaussian distribution is obtained with an associated standard deviation increasing with the square root of time. The effect is the dilution and, as a consequence, an increase of the entropy of the system. In the presence of a continuous input, with pure advection one gets a step function, with constant height and increasing length and in this case a steady-state distribution can easily be reached by setting the same velocity also at the open boundaries. On the contrary, the transfer function for the diffusive case behaves similarly to a δ function, continuously increasing in the neighbourhood of the source point (Diachisin, 1963) but also enlarging on the whole length with the partial output on the open boundaries. Further, a two-dimensional system also diverges and a steady state can be reached only in a 3D domains. With three dimensions, the advective contribution is to transport a pollutant patch from one point to another, without changing its shape, unless undesired numerical diffusion occurs. (Dejak et al., 1987a). The enlargement of the patch can only be described through an eddy diffusivity tensor. In this case in a sphere of ray Δr around the source, the amount of pollutant concentration accumulated does not increase infinitely, but tends to be constant, as it approaches the steady state:

$$q_0 = \dot{Q}t\left[-\frac{\Delta r}{\sqrt{\pi K t}}\exp\left(-\frac{\Delta r^2}{4Kt}\right)+\left(1-\frac{\Delta r^2}{2Kt}\right)\mathrm{erf}\left(\frac{\Delta r}{2\sqrt{Kt}}\right)+\frac{\Delta r^2}{2Kt}\right]=$$

$$=\frac{\dot{Q}\Delta r^2}{2K}\left[1-\frac{2\Delta r}{3\sqrt{\pi K}}\cdot\frac{1}{\sqrt{t}}-\frac{\Delta r^3}{48\sqrt{\pi K}}\cdot\frac{1}{\sqrt{t^3}}+\ldots\right]$$

$$\lim_{t\to+\infty} q_0 = \frac{\dot{Q}\Delta r^2}{2K} \tag{4}$$

These considerations, derived from the analysis of simple systems, show that steady states or regime conditions are not easy to model and show the need of theoretical investigations, which could lead to counter-intuitive results. At steady-state, the entropy production of the system is at its minimum and it is continuously exported through the open boundaries. This situation is typical of irreversible phenomena: the internal processes take place much more rapidly than the chemical reactions or the macroscopic processes driven by the slow variations of the forcing functions. In fact the modeling effort is aimed to reproduce a dissipative structure, in which the regime steady-state is maintained by internal microscopic perturbations that counteract the macroscopic ones caused by temporal variations of external forcing functions. According with the definition of local equilibrium (Prigogine, 1962; Lvov et al., 1996) there is a strong analogy between the relationship of regime states and steady states, from one side, and steady states and equilibrium states, on the other.

To validate this method in the simplest way, it is better to apply it first to ecosystems only remixed by periodical phenomena, like the tide in the Venice Lagoon. It is easier to describe interactions among chemical, biological, thermal and diffusive phenomena, in such ecosystems than in waterbodies characterized also by residual currents. Transport equations, currently adopted in hydrodynamic models, can not be straightforwardly applied to a combined model like the one here presented, but the velocity fields elaborated by them represent a valuable source of information for the definition of a diffusivity field on the horizontal layers and permit to extend the treatment of molecular diffusion in the statistical thermodynamics to the turbulent one. This alternative approach does not take into account the structure of the model used for deriving the velocity fields, or its closure equations, but it only makes use of its results.

It has been shown by both field measurements and modellistic studies, that in the case of the lagoon of Venice residual currents are negligible, as it may be deduced by observing the tidally averaged Eulerian velocities, displayed as arrows at each gird point in Figure 1a, compared with the actual velocities, Figure 1b. The residuals are two orders of magnitude smaller than the average of the absolute values, and therefore irrelevant if compared with the stochastic fluctuations and numerical errors. This finding has been checked by performing the statistical analysis of the velocity fields in Lagrangian reference systems. Nevertheless, in this case the average ratio between the velocity and its absolute value is not significative and one has to compute, for each trajectory, the ratio between the autocovariance in respect of its initial velocity and the average variance. This ratio gives the Lagrangian correlation coefficient (the Eulerian coefficient substantially reproduce the ratios outlined above), which is about 1/100 for the lagoon of Venice. By integrating such coefficient (Csanady 1980, Monin e Yaglom, 1975), one obtain the Lagrangian time scale, which is important for assuring the validity of Fick's law (Dejak et al., 1996). These and the subsequent elaboration presented in this paper are based on a set of 2D velocity fields elaborated by the Danish Hydraulic Institute and provided by the Consorzio Venezia Nuova only for testing the methodology. The method is then applied to two more reliable available 3D velocity fields, calculated from a model which uses a more detailed bathymetry and includes realistic buoyancy terms (Casulli 1994).

Statistical Thermodynamic Approach to Turbulent Diffusivity

The molecular diffusivity of a perfect gas, in the simplest statistical thermodynamic treatment, as said above, is given by the product of the molecular velocity and the mean free path. The equivalent of the mean free path in turbulent diffusion may be assumed to be the size of the vortexes produced by tidal agitation that cause the deviations from the inertial trajectories of particles, whereas the molecular velocities are replaced by hourly

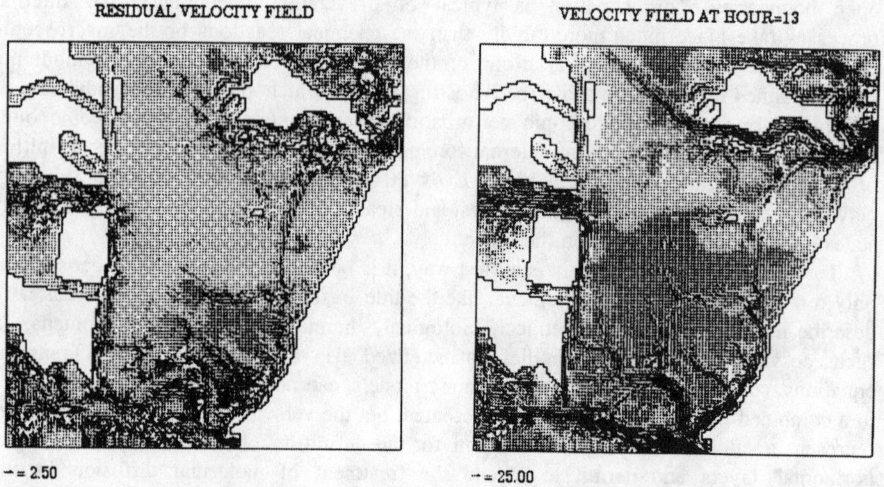

Figure 1 a, b. Tidally averaged velocities (a) and example of velocity field (b). To be able to represent the small residual velocities, the scale of Figure a is 10 times smaller than the scale of Figure b.

averaged particle velocity vectors. Analogously to the mean free path, the sizes of vortices are independent from the time variation of forcing functions and therefore they could be calculated by a long time average, being one hour the time step, depending only from local morphology and the mean tidal energy input. The size of two-dimensional vortexes may be estimated for each point by a statistical analysis of the distribution of the points covered by any trajectory designed, during the tidal excursion, by pollutant particles released at one grid point at different tidal moments, or by computing the deterministic envelope of a family of the trajectories. Of the two methods, the second is affected by strong statistical fluctuations, and so the first is preferred.

Since residual currents are negligible, it should be expected that almost all the trajectories are closed curves. Figure 2 shown an example of an ensemble of these trajectories which starts from a given point at given time within a tidal cycle: the bathymetry is also shown in the background. This ensemble can be statistically described with a dispersion ellipse that includes 95% of all the trajectories points. The ellipse at each grid point, see Figure 2, is characterized by the length of the major axis (Figure 3) and minor axis (Figure 4), by the eccentricity (Figure 5), and by the angle between the major axis and the ordinate axis of the Cartesian mesh (N-S direction). The parameters of the ellipses are estimated as in (Lindler 1964) and are evidenced in Figure 2 by using different grey tones (see the scale below the figure). As one can see, the eccentricity is markedly different from 1 along the deepest channels, whereas in shallower areas the ellipse is less eccentric: this substantiates the need of a tensor reproducing the concentration patterns in all zones of the modelled area. The presence of important transversal channels is reflected by the fact that the length of the major axis of the ellipses along channels oriented NW-SE or NE-SW is comparable with the one along the main N-S and E-W channels.

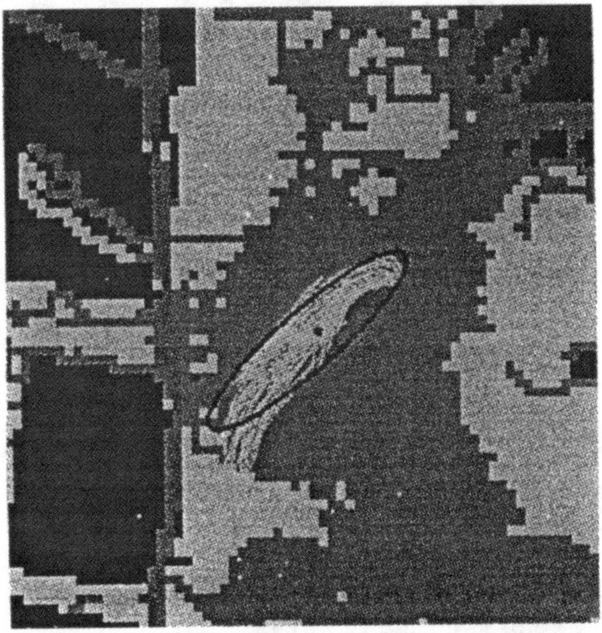

Figure 2. Envelope of tidal trajectories .

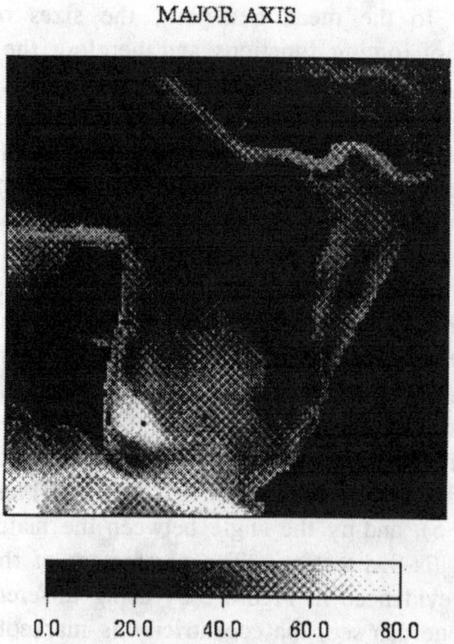

Figure 3. Major axis of the ellipses

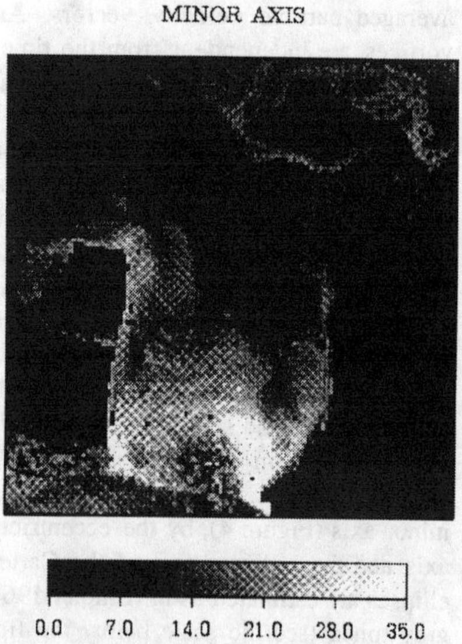

Figure 4. Minor axis of the ellipses.

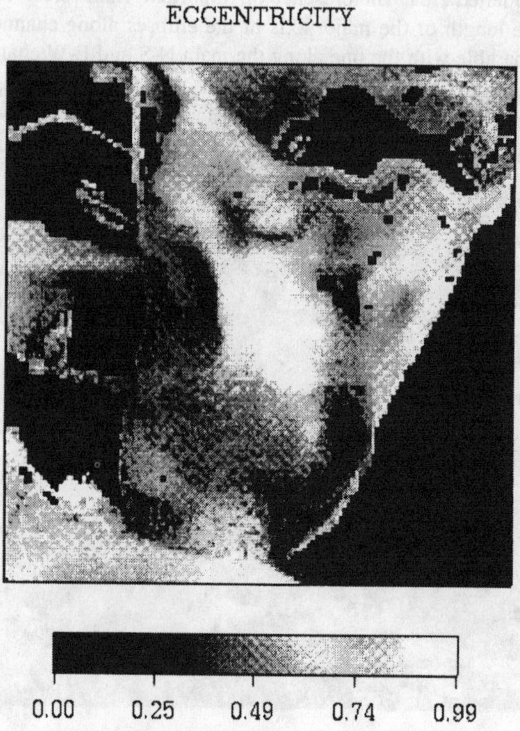

Figure 5. Eccentricity of the ellipses.

The definition of the ellipse close to the edges of the modelled area has requested the introduction of reasonable approximations of trajectories at the boundaries: for this purpose the velocity component perpendicular at a closed boundary is reflected, while outside the open boundaries the particle is moved by a velocity field, assumed locally symmetrical to the actual one. In this way, most of the trajectories departing from points near the open boundaries are closed. These conditions are general and make the method readily applicable to other water bodies hydrodynamics similar to the lagoon of Venice and it is possible to extend this also to systems with not negligible residual currents, by the separation of the two effects.

By comparing the integral of the Fourier–Fick equations with the statistical distribution in a Lagrangian reference system, one can express the diffusivity in term of the parameters of the distribution:

$$K_{ii} = \frac{1}{2}\frac{d\sigma_i^2}{dt} = \sigma_i \frac{d\sigma_i}{dt} = \sigma_i v_i = \overline{\sigma}_i \overline{v}_i \tag{5}$$

where the Lagrangian derivative of σ_i is v_i and the two averages are decoupled, as in the molecular case, as the first is averaged over a long time span, in order to smooth the fluctuations, and the second over a short time, much shorter than a tidal period.

Carrying on the analogy with molecular diffusivity, the eddy diffusivities along the canonical axes of each ellipse are determined by a Lagrangian vector: its components are represented by the products between the two ellipse semiaxes and the absolute value of the projections of hourly averaged velocities on them. The Lagrangian diffusivity tensor is then given by a matrix, having in the principal diagonal the two components of this vector and, outside, zero value, so that it results locally diagonal in the reference system of each ellipse. Even though the components of the tensor are averaged over a few tidal cycle, they are not negligible, as they are given by the absolute value of a scalar product of two vectors with the same origin: one of them is fixed (σ) and the other varies its direction (v).

In order to obtain the Eulerian tensor, the Lagrangian one is rotated according with the orientation of the fixed Cartesian axes by applying a linear transformation and its inverse, which lead to a non diagonal but still symmetrical space and time varying diffusivity tensor. Finally, the tensor is rescaled, by comparing its space and time averaged value with a constant eddy diffusivity of $22 m^2 s^{-1}$ which include tidal agitation and was evaluated for the water body as a whole (Dejak et al., 1987b).

Results are summarised in Figures 6a, 6b, 6c, which are similar to the previous figure but show respectively the daily averaged diagonal components, K_{xx} and K_{yy} and the cross term, K_{xy} of the tensor. The absolute values of the cross terms, Figure 6c, are realistic and consistent with the bathymetry shown in Figure 2: as one can see from Figures 6a and b, they are negative, coloured in white, in the channels oriented NE-SW and positive, coloured in black, on the NW-SE direction. The two diagonal components in the tensors are dominant and assume their highest value on the channel oriented S-N and N-E.

Computational Aspects and Numerical Integration

The lagoon of Venice is a rather complex network of deep channels (maximum depth: 20m) and shallow areas joined to the Adriatic sea through three inlets and characterized by the incoming of sea water as well as of fresh water from 20 rivers. The computational domain covers an area of about 180 km^2 and it is discretised by a fine 3D mesh of size $\Delta x = \Delta y =$ 100m in the horizontal layers and $\Delta z = 1$m along the vertical axis. This yields to 128×128

× 20=327680 cells, only a tenth of which belongs to the water body itself as the average depth is about 1 m, as shallow areas cover most of the basin. The time step is $\Delta t = 1$ hour.

The numerical integration of the diffusion equation in presence of a continuous point source has been performed with a finite difference explicit method for each horizontal layer and with an implicit Laasonen scheme along each vertical column (Pastres et al., 1995a). The explicit method adopted to solve the horizontal diffusion second order partial differential equation with space and time varying diffusivity tensors, permits one to check directly non only the convergence, consistency, and stability but also the conservativity of the solutions. The last property is crucial in the simulation of reactive-diffusive systems because a non conservative scheme would lead to mass balance alterations. In scalar terms, the horizontal diffusion equation can be written as:

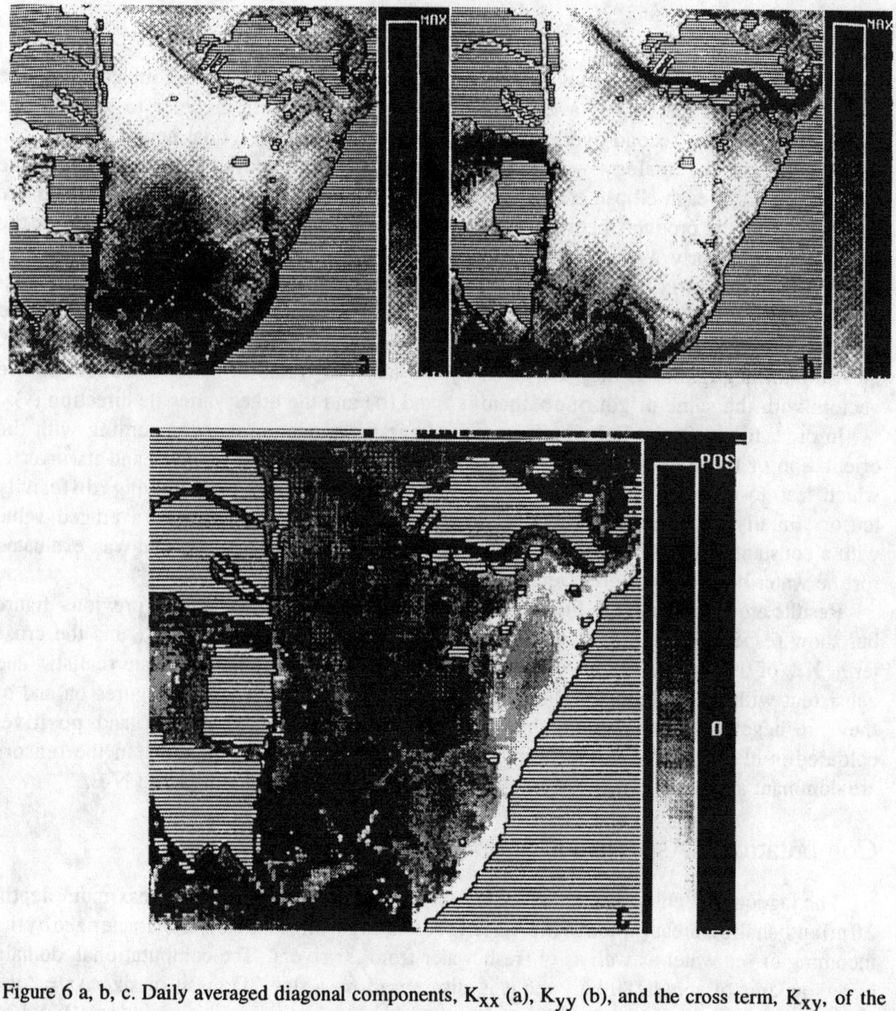

Figure 6 a, b, c. Daily averaged diagonal components, K_{xx} (a), K_{yy} (b), and the cross term, K_{xy}, of the tensor.

$$\frac{\partial c}{\partial t}=\frac{\partial}{\partial x}\left(K_{xx}\frac{\partial c}{\partial x}\right)+\frac{\partial}{\partial y}\left(K_{yy}\frac{\partial c}{\partial y}\right)+\frac{\partial}{\partial x}\left(K_{xy}\frac{\partial c}{\partial y}\right)+\frac{\partial}{\partial y}\left(K_{xy}\frac{\partial c}{\partial x}\right)=$$
$$K_{xx}\frac{\partial^2 c}{\partial x^2}+K_{yy}\frac{\partial^2 c}{\partial y^2}+2K_{xy}\frac{\partial^2 c}{\partial x\partial y}+\frac{\partial K_{xx}}{\partial x}\frac{\partial c}{\partial x}+\frac{\partial K_{yy}}{\partial y}\frac{\partial c}{\partial y}+\frac{\partial K_{xy}}{\partial x}\frac{\partial c}{\partial y}+\frac{\partial K_{xy}}{\partial y}\frac{\partial c}{\partial x} \tag{6}$$

The scheme adopted is forward in time and centred in space except for the last two terms that are forward in space in order to guarantee the conservativity of the method:

$$\begin{aligned}c_{i,j}^{t+1}&=c_{i,j}^{t}+D_{xx(i,j)}^{t}\cdot\left(c_{(i+1,j)}^{t}-2c_{(i,j)}^{t}+c_{(i-1,j)}^{t}\right)+D_{yy(i,j)}^{t}\cdot\left(c_{(i,j+1)}^{t}-2c_{(i,j)}^{t}+c_{i,j-1}^{t}\right)+\\&\frac{D_{xy(i,j)}^{t}}{2}\cdot\left(c_{(i+1,j+1)}^{t}-c_{(i-1,j+1)}^{t}-c_{(i+1,j-1)}^{t}+c_{(i-1,j-1)}^{t}\right)+\\&\frac{D_{xy(i+1,j)}^{t}-D_{xy(i-1,j)}^{t}}{4}\cdot\left(c_{(i,j+1)}^{t}-c_{(i,j-1)}^{t}\right)+\frac{D_{xy(i,j+1)}^{t}-D_{xy(i,j-1)}^{t}}{4}\cdot\left(c_{(i+1,j)}^{t}-c_{(i-1,j)}^{t}\right)+\\&\left(D_{xx(i+1,j)}^{t}-D_{xx(i,j)}^{t}\right)\cdot\left(c_{(i+1,j)}^{t}-c_{(i,j)}^{t}\right)+\left(D_{yy(i,j+1)}^{t}-D_{yy(i,j)}^{t}\right)\cdot\left(c_{(i,j+1)}^{t}-c_{(i,j)}^{t}\right)\end{aligned} \tag{7}$$

where D_{xx} , D_{xy} , D_{yy} are the dimensionless diffusion numbers: $D^t_{xx} = K^t_{xx} \cdot \Delta t/(\Delta x)^2$ and analogously for the other two components. The most important advantage of this discretization and its originality, lie in the fact that, as easily proved by algebraic manipulation, the scheme unconditionally assures the conservativity property, a fundamental requirement for reproducing chemical and biological phenomena that require mass balances. Open boundary conditions are set according to a delocalized Gaussian extrapolation based on the values of the concentration at three grid points adjacent to the boundary (Dejak et al., 1987b). These conditions assure the continuity of the solution and more realistic outward fluxes. The stability is assured when the following inequality is satisfied (Richtmyer and Morton, 1967): $\max(D^t_{xx}) + \max(D^t_{yy}) \leq 1/2$. The instabilities initially caused by the cross term (especially in the few grid points where D_{xy} is greater than one of the diagonal components), are smoothed out as the numerical integration proceeds, as one can see from Figure 7a, 7b, 7c, 7d, where the results are presented by using a scale of grey tones.

The extension of the method to a three-dimensional grid is straightforward, as vertical diffusion is assumed to be independent from the other ones. Under this hypothesis, one can use the fractional step method for separately treating the diffusion on the xy plane and along the vertical. The integration on the third dimension is carried out with the implicit scheme, (Laasonen, 1949) already implemented in the previous version of the model. Depth varying vertical diffusivities were used, estimated by statistical examination of temperature vertical profiles, as explained above; these enable, without the assumption of instantaneous mixing, the reproduction of heat diffusion from the surface to the bottom water cells (Dejak et al., 1991).

The results are displayed in Figure 8a, 8b, 8c, 8d, which are similar to the previous illustrations. They are realistic and physically meaningful, as the evolution of the spatial patterns of the concentration integrated along the vertical axis shows the appearance of the network of channels, along which the vertically integrated concentrations are higher. This highly anisotropic distribution is typical of this environment and could never be modelled by using only a spatial and temporal constant diffusivity, even though this simplified approach has been successfully applied. Nevertheless, the improvements here presented will make possible a more detailed modeling of biological, chemical and physical processes occurring in the waterbody.

Conclusions

The semiempirical method here described for estimating the diffusivity tensors in a real water body, governed by the tidal agitation, is different from the Eulerian approaches to the estimation of eddy diffusivity, being based only on the statistical analysis of the velocity fields given by a hydrodynamic model. With this approach, which involves the evaluation of ensemble averages in Lagrangian reference systems, the quality of the results obtained, greatly depends on the features of the model used for the calculation of the diffusivity tensors and on the accuracy of the spatial discretization of the morphology included in the starting hydrodynamic model.

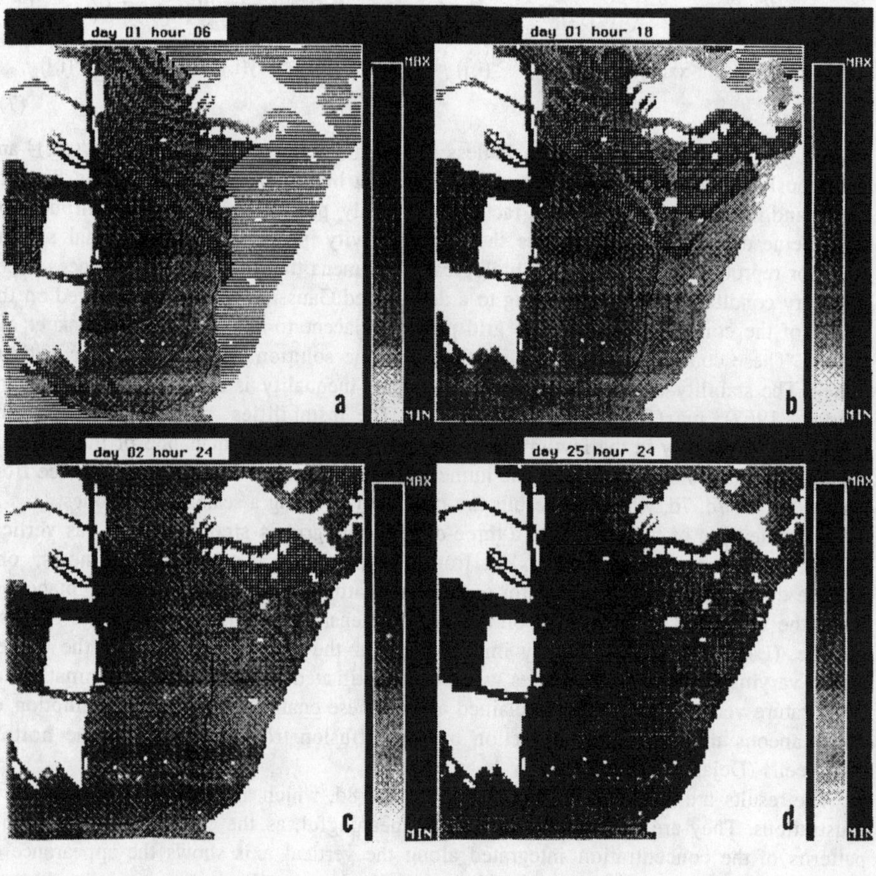

Figure 7 a, b, c, d. Results of the integration (surface layer): instabilities are clearly visible 6 hours (a) after the simulation to the release as a point source of a persistent pollutant in the industrial area of the Venice Lagoon; instabilities are smoothed out as the integration proceeds (b, c and d). (Concentration values are shown in the colour scale in the right; instability: ▦ sea: ▤ .)

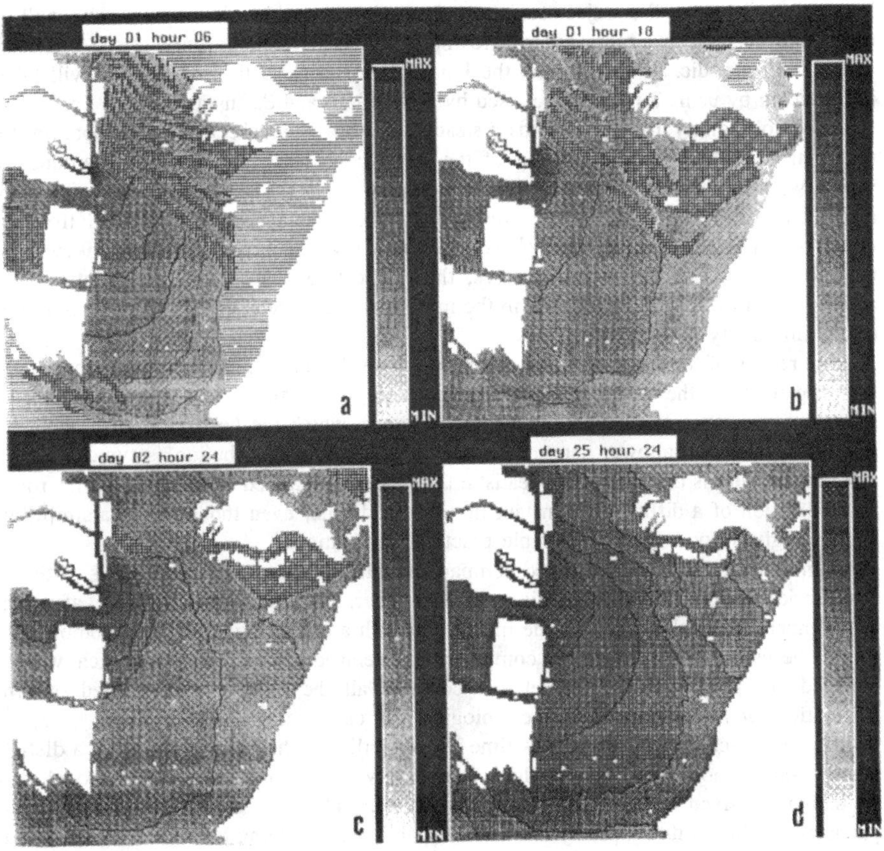

Figure 8 a, b, c, d. Vertically integrated concentration of a persistent pollutant after release as a source point in the industrial area of the Venice Lagoon.

Shear stress too can be calculated using eddy diffusivities. It does not require the direct knowledge of the Lagrangian vector field that was used to compute the diagonal tensor but of the tensor itself. In this way von Kàrmàn's dimensional analysis could be simplified using only first derivatives and there is no need to calculate less precise second derivatives.

Some doubts might arise about the ergodicity of the statistical treatment used to determine the ellipses, because the points of the trajectories do not seem to have a Gaussian distribution, but a toroidal one. The resolution of minor vortexes, originating from the turbulent energy cascade of bigger eddies, is not possible to be simulated with the presently available advective models, and so it is not possible to smooth out the toroidal distribution set of curves. Nevertheless, the solution adopted theoretically guarantees the statistical-thermodynamic ergodicity since the whole reachable phase space is really reached for both the position and momentum subspace, at least in the horizontal layers; as long as the vertical direction is concerned, it would be necessary to consider the shear stress and in general a better knowledge of all the concomitant phenomena. The interactions between the trajectories, not taken into consideration up to now, will be taken into account in a more general treatment. In fact, (5) holds only if the ensemble averages are ergodic. To this regard, the longitudinal terms of the shear stress will taken into account by using a method proposed by Aris (1956) and Saffman (1962).

The evolutionary pattern towards a steady state has been checked by comparing the principal part of the time derivative of the mean concentration in a sphere around the source with the theoretical value that is proportional to $t^{3/2}$, as one can see by deriving the right hand side of q_0 in (4). In the present case, the time derivative computed from the model output and multiplied by $t^{-3/2}$ increases first towards a constant value, likewise in the theoretical case, but then decreases, thus indicating that, in this application, the achievement of steady state is faster in the more realistic system than in the 3D case with costan diffusivity (Figure 9).

The results of the integration, displayed in Figure 7 a, b, c, d, and the evolution of the spatial patterns of the concentration integrated along the vertical axis (Figure 8 a, b, c, d) show the appearance of a network of channels, along which the horizontal displacements are higher than in shallower zones. This highly anisotropic distribution is typical of this environment and is certainly more realistic than the one obtained by using the first rough approximation of a diffusivity constant in space and time, even though such a simplified approach gives good results in simple practical applications.

A more quantitative and detailed comparison with the observations, that is with the time series quoted at the beginning of the paper, will be possible only after the development of an algorithm for the integration with a conservative implicit method. This step is necessary for reducing the computational demand of the program, which will be included in a more complex model that contains all the routines that carried out the calculation for the description of the biological and chemical processes.

This approach already introduces time irreversibility in the transport term of a distributed parameter model and preserve the conservativity, which in a water quality model must be satisfied, given the reactive coupling between physical and chemical-biological phenomena. Due to its simplicity and to its capability of evolving towards a regime steady state and of maintaining for a simulated year at a reasonable computational cost, this 3D diffusion model offers a great advantage in respect to purely hydrodynamic models, that hardly permit to explain the dilution of a dissolved substance and thus the extension of the method to even more complex physical, chemical, biochemical and ecological processes, as well as their interactions, which involve simultaneous dispersion of all related variables.

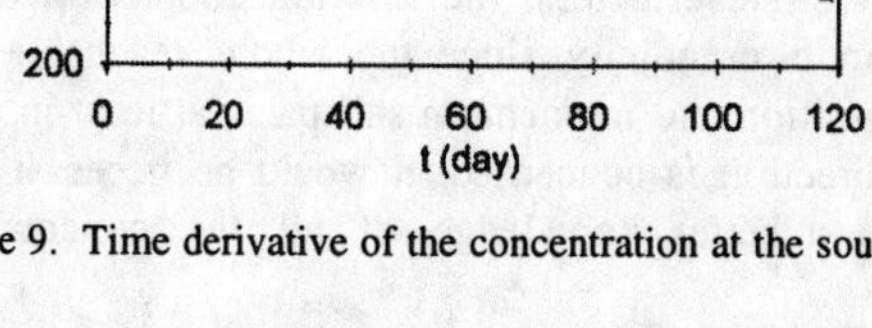

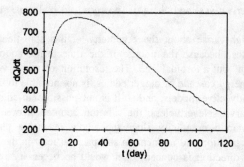

Figure 9. Time derivative of the concentration at the source.

References

Aris, R., On the dispersion of a solute in a fluid flowing through a tube. *Proc. Roy. Soc. London A235,* 67-77, 1956.

Bertonati, M., C. Dejak, L. I. Mazzei, and G. Pecenik, Eutrophication model of the Venice Lagoon: statistical treatment of "in situ" measurements of phytoplankton growth parameters. *Ecol. Model., 37,* 391-404, 1987.

Casulli, V. and E. Cattani, Stability, accuracy and efficiency of a semi-implicit method for three-dimensional shallow water flow. *Computers Math. Applic.* 27, 99-112, 1994.

Cheng, R. T., T. M. Powell and T. M. Dillon, Numerical models of wind-driven circulation in lakes. *Appl. Math. Model., 1,* 141-159, 1967.

CRTSF (Commissione Regionale Tecnico-Scientifica di controllo dell'impianto della centrale termica ENEL di Fusina), Venice: *Technical Report,* Vol III: experimental measurements, 1983.

CRTSF (Commissione Regionale Tecnico-Scientifica di controllo dell'impianto della centrale termica ENEL di Fusina), Venice: *Technical Report,* Vol III: experimental measurements, 1989.

Csanady, G. T., Turbolent diffusion in the Environment. Reidel Publ. Comp. Dordrecht Holl. 1980.

Dejak, C., Ricerca sulla teoria delle soluzioni concentrate di elettroliti forti: XXVIII impostazione dellateoria con raggio ionico variabile. *Annali di chimica 30,* 971-978, 1960.

Dejak, C., Licheri and G. Piccalunga, On the effect of termination error in Fourier analysis of X-ray diffraction data for liquid water. *J. Appl. Cryst., 3,* 183-185, 1970.

Dejak, C., L. I. Mazzei, and G. Cocco, Research on the primary kinetic salt effect. XII The influence of tetra-alkyl-ammonium salts on the iodide-persulfate reaction. *Gazz. Chim. Ital., 101,* 606-624, 1971.

Dejak, C., L. I. Mazzei, O. Devoto, and G. Cocco, Colligative properties of solution of strong electrolytes and of sea and brackish water. Theoretical calculation of the osmotic coefficient. *Desalination, 10,* 263-272, 1972.

Dejak, C., L. I. Mazzei, I. Lavagnini, and G. Saba, Diffusion models for chemical and thermal Pollution. *Proceeding of the 3rd Int. Congress of Iranian Chem. Soc., Shiraz,* 261-262, 1975a.

Dejak, C., L. I. Mazzei, G. Paschina, and G. Saba, Analysis of the periodical behaviour of pollution. *La Chimica e l'Industria, 57,* 5-9, 1975b.

Dejak, C., L. I. Mazzei, and L. Meregalli, Vertical averages in a three-dimensional diffusion of pollutants. *Il Nuovo Cimento, 4c,* 493-510, 1981.

Dejak, C., L. I. Mazzei, E. Messina, and G. Pecenik, Steady state achievement by introduction of true tidal velocities in a pollution model of the Venice Lagoon. *Ecol. Model., 37,* 59-79, 1987a.

Dejak, C., L. I. Mazzei, M. Molin, and G. Pecenik, Tidal three-dimensional diffusion in a model of the lagoon of Venice and reliability conditions for its numerical integration. *Ecol. Model., 37,* 81-101, 1987b.

Dejak, C., L. I. Mazzei, E. Messina, and G. Pecenik, An advective-diffusive pollution model of the lagoon of Venice. *Ecol. Model., 37,* 47-57, 1987c.

Dejak, C., L. I. Mazzei, L. Meregalli, and E. Messina, A two-dimensional diffusive model of the Venice Lagoon and relative open boundaries conditions. *Ecol. Model., 37,* 21-45, 1987d.

Dejak, C. and G. Pecenik. A trophic-diffusion 3D model of the Venice Lagoon. *Elsevier Oceanography Series, 43,* 301-309, 1987.

Dejak, C., D. Franco, R. Pastres, and G. Pecenik, A steady state achieving 3D eutrophication diffusion submodel. *Environmental Software, 4,* 94-101, 1989.

Dejak, C., D. Franco, R. Pastres, and G. Pecenik, A 3-D Eutrophication-Diffusion Model of the Venice Lagoon: Some Applications. In *Residual Currents and Long-term Transport", Coastal and Estuarine Studies,* edited by R. T. Cheng, pp. 38, 526-538, Springer Verlag, N. Y. Inc. New York, 1990.

Dejak, C., D. Franco, R. Pastres, G. Pecenik and C. Solidoro, Thermal exchanges at air-water interfaces and reproduction of temperature vertical profiles in water columns. *Journal of Marine Systems, 3,* 465-476, 1991.

Dejak, C., C. Solidoro, and G. Pecenik, The Boltzmann's H-Theorem and the arrow of time: consideration about biological and technological evolution. *Ecol. Phys. Chemistry,* 2, 85-96, 1992.

Dejak, C., D. Franco, R. Pastres, G. Pecenik and C. Solidoro, An informational approach to model time series of environmental data though negentropy estimation. *Ecol. Model., 67,* 199-220, 1993.

Dejak, C., R. Pastres, G. Pecenik and C. Solidoro, Conditions for the applications of transport models to problems of environmental chemistry. *Annali di Chimica. J. Analyt. Environ. Chemistry 87,* 617-634, 1996.

Diachishin, A. N., Waste disposal in tidal waters. *J. San Eng. Div., 89*(SA4), 23-43, 1963.

Gray, W. G., Physics based modelling of lakes, reservoirs and inpoundmens *ASCE Publ.* NY USA pp. 95-97, 1986.
Hirshfelder, J., C. F. Curtiss, and R. B. Bird, *Molecular Theory of Gases and Liquids*. J. Wiley & Sons Inc. N. Y. Edition 3rd, 1966.
Linder, A., *Statistische Methoden* IV Ed. *Birkhauser Verl. Basel.* p. 179, 1964.
Kantha, L., O. M. Phillip, and R. S. Arad, On turbolent entrainment at a stable density interface. *J. Fluid Mech. 79*, 753-768, 1977.
Kàrmàn, T. von, Mecanische Ähnlichkeit und Turbolenz. *Nach Ges. Wiss., Göttingen, Math. Phys. Klasse*, 58, 1930.
Laasonen, G., Über eine Methode zur Lösung der Wärmegleichung. *Acta Math, 81*, 309, 1949.
Lvov, S. N., R. Pastres and A. Marcomini, Thermodynamical stability analysis of the carbon biogeochemical cycle in aquatic shallow environment. *Geochimica et Cosmochimica Acta 60* (1996) 3569-3579, 1996.
Mariotti, M. and C. Dejak, Valutazioni dei flussi energetici all'interfaccia acqua-aria nella laguna di Venezia. *Ambiente e Risorse,10*, 73, 1982.
Monin, A. S. and A. M. Yaglom, *Statistical Fluid Mechanics of Turbulence*. I vol. MIT Press, Mass. V Ed. pp. 364-373, 1987.
Okubo, A., Oceanic diffusion diagrams. *Deep sea results*. *18*, 789-204, 1971.
Pastres, R, D. Franco, G. Pecenik, C. Solidoro, and C. Dejak, Using parallel computer in environmental modelling: a working example. *Ecol. Model. 80*, 69-85, 1995a.
Pastres, R, D. Franco, G. Pecenik, C. Solidoro, and C. Dejak, First order sensitivity analysis of a distributed parameters ecological model. *SAMO 95 International Symposium*, 1995b.
Pecenik, G., R. Pastres, S. Righetto, D. Franco, and C. Dejak, Analisi di sensitività di modelli ambientali: applicazione all'ecosistema lagunare. *Ambiente e Risorse*, *4*, 26-38, 1990.
Pecenik, G., C. Solidoro, D. Franco, C. Dejak, and D. Franco, Modellazione macroalgale per un macromodello eutrofico diffusivo della laguna di Venezia. *SITE* , *12*, 881-890, 1991.
Prigogine, I., Non equilibrium statistical mechanics. *Interscience Pub., N. Y.*, XVI, 1962.
Prandtl, L., *Z. Angew Math. Mech.*, *5*,136, 1925.
Richtmyer, R. D. and R. W. Morton, Difference methods for initial value problems, (2nd edition) *Wiley-Interscience, London*, 405, 1925.
Rossi, G. and Stoicovic, Misure di corrente nel canale dei petroli nella laguna di Venezia in località Fusina. ENEL, Technical; Report, 320-374, 1974.
Saffman, P. G., The effect of windshear on horizontal spread from an istantaneous ground source. *Quant. J. Roy. Soc. 88*, 382-393, 1962.
Smagorinsky, J., S. Manabe and J. L. Holloway, Numerical results from a nine-level general circulation model of the atmosphere. *Mon. Wea. Rev. 93*, 727-768, 1965.
Snedecor, G. W. Statistical Methods V Ed. *Lower St. Univ. Press*, p 164, 1956.
Solidoro, C., C. Dejak, D. Franco, R. Pastres, and G. Pecenik, A model for macroalgae and phytoplancton growth in the Venice lagoon. *Environmental Intnl.*, 21, 619-626, 1995.
Solidoro, C, G. Pecenik, R. Pastres, D. Franco, and C. Dejak, Modelling macroalgae (ulva rigida) in the Venice lagoon: Model structure identifiaction and first parameter estimation, *Ecol. Model.*, *94*, 191-206, 1997.
Taylor, G. I., Scientific Papers Vol 2 *Cambridge Univ. Press,* 1960.

List of Contributors

K. Aagaard
Polar Science Center
University of Washington
Seattle WA 98195
USA

A. Anis
Dept. of Geophysics and Planetary Science
Faculty of Science
The Yigal Allon Kinneret Limnological Lab.
PO Box 345
Tiberias 14102
ISRAEL

Y. Asada
Lake Biwa Research Institute
1-10 Uchide-hama
Otsu 520
JAPAN

P. G. Baines
CSIRO Division of Atmospheric Research
Private Bag No 1
Aspendale Victoria 3195
AUSTRALIA

G. Bauer
Institut fur Mechanik (AG III)
Technische Hochschule Darmstadt
Hochschulstr. 1
64289 Darmstadt
GERMANY

E. Bäuerle
Institut fur Umweltphysik Universitat Heidelberg
Im Neuenheimer Feld 366
Heidelberg 69120
GERMANY

A. L. Berestov
Dept. of Mechanical and Aerospace Eng.
Arizona State University
PO Box 876106
Tempe Arizona AZ 85287-6106
USA

E. C. Carmack
Institute of Ocean Sciences
P. O. Box 600
Sidney BC V8L 4B2
CANADA

X. Chen
Dept. of Coastal & Oceanographic Engineering Department
University of Florida
Gainesville Florida 32611
USA

C. Dejak
Department of Physical Chemistry
Environmental Section
University of Venice
Dorsoduro 2137 Venice
ITALY

I. P. D. De Silva
Centre for Water Research
Department of Environmental Engineering
The University of Western Australia
Nedlands Western Australia 6907
AUSTRALIA

M. A. Donelan
University of Miami
RSMAS/Applied Marine Physics
4600 Rickenbacker Causeway
Miami Florida 33149
USA

C. C. Eriksen
School of Oceanography
University of Washington
Box 357940
Seattle WA 98195-7940
USA

D. Farmer
Institute of Ocean Sciences
9860 West Saanich Road
Sidney BC V8L 4B2
CANADA

A. V. Fedorov
Scripps Institution of Oceanography
University of California San Diego
La Jolla CA 92093-0230
USA

H. J. S. Fernando
Dept. of Mechanical and Aerospace Eng.
Arizona State University
PO Box 876106
Tempe Arizona AZ85287-6106
USA

J. H. Ferziger
Environmental Fluid Mechanics Laboratory
Department of Civil Engineering
Stanford University
Stanford CA 94305-4020
USA

C. Garrett
Centre for Earth and Ocean Research
University of Victoria
Victoria BC V8W 2Y2
CANADA

J. Gemmrich
Institute of Ocean Sciences
9860 West Saanich Road
Sidney BC V8L 4B2
CANADA
and
Department of Physics and Astronomy
University of Victoria
Victoria BC V8W 2Y2
CANADA

C. H. Gibson
Departments of Applied Mechanics and
Engineering Sciences and Scripps
Institution of Oceanography
University of California at San Diego
La Jolla California 92093-0411
USA

M. Gloor
Swiss Federal Institute for Environmental
Science and Technology (EAWAG) and
Swiss Federal Inst. of Technology (ETH)
CH-8600 Dübendorf
SWITZERLAND

M.C. Gregg
Appl. Phys. Lab. and Sch. of Oceanography
University of Washington
522 Henderson Hall
Seattle WA 98105
USA

R. Grimshaw
Department of Mathematics
Monash University
Clayton Victoria 3168
AUSTRALIA

P. Güting
Institut fur Mechanik (AG III)
Technische Hochschule Darmstadt
Hochschulstr. 1
64289 Darmstadt
GERMANY

S. P. Haigh
Department of Mathematics
University of British Columbia
Vancouver BC V6T 1Z2
CANADA

P. F. Hamblin
Aquatic Ecosystems Restoration Branch
National Water Research Institute
867 Lakeshore Rd.
Burlington Ontario L7R 4A6
CANADA

E. Hollan
Institut für Seenforschung
Landesanstalt für Umweltschutz
Baden-Württemberg
D-88085 Langenargen
GERMANY

J. C. R. Hunt
Meteorological Office
London Road
Bracknell
Berkshire RG12 2SZ
UNITED KINGDOM

K. Hutter
Institut fur Mechanik (AG III)
Technische Hochschule Darmstadt
Hochschulstr. 1
64289 Darmstadt
GERMANY

J. Imberger
Department of Environmental Engineering
Centre for Water Research
University of Western Australia
Nedlands Western Australia 6907
AUSTRALIA

D. M. Imboden
Swiss Federal Institute of Technology
ETH, 8092 Zürich, Switzerland
and Swiss Federal Institute of
Environmental Science and Technology
EAWAG 8600 Dübendorf
SWITZERLAND

G. N. Ivey
Department of Environmental Engineering
Centre for Water Research
University of Western Australia
Nedlands Western Australia 6907
AUSTRALIA

A. Javam
Department of Environmental Engineering
Centre for Water Research
University of Western Australia
Nedlands Western Australia 6907
AUSTRALIA

R. Jiang
Laboratoire de recherches hydrauliques
Ecole Polytechnique Fédérale de Lausanne
CH-1015 Lausanne
SWITZERLAND

E. P. Jones
The Bedford Institute of
Oceanography (BIO)
Dartmouth NS B2Y 4A2
CANADA

K. B. Katsaros
National Oceanic & Atmospheric
Administration
Atlantic Oceanographic and Meteorological
Laboratory
4301 Rickenbacker Causeway
Miami Florida 33149
USA

J. R. Koseff
Environmental Fluid Mechanics Laboratory
Department of Civil and Environmental
Engineering
Stanford University
Stanford CA 94305-4020
USA.

A. S. Ksenofontov
Kabardinsky-Balkarsky State University
Nalchik
RUSSIA

M. Kumagai
Lake Biwa Research Institute
1-10, Uchidehama, Otsu 520
JAPAN

G. A. Lawrence
Civil Engineering Department
The University of British Columbia
2324 Main Mall
Vancouver BC V6T 1Z4
CANADA

C. Lemckert
School of Engineering
Gold Coast Campus Griffin University
PMB 50 Gold Coast Mail Centre
Queensland 9726
AUSTRALIA

U. Lemmin
Laboratoire de recherches hydrauliques
Ecole Polytechnique Fédérale de Lausanne
CH-1015 Lausanne
SWITZERLAND

M. Li
Centre for Earth and Ocean Research
University of Victoria
Victoria BC V8W 2Y2
CANADA
and
Institute of Ocean Sciences
PO Box 6000
9800 West Saanich Road
Sidney BC V8L 4B2
CANADA

C-L. Lin
Environmental Fluid Mechanics Laboratory
Department of Civil Engineering
Stanford University
Stanford CA 94305-4020
USA

P. F. Linden
Department of Applied Mathematics and Theoretical Physics
University of Cambridge
Cambridge CB3 9EW
UNITED KINGDOM

I. D. Lozovatsky
Department of Mechanical and Aerospace Engineering
Arizona State University
Tempe AZ 85287-6106
USA
and
P.P. Shirshov Institute of Oceanology
Russian Academy of Sciences
117851 Moscow
RUSSIA

R. W. Macdonald
Institute of Ocean Sciences
PO Box 600
Sidney BC V8L 4B2
CANADA

S. MacIntyre
Marine Science Institute and
Institute for Computational Earth System Science
University of California
Santa Barbara CA 93106-6150
USA

J. J. M. Magnaudet
Institut de Mecanique des Fluides
CNRS URA 005
2 allee Professeur C. Soula
Toulouse 31400
FRANCE

T. Maxworthy
Depts .of Aerospace and Mechanical Eng.
University of Southern California
Los Angeles CA 90089-1191
USA

F. A. McLaughlin
Institute of Ocean Sciences
PO Box 600
Sidney BC V8L 4B2
CANADA

W. K. Melville
Scripps Institution of Oceanography
University of California San Diego
9500 Gilman Drive
La Jolla CA 92093-0230
USA

S. G. Monismith
Environmental Fluid Mechanics Laboratory
Stanford University
Stanford Ca. 94305-4020
USA
and
Institut de Mecanique des Fluides CNRS URA 005
2 allee Professeur C. Soula
Toulouse 31400
FRANCE

S. Nakano
Lake Biwa Research Institute
1-10 Uchidehama Otsu 520
JAPAN

S. Nishida
Department of Civil Engineering
Hachinohe Institute of Technology
Hachinohe 031
JAPAN

M. Ohtani
Department of Engineering Science
Faculty of Engineering
Hokkaido University
Sapporo 060
JAPAN

R. L. Oliver
Murray-Darling Freshwater Res. Centre
Co-operative Research Centre for
Freshwater Ecology
PO Box 921
Albury NSW 2640
AUSTRALIA

R. G. Perkin
Institute of Ocean Sciences
PO Box 600
Sidney BC V8L 4B2
CANADA

R. Pastres
Department of Physical Chemistry
Environmental Section
University of Venice
Dorsoduro 2137 Venice
ITALY

G. Pecenik
Department of Physical Chemistry
Environmental section
University of Venice
Dorsoduro 2137 Venice
ITALY

I. Polenghi
Department of Physical Chemistry
Environmental section
University of Venice
Dorsoduro 2137 Venice
ITALY

V. Polonichko
Institute of Ocean Sciences
9860 West Saanich Road
Sidney BC V8L 4B2
CANADA
and
School of Earth & Ocean Sciences
University of Victoria
Victoria BC V8W 2Y2
CANADA

D. Ramsden
Can. Ctr. for Clim. Modeling and Analysis
University of Victoria
PO Box 1700 MS3339
Victoria BC V8W 2Y2
CANADA

C. S. Reynolds
Freshwater Biological Association
NERC Institute of Freshwater Ecology
Ambleside Cumbria GB-LA22 0LP
UNITED KINGDOM

P. B. Rhines
School of Oceanography
University of Washington
Box 357940
Seattle Washington 98195
USA

A. Saggio
Department of Environmental Engineering
Centre for Water Research
University of Western Australia
Nedlands Western Australia 6907
AUSTRALIA

S. Schofield
Dept. of Coastal & Oceanogr. Eng. Dept.
University of Florida
Gainesville Florida 32611
USA

Y. P. Sheng
Department of Coastal & Oceanographic Engineering Department
University of Florida
Gainesville Florida 32611
USA

C. Solidoro
Department of Physical Chemistry
Environmental section
University of Venice
Dorsoduro 2137 Venice
ITALY

E. J. Strang
Department of Mechanical and Aerospace Engineering
Arizona State University
PO Box 876106
Tempe Arizona AZ85287-6106
USA

M. Straskraba
Biomath. Lab. Entomological Institute
Acad. of Sciences of the Czech Republic and Faculty of Biological Sciences
University of South Bohemia
Ceské Budejovice
CZECH REPUBLIC

J. H. Swift
Scripps Institution of Oceanography
University of California
La Jolla CA 92037
USA

S. G. Teoh
Arup Jururunding Sdn Bhd
Industrial Engineering
No. 25-30 Jalan Ara 7/3B
Bandar Sri Damansara
52200
KUALA LUMPUR

S. A. Thorpe
Department of Oceanography
Southampton Oceanography Centre
Empress Dock
Southampton SO14 3ZH
UNITED KINGDOM

Y. Wang
Institut fur Mechanik (AG III)
Technische Hochschule Darmstadt
Hochschulstr. 1
64289 Darmstadt
GERMANY

J. Weinstock
Environmental Research Laboratories
National Oceanic and Atmospheric Administration
325 Broadway
Boulder CO 80303
USA

J. Whittington
Murray-Darling Freshwater Research Centre
Co-operative Research Centre for Freshwater Ecology
PO Box 921
Albury NSW 2640
AUSTRALIA

A. Wüest
Swiss Federal Inst. for Environ. Sci. and Tech. (EAWAG) and
Swiss Fed. Institute of Technology (ETH)
CH-8600 Dübendorf
SWITZERLAND

H. Yamazaki
Department of Ocean Sciences
Tokyo University of Fisheries
4-5-7 Konan Minato-ku Tokyo 108
JAPAN

S. Yoshida
Department of Engineering Science
Faculty of Engineering
Hokkaido University
Sapporo 060
JAPAN

Z. Zhu
Civil Engineering Department
The University of British Columbia
2324 Main Mall
Vancouver BC V6T 1Z4
CANADA